# Heat Transfer

**Jean Taine**
*Ecole Centrale Paris*

and

**Jean-Pierre Petit**
*Ecole Centrale Paris*

**Prentice Hall**
New York London Toronto Sydney Tokyo Singapore

First published 1993 by
Prentice Hall International (UK) Ltd
Campus 400, Maylands Avenue
Hemel Hempstead
Hertfordshire, HP2 7EZ
A division of
Simon & Schuster International Group



Typeset in 10/12pt Times
by P&R Typesetters Ltd, Salisbury

Printed and bound in Great Britain by
Redwood Books, Trowbridge, Wiltshire

Library of Congress Cataloging-in-Publication Data

Taine, J. (Jean), 1949–
[Transferts thermiques. English]
Heat transfer/Jean Taine, Jean-Pierre Petit: English translation by J. Roger Calvert.
p. cm.
Translation of: Transferts thermiques.
Includes bibliographical references and index.
ISBN 0-13-387994-1
1. Heat—Transmission. I. Petit, Jean-Pierre. II. Title.
QC320.T3513 1992 92-16861
536′.2—dc20 CIP

British Library Cataloguing in Publication Data

A catalogue record for this book is available from the British Library

ISBN 0-13-387994-1 (pbk)

1 2 3 4 5 97 96 95 94 93

# Contents

*Foreword by Michel Combarnous* vii

*Preface* ix

*Nomenclature* xi

**Part I Introduction to Steady Energy Transfer** **1**

1 The principal modes of energy transfer 3
2 Linear conduction 25
3 First approach to convection. Applications 51
4 Introduction to radiation 78

**Part II Non-steady Conduction** **113**

5 Scope. General theorems 115
6 Semi-infinite geometry. Response after a short time interval 124
7 Finite geometry. Response at an arbitrary time 135
8 Time and length scales 140

**Part III Radiation** **169**

9 Radiative properties of opaque bodies 171
10 Radiative transfer between opaque bodies 186
11 Semi-transparent media 215

**PART IV Mechanics of Non-isothermal Fluids. Convection** **247**

12 Dimensional approach to convection 249
13 Balance equations 260
14 External laminar convection. Boundary layers 268
15 Internal laminar convection. Development of flow conditions in forced convection 282

16 Turbulent forced convection 292
17 Empirical correlations in turbulent flow 325
18 Coupling between convection and radiation 330

**Part V Basic Data** **345**

1 Forced internal convection 347
2 Forced external convection 386
3 Free external convection 391
4 Free internal convection 397
5 Radiation 409
6 Conversion factors, temperature scales 420
7 Thermophysical properties 424
8 Convection equations in various coordinate systems 433

**Part VI Applications** **437**

1 Insulation of a cryogenic container 439
2 Principles of infra-red long-range detection 446
3 Temperature of a material exposed to solar radiation 450
4 Heating a metal ingot in a furnace 455
5 Temperature measurement 458
6 Is the wine cool? The earthenware jar 478
7 Industrial heat recovery (controlled forced ventilation with two flows) 485
8 Supplementary heating using solar radiation 492
9 Thermal study of a nuclear reactor 504
10 Heat treatment in a tube furnace 510
11 Laser heat treatment of steel 515
12 Thermal inertia of a building 523
13 Temperature measurement by laser-induced photothermal effect 528
14 Window for a domestic oven 535
15 Combustion chamber of an aircraft engine (greatly simplified) 546
16 Control of an electric furnace 553
17 Insulating windows 558
18 Environmental effects on a photovoltaic solar cell 567

*Bibliography* 576

*Subject index* 583

# Foreword

For about fifteen years, Jean Taine has been working in the field of heat transfer as a Professor at the Ecole Centrale Paris, which is a well-known general engineering school in France. Founded in 1829 by a group of scientists and engineers, among whom was the first professor of the school in physics and heat transfer, Eugène Péclet (1793–1893), it produced the scientific and technical international society of engineers whose members included Gustave Eiffel, Louis Blériot, Jean-Pierre Peugeot, André Michelin and, more recently, Marcel Véron (1900–1984), another impressive professor of heat transfer at the Ecole Centrale.

Jean Taine has spent most of the last fifteen years teaching students and doing research, mainly in radiative heat transfer, a field where both scales for the description of the transport phenomena must necessarily be used – a microscopic scale in which the physics of the phenomena have to be carefully understood and described, and a macroscopic scale, more familiar to the engineer, in which the general rules for modelling must be defined.

Very active in both research and teaching, Jean Taine wrote a general book on the field, with a contribution from Dr Jean-Pierre Petit for 'Basic Data', Part V. Published in France in 1989, this book was, and still is, very successful. In 1991, it was followed by the publication of a collection of practical examples of application, written by the same authors.

The present edition is not only based on a brilliant translation by Dr Roger Calvert of the original French publication, but it has also been improved on many points. In particular, it has been enlarged upon by the inclusion of a thorough description of some applications.

There is no doubt that this book will be an efficient tool for advanced students, scientists and engineers who are interested in the field of heat transfer, as well as for numerous colleagues in the international community of heat transfer, within which Jean Taine is gradually acquiring considerable repute.

*Michel Combarnous*

*Professor at the University of Bordeaux*

*Corresponding Member of the French Academy of Sciences*

# Preface

This work has two aims:

- to provide a progressive and complete textbook of heat transfer, written from the physicist's point of view;
- to be a general reference work for the student, engineer or researcher.

It is arranged in six parts. The first four (written by Jean Taine) constitute the **course** proper, while the fifth part entitled 'Basic Data' (author Jean-Pierre Petit) is a **tool** for tackling a large number of real applications. The sixth part, 'Applications', is a set of daily or industrial problems, where modelling based on realistic data is emphasized. The combination makes a coherent and indivisible whole.

The authors have not hesitated to introduce new terminologies or theories relative to key concepts (equilibrium radiation, conducto-convective flux, etc.). The choices were dictated by two essentials: simplicity of language and scientific rigour.

A progressive approach to energy transfer necessitates the fundamental study in the first part, entitled 'Introduction to Steady Energy Transfer', before the ideas and methods relevant to the various transfer modes are tackled (Parts II–IV). The approach chosen has the advantage of giving a first presentation of the whole discipline and allowing basic concepts of other modes of heat transfer to be used during the study of one particular mode. The ideas developed in the final chapters (11, 16 and 18) of the various parts, although already applied in certain design offices, are still the subject of research.

The authors would like to thank

- for fruitful discussions and communications: Serge Bories, Sébastien Candel and Jean-Marc Delhaye;
- for continual exchange of ideas, significant contributions and efficient help in the preparation of the work: all the members of the teaching team in Heat Transfer at E.C.P. and in particular Chidiac Chidiac, Jean-Jacques Greffet, Jean-Pierre Martin,Philippe Mignon, Marie-Yvonne Perrin, Louis Philippe and Anouar Soufiani.

Finally, they offer their thanks to Marie-France Martin who readily and efficiently handled the complex physical preparation of this work.

*Jean Taine, Jean-Pierre Petit*
*October 1991*

# Nomenclature

| | |
|---|---|
| $a$, ($a_t$) | thermal diffusivity (turbulent) |
| $A_\lambda$, $A_\nu$, $A$ | global absorption factor (semi-transparent medium) |
| $A$, $dA$, etc. | area |
| $b$ | thermal effusivity |
| $c$, ($c_o$) | speed of light (in free space) |
| $c$, $c_p$, $c_v$ | specific heat, at constant pressure, at constant volume |
| $Cr$ | capacity rate |
| $\mathbf{d}$, $d_{ij}$ | strain rate tensor |
| $D$, ($D_h$) | diameter (hydraulic diameter) |
| $e$, $E$, $\mathscr{E}$ | internal energy: per unit mass, per unit volume, total |
| $E$ | heat exchanger efficiency |
| $\mathbf{E}$ | electric field vector |
| $E_\lambda$, $E_\nu$, $E$ | global emission factor (semi-transparent medium) |
| $e_{kin}$, $E_{kin}$, $\mathscr{E}_{kin}$ | kinetic energy: per unit mass, per unit volume, total |
| $f$, $f_{ij}$, $df$, etc. | radiative shape factors |
| $g$ | resultant body force per unit mass (sometimes acceleration due to gravity) |
| $h$, $\hbar$ | Planck's constants |
| h, H, $\mathscr{H}$ | enthalpy: per unit mass, per unit volume, total for system |
| $h$, $h(x)$, $\bar{h}$, $h_R$ | heat transfer coefficient: convective, local, mean, radiative |
| $h_g$ | global heat transfer coefficient |
| $I'_\lambda$, $I_\lambda$, ($I^o_\lambda$), $I$, $I'_\nu$, $I_\nu$, ($I^o_\nu$) | intensity: directional, isotropic (equilibrium) total |
| $I$ | electric current intensity |
| $\mathbf{j}$ | electric current density vector |
| $k$ | mean turbulent kinetic energy |
| $k$, $k'$ | von Karman constants |
| $k_B$ | Boltzmann's constant |
| $L_m$, $L_{th}$ | entry length: mechanical, thermal |
| $l$, $L$ | lengths |
| $\mathscr{L}_v$, $\mathscr{L}_f$ | latent heat per unit mass: vaporization, fusion |
| $M$ | molar mass |

| | |
|---|---|
| $m$, d$m$, $\partial m$ | mass |
| $\dot{m}$, d$\dot{m}$ | mass flux |
| $\hat{n}$, $n$ | complex, real indices |
| $\mathbf{n}$, $n_i$ | unit normal vector |
| $p$ | pressure |
| $p_v$ | phase diffusion function |
| $P$ | power dissipated per unit volume |
| $\mathscr{P}$ | wetted perimeter |
| $\mathbf{q}$, $\mathbf{q}^{cd}$, $\mathbf{q}^{R}$, $\mathbf{q}^{cv}$ | flux density vector: conductive, radiative, convective |
| Q, $\dot{Q}$ | quantity of heat, heat flow rate |
| $R_\lambda$, $R_\nu$, $R$ | global reflection factor (semi-transparent medium) |
| $R$, $R_h$ | radius, hydraulic radius |
| $r$, $R$ | perfect gas constant per unit mass, molar |
| $\mathscr{R}$ | thermal resistance |
| $\mathbf{r}$ | position vector |
| $S$, d$S$, (d$S_a$, d$S_n$) | surface, element of surface (open system, material) |
| $t$ | time |
| $T$ | temperature (usually absolute) |
| $T_\lambda$, $T_\nu$, $T$ | global transmission factor (semi-transparent medium) |
| $T_{ij}$, $\underline{\mathbf{T}}$ | stress tensor |
| $\mathbf{v}$ | velocity vector |
| $u$, $v$, $w$ | velocity components |
| or $v_i$, $v_j$, $v_k$ | (tensor notation) |
| $V$, $V_a$, $V_m$, $dV$, d$V_a$, d$V_m$ | volume: general, open, material |
| $\mathbf{w}$ | velocity of an element of a control surface |
| $W$, $\dot{W}$ | work, mechanical power |
| $x$, $y$, $z$ | Cartesian coordinates |
| | |
| $\alpha'_\lambda$, $\alpha_\lambda$, $\alpha$ | absorptivity |
| $\beta$ | coefficient of thermal expansion |
| $\beta_v$ | extinction coefficient |
| $\Gamma$, d$\Gamma$ | thermal conductance or velocity constant (Chapter 9) |
| $\delta$, $\delta_{th}$, $\delta_m$ | boundary layer thickness, thermal, velocity |
| $\delta_{ij}$ | Kronecker tensor (unit tensor) |
| $\varepsilon$ | turbulent dissipation rate |
| $\varepsilon'_\lambda$, $\varepsilon_\lambda$, $\varepsilon$ | emissivity |
| $\hat{\varepsilon}$, $\varepsilon_1$, $\varepsilon_2$ | relative permittivity: complex, real, imaginary |
| $\kappa$, $\kappa_\nu$, $\kappa_\lambda$ | absorption coefficient |
| $\lambda$, ($\lambda_t$) | wavelength or thermal conductivity (turbulent) |
| $\mu_F$ | chemical potential (Fermi level) |
| $\mu$, ($\mu_T$) | dynamic viscosity (turbulent) |
| $\nu$, ($\nu_T$) | kinematic viscosity (turbulent) |
| $\nu$ | frequency |
| $\Pi$ | turbulence production |

| | |
|---|---|
| $\Pi_i$ | dimensionless group |
| $\rho_\lambda$, $\rho_\nu$, $\rho$, etc. | reflectivity |
| $\sigma$ | Stefan's universal constant |
| $\sigma_v$ | diffusion coefficient |
| $\tau_\lambda$, $\tau_\nu$, $\tau$, etc. | transmissivity |
| $\tau_{ij}$, $(\tau^t_{ij})$, $\tau$, $(\tau^t)$ | viscous stress tensor (turbulent) |
| $\tau_w$ | wall shear stress |
| $\omega$ | circular frequency |
| $\Omega$, $d\Omega$ | solid angle |
| $\Phi$ | flux (in W) |
| $\varphi$, $\varphi^{cd}$, $\varphi^{cc}$, $\varphi^R$, $\varphi^{cv}$ | flux density (in W m$^{-2}$), conductive, conducto-convective, radiative, convective |

**Common dimensionless groups**

| | |
|---|---|
| $Bi$ | Biot number |
| $c_F$ | friction coefficient |
| $Ec$ | Eckert number |
| $Fo$ | Fourier number |
| $Gr_x$, $Gr_L$, $Gr_D$, etc. | Grashof number |
| $Nu$ | Nusselt number |

*External convection*:

| | |
|---|---|
| $Nu_x(x)$ | Nusselt number based on $x$ at position $x$ |
| $Nu_x(\mathbf{r})$ | Nusselt number based on $x$ at position $\mathbf{r}$ |

$Nu_x(\mathbf{r})$ is related to $h(\mathbf{r})$ by $Nu_x(\mathbf{r}) = h(\mathbf{r})x/\lambda_f$

$Nu_L$ is related to $\bar{h}$, the mean over $L$, by $Nu_L = \bar{h}L/\lambda$

*Internal convection*:

| | |
|---|---|
| $Nu_{D_h}(L)$ | Nusselt number based on $D_h$ at point $L$ |
| $Nu_{D_h}(x)$ | Nusselt number based on $D_h$ at point $x$ |
| $Nu_{D_h}(\mathbf{r})$ | Nusselt number based on $D_h$ at point $\mathbf{r}$ |

$Nu_{D_h}(\mathbf{r})$ is related to $h(\mathbf{r})$ by $Nu_{D_h}(\mathbf{r}) = h(\mathbf{r})D_h/\lambda$

$Nu_{D_h}$ is related to $\bar{h}$, the mean over $L$, by $Nu_{D_h} = \bar{h}L/\lambda$

in fully developed internal convection (if $h$ is independent of $\mathbf{r}$): $Nu_{D_h} = hD_h/\lambda$

| | |
|---|---|
| $Pe$ | Péclet number |
| $Pr$, $(Pr_t)$ | Prandtl number (turbulent) |
| $Ra_x$, $Ra_L$, etc. | Rayleigh number |
| $Re_x$, $Re_D$, $Re_L$, etc. | Reynolds Number |
| $Ri$ | Richardson number |
| $St$ | Stanton number |

PART I

# Introduction to Steady Energy Transfer

This first part of the work is deliberately restricted to geometrically simple (one-dimensional) systems and to the equally restricting steady-state case. The principal modes of heat transfer (conduction, convection and radiation) are introduced at the beginning of this part, and a first overview of the horizons of the extremely diverse subject is given. By eliminating, initially, the complications which usually arise from complex geometry and unsteadiness, we allow the **physical approach to the phenomena to be given prominence**.

Chapter 1 aims to define the general scope of the subject and to analyse the different types of heat transfer qualitatively. From Chapter 2, which is concerned with the linear approach to conduction, the complex problem of **modelling all the transfers for a real system** is posed and a methodical system is suggested. Numerous examples of phenomena or systems in which the three transfer modes appear are similarly treated in Chapters 3 and 4, dedicated to a first approach to convection and to thermal radiation respectively.

# 1

# The Principal Modes of Energy Transfer

## 1.1 Physical limitations of the study

### 1.1.1 The system

Every physical system is made up of two continuously interacting sub-systems: a material system and a radiation field.

- The **material system** is generally considered to be a continuous medium [1]. It is actually made up, on the elementary scale, of molecules (including atoms and ions), electrons, fictitious particles such as phonons (which represent quanta of vibrational energy in a solid), etc.

  The assumption of a continuous medium involves restricting analysis to a material system made up of elements of arbitrarily small volume on the macroscopic scale but sufficiently large for the number of molecules in them to be extremely high (of the order $10^{15}$–$10^{20}$, to give a general idea). Such elements may, under certain physical conditions, be characterized statistically by macroscopic physical properties averaged over all the molecules they contain (average mass, velocity, pressure or temperature). A simple example of the transition from the elementary scale to that of the continuous medium with regard to temperature is given in Section 1.1.3.
- The electromagnetic **radiation field** is characterized on the macroscopic scale by the definition at every point $\mathbf{r}$ in space and for every direction $\Delta$, of a quantity $I'_\nu$, the monochromatic intensity, related to the frequency $\nu$, which will be defined in Chapter 4. The radiation field results from a distribution of photons, each characterized by a frequency $\nu$, a momentum $\mathbf{p}$ and a spin $s$.

### 1.1.2 Perfect thermodynamic equilibrium and thermal equilibrium

A fundamental difficulty in the study of energy transfer lies in the transition from the elementary scale to that of the continuous medium: this is the realm of statistical mechanics [2]. The simplest case to handle, which serves as a reference, is that of **thermodynamic equilibrium**. This is characterized in statistical mechanics by statistical laws for the distribution of the energies of the molecules (or other particles: photons, phonons, etc.), which depend on the type of particles: Fermi–Dirac distributions (fermions), Bose–Einstein distributions (bosons), Bose–Einstein with indeterminate numbers (photons) and, in the macroscopic limit, the **Maxwell–Boltzmann** distribution (the usual case for molecules) [2]. In these laws, the (absolute) thermodynamic temperature $T$ appears as a sole parameter. When one moves to the realm of continuous media, thermodynamic equilibrium (also known as **perfect thermodynamic equilibrium PTE**) is characterized by the uniformity of a set of physical properties which describe its state: temperature $T$ but also chemical potential $\mu$, pressure $p$, etc. More restrictively, one can characterize **thermal equilibrium** by uniformity of temperature throughout the system. We may note in this connection that, in thermal equilibrium, the following apply:

- The material system is isothermal at a temperature $T$.
- The radiation field has a uniform distribution of intensity, which depends only on $T$ (the expression for the intensity of equilibrium radiation derives from statistical mechanics [2], and will be covered in Chapter 4).
- The material system and the radiation field are in mutual equilibrium (at the same temperature).

*Important note*
In certain particular cases the radiation field is in thermal equilibrium (at a given temperature $T_R$) although within a medium which is not in thermal equilibrium, but contains temperature gradients: an example appears in Section 4.8. The opposite situation can also occur.

### 1.1.3 The concept of the continuous medium

Let us consider the situation of a group of identical atoms of mass $m$, in thermal equilibrium. The thermodynamic temperature of the system may be calculated by the classical equation:

$$\frac{3}{2} N k_B T = \sum_{s=1}^{N} \frac{m v_s^2}{2} \tag{1.0}$$

where $N$ is the number of atoms in the element of volume $dV$ under consideration, $k_B$ is Boltzmann's constant, $v_s$ is the speed of an atom relative to $dV$. The distribution of the speeds of the atoms obeys Maxwell–Boltzmann statistics, arising from collisions between atoms (making allowance for the balance between the atoms entering $\mathrm{d}V$ and those leaving). If the volume concerned contains only a few atoms, the pseudo-temperature defined by equation (1.0) will fluctuate very strongly as $\mathrm{d}V$ is increased. When the number of atoms contained in volume $\mathrm{d}V$ is sufficient to obtain a Boltzmann distribution ($N \sim 10^{15}$ to $10^{20}$), an increase in $\mathrm{d}V$ cannot significantly affect the temperature calculated by the above equation. The medium has become continuous on this scale. The volume element $\mathrm{d}V$ under consideration is nevertheless arbitrarily small by comparison with the dimensions of the system.

A similar argument may be followed in respect of all other macroscopic physical quantities (speed, pressure, density), which contribute to the lack of equilibrium in the system.

### *1.1.4 Thermal non-equilibrium*

The systems considered in this work are in thermal non-equilibrium, for example due to constraints imposed by the surroundings. Numerous mechanisms on the elementary scale, however, tend to return the medium to a state of thermal equilibrium:

- collisions between molecules or between a molecule and a wall;
- molecule/photon interactions (absorption, spontaneous emission, stimulated emission, etc.);
- interactions among phonons, between phonons and electrons, electrons and photons, etc.;
- other interactions.

The energy changes of the physical system result from the combination at the macroscopic scale of all these elementary mechanisms of energy transfer and also transfers of energy associated with macroscopic mass transfer (convection).

In the case of a **significant departure from thermal equilibrium** the material system can only be described, at a point $\mathbf{r}$ and an instant $t$, by statistical distributions of the populations $N_{ij}$ of the various species $j$ (molecules and other particles) at the various energy levels $i$, which may be continuous or discrete (see, for example, references [3], [4] and [5]). The mechanisms mentioned above, collisions, for example, lead to relaxation equations [3] which govern the evolution of the populations $N_{ij}$, analogous to the equations of chemical kinetics (the medium is in fact a **reactive medium** in both cases). In the case of a small departure from thermal equilibrium a powerful method for describing the evolution of the system is the Boltzmann formulation (references [4], [5], [6], etc.). Such an approach, which makes use of statistical mechanics, is not developed in this work, which is limited to the basics of the discipline. However, it is necessary for the study of heat transfer in numerous applications:

- in LASERS;
- in *cold* plasmas used as an energy source or for the production of composites (welding torches, etc.) and in fusion plasmas;

- in atmospheric re-entry phenomena for space vehicles (shuttles, missiles, etc.);
- in highly rarefied materials,
- etc.

### 1.1.5 The hypothesis of local thermodynamic equilibrium (LTE)

The physics of transfer in continuous media rests on the hypothesis of local thermodynamic equilibrium (LTE) which corresponds to a situation of **mild non-equilibrium**.

Under this hypothesis, it is assumed that during an interval of time $dt$ and in an arbitrarily small (but macroscopic) element of volume $dV$ the material system is locally infinitely close to a state of equilibrium, describable by a group of intensive and extensive physical properties.

Practically, the hypothesis of LTE means that it is possible to define the normal physical variables $T(\mathbf{r}, t)$, $p(\mathbf{r}, t)$, $\mu(\mathbf{r}, t)$, etc., at any instant $t$ and for every point $\mathbf{r}$. In so far as temperature is concerned, the hypothesis is equivalent to assuming that the internal degrees of freedom of the material are thermalized (i.e. the populations of the energy levels follow the Maxwell–Boltzmann distribution). Since under normal conditions the characteristic times for the return to thermal equilibrium by the mechanisms seen above are of the order of $10^{-9}$–$10^{-12}$ s, **the hypothesis of LTE is generally justified**, unless the causes of the non-equilibrium are maintained (for example, optical pumping in a laser). An alternative, equivalent, approach to the realization of LTE in a small macroscopic element of volume $dV$ is that the mean free paths of the carriers of thermal energy shall be small in comparison to the dimensions of the element.

We will adopt the hypothesis of LTE for the remainder of this work. The physical system considered is then the site of the following irreversible macroscopic processes with which flux is associated:

- Relative to an element of matter, the cumulative effect on the macroscopic scale of transport of the various physical quantities (electric charge, number of molecules of a given type, energy) by particles (molecules, electrons, phonons, etc.) translates to macroscopic **diffusion** fluxes: electrical conduction, diffusion of one type of species into another, **thermal conduction**, etc.
- Similarly, associated with every macroscopic mass transfer caused by the global motion of part of the material system, there are associated macroscopic fluxes of electric charge, energy, etc. These are the convective phenomena: electrical convection, **thermal convection**, etc.
- Interactions between the molecules of the material system and the photons of the radiation field lead, in cases where the material system and the radiation field are not in equilibrium, to macroscopic energy fluxes in the form of **radiation**.
- Numerous other transport phenomena, more or less complex, also cause energy fluxes; we may mention the case of thermoelectric effects [2,7].

### 1.1.6 Objectives and conventions

The objective of this work is to determine, for any physical system in LTE, the evolutions of the temperature $T(\mathbf{r}, t)$ and the energy flux (for all forms of energy) which is necessary in order to control the process.

A process will be termed **unsteady** if, by comparison with a given frame of reference, the physical quantities, $A$, considered (scalars, vectors or tensors) depend explicitly on time:

$$\frac{\partial A(\mathbf{r}, t)}{\partial t} \neq 0 \tag{1.1}$$

A process will be termed **steady** in the opposite case:

$$\frac{\partial A}{\partial t} = 0 \tag{1.2}$$

**In Part I of this book** (Chapters 1–4), **we will confine ourselves to steady processes.**

We will adopt the following conventions. The power $d\Phi$ (expressed in watts) algebraically crossing an elementary surface $dS$, whose normal is $\mathbf{n}$ (Figure 1.1), and called the energy flux through $dS$, is given as a function of the energy flux density vector $\mathbf{q}$:

$$d\Phi = \mathbf{q} \cdot \mathbf{n} \, dS \tag{1.3}$$

The quantity

$$\varphi = \mathbf{q} \cdot \mathbf{n} \tag{1.4}$$

is called the (algebraic) flux density through the surface $dS$ (expressed in W m$^{-2}$).

We shall define the quantities $d\Phi$, $\mathbf{q}$ and $\varphi$ for each of the modes of transfer listed above (conduction, convection and radiation).

*Note: convention for open surfaces*

**In most cases it is possible to use a frame of reference with a system of axes normal to the surfaces under consideration. By convention, the normals to the surfaces are in the positive sense of the axes.**

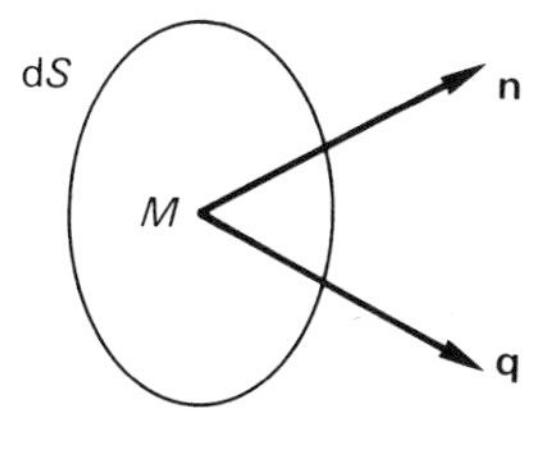


**Figure 1.1**

## *1.2 Introduction to radiative flux*

Chapter 4 in this introductory part is devoted to the study of thermal radiation in simple cases. In this section we will limit ourselves to introducing the idea of radiative flux. Energy is permanently exchanged between a material system and a radiation field by the following processes:

- The spontaneous **emission** of radiation is a conversion of thermal energy (energy of vibration or rotation, electronic energy, energy of phonons, etc.) to a radiative energy (of photons).
- **The absorption** of radiation is an inverse conversion, of radiative energy to thermal energy.

*Note*

(1) We will not consider here the opposite phenomenon to absorption, called induced emission or negative absorption, which plays an essential rôle in lasers [8], which are not in LTE. This phenomenon is included (negatively) within absorption.

(2) We will not take into account either complex multi-photon phenomena or diffusion with change of frequency.

(3) Other processes (**scattering without frequency change**, **reflection**) correspond to changes in direction of an incident ray, but do not affect the energy of the material system.

We distinguish three types of medium, so far as radiation is concerned: the two important limiting cases of **transparent** bodies and **opaque** bodies, and the general case of the **semi-transparent** medium.

- A **transparent** medium does not interact with the radiation field. It does not emit, absorb, reflect or diffuse; all incident radiation is transmitted, whatever its direction and frequency (or wavelength).
- An **opaque body** does not transmit any portion of the incident radiation (I), which may thus be absorbed (A) or reflected (R) (Figure 1.2). More precisely, we define an opaque body as one where the depth of penetration of radiation $\eta$ is small in

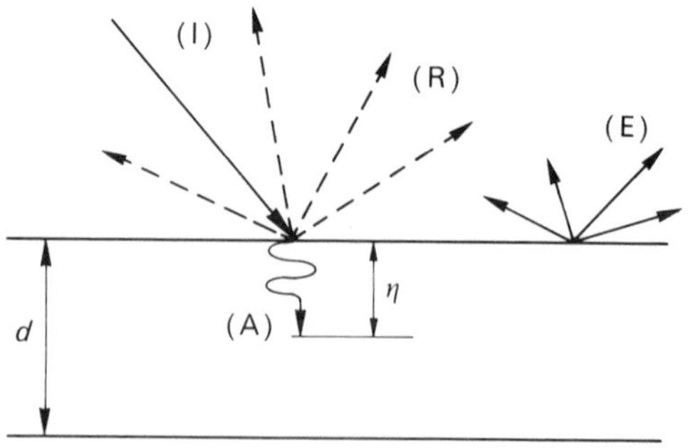


**Figure 1.2**

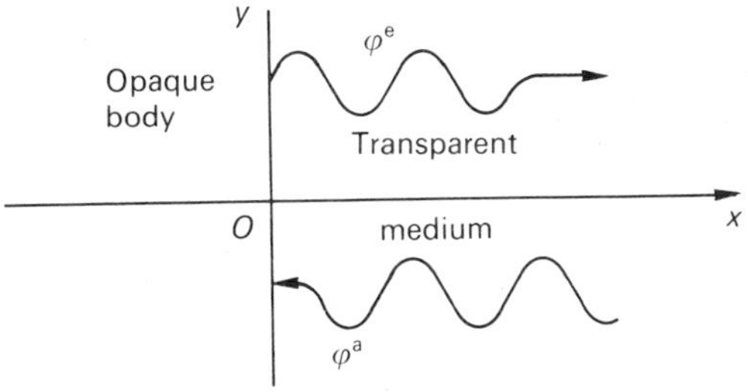


**Figure 1.3**

comparison to a typical dimension $d$ of the system. In so far as an opaque body absorbs radiation (A), it is likely to emit it (E) at every point on its boundary.

Consider an opaque body whose boundary surface is normal at $O$ to the axis $Ox$ (Figure 1.3). The **radiative energy flux density** $\varphi^R$ exchanged between the body and the radiation field is written (taking flux in the $Ox$ direction as positive) as:

$$\varphi^R = \varphi^e - \varphi^a \tag{1.5}$$

where $\varphi^e$ and $\varphi^a$ are the magnitudes of the **flux density emitted and absorbed**, respectively, by the body at the point $O$. In the case of a solid body, for example, the first term in equation (1.5), $\varphi^e$, represents the loss of thermal energy (phonons, etc.) by emission; the second, $\varphi^a$, the gain of energy at the boundary of the body by absorption. The quantities $\varphi^e$ and $\varphi^a$ should obviously be integrated over the whole spectrum of wavelengths (frequencies) and over all directions in the hemisphere external to the opaque body.

*Note*. In thermal equilibrium (see Section 1.1.2) the radiative flux density $\varphi^R$ is zero. When not in equilibrium, $\varphi^R$ represents an energy flux which propagates in the form of radiation in a transparent medium and diffuses by conduction in an opaque solid.

- A **semi-transparent medium** reflects, absorbs or diffuses the incident radiation, or transmits it for a finite distance. It also emits radiation. Whereas in the case of an opaque body the interactions between the radiation and the material may, as a first approximation, be treated as a surface phenomenon, they must certainly be treated as a volume one in the case of a semi-transparent medium. We will mainly consider transfers between opaque bodies through transparent media in this work. The case of semi-transparent media is, however, considered in Chapter 11.

## *1.3 Conduction*

The transfer of energy by conduction occurs **through an element of material** across which there exists a temperature gradient. It represents the global effect of energy transport by elementary carriers (molecules, phonons, electrons, etc.).

In the case of a **fluid** the elementary carriers (molecules, atoms, ions, etc.) are characterized by energies of translation, possibly vibration and rotation, electronic energy, etc. Enskog's formulation [4–6,9], derived from Boltzmann's equation, allows this phenomenon to be modelled statistically, taking account of the intermolecular collisions depending on the interaction potentials. The thermal conductivity of a medium may be calculated to a certain degree of accuracy by this method, even for mixtures of species.

In the case of a **solid** the atoms are arranged in a more or less perfect crystal lattice. The elementary energy vectors are phonons (quanta of lattice vibration) and perhaps free electrons (both electrical and thermal conduction). The modelling of electrical and thermal conduction uses the methods of solid state physics, again having recourse to Boltzmann's equation [10,11].

### *1.3.1 Conductive flux*

For isotropic homogeneous materials (fluids or solids), the flux density vector $\mathbf{q}^{cd}$ is proportional to the local temperature gradient $\nabla T$:

$$\mathbf{q}^{cd} = -\lambda(T)\nabla T \tag{1.6}$$

The coefficient of proportionality $\lambda(T)$ is called the **thermal conductivity** and is, in general, strongly dependent on temperature. The negative sign in equation (1.6) is required by the second law of thermodynamics. Equation (1.6) is generally called **Fourier's law**. It is an approximation to the first-order response of a system, and is analogous to many other physical laws concerned with similar transport phenomena, producing fluxes of charge, concentration, etc.:

- Ohm's law, in its vector form:

$$\mathbf{j} = \sigma\mathbf{E} = -\sigma\nabla V_{el} \tag{1.7}$$

  where $\mathbf{j}$, $\mathbf{E}$, $\sigma$ and $V_{el}$ represent the current density and electric field vectors, the conductivity and the electrical potential respectively.
- Fick's law, for the diffusion of a very dilute substance through another:

$$\mathbf{d}_\alpha = -D_\alpha\nabla(n_\alpha/n) \tag{1.8}$$

  where $\mathbf{d}_\alpha$ is the diffusion rate of the substance $\alpha$, $D_\alpha$ the diffusivity and $n_\alpha/n$ the molar fraction.

*Notes*

(1) Many bodies cannot be considered homogeneous and isotropic. Composites, fibrous insulators, rubbers and flaky materials have different conductivities along, and perpendicular to, the plane of cleavage. Fourier's law may then be generalized by treating $\lambda$ as a conductivity tensor. An example of a material of this type is shown in Figure 1.4. We see that the conductivity depends in a very critical manner on the direction.

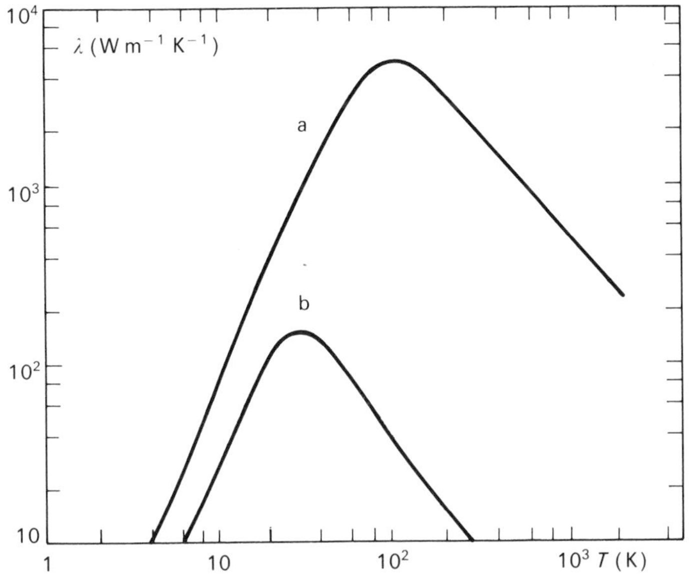


**Figure 1.4** Conductivity of pyrolytic graphite [12]: (a) parallel to the cleavage plane; (b) perpendicular to the cleavage plane.

(2) Fourier's law does not explicitly include time. It assumes an instantaneous response throughout the medium to a temperature change occurring at a point M. This assumption is valid so long as the time scales considered are long in comparison to those typical of transfer by collisions between the elementary carriers (the relaxation time). In practice, Fourier's law is valid for virtually all applications. A more general discussion appears in Part II of this work (Appendix 1).

### *1.3.2 Orders of magnitude of thermal conductivities*

We will use SI units exclusively. However, English-speaking authors, despite recent efforts, still generally manage with non-rational units.

Conversion factors are given in Part V, 'Basic Data'. The range of thermal conductivities is much less than for electrical conductivities (a ratio of $1\text{–}5 \times 10^4$, against $1\text{–}10^{25}$). The thermal conductivities of various materials are shown in Figure 1.5, as functions of temperature. The distinction between thermal conductors and insulators is somewhat arbitrary; there is a certain correspondence, as follows, between electrical conductors and insulators:

- Among the good conductors, we may cite metals in general, and copper ($\lambda(300\ \text{K}) = 400\ \text{W m}^{-1}\ \text{K}^{-1}$) and aluminium ($\lambda(300\ \text{K}) = 210\ \text{W m}^{-1}\ \text{K}^{-1}$) (and their alloys) in particular.

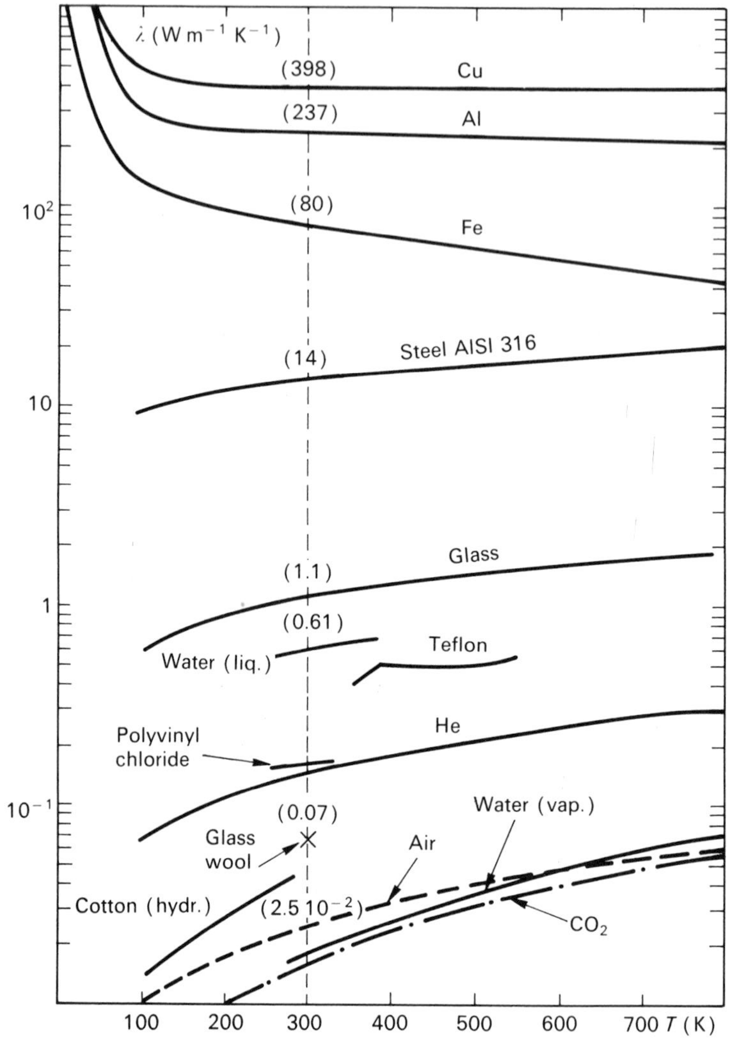


**Figure 1.5** Variations of conductivity of some materials with temperature.

- Steels are mediocre conductors of heat; for a typical stainless steel the conductivity does not rise above 14 $W\,m^{-1}\,K^{-1}$ at 300 K. Glasses have conductivities around 1 $W\,m^{-1}\,K^{-1}$.
- Gases have low conductivities (of the order of a few times $10^{-2}\,W\,m^{-1}\,K^{-1}$). They increase very roughly as absolute temperature, $T^{0.8}$, and are, to a first approximation, independent of pressure. The conductivity of a mixture of gases is a complex function of the conductivities of its constituents [4,8]; account must be taken of the different types of interaction at the molecular scale. Conductivity

tables exist for certain common mixtures [11]. At first sight, gases make excellent insulators. But a fundamental restriction on their use as thermal insulators quickly appears: the phenomenon of natural convection discussed in the next section. However, one can make insulators with dispersed materials (fibrous or porous), filled with air. The interstices are then so small that natural convection is unable to develop, due to viscous effects. Glass wool is an example.

## 1.4 Convective and conducto-convective fluxes at a wall

### 1.4.1 The phenomenon of convection

Thermal convection is a **transfer of energy relative to a given system of axes**, consequent on a macroscopic mass transfer at that system.

Let us consider a fluid flow, or, more precisely, an elementary stream tube, that is, a tube to which the streamlines are tangential (Figure 1.6). This stream tube intersects some plane (for example, $Oyz$), defining an elementary region $d\Sigma$ around a point M, whose normal is $\mathbf{n}$. Leaving $d\Sigma$ there is a mass flow rate $d\dot{m}$, given by:

$$d\dot{m} = \rho(\mathbf{r})\mathbf{v}\cdot\mathbf{n}\,d\Sigma \tag{1.9}$$

where $\rho(\mathbf{r})$ denotes the density and $\mathbf{v}$ the local velocity of the fluid relative to the element $d\Sigma$ of $Oyz$. Associated with this macroscopic mass transfer is an enthalpy transfer, specified by a convective flux $d\Phi^{cv}$ at $M$, relative to $Oyz$:

$$d\Phi^{cv} = \rho(\mathbf{r})\mathbf{v}\cdot\mathbf{n}h(T(\mathbf{r}))\,d\Sigma = d\dot{m}h \tag{1.10}$$

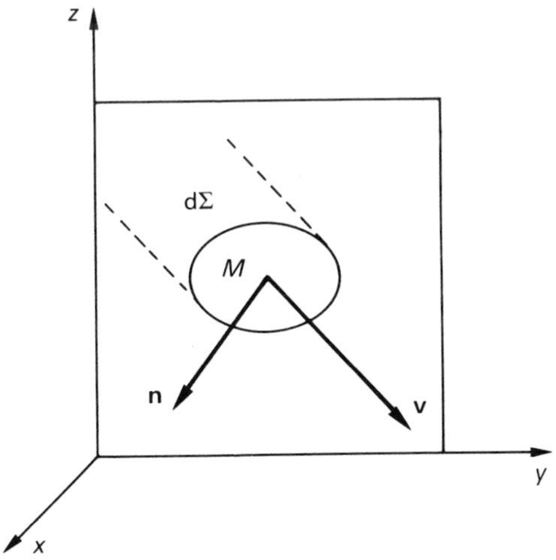


**Figure 1.6**

and a convective flux density vector:

$$\mathbf{q}^{cv} = \rho \mathbf{v} \mathrm{h} \qquad \text{such that} \qquad \varphi^{cv} = \rho \mathbf{v} \cdot \mathbf{n} \mathrm{h} \tag{1.11}$$

where h is the local specific enthalpy of the fluid at $M$ relative to an arbitrary temperature datum. Traditionally, three types of fluid convection are distinguished:

- **Forced convection** occurs when the fluid motion is imposed from outside the system. This is the case with industrial heat exchangers or vehicle 'radiators' (which are in fact principally convectors). Two fluids in motion exchange energy through a wall whose temperature is different from theirs. Since the flow velocities can reach very high values, heat transfer by forced convection is often extremely effective (see Section 1.4.2).
- **Natural (or free) convection** occurs spontaneously, in certain conditions, in a fluid in which a temperature gradient, generally imposed externally, exists. Let us take as an example a room whose floor is hotter than its ceiling (underfloor heating). The warm air near the floor, which is less dense than the cold air near the ceiling, will rise due to buoyancy forces, while the cold air will fall (Figure 1.7).

  The process will be slowed down by viscous friction forces, and the temperature difference, which causes the phenomenon, will be reduced by thermal conduction. Under the opposing forces (buoyancy and friction) the fluid can reach a limiting velocity in some conditions, and there is steady flow throughout. This flow transfers energy from floor to ceiling. This mode of transfer is clearly much more important than the purely conductive transfer through still air (since the thermal conductivity is very low). We note that conduction is a dissipative phenomenon which opposes natural convection. Natural convection is usually a lot less effective than forced convection (which would occur in our example if we used a fan in the room). One is, however, often obliged to rely solely on natural convection in many applications, for technical or economic reasons such as cost or reliability. One then uses fins, which increase the area of the heat transfer surface between the fluid and the system. Examples are cooling of compressors, large electrical power transformers, electronic components (see Section 2.3).
- **Mixed convection** is a combination of the two preceding phenomena, natural and forced convection. It occurs when the (imaginary) flow speeds due to the two types of convection, considered separately, are of the same order of magnitude.

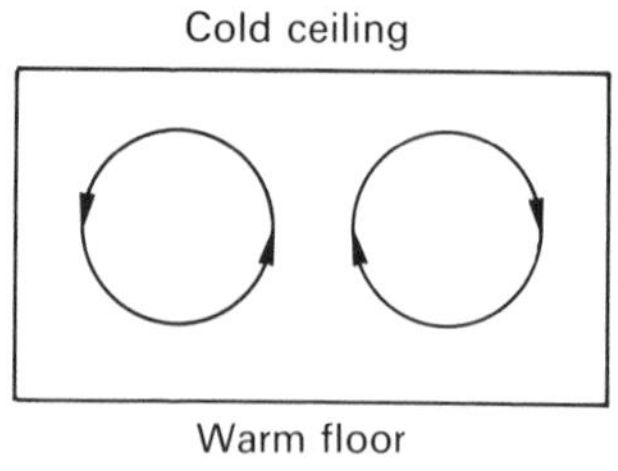


**Figure 1.7**

Whatever the type of convection, the fluid flow can be divided into two régimes, according to velocity: laminar and turbulent. We will not study these régimes in detail in this introductory section; we will limit ourselves to a short qualitative description.

Let us consider a flow in a transparent tube of constant cross-section, and suppose that we can use tracers (particles in suspension or ink streaks) to follow the movement of fluid particles.

- At low speed, the flow is ordered, parallel to the axis of the tube, and without mixing in the fluid. A thin thread of ink does not spread out appreciably. The flow is **laminar**. At every point, the speed and temperature of the fluid can be defined precisely.
- As the speed rises, the type of flow changes completely beyond a critical speed $v_c$ (which depends on the fluid properties, the tube diameter and the roughness of the tube walls). The ink threads are moved around by the random turbulent motions. The flow is three-dimensional, with a mixing which enhances heat transfer just as it speeds the distribution of colour throughout the fluid. The régime is **turbulent**. It is not possible to predict the velocity or the temperature of a fluid element at a given moment. Generally, only time mean values of temperature and velocity are known.

In a convection problem, it is necessary to determine the velocity and temperature fields simultaneously in order to calculate the flux transfers in the system. Such an approach will be used in the Part IV of this book. We will now, as a first stage, analyse a large number of applications by making use of the very simple model of the convective heat transfer coefficient. This will be introduced in the next section by means of an important example.

### 1.4.2 Combined conduction and convection at a wall

Let us consider a solid wall in contact with a flowing fluid, and assume for the time being, for simplicity, that radiative transfer is negligible. The conductive flux density at the wall, but inside the solid $S$, is a function of the conductivity of the solid $\lambda_S$ and the temperature field $T_S$. If the wall is normal to $Oy$,

$$\varphi^{cd}|_{wS} = -\lambda_S \left.\frac{\partial T_S}{\partial y}\right|_w \tag{1.12.1}$$

The heat transfer in the fluid at the wall is purely conductive, since the velocity of the fluid must be zero there. The expression for the flux density on the fluid side is

$$\varphi^{cd}|_{wf} = -\lambda_f \left.\frac{\partial T_f}{\partial y}\right|_w \tag{1.12.2}$$

where $\lambda_f$ and $T_f$ represent the conductivity and the temperature field of the fluid. It follows, therefore (in the absence of radiation), that

$$-\lambda_S \frac{\partial T_S}{\partial y}\bigg|_{wS} = -\lambda_f \frac{\partial T_f}{\partial y}\bigg|_{wf} \tag{1.12.3}$$

Furthermore, the **temperature is continuous at the wall**. In the expression for the conductive flux on the fluid side, the awkward problem is to find the wall temperature gradient $\partial T_f/\partial y|_w$, which depends on convection (and is sometimes, incorrectly, called the convective flux at the wall). We will call it the **conducto-convective flux** at the wall, denoted by $\varphi^{cc}$, since it results from a combination of conduction perpendicular and convection parallel to the flow direction. To calculate it exactly it is necessary to solve the coupled thermal and flow problems. Actually, we will adopt a simplified model, which we will introduce by means of an example. Let us consider a fluid flowing steadily between two plates (of infinite extent in the $x$ and $y$ directions) maintained at temperatures $T_1$ and $T_2$ respectively (Figure 1.8).

Consider a position far from the entry to the plates, so that the fluid properties no longer depend on its entry temperature. The flow is assumed to be very turbulent in the region of interest (see Section 1.4.1), and the mean velocity is everywhere parallel to $Ox$. In the central part of the flow, the mixing is vigorous, favouring heat transfer. The temperature profile, averaged over a long time period, compared with the turbulent fluctuations, is then uniform at a value $T_m$ in the central region between the plates. On the other hand, in the immediate vicinity of the walls, the fluid velocity is low (zero at the wall) because of the effects of viscous friction. In a boundary sub-layer of thickness $\zeta$ the heat transfer in the $y$ direction is more or less conductive (and linear on the assumption that the fluid conductivity $\lambda_f$ is constant). The mean temperature profile in the $y$ direction is sketched for this limiting case in Figure 1.9.

With the temperature profile of Figure 1.9, the conducto-convective flux density at the wall $y = 0$ is given by:

$$\begin{aligned}\varphi^{cc}|_{y=0} &= -\lambda_f \frac{\partial T_f}{\partial y}\bigg|_{y=0} \\ &= -\lambda_f\left(\frac{T_m - T_1}{\zeta}\right) \\ &= h(T_1 - T_m)\end{aligned} \tag{1.13}$$

$h$ is called the **convective heat transfer coefficient at the wall**. It does not generally depend on the nature of the wall, but solely on the fluid properties and the nature of the flow. This quantity can in many cases be assumed constant, so long as the fluid temperature does not vary too much. In a closed geometry (pipes), this approximation is justified far from the pipe entry. In an open geometry (flat plate), an average value over the length of the system is used for the coefficient $h$. (A systematic study of this quantity is undertaken in Part IV of this work.)

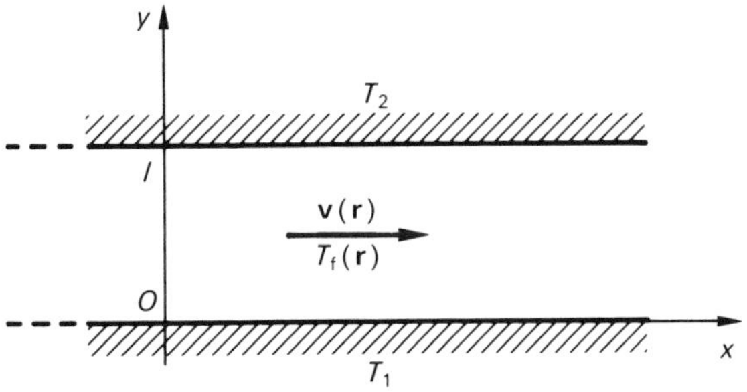


**Figure 1.8**

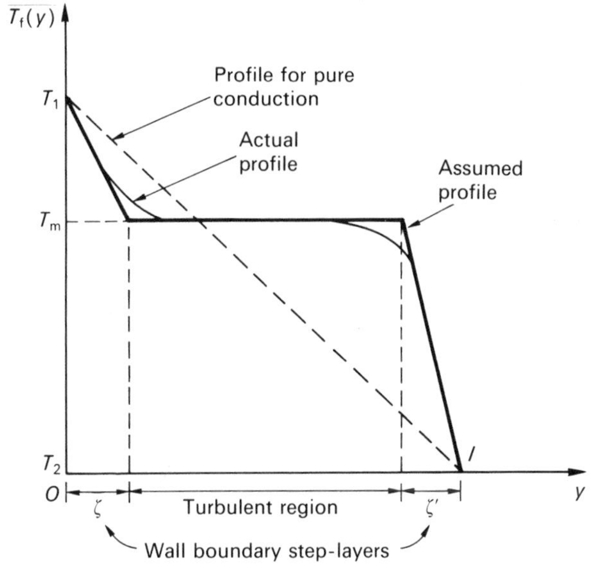


**Figure 1.9**

Equation (1.13) may be generalized to calculate the local conducto-convective flux density, $\varphi^{cc}$ at a wall in the form

$$|\varphi^{cc}| = h|T_w - T_c| \tag{1.14}$$

where $h$ is the convective heat transfer coefficient, $T_w$ the local wall temperature and $T_c$ a characteristic local temperature of the fluid. This could be the temperature of a fluid at rest with natural convection or the temperature averaged over the cross-section of a pipe (as defined in Section 3.2), known as the mixing temperature. Equation (1.14) is written in terms of absolute values; the sign depends on the choice of axes. In this introductory treatment, we will assume that $h$ and $T_c$ are known.

**Table 1.1** Order of magnitude of convective heat transfer coefficients

| *Type* | *Fluid* | $h\ (W\,m^{-2}\,K^{-1})$ |
|---|---|---|
| Natural convection | gas | 5–30 |
| | water | 100–1000 |
| Forced convection | gas | 10–300 |
| | water | 300–12 000 |
| | oil | 50–1700 |
| | liquid metal | 6000–110 000 |
| Phase change | water (boiling) | 3000–60 000 |
| | water (condensing) | 5000–110 000 |

The effectiveness of convective transfer is obvious if we consider Figure 1.9. The turbulent mixing of the fluid makes the temperature uniform in the central region and increases the temperature gradients in the vicinity of the two walls, in the boundary sub-layer of thickness $\zeta$. The mixing zone is more important, and the boundary layers thinner, when the mean velocity of the fluid is higher.

The flux density at the wall thus rises. From (1.13):

$$\varphi_{\mathrm{w}}^{\mathrm{cc}}\big|_{y=0} = -\lambda_{\mathrm{f}}\frac{T_{\mathrm{m}} - T_1}{\zeta} \tag{1.15}$$

If we suppose the fluid to be completely still, with the same conditions at the walls, the transfer is purely conductive and the flux density at the wall $y = 0$ may be written:

$$\varphi_{\mathrm{w}}^{\mathrm{cd}}\big|_{y=0} = -\lambda_{\mathrm{f}}\frac{T_2 - T_1}{l} \tag{1.16}$$

where $l$ is the distance between the walls.

If $(T_2 - T_1)$ and $(T_{\mathrm{m}} - T_1)$ are the same order of magnitude, the ratio of the conductive flux density without and with convection is of the order $\zeta/l$. This may be very small, thus demonstrating the effectiveness of convective heat transfer.

In Table 1.1 we have, as a guide, grouped together typical values for the convective heat transfer coefficient $h$ according to the type of convection, the type of fluid and the flow régime. Some two-phase cases are also included.

## 1.5 Standard boundary conditions

**Temperature is generally a continuous quantity**. The same applies to energy flux density (taking account of all forms of energy). In this section we will tackle a number of examples (not exhaustive) relating to boundary conditions for thermal flux.

### 1.5.1 Example 1

Consider an opaque solid of conductivity $\lambda_S$, in the region $x < 0$, (bounded by the plane $x = 0$), in contact with a transparent fluid at an initial temperature $T_0$ in which a natural convection develops. The convective heat transfer coefficient is $h$ at the wall, and the solid exchanges a radiative flux $\varphi^R$ with its surroundings, through the fluid (Figure 1.10).

The temperature distribution is given by $T_S$ in the solid. Continuity of energy flux in the neighbourhood of $x = 0$, with the sign convention of positive flux in the positive $x$ direction, leads to

$$\varphi^{cd} = \varphi^{cc} + \varphi^R$$

$$-\lambda_S \left.\frac{\partial T_S}{\partial x}\right|_{x=0} = h[T_S(0) - T_0] + \varphi^R \tag{1.17}$$

and

$$\varphi^R = \varphi^e - \varphi^a \tag{1.18}$$

where $\varphi^a$ and $\varphi^e$ represent the **magnitudes** of the emitted and absorbed radiation flux densities, over all wavelengths, in the neighbourhood of $x = 0$.

*Note*
Within the fluid, where there is convective transfer the expression $h(T_S(0) - T_0)$ is a phenomenological expression for the conductive flux at the wall. If the region $x > 0$ was a transparent solid conductor of conductivity $\lambda'$ and temperature $T'_S$, the continuity condition would become:

$$-\lambda_S \left.\frac{\partial T_S}{\partial x}\right|_{x=0} = -\lambda'_S \left.\frac{\partial T'_S}{\partial x}\right|_{x=0} + \varphi^R \tag{1.19}$$

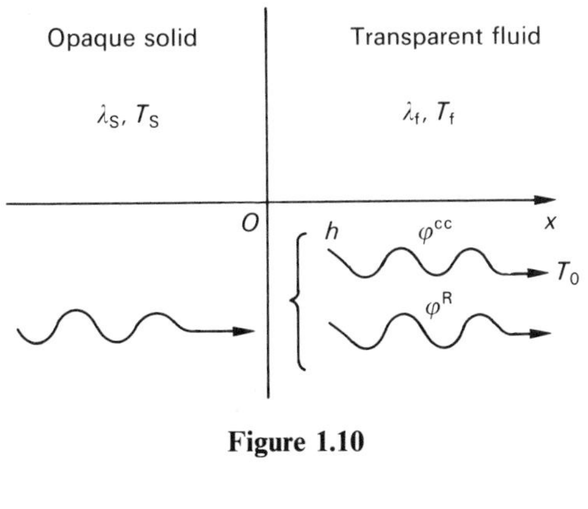


**Figure 1.10**

### 1.5.2 Example 2

We will consider the same problem as in Section 1.5.1, but replace the opaque solid in $x < 0$ with a solid which is semi-transparent to radiation (Figure 1.11).

In these circumstances the exchanges between the solid and the radiation can no longer be considered to be taking place (to a first approximation) at the suface, but, rather, they take place throughout the volume. In particular, the absorption of radiation at the fluid/solid interface is zero. There is continuity of radiative flux at the wall $x = 0$:

$$\varphi_S^R(0^-) = \varphi_f^R(0^+) \tag{1.20}$$

The radiation flux densities in the fluid $\varphi_f^R(0^+)$ and the solid $\varphi_S^R(0^-)$ can be expressed in terms of the local properties of the radiation field. Furthermore, the continuity of conductive flux density at $x = 0$ can be expressed for the material as:

$$-\lambda_S \left.\frac{\partial T_S}{\partial x}\right|_{x=0} = h(T_S(0) - T_0) \tag{1.21}$$

The problem to be solved is then generally much more complex than the preceding one, since the temperature function in the semi-transparent medium, $T_S(x)$, is very tightly coupled to the quantity characterizing the radiation field (the intensity); see Chapter 11.

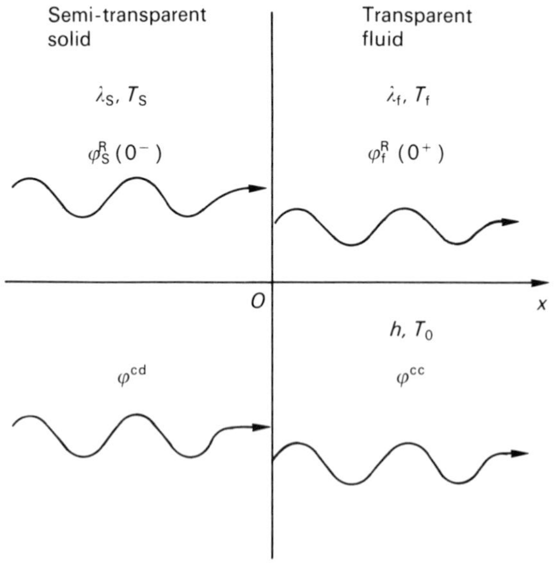


**Figure 1.11**

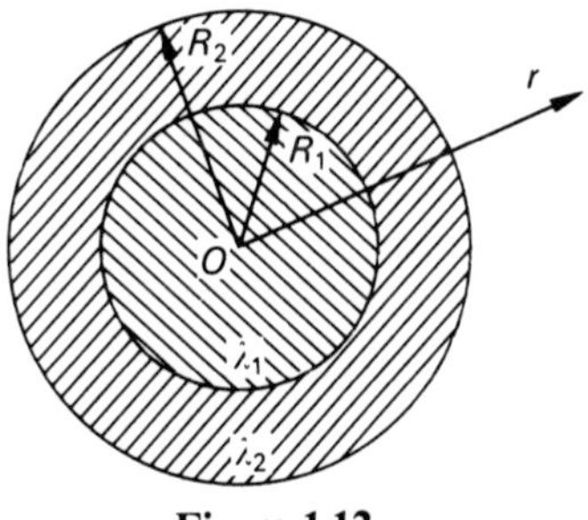


**Figure 1.12**

### 1.5.3 Example 3. Thermal contact

Let us consider a cylindrical element of conductivity $\lambda_1$ enclosed in a sleeve of conductivity $\lambda_2$ (Figure 1.12). Whatever the precision of manufacture of the faces of the cylinder and the sleeve at $r = R_1$, the contact between these two components will never be perfect because of surface roughness, different coefficients of expansion, etc. In practice, the contact will be discontinuous and air will be trapped in the interstices. As the dimensions of these are very small (of the order of $\mu$m), convection cannot develop and **transfer will be by both conduction and radiation simultaneously**. Since air is a very poor conductor, the temperature drop between the two components will not be negligible:

$$T_1(R_1) \neq T_2(R_1) \tag{1.22}$$

The energy flux density is clearly continuous since the corresponding surfaces in medias 1 and 2 are very close together. We need to use a very precise model for thermal contact since the geometry of the air film is difficult to specify. In steady conditions, it is possible to introduce the concept of thermal contact resistance (see Section 2.2). In unsteady conditions, the problem is much more recalcitrant.

### 1.5.4 Example 4. Moving boundary between phases

For this relatively complex case, refer to the examples presented in Sections 3.2.3 and 3.3.1.2.

## 1.6 Energy balance for steady-state transfers without flow

### 1.6.1 General balance equations

We will consider rigid, fixed, continuous material systems. The control volume $V$ is bounded by a surface $S$. In the steady state, the total internal energy of the system $\mathscr{E}$ is invariant.

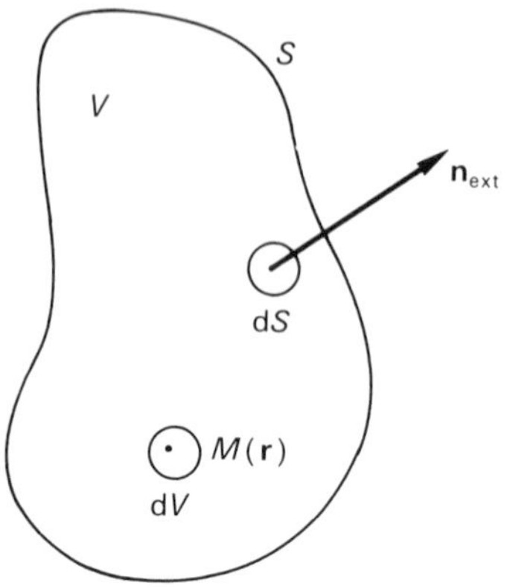


**Figure 1.13**

In the absence of motion, and thus work, the energy balance for unit time is:

Net flux crossing the surface $S$ + power generated within $V = 0$ (1.23)

In integral form, the energy equation becomes:

$$\frac{\partial \mathscr{E}}{\partial t} = \int_S -\mathbf{q} \cdot \mathbf{n}_{ext} \, dS + \int_V P \, dV = 0 \qquad (1.24)$$

See Figure 1.13, where $\mathbf{n}_{ext}$ is the outwards normal of an element of surface d$S$, $\mathbf{q}$ represents the energy flux density, with both conductive and radiative contributions. In the absence of motion, convection does not contribute to the balance. The negative sign on the first term allows for the thermodynamic convention for the direction of the normal, which is opposite to the mathematical convention. $P$, the power per unit volume released at a point $M$, is considered positive when it corresponds to the heating of the system (e.g. Joule effect), which is generally the case.

The energy equation (1.24) may be written either for a finite or an elementary system. In the latter case, it often takes the mathematical form of a differential or partial differential equation in temperature. The constants of integration in this equation are evaluated by considering the **thermal boundary conditions**, which are generally similar to those considered in Section 1.5.

### 1.6.2 *One-dimensional example*

Consider a uniform plate of thickness $e$ and constant conductivity $\lambda$, normal to the $x$ axis and infinite in the $y$ and $z$ directions (Figure 1.14).

The surface $x = +e$ is fixed at temperature $T_0$. At $x = 0$, the plate is subjected to a conducto-convective flux in a fluid with a characteristic temperature $T_f$, with a convective heat transfer coefficient $h$. A power $P(x)$ per unit volume is dissipated within the plate. And, finally, there is a radiative flux in the positive $x$ direction,

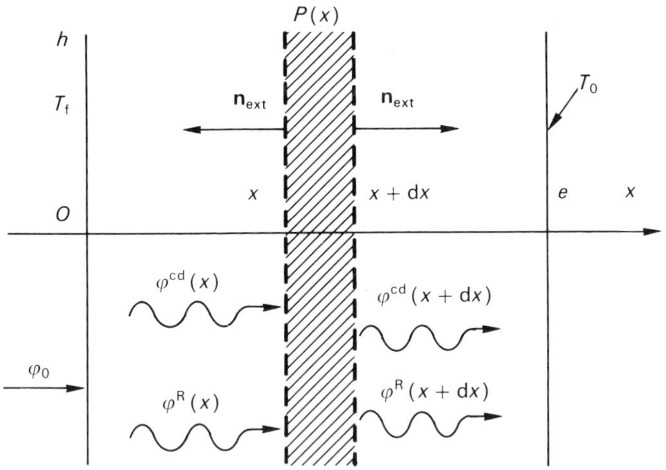


**Figure 1.14**

which is attenuated according to the law (supposed known) which describes the case where **emission from the plate is negligible**:

$$\varphi^R(x) = \varphi_0 \, e^{-\kappa x}, \qquad \text{where } \kappa e \gg 1 \tag{1.25}$$

Clearly, this problem is unaffected by movement in the $y$ or $z$ directions, and can be treated as one-dimensional, in the $x$ direction.

The fundamental point is to define the control volume (here an elementary slice of the plate between $x$ and $x + \delta x$ of **unit area** in the $yz$ plane), and apply an energy balance to it. This material system is subject to two types of flux, radiative and conductive, at its boundaries. The corresponding flux densities (positive in the positive $x$ direction) are given by equation (1.25) and by Fourier's law:

$$\varphi^{cd}(x) = -\lambda \frac{dT}{dx} \tag{1.26}$$

The energy flux across closed surface $S$ in the system, in the outwards direction, then becomes:

$$\int_S -\mathbf{q}\cdot\mathbf{n}_{ext}\, dS = \{\varphi^R(x) + \varphi^{cd}(x)\} - \{\varphi^R(x + dx) + \varphi^{cd}(x + dx)\} \tag{1.27}$$

$$= -\left\{\frac{d\varphi^R}{dx} + \frac{d\varphi^{cd}}{dx}\right\} dx \tag{1.28}$$

The power generated in the control volume is $P(x)\,dx$. It then follows from equations (1.24)–(1.28) that:

$$\lambda \frac{d^2 T}{dx^2} + \kappa \varphi_0 \, e^{-\kappa x} + P(x) = 0 \tag{1.29}$$

This is the control volume energy equation, whose solution depends on two constants, determined from the boundary conditions:

$$T(e) = T_0 \tag{1.30}$$

$$-\lambda \frac{dT}{dx}(0) = h[T_f - T(0)] \tag{1.31}$$

Equation (1.31) arises from the fact that the medium is considered to be semi-transparent and that there is (separate) continuity of thermal flux density (1.31) and radiative flux density $\varphi_0$ at $x = 0$ (see Section 1.5.2).

*Notes*

1. In equation (1.29), the term $\kappa\varphi_0 e^{-\kappa x}$ is apparently a power density generated in the material (just like $P(x)$). The explanation is obvious – this term represents the fraction of radiative energy which is converted to thermal energy in the control volume.
2. If for any $x$, the term in $e^{-\kappa x}$ is negligible on the macroscopic scale in the neighbourhood of $x = 0$, the radiative term disappears from equation (1.29), which becomes:

$$\lambda \frac{d^2T}{dx^2} + P(x) = 0 \tag{1.32}$$

with boundary conditions:

$$T(e) = T_0 \tag{1.33}$$

$$-\lambda \frac{dT}{dx}(0) = h(T_f - T(0)) + \varphi_0 \tag{1.34}$$

Equation (1.34) demonstrates that the radiative flux $\varphi^R(0)$ is converted to energy of the material system at the surface; we recall that we have neglected emitted flux in this example. Under these conditions, the plate is an **opaque body**, which corresponds to the limiting case where $\kappa e$ is very large compared to 1 (see Section 1.2).

# 2

# Linear Conduction

The objective of this chapter is to apply the energy conservation equation to the simple case of a stationary medium in which conduction takes place. Conducto-convective or radiative transfers which appear in the boundary conditions will be treated as linearizeable (this will be discussed in Section 4.6). Since the conductivities and convective heat transfer coefficients are, in this first instance, assumed to be independent of temperature, we speak of linear conduction.

A model based on an electrical analogy is developed in Section 2.1, the classical treatment of cooling fins is the subject of Section 2.2, while Section 2.3 discusses the general structure of steady conduction problems. A synthesis of these different models and an initial solution method are proposed in Section 2.4, through their application to the simplified steady energy balance of an apartment.

## *2.1 The electrical analogy and its limitations*

### *2.1.1 Principles*

We have seen in Chapter 1 that Fourier's law for conductive flux

$$\mathbf{q}^{cd} = -\lambda(T)\nabla T \tag{2.1}$$

is analogous to Fick's law and, in particular, to Ohm's law in its local form (equation (1.7)). We note the obvious correspondence between the electrical and thermal quantities:

| | | | | |
|---|---|---|---|---|
| Conductivity | $\sigma(T)$ | $\leftrightarrow$ | $\lambda(T)$ | |
| Potential | $V_{el}$ | $\leftrightarrow$ | $T$ | Temperature |
| Current density | $\mathbf{j}$ | $\leftrightarrow$ | $\mathbf{q}^{cd}$ | Conductive flux density |
| Current | $I$ | $\leftrightarrow$ | $\Phi^{cd}$ | Conductive flux |

It is evident that in both cases the isotherms (equipotentials) are normal to the flux lines and tubes (current lines and tubes).

The purpose of the electrical analogy, in the steady-state situation, is to make use of the simple and well-known techniques of linear steady electrokinetics: ideas of resistance and conductance, series and parallel connection, Thévenin and Norton's theorems, network theory, etc. However, the utility of these methods is much more limited than in electrokinetics for the following two basic reasons:

- Electrical and thermal conductivities depend in general on the temperature $T$. While this causes some minor inconvenience in electrical conduction, it makes problems non-linear in $T$ for thermal conduction. **The thermal conductivity will be assumed uniform, isotropic and independent of** $T$, at a value appropriate to the general temperature level for the application considered.
- Conduction often occurs in association with radiation, which is seldom linearizeable. When the radiation is not linear, the electrical analogy cannot be used (see Section 4.6 in this connection).

Subject to these two reservations, it is possible to define thermal resistance and conductance. Consider an elementary flux tube between two isothermal sections $dS_1$ and $dS_2$, at temperatures $T_1$ and $T_2$ respectively (Figure 2.1). In the steady state, the conductive flux $d\Phi$ through an isothermal section, in terms of the curvilinear coordinate $s$, is:

$$d\Phi = -\lambda \frac{\partial T}{\partial s} dS(s) \tag{2.2}$$

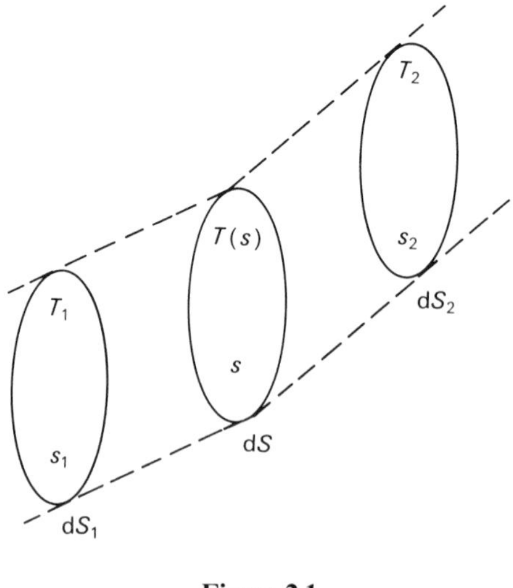


**Figure 2.1**

After integration from $s_1$ to $s_2$, and using the electrical convention of considering the voltage drop, we obtain the temperature drop:

$$T_1 - T_2 = \frac{1}{\lambda}\left[\int_{s_1}^{s_2} \frac{\mathrm{d}s}{dS(s)}\right] \mathrm{d}\Phi = \mathscr{R}\,\mathrm{d}\Phi \qquad (2.3)$$

We may note that $\mathscr{R}$, which is the thermal resistance of the tube between $s_1$ and $s_2$, is the inverse of a differential element. Equation (2.3) is analogous to the common form of Ohm's law:

$$V_1 - V_2 = RI \qquad (2.4)$$

$R$ being the electrical resistance. We can similarly introduce thermal conductance $\mathrm{d}\Gamma$:

$$\mathrm{d}\Gamma = \lambda\left[\int_{s_1}^{s_2} \frac{\mathrm{d}s}{\mathrm{d}S(s)}\right]^{-1} \qquad (2.5)$$

It is obvious that, starting from equation (2.3), and by considering tubes in series or parallel, we can obtain the thermal resistance of a finite tube between two arbitrary isothermal surfaces. The basic idea is, as in steady electrokinetics, that the **energy flux within a tube is conserved** (and the energy flux density clearly is not!).

### *2.1.2 Simple examples of applications*

These, non-exhaustive, examples are by way of illustration.

- *Resistance of an infinite cylindrical sleeve with isothermal faces.* The system is shown in cross-section in Figure 2.2. We consider a slice of thickness d$z$ in cylindrical coordinates. The coaxial cylinders of radius $r$ are isothermal at a temperature $T(r)$. With the usual notation we have, for any $r$:

$$\mathrm{d}\Phi = -\lambda \frac{\mathrm{d}T}{\mathrm{d}r} 2\pi r \mathrm{d}z \qquad \text{so that:} \qquad T_1 - T_2 = \frac{\mathrm{d}\Phi}{2\pi\lambda\,\mathrm{d}z} \ln\left[\frac{R_2}{R_1}\right] \qquad (2.6)$$

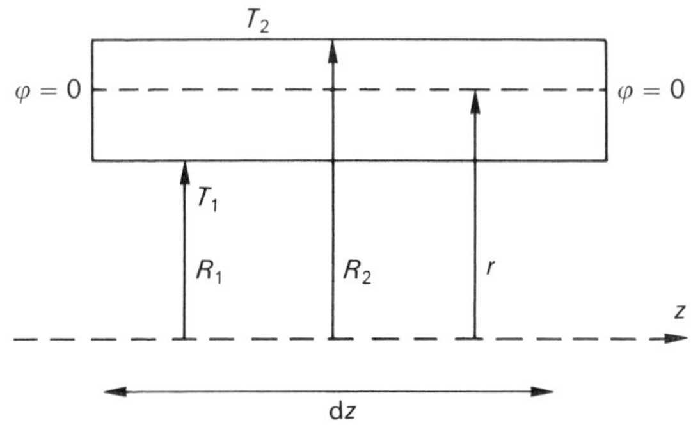


**Figure 2.2**

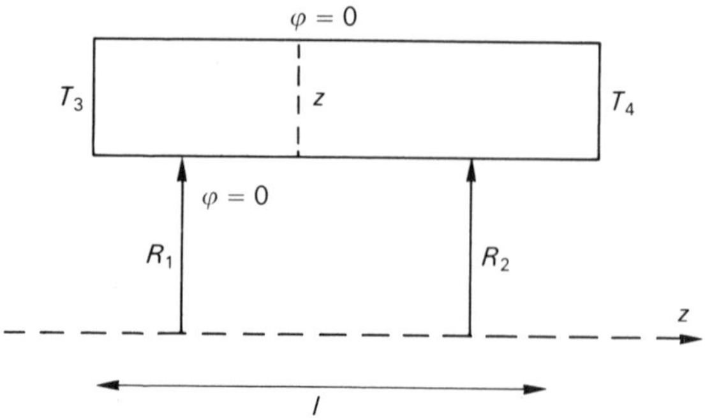


**Figure 2.3**

The required thermal resistance is thus:

$$\mathscr{R} = \frac{1}{2\pi\lambda dz} \ln\left[\frac{R_2}{R_1}\right] \tag{2.7}$$

- *Resistance of a cylindrical sleeve of finite length l whose cylindrical faces are insulated and whose cross-sections are isothermal (Figure 2.3).* In Cartesian coordinates, we obtain:

$$\mathscr{R} = \frac{l}{\pi[R_2^2 - R_1^2]\lambda} \tag{2.8}$$

- *Resistance of the volume contained between two concentric isothermal spherical surfaces and an insulated conical surface (Figure 2.4).* We obtain:

$$\mathscr{R} = \frac{1}{\lambda\Omega}\left[\frac{1}{R_1} - \frac{1}{R_2}\right] \tag{2.9}$$

where $\Omega$ is the solid angle of the cone.

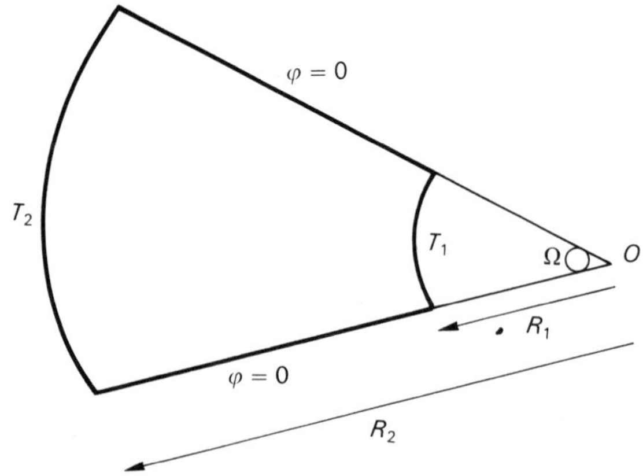


**Figure 2.4**

- *Thermal resistance for conducto-convective transfer*. The conducto-convective flux at a wall is linear; it is therefore an isothermal wall of temperature $T_w$, of area $\Sigma$, which receives a flux described by a constant heat transfer coefficient $h$ from a fluid of characteristic temperature $T_c$. The flux at the wall will be:

$$\Phi^{cc} = \Sigma h(T_w - T_c) \tag{2.10}$$

The corresponding thermal resistance is thus:

$$\mathscr{R} = \frac{1}{h\Sigma} \tag{2.11}$$

and the thermal conductance is:

$$\Gamma = \Sigma h \tag{2.12}$$

$h$ **has the appearance of a thermal conductance per unit area**. We note that in the very specific conditions of linear radiative flux, we define a thermal conductance associated with the radiative transfer $h_R$ in the same way (see Section 4.6).

- *The paradox of cylindrical insulation*. It is proposed to insulate a cylindrical pipe whose external temperature at radius $r = R_1$ is held at $T_1$ (Figure 2.5), using an insulating sleeve of conductivity $\lambda$ between $r = R_1$ and $r = R_2$, whose external face is cooled by natural convection in air at rest of temperature $T_a$ and by linear radiation, with heat transfer coefficient $h$. The problem is to calculate the rate of heat lost, in the steady state, as a function of the thickness of the sleeve (and thus of $R_2$).

  Between $r = R_1$ and air, we have resistances in series, as shown schematically in Figure 2.6:

  Using equations (2.7) and (2.11), we can obtain an overall thermal resistance for a length d$z$:

$$\mathscr{R} = \mathscr{R}^{cd} + \mathscr{R}^{cc} = \frac{1}{2\pi\lambda dz}\ln\left[\frac{R_2}{R_1}\right] + \frac{1}{2\pi R_2 h dz} \tag{2.13}$$

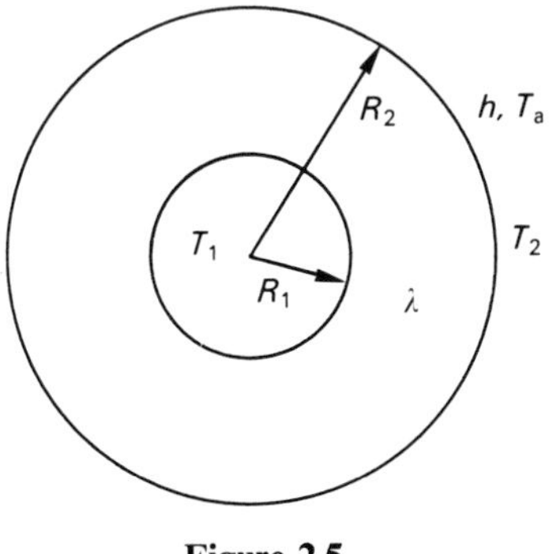


**Figure 2.5**

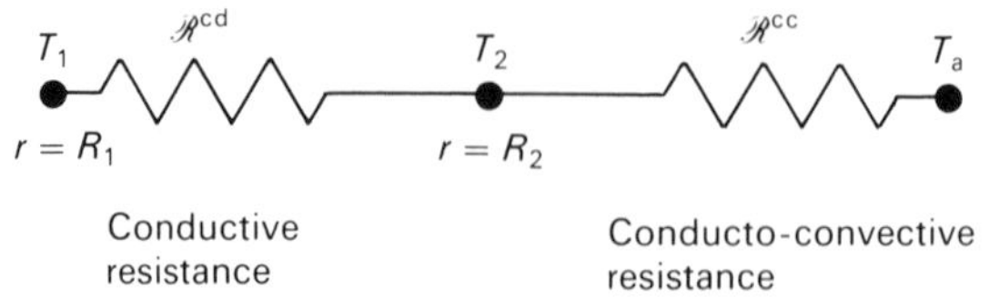


**Figure 2.6**

This has a minimum value when $R_2 = \lambda/h$ (so long as $R_1$ is less than $\lambda/h$). This means that the flux $\Phi$ dissipated is a maximum at $R_2 = \lambda/h$, and is greater than the flux $\Phi_0$ dissipated in the absence of the insulating sleeve (Figure 2.7). This property is used to increase the thermal dissipation of electrical conductors (sheathed copper wires) in order to carry more current. $T_1$ is the maximum admissible temperature at $r = R_1$. If $R_1$ is larger than $\lambda/h$, the flux dissipated falls steadily as the insulation thickness increases.

It is clear, on the other hand, that the fact that the flux dissipated passes through a maximum for $R_2$ equal to $\lambda/h$ is catastrophic in certain applications. Suppose, for example, that we wish to insulate a hot water pipe, and that the insulation, affected by humidity, has a conductivity of 0.5 W m$^{-1}$ K$^{-1}$. The flux will be maximum for an external insulation diameter of 0.1 m, if one adopts a realistic value, $h = 10$ W m$^{-2}$ K$^{-1}$.

This well-known fact constitutes the paradox of the insulator, and is typical of cylindrical geometries. Its explanation is simple: the effect of increasing the external surface area (which tends to increase the flux dissipated) is greater than the increase of the thermal resistance due to the thickness.

- *Thermal resistance of a plate heat exchanger element: overall coefficient.* We consider an exchanger between two fluids circulating between parallel plates. A wall of

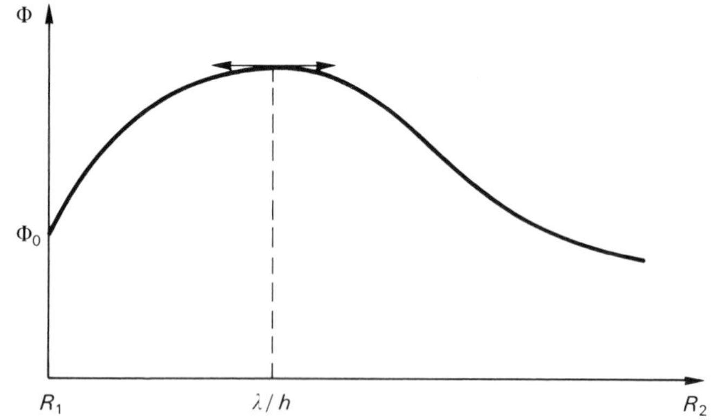


**Figure 2.7**

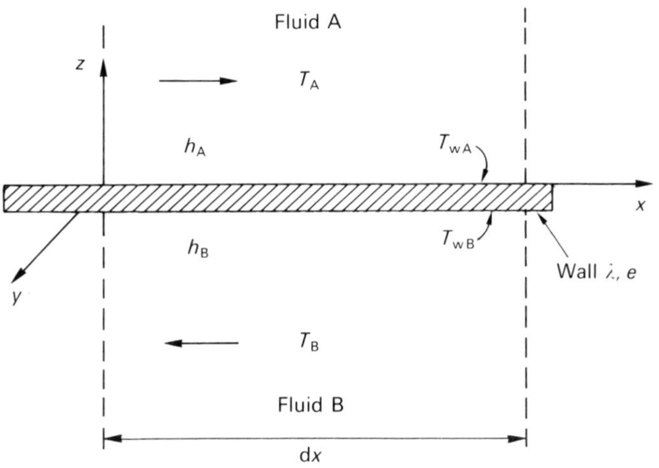


**Figure 2.8**

thickness $e$ separates the two fluids (Figure 2.8). It has a conductivity $\lambda$ and heat is exchanged through it.

We look at an element of the exchanger of length d$x$, sufficiently short for the temperatures $T_A$ and $T_B$ of the two fluids A and B and the temperatures of the two faces of the dividing wall $T_{wA}$ and $T_{wB}$ to be considered constant in the $x$ direction. The width of the exchanger (in the $y$ direction) is $L$, and is sufficiently large for edge effects to be negligible. We neglect axial conduction (in the $x$ direction) in the plate in comparison with the transfer in the $z$ direction. The fluids only exchange heat with the dividing plate through conducto-convection, for which the coefficients are $h_A$ and $h_B$. The equivalent electrical circuit is shown in Figure 2.9.

The energy flux exchanged between the two fluids in the element d$x$ is thus:

$$d\Phi = \frac{T_A - T_B}{\mathcal{R}} = d\Gamma[T_A - T_B] \tag{2.14}$$

with the thermal conductance given by:

$$d\Gamma = L\,dx\left[\frac{1}{h_A} + \frac{e}{\lambda} + \frac{1}{h_B}\right]^{-1} = L\,dx h_g \tag{2.15}$$

In (2.15), $h_g$ represents a conductance per unit area, and is called the **global heat transfer coefficient** between the two fluids. This notion is essential to characterize a heat exchanger (see Section 3.5).

**Figure 2.9**

*Notes*

1. If the plate is a good conductor, the term $e/\lambda$ is often negligible compared to the terms $1/h_A$ and $1/h_B$. This is usually the case for gases separated by a metal plate.
2. If one of the conducto-convective transfers is poor, it is useful to increase the heat exchange area for the corresponding fluid by fitting fins (see Section 2.2). This is done on the air/surface side of an automobile radiator, for example.
3. It is generally necessary to allow for fouling of the surface, in the medium to long term, which adds an external thermal resistance. Specialist works give empirical expressions for the relative resistance of fouled surfaces depending on the nature of the fluids and the duty (soot, mussels in the cooling circuits of nuclear power stations, etc.).
4. References provide analytical expressions (or the results of numerical modelling) for the thermal resistance of various geometrical configurations [13,14]. This method is, however, generally limited to relative simple geometries. It often allows approximate results to be obtained for steady-state systems.

## 2.2 Fins and the fin approximation

We have seen that the energy flux removed from a component or system by conducto-convective transfer is strongly dependent on the type of fluid concerned (air or water are the most common) and on whether the convection is forced or free. Refer to Table 1.1 in Section 1.4 for the range of variation of the heat transfer coefficient $h$. The case of natural convection in air is doubly unfavourable in this respect. But for reliability reasons, it is the only type of heat exchange possible in numerous industrial applications (free convection is self-powered, while forced convection is dependent on, for example, the operation of mechanical components or the continuity of cooling water supplies). As well as reliability, there are obviously also economic advantages.

In such conditions a simple way of increasing the flux transferred between a system and a fluid is to **artificially increase the heat exchange surface between the system and the fluid**: we attach **fins** to the surface of the basic component. Well-known cases are the fins on central heating 'radiators' or the cooling fins on a thyristor (a power electronic component) shown in plan view in Figure 2.10.

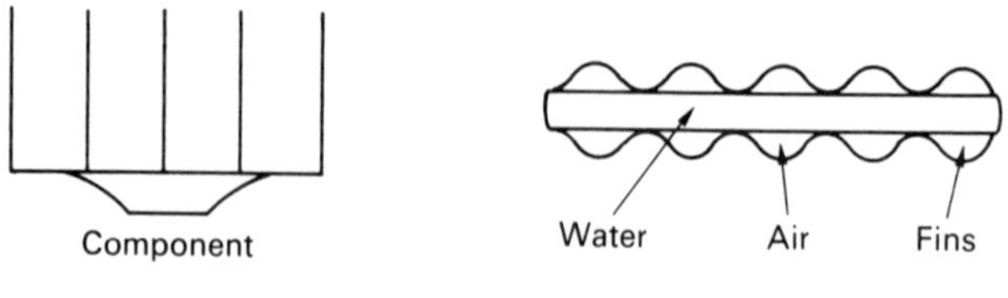


**Figure 2.10**

It goes without saying that the dimensions of the fins must be optimized with regard to economic considerations (improvement of heat transfer relative to increased cost). Fins are also used in forced convection: for example in an automobile radiator, where they appear on the side of the fluid with the poorer value of $h$ – the air side.

An important concept is the efficiency of a fin (or an array of fins). There are two definitions in the literature according to the reference used. We will generally adopt the following definition of fin efficiency:

$$\eta = \frac{\Phi_f}{\Phi_0} \tag{2.16}$$

which is the ratio of the flux $\Phi_f$ dissipated by a fin and the flux $\Phi_0$ which would be dissipated in the same external conditions by the plane surface (without the fin). By the same conditions, we mean that the exterior fluid remains at the same reference temperature $T_f$ and has the same heat transfer coefficient $h$. In fact, the convective heat transfer coefficients between the fluid and the plane component or the fin system may be different, in the same way as the equivalent radiative heat transfer coefficients $h_R$, on the assumption that radiative transfers are linear (see Section 4.6). We wish to make $\eta$ as large as possible (and greater than one), at minimum cost. But it rapidly becomes difficult to maintain the value of $h$ if the number of fins increases.

The other definition of efficiency often used in heat exchangers is:

$$\eta' = \frac{\Phi_f}{\Phi_f(T_0)} \qquad 0 \leqslant \eta' \leqslant 1 \tag{2.17}$$

where $\Phi_f(T_0)$ is the flux which would be dissipated by a fin if the whole of its surface was at temperature $T_0$ of the substrate in the absence of the fin (i.e. an ideal fin of infinite conductivity).

### *2.2.1 The fin approximation*

We will restrict ourselves here to the simple case of a single fin of arbitrary cross-section $A$ in free air of temperature $T_a$, cooling a plane substrate which is at temperature $T_0$ (Figure 2.11).

We assume that the transfer between the fin and the surroundings is solely convective (although any radiative transfers may in certain cases be linearizeable – see Section 4.6). The convective heat transfer coefficient is assumed to have a constant value $h$ on all surfaces of the fin exposed to the air. The problem is to find the energy flux conducted through the base of the fin ($z = 0$): **in the steady state, this flux is equal to the total dissipation through all the faces of the fin which are immersed in the fluid**. In theory, it is necessary to determine first the three-dimensional temperature distribution throughout the fin; this calculation is extremely complex. In practice, the calculation can often be considerably simplified by using a well-known approximation: the fin approximation.

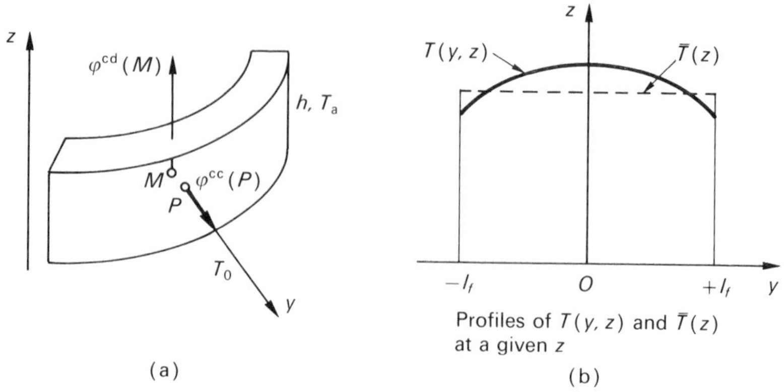


**Figure 2.11**

The conducto-convective flux density exchanged by an element of the side of the fin between $z$ and $z + dz$, around the point $P$, whose outward normal is in the $y$ direction (see Figure 2.11(a)) is given by:

$$\varphi^{cc}(P) = h[T(P) - T_a] = -\lambda \frac{\partial T}{\partial y}(P) \tag{2.18}$$

This flux density, associated with a low convective heat transfer coefficient $h$ is assumed to be small in comparison to the conductive flux density $-\lambda \partial T/\partial z(M)$ in the fin crossing an element of area normal to the $z$ direction, around $M$ (Figure 2.11(a)). Thus:

$$|h[T(P) - T_a]| \ll \left| -\lambda \frac{\partial T}{\partial z}(M) \right| \tag{2.19.1}$$

or, more simply, using equation (2.18):

$$\left| \frac{\partial T}{\partial y}(P) \right| \ll \left| \frac{\partial T}{\partial z}(M) \right| \tag{2.19.2}$$

These apply wherever points $P$ and $M$ are within the slice $[z, z + dz]$ of the fin. It follows from equations (2.19) that whatever direction normal to the $z$ axis is considered, the variations of temperature within the fin are smaller in that direction than in the $z$ direction.

The **fin approximation** comprises the supposition that, as a first approximation, the isotherms in the fin are normal to the $z$ axis (which is equivalent to neglecting variations in temperature normal to the $z$ axis). **The temperature field is then one-dimensional, in the $z$ direction**, which simplifies calculations considerably. In practice, we replace the two- or three-dimensional temperature profile ($T(y, z)$ or $T(x, y, z)$) with the average of the profile at a cross-section $z = \text{const}$, $\bar{T}(z)$: see Figure 2.11(b).

*Notes*

1. The fin considered in Figure 2.11(a) is solid. It is equally possible to envisage hollow fins with perhaps another fluid inside characterized by different reference temperature $T'_a$ and heat transfer coefficient $h'$. The fin approximation may be generalized to this case without difficulty.
2. The validation of this approximation and its limitations will be discussed below, Section 2.3.

### *2.2.2 Calculation of the efficiency of a fin*

To obtain linear equations, it is necessary to make the following assumptions:

- The conductivity of the material is independent of temperature.
- The radiative flux is negligible or linearizeable; there will be a unique heat transfer coefficient $h$, assumed constant.

In the absence of any power dissipation within the material, the energy balance is written:

$$\frac{\partial \mathscr{E}}{\partial t} = \int_S (-\mathbf{q}) \cdot \mathbf{n}_{\text{ext}}\, dS = 0 \tag{2.20}$$

we will apply it to a layer of the fin between $z$ and $z + dz$ (Figure 2.12). If $A$ represents the horizontal cross-sectional area of the fin and $\mathscr{P}$ the '**wetted perimeter**' (perimeter immersed in fluid), we obtain:

$$A[\varphi^{cd}(z) - \varphi^{cd}(z + dz)] + \mathscr{P}\, dz h[T_a - \bar{T}(z)] = 0 \tag{2.21}$$

with:

$$\varphi^{cd} = -\lambda \frac{d\bar{T}(z)}{dz} \tag{2.22}$$

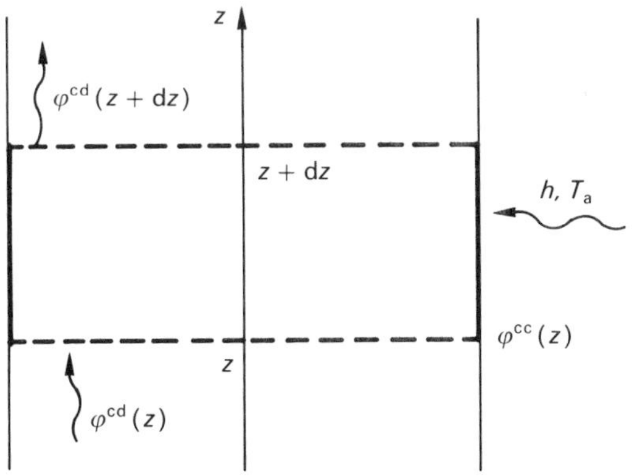


**Figure 2.12** Cross-section of a fin.

we obtain:

$$\frac{d^2\bar{T}}{dz^2} - \frac{h}{\lambda}\frac{\mathscr{P}}{A}[\bar{T}(z) - T_a] = 0 \tag{2.23}$$

Two boundary conditions are required for this equation, for example:

- at $z = 0$

$$\bar{T}(0) = T_0 \tag{2.24}$$

- at $z = l_c$ (the tip of the fin)

$$-\lambda\frac{d\bar{T}}{dz}(l_c) = h[\bar{T}(l_c) - T_a] \tag{2.25}$$

Examples of more complex boundary conditions appear in Section 2.4. Equations (2.23)–(2.25) can be solved by substituting:

$$m^2 = \frac{\mathscr{P}h}{A\lambda} \tag{2.26}$$

which gives:

$$\frac{\bar{T} - T_a}{T_0 - T_a} = \frac{\exp[m(l_c - z)](1 + m\lambda/h) - \exp[-m(l_c - z)](1 - m\lambda/h)}{\exp(ml_c)(1 + m\lambda/h) - \exp(-ml_c)(1 - m\lambda/h)} \tag{2.27}$$

The temperature at the tip of the fin is then:

$$\bar{T}(l_c) = T_a + [T_0 - T_a]\frac{2m\lambda/h}{\exp(ml_c)(1 + m\lambda/h) - \exp(-ml_c)(1 - m\lambda/h)} \tag{2.28}$$

This temperature clearly tends to $T_a$ for a sufficiently long fin, described as **infinite** (in the thermal sense), satisfying the criterion:

$$e^{ml_c} \gg 1 \tag{2.29}$$

In practice, this is valid when $l_c$ is greater than $3/m$ (precision $4 \times 10^{-2}$) or $5/m$ (precision better than $10^{-2}$).

The flux dissipated by the fin crosses its base ($z = 0$), and is given by:

$$\Phi_f = A\left\{-\lambda\frac{d\bar{T}}{dz}(0)\right\} \tag{2.30}$$

This flux should be compared to the conducto-convective flux exchanged by the base alone, in the absence of the fin (assuming the same heat transfer coefficient $h$):

$$\Phi_0 = Ah[T_0 - T_a] \tag{2.31}$$

The efficiency of the fin, as defined in equation (2.16), is then:

$$\eta = \frac{\lambda}{h}m\frac{[1 + m\lambda/h]\exp(ml_c) + [1 - m\lambda/h]\exp(-ml_c)}{[1 + m\lambda/h]\exp(ml_c) - [1 - m\lambda/h]\exp(-ml_c)} \tag{2.32}$$

This efficiency should clearly be greater than 1 for the system to be of interest. It follows from (2.32), that, for an infinite fin (satisfying (2.29)), the **gain in efficiency is effectively zero after a certain fin length**; no more flux is dissipated at the tip of the fin. This follows from the fact that after a certain length the temperature of the fin is close to the reference temperature of the fluid, and the flux exchanged tends to zero.

We can now begin to discuss the validity of the fin approximation. Starting from the temperature field obtained by assuming the validity of equation (2.27), it is possible to study the validity of criterion (2.29). Remembering (2.18), we may just calculate the ratio:

$$\frac{-\lambda\,\mathrm{d}\bar{T}/\mathrm{d}z}{[h(\bar{T}-T_a)} = \left(\frac{\lambda m}{h}\right)\frac{\exp[m(l_c - z)](1 + m\lambda/h) + \exp[-m(l_c - z)](1 - m\lambda/h)}{\exp[m(l_c - z)](1 + m\lambda/h) - \exp[-m(l_c - z)](1 - m\lambda/h)} \tag{2.33}$$

and compare it with 1.

If we consider an infinite fin (satisfying condition (2.29)), we see that so long as the quantity $\exp[m(l_c - z)]$ is large compared to 1 (that is, at a certain distance from the fin tip) the ratio given by (2.33) tends towards $\lambda m/h$. Criterion (2.19) is satisfied if:

$$\frac{\lambda}{h}m = \frac{\lambda}{h}\sqrt{\frac{h\mathscr{P}}{\lambda A}} = \left(\frac{h}{\lambda}\left(\frac{A}{\mathscr{P}}\right)\right)^{-1/2} \gg 1 \tag{2.34}$$

The dimensionless quantity $h/\lambda(A/\mathscr{P})$ is called the Biot number. $A/\mathscr{P}$ represents a characteristic dimension $D_t$ of the fin (the ratio of its cross-section to its wetted perimeter). The Biot number is written:

$$Bi = \frac{hD_t}{\lambda} = \frac{h}{\lambda}\left(\frac{A}{\mathscr{P}}\right) \tag{2.35}$$

A criterion for the validity of the fin approximation for the infinite (in the sense of (2.29)) fin model is thus:

$$Bi \ll 1 \tag{2.36}$$

In fact, we will see in the next section that a practical condition for the validity of this approximation is less restrictive (if one is only interested in the total flux exchanged by the whole fin).

Let us look at the phenomena at the tip of the fin (not necessarily 'infinite' in the thermal sense). From (2.33), with $z = l_c$:

$$\left[\frac{-\lambda\,\mathrm{d}\bar{T}/\mathrm{d}z}{h(\bar{T}-T_a)}\right]_{z=l_c} = 1$$

The problem should normally be considered in two or three dimensions in the neighbourhood of the fin tip; the fin approximation is no longer valid. We note that this is of little interest in practice because, for a well-designed (and thus thermally infinite) fin, there is very little energy exchange at $x = l_c$ because the fin temperature is very close to that of the fluid.

### 2.2.3 Non-dimensionalization of the problem

Dimensional analysis is used systematically in fluid mechanics and heat transfer to identify the relevant parameters in a problem or to simplify the display of results. The very simple example of fins considered above provides an illustration. The basic point is to choose reference values (temperature differences, length, etc.) to obtain reduced quantities, indicated by the symbol +, and preferably normalized to lie between 0 and 1.

- For temperature differences, the reference is clearly $T_0 - T_a$:

$$T^+ = \frac{\bar{T} - T_a}{T_0 - T_a} \tag{2.37}$$

- For length, we select a distance characteristic of the transverse dimension of the fin, say $D_t = A/\mathscr{P}$; then:

$$z^+ = \frac{\mathscr{P} z}{A} \tag{2.38}$$

Equations (2.23)–(2.25) become:

$$\frac{\overline{\mathrm{d}^2 T^+}}{\mathrm{d}z^{+2}} - Bi\, T^+ = 0 \tag{2.39}$$

with $\bar{T}^+(0) = 1$ and $\dfrac{\overline{\mathrm{d}T^+}}{\mathrm{d}z^+}(z_0^+) + Bi\, \bar{T}^+(z_0^+) = 0.$

The dimensionless quantity $Bi$, the **Biot Number**, which was introduced in the last section, appears:

$$Bi = \frac{hD_t}{\lambda} \tag{2.40}$$

and also the non-dimensionalized length of the fin $z_0^+$:

$$z_0^+ = \frac{\mathscr{P} l_c}{A} \tag{2.41}$$

The new equation (2.39) is much simpler than previously, and the temperature differences are normalized to lie between 0 and 1, which generally simplifies the application of numerical methods. Above all, it emerges that the thermal problem depends on only two dimensionless quantities $Bi$ and $z_0^+$, so that the non-dimensional solutions must be of the form:

$$\bar{T}^+ = f(z^+, Bi, z_0^+) \tag{2.42}$$

$$\eta = g(Bi, z_0^+) \tag{2.43}$$

A class of **thermally similar** problems may be defined, characterized by geometric similarity, the same type of boundary conditions and the same values of $Bi$ and $z_0^+$. This simplifies the compilation of results.

*Note*
The Biot number is only meaningful for a linear problem in which radiative transfers are linearized (see Section 4.6) and $h$ and $\lambda$ are independent of temperature.

## 2.3 Steady conduction in several dimensions

Because the system geometry and the boundary conditions become somewhat complex, the solution of even a steady conduction problem requires, in practice, recourse to numerical methods. And if the conductivity depends on temperature and the boundary conditions include radiative terms, these problems may be very difficult.

We will limit ourselves in this section to a simple two-dimensional problem in order to show the structure of a multi-dimensional conduction problem.

### 2.3.1 Two-dimensional analysis of a fin

We propose to calculate, for a simple geometry, the exact flux dissipated by a fin, without making the approximation that temperature is a function of $z$ only, and compare the result to that obtained using the fin approximation.

Let us consider a fin which is **semi-infinite in the $z$ direction (in the sense of (2.29))**, infinite in the $y$ direction and of width $2l_a$ in the $x$ direction. The base ($z = 0$) is held at $T_0$; the fin is cooled by convection in air at $T_a$ (with a heat transfer coefficient $h$ assumed constant). This problem is a simplification of that in Section 2.2.2. The problem is invariant in the $y$ direction, and may be treated as two-dimensional in the $x$ and $z$ directions (Figure 2.13).

For a system of volume $V$, bounded by a surface $S$, the energy equation for a purely conductive transfer without heat generation is written:

$$\frac{\partial \mathscr{E}}{\partial t} = \int_S -\mathbf{q}^{cd} \cdot \mathbf{n}_{ext} \, dS = 0 \tag{2.44}$$

At every point in the system let:

$$\nabla \cdot [-\lambda(T) \nabla T] = 0 \tag{2.45}$$

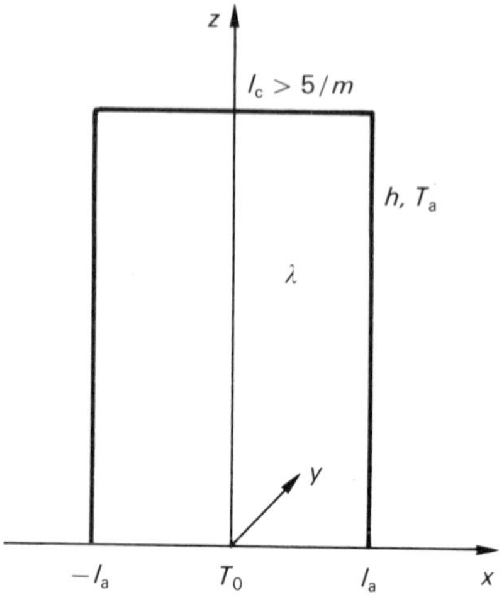


**Figure 2.13**

**If we assume that the conductivity is independent of temperature**, the energy equation takes the particularly simple form:

$$\nabla^2 T = 0 \tag{2.46}$$

For the fin, we obtain in Cartesian coordinates:

$$\frac{\partial^2 T}{\partial x^2} + \frac{\partial^2 T}{\partial z^2} = 0 \tag{2.47.1}$$

with the four boundary conditions:

$$T(x, 0) = T_0 \tag{2.47.2}$$

$$T(x, \infty) = T_a \tag{2.47.3}$$

$$-\lambda \frac{\partial T}{\partial x}(l_a, z) = h[T(l_a, z) - T_a] \tag{2.47.4}$$

$$-\lambda \frac{\partial T}{\partial x}(0, z) = 0 \tag{2.47.5}$$

*Notes*

1. Equation (2.47.3) takes the place of the fin tip condition for a finite fin.
2. Equation (2.47.5), expressing the condition that the flux across the *yz* plane is zero by symmetry, could be replaced by a condition analogous to (2.47.4):

$$-\lambda \frac{\partial T}{\partial x}(-l_a, z) = -h[T(-l_a, z) - T_a]$$

but equation (2.47.5) simplifies the calculations.

We compose non-dimensionality on the system as before (Section 2.2.3):

$$T^+(x^+, z^+) = \frac{T - T_a}{T_0 - T_a}, \qquad x^+ = \frac{x}{l_a}, \qquad z^+ = \frac{z}{l_a} \tag{2.48}$$

The choice of $l_a$, the fin half-width, as the reference length is consistent with equation (2.38). Equations (2.47.1)–(2.47.5) become:

$$\frac{\partial^2 T^+}{\partial x^{+2}} + \frac{\partial^2 T^+}{\partial z^{+2}} = 0 \tag{2.49.1}$$

with the four boundary conditions:

$$T^+(x^+, 0) = T_0 \tag{2.49.2}$$

$$T^+(x^+, \infty) = 0 \tag{2.49.3}$$

$$\frac{\partial T^+}{\partial x^+}(1, z^+) + Bi\, T^+(1, z^+) = 0 \tag{2.49.4}$$

$$\frac{\partial T^+}{\partial x^+}(0, z^+) = 0 \tag{2.49.5}$$

This system of equations depends on only one physical magnitude, the Biot number $Bi$ (for a finite fin there would also be a second parameter, for example $z_0^+$):

$$Bi = \frac{hl_a}{\lambda} \tag{2.50}$$

These equations are readily solved, for example by the method of separation of variables (see Appendix 4 to Part II).

### 2.3.2 *Validity of the fin approximation ('infinite' fin) as regards flux dissipated*

Given the solution $T^+(x^+, z^+)$, we can obtain the exact flux dissipated by the fin in terms of dimensional quantities:

$$\Phi_f^{ex} = 2\,dy \int_0^{l_a} \left\{ -\lambda \frac{\partial T}{\partial z}(x, 0) \right\} dx \tag{2.51}$$

An approximation $\Phi_f^{ap}$ to this flux can be obtained using the fin approximation. From (2.27) and (2.30), if $l_c$ tends to infinity:

$$\Phi_f^{ap} = 2\,dy[T_0 - T_a]\sqrt{\lambda h l_a} \tag{2.52}$$

The quantity $(\Phi_f^{ap} - \Phi_f^{ex})/\Phi_f^{ex}$ is plotted as a function of $Bi$ in Figure 2.14.

**So long as the Biot number is less than 1, the error in the flux is less than 12%.** This order of magnitude for the critical Biot number is valid for other fin geometries,

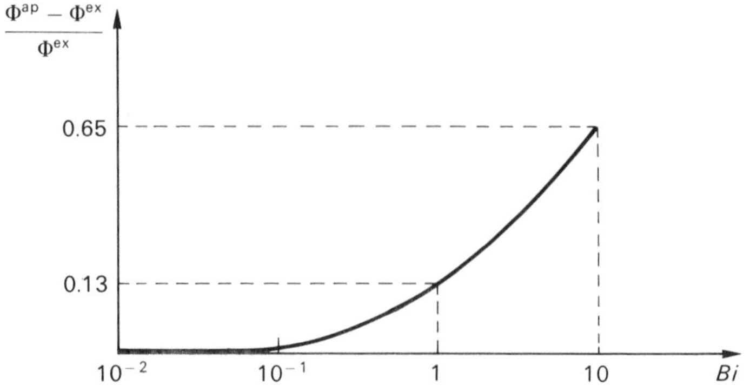


**Figure 2.14**

and provides an easy test as to whether the fin approximation may be used. In the case of a finite fin, $z_0^+$ must also be considered (see Section 2.2.3). We note that the criterion $Bi < 1$ obtained here is much less severe than the local criterion $Bi \ll 1$ obtained in Section 2.2.2. The reason is that here we are only interested in the global flux dissipated by the fin.

**Example of application**

Let us consider an 'infinite' fin made of stainless steel ($\lambda \simeq 15$ W m$^{-1}$ K$^{-1}$), whose thickness $2l_f$ is $2 \times 10^{-3}$ m, and use the table of $h$ values (Section 1.4.2) to investigate in which cases the fin approximation is valid. The Biot number is:

$$Bi = \frac{hl_f}{\lambda} \simeq 7 \times 10^{-5}h \tag{2.53}$$

The approximation is justified on the preceding criterion ($Bi < 1$) so long as $h$ is less than 15 000 W m$^{-2}$ K$^{-1}$ – for natural convection with any fluid and forced convection for common fluids (gas, water, oil). On the other hand, it will not be (with the values given) in forced convection with a liquid metal ($h$ ranging from 6000 to 10.000) or if there is a phase change (boiling or condensation) on the surface of the fin.

*Note*

The steel selected in our example represents an unfavourable case from the point of view of the fin approximation, in comparison with aluminium (fifteen times better conductor) or copper (twenty-five times better).

## 2.4 Example of an application. Simplified energy balance for an apartment

An apartment is situated on the third floor of a five-storey building, between identical apartments, all heated to a temperature $T_i$ (Figures 2.15 and 2.16). This apartment opens off the landing, at the same temperature $T_i$.

The three structural walls, running north–south (two side walls and a central wall), and the slabs making up the ceiling and floor, are concrete of thickness $e_w$ and conductivity $\lambda_w$ (Figure 2.15). These three walls terminate at the south façade; the central wall stops at the north wall, the two side walls flank the loggia, which is only open on one side, to the north. **Perfectly insulating** joints connect the three walls to the floor and ceiling slabs (Figure 2.16). (This heuristic property of the joints has no purpose except to simplify this particular problem.)

The walls of the façades, running east–west, are of lightweight concrete, of thickness $e_l$ and conductivity $\lambda_l$. The south façade is pierced by two glass windows, each of area $S_g$, thickness $e_g$ and conductivity $\lambda_g$; the north façade by two glass doors each of area $S_d$, made of the same glass. The glass is, in all cases, perfectly insulated from the walls. Similarly, the two façade walls are fixed to the load-bearing structure by 'perfectly insulating' joints.

The internal partitions, of thickness $e_p$ and conductivity $\lambda_p$ are shown on Figure 2.15; the existence of doors is ignored.

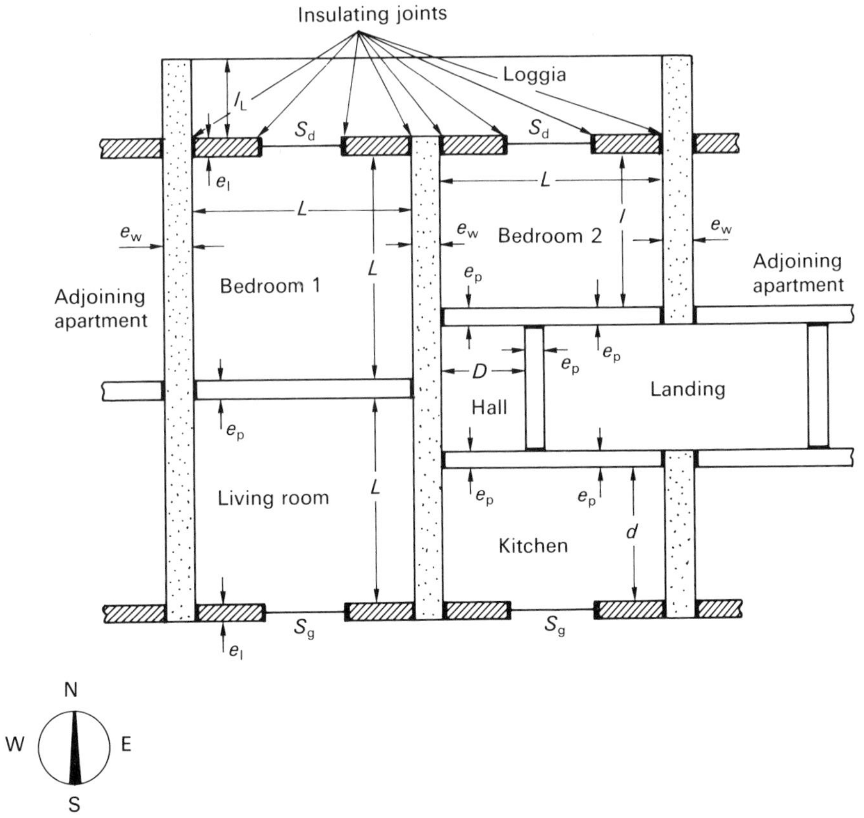


**Figure 2.15** Plan view on section AA.

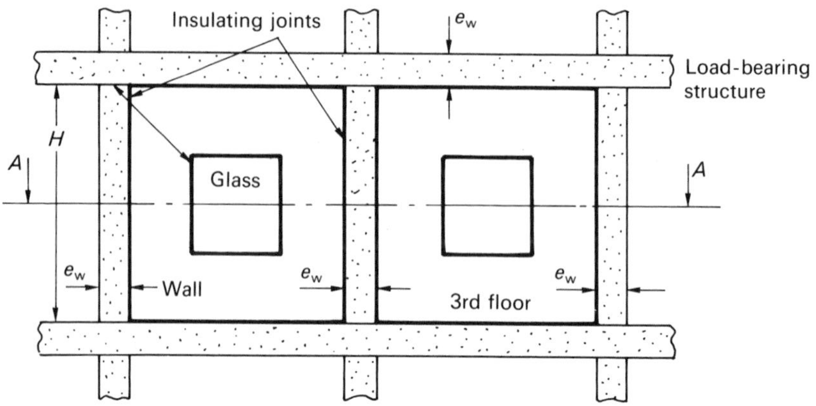


**Figure 2.16** Vertical section.

The air inside the apartment and landing is, a certain distance from the external walls, at a temperature $T_i$, and the external surroundings are at a temperature $T_e$. The heat transfer coefficient is taken as $h_i$ for all the internal walls, and $h_e$ for the external walls. It is assumed that these values take account of both linearized radiation (see Section 4.6) and conducto-convective transfers.

The air in the apartment is replaced once per hour by the introduction of external air at $T_e$, by convection.

### 2.4.1 The problem

Calculate the power necessary to maintain the temperature $T_i$ inside the apartment; under steady conditions.

Use the following numerical values:

| | | | | | |
|---|---|---|---|---|---|
| $l_l = 2$ m | $H = 2.5$ m | $L = 4.8$ m | $d = 2.5$ m | $D = 1.5$ m | |
| $e_w = 0.30$ m | $e_l = 0.20$ m | $e_p = 0.15$ m | $e = 4 \times 10^{-3}$ m | | |
| $\lambda_w = 5$ | $\lambda_l = 0.2$ | $\lambda_p = 1.5$ | $\lambda_g = 1$ | | $(\mathrm{W\,m^{-1}\,K^{-1}})$ |
| $S_g = 1\ \mathrm{m}^2$ | $S_d = 2.5\ \mathrm{m}^2$ | $h_i = 10$ | $h_g = 30$ | | $(\mathrm{W\,m^{-2}\,K^{-1}})$ |
| $T_e = 0$ °C | $T_i = 19$ °C | | | | |

Properties of air:

$\lambda_g = 2 \times 10^{-2}\ \mathrm{W\ m^{-1}\ K^{-1}}$; $c_{pg} = 1000\ \mathrm{J\ K^{-1}\ kg^{-1}}$; $\rho_g = 1.3\ \mathrm{kg\ m^{-3}}$

### 2.4.2 Approach

The problem appears, at first sight, to be complex. A systematic approach is desirable to simplify it. This consists of the following:

1. Specification of the essential **parameters** which govern the thermal problem.
2. Precise definition of the **system** to be studied, and its sub-division into linked or independent sub-systems. This is the most important part of the modelling. We must:
   (a) exploit the **symmetries** of the system to a maximum, to simplify the solution;
   (b) take account of the parameters defined in (1);
   (c) define accurately the links between the system (or a sub-system) and its surroundings – the **boundary conditions**;
   (d) **model** the heat transfer as simply as possible, relying on assumptions which will be confirmed after solution, or on **justified approximations**. Thermal problems are nearly always unique and original. A study of the orders of magnitude involved in the **particular problem** concerned will often allow a much simpler model to be adopted.
3. Writing the **balance** equations for the sub-systems identified, with their various boundary conditions.
4. Checking, after solution, the validity of any assumptions made in (2d) above.

### 2.4.3 Solution

We will follow the procedure defined above rigorously.

1. THE PARAMETERS

The present case is trivial: the source is the interior of the apartment at $T_i$, and energy is lost to the surroundings at $T_e$. A first question arises, however – exactly which part of the structure of the apartment is at $T_i$? The modelling will settle this point.

2. THE SYSTEM

**(a) The parameters**

If we consider a vertical cross-section of the apartment (Figure 2.17), perpendicular to the north–south axis, it is obvious by symmetry of geometrical and thermal conditions that the fluxes exchanged with the apartments above, below and each side are zero. The system is thus bounded by the planes of symmetry in Figure 2.17. We notice that the hall has the same symmetry.

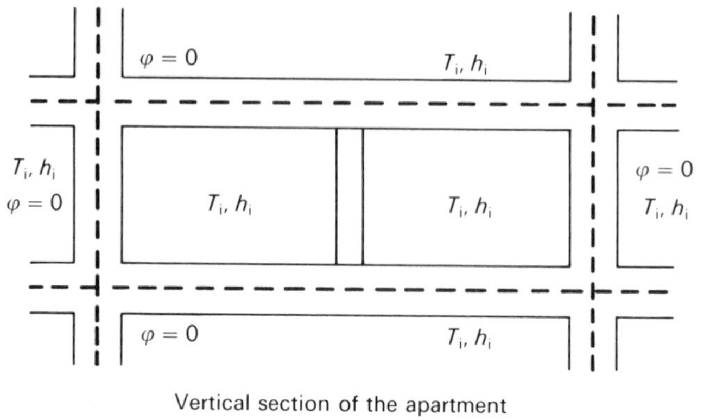


**Figure 2.17** Vertical section of the apartment.

**(b, c) Parameters and boundary conditions**

The fact that there are only two parameters suggests the existence of 'parallel' thermal transfers in several sub-systems:

- Convective air replacement. Fresh air enters at $T_e$; stale air leaves at $T_i$, with the same mass flow rate;
- Transfers through the north and south façades (light walls and windows).

  The heuristic existence of perfectly insulating joints decouples these two problems. These two separate, very simple sub-systems, are shown in Figure 2.18. We should note that the temperature $T_{we}$ and $T_{wi}$ are different for the two sub-systems.
- Transfers through the external surfaces of the structural walls, floor and ceiling. This calculation appears complex due to the existence of internal structures in the apartment. It is here that the modelling is crucial.

**(d) Modelling of the structural walls**

It is intuitive that:

- the fin approximation may be used for the structural walls. In fact, the Biot numbers for walls of thickness $e_w/2$, insulated on one face and exchanging heat

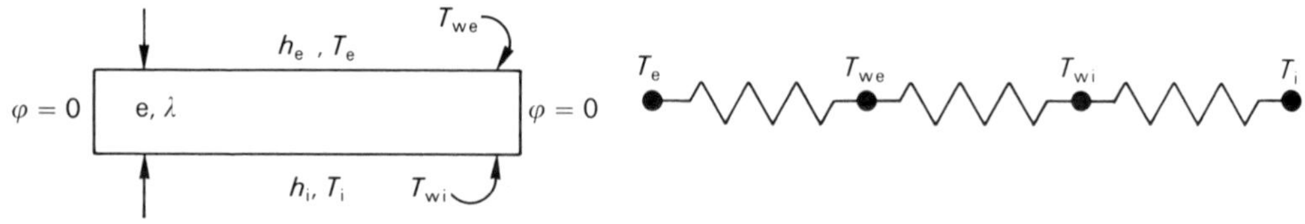


**Figure 2.18**

conducto-convect · ($h_i$ and $h_e$) on the other are:

$$Bi_{\text{int}} = \frac{h_i e_w}{2\lambda_w} \simeq 0.83$$

$$Bi_{\text{ext}} = \frac{h_e e_w}{2\lambda_w} \simeq 0.41$$

The c· or the validity of the fin approximation, to around 10%, is $Bi < 1$. This crit . is satisfied so long as $h_i$ and $h_e$ are known to this same precision.

The structural walls behave as 'infinite' fins inwards as well as outwards (see Section 2.2.2). The criterion is that the length of the fin is greater than an effective length $L_e = 5/m$. Thus:

$$L_{\text{e ext}} = 5\left(\frac{\lambda_w e_w}{2h_e}\right)^{0.5} \simeq 0.7 \text{ m}$$

$$L_{\text{e int}} = 5\left(\frac{\lambda_w e_w}{2h_i}\right)^{0.5} \simeq 1 \text{ m}$$

It is thus clear that more than 1 m from the façade, all the internal partitions (including the hall) are isothermal at $T_i$. **The internal partitions in the apartment play no part in the problem**. Similarly, the loggia walls are at a temperature $T_e$ at 0.7 m from the façade.

## 3. ENERGY BALANCE

### (a) Air replacement

The convective fluxes entering and leaving are $\dot{m}\text{h}_e$ and $\dot{m}\text{h}_i$, where $\dot{m}$, $\text{h}_e$ and $\text{h}_i$ represent the mass flow and specific enthalpies respectively. The heat flux due to air replacement is thus:

$$\Phi_{\text{rep}} = \dot{m}(\text{h}_e - \text{h}_i) = \rho \frac{V}{\Delta t} c_p (T_e - T_i)$$

where $V/\Delta t$ is the volume flow rate. Whence:

$$\Phi_{\text{rep}} = -1500 \text{ W}$$

### (b) The north and south façades

In a similar way to section (2b), the flux crossing the north façade is, with the notation given:

$$\Phi_{\text{fac}}^{\text{N}} = (T_e - T_i)\left[2S_d\left(\frac{1}{h_i} + \frac{e_g}{\lambda_g} + \frac{1}{h_e}\right)^{-1} + S_I\left(\frac{1}{h_i} + \frac{e_I}{\lambda_I} + \frac{1}{h_e}\right)^{-1}\right]$$

whence $\Phi_{\text{fac}}^{\text{N}} \simeq -1000$ W. Similarly for the south façade, $\Phi_{\text{fac}}^{\text{S}} \simeq -650$ W.

**(c) The structural walls, floor and ceiling**

- Central wall. Figure 2.19 shows the model adopted following the analysis in Section (2d). The isotherms are perpendicular to the $x$ axis:

  in the conductive wall between $x = 0$ and $x = e_I$; in the fin: $x > e_I$;

  The flux $\Phi$ crossing the face $x = 0$ is given by:

$$\Phi = He_w h_e (T_e - T_{we})$$

$$= He_w \lambda_w \frac{T_{we} - T(e_I)}{e_I}$$

$$= He_w \left[ -\lambda_w \frac{\partial T'}{\partial x}(e_I) \right]$$

The temperature field in the fin, $T'(x)$, is the solution of:

$$\frac{d^2 T'}{dx^2} - \frac{2h_i}{\lambda_w e_I}(T' - T_i) = 0$$

$$T'(\infty) = T_i$$

$$T'(e_I) = T(e_I)$$

Thus:

$$\frac{T'(x) - T_i}{T(e_I) - T_i} = \exp\left( -\left( \frac{2h_i}{\lambda e_{\mp}} \right)^{0.5} x \right)$$

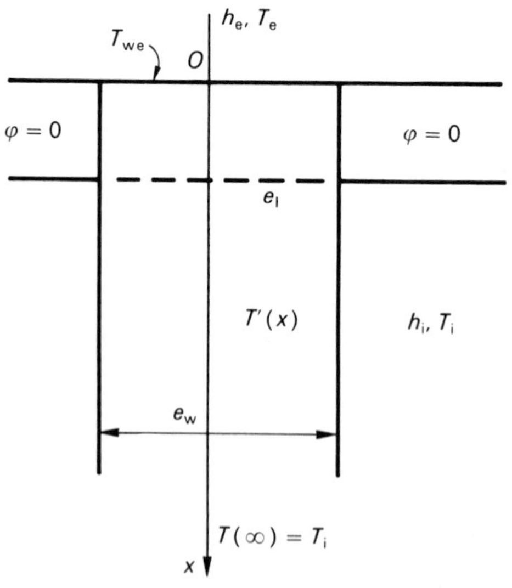


**Figure 2.19**

We then obtain, for the central fin:

$$\Phi_{cf} = \lambda_w e_w H \frac{(T_e - T_i)}{e_I + \sqrt{\lambda_w e_I/(2h_i)} + \lambda_w/h_e}$$

and:

$$\Phi_{cf} = -75 \text{ W}$$

- External walls, floor and ceiling. The model used is shown in Figure 2.20. There are two infinite fins, joined by a conductor of thickness $e_I$. The flux across the face $x = 0$ is:

$$\Phi = He_w\left[-\lambda_w \frac{\partial T_1}{\partial x}(0)\right]$$

$$= He_w\lambda_w \frac{T_1(0) - T_2(e_I)}{e_I}$$

$$= He_w\left[-\lambda_w \frac{\partial T_2}{\partial x}(e_I)\right]$$

The temperature fields $T_1(x)$ and $T_2(x)$ are calculated as before. We obtain, for the external walls:

$$\Phi_{ew} = (H + 2L + 2e_w)\frac{e_w\lambda_w(T_e - T_i)}{e_I + \sqrt{\lambda_w e_I/(2h_i)} + \sqrt{\lambda_w e_I/(2h_e)}}$$

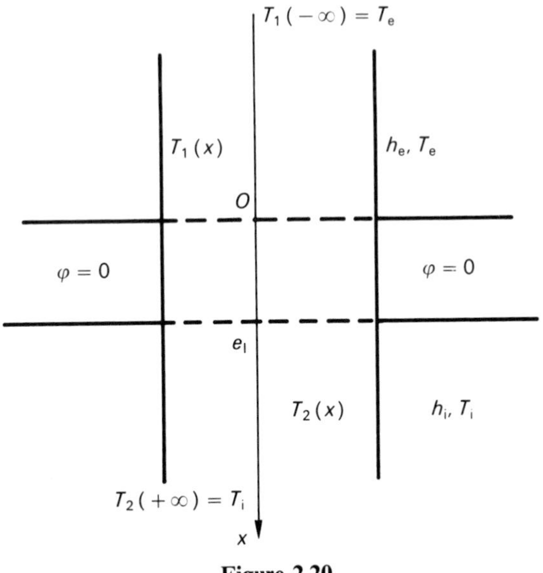


**Figure 2.20**

and

$$\Phi_{ew} = -300 \text{ W}$$

CONCLUSION

The steady-state heat losses for this apartment for an interior temperature of 19 °C and an exterior temperature of 0 °C are around 3500 W. The majority of the loss is associated with the air replacement. It could be of interest to use a regenerative heat exchanger: the incoming air would be heated by the stale air leaving. The losses through the façades could be usefully reduced by double glazing and wall insulation. All these solutions would incur non-trivial investments, and would require an economic analysis which is outside the scope of this course. The losses due to the fin effects of the structure of the building are unavoidable. They would not be negligible once the other losses had been reduced.

*Notes*

1. We notice that the various fluxes exchanged by the apartment are proportional to $T_e - T_i$ (i.e. they are linear). This is (for the steady state) the justification for the (French) building industry standards (numbers K, G, B and (shortly) C) which are based on work of this kind (see Section 4.6).
2. Another very important impact aspect, the unsteady behaviour of the structure, may be studied using the same approach (see Part VI, Application 12).

# 3

# First Approach to Convection. Applications

In Chapter 2 we limited ourselves to the analysis of rigid systems. Convective heat transfer is not present within such a system. However, movement does often occur in this case in a surrounding fluid. The convective transfer associated with this movement is linked to conduction in the fluid at the boundary of the system under consideration, and modifies the temperature field in the fluid, $T_f(y)$, near the wall (Figure 3.1). The conductive flux in the fluid at the surface $y = 0$ of the system, called conducto-convective flux (see Section 1.4.2), is expressed in terms of the convective heat transfer coefficient $h$ and a characteristic temperature $T_c$ by the relationship:

$$\varphi^{cc}(0) = -\lambda_f \frac{\partial T_f}{\partial y}(0) = h[T(0) - T_c] \tag{3.1}$$

In this chapter we will, on the contrary, consider the possibility of movement of a fluid (or a solid) within the system. It is now necessary to take into account the convective heat transfer associated with the mass transfer relative to the selected reference. The flux density $\varphi^{cv}$ across a surface $d\Sigma$, whose normal is in direction $\mathbf{n}$, is given by the equation (see 1.11):

$$\varphi^{cv} = \rho \mathrm{h}(T)\, \mathbf{v} \cdot \mathbf{n} \tag{3.2}$$

where $\mathrm{h}(T)$, $\rho$ and $\mathbf{v}$ are the specific enthalpy, density and **fluid velocity relative to** $d\Sigma$ (Figure 3.2).

The study of the phenomenon of convection is approached progressively in this chapter. In Sections 3.1 and 3.2 we will treat the system as a **material system** and an **open system** in simple cases referring to solids and liquids. We will then, in Section 3.3, introduce the general theory applicable to deformable and compressible fluids, and establish the corresponding energy equation. The principles of the study of energy transfer based on the **mechanics of non-isothermal fluids** are then established (this analysis is the subject of Part IV of this work). The theory introduced is applied as a first approach to simple cases relating to heat exchangers in Sections 3.4 and 3.5.

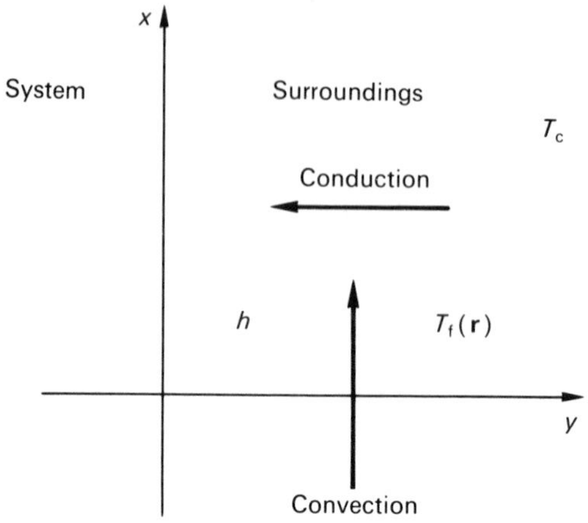


Figure 3.1

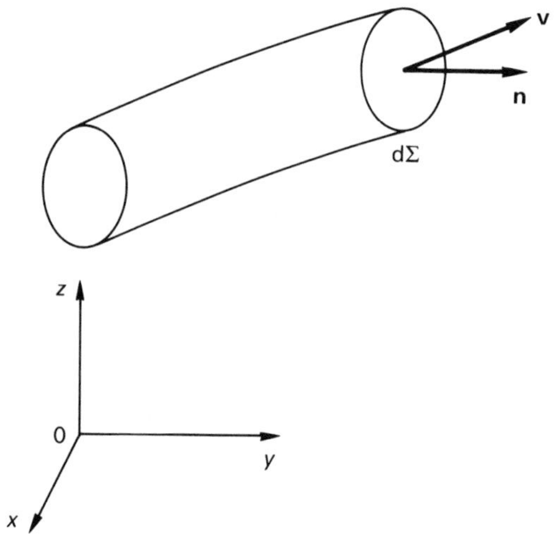


Figure 3.2

## *3.1 Treatment as a material system*

### *3.1.1 Theory for a material system*

We define a **material system** as a continuous system bounded by a closed surface $S_m(t)$, which may be mobile and deformable, across which at any point and at any instant $t$ there is no macroscopic mass flux. In other words, it is a macroscopic quantity of matter whose evolution over time we will follow.

*Note*
This approach clearly belongs to the theory of continuous media (see Sections 1.1.1 and 1.1.3). At the molecular scale there are a considerable number of molecules entering and leaving the system across its surface $S_m(t)$ in a time interval $dt$. We assume, in fact, by defining the system in this manner that at any instant the mass carried into the system by molecules entering compensates exactly for the deficit caused by those leaving. Here we make the assumption of local thermodynamic equilibrium (Section 1.1.4); the system may be broken down into a very large number of elementary material systems of volume $dV_m(t)$, bounded by closed surfaces, deformable and moving. Within each of these elementary material systems all the molecules which entered during $dt$ have their velocities redistributed by collisions to the Maxwell–Boltzmann distribution corresponding to the temperature $T(t)$ of the element concerned. The same analysis can be made for pressure $p$ or chemical potential $\mu$.

We apply **the generalized first law of thermodynamics** to such a system, in conditions of LTE, in the form:

$$\frac{d}{dt}(\mathscr{E} + \mathscr{E}_k) = \dot{W} + \dot{Q} \tag{3.3}$$

where $\mathscr{E}$ and $\mathscr{E}_k$ represent the internal and kinetic energies of the whole system, and $\dot{W}$ and $\dot{Q}$ are the rates of energy exchange with the surroundings in the form of reversible work and heat respectively. The only exchanges of heat to be taken into account at the boundaries of a material medium with its surroundings are:

- conductive transfers (including perhaps conducto-convective fluxes);
- radiative transfers.

There is obviously no convective exchange at the boundary $S_m(t)$ of a material medium, by its very definition.

### *3.1.2 Application example 1. Wire drawing*

By way of example, let us study the heat balance of a metal wire of section $\Sigma$ drawn at a very high speed $u$ (in the $x$ direction) (Figure 3.3). The wire is cooled by radiation and by conducto-convection to the ambient air at temperature $T_a$, with heat transfer coefficient $h$. The system considered is the slice of material which at the instant $t$ lies in the interval $[x, x + dx]$. We assume the temperature is constant across each cross-section.

The external fluxes which must be taken into account for this system are as follows:

- The conductive fluxes $\Sigma\varphi^{cd}(x)$ and $\Sigma\varphi^{cd}(x + dx)$ across the cross-sections of the wire and the conducto-convective flux $\mathscr{P}\,dx\,h(T(x) - T_a)$ across the sides, where $\mathscr{P}$ is the perimeter of the wire. (It is the external air which is in motion, in forced convection, relative to the wire.)

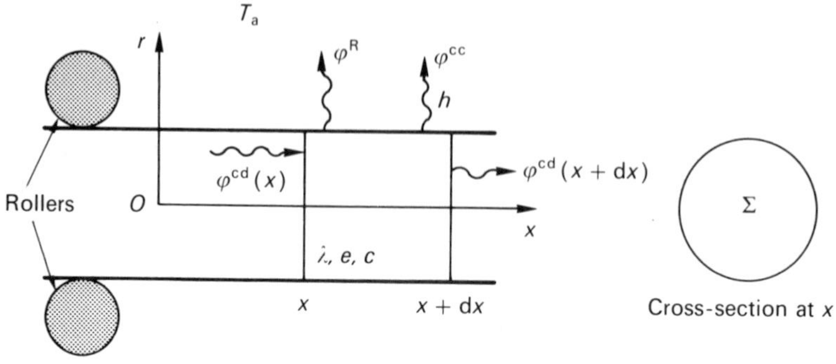


**Figure 3.3**

- A radiative flux $P\,dx\varphi^R$ (treated as positive in the $r$ direction) between the side of the slice and the surroundings.

For the wire, the kinetic energy $\mathscr{E}_k$ of the material system is constant in the frame of reference chosen (relative to the factory), and the term $\dot{W}$ in (3.3) is negligible. It follows that:

$$\frac{d\mathscr{E}}{dt} = C\frac{dT}{dt} = (\rho c\Sigma\,dx)\left(\frac{\partial T}{\partial t} + u\frac{\partial T}{\partial x}\right) \tag{3.4}$$

where $C$ and $c$ represent the thermal capacity and specific heat respectively.

The first (bracketed) term in (3.4), $\partial T/\partial t$, corresponds to an **unsteady** régime (variation of temperature with time at a fixed point $x$ **relative to the factory**); in this part of the course we are restricting ourselves to steady régimes, characterized by:

$$\frac{\partial T}{\partial t} = 0 \tag{3.5}$$

The second (bracketed) term in (3.4), $u\partial T/\partial x$, exists in a steady régime; it represents the variation of temperature per unit time of an element of material followed during its motion relative to the factory.

Taking account of the relevant transfers, we obtain from the above:

$$\dot{Q} = \Sigma[\varphi^{cd}(x) - \varphi^{cd}(x + dx)] + \mathscr{P}\,dx[h(T_a - T(x)) - \varphi^R] \tag{3.6}$$

Thus, from (3.3) to (3.6):

$$\frac{d\mathscr{E}}{dt} = \rho\Sigma cu\frac{\partial T}{\partial x}dx = \dot{Q} = \lambda\Sigma\frac{d^2T}{dx^2}dx + \mathscr{P}\,dx[h(T_a - T) - \varphi^R(T_a, T)] \tag{3.7.1}$$

with the boundary conditions:

$$T(0) = T_0; \qquad T(\infty) = T_a \tag{3.7.2}$$

## 3.2 Treatment as an open system with fixed boundaries

### 3.2.1 Theory for an open system

An **open system** has a boundary which is permeable to macroscopic mass flow. In this section, we will limit ourselves to open systems whose boundaries are fixed in the chosen frame of reference; the general case will be treated in Section 3.3.

In the steady state, the physical quantities considered are invariant with respect to time in this frame of reference:

$$\frac{\partial \mathscr{E}_{\mathrm{k}}}{\partial t} = 0; \qquad \frac{\partial \mathscr{E}}{\partial t} = 0; \qquad \frac{\partial \mathscr{H}}{\partial t} = 0 \tag{3.8}$$

where $\mathscr{H}$ represents the enthalpy of the system.

The heat transfers to be considered are:

- conductive $\varphi^{\mathrm{cd}}$ or $\varphi^{\mathrm{cc}}$;
- radiative $\varphi^{\mathrm{R}}$;
- convective $\varphi^{\mathrm{cv}}$, associated with the macroscopic flux crossing the system boundaries.

In the case of a solid or a liquid (incompressible in practice), there is a quasi-equivalence between internal energy $\mathscr{E}$ and enthalpy $\mathscr{H}$. We will generally work in terms of enthalpy, which is equally applicable to the case of a gas (see Section 3.3).

If we neglect variations in kinetic energy and power exchanged in the form of work, the energy conservation equation for a fixed open system in steady state becomes:

$$\mathbf{q}^{\mathrm{en}} = \mathbf{q}^{\mathrm{cd}} + \mathbf{q}^{\mathrm{R}} + \mathbf{q}^{\mathrm{cv}} \tag{3.9.1}$$

where $\mathbf{q}^{\mathrm{en}}$ is the energy flux density vector (the radiative flux density vector $\mathbf{q}^{\mathrm{R}}$ will be defined in Section 4.2.3).

$$\frac{\partial \mathscr{H}}{\partial t} = \int_{\mathrm{S}} -\mathbf{q}^{\mathrm{en}} \cdot \mathbf{n}_{\mathrm{ext}}\, \mathrm{d}S + \int_{\mathrm{V}} P\, \mathrm{d}V = 0 \tag{3.9.2}$$

where $P$ is the power dissipated per unit volume.

The validity of this equation when extended to a perfect gas is discussed in Section 3.3 in terms of general transport theories. This conservation is illustrated by two examples (Section 3.2.2 and 3.2.3).

### 3.2.2 Example 1. Wire drawing

We will re-work the example from Section 3.1.2 in terms of the open system which is between $x$ and $x + \mathrm{d}x$ at whatever instant is considered. Equation (3.9.2) becomes:

$$\frac{\partial \mathscr{H}}{\partial t} = \dot{Q} = \lambda \Sigma \frac{\mathrm{d}^2 T}{\mathrm{d}x^2}\mathrm{d}x + \mathscr{P}\,\mathrm{d}x[-\varphi^{\mathrm{R}} + h(T_{\mathrm{a}} - T)] - \Sigma \rho c u \frac{\mathrm{d}T}{\mathrm{d}x}\mathrm{d}x = 0 \tag{3.10}$$

The first term $\lambda\Sigma\,\mathrm{d}^2T/\mathrm{d}x^2\,\mathrm{d}x$ represents the overall effect of the conductive contributions through the planes $x$ and $x+\mathrm{d}x$; the second term $\mathscr{P}\,\mathrm{d}x[-\varphi^{\mathrm{R}}+h(T_{\mathrm{a}}-T)]$ represents the radiative and conductive contributions through the sides of the wire; and the last term $-\Sigma\rho cu\,\mathrm{d}T/\mathrm{d}x$ represents the sum of the convective contributions $\Sigma\rho u\mathrm{h}(T(x))$ at $x$ and $-\Sigma\rho u\mathrm{h}(T(x+\mathrm{d}x))$ at $x+\mathrm{d}x$. The conservation equation (3.10) is obviously identical to equation (3.7.1), obtained by consideration of the material system.

Considering the solution of equation (3.7.1) and (3.7.2):

$$\frac{T-T_{\mathrm{a}}}{T_0-T_{\mathrm{a}}}=\exp\left\{-\frac{\rho cux}{2\lambda}\left[\left(1+\frac{4\mathscr{P}h\lambda}{\Sigma\rho^2c^2u^2}\right)^{1/2}-1\right]\right\} \tag{3.11}$$

We see that the contribution of axial conduction (in the $x$ direction), represented by $\lambda$, is negligible in comparison with the convective contribution (for an open system) if:

$$u \gg \left(\frac{4\lambda\mathscr{P}h}{\Sigma\rho^2c^2}\right)^{1/2} \tag{3.12}$$

In practically all common applications, this condition is satisfied, whether we are concerned with a solid (the wire, for example), a liquid or even a gas (we will see in Section 3.4 that for a dilatable and compressible gas the overall conservation equation has a structure analogous to equation (3.7.1)). To convince ourselves, we need only consider the values of $\lambda$, $\rho$ and $c$ for the application concerned, with realistic values of $\mathscr{P}$ and $\Sigma$ and the range of variation of the convective heat transfer coefficient $h$ (see Table 1.1 in Section 1.4.2).

We will assume in the following that **axial conductive heat transfer (for a solid, a liquid or a gas) is in practice negligible compared to the convective heat transfer in the same direction.**

*Notes*

1. This hypothesis will be demonstrated in Part IV of this course by consideration of orders of magnitude in the energy equation.
2. We should note that the most likely exception to this hypothesis is natural convection at very low velocities (and thus low $h$) of liquid sodium (high value of $\lambda$).

### 3.2.3 *Example 2. Solid–liquid interface, fusion front*

One method for the storage of energy for future use is to use the considerable latent heat of the liquid–solid change of state. We use salts whose melting point can be as high as 900 °C, so that the heat quality is not degraded. Normal plate heat exchangers may be used, but are not very effective. Better are direct contact exchangers. One of the problems to be solved is to relate the speed of movement of the fusion front to the energy flux at the interface on both the liquid and solid sides.

If we make the simplifying assumption that **the densities $\rho_{\mathrm{s}}$ and $\rho_{\mathrm{L}}$ of the solid and liquid phases are equal**, with the value $\rho$, we can model the fusion front as a **rigid**

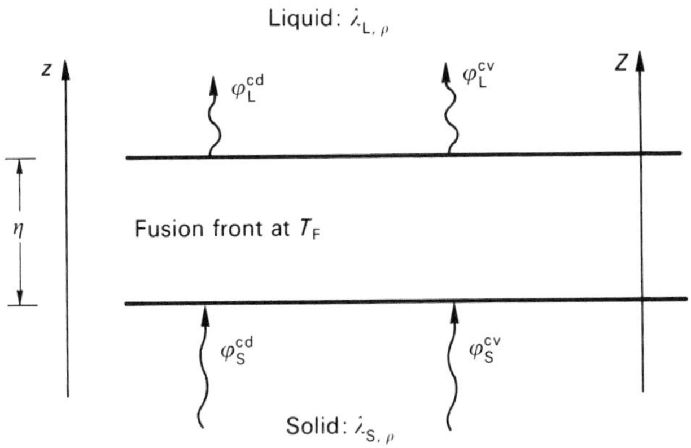


**Figure 3.4**

layer of thickness $\eta$, at a uniform temperature $T_F$ consisting of two intimately mixed phases (see Figure 3.4). The simplest control system to choose is the fusion front of thickness $\eta$ itself. It is in motion at a speed $\dot{z}$ in the $z$ direction, and is an **open** system. We apply the conservation equations to this fusion front, in the $z$ direction.

There are obviously convective and conductive transfers across both faces in this case (see Figure 3.4):

$$\Sigma\varphi_S^{cd} + \Sigma\varphi_S^{cv} - \Sigma\varphi_L^{cd} - \Sigma\varphi_L^{cv} = \Sigma\varphi_S^{cd} + \dot{m}h_S(T_F) - \Sigma\varphi_L^{cd} - \dot{m}h_L(T_F) = 0 \quad (3.13)$$

where $\Sigma$, $h_S(T_F)$ and $h_L(T_F)$ are the area of the front and the specific enthalpies of the solid and liquid phases, $\dot{m}$ is the mass flux (obviously constant) of the liquid and solid across the faces of the fusion front:

$$\dot{m} = -\rho\Sigma\dot{z}_F \quad (3.14)$$

The conservation equation becomes:

$$-\lambda_S \frac{\partial T_S}{\partial z} + \lambda_L \frac{\partial T_L}{\partial z} = \rho\dot{z}_F(h_S - h_L) = -\rho\dot{z}_F\mathscr{L} \quad (3.15)$$

where $\mathscr{L}$ is the specific latent heat of the change of state.

This result is clearly independent of the geometry; it can be generalized to a liquid–vapour change of state (see Section 3.3.1.2), which is characterized by a very large variation in the density $\rho$.

## 3.3 General case of a deformable medium

The problem of transfer in a moving fluid is basic. As for a solid, it is necessary to establish the energy law of the system. The problem is complicated by the fact that **a fluid is deformable**, and also often **dilatable** ($\partial\rho/\partial T \neq 0$), particularly a gas when significant energy is put into or extracted from it; similarly if may be **compressible** ($\partial\rho/\partial p \neq 0$).

### 3.3.1 Preliminary (transport) theorems

3.3.1.1 TRANSPORT THEOREM FOR AN OPEN SYSTEM WITH MOVING BOUNDARIES

Let $A$ and $a$ be symbols for a physical quantity (scalar, vector or tensor), relative to unit volume and unit mass respectively:

$$A = \rho a \tag{3.16}$$

Note that the density $\rho$ is an exception to this scheme.

Let us consider a mobile, deformable, continuous system of volume $V(t)$, bounded by a surface $S(t)$ (Figure 3.5). We will assume, without proof the following theorem [15]:

$$\frac{\mathrm{d}}{\mathrm{d}t}\int_{V(t)} A\,\mathrm{d}V = \int_{V(t)} \frac{\partial A}{\partial t}\,\mathrm{d}V + \int_{S(t)} A\mathbf{w}\cdot\mathbf{n}_{\mathrm{ext}}\,\mathrm{d}S \tag{3.17}$$

In this equation $\mathbf{n}_{\mathrm{ext}}$ represents the outward normal of an element of the surface $S$ and $\mathbf{w}$ **the vector velocity of element** $\mathrm{d}S$ **of the surface** (as opposed to that of the element of material which is at this point at the instant considered, which is denoted by $\mathbf{v}$).

3.3.1.2 APPLICATION EXAMPLE. VAPORIZATION OF A DROP

Two-phase liquid–vapour transfers play a very considerable rôle, as much in large power stations (whether nuclear or conventional) as in combustion (vaporization of atomized fuel). We propose to establish the expression for the continuity of energy flux at the phase front for this case. The expression is a generalization of that

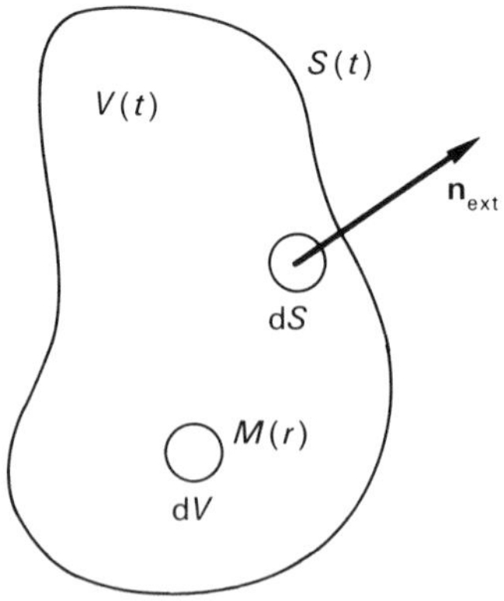


**Figure 3.5**

introduced in Section 3.2.3 for the case of liquid–solid phase change at constant density.

We will consider the heat transfer at the boundary of a spherical liquid drop of radius $R(t)$ (Figure 3.6), characterized by conductivity, density, specific heat and specific enthalpy $\lambda_L$, $\rho_L$, $c_{pL}$ and $h_L$, which evaporates in a surrounding medium at temperature $T_a$ far away, consisting of pure water vapour (with very different conductivity, density, specific heat and specific enthalpy $\lambda_v$, $\rho_v$ $c_{pv}$ and $h_v$). The temperature profile in the liquid phase is denoted by $T_L$ and that in the vapour phase by $T_v$. The origin is taken at the centre of the drop.

For simplicity, we will assume that the drop is opaque to radiation and the vapour is transparent. (The result to be obtained can be generalized to two semi-transparent media.)

The drop is an open system with variable volume $V(t)$. In the energy equation, we neglect the power exchanged by the drop in the form of work. The equation becomes, in our usual notation, and with $\mathbf{q}^R$ to be defined in Section 4.2.3:

$$\frac{d\mathscr{H}_L}{dt} = \int_{S(t)} -(\mathbf{q}^{cd} + \mathbf{q}^R + \mathbf{q}^{cv})\cdot\mathbf{n}_{ext}\, dS + \int_{V(t)} P\, dV \qquad (3.18)$$

The first term in (3.18) becomes, using the transport theorem:

$$\frac{d\mathscr{H}_L}{dt} = \frac{d}{dt}\int_{V(t)} H_L\, dV = \int_{V(t)} \frac{\partial H_L}{\partial t}\, dV + \int_{S(t)} H_L\, \dot{R}\, dS \qquad (3.19.1)$$

making use of the relation:

$$\mathbf{w}\cdot\mathbf{n}_{ext} = \dot{R} \qquad (3.19.2)$$

where $H_L$ and $\dot{R}$ represent the enthalpy per unit volume and the radial velocity of the boundary relative to the specified origin. Further:

$$\frac{d\mathscr{H}_L}{dt} = \int_0^{R(t)} \frac{\partial H_L}{\partial t} 4\pi r^2\, dr + 4\pi R^2 \dot{R} \rho_L h_L(T_F) \qquad (3.20)$$

where $T_F$ is the temperature of the phase front, assumed constant.

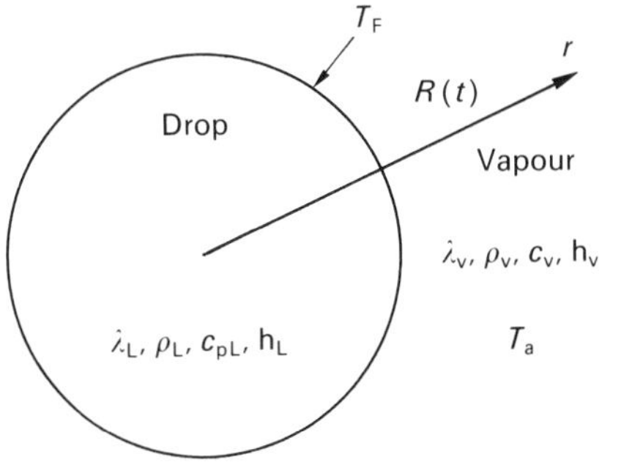


**Figure 3.6**

The first term on the right-hand side of equation (3.20) represents the increase in enthalpy of the system between $t$ and $t + \mathrm{d}t$ due to its **time dependent** nature (considering the volume $V(t)$ at the instant $t$). The second term represents the increase in enthalpy of the system between $t$ and $t + \mathrm{d}t$ due to its **increase in volume** $4\pi R^2 \dot{R}$ (considering the enthalpy value on the boundary at the instant $t$).

In this particular application, the second integral in (3.18) is zero, and the above equation becomes:

$$\frac{\mathrm{d}\mathscr{H}}{\mathrm{d}t} = 4\pi R^2 \left[ \lambda_{\mathrm{v}} \frac{\partial T_{\mathrm{v}}}{\partial r}(R(t), t) - \varphi^{\mathrm{R}}(T_{\mathrm{F}}, T_{\mathrm{a}}) \right] + 4\pi R^2 \rho_{\mathrm{v}} \mathbf{v}_{\mathrm{v}} \cdot (-\mathbf{n}_{\mathrm{ext}}) \mathrm{h}_{\mathrm{v}}(T_{\mathrm{F}}) \quad (3.21)$$

The last term represents the convective flux of vapour across the front; $\mathbf{v}_{\mathrm{v}}$ **is the local velocity of the vapour relative to the element of surface area**, which is related to the local velocity of the liquid phase $\mathbf{v}_{\mathrm{L}}$ by the conservation equation:

$$4\pi R^2 \rho_{\mathrm{v}} \mathbf{v}_{\mathrm{v}} = 4\pi R^2 \rho_{\mathrm{L}} \mathbf{v}_{\mathrm{L}} \quad (3.22)$$

and, in addition, since velocities are positive in the $r$ direction:

$$\mathbf{v}_{\mathrm{L}} \cdot \mathbf{n}_{\mathrm{ext}} = -\dot{R} \quad (3.23)$$

Finally, because the drop is an opaque body, the time variation of enthalpy of the drop is due only to conduction in the liquid phase (provided $\rho_{\mathrm{L}}$ is constant inside the drop):

$$\int_0^{R(t)} \frac{\partial H_{\mathrm{L}}}{\partial t} 4\pi r^2 \, \mathrm{d}r = 4\pi R^2 \lambda_{\mathrm{L}} \frac{\partial T_{\mathrm{L}}}{\partial r}(R(t), t) \quad (3.24)$$

Making use of equations (3.18)–(3.24), the interface condition becomes:

$$-\lambda_{\mathrm{L}} \frac{\partial T_{\mathrm{L}}}{\partial r}(R(t), t) = -\lambda_{\mathrm{v}} \frac{\partial T_{\mathrm{v}}}{\partial r}(R(t), t) + \varphi^{\mathrm{R}}(T_{\mathrm{F}}, T_{\mathrm{a}})) - \dot{R} \rho_{\mathrm{L}}(\mathrm{h}_{\mathrm{v}}(T_{\mathrm{F}}) - \mathrm{h}_{\mathrm{L}}(T_{\mathrm{F}})) \quad (3.25)$$

This expression is a generalization of (3.15); the enthalpy of $h_{\mathrm{v}}$ to $h_{\mathrm{L}}$ is of course the specific latent heat of vaporization $\mathscr{L}_{\mathrm{v}}$.

### 3.3.1.3 TRANSPORT THEOREM FOR A MATERIAL SYSTEM

The application of the mass transport theorem to a material system of volume $V_{\mathrm{m}}(t)$ and surface $S_{\mathrm{m}}(t)$ is obvious once we set:

$$\mathbf{w}(\mathbf{r}, t) = \mathbf{v}(\mathbf{r}, t) \quad (3.26)$$

i.e. the velocity of a material element at a point on the surface, $\mathbf{v}$, is the same as the velocity of the surface, $\mathbf{w}$. We write the conservation of mass:

$$\frac{\mathrm{d}m}{\mathrm{d}t} = \int_{V_{\mathrm{m}}(t)} \frac{\partial \rho}{\partial t} \mathrm{d}V_{\mathrm{m}} + \int_{S_{\mathrm{m}}(t)} \rho \mathbf{v} \cdot \mathbf{n}_{\mathrm{ext}} \, \mathrm{d}S_{\mathrm{m}} = 0 \quad (3.27)$$

which becomes, after obvious manipulation, in local form:

$$\frac{\partial \rho}{\partial t} + \nabla \cdot (\rho \mathbf{v}) = 0 \tag{3.28}$$

the **mass conservation equation**, known as the **equation of continuity**.

In the case of a material system, the form (3.17) of the transport theorem can be transformed from an integral over $S_m$ to an integral over $V_m$, using the notation of equation (3.16):

$$\frac{d\mathscr{A}}{dT} = \frac{d}{dt}\int_{V_m(t)} \left[ \rho \frac{\partial a}{\partial t} + a \frac{\partial \rho}{\partial t} + a\nabla \cdot (\rho \mathbf{v}) + \rho \mathbf{v} \cdot \nabla a \right] dV_m \tag{3.29}$$

which becomes, after simplification using (3.28):

$$\frac{d}{dt}\int_{V_m(t)} \rho a \, d\dot{V}_m = \int_{V_m(t)} \rho \frac{da}{dt} dV_m \tag{3.30}$$

This equation is valid whether $a$ is a scalar, a vector or a tensor, and is often called Reynolds theorem.

### 3.3.2 Deviation of the balance equations for a fluid element[1]

#### 3.3.2.1 EQUATION OF STATE

We recall that a fluid is characterized by an **equation of state**, relating pressure $p$, density $\rho$ and absolute temperature $T$:

$$f(p, \rho, T) = 0 \tag{3.31}$$

For a perfect gas, this equation takes the particularly simple form:

$$p = \rho r T \tag{3.32}$$

where $r$ is the specific gas constant; more general equations may be set up for real gases using virial coefficients (van der Waals, etc.). For a liquid, which is, to a first approximation, constant volume (i.e. indilatable $\partial\rho/\partial T = 0$ and incompressible $\partial\rho/\partial p = 0$) the equation of state reduces to

$$\rho = \rho_0 \tag{3.33}$$

Many balance equations can be written for a fluid element; we will limit ourselves to the most common: balance of mass, of momentum and of enthalpy.

#### 3.3.2.2 BALANCE OF MASS

The equation of **balance of mass** was derived above (Section 3.3.1.3):

$$\frac{\partial \rho}{\partial t} + \nabla \cdot (\rho \mathbf{v}) = 0 \tag{3.34.1}$$

[1] Section 3.3.2 may be skipped on a first reading.

In the case of a constant density fluid (as in (3.33)) it becomes:

$$\nabla \cdot \mathbf{v} = 0 \tag{3.34.2}$$

3.3.2.3 BALANCE OF MOMENTUM

The equation of **balance of momentum** arises from the application of basic principles to a material system:

$$\frac{\mathrm{d}}{\mathrm{d}t}\int_{V_{\mathrm{m}}(t)} \rho\mathbf{v}\,\mathrm{d}V_{\mathrm{m}} = \int_{V_{\mathrm{m}}(t)} \rho\frac{\mathrm{d}\mathbf{v}}{\mathrm{d}t}\,dV_{\mathrm{m}} = \int_{V_{\mathrm{m}}(t)} \rho\mathbf{g}\,\mathrm{d}V_{\mathrm{m}} + \int_{S_{\mathrm{m}}(t)} \mathbf{t}(\mathbf{n}_{\mathrm{ext}})\,\mathrm{d}S_{\mathrm{m}} \tag{3.35}$$

in which appear the resultant of the body forces per unit mass $\mathbf{g}$ and the stresses $\mathbf{t}(\mathbf{n}_{\mathrm{ext}})$ acting on an element $\mathrm{d}S_{\mathrm{m}}$ of the system surface, whose outward normal is $\mathbf{n}_{\mathrm{ext}}$.

*Important note*

The stresses on a surface depend on the sense chosen for the normal. We consider the force per unit area exerced by the fluid adjacent to the normal on the (usually imaginary) element of surface. By the principle of action and reaction, the fluid on the other side of the surface exerts an opposite force on it:

$$\mathbf{t}(-\mathbf{n}) = -\mathbf{t}(\mathbf{n}) \tag{3.36}$$

A demonstration of this is given in reference [1].

The stresses $\mathbf{t}(\mathbf{n}_{\mathrm{ext}})$ include pressure and viscous forces (we restrict ourselves here to common Newtonian fluids):

- The **pressure forces** represent the cumulative effect on the surface of the momenta of the molecules in the exterior half-space bounded by the surface. It is independent of elementary interactions between molecules, and exists even in a gas with 'point' particles with no interactions. The expression for the pressure stress is:

$$\mathbf{t}_{\mathrm{p}}(\mathbf{n}_{\mathrm{ext}}) = -p\mathbf{n}_{\mathrm{ext}} \tag{3.37}$$

- The **viscous stresses** on the other hand, arise from the cumulative effects of interactions between the molecules of the fluid (governed by inter-molecular potentials) when **deformations** are imposed on a fluid element, considered as a continuous medium. The idea of deformation, as opposed to other transformations of a material element (translation or notation in a given coordinate system), is defined in Appendix 2 to Part I. As a consequence of deformations imposed on a material element, a vector field of viscous stress (in the case of a Newtonian fluid) appears at all points on the surface of the element. This is characterized by the stress vector $\mathbf{t}_{\mathrm{v}}(\mathbf{n}_{\mathrm{ext}})$, given by:

$$\mathbf{t}_{\mathrm{v}}(\mathbf{n}_{\mathrm{ext}}) = \mathbf{n}_{\mathrm{ext}} \cdot \underline{\tau} \tag{3.38}$$

where $\underline{\tau}$ is the viscous stress tensor, which is a function only of the fluid rate of deformation tensor $\underline{\mathbf{d}}$ (see Appendix 2 to Part 1). Note that the scalar product of the vector $\mathbf{n}_{\mathrm{ext}}$ and the symmetric (second-order) tensor $\underline{\tau}$ gives the viscous stress vector $\mathbf{t}_{\mathrm{v}}(\mathbf{n}_{\mathrm{ext}})$. In tensor notation (see Appendix 1 of this part for notational conventions), equation (3.38) becomes:

$$t_{vj}(n_{ext_i}) = n_{ext_i}\tau_{ij} \tag{3.39}$$

where $t_{vj}$ and $n_{ext_i}$ are the viscous stress and outwards normal vectors and $\tau_{ij}$ is the viscous stress tensor given by the formula (with $\kappa = 0$):

$$\tau_{ij} = \mu\left[\left(\frac{\partial v_i}{\partial x_j} + \frac{\partial v_j}{\partial x_i}\right) - \frac{2}{3}\frac{\partial v_k}{\partial x_k}\delta_{ij}\right] \tag{3.40}$$

where $\delta_{ij}$ is the Kronecker function and the repeated appearance of the subscript $k$ in a term implies summation over all values of $k$ (see notational conventions in Appendix 1 of this part).

*Notes*

1. In tensor notation the pressure stress given by equation (3.37) becomes:

$$t_{pj}(n_{ext_i}) = -pn_{ext_i}\delta_{ij} \tag{3.41}$$

The total stress tensor for a Newtonian fluid is thus:

$$T_{ij} = \tau_{ij} - p\delta_{ij}$$

with: (3.42)

$$t_j(n_{ext_i}) = n_{ext_i}T_{ij}$$

2. Note that the stress tensors $\tau_{ij}$ and $T_{ij}$ are symmetrical with respect to $i$ and $j$.

The second term in the second part of equation (3.35) becomes, in tensor notation:

$$\int_{S_m(t)} t_i(n_{ext_j})\,dS_m = \int_{S_m} n_{ext_j}T_{ji}\,dS_m = \int_{V_m} \frac{\partial}{\partial x_j}T_{ji}\,dV_m \tag{3.43}$$

The balance of momentum equation (3.35) can then be written in local form:

$$\rho\frac{dv_i}{dt} = \rho g_i - \frac{\partial}{\partial x_i}p + \frac{\partial}{\partial x_j}\tau_{ij} \tag{3.44}$$

and, assuming that the **dynamic viscosity** $\mu$ is independent of temperature and, more generally, **constant**:

$$\rho\frac{dv_i}{dt} = \rho g_i - \frac{\partial}{\partial x_i}p + \mu\frac{\partial^2}{\partial x_j^2}v_i + \frac{\mu}{3}\frac{\partial}{\partial x_i}\left(\frac{\partial v_k}{\partial x_k}\right) \tag{3.45}$$

*Notes*

1. In the case of constant volume transformations of the fluid (equation (3.34.2)), the last term is zero.
2. The term $\partial^2 v_i/\partial x_j^2$ (here given in tensor notation) expands in Cartesian coordinates to: $\partial^2 v_i/\partial x_1^2 + \partial^2 v_i/\partial x_2^2 + \partial^2 v_i/\partial x_3^2$.

3. A more complete statement of the momentum balance equation will be found in Chapter 13 (specifically, with $\mu$ dependent on temperature). This equation is usually called the **Navier–Stokes equation.**

#### 3.3.2.4 BALANCE OF INTERNAL ENERGY, KINETIC ENERGY AND ENTHALPY

The balance equation for internal energy for an element of a material system derives from the generalized form of the **first law of thermodynamics**. If $e$ is the specific internal energy of the fluid, the latter states:

$$\frac{\mathrm{d}}{\mathrm{d}t}(\mathscr{E} + \mathscr{E}_{\mathrm{k}}) = \frac{\mathrm{d}}{\mathrm{d}t}\int_{V_{\mathrm{m}}(t)} \rho\left(e + \frac{v_i^2}{2}\right)\mathrm{d}V_{\mathrm{m}} = \int_{V_{\mathrm{m}}(t)} \rho\left(\frac{\mathrm{d}e}{\mathrm{d}t} + v_i\frac{\mathrm{d}v_i}{\mathrm{d}t}\right)\mathrm{d}V_{\mathrm{m}} = \dot{W} + \dot{Q} \tag{3.46}$$

where $\mathscr{E}$ and $\mathscr{E}_{\mathrm{k}}$ are the internal and kinetic energies of the system, $\dot{W}$ and $\dot{Q}$ the rate of energy transferred by it in the form of work and heat, i.e.:

$$\dot{W} = \int_{V_{\mathrm{m}}(t)} \rho g_i v_i \,\mathrm{d}V_m + \int_{S_{\mathrm{m}}(t)} t_i(n_{\mathrm{ext}_j})v_i \,\mathrm{d}S_{\mathrm{m}} \tag{3.47}$$

is the sum of the powers associated with body forces and stresses, and:

$$\dot{Q} = \int_{V_{\mathrm{m}}(t)} P\,\mathrm{d}V_{\mathrm{m}} + \int_{S_{\mathrm{m}}(t)} -(q_i^{\mathrm{cd}} + q_i^{\mathrm{R}})n_{\mathrm{ext}_i}\,\mathrm{d}S_{\mathrm{m}} \tag{3.48}$$

is the sum of the powers generated within the material element and crossing its surface. We note that for a material system there is no convection term in the balance. The second term in (3.47) becomes (using (3.39)–(3.42)):

$$\begin{aligned}\int_{S_{\mathrm{m}}(t)} (t_{\mathrm{p}j} + t_{\mathrm{v}j})v_j\,\mathrm{d}S_{\mathrm{m}} &= \int_{S_{\mathrm{m}}(t)} n_{\mathrm{ext}_i}(-p\delta_{ij} + \tau_{ij})v_j\,\mathrm{d}S_{\mathrm{m}} \\ &= \int_{V_{\mathrm{m}}(t)} \frac{\partial}{\partial x_i}[-p\delta_{ij}v_j + \tau_{ij}v_j]\,\mathrm{d}V_{\mathrm{m}} \\ &= \int_{V_{\mathrm{m}}(t)} \left(-\frac{\partial}{\partial x_i}(pv_i) + \frac{\partial}{\partial x_i}(\tau_{ij}v_j)\right)\mathrm{d}V_{\mathrm{m}}\end{aligned} \tag{3.49}$$

The generalized first law in local form is thus:

$$\rho\left(\frac{\mathrm{d}e}{\mathrm{d}t} + v_i\frac{\mathrm{d}v_i}{\mathrm{d}t}\right) = \rho g_i v_i - \frac{\partial}{\partial x_i}(pv_i) + \frac{\partial}{\partial x_i}(\tau_{ij}v_j) - \frac{\partial}{\partial x_i}(q_i^{\mathrm{cd}} + q_i^{\mathrm{R}}) + P \tag{3.50}$$

The terms on the right-hand side of this equation are, respectively, the powers associated with body forces, pressure forces, viscous forces and thermal effects.

The energy balance equation is obtained from equation (3.50) by substitution, term by term, from the kinetic energy balance equation. The **kinetic energy balance equation** is obtained by taking the scalar product of the momentum balance equation

(3.44) with $v_i$:

$$\rho v_i \frac{\mathrm{d}v_i}{\mathrm{d}t} = \rho g_i v_i - v_i \frac{\partial}{\partial x_i} p + v_i \frac{\partial}{\partial x_i} \tau_{ji} \tag{3.51}$$

The internal energy balance equation follows by subtraction of (3.51) and (3.50), term by term. After some simple manipulation:

$$\rho \frac{\mathrm{d}e}{\mathrm{d}t} = -p\left(\frac{\partial}{\partial x_i} v_i\right) + \tau_{ij} \frac{\partial}{\partial x_i} v_j + P - \frac{\partial}{\partial x_i} (q_i^{\mathrm{cd}} + q_i^{\mathrm{R}}) \tag{3.52}$$

Note that to obtain (3.52), we have made use of the fact that $v_i \partial/\partial x_j \tau_{ji}$ is equal to $v_j \partial/\partial x_i \tau_{ij}$ because a double summation has been made over $i$ and $j$, and the tensor $\tau_{ij}$ is symmetric.

Equation (3.52) is, in fact, less interesting in non-adiabatic processes. It is preferable to introduce the enthalpy balance equation by using:

$$e = \mathrm{h} - \frac{p}{\rho} \tag{3.53}$$

where h is the specific enthalpy. This leads (using the continuity equation (3.34)) to:

$$\frac{\mathrm{dh}}{\mathrm{d}t} = \frac{\mathrm{d}e}{\mathrm{d}t} + \frac{1}{\rho}\frac{\mathrm{d}p}{\mathrm{d}t} + \frac{p}{\rho}\frac{\partial}{\partial x_i} v_i \tag{3.54}$$

The enthalpy balance equation is then:

$$\rho \frac{\mathrm{dh}}{\mathrm{d}t} = -\frac{\partial}{\partial x_i} (q_i^{\mathrm{cd}} + q_i^{\mathrm{R}}) + P + \tau_{ij} \frac{\partial}{\partial x_i} v_j + \frac{\mathrm{d}p}{\mathrm{d}t} \tag{3.55}$$

The first three terms on the right-hand side correspond to irreversible heat exchanges; the third, which is always positive, corresponds to heating of the fluid by irreversible viscous forces. The last term is reversible, and is associated with pressure forces. For a liquid, the difference between enthalpy and internal energy is often academic. We usually assume:

$$\frac{\mathrm{dh}}{\mathrm{d}t} = c \frac{\mathrm{d}T}{\mathrm{d}t} \tag{3.56}$$

where $c$ is the specific heat of the liquid. For a real gas, the specific enthalpy is given by:

$$\mathrm{dh} = c_p \, \mathrm{d}T + \frac{1}{\rho} (1 - T\beta) \, \mathrm{d}p \tag{3.57}$$

where $\beta$ is the coefficient of thermal expansion, whose derivative is equal to $-1/\rho(\partial\rho/\partial T)_{\mathrm{p}}$. The general energy equation then takes the form, derived from (3.55):

$$\rho c_{\mathrm{p}} \frac{\mathrm{d}T}{\mathrm{d}t} = \rho c_{\mathrm{p}} \left(\frac{\partial T}{\partial t} + v_i \frac{\partial T}{\partial x_i}\right) = -\frac{\partial}{\partial x_i} (q_i^{\mathrm{cd}} + q_i^{\mathrm{R}}) + P + T\beta \frac{\mathrm{d}p}{\mathrm{d}t} + \tau_{ij} \frac{\partial}{\partial x_i} v_j \tag{3.58}$$

The scalar quantity representing the heating due to viscous stresses (proportional to $\mu$, if $\mu$ is constant) is generally written:

$$\mu\Phi_D = \tau_{ij} \frac{\partial}{\partial x_i} v_j \tag{3.59}$$

$\Phi_D$ is called the dissipation function. In practical cases where there is significant energy transfer by conduction or radiation, the last two terms of (3.58) (the dissipation function and the effect of pressure forces) are negligible. A detailed discussion of this point is given in reference [15]. These terms only play an important role in compressible flow (velocities close to the speed of sound). **In most common applications, the energy balance equation** (in integral form and vector notation) **is**:

$$\int_{V_m(t)} \rho(T)c_p(T)\frac{dT}{dt}\,dV_m = \int_{V_m(t)} \rho c_p\left(\frac{\partial T}{\partial t} + \mathbf{v}\cdot\boldsymbol{\nabla}T\right)dV_m$$

$$= \int_{V_m(t)} P\,dV_m + \int_{S_m(t)} -(\mathbf{q}^{cd} + \mathbf{q}^R)\cdot\mathbf{n}_{ext}\,dS_m \tag{3.60}$$

Note that for a solid, a liquid or a perfect gas, the first term in this equation represents the time derivative of the enthalpy $d\mathscr{H}/dt$.

*Note*

For a perfect gas ($\beta = 1/T$) and for an adiabatic process (no heat transfer) the local energy equation reduces to:

$$\rho c_p\,dT = dp \tag{3.61}$$

Combining (3.61) with the equation of state ($p = \rho r T$) and the relation $c_p - c_v = r$, it becomes:

$$\frac{dp}{p} = \frac{c_p}{c_p - c_v}\frac{dT}{T} = \frac{\gamma}{\gamma - 1}\frac{dT}{T} \tag{3.62}$$

the well-known equation for an isentropic process in a perfect gas.

In the general case of transfer in systems involving fluids, obtaining temperature fields and heat fluxes involves the solution of the energy equations (3.58) and the set of dynamic equations which are coupled to it (via the term $\mathbf{v}\cdot\boldsymbol{\nabla}T$) representing the balance of mass, momentum, kinetic energy, etc. **This approach is developed in Part IV of this work (Chapters 13–16).**

## *3.4 A direct application. Convection in a duct*

We will consider steady flow of a fluid, at a constant mass flow rate $\dot{m}$, in a cylindrical duct with parallel generators, of constant cross-section $\Sigma$.

### 3.4.1 Simplifying assumptions

The thermal problem is separated from the mechanical problem, by means of a number of assumptions:

- The thermophysical properties of the fluid ($\lambda$, $c_p$, etc.) are assumed constant, and, in particular, are assumed independent of temperature. We will take, in effect, their mean values over the length of the duct. **The density $\rho$ may well vary appreciably with $T$** (particularly for a gas).
- The fluid velocity **v** is everywhere parallel to the generators, i.e. in the $x$ direction (Figure 3.7). It has a magnitude $u$; this implies that we are far from the entrance of the tube. In steady flow, the balance of mass in a flux tube of area $d\Sigma$ between sections $x$ and $x + dx$ is:

$$(\rho u)(x, y, z)\, d\Sigma = (\rho u)(x + dx, y, z)\, d\Sigma \tag{3.63}$$

  whence:

$$\frac{\partial}{\partial x}(\rho u) = 0 \tag{3.64}$$

  We note that the axial velocity $u$ is not constant along a flux tube in so far as $\rho$ varies (notably for a gas).
- The pressure gradient which creates the flow is sufficiently weak that we can consider the heat exchange to be at constant pressure and characterized by $c_p$ (see the conclusions of Section 3.3.2.4).
- If the fluid is a gas, it is assumed perfect. Then, whether we are concerned with a liquid or a gas, equation (3.60) gives the derivative of the enthalpy of the system, $\mathscr{H}$, with respect to time ($d\mathscr{H}/dt$).

### 3.4.2 Energy balance

The fluid temperature varies, in general, with all three space coordinates. We will adopt a simplified one-dimensional model, similar to the fin approximation, by introducing the concept of **mixed fluid temperature**.

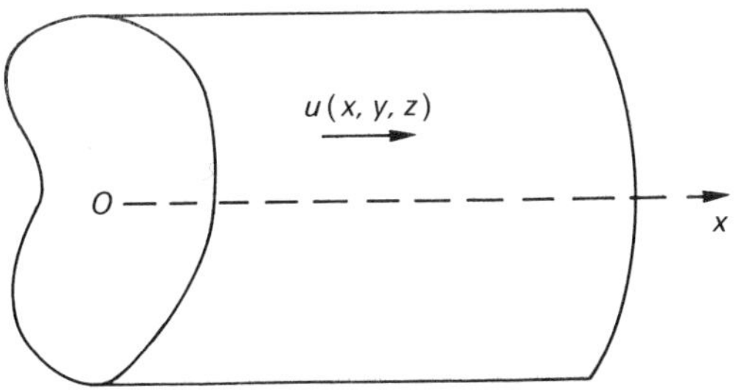


**Figure 3.7**

3.4.2.1 THE CONCEPT OF MIXED FLUID TEMPERATURE

Consider the material system comprising the fluid which, at instant $t$, lies between $x$ and $x + \mathrm{d}x$, as shown in Figure 3.8(a).

An instant later, the system is displaced and deformed (see Figure 3.8(b)). The deformation is important because the flow lines in the immediate vicinity of the wall are practically undisplaced (no velocity at the wall), in contrast to the flow lines in the central part of the tube.

The change of enthalpy per unit time in the system under steady conditions is given by:

$$\frac{\mathrm{d}\mathscr{H}}{\mathrm{d}t} = \mathrm{d}x \int_{\Sigma} \rho c_{\mathrm{p}} u \frac{\partial T}{\partial x} \mathrm{d}y \, \mathrm{d}z \tag{3.65}$$

in which $\rho$, $u$ and $T$ depend on $x$, $y$, $z$, while $c_{\mathrm{p}}$ is constant. After transformation, using (3.64), we obtain:

$$\frac{\mathrm{d}\mathscr{H}}{\mathrm{d}t} = c_{\mathrm{p}} \, \mathrm{d}x \left[ \frac{\mathrm{d}}{\mathrm{d}x} \int_{\Sigma} \rho u T \, \mathrm{d}y \, \mathrm{d}z \right] \tag{3.66}$$

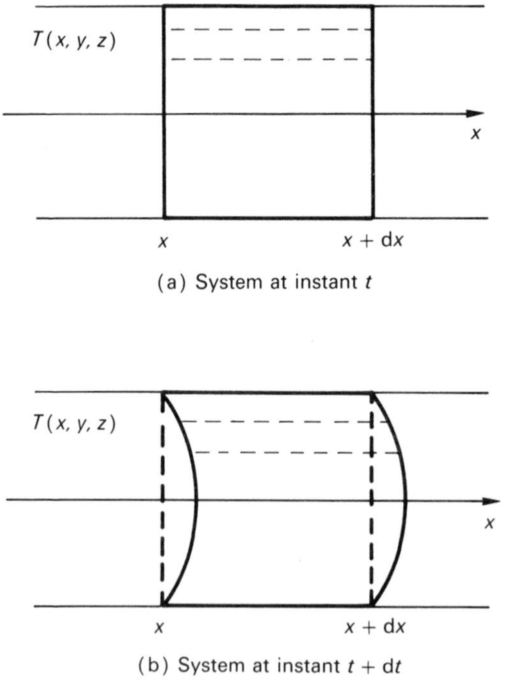


**Figure 3.8**

The integral over $\Sigma$ can be interpreted as an average of temperature weighted by mass flow: in fact enthalpy. The quantity:

$$T_m(x) = \frac{\displaystyle\int_\Sigma \rho u T \,\mathrm{d}y\,\mathrm{d}z}{\displaystyle\int_\Sigma \rho u \,\mathrm{d}y\,\mathrm{d}z} = \frac{\displaystyle\int_\Sigma \rho u T \,\mathrm{d}y\,\mathrm{d}z}{\dot{m}} \tag{3.67}$$

is called the **mixed fluid temperature**. This would be the temperature of the cross-section $\Sigma$ if thermal equilibrium were reached at it, whence the term of mixed fluid temperature. It is an overall temperature at the cross-section, associated with the mass flow $\dot{m}$. Using this, equation (3.66) becomes:

$$\frac{\mathrm{d}\mathscr{H}}{\mathrm{d}t} = \dot{m} c_\mathrm{p}\,\mathrm{d}T_\mathrm{m} \tag{3.68}$$

This new rigorous expression is formally identical to that appropriate to a solid body motion (see equation (3.4)).

*Note*
If $c_\mathrm{p}$ depends on $T(x, y, z)$, the expression for the mixed fluid temperature (3.67) can be generalized by replacing $\rho u$ by $\rho u c_\mathrm{p}$.

### 3.4.2.2 ENERGY BALANCE FOR A THIN SLICE

Suppose, for simplicity, that radiative exchanges are negligible. Furthermore, **axial conduction is assumed negligible** (see the conclusions to Section 3.2.2). This assumption will similarly be justified in Chapter 14 by consideration of orders of magnitude.

With conducto-convective flux at the wall, the energy balance for a layer of material $[x, x + \mathrm{d}x]$ is:

$$\frac{\mathrm{d}\mathscr{H}}{\mathrm{d}t} = \dot{m} c_\mathrm{p}\,\mathrm{d}T_\mathrm{m} = \mathscr{P}\,\mathrm{d}x h(T_\mathrm{w}(x) - T_\mathrm{m}(x)) \tag{3.69}$$

where $\mathscr{P}$ is the wetted perimeter of the duct, and $T_\mathrm{w}(x)$ the wall temperature, assumed constant at any cross-section (perhaps averaged around the wetted perimeter $\mathscr{P}$). In the very special case where the wall temperature $T_\mathrm{w}$ of the duct is also constant with $x$, the fluid temperature tends exponentially to $T_\mathrm{w}$ according to:

$$\frac{T_\mathrm{m} - T_\mathrm{w}}{T_0 - T_\mathrm{w}} = \exp\left[-\frac{\mathscr{P} h x}{\dot{m} c_\mathrm{p}}\right] \tag{3.70}$$

where $T_0$ is the temperature at the duct entrance. This situation can occur when there is a two-phase system on the other side of the wall throughout its length – an ideal condenser or evaporator (Figure 3.9).

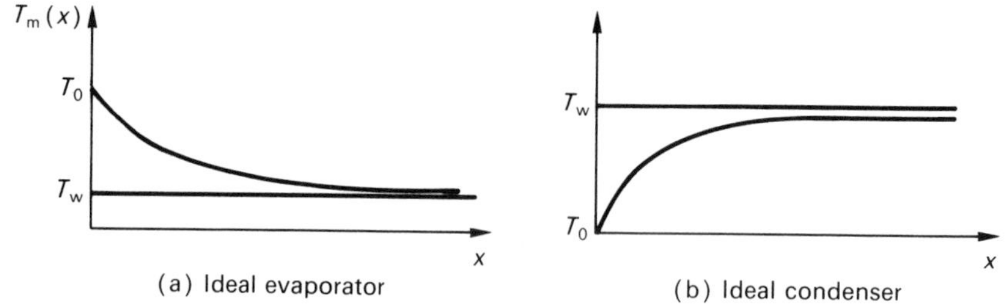


**Figure 3.9**

## *3.5 Ideas on heat exchangers*

Consider a plane heat exchanger with two fluids, 'hot' $h$ and 'cold' $c$, separated by a conducting wall of thickness $e$ and conductivity $\lambda$, across which heat is exchanged.

The wall is impermeable to matter and permeable to heat; a generalization to porous walls can be envisaged in chemical engineering. The two other walls, parallel to the median wall, are perfectly insulating and are at distance $d$ from the median wall. This setup is realized in practice by a stack of heat exchanger elements of the type described here. We assume that the extent of the heat exchanger $L$ in the $y$ direction is large enough compared to $d$ for end effects to be neglected.

Such an exchanger is called **co-flow** if the fluids circulate in the same sense in the $x$ direction, and **counter-flow** if they circulate in opposite senses (as in Figure 3.10). **We will adopt the hypotheses and conclusions of Section 3.4** and will not consider radiative transfers (or linearized transfers taken into account in the transfer coefficients adopted).

Let $\dot{m}_h$ and $\dot{m}_c$ be the constant mass flows of the hot and cold fluids, **taken as positive in the positive $x$ direction**; $\dot{m}_h$ will always be positive, $\dot{m}_c$ may be positive or negative according to the type of exchanger. We will use $T_{mh}(x)$, $T_{mc}(x)$, $T_{wh}(x)$, $T_{wc}(x)$ to represent the temperatures of the 'hot' and 'cold' fluids and corresponding walls, at position $x$. The energy balances for the slices of fluid between $x$ and $x + dx$ at instant $t$ (see Figure 3.10) are then:

$$-\mathrm{d}\Phi = \dot{m}_h c_{ph}\,\mathrm{d}T_{mh} = -L\,\mathrm{d}x h_h (T_{mh} - T_{wh}) \qquad (3.71)$$

$$\mathrm{d}\Phi = \dot{m}_c c_{pc}\,\mathrm{d}T_{mc} = -L\,\mathrm{d}x h_c (T_{mc} - T_{wc}) \qquad (3.72)$$

where $h_h$ and $h_c$ are the **wall heat transfer coefficients**, **averaged over the length of the exchanger** on the hot and cold sides. Obviously, in this perfect heat exchanger, the heat lost by the hot fluid, $|\mathrm{d}\Phi|$, is gained by the cold fluid.

The temperature differences $T_{mh} - T_{wh}$ and $T_{mc} - T_{wc}$ can be calculated using the global heat transfer coefficient $h_g$ (called $U$ is some literature), appropriate to the conductance per unit area of an element $\mathrm{d}x$ of the exchanger (see Section 2.1.2).

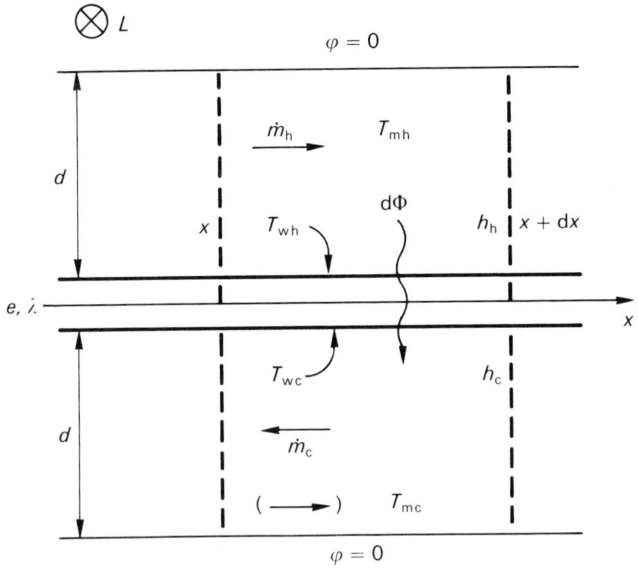


**Figure 3.10** Section of plane heat exchanger.

Therefore:

$$d\Phi = -L\,dx h_c(T_{mc} - T_{wc}) = L\,dx h_h(T_{mh} - T_{wh}) = -L\,dx h_g(T_{mc} - T_{mh}) \tag{3.73.1}$$

whence:

$$T_{mc} - T_{wc} = \frac{h_g}{h_c}(T_{mc} - T_{mh}) \tag{3.73.2}$$

and

$$T_{mh} - T_{wh} = \frac{h_g}{h_h}(T_{mh} - T_{mc}) \tag{3.73.3}$$

in which $h_g$ represents[2]:

$$h_g = \left(\frac{1}{h_h} + \frac{e}{\lambda} + \frac{1}{h_c}\right)^{-1} \tag{3.74}$$

This simplifies to:

$$\dot{m}_h c_{ph}\,dT_{mh} = -L\,dx h_g(T_{mh} - T_{mc}) \tag{3.75.1}$$

$$\dot{m}_c c_{pc}\,dT_{mc} = +L\,dx h_g(T_{mh} - T_{mc}) \tag{3.75.2}$$

[2] One could add fouling resistance (see Section 2.2.2), equivalent to adding extra conducting layers, whose properties would have to be found.

By subtracting (3.75.1) and (3.75.2) term by term, integrating and letting the entrance to the exchanger at $x = 0$, we obtain:

$$T_{mh}(x) - T_{mc}(x) = (T_{mh}(0) - T_{mc}(0)) \exp\left[-h_g L x \left(\frac{1}{\dot{m}_h c_{ph}} + \frac{1}{\dot{m}_c c_{pc}}\right)\right] \quad (3.76)$$

Substituting (3.76) into (3.75.1) and (3.75.2), and writing $S_{max}$ for the quantity $Ll$, we obtain:

$$\frac{T_{mh}(0) - T_{mh}(l)}{T_{mh}(0) - T_{mc}(0)} = \frac{\dot{m}_c c_{pc}}{\dot{m}_c c_{pc} + \dot{m}_h c_{ph}} \left\{1 - \exp\left[-h_g S_{max}\left(\frac{1}{\dot{m}_h c_{ph}} + \frac{1}{\dot{m}_c c_{pc}}\right)\right]\right\} \quad (3.77)$$

$$\frac{T_{mc}(l) - T_{mc}(0)}{T_{mh}(0) - T_{mc}(0)} = \frac{\dot{m}_h c_{ph}}{\dot{m}_c c_{pc} + \dot{m}_h c_{ph}} \left\{1 - \exp\left[-h_g S_{max}\left(\frac{1}{\dot{m}_h c_{ph}} + \frac{1}{\dot{m}_c c_{pc}}\right)\right]\right\} \quad (3.78)$$

Equations (3.77) and (3.78) are valid for both co-flow ($\dot{m}_h, \dot{m}_c > 0$) and counter-flow ($\dot{m}_h > 0, \dot{m}_c < 0$) heat exchangers, whatever their cross-sectional shape: cylindrical, triangular, etc. In general in a more complex industrial heat exchanger, the heat transfer surface $S_{max}$ offered to one fluid is greater than that offered to the other. In such a case, the **global heat transfer coefficient $h_g$ is conventionally based on** $S_{max}$.

In the particular case where $\dot{m}_c c_{pc}$ (or $\dot{m}_h c_{ph}$) tends to infinity, so that $T_{mc}(x)$ is constant at $T_{mc}(0)$ (or $T_{mh}(x)$ is constant at $T_{mh}(0)$), we come back to a tube with constant wall temperature, as already considered in Section 3.4.2.2 (equation (3.70)).

The temperatures and temperature differences of the two fluids vary exponentially. The cases of co-flow and counter-flow exchangers are very different:

- For co-flow heat exchangers, $\dot{m}_h$ and $\dot{m}_c$ are positive; the fluid temperatures tend to converge at the common exit of the exchanger (Figure 3.11).
- For counter-flow heat exchangers, $\dot{m}_h$ is positive, and $\dot{m}_c$ negative; the exponent in (3.76) tends to zero with $\dot{m}_c c_{pc} + \dot{m}_h c_{ph}$. We can distinguish various cases of variation shown in Figures 3.12(a, b and c).

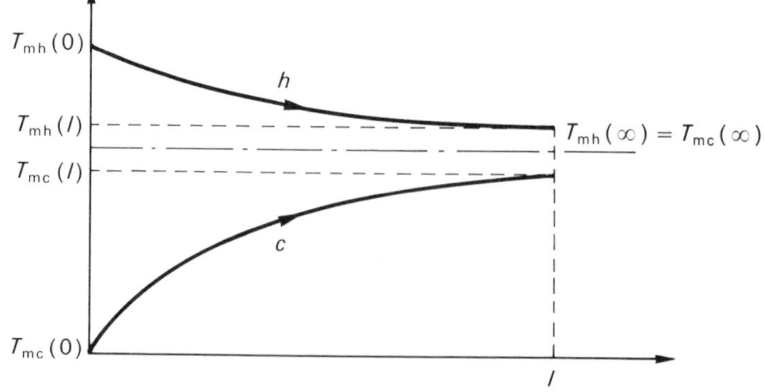


**Figure 3.11** Co-flow heat exchanger.

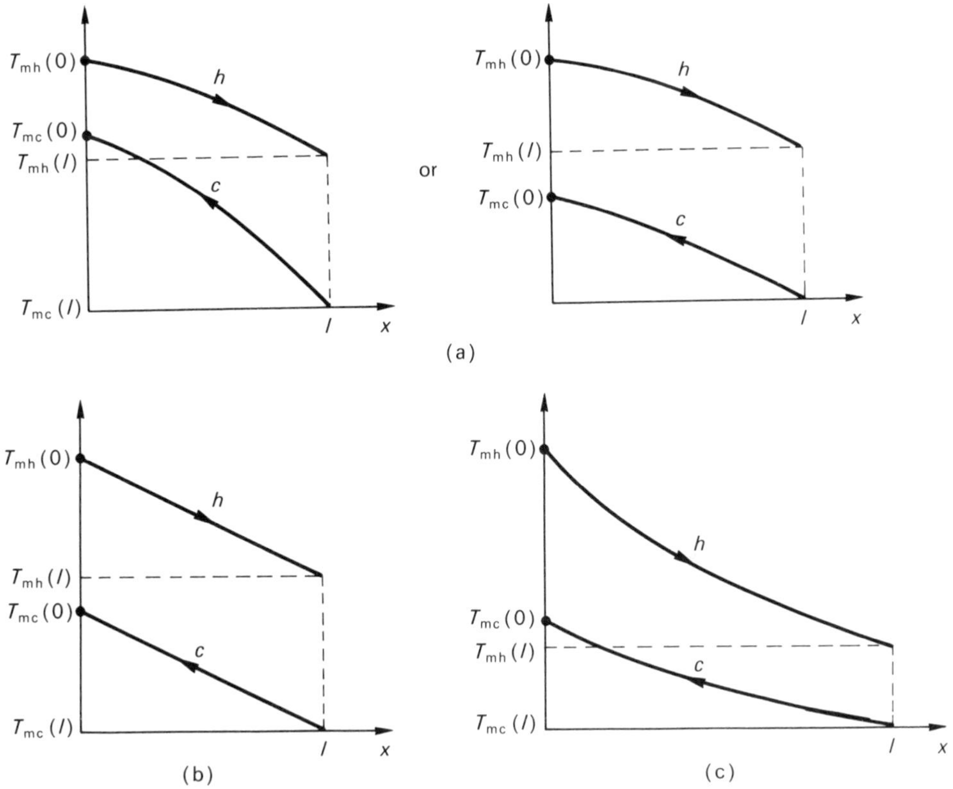


**Figure 3.12** (a) Counter-flow heat exchanger ($\dot{m}_h c_{ph} + \dot{m}_c c_{pc} > 0$). (b) Counter-flow heat exchanger ($\dot{m}_h c_{ph} + \dot{m}_c c_{pc} = 0$). (c) Counter-flow heat exchanger ($\dot{m}_h c_{ph} + \dot{m}_c c_{ph} < 0$).

We see that for a counter-flow heat exchanger, the 'cold' fluid may leave at a higher temperature than the 'hot' fluid ($T_{mc}(0) > T_{mh}(l)$ – Figure 3.12(a)), which is impossible for a co-flow heat exchanger. The former type thus appears to give a better performance than the latter. We will look at this in the next paragraph.

### 3.5.1 *Efficiency and number of transfer units*

A heat exchanger can be considered as a quadrapole (the poles being the inlet and outlets of the hot and cold fluids). From now on, we will designate temperatures with superscripts *i* for inlet and *o* for outlets. $\dot{m}c_p$ will be called the capacity rate $Cr$:

$$Cr = |\dot{m}c_p| \tag{3.79}$$

$$\begin{aligned} Cr_{min} &= \min(Cr_h, Cr_c) \\ Cr_{max} &= \max(Cr_h, Cr_c) \end{aligned} \tag{3.80}$$

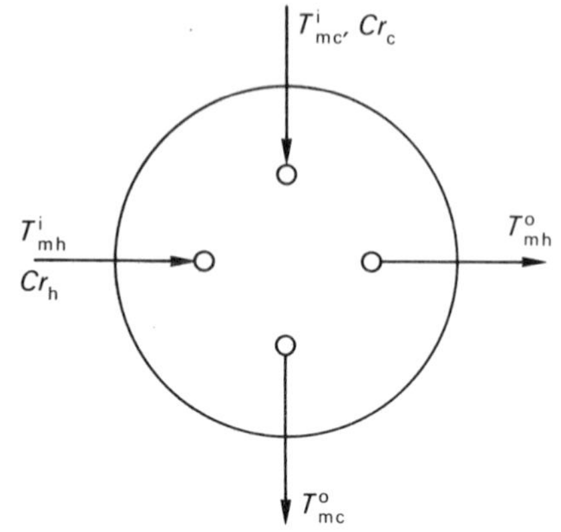


**Figure 3.13**

The general scheme of a heat exchanger is shown in Figure 3.13. There are many different types of heat exchanger (co-flow, counter-flow, cross-flow, mixed flow, etc.), but they can all be described by this quadrapole model.

The fundamental equation for a perfect heat exchanger (without losses to the surroundings) is:

$$|\Phi| = Cr_h(T^{i}_{mh} - T^{o}_{mh}) = Cr_c(T^{o}_{mc} - T^{i}_{mc}) \tag{3.81}$$

When we recall that industrial heat exchangers are constructed of compact stacks of basic modules, we can see that the losses are in practice negligible. In this equation, $|\Phi|$ represents the flux exchanged by the two fluids over the whole length of the exchanger.

To define the concept of heat exchanger performance, we introduce a dimensionless number $E$ called exchanger efficiency, defined as the ratio of the flux transferred from the hot to the cold fluid to the maximum theoretically transmissible flux $\Phi_{th}$:

$$\Phi_{th} = Cr_{min}(T^{i}_{mh} - T^{i}_{mc}) \tag{3.82}$$

To calculate $\Phi_{th}$, we assume that the fluids undergo the maximum change in temperature allowed by the second law of thermodynamics, $(T^{i}_{mh} - T^{i}_{mc})$. This could be achieved, for example, by a counter-flow heat exchanger of infinite length. It is easy to see from (3.81) that only the fluid with the lower capacity rate $Cr_{min}$ can actually undergo this temperature change. Therefore:

$$E = \frac{\Phi}{\Phi_{th}} = \frac{Cr_h(T^{i}_{mh} - T^{o}_{mh})}{Cr_{min}(T^{i}_{mh} - T^{i}_{mc})} = \frac{Cr_c(T^{o}_{mc} - T^{i}_{mc})}{Cr_{min}(T^{i}_{mh} - T^{i}_{mc})} \tag{3.83}$$

By way of example, let us consider the efficiencies of the co-flow and counter-flow heat exchangers in the preceding section. For the co-flow heat exchanger, we obtain:

$$E = \left(1 + \frac{Cr_{\min}}{Cr_{\max}}\right)^{-1} \left\{1 - \exp\left[-\frac{h_g S_{\max}}{Cr_{\min}}\left(1 + \frac{Cr_{\min}}{Cr_{\max}}\right)\right]\right\} \tag{3.84}$$

and for the counter-flow exchanger:

$$E = \left\{1 - \exp\left[-\left(1 - \frac{Cr_{\min}}{Cr_{\max}}\right)\frac{h_g S_{\max}}{Cr_{\min}}\right]\right\} \times \left\{1 - \left(\frac{Cr_{\min}}{Cr_{\max}}\right)\exp\left[-\left(1 - \frac{Cr_{\min}}{Cr_{\max}}\right)\frac{h_g S_{\max}}{Cr_{\min}}\right]\right\}^{-1} \tag{3.85}$$

It is clear from (3.84) and (3.86) that the efficiencies of the two types of heat exchangers are increased if the maximum heat exchange area increases (which should be obvious). For the co-flow heat exchanger, $E$ is limited by $(1 + Cr_{\min}/Cr_{\max})^{-1}$, while **for a counter-flow heat exchanger**, $E$ **may approach** 1 (but only for an 'infinite' length exchanger, which will be defined below). Increase in the area available for heat exchange between the two fluids, and thus in the size and, above all, the cost of the heat exchanger, is clearly limited by economic factors.

The efficiency of the heat exchanger is a function of two dimensionless quantities: the ratio $Cr_{\min}/Cr_{\max}$ and the group $h_g S_{\max}/Cr_{\min}$. This latter group can be considered as the ratio between the maximal exchange area $S_{\max}$ to a reference area $Cr_{\min}/h_g$, which is called the transfer unit; $h_g$ represents, we recall, the average overall heat transfer coefficient referred to the maximal exchange area, and $Cr_{\min}$ the lesser of the two fluids' 'capacity discharge rates'; the reference area represents the area needed to increase the temperature of one of the fluids by about one degree when the typical temperature difference between the two fluids is about one degree.

We will give the name *number of transfer units* (NTU) for a given type of heat exchanger to the quantity:

$$NTU = \frac{h_g S_{\max}}{Cr_{\min}} \tag{3.86}$$

where $S_{\max}$ is the larger exchange area, $h_g$ the overall average heat transfer coefficient (based on $S_{\max}$) and $Cr_{\min}$ the lower capacity rate. We then have the general characteristic relationship for any type of heat exchanger:

$$E = f\left(NTU, \frac{Cr_{\min}}{Cr_{\max}}\right) \tag{3.87}$$

For every type of heat exchanger, characteristic curves giving $E$ as a function of $NTU$, for various values of $Cr_{\min}/Cr_{\max}$ can be drawn (see, for example, Figure 3.14).

The efficiency is generally increased when:

- NTU (and thus the heat transfer area) is increased. For NTU of order 4, the efficiency tends asymptotically to a limit. Increasing the area produces only a small increase in efficiency.

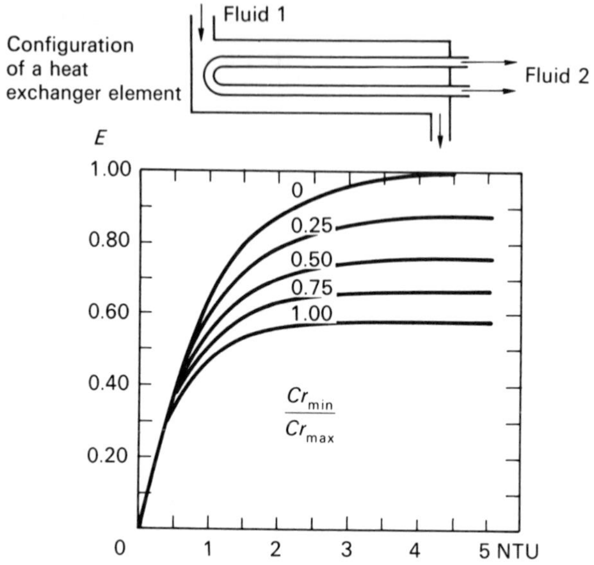


**Figure 3.14** Efficiency of a basic heat exchanger [16].

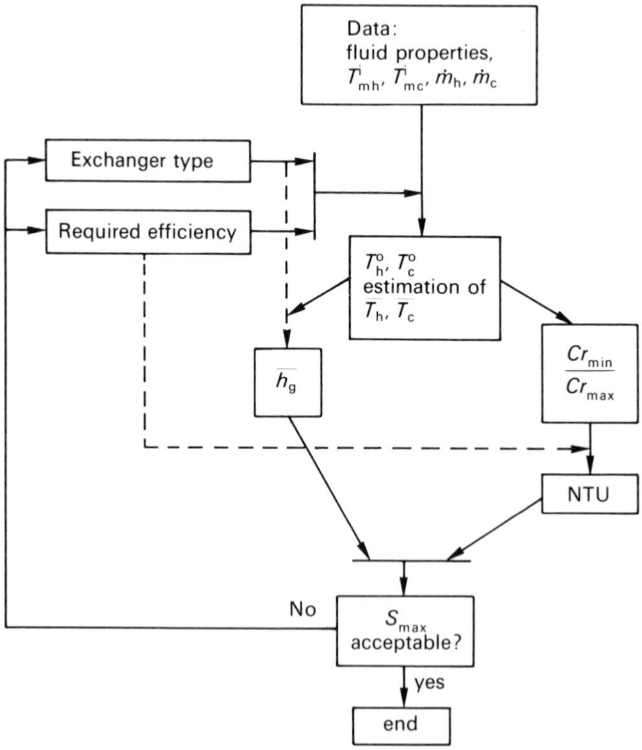


**Figure 3.15** Heat exchanger calculation method.

- The ratio $Cr_{min}/Cr_{max}$ is lower. But in the limit, the flow of one of the fluids becomes negligible, which is of little interest.

### 3.5.2 Heat exchanger calculation method

Suppose that we wish, in an industrial installation, to preheat one fluid with another (e.g. combustion mixture preheating using the products of the combustion). The entry and exit temperatures are specified. A type of heat exchanger is chosen to match the particular problem. We must determine the heat transfer area $S_{max}$ (and thus the volume of the exchanger, taking account of its compactness – square metres of heat exchange area per cubic metre). It is possible to do this iteratively, as shown in Figure 3.15.

*Notes*

1. There are innumerable types of exchangers and working fluids. For particular applications, it is useful to refer to specialist publications (e.g. reference [17]).
2. The problems of cost, fouling, leakage, thermal stress, vibrations, etc., must be taken into account.

# 4

# Introduction to Radiation

If we restrict ourselves to radiative energy transfers between **opaque bodies** through **transparent media** (see definitions in Section 1.1), radiative transfers only appear in the thermal balance of a system at the level of the boundary conditions (see Section 1.4). We will take this point of view in this chapter and only consider the configurations which are geometrically and physically simple. Part III of this work (Chapters 9–11) will, on the other hand, be devoted to a generalization of the theory of these transfers to arbitrary geometries and to semi-transparent media.

The basic problem is to evaluate the radiative flux density $\varphi^{R}$, which appears in the expression for energy flux at the boundary of an opaque body.

For example, if $T_s$ and $\lambda_s$ are the temperature field and conductivity of an opaque solid exchanging energy through a transparent fluid by radiation and with the fluid (of characteristic temperature $T_a$) by conducto-convective transfer with heat transfer coefficient $h$ (Figure 4.1), the continuity of energy flux requires:

$$-\lambda_s \frac{\partial T_s}{\partial x}(0) = \varphi^{R} + h(T_s(0) - T_a) \tag{4.1}$$

The field of thermal radiation is split into elementary rays, characterized by frequencies $\nu$ or wavelengths $\lambda$ (Section 4.1).

**The radiative flux density** $\varphi^{R}$ at a point $O$, positive in the positive $x$ direction, is the **sum over all wavelengths** ($\lambda \in [0, \infty]$) **of the monochromatic flux density** $d\varphi_{\lambda}^{R}$, **defined for the spectral interval** $[\lambda, \lambda + d\lambda]$:

$$\varphi^{R} = \int_{\lambda=0}^{\lambda=\infty} d\varphi_{\lambda}^{R} \tag{4.2}$$

All problems in thermal radiation start with, firstly, the determination of $d\varphi_{\lambda}^{R}$, and, secondly, the integration of equation (4.2). If the radiative flux densities $\varphi^{R}$ and $d\varphi_{\lambda}^{R}$ are algebraic values, the convention is to express $d\varphi_{\lambda}^{R}$ in terms of the particular monochromatic flux densities defined in Section 1.1 (see Figure 4.2(a and b)), which

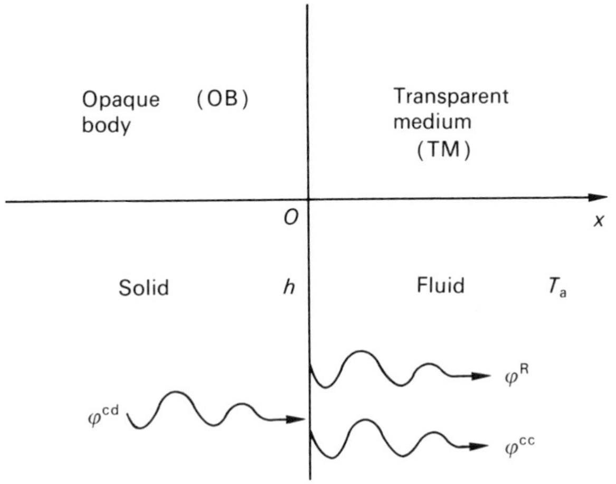


**Figure 4.1**

are arithmetic quantities:

- an **emitted** monochromatic flux density $d\varphi_\lambda^e$;
- a monochromatic flux density which will be **absorbed** $d\varphi_\lambda^a$;
- a **reflected** monochromatic flux density $d\varphi_\lambda^r$;
- an **incident** monochromatic flux density $d\varphi_\lambda^i$;
- an **emergent** (or **leaving**) monochromatic flux density $d\varphi_\lambda^l$;

From the point of view of the opaque body, the monochromatic radiation flux density $d\varphi_\lambda^R$ at $x = 0^-$ is:

$$d\varphi_\lambda^R = d\varphi_\lambda^e - d\varphi_\lambda^a \tag{4.3}$$

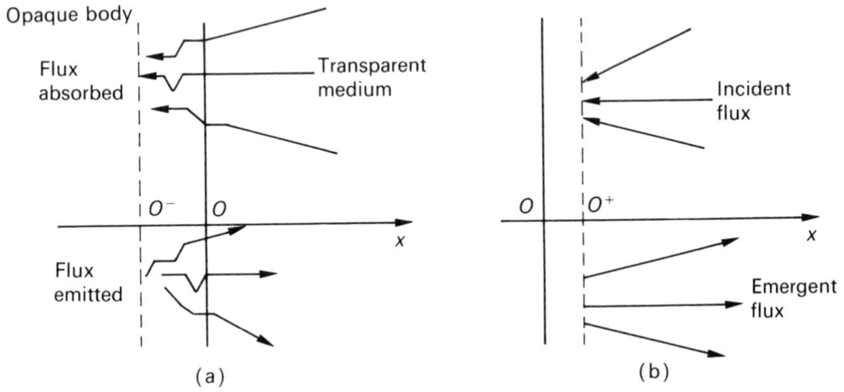


**Figure 4.2** (a) Seen from the opaque body. (b) Seen from the transparent medium.

From the point of view of the transparent medium, $d\varphi_\lambda^R$ at $x = 0^+$ is:

$$d\varphi_\lambda^R = d\varphi_\lambda^l - d\varphi_\lambda^i \tag{4.4}$$

It is obvious that (4.3) and (4.4) are equivalent for an opaque body (which does not transmit any part of the incident radiation). In fact, the incident monochromatic flux will be partly absorbed and partly reflected:

$$d\varphi_\lambda^i = d\varphi_\lambda^a + d\varphi_\lambda^r \tag{4.5}$$

while the emergent flux will be made up of the emitted and reflected fluxes:

$$d\varphi_\lambda^l = d\varphi_\lambda^e + d\varphi_\lambda^r \tag{4.6}$$

Using (4.4), (4.5) and (4.6), we come back to (4.3).

## *4.1 Domain of thermal radiation*

Electromagnetic radiation (EMR) is made up of quanta called photons, with energy $h\nu$, which are ultra-relativistic particles whose study is in the realm of statistical and quantum physics (Bose–Einstein statistics of indeterminate number of particles [2]). This study does not come within the purview of this course; we will, however, note that the laws of thermal radiation have a rigorous basis in this theory.

Here we will consider EMR propagating in transparent, non-dispersive media whose absolute refractive index $n$ is equal to 1 (free space and, to an excellent approximation, transparent gases such as $O_2$, $N_2$). The basic quantity characterizing an EMR is its frequency $\nu$, which is unchanged by reflection or passage through any medium. With the assumptions just made, the wavelength $\lambda$ of the radiation is also invariant:*

$$\lambda = \frac{c_0}{\nu} \tag{4.7}$$

where $c_0$ is the speed of the radiation in free space. Following common usage, we will generally use the wavelength expressed in $\mu$m ($10^{-6}$ m).

In Section 4.4 we will demonstrate that the most 'representative' wavelength $\lambda_m$ in a transparent medium, at temperature $T$ and in thermodyanmic equilibrium with the opaque bodies which surround it, obeys the remarkable formula:

$$T\lambda_m(T) = 2898\ \mu\text{m K} \tag{4.8}$$

From (4.8), we can deduce the characteristic wavelengths of radiation in equilibrium with sources at various temperatures (Figure 4.3).

* $\lambda$ is not invariant in a semi-transparent medium. In fact, $n$ and $c$ can depend on the local properties of the medium, have an imaginary part of even depend on $\nu$. The spectrum will usually be in terms of frequency (Hz) or more generally of wave number in free space (c m$^{-1}$).

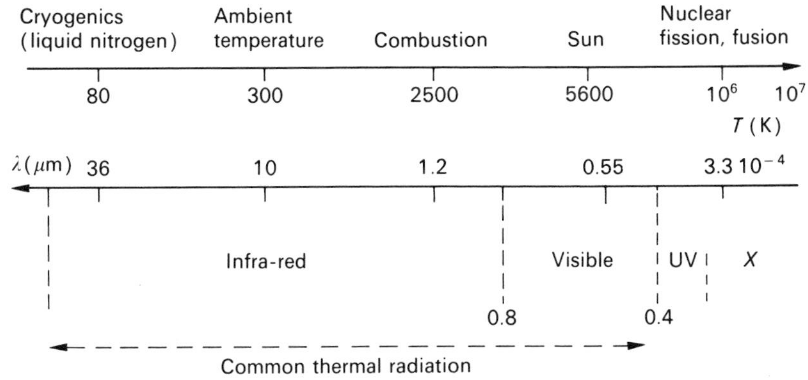


**Figure 4.3** The various electromagnetic radiations.

The normal domain of thermal radiation covers the near infra-red (0.8–70 μm) and the visible (0.4–0.8 μm). It is clear from equation (4.8) that **radiation is as much a thermodynamic as an electromagnetic concept**. The only available methods of measuring high temperatures are radiative (difficult to implement, generally making use of lasers, they are precise because they do not disturb the system being studied).

## 4.2 Theory of monochromatic flux

### 4.2.1 Directional monochromatic flux

A directional quantity is related to an elementary solid angle $d\Omega$, subtended around a direction defined by a unit vector $\mathbf{u}$. It may be marked with the symbol $'$.

Let us consider an element of surface $dS_1$ around a point $O_1$ which exchanges radiation with an element of surface $dS_2$ around a point $O_2$ (Figure 4.4). Let $\mathbf{n}_1$ and $\mathbf{n}_2$ be the normals to $dS_1$ and $dS_2$, $\theta_1$ and $\theta_2$ the angles $(\mathbf{n}_1, \mathbf{u}_1)$ and $(\mathbf{n}_2, \mathbf{u}_2)$ and $\mathbf{u}_1$ the unit vector from $O_1$ towards $O_2$. The directional monochromatic flux exchanged between $S_1$ and $S_2$ is proportional to:

- **the projected area of** $dS_1$ **in the direction** $O_1O_2$, i.e. $dS_1 \cos\theta_1$;
- the solid angle $d\Omega_1$ through which one sees $dS_2$ from $O_1$, i.e. $dS_2 \cos\theta_2 / O_1O_2^2$;
- the size $d\lambda$ of the spectral interval $[\lambda, \lambda + d\lambda]$.

We finally obtain:

$$d^5\Phi'_\lambda = I'_\lambda(O_1, \mathbf{u}) \frac{dS_1 \cos\theta_1 \, dS_2 \cos\theta_2}{O_1O_2^2} d\lambda \tag{4.9}$$

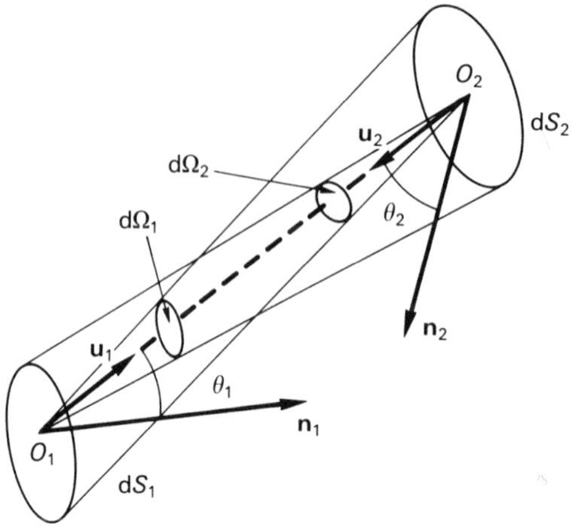


**Figure 4.4**

Equation (4.9) defines the **monochromatic intensity $I'_\lambda(O_1, \mathbf{u}_1)$ of the radiation considered**. It is a directional quantity which is invariant at all points on the line $O_1O_2$ for the type of radiation concerned, for the case where the medium between $O_1$ and $O_2$ is transparent. The geometric element of equation (4.9) is symmetrical with respect to the surfaces $dS_1$ and $dS_2$; (4.9) may be written in two different forms. Using the solid angle $d\Omega_1$, we obtain:

$$d^5\Phi'_\lambda = I'_\lambda(O_1, \mathbf{u}_1)\, dS_1 \cos\theta_1\, d\Omega_1\, d\lambda \tag{4.10}$$

If we use the solid angle $d\Omega_2$ through which one sees $dS_1$ from $O_2$, i.e. $dS_1 \cos\theta_1/O_1O_2^2$, we obtain:

$$d^5\Phi'_\lambda = I'_\lambda(O_1, \mathbf{u}_1)\, dS_2 \cos\theta_2\, d\Omega_2\, d\lambda \tag{4.11}$$

The intensity, the energy quantity, is additive. For example, the directional monochromatic flux vector leaving the surface $dS_1$, assumed opaque, $d^5\Phi'^{l}_{1\lambda}$ is made up of an emitted flux $d^5\Phi'^{e}_{1\lambda}$ and a reflected flux $d^5\Phi'^{r}_{1\lambda}$:

$$d^5\Phi'^{l}_{1\lambda} = d^5\Phi'^{e}_{1\lambda} + d^5\Phi'^{r}_{1\lambda} \tag{4.12}$$

Using equations analogous to (4.9), we obtain the relationship between the corresponding intensities:

$$I'^{l}_{1\lambda} = I'^{e}_{1\lambda} + I'^{r}_{1\lambda} \tag{4.13}$$

If $I'^{i}_{2\lambda}$ is the directional monochromatic flux incident on $dS_2$, we have:

$$I'^{l}_{1\lambda} = I'^{i}_{2\lambda} = I'^{a}_{2\lambda} + I'^{r}_{2\lambda} \tag{4.14}$$

where $I'^{r}_{2\lambda}$ is the intensity of the radiation incident on $dS_2$ and reflected into half-space, and $I'^{a}_{2\lambda}$ is the intensity of the radiation absorbed by $dS_2$.

*Important note*
The concept of monochromatic intensity $I'_\lambda$ is linked to the spectrum quantity chosen as a reference: in this case, the wavelength $\lambda$. It is possible to define **other quantities** which are monochromatic intensities referenced to the frequency $\nu$ ($I'_\nu$) or wave number $\sigma$ ($I'_\sigma$), which is the usage for semi-transparent media. The relationships between the three quantities are:

$$I'_\lambda \, \mathrm{d}\lambda = -I'_\nu \, \mathrm{d}\nu = -I'_\sigma \, \mathrm{d}\sigma \tag{4.15.1}$$

The minus signs appearing in (4.15.1) appear because the differentials $\mathrm{d}\lambda$ and $\mathrm{d}\nu$ are of opposite sign. Recall the expression for $\mathrm{d}\lambda$:

$$\mathrm{d}\lambda = (-c_0/n\nu)\left[\frac{(\mathrm{d}n/\mathrm{d}\nu)}{n} + \frac{1}{\nu}\right]\mathrm{d}\nu \tag{4.15.2}$$

## 4.2.2 Hemispherical monochromatic flux

### 4.2.2.1 THE GENERAL CASE

To calculate the radiative flux, it is necessary to obtain, as a preliminary, the monochromatic fluxes (emitted, reflected, absorbed, etc.) in all directions in the half-space of the transparent medium bounded by the surface element of the opaque body considered. Such fluxes are termed **hemispherical**. As before, we can use equation (4.10) or (4.11).

For example, the monochromatic flux leaving $\mathrm{d}S_2$ (emitted and reflected) is (without the $'$ symbol, since it is hemispherical flux):

$$\mathrm{d}^3\Phi^1_\lambda = \mathrm{d}S_1 \, \mathrm{d}\lambda \int_{2\pi} I'^1_\lambda(O_1, \mathbf{u}_1) \cos\theta_1 \, \mathrm{d}\Omega_1 \tag{4.16}$$

If the intensity is symmetrical about $\mathbf{n}_1$ (as in Figure 4.5), this last expression becomes (substituting dependence on $\theta_1$ for dependence on $u_1$):

$$\mathrm{d}^3\Phi^1_\lambda = \mathrm{d}S_1 \, \mathrm{d}\lambda \int_0^{\pi/2} I'^1_\lambda(O_1, \theta_1) \cos\theta_1 2\pi \sin\theta_1 \, \mathrm{d}\theta_1 \tag{4.17}$$

Equations analogous to (4.16) and (4.17) can be written for flux emitted or reflected by $\mathrm{d}S_1$, incident on $\mathrm{d}S_1$, etc., by substitution of the intensities corresponding to $I'^1_\lambda(O_1, \theta_1)$.

### 4.2.2.2 ISOTROPIC RADIATION

If the intensity of the radiation considered is independent of direction (we will represent this case by $I_\lambda$, omitting the prime $'$), equations (4.16) and (4.17) take a special form. Using the identity:

$$\int_0^{\pi/2} \cos\theta_1 (2\pi \sin\theta_1 \, \mathrm{d}\theta_1) = \pi \tag{4.18}$$

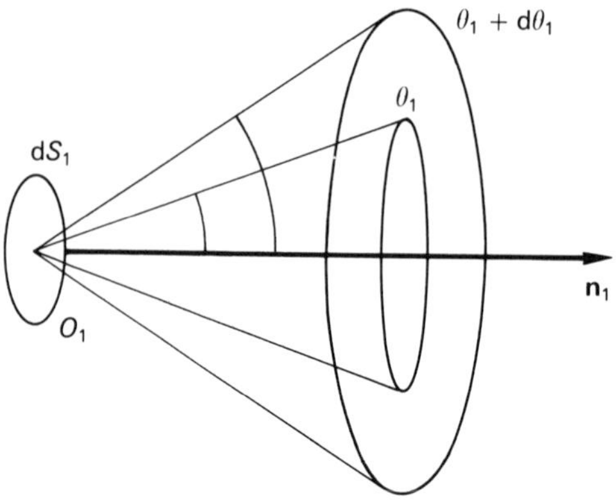


**Figure 4.5**

we obtain:

$$d^3\Phi^l_\lambda = \pi I^l_\lambda(O_1)\, dS_1\, d\lambda \tag{4.19}$$

The emitted, reflected, emergent and incident monochromatic flux densities then take the very simple forms:

$$\begin{aligned} d\varphi^e_\lambda &= \pi I^e_\lambda(O_1)\, d\lambda \\ d\varphi^r_\lambda &= \pi I^r_\lambda(O_1)\, d\lambda \\ d\varphi^l_\lambda &= \pi I^l_\lambda(O_1)\, d\lambda \\ d\varphi^i_\lambda &= \pi I^i_\lambda(O_1)\, d\lambda \end{aligned} \tag{4.20}$$

### *4.2.3 Radiative flux and the radiative flux density vector*

Let **n** be the unit vector of the outwards directed normal at a point M of an opaque body (Figure 4.6), and $\theta$ the angle $(\mathbf{u}, \mathbf{n})$. The monochromatic radiative flux density

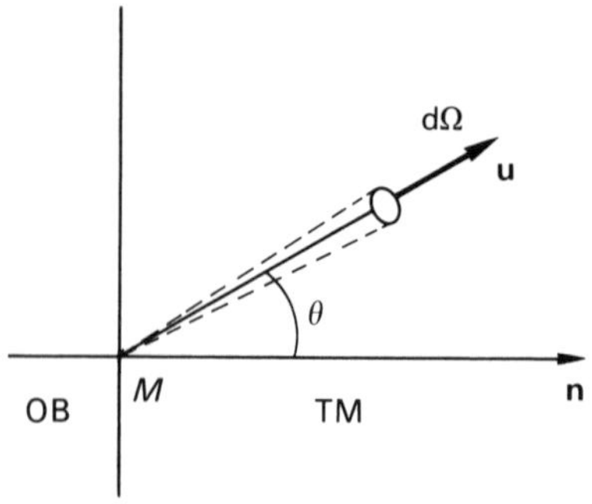


**Figure 4.6**

at M, in our usual notation, is:

$$d\varphi_\lambda^R = d\varphi_\lambda^e - d\varphi_\lambda^a = d\varphi_\lambda^l - d\varphi_\lambda^i \tag{4.21}$$

The emergent flux density is in the half space $\mathbf{u}\cdot\mathbf{n} = \cos\theta > 0$, the incident flux density in the half-space $\mathbf{u}\cdot\mathbf{n} = \cos\theta < 0$.

Thus:

$$d\varphi_\lambda^l = d\lambda \int_{2\pi(\cos\theta>0)} I_\lambda'^l \cos\theta \, d\Omega; \qquad d\varphi_\lambda^i = d\lambda \int_{2\pi(\cos\theta<0)} I_\lambda'^i |\cos\theta| \, d\Omega \tag{4.22}$$

The distribution of intensity (for all radiations together) at point M is $I'_\lambda(\mathrm{M}, \mathbf{u})$ defined by:

$$\begin{aligned} \cos\theta > 0 \qquad & I'_\lambda(\mathrm{M}, \mathbf{u}) = I_\lambda'^l(\mathrm{M}, \mathbf{u}) \\ \cos\theta < 0 \qquad & I'_\lambda(\mathrm{M}, \mathbf{u}) = I_\lambda'^i(\mathrm{M}, \mathbf{u}) \end{aligned} \tag{4.23}$$

Then, from (4.21) and (4.23),

$$d\varphi_\lambda^R = \mathbf{n}\cdot\left(\int_{4\pi} I'_\lambda(\mathrm{M}, \mathbf{u})\mathbf{u} \, d\Omega\right) \tag{4.24}$$

The quantity in parentheses in equation (4.24) is called the **monochromatic radiation flux density vector**. By integration over the spectrum, we define a radiative flux density vector $\mathbf{q}^R$, analogous to $\mathbf{q}^{cd}$:

$$\mathbf{q}^R = \int_0^\infty d\lambda \int_{4\pi} I'_\lambda(\mathrm{M}, \mathbf{u})\mathbf{u} \, d\Omega \tag{4.25}$$

such that:

$$\varphi^R = \mathbf{q}^R \cdot \mathbf{n} \tag{4.26}$$

We notice a marked similarity at the elementary scale between the phenomena of radiative and conductive heat transfer:

- The radiative flux density vector arises from the contributions of the energies of all the photons crossing a unit area of surface in unit time, in any direction.
- The conductive flux density arises, in the simple case of a monatomic gas, from the contributions of the kinetic energies of all the molecules crossing a unit area of surface in unit time, in any direction.

## *4.3 Thermal equilibrium and the directional monochromatic factors characterizing an opaque body*

The phenomena of absorption, emission and reflection are characterized by **directional monochromatic quantities** called **absorbtivity** (absorbtion factor), **emissivity** (emission factor) and **reflectivity** (reflection factor), which depend only on the opaque body concerned and the refractive index $n$ of the associated transparent medium. Below, we will introduce the transmissivity, for non-opaque bodies. In thermal equilibrium, these factors are linked by simple relationships which can be generalized when the system is not in equilibrium (but is in LTE).

### *4.3.1 Directional monochromatic absorptivity and reflectivity*

Consider radiation characterized by the incident directional monochromatic flux $d^5\Phi_\lambda'^{i}$, incident on the opaque surface $dS_1$, in the solid angle $d\Omega$ (Figure 4.7).

This flux will in part be absorbed (fraction $d^5\Phi_\lambda'^{a}$) and in part reflected in $2\pi$ steradians ($d^5\Phi_\lambda'^{\bullet r}$). Whence:

$$d^5\Phi_\lambda'^{i} = d^5\Phi_\lambda'^{a} + d^5\Phi_\lambda'^{\bullet r} \tag{4.27}$$

From the definition of the directional monochromatic absorbtivity of $dS_1$, $\alpha_\lambda'$:

$$\alpha_\lambda'(O_1, \theta_1) = \frac{d^5\Phi_\lambda'^{a}}{d^5\Phi_\lambda'^{i}} \tag{4.28}$$

and that of the directional monochromatic hemispheric reflectivity of $dS_1$, $\rho_\lambda'^{\bullet}$:

$$\rho_\lambda'^{\bullet}(O_1, \theta_1) = \frac{d^5\Phi_\lambda'^{\bullet r}}{d^5\Phi_\lambda'^{i}} \tag{4.29}$$

we can deduce the relationship:

$$\alpha_\lambda'(O_1, \theta_1) + \rho_\lambda'^{\bullet}(O_1, \theta_1) = 1 \tag{4.30}$$

The two factors defined by (4.28) and (4.29) clearly lie in the interval [0, 1]. Equation (4.30) is valid even when not in equilibrium.

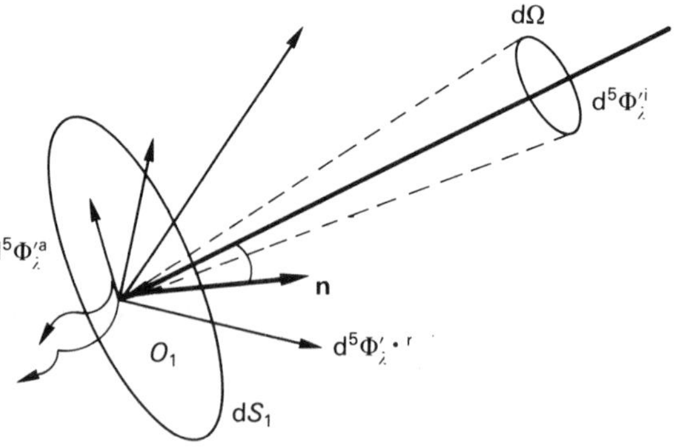


**Figure 4.7**

### 4.3.2 *Equilibrium radiation*

It can be shown in statistical physics [2] that in a transparent, non-dispersive medium with refractive index of 1, in thermal equilibrium with adjacent opaque bodies, an **equilibrium radiation** exists at every point. The intensity of this radiation is independent of direction (isotropic radiation) and depends only on the wavelength and the temperature. The intensity $I_\lambda^0(T)$ is given by Planck's equation:

$$I_\lambda^0 = 2hc_0^2\lambda^{-5}\left[\exp\left(\frac{hc_0}{k_B\lambda T}\right) - 1\right]^{-1} \tag{4.31}$$

in which $h$ and $k_B$ denote Planck and Boltzmann's universal constants:

$$\begin{aligned} k_B &= 1.3805\ 10^{-23}\ \mathrm{J/K} \\ h &= 6.626\ 10^{-34}\ \mathrm{Js} \\ c_0 &= 2.998\ 10^8\ \mathrm{m/s} \end{aligned} \tag{4.32}$$

The predictions of (4.31) agree with experiment to a remarkable degree of accuracy; this was historically one of the first successes of statistical physics. **No assumptions save that of a constant temperature $T$ are needed to obtain this result**. Equilibrium radiation will serve as a reference for all flux calculations, as we will show in the next section. Its basic properties are demonstrated in Section 4.4.

### 4.3.3 *Directional monochromatic emissivity*

Considering a system in thermal equilibrium at temperature $T$, the **emergent** monochromatic flux in the solid angle $d\Omega_1$ of an opaque surface element $dS_1$ is given (after (4.10) and (4.31)) by:

$$(d^5\Phi_\lambda'^{l})^{eq} = I_\lambda^0(T)\, dS_1 \cos\theta_1\, d\Omega_1\, d\lambda \tag{4.33}$$

This flux is made up of an **emitted** flux characterized by the intensity $I_\lambda'^{e}(O_1, \theta_1, T)$:

$$(d^5\Phi_\lambda^{\bullet\prime r})^{eq} = I_\lambda'^{e}(O_1, \theta_1, T)\, dS_1 \cos\theta_1\, d\Omega_1\, d\lambda \tag{4.34}$$

and a **reflected** flux $(d^5\Phi_\lambda^{\bullet\prime r})^{eq}$ in the solid angle $d\Omega_1$, originating from all the directions of the half-space visible to the opaque body (Figure 4.8).

These various fluxes satisfy the relation:

$$(d^5\Phi_\lambda'^{l})^{eq} = (d^5\Phi_\lambda'^{e})^{eq} + (d^5\Phi_\lambda^{\bullet\prime r})^{eq} \tag{4.35}$$

From (4.33)–(4.35), we can draw a fundamental conclusion:

$$(I_\lambda'^{e})^{eq}(O_1, \theta_1, T) \leqslant I_\lambda^0(T) \tag{4.36}$$

This result, which has been derived for equilibrium conditions, can be generalized to the case where there are temperature gradients in the system (by the hypothesis of local thermodynamic equilibrium LTE).

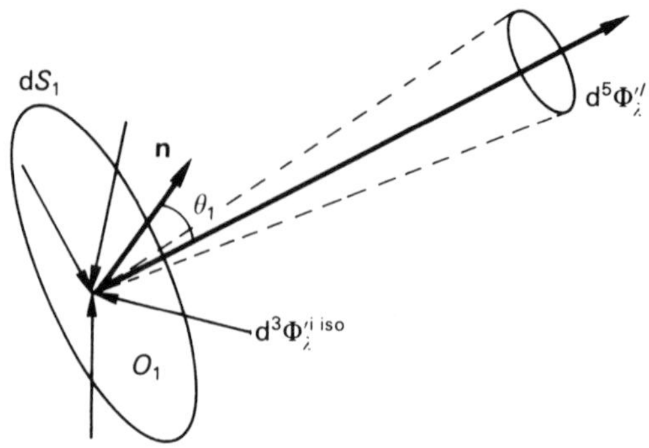


**Figure 4.8**

**The monochromatic intensity, $I'^e_\lambda(O_2, \theta_1, T)$, of the radiation emitted by an opaque body at temperature $T$ is always less than or equal to that of the equilibrium radiation at the same temperature $I^0_\lambda(T)$.**

The intensity of equilibrium radiation thus provides a reference and a limit to characterize the monochromatic intensity of the radiation emitted by an opaque body. We thus define the **directional monochromatic emissivity**, $\varepsilon'_\lambda$, of an opaque body by the equation:

$$I'^e_\lambda(O_1, \theta_1, T_1) = \varepsilon'_\lambda(O_1, \theta_1, T_1) I^0_\lambda(T_1) \tag{4.37}$$

$\varepsilon'_\lambda$ satisfies the inequality:

$$0 \leqslant \varepsilon'_\lambda \leqslant 1 \tag{4.38}$$

### *4.3.4 The fundamental law of thermal radiation (Draper's or Kirchoff's law)*

It can be shown [18], starting from a general principle of radiative diffusion known as Helmholtz's reciprocity principle, that in thermal equilibrium **the directional monochromatic absorbitivity, $\alpha'_\lambda$ of an opaque body is equal to its directional monochromatic emissivity $\varepsilon'_\lambda$:**

$$\varepsilon'_\lambda(O_1, T_1, \theta_1) = \alpha'_\lambda(O_1, T_1, \theta_1) \tag{4.39}$$

This relationship will be assumed valid outside equilibrium (but in LTE).

### *4.3.5 Common special cases*

#### 4.3.5.1 THE GREY BODY

A grey body is an opaque body whose radiative properties ($\alpha'$, $\varepsilon'$, $\rho'^{\bullet}$) are independent of wavelength but depend *a priori* on the direction $\theta_1$. In this very special case, (4.39) becomes:

$$\varepsilon'(O_1, T_1, \theta_1) = \alpha'(O_1, T_1, \theta_1) \tag{4.40}$$

This approximation is often made in a careless manner. A body can generally be considered as grey in a certain range of wavelengths. For example, in the middle infra-red (around 10 $\mu$m), dry plaster has an emissivity in the normal direction of around 0.9. Referring to Figure 4.3, it could be considered grey for radiation from low temperature sources (walls, ground, atmosphere, people, . . .). But plaster is much more highly reflecting for solar radiation (at 5600 K). Its absorbtivity is then no more than about 0.2. It could still be considered as grey, but with a different value of $\varepsilon$ for the latter radiation. Glass is another example of the limitations of the idea of a grey body (see Section 14.7).

#### 4.3.5.2 BODY WITH ISOTROPIC RADIATIVE PROPERTIES

These bodies are such that the monochromatic emissivity, absorbtivity and reflectivity are independent of direction. They thus depend only on wavelength; equation (4.39) becomes:

$$\varepsilon_\lambda(O_1, T_1, \theta_1) = \alpha_\lambda(O_1, T_1, \theta_1) = 1 - \rho_\lambda(O_1, T_1, \theta_1) \tag{4.41}$$

This approximation is often justified by the fact that the directional emissivity and absorbtivity factors are stationary in the neighbourhood of the normal direction (Figure 4.9). For example, $\varepsilon'_\lambda$ and $\alpha'_\lambda$ vary little between 0 and 60°; these incidences are dominant in the expression for flux because of the weighting by $\cos\theta_1$ in equation (4.10). The interest in the isotropic approximation is obviously to simplify the calculations considerably. A systematic study of these properties is undertaken in Part III: Radiation, Chapter 9.

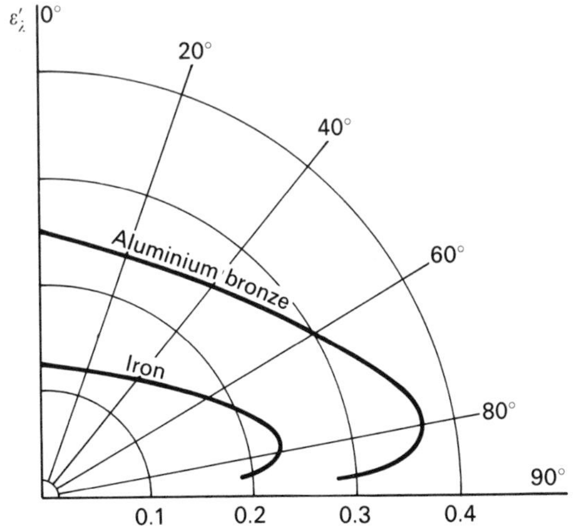


**Figure 4.9**

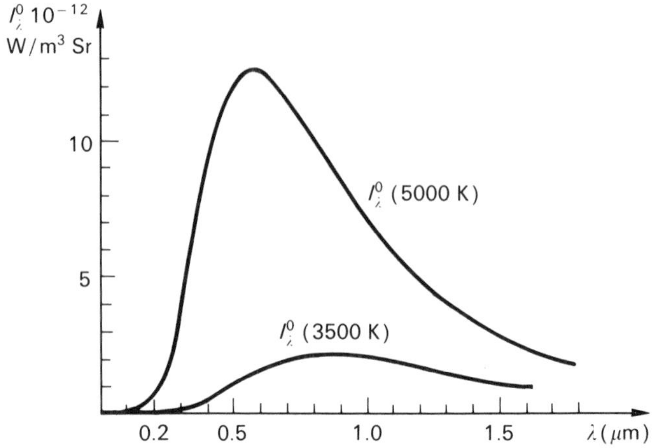


**Figure 4.10**

4.3.5.3 THE BLACK BODY

A black body, by definition, absorbs all radiation of whatever wavelength and direction:

$$\alpha'_\lambda = 1 \qquad \forall\ \lambda \text{ and } \theta_1 \tag{4.42}$$

We can deduce two fundamental properties:

- From (4.30), a black body does not reflect any radiation;
- From (4.37) and (4.39), we obtain:

$$(I'^{\mathrm{e}}_\lambda)^{\mathrm{BB}}(T) = I^0_\lambda(T) \qquad \forall\ \theta_1 \tag{4.43}$$

**The monochromatic intensity of radiation emitted by a surface element of a black body of temperature $T$ is equal to the intensity of equilibrium radiation at temperature $T$.**

In the following, $I^0_\lambda(T)$ will denote simultaneously the intensity of equilibrium radiation and that of the radiation emitted by a black body, locally at temperature $T$.

## *4.4 Properties of equilibrium radiation*

The intensity of equilibrium radiation is given (for a transparent medium of refractive index 1 – see (4.31)) by:

$$I^0_\lambda(T) = \frac{2hc_0^2\lambda^{-5}}{\exp\left(\dfrac{hc_0}{\lambda k_\mathrm{B} T}\right) - 1} \tag{4.44}$$

- This function has the characteristic shape of Figure 4.10.

- It has a maximum (for given $T$) for $\lambda_m(T)$ satisfying:

$$T \cdot \lambda_m(T) = 2898 \ \mu\text{mK} \tag{4.45}$$

which associates a characteristic wavelength with any specified temperature (see Section 4.1).

- The intensity curves corresponding to different temperatures do not cross (Figure 4.10). If the wavelength $\lambda$ is fixed:

$$T' > T \Rightarrow I_\lambda^0(T') > I_\lambda^0(T) \tag{4.46}$$

- Using normalized coordinates, defined by:

$$x = \frac{\lambda}{\lambda_m(T)}, \qquad y = \frac{I_\lambda^0(T)}{I_{\lambda_m(T)}(T)} \tag{4.47}$$

we can see from Figure 4.11 that $y$ is only significant when:

$$0.5 < x < 8 \tag{4.48}$$

**In practice, the range from $\lambda_m/2$ to $8\lambda_m$ corresponds to 98% of the energy equilibrium radiation.** This is also (from (4.37)) the range of wavelengths of radiation **emitted** by a body in LTE.

- The **total intensity** of equilibrium radiation is defined by:

$$I^0(T) = \int_0^\infty I_\lambda^0(T)\,\mathrm{d}\lambda = \frac{\sigma}{\pi} T^4 \tag{4.49}$$

In this important relationship, $\sigma$ represents a universal constant, Stefan's constant, which is a function of $h$, $k_B$ and $c_0$:

$$\sigma = \frac{2\pi^5 k_B^4}{15 c_0^2 h^3} = 5.67 \times 10^{-8} \ \text{Wm}^{-2}\,\text{K}^{-4} \tag{4.50}$$

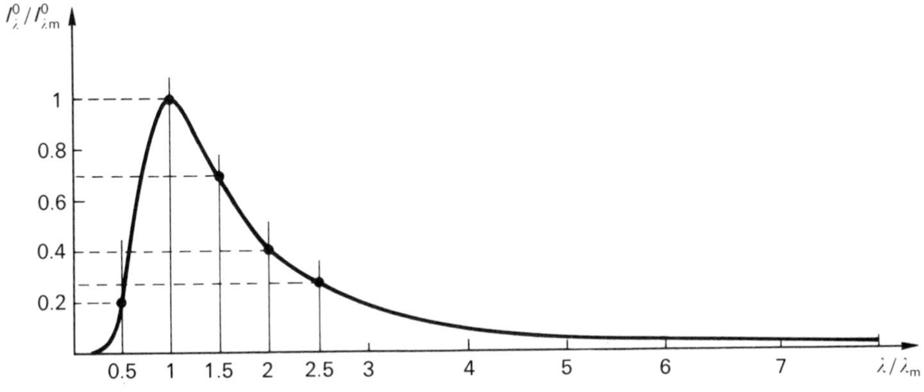


**Figure 4.11**

The total flux density of incident or emergent isotropic radiation, in equilibrium at temperature $T$, on an element of surface is, from (4.20):

$$\varphi^{\mathrm{i}} = \varphi^{\mathrm{l}} = \int_0^\infty \pi I_\lambda^0(T)\,\mathrm{d}\lambda = \sigma T^4 \tag{4.51}$$

(Equation (4.51) also gives the total flux **emitted** by a black body.)

- In practice, for the calculation of radiative heat transfer we use the quantity:

$$z\left(\frac{\lambda_1}{\lambda_{\mathrm{m}}(T)}, \frac{\lambda_2}{\lambda_{\mathrm{m}}(T)}\right) = \frac{\displaystyle\int_{\lambda_1}^{\lambda_2} I_\lambda^0(T)\,\mathrm{d}\lambda}{\displaystyle\int_0^\infty I_\lambda^0(T)\,\mathrm{d}\lambda} = \frac{\displaystyle\int_{\lambda_1}^{\lambda_2} \pi I_\lambda^0\,\mathrm{d}\lambda}{\sigma T^4} \tag{4.52}$$

which represents the ratio of the equilibrium radiation flux in the spectral band $[\lambda_1, \lambda_2]$ to the total flux for the same equilibrium radiation integrated over the whole spectrum. Values of $z(0, \lambda/\lambda_{\mathrm{m}}(T))$ as a function of $x = \lambda/\lambda_{\mathrm{m}}(T)$ are given in 'Basic Data' (Part V of this work).

## 4.5 Examples of direct application

The examples developed in this section correspond to particularly simple cases of radiative transfer. Complex three-dimensional applications will be treated in Part III of this work.

### 4.5.1 Opaque convex body surrounded by a black body

We consider a **convex** body of surface $S_1$, isothermal at temperature $T_1$, of directional monochromatic emissivity $\varepsilon'_\lambda(\theta, \phi, \lambda, T_1)$, completely surrounded by a black body, of some surface $S_2$, possibly concave, isothermal at $T_2$ (Figure 4.12). The medium

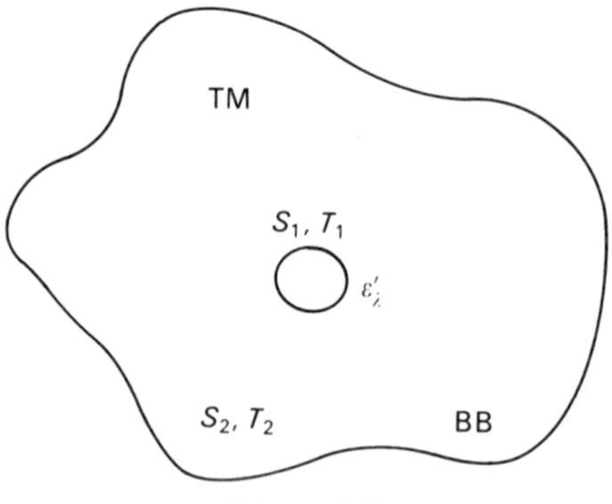


**Figure 4.12**

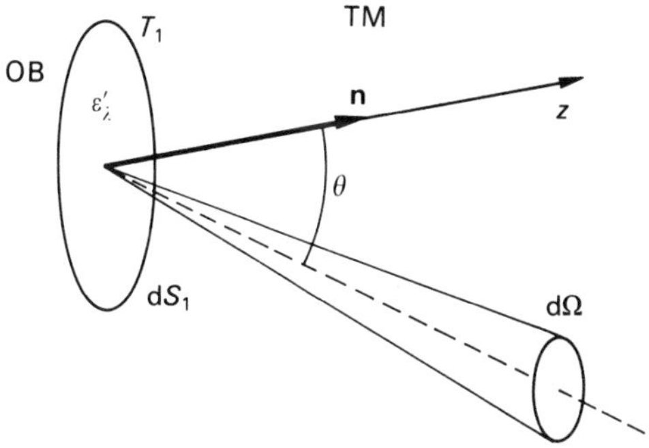


**Figure 4.13**

separating the two bodies is perfectly transparent. What is the radiative flux $\Phi^R$ crossing the surface $S_1$ and $S_2$?

Consider an **arbitrary** element $dS_1$ of the surface $S_1$ (Figure 4.13). If we choose an axis $Oz$ oriented towards the transparent medium, the flux across $dS_1$ is:

$$d\Phi^R = d\Phi^e - d\Phi^a \tag{4.53}$$

where $d\Phi^e$ and $d\Phi^a$ represent the flux emitted and absorbed. We have:

$$d\Phi^e = dS_1 \int_0^\infty d\lambda \int_{\Omega=2\pi} \varepsilon'_\lambda I^0_\lambda(T_1) \cos\theta \, d\Omega \tag{4.54}$$

and:

$$d\Phi^a = dS_1 \int_0^\infty d\lambda \int_{\Omega=2\pi} \varepsilon'_\lambda I'^i_\lambda \cos\theta \, d\Omega \tag{4.55}$$

where $I'^i_\lambda$ represents the intensity of the radiation incident on $dS_1$ in the solid angle $d\Omega$. If $S_2$ is a black body at temperature $T_2$, $I'^i_\lambda$ is independent of direction and is given by:

$$I'^i_\lambda = I^0_\lambda(T_2) \tag{4.56}$$

We then deduce:

$$d\Phi^R = dS_1 \int_0^\infty d\lambda \int_0^{\pi/2} \varepsilon'_\lambda (I^0_\lambda(T_1) - I^0_\lambda(T_2)) \cos\theta \, 2\pi \sin\theta \, d\theta \tag{4.57}$$

**The radiative flux does not depend in any way on the geometry of $S_2$, which may be concave!** This classical result has applications (e.g. ovens with 'black' walls). In the particular case of a grey body with, in addition, isotropic emission, the radiative flux becomes (since $\varepsilon'_\lambda$ is independent of $\lambda$ and $\theta$):

$$\varepsilon'_\lambda = \varepsilon(T_1) \tag{4.58}$$

$$d\Phi^R = dS_1 \varepsilon_1(T_1) \int_0^\infty (I^0_\lambda(T_1) - I^0_\lambda(T_2)) \, d\lambda \int_0^{\pi/2} \cos\theta \, 2\pi \sin\theta \, d\theta \tag{4.59}$$

so that:

$$\mathrm{d}\Phi^{\mathrm{R}} = \varepsilon_1(T_1)\,\mathrm{d}S_1\sigma(T_1^4 - T_2^4) \tag{4.60}$$

### *4.5.2 Opaque convex body, of small dimensions, in an enclosure in thermal equilibrium*

Let us consider first a closed region at a uniform temperature $T_2$ (Figure 4.14). **The walls are opaque with arbitrary thermal properties**. The radiation within this region is characterized for every point and every direction by the intensity of equilibrium radiation $I_\lambda^0(T_2)$. If we introduce a **very small** opaque body into this region, held by some means at a temperature $T_1$, we may make the following hypotheses:

- The body at temperature $T_1$ makes a negligible perturbation to the radiation in the region;
- The radiation emitted by the body at temperature $T_1$ and reflected back to it from the surroundings is negligible.

In these conditions, the radiative flux exchanged is given by:

$$\mathrm{d}\Phi^{\mathrm{R}} = \mathrm{d}\Phi^{\mathrm{e}} - \mathrm{d}\Phi^{\mathrm{a}} \tag{4.61}$$

with, as in Section 4.5.1:

$$\mathrm{d}\Phi^{\mathrm{e}} = \mathrm{d}S_1 \int_0^\infty \mathrm{d}\lambda \int_0^{\pi/2} \varepsilon'_\lambda I_\lambda^0(T_1)\cos\theta\, 2\pi \sin\theta\, \mathrm{d}\theta \tag{4.62}$$

On the other hand, if the incident flux is characterized by the intensity $L_\lambda^0(T_2)$, the flux absorbed is:

$$\mathrm{d}\Phi^{\mathrm{a}} = \mathrm{d}S_1 \int_0^\infty \mathrm{d}\lambda \int_0^{\pi/2} \varepsilon'_\lambda I_\lambda^0(T_2)\cos\theta\, 2\pi \sin\theta\, \mathrm{d}\theta \tag{4.63}$$

$\mathrm{d}\Phi^{\mathrm{R}}$ thus has exactly the same form as in Section 4.5.1 above.

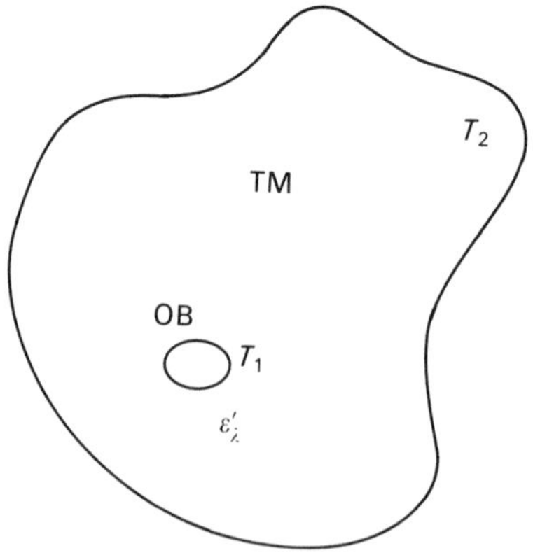


**Figure 4.14**

### 4.5.3 Principles of bichromatic pyrometry

A classical method of measuring the temperature $T$ of an element of an opaque body is based on measuring the flux received by an optical detector (germanium–gold, indium antimonide, etc.) at two different wavelengths, from where it gets the same bichromatic pyrometry.

We use the configuration of Figure 4.15, with the following hypotheses:

(H1) The flux reflected by an element $dS$ of the surface of which we wish to measure the temperature is negligible compared to the flux emitted by that element.
(H2) The medium between $dS$ and the detector $d\Sigma$ is transparent.
(H3) The aligned optical system which focuses the radiation at a point on the detector is perfectly defined.
(H4) An interference filter (single-mode Fabry–Perot type) is inserted near the detector. Its transmission and reflection characteristics are shown schematically in Figure 4.16.

For a given interference filter, the detector will only receive incident flux $\Delta\Phi^{i}_{1\lambda}$ in the spectral interval $[\lambda_1, \lambda_1 + \Delta\lambda_1]$. This flux gives rise to the measured voltage; $\Delta\Phi^{i}_{1\lambda}$ is given by:

$$\Delta\Phi^{i}_{1\lambda} = d\Sigma \int_{d\omega} \cos\theta \, d\Omega \int_{\lambda_1}^{\lambda_1+\Delta\lambda_1} \tau^{OS}_{\lambda} \varepsilon'_{\lambda} I^{0}_{\lambda}(T)\, d\lambda \qquad \text{with } \cos\theta \simeq 1 \quad (4.64)$$

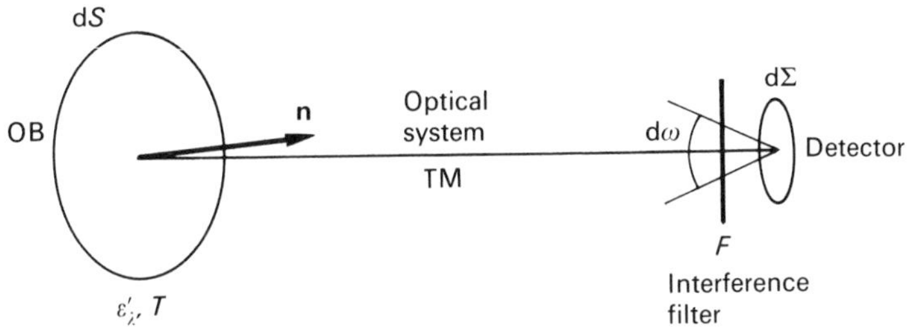


**Figure 4.15**

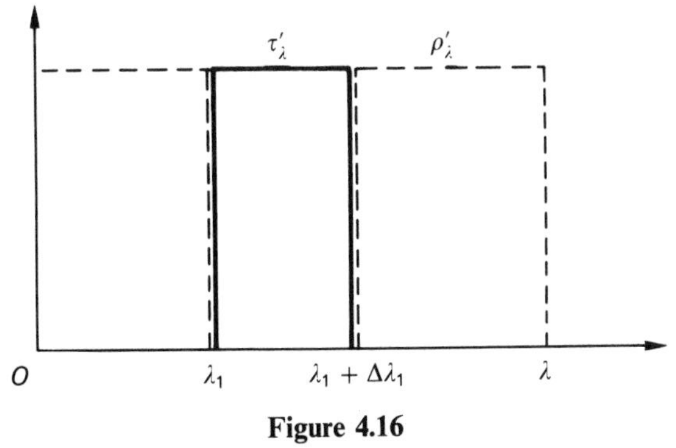


**Figure 4.16**

The spatial integral is made over the solid angle $d\omega$ through which the element of surface $dS$ is seen from the detector through the optical system of transmissivity $\tau_\lambda^{OS}$ and the interference filter (see Figure 4.15). The angle $\theta$ is clearly practically zero to a good approximation. We make the supplementary hypothesis:

(H5) The emissivity $\varepsilon'_{\lambda 1}$ of the body $dS$ is invariant over $\Delta\lambda_1$ and in the directions making up the solid angle $d\omega$; similarly $\tau_{\lambda_2}^{OS}$ is invariant over $\Delta\lambda_1$. Under these conditions,

$$\Delta\Phi^i_{1\lambda} = d\Sigma\tau_{\lambda_1}^{OS}\varepsilon'_{\lambda_1}\int_{\lambda_1}^{\lambda_1+\Delta\lambda_1} I_\lambda^0(T)\,d\lambda \int_\omega \cos\theta\, d\Omega \qquad (4.65)$$

Similarly, we can use a second filter $F_2$, passing the interval $[\lambda_2, \lambda_2 + \Delta\lambda_2]$, for which the detector receives flux:

$$\Delta\Phi^i_{2\lambda} = d\Sigma\tau_{\lambda_2}^{OS}\varepsilon'_{\lambda_2}\int_{\lambda_2}^{\lambda_2+\Delta\lambda_2} I_\lambda^0(T)\,d\lambda \int_\omega \cos\theta\, d\Omega \qquad (4.66)$$

The ratio of the two fluxes:

$$\frac{\Delta\Phi^i_{1\lambda}}{\Delta\Phi^i_{2\lambda}} = \frac{\tau_{\lambda_1}^{OS}}{\tau_{\lambda_2}^{OS}}\frac{\varepsilon'_{\lambda_1}}{\varepsilon'_{\lambda_2}}\frac{\int_{\lambda_1}^{\lambda_1+\Delta\lambda_1} I_\lambda^0(T)\,d\lambda}{\int_{\lambda_2}^{\lambda_2+\Delta\lambda_2} I_\lambda^0(T)\,d\lambda} \qquad (4.67)$$

depends on the following:

1. $T$, the quantity to be determined, by way of the ratio of the two integrals (an implicit function of $T$, which can be precisely evaluated numerically);
2. the known ratio of the transmissivities of the optical system;
3. the unknown (in industrial conditions) ratio of the emissivities of the body concerned at the two wavelengths $\lambda_1$ and $\lambda_2$. In practice, much depends on the choice of $\lambda_1$ and $\lambda_2$, which depends on two opposing considerations:
   - $\lambda_1$ and $\lambda_2$ should be sufficiently far apart for the ratio of the integrals (for the same $\Delta\lambda$) to be significantly different from 1;
   - $\lambda_1$ and $\lambda_2$ should be sufficiently close together for $\varepsilon'_{\lambda_1}$ and $\varepsilon'_{\lambda_2}$, which are generally unknown, to be reasonably assumed to be equal.

It is usually appropriate to select $\lambda_1$ and $\lambda_2$ in the range $[0.6\lambda_m(T), 0.9\lambda_m(T)]$, i.e. where the function $I_\lambda^0(T)$, which has a maximum at $\lambda_m(T)$, varies most rapidly.

*Notes*

1. The validity of the assumption of constant $\varepsilon'_\lambda$ can be tested by using a third wavelength $\lambda_3$ and comparing the temperatures $T_{12}$, $T_{13}$ and $T_{23}$ found for the element $dS$ from the pairs $(\lambda_1, \lambda_2)$, $(\lambda_1, \lambda_3)$ and $(\lambda_2, \lambda_3)$ respectively.
2. Using a sophisticated laser technique one can separate the emitted radiation from the radiation reflected by $dS$, in cases where the latter is not negligible.

## 4.6 Conditions for linearization of radiative flux

In numerous cases of practical interest, involving a grey body of emissivity $\varepsilon$ which is completely surrounded, say by an isothermal black body at $T_a$ (Section 4.5.1), or even more often by a medium in equilibrium at $T_a$ (Section 4.5.2), the radiative flux at the grey body at temperature $T$ is given by:

$$\varphi^R = \varepsilon\sigma(T^4 - T_a^4) \tag{4.68}$$

In general, this will occur in parallel with a conducto-convective flux given by:

$$\varphi^{cc} = h(T - T_a) \tag{4.69}$$

from either forced or natural convection. The accuracy of the estimate of $h$ is roughly 10–30%. When $\varphi^{cc}$ and $\varphi^R$ are the same order of magnitude, it is pointless to calculate $\varphi^R$ to a high accuracy, the more so as $\varepsilon$ is also estimated to about 10% accuracy.

If the temperatures $T$ and $T_a$ are relatively close together, i.e.:

$$\frac{T - T_a}{T} \ll 1 \tag{4.70.1}$$

we can linearize (4.68):

$$(T^4 - T_a^4) = (T - T_a)(T + T_a)(T^2 + T_a^2) \simeq 4T_m^3(T - T_a) \tag{4.70.2}$$

where $T_m$ is an intermediate temperature which, to a first approximation, may be taken as $(T + T_a)/2$. We can re-write (4.68) in the form:

$$\varphi^R = h_R(T - T_a) \tag{4.71}$$

analogous to (4.69), where the radiative heat transfer coefficient $h_R$ is given by $4\varepsilon\sigma T_m^3$. Obviously, the linearization condition (4.70) must be checked in every case, depending on the temperature levels concerned and the required accuracy.

**Numerical example**

Consider a building: the temperatures are around 300 K; then $h_R \simeq 6\ \mathrm{W\,m^{-2}\,K^{-1}}$ for $\varepsilon = 1$; this is not very different from $h_{cc} \simeq 4\ \mathrm{W\,m^{-2}\,K^{-1}}$, often used as a first estimate for natural convection. We thus obtain the notorious $h_{TOTAL} \simeq 10\ \mathrm{W\,m^{-2}\,K^{-1}}$, generally attributed to convection, although radiation is dominant!

All the thermal transfers (conductive, convective, conducto-convective and radiative) involved in the building are thus linearizable. This explains the appearance in the industry standards of a global coefficient $K$ (or $U$), analogous to a thermal conductance, such that the losses from the building are given by:

$$\Phi = KS(T_{int} - T_{ext}) \tag{4.72}$$

## 4.7 Extension to media transparent in bands

Up to now, we have considered transfers between opaque bodies through a transparent medium. But many substances (glass, atmospheric air, water, etc.) are semi-transparent (see Section 1.1). In this section, we will limit ourselves to the special cases of bodies which are transparent or opaque in bands of wavelength. Semi-transparent media will be studied in Chapter 11.

The directional transmissivity (perhaps monochromatic) of a material is defined as the ratio of the directional transmitted flux to the directional flux incident on the material:

$$\tau'_\lambda = \frac{d\Phi'^t_\lambda}{d\Phi'^i_\lambda} \tag{4.73}$$

Incident radiation may be transmitted, reflected or absorbed. The conservation of energy becomes (see Figure 4.17):

$$\alpha'_\lambda + \rho'^{\bullet}_\lambda + \tau'_\lambda = 1 \tag{4.74}$$

Since we are restricting ourselves to media which are either transparent or opaque, $\tau'_\lambda$ is either 0 or 1. For example, the radiative properties of a glass window at normal incidence may be taken as a first approximation as shown in Figure 4.18.

The cut-off wavelength $\lambda_c$ varies from 3 to 5 $\mu$m according to the type of glass. The actual material properties are shown by the thick lines.

To calculate the radiative flux, we divide the spectrum into bands of wavelength. We then treat the body as either opaque or transparent. For the latter, we recall, no radiative energy is exchanged with the material system.

The atmosphere is another example of a medium of this type. The main constituents of air ($N_2$ and $O_2$) are not absorbent in the infra-red. Rather, it is the trace constituents $CO_2$ (concentration $\simeq 3.5 \times 10^{-4}$) and $H_2O$ (a few %, depending

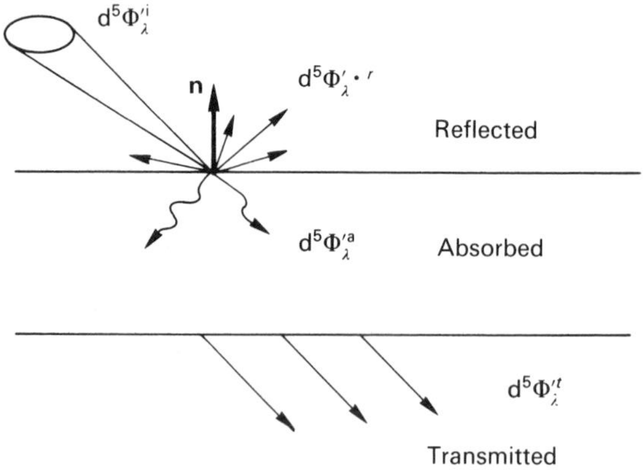


**Figure 4.17**

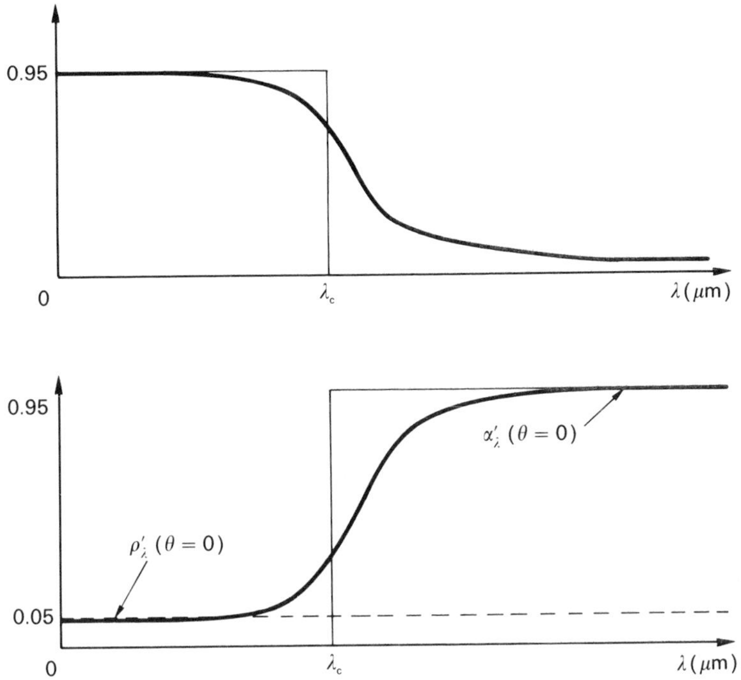


**Figure 4.18**

on the degree of humidity) which are absorbent. Air can be considered as transparent over a few centimetres, but absorbent in bands over a few metres. In particular, the atmosphere, which represents tens of kilometres of air, can be crudely modelled in the infra-red as shown in Figure 4.19.

We notice in particular the existence of a window of transparency in the atmosphere between 9 and 12 $\mu$m in which a body below the atmosphere emits radiation without receiving any in return (the atmospheric window). This means that the earth is continuously dissipating energy between 9 and 12 $\mu$m through the

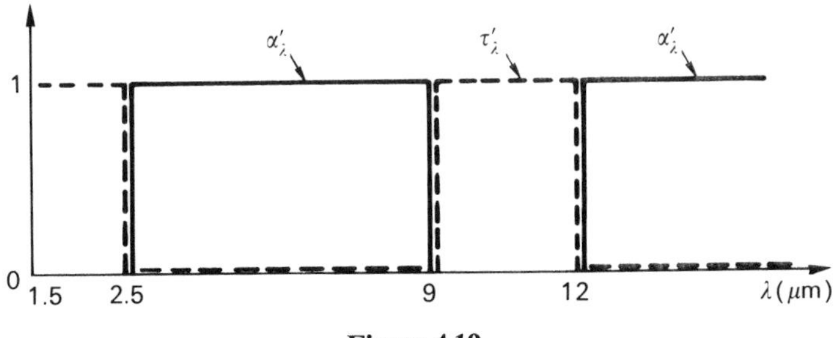


**Figure 4.19**

atmosphere, without the latter on its part emitting any towards the earth. This loss is important since it is in the part of the spectrum where the equilibrium intensity at 300 K is maximum.

*Note*

In the visible region, an important fraction of solar radiation is transmitted by the atmosphere. As a very rough approximation, this radiation is equivalent under the atmosphere to that from a point black body source at 5700 K. In the sun, we receive:

- a directional contribution from the point source concerned – the sun;
- a diffuse contribution, approximately isotropic, due to the diffusion of solar radiation by molecules and aerosols. The blue colour of a clear sky is explained by the selective diffusion of the higher spectral frequencies to which the eye is sensitive. The red of a beautiful sunset arises for the same reason – we see that which has not been diffused.

## 4.8 Application examples

### 4.8.1 Using the concept of equilibrium radiation

4.8.1.1 MEASURING THE TEMPERATURE OF A GAS WITH A THERMOCOUPLE

Consider a transparent gas in turbulent flow in a duct of diameter $D = 0.6$ m, which may be considered to be a grey body with emissivitity $\varepsilon_w = 0.5$ (Figure 4.20).

We propose to measure the temperature of the gas with a thermocouple of diameter $d = 0.5$ mm positioned on the duct axis. For simplicity, this thermocouple is assumed to be insulated after its end face; the thermocouple is a grey body of

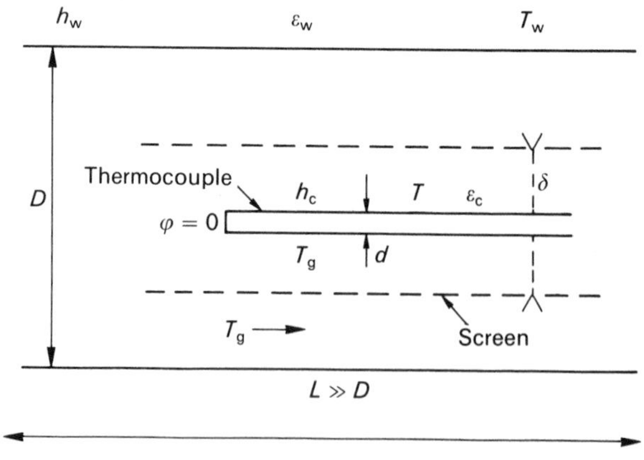


**Figure 4.20**

emissivity $\varepsilon_c = 0.8$. The characteristic temperature of the gas, $T_g$, and the wall temperature are assumed to be constant over a length $L$ which is large compared to $D$; similarly the temperature $T$ of the thermocouple. The convective heat transfer coefficients are $h_c = 100$ W m$^{-2}$ K$^{-1}$ and $h_w = 30$ W m$^{-2}$ K$^{-1}$ on the surface of the thermocouple and the duct respectively. What difference will there be between the temperature $T$ measured by the thermocouple and that of the gas $T_g$?

Numerical values: $T_w = 200°C$, $T_g = 600°C$

With what modification could we obtain a more accurate measurement of the temperature?

4.8.1.2 SOLUTION

We will follow the usual procedure (see Section 2.3):

1. The temperature $T$ of the thermocouple depends on the gas temperature $T_g$ and also, by radiation, on the wall temperature $T_w$.
2. The crucial point is to model the radiative transfer between the thermocouple and the wall. The transparent gas has no effect on this transfer, and **its temperature is irrelevant**. We therefore have a thermocouple of diameter $d = 5 \times 10^{-4}$ m in an isothermal tube of diameter $D = 0.6$ m, at a temperature $T_w$. If we ignore the presence of the thermocouple, the whole of the inside of the tube is filled with an equilibrium radiation in all directions with an intensity $I_\lambda^0(T_w)$. The perturbation introduced by the thermocouple is related to the ratio of the flux emitted by the thermocouple to that emitted by the tube, per unit length: $\Phi_{th}/\Phi_{tube} = \varepsilon_c \pi d dz \sigma T^4 / \varepsilon_c \pi D\, dz \sigma T_w^4$. If as a first approximation (which is an overestimate) we put $T = T_g$, this ratio is $1.5 \times 10^{-3}$. Since the error in the heat transfer coefficient $h_c$ is of the order of 10%, it is justifiable to ignore the disturbance introduced by the thermocouple. **This will receive an equilibrium radiation at temperature $T_w$, whatever the value of $\varepsilon_w$!**
3. The only **balance** to be done is that of the thermocouple. In steady state we have:

$$\int_S (\mathbf{q}^{cd} + \mathbf{q}^R)\cdot(-\mathbf{n}_{ext})\, dS = 0 \tag{4.75}$$

so:

$$\pi\, dL[h_c(T_g - T) + (\varphi^a - \varphi^e)] = 0 \tag{4.76}$$

The flux absorbed $\varphi^a$ is:

$$\varphi^a = \varepsilon_c \varphi^i = \varepsilon_c \sigma T_w^4 \tag{4.77}$$

and that emitted $\varphi^e$ is:

$$\varphi^e = \varepsilon_c \sigma T^4 \tag{4.78}$$

Whence the equation:

$$h_c(T_g - T) + \varepsilon_c \sigma (T_w^4 - T^4) = 0 \tag{4.79}$$

which may be solved by iteration. The result is $T \simeq 760$ K, to be compared with $T_g = 873$ K and $T_w = 473$ K.

4.8.1.3 COMMENTS

1. The radiation incident on the thermocouple is made up of the flux emitted by the tube, $\varepsilon_w \sigma T_w^4$, and that **reflected** by the tube $(1 - \varepsilon_w)\sigma T^4$.
2. The temperature difference $T_g - T$ is considerable! The thermocouple indicates its own temperature, not that of the gas. What experimental precaution should we take to remedy this?

We put a screen of diameter $\delta$ (small compared with $D$) around the thermocouple. The screen will reach a temperature close to $T$ calculated above; actually its temperature $T_s$ is much closer to $T_g$ because it exchanges heat with the gas on both sides.

The perturbing effect on the new temperature $T'$ of the thermocouple is now that of the screen rather than the tube wall. The effect is reduced for two reasons:

(a) the temperature difference $T_s - T'$ is small;
(b) the ratio $d/\delta$ is not very small: the screen does not impose an equilibrium radiation on the thermocouple.

The correction is not, however, exact [19].

## 4.8.2 Thermal study of an incandescent light bulb

4.8.2.1 SPECIFICATION

A lamp of power $\mathscr{P}$ is sketched below. It comprises the following (Figure 4.21):

1. A spherical glass envelope of radius $R$, thickness $e$, conductivity $\lambda_g$, emissive $\varepsilon_{g\lambda}$ and is assumed to be at constant temperature $T_g$.

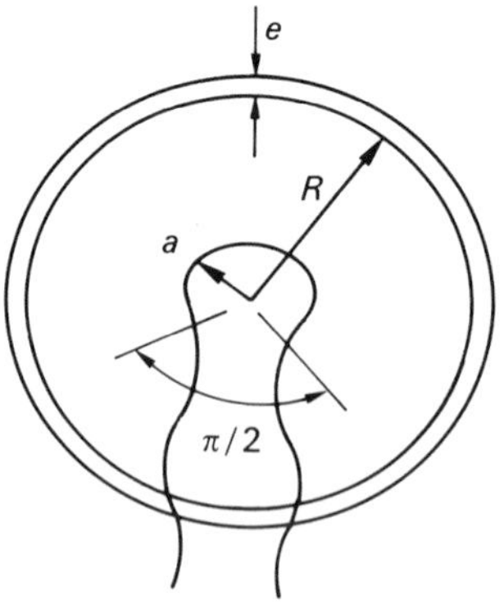


**Figure 4.21**

2. A tungsten filament, at a uniform temperature $T_F$, surface area $S_F$, equivalent to $\frac{3}{4}$ of a turn of radius $a$ and the same centre $O$ as the bulb. The tungsten is opaque with emissivity $\varepsilon_{F\lambda}$. We make the following approximations:
   - we assume that all the radiation emitted by the filament arrives at the glass envelope, and not at another point on the filament;
   - the presence of the electrical conductors connected to the filament is negligible;
   - the power $\mathscr{P}$ is dissipated in the filament;
   - the heat transfer inside the bulb is purely radiative.

1. The surroundings are at temperature $T_a$; the natural convective heat transfer coefficient **on the outside** is $h$. Determine the output of the bulb (the ratio of the light output between 0.4 and 0.8 $\mu$m to the power $\mathscr{P}$). Determine $T_g$. Verify that the temperature of the glass is uniform in the radial direction.
2. The tungsten has in fact an emissivitity as shown in Figure 4.22. What consequences would this have on the answers to 1? (Limit discussion to semi-quantitative arguments.)

Numerical data:

$\mathscr{P} = 60$ W; $R = 2.5 \times 10^{-2}$ m; $e = 5 \times 10^{-4}$ m; $S_F = 2 \times 10^{-5}$ m$^2$;
$T_a = 20°$C; $h = 5$ W m$^{-2}$ K$^{-1}$; $a = 1.5 \times 10^{-2}$ m;

Glass: $\lambda_g = 1\ Wm^{-2}\ K^{-1}$

$$\varepsilon_{g\lambda}: \begin{cases} \lambda \geqslant 3.5\ \mu\text{m}; & \tau_{g\lambda} = 1; \quad \varepsilon_{g\lambda} = 0 \\ \lambda > 3.5\ \mu\text{m}; & \tau_{g\lambda} = 0; \quad \varepsilon_{g\lambda} = 1 \end{cases}$$

Tungsten:

- $\varepsilon_F = 0.4$ (independent of $\lambda$ – question 1)
- see Figure 4.22 (question 2)

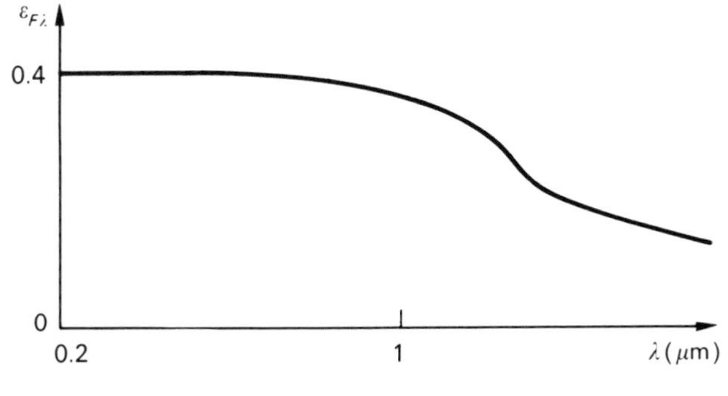


**Figure 4.22**

4.8.2.2 SOLUTION

Using the normal sequence:

1. The parameters affecting the thermal study seem to be the power $\mathscr{P}$ of the bulb, the glass temperature $T_g$ and the temperature of the surroundings $T_a$. In fact, it is also necessary to know the spectral range of the light emitted by the lamp, for which the filament temperature $T_F$ is the parameter.
2. The sub-systems to be studied are:
   - the filament, whose heat balance will give the temperature $T_F$, allowing the output of the lamp to be determined;
   - the glass envelope, whose heat balance allows the temperature of the glass to be determined.

   *Note.* The glass is treated as a black body for all sources whose temperatures are below 400 K (range $[\lambda_m/2; 8\lambda_m]$).
3. Heat balance for the filament. In the steady state:

$$\int_V P\,dV + \int_S \mathbf{q}^{en}\cdot(-\mathbf{n}_{ext})\,dS = 0 \tag{4.80}$$

If we only consider radiative transfer, with a glass temperature assumed to be below 400 K:

$$\mathscr{P} + \Phi_a - \Phi_e = 0 \Rightarrow \mathscr{P} = \varepsilon_F S_F \sigma(T_F^4 - T_g^4) \tag{4.81}$$

$T_g$ is clearly negligible compared to $T_F$: $T_F = 3400$ K. The peak emission will be at $\lambda_m(T_F) = 0.85\ \mu$m, which is at the limit of the visible and the infra-red.

Output of the lamp: This is the ratio of the flux emitted in the visible and transmitted through the glass to the flux $\mathscr{P}$ dissipated by the lamp:

$$R = \frac{S_F}{\mathscr{P}}\int_{0.4}^{0.8\,\mu m} \varepsilon_F \pi I_\lambda^0(T_F)\,d\lambda = z\left(0, \frac{0.8}{\lambda_m(T_F)}\right) - z\left(0, \frac{0.4}{\lambda_m(T_F)}\right) \simeq 0.2 \tag{4.82}$$

4. Heat balance for the glass:
   - On the outside, the glass exchanges energy both by conducto-convection and by radiation. The glass receives **equilibrium radiation** (with flux density $\varphi^i$) at temperature $T_a$ and emits radiation. So:

$$\varphi^R = \varphi^a - \varphi^e = \varepsilon_g \varphi^i - \varepsilon_g \sigma T_g^4 \tag{4.83.1}$$

   Since the glass behaves like a black body for all radiation emitted by a source whose temperature is below 400 K, we obtain:

$$\varphi^R = \varphi^i - \sigma T_g^4 = \sigma(T_a^4 - T_g^4) \tag{4.83.2}$$

   - Inside the bulb, the glass absorbs the radiation emitted by the filament beyond 3.5 $\mu$m. It also receives radiation emitted by itself (autoillumination), which it absorbs completely. This radiation is exactly balanced by the radiation it emits.

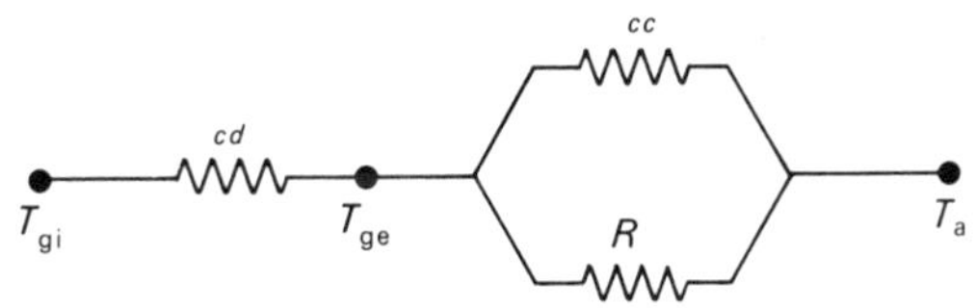


**Figure 4.23**

The balance may thus be written using (4.52):

$$S_g[h(T_a - T_g) + \sigma(T_a^4 - T_g^4)] + \mathscr{P}\left[1 - z\left(0, \frac{3.5}{\lambda_m(T_F)}\right)\right] = 0$$

We obtain, by iteration, $T_g \simeq 330$ K (or 57°C).

*Notes*:

($\alpha$) Having regard to the low temperature of the glass, it is possible to linearize the balance expression:

$$h(T_a - T_g) + \sigma(T_a^4 - T_g^4) = h(T_a - T_g) + 4\sigma T_m^3(T_a - T_g) = h_{eq}(T_a - T_g)$$

which gives $h_{eq} \simeq 12$ W m$^{-2}$ K$^{-1}$.

($\beta$) The temperature of the glass is clearly uniform in the radial direction. A very simple thermal equivalent circuit allows this to be demonstrated (Figure 4.23). From this:

$$\frac{T_{gi} - T_{ge}}{T_{ge} - T_a} = \frac{eh_{eq}S_g}{\lambda_g S_g} = Bi = 6 \times 10^{-3} \ll 1$$

($\gamma$) With the more realistic data for the emissivity of tungsten, the results are modified in the following manner:

- when the emissivity is less than or equal to 0.4, and the electric power dissipated by the lamp is unchanged, the temperature $T_F$ of the filament is increased;
- in the visible range, $\varepsilon_{F\lambda}$ is unchanged, but $T_F$ is increased, so the output $R$ is higher;
- finally, the function $z(0, 3.5/\lambda_m(T_F))$ increases when $T_F$ increases, so that the temperature $T_g$ falls.

In practice, it is necessary to introduce an inert gas into the bulb, at a fairly low pressure, to prevent volatilization of the filament. Within this gas, complex but relatively weak natural convection currents appear, whose effect is:

- to cool the filament;
- to reduce the output.

# Conventions: Vector and Tensor Notation

In simple expressions, **vector notation** is usually used in this work. On the other hand, in more complex calculations or expressions **tensor notation conventions** are more often used, since they are the **clearest** and most concise.

The mathematical objects to be found in this work are as follows:

- **Scalars**: pressure, temperature, etc.
- **Vectors**: velocity, normal direction, vector flux density, etc.
- **Tensors**: stress, rate of deformation, etc.
- **Operators** acting on scalars, vectors or tensors: gradient, divergence, Laplacian, etc.

*Note*
The components of a vector, a tensor or an operator take different forms in different systems of orthogonal coordinates (Cartesian, cylindrical polar, spherical polar, etc.), but the mathematical object is invariant. **We will use representations** (vector or tensor) **which are independent of the coordinate system chosen**.

## 1. Vector notation

- A scalar is represented by a non-bold letter: $p$, $T$, $a$, $\lambda$, etc.
- A vector is represented by a bold letter: $\mathbf{v}$, $\mathbf{w}$, $\mathbf{q}$, etc.
- A tensor is represented by a bold and underlined letter $\underline{\boldsymbol{\tau}}$, $\underline{\mathbf{d}}$.
- The components of a vector or a tensor in vector notation have specific symbols, for example $\mathbf{v}$ $(u, v, w)$, or are given subscripts appropriate to the coordinate axes: $(v_x, v_y, v_z)$ or $(v_r, v_\theta, v_z)$ etc., $\tau_{xx}$, $\tau_{xy}$, etc., but never integer subscripts such as $i$, $j$, $k$, etc.

- Operators are represented as vectors:
  $\nabla$ gradient operator;
  $\nabla\cdot$ divergence operator (scalar product by $\nabla$);
  $\nabla^2$ Laplacian operator (or $\nabla\cdot\nabla$).

## 2. Tensor notation

- A scalar (a tensor of order 0) has no subscript: $p$, $T$ etc.
- A vector (tensor of order 1, with two or three dimensions in this work) is represented by a running subscript. For example, $v_i$, $q_k$ represent the vectors of velocity and flux density. The running subscript can take two or three values (for two- and three-dimensional problems), which depend on the type of orthogonal reference chosen (Cartesian, cylindrical, spherical, etc.). For example, in Cartesian coordinates, the components of the velocity vector are $(v_x, v_y, v_z)$. In tensor notation, it is represented by $v_j$; this notation is obviously independent of the type of reference chosen.
- A tensor (restricting ourselves to second-order tensors) is an object defined by two running subscripts, for example $i$ and $j$. The quantity $\partial v_i/\partial x_j$ is a tensor of this type which can be represented by a 3 × 3 table (if the velocity vector $v_i$ and the position vector $x_j$ are three-dimensional). In Cartesian coordinates this tensor has nine components;

$$\begin{matrix} \partial v_x/\partial x & \partial v_x/\partial y & \partial v_x/\partial z \\ \partial v_y/\partial x & \partial v_y/\partial y & \partial v_y/\partial z \\ \partial v_z/\partial x & \partial v_z/\partial y & \partial v_z/\partial z \end{matrix}$$

  where the Cartesian components of $v_i$ and $x_j$ are $(v_x, v_y, v_z)$ and $(x, y, z)$ respectively. Note that the rows refer to the first subscript given $(i)$ and the columns to the second $(j)$.

*Note*
One tensor plays a particularly important role: the Kronecker tensor:

$$\delta_{ij}\begin{cases} =0 & \text{if } i \neq j \\ =1 & \text{if } i = j \end{cases}$$

*Basic rule for tensor notation:*

**The repetition within a term (a monomial which may contain differentials) of the same running subscript $i$, $j$ or $k$ represents an implicit summation over all the values taken by that subscript.**

EXAMPLES:

- $\partial/\partial x_j\,\partial/\partial x_j$ or $\partial^2/\partial x_j^2$ represents the Laplacian operator: $\partial^2/\partial x^2 + \partial^2/\partial y^2 + \partial^2/\partial z^2$ in Cartesians, $\partial^2/\partial r^2 + 1/r\,\partial/\partial r + 1/r^2\,\partial^2/\partial\theta^2 + \partial^2/\partial z^2$ in cylindrical coordinates, etc.
- $\partial v_k/\partial x_k$ represents the divergence of velocity: ($\partial v_x/\partial x + \partial v_y/\partial y + \partial v_z/\partial z$ in Cartesians).
- $\tau_{ij}\,\partial v_j/\partial x_i$ represents a double summation over $i$ and over $j$ (nine terms), etc.
- $\tau_{ii}$ represents the trace of the tensor $\tau_{ij}$ ($\tau_{xx} + \tau_{yy} + \tau_{zz}$ in Cartesians).
- $\delta_{ij}\,v_j = v_i$.

## 3. *Some equivalent notation*

| Vector notation | Tensor notation | Expansion (*Cartesian coordinates*) |
|---|---|---|
| $\mathbf{v}$ | $v_j$ | $(v_x, v_y, v_z)$ |
| $\nabla T$ | $\partial T/\partial x_i$ | $(\partial T/\partial x, \partial T/\partial y, \partial T/\partial z)$ gradient |
| $\nabla\cdot\mathbf{v}$ | $\partial v_k/\partial x_k$ | $\partial v_x/\partial x + \partial v_y/\partial y + \partial v_z/\partial z$<br>Divergence |
| $\nabla^2\mathbf{v}$ | $\partial^2 v_k/\partial x_j^2$ | $\partial^2 v_x/\partial x^2 + \partial^2 v_x/\partial y^2 + \partial^2 v_x/\partial z^2$<br>$\partial^2 v_y/\partial x^2 + \partial^2 v_y/\partial y^2 + \partial^2 v_y/\partial z^2$<br>$\partial^2 v_z/\partial x^2 + \partial^2 v_z/\partial y^2 + \partial^2 v_z/\partial z^2$ |
| $\nabla\cdot\underline{\tau}$ | $\partial\tau_{ij}/\partial x_i$ | $\partial\tau_{xx}/\partial x + \partial\tau_{yx}/\partial y + \partial\tau_{xz}/\partial z$<br>$\partial\tau_{xy}/\partial x + \partial\tau_{yy}/\partial y + \partial\tau_{zy}/\partial z$<br>$\partial\tau_{xz}/\partial x + \partial\tau_{yz}/\partial y + \partial\tau_{zz}/\partial z$ |

APPENDIX 2

# The Rate of Strain Tensor $d_{ij}$ and Viscous Stress Tensor $\tau_{ij}$

A more detailed treatment of this subject is to be found in *Mécanique des Fluides* by S. Candel [1] (in French).

## *1. The rate of strain tensor*

We will follow the evolution of a **continuous material** system (see the definitions of these terms in Sections 1.1 and 3.1) between times $t$ and $t + \mathrm{d}t$. For convenience, we will use **tensor notation (see Appendix 1 – Conventions)**.

In some orthogonal coordinate system $Ox_1x_2x_3$ we consider at some instant $t$ the position of a reference point $A$ (position vector $x_i$) and a moving point $G$ (position vector $x_i + \mathrm{d}x_i$).

At the instant $t + \mathrm{d}t$, the point $A$, which has a velocity $v_i$ at instant $t$, is at a position given by:

$$x_i' = x_i = v_i \,\mathrm{d}t \tag{1}$$

while the point $G$ will be at the position given by the vector:

$$x_i' + \mathrm{d}x_i' = x_i + \mathrm{d}x_i + v_i \,\mathrm{d}t + \frac{\partial v_i}{\partial x_j}\mathrm{d}x_j \,\mathrm{d}t \tag{2}$$

In this expression, $\partial v_i/\partial x_j$ is a second-order tensor. A simple transformation of (2) leads to:

$$x_i' + \mathrm{d}x_i' = x_i + \mathrm{d}x_i + v_i \,\mathrm{d}t + \left[\frac{1}{2}\left(\frac{\partial v_i}{\partial x_j} + \frac{\partial v_j}{\partial x_i}\right) + \frac{1}{2}\left(\frac{\partial v_i}{\partial x_j} - \frac{\partial v_j}{\partial x_i}\right)\right]\mathrm{d}x_j \,\mathrm{d}t \tag{3}$$

In this result:

- $dx_i$ represents the **initial position vector** of $G$ relative to $A$.
- $v_i\,dt$ represents a vector **translation** of A.
- $\frac{1}{2}(\partial v_i/\partial x_j - \partial v_j/\partial x_i)\,dx_j$ represents the vector resulting from the scalar product of the antisymmetric second-order tensor $r_{ij}$ (defined below, whose diagonal elements are zero) with the initial position vector $dx_j$. The tensor $r_{ij}$ given by:

$$r_{ij} = \frac{1}{2}\left(\frac{\partial v_i}{\partial x_j} - \frac{\partial v_j}{\partial x_i}\right) \tag{4}$$

  is called the **rate of rotation tensor** of the material system. We can recognize the components of $\nabla \times \mathbf{v}$ in its components; it defines the rotation given to a material element $dx_j$ of the system in unit time.
- $\frac{1}{2}(\partial v_i/\partial x_j + \partial v_j/\partial x_i)\,dx_j$ represents the vector resulting from the scalar product of the symmetric second-order tensor $d_{ij}$ (defined below) with the initial position vector $dx_j$. The tensor $d_{ij}$ given by:

$$d_{ij} = \frac{1}{2}\left(\frac{\partial v_i}{\partial x_j} + \frac{\partial v_j}{\partial x_i}\right) \tag{5}$$

  is called the **rate of strain tensor** of the material system. This part of the transformation, in contrast to the preceding ones (translation and rotation), does not preserve angles or distances: it is a **deformation**.

We can consider two classes of deformations:

- Those which correspond to the diagonal elements of $d_{ij}$; for example (in Cartesian coordinates) $\partial v_x/\partial x$, $\partial v_y/\partial y$ and $\partial v_z/\partial z$. Their interpretation is straightforward: a positive increment in axial speed produces a **dilation** of the material element, a negative increment produces a **contraction** (see the example in Figure 1). We speak of normal or axial deformation of the material system.
- The deformations corresponding to the non-diagonal elements of the tensor $d_{ij}$ are called shearing or angular deformation. For example, in Cartesian coordinates, $\frac{1}{2}(\partial v_x \partial y + \partial v_y/\partial x)$. This case is shown in Figure 2.

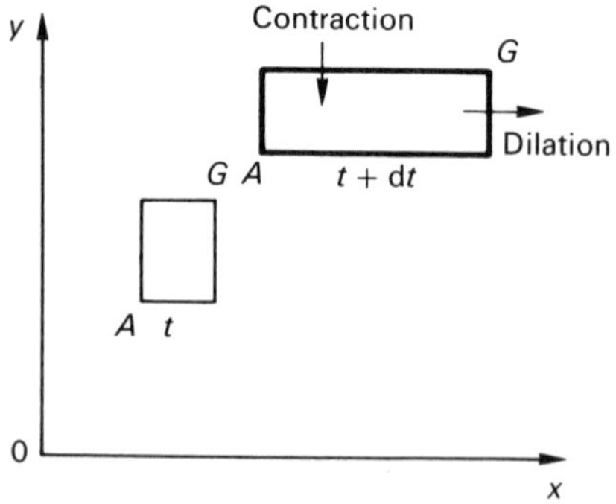


**Figure 1** Normal deformation.

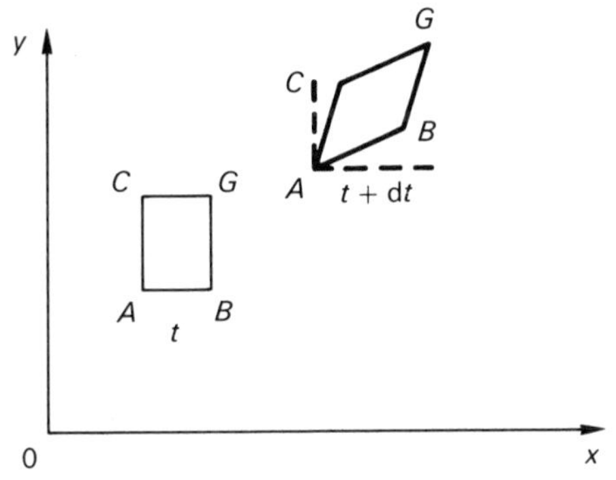


**Figure 2** Angular deformation (pure shear without rotation).

The phenomenon of angular deformation appears well in $B$, $C$ and $G$. Note that we are looking at a case of deformation by pure shear without rotation; that is, the corresponding components of the rate of rotation tensor are zero:

$$\frac{\partial v_x}{\partial y} = \frac{\partial v_y}{\partial x}$$

*Note*

To convince ourselves of the power of tensor notation, let us express equations (1), (2) and (3) in Cartesian coordinates:

$$\begin{aligned} x' &= x + v_x\,\mathrm{d}t \\ y' &= y + v_y\,\mathrm{d}t \\ z' &= z + v_z\,\mathrm{d}t \end{aligned} \tag{1a}$$

$$x' + \mathrm{d}x' = x + \mathrm{d}x + v_x\,\mathrm{d}t + \left(\frac{\partial v_x}{\partial x}\,\mathrm{d}x + \frac{\partial v_x}{\partial y}\,\mathrm{d}y + \frac{\partial v_x}{\partial z}\,\mathrm{d}z\right)\mathrm{d}t \tag{2a}$$

$$y' + \mathrm{d}y' = \text{etc.}$$

$$z' + \mathrm{d}z' = \text{etc.}$$

$$\begin{aligned} x' + \mathrm{d}x' = x + \mathrm{d}x + v_x\,\mathrm{d}t + \left(\frac{\partial v_x}{\partial x}\,\mathrm{d}x\,\mathrm{d}t\right) + \frac{1}{2}\left(\frac{\partial v_x}{\partial y} + \frac{\partial v_y}{\partial x}\right)\mathrm{d}y\,\mathrm{d}t \\ + \frac{1}{2}\left(\frac{\partial v_x}{\partial z} + \frac{\partial v_z}{\partial x}\right)\mathrm{d}z\,\mathrm{d}t + \frac{1}{2}\left(\frac{\partial v_x}{\partial y} - \frac{\partial v_y}{\partial x}\right)\mathrm{d}y\,\mathrm{d}t + \frac{1}{2}\left(\frac{\partial v_x}{\partial z} - \frac{\partial v_z}{\partial x}\right)\mathrm{d}z\,\mathrm{d}t \end{aligned} \tag{3a}$$

$$y' + \mathrm{d}y' = \text{etc.}$$

$$z' + \mathrm{d}z' = \text{etc.}$$

The components of the rate of strain tensor as a Cartesian matrix are:

$$\underline{\mathbf{d}} = \begin{pmatrix} \partial v_x/\partial x & \frac{1}{2}(\partial v_x/\partial y + \partial v_y/\partial x) & \frac{1}{2}(\partial v_x/\partial z + \partial v_z/\partial x) \\ \frac{1}{2}(\partial v_y/\partial x + \partial v_x/\partial y) & \partial v_y/\partial y & \frac{1}{2}(\partial v_y/\partial z + \partial v_z/\partial y) \\ \frac{1}{2}(\partial v_z/\partial x + \partial v_x/\partial z) & \frac{1}{2}(\partial v_z/\partial y + \partial v_y/\partial z) & \partial v_z/\partial x \end{pmatrix} \tag{5a}$$

All these laborious expressions and calculations are pointless if tensor notation is used.

## 2. The viscous stress tensor

It is obvious that an element of fluid in simple rotation or simple translation is not subject to stresses other than those of pressure.

On the other hand, deformation of a fluid element (normal or angular, or, alternatively, dilation or shear) will generate stresses which will tend to oppose the deformation. For most fluids (known as Newtonian) the hypothesis of a viscous stress tensor which is proportional to the rate of strain tensor is valid for constant volume transformations. The coefficient of proportionality is conventionally twice the fluid dynamic viscosity $\mu$. For the general case [20]:

$$\tau_{ij} = 2\mu d_{ij} + \left(\kappa - \frac{2}{3}\mu\right) d_{kk}\delta_{ij} \tag{6}$$

where $\kappa$ is the second coefficient of viscosity – the bulk viscosity.

The second term on the right-hand side of (6) will be zero when $d_{kk}$, the trace of the rate of strain tensor $(\partial v_x/\partial x + \partial v_y/\partial y + \partial v_z/\partial z)$, is zero, that is, for a constant volume or constant density ($\rho_0 = \text{const}$) transformation. When the contractions and dilations do not cancel out, the volume of an element will not be constant and we must include $\kappa$ [20]. This last reference gives a physical interpretation of bulk viscosity: it is related to the relaxation times between the translational and internal (rotation, vibration, etc.) degrees of freedom of the molecules. In consequence, it is zero for monatomic gases.

PART II

# Non-steady Conduction

The physical aspects of phenomena are highlighted in this part. The various properties of non-steady conduction are illustrated by two-dimensional $(x, t)$ or $(r, t)$ problems, to simplify them. Generalization to four-dimensional problems $(\mathbf{r}, t)$ with boundary or initial conditions which are time or space dependent is introduced here, using **standard mathematical techniques** in the **various appendices to this part**. The common numerical methods (finite element, finite difference, etc.) are not covered in this work.

# 5

# Scope. General Theorems

## 5.1 The general problem

In this chapter, we will restrict ourselves to **fixed**, **rigid systems**, otherwise convection will appear. The general energy equation, derived in Chapter 3 (equation (3.60)) for such a system of volume $V$, bounded by a surface $S$ (see Figure 5.1) is:

$$\int_V \rho c_p \frac{\partial T}{\partial t} \, \mathrm{d}V = \int_S -(\mathbf{q}^{\mathrm{cd}} + \mathbf{q}^{\mathrm{R}}) \cdot \mathbf{n}_{\mathrm{ext}} \, \mathrm{d}S + \int_V P \, \mathrm{d}V \tag{5.1}$$

In this equation, $T(\mathbf{r}, t)$, $c_p(\mathbf{r}, t)$, $\rho(\mathbf{r}, t)$ and $P(\mathbf{r}, t)$ represent the temperature, specific heat (at constant pressure, in the case of a gas), density and power per unit volume dissipated at the point $M(\mathbf{r})$ at the instant $t$; $\mathbf{q}^{\mathrm{cd}}$ and $\mathbf{q}^{\mathrm{R}}$ are, respectively, the flux density vectors for conduction and radiation at $\mathbf{r}$ at instant $t$. In local form, and after transformation of the first integral on the right-hand side, equation (5.1) becomes:

$$\rho c_{\mathrm{p}} \frac{\partial T}{\partial t} = -\nabla \cdot (\mathbf{q}^{\mathrm{cd}} + \mathbf{q}^{\mathrm{R}}) + P \tag{5.2}$$

This is the form of the energy equation which we will use. The conductive flux density vector $\mathbf{q}^{\mathrm{cd}}$ is usually given, for an isotropic medium, by the linear law known as Fourier's law:

$$\mathbf{q}^{\mathrm{cd}} = -\lambda(\mathbf{r}, t) \nabla T \tag{5.3}$$

The linear approximation is only valid if the conductivity $\lambda$ is independent of temperature, which is not usually the case. If the medium is not isotropic, (5.3) can be generalized in tensor form. This makes the solution of the problem much harder numerically, though not necessarily physically. Because of this, we will not consider this situation further, although it arises in the field of textiles (tyres, for example) or composite materials whose use is becoming general.

Law (5.3) also does not take account of any delay phenomenon in the response of the flux to a sudden perturbation in the temperature field. If we consider very

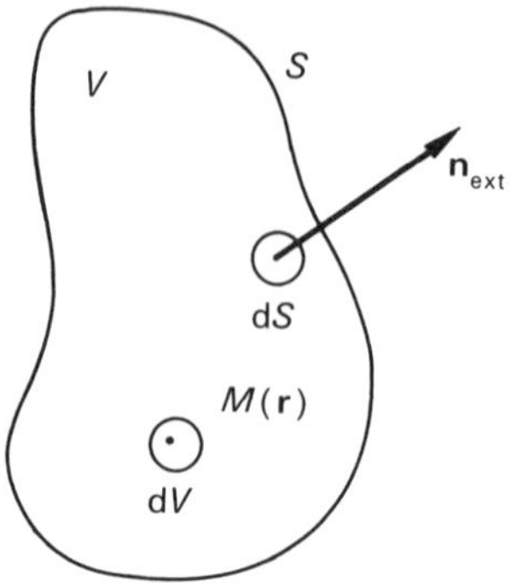


**Figure 5.1**

rapid phenomena (for example, the effect of the triggering of a phase blocked mode YAG laser, whose duration would not exceed $10^{-12}$ s), this law is clearly unsatisfactory. Conduction is basically due to transfer of energy by carriers (interactions between molecules in a gas or a liquid, or interactions involving phonons and electrons in crystalline materials): the characteristic times for the transfers (or **relaxation times**) are of the order $10^{-9}$–$10^{-11}$ s. At these scales of time, Fourier's law is not valid, and must be generalized (see Appendix 1 of this part). But in most engineering applications, a response of around 1 ms or a frequency below 1 kHz is sufficient; equation (5.3) is then adequate.

The introduction of the radiative flux density vector $\mathbf{q}^R$ (see Section 4.23) is only of interest in the study of semi-transparent media. Inside an opaque body, and away from its surface, the radiative flux is clearly zero. Similarly, for the other extreme case of a perfectly transparent medium which does not interact with the radiation field, we obtain:

$$\nabla \cdot \mathbf{q}^R = 0 \tag{5.4}$$

This last relation expresses the fact that whatever radiative energy enters a transparent system of any kind leaves it in radiative form. Thus so long as the system under consideration only comprises **opaque bodies within transparent surroundings**, the energy equation takes the simpler form:

$$\rho c_p \frac{\partial T}{\partial t} = +\nabla \cdot [\lambda(\mathbf{r}, t) \nabla T] + P(\mathbf{r}, t) \tag{5.5}$$

The radiative flux only appears in the expression for the boundary conditions at the surface of the opaque bodies. A problem of unsteady conduction is, under these conditions, defined by a system comprising the following:

- **The energy equation** (or several if the system can be divided into several sub-systems).
- **The thermal boundary conditions** on all surfaces. The different types of standard boundary condition have been described in Chapter 1.

- **Initial conditions**, defining the system for all points **r** at $t = 0$. For example, we might set:

$$T(\mathbf{r}, t = 0) = T_0(\mathbf{r}) \tag{5.6}$$

  The initial condition could also introduce a flux distribution. This would not necessarily correspond to a state of thermal equilibrium, nor even to a steady state. All the same, the final solution may correspond to a state which is an asymptotic function of time or an arbitrary state.

The preceding system of equations, which will involve at least a partial differential equation in three space variables and one of time, will not in general have a simple analytic solution. The solution is complicated by the following:

- Problems associated with the system geometry, often three-dimensional.
- The strong dependence of certain quantities ($\lambda$, $\rho$, $c_p$) on local temperature $T(\mathbf{r}, t)$. If the conductivity $\lambda$ depends on $T$, the energy equation is no longer linear; similarly if in the conducto-convective boundary conditions the heat transfer coefficient $h$ depends on $T$. The situation is even more complex when the radiative boundary conditions cannot be linearized in $T$.
- Anisotropic or heterogeneous behaviour of the materials considered.

It is not one of the objectives of this work to introduce the numerical methods usually applied. We limit ourselves here to linear problems, through which we will bring out techiques which are common to all conduction problems.

*Note: semi-transparent media*
If the system concerned is a semi-transparent medium, as defined in Section 1.2, an extra non-zero term $-\nabla \cdot \mathbf{q}^R$ appears in equation (5.2). This term represents the power absorbed (per unit volume) less that emitted, and can be treated as a power per unit volume $P_R$ dissipated at point **r**, and incorporated into $P(\mathbf{r}, t)$. Equation (5.5) is still applicable, and is therefore quite general. The solution of non-steady conduction problems is complicated by the tight coupling between conduction and radiation. The system to be solved in this case comprises the following:

- one or more energy equations;
- the thermal boundary conditions (see the semi-transparent case in Section 1.5);
- the initial conditions;
- the radiation transfer equation which gives the law for the development of the monochromatic intensity $I'_\nu$ for every direction at point **r** (see Chapter 11);
- the radiative boundary conditions (values of directional monochromatic intensity $I'_\nu$ at the interfaces).

The coupling is two-fold: the determination of $I'_\nu(\mathbf{r}, t)$ at **r** for one direction (radiation transfer equation) requires knowledge of the temperature field $T(\mathbf{r}, t)$ at every point; the source term $P_R(\mathbf{r}, t)$ in the equation requires knowledge of all the local directional intensities at **r**.

## 5.2 Scope

We will assume for the remainder of this part that the conductive transfers are represented by systems which are linear in $T$, which is equivalent to saying that the **conductivity** $\lambda$, **the density** $\rho$, **the specific heat** $c_p$ **and the convective heat transfer coefficient** $h$ **are constant**, and that any possible radiative transfers appearing in the boundary conditions are linearizable. Under these assumptions, the energy equation (5.6) becomes:

$$\rho c_p \frac{\partial T}{\partial t} = \lambda \nabla^2 T + P(\mathbf{r}, t) \tag{5.7}$$

If we introduce a thermal **diffusivity** $a$ defined by:

$$a = \frac{\lambda}{\rho c_p} \tag{5.8}$$

(whose units are $m^2/s$), this equation becomes:

$$\frac{\partial T}{\partial t} = a\nabla^2 T + \frac{P}{\rho c_p}(\mathbf{r}, t) \tag{5.9}$$

which is a classical diffusion equation, comparable to the Navier–Stokes equation which represents the diffusion of momentum. Versions of this equation in various coordinate systems appear in Part V.

*Note*
It is important to note that a linear equation like (5.9) is also satisfied by the components of the flux density vector. This can, in some cases, simplify the solution of a problem (see, for example, Section 6.2.2).

## 5.3 The technique of superposition

The solution of a linear system of partial differential equations consisting of the energy equation and the boundary and initial conditions is unique. It is often useful to split a linear system of equations of this type, $S^{(0)}$, into sub-systems $S^{(1)}$, $S^{(2)}$, . . ., $S^{(n)}$, which are also linear, in such a way that their sum term-for-term gives the original linear system $S^{(0)}$. The unique solution $T^{(0)}$ of the system $S^{(0)}$ is then the sum of the solutions $T^{(i)}$ of all the sub-systems $S^{(i)}$:

$$T^{(0)} = \sum_{i=1}^{n} T^{(i)} \tag{5.10}$$

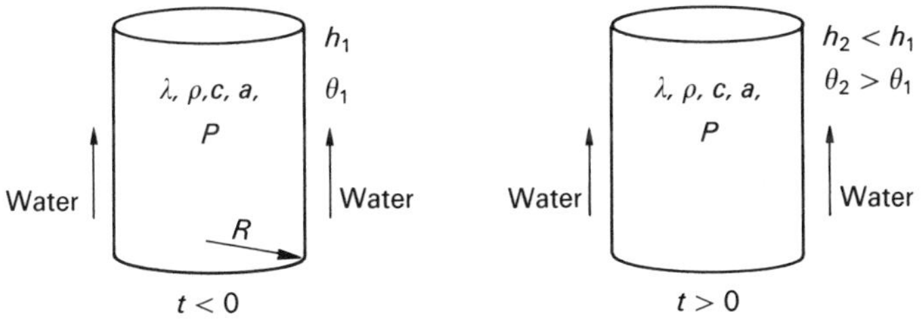


**Figure 5.2**

To make this theorem clear, we will illustrate it with a practical example. The generalization to other cases (different geometry, boundary and initial conditions) is obvious.

We consider a very long cylinder, of radius $R$, of a fissile material with constant conductivity $\lambda$ and diffusivity $a$. A power $P$ per unit volume (assumed constant) is dissipated within this rod (Figure 5.2). In normal operation, it is in a steady state; the power produced in this cylinder by the fission reaction is removed by a coolant (pressurized water), whose mixed fluid temperature is taken as constant at $\theta_1$. The heat transfer has a constant convective coefficient $h_1$. To simplify things, we will ignore the stainless steel sheath between the fissile material and the water. At time $t = 0$ an incident occurs in the cooling circuit. In a very schematic way, we will assume that the cooling water temperature changes immediately to a value $\theta_2$ ($\theta_2 > \theta_1$), while the heat transfer coefficient becomes $h_2$ ($h_2 < h_1$).

The problem (crucial in industry) is to find the time interval $\Delta t$ after which the fissile material has reached a critical temperature $T_c$, if the system is left to itself. In brief, we have to know how long we have to slow down the reaction (to reduce $P$ as far as possible) by dropping the fission reaction moderator rods.

The temperature field $T_2(\mathbf{r}, t)$ is, for $t > 0$, the solution of the system comprising the following:

- the energy equation (in infinite cylindrical coordinates):

$$\frac{\partial T_2}{\partial t} = a\left(\frac{\partial^2 T_2}{\partial r^2} + \frac{1}{r}\frac{\partial T_2}{\partial r}\right) + \frac{P}{\rho c} \tag{5.11.1}$$

- the boundary conditions:

$$-\lambda \frac{\partial T_2}{\partial r}(R, t) = h_2(T_2(R, t) - \theta_2) \tag{5.11.2}$$

$$-\lambda \frac{\partial T_2}{\partial r}(0, t) = 0 \tag{5.11.3}$$

- the initial condition:

$$T_2(r, t = 0) = T_1^s(r) \tag{5.11.4}$$

$T_1^s(r)$ represents the initial steady state solution associated with the pair of values $(h_1, \theta_1)$, which is a solution of the system (5.12), with the value $i = 1$:

$$i = 1, 2 \qquad a\left(\frac{\partial^2 T_i^s}{\partial r^2} + \frac{1}{r}\frac{\partial T_i^s}{\partial r}\right) + \frac{P}{\rho c} = 0 \tag{5.12.1}$$

$$-\lambda \frac{\partial T_i^s}{\partial r}(R) = h_i(T_i^s(R) - \theta_i) \tag{5.12.2}$$

$$-\lambda \frac{\partial T_i^s}{\partial r}(0, t) = 0 \tag{5.12.3}$$

whose solution is clearly:

$$T_i^s = \theta_i + P\left[\frac{R^2 - r^2}{4\lambda} + \frac{R}{2h_i}\right] \tag{5.13}$$

We may comment that $T_2^s$ is the steady state solution of the system for $i = 2$, that is, for the pair of values $(h_2, \theta_2)$ after an 'infinite' time, a notion which we will make more precise below.

The direct solution of equations (5.11) is theoretically possible but very laborious. It is much simpler to use the superposition technique described at the beginning of this section, by putting*:

$$T_2(r, t) = T_2^s(r) + T(r, t) \tag{5.14}$$

Equations (5.11) are replaced by the system (5.12), with $i = 2$, for which the solution $T_2^s(r)$ is known, and a set of equations in $T(r, t)$, which is the term-by-term difference between (5.11) and (5.12):

$$\frac{\partial T}{\partial t} = a\left(\frac{\partial^2 T}{\partial r^2} + \frac{1}{r}\frac{\partial T}{\partial r}\right) \tag{5.15.1}$$

$$-\lambda \frac{\partial T}{\partial r}(R, t) = h_2 T(R, t) \tag{5.15.2}$$

$$-\lambda \frac{\partial T}{\partial r}(0, t) = 0 \tag{5.15.3}$$

$$T(r, 0) = T_1^s(r) - T_2^s(r) = \theta_1 - \theta_2 + \frac{PR}{2}\left(\frac{1}{h_1} - \frac{1}{h_2}\right) = T_0 < 0 \tag{5.15.4}$$

It is obviously much simpler to solve (5.15) than (5.11) due to the following:

- The power term $P$ has disappeared.
- Equation (5.15.2) has become homogeneous.
- The initial condition is now a constant.

* An alternative, equivalent, choice could be: $T_2(r, t) = T_1^s(r) + T'(r, t)$.

The mathematical solution of (5.15) is standard. It is given in Appendix 3 to this second part, to lighten the discussion here. A numerical application based on this example is also given there.

It is possible to put a physical interpretation on (5.15). It represents the diffusion of heat in an infinitely long cylinder of radius $R$, initially at a uniform negative temperature $T_0$, which is heated by conducto-convection at its surface, with heat transfer coefficient $h$ and mixed fluid temperature $T_m = 0$. This is a classical type of problem and tabulations are available for various one-dimensional geometries and boundary conditions.

In the tables, it is preferable to present the solution in terms of a minimal number of dimensionless groups. These groups in fact represent the true physical parameters which control the problem. The object of the next section is to present a systematic method of dimensional analysis, generalizing the simple approach of Section 2.2.

## 5.4 Dimensional analysis – the Π theorem

Dimensional analysis is based on a classical theorem, the Vaschy–Buckingham or Π theorem, which we will simply state and apply to various cases:

**If a physical phenomenon is described by $n$ physical quantities $p_i$ $(i = 1, n)$, which are expressible in terms of $k$ basic independent physical dimensions, it is possible to group the $n$ quantities into $n - k$ dimensionless ratios $\Pi_j$ $(j = 1, n - k)$, and the solution of the problem may be written:**

$$f(\Pi_1, \Pi_2, \ldots, \Pi_{n-k}) = 0 \tag{5.16}$$

Let us apply this theorem to two examples, in order to understand its power and its limitations:

**Example 1**

Let us look at the phenomena described by equations (5.15). The problem depends on eight physical quantities ($t$, $T$, $T_0$, $a$, $r$, $\lambda$, $2R$, $h_2$). We note that the quantities $T$ and $T_0$ actually represent increments of temperature, relative to the mixed fluid temperature $T_m = 0$. The SI units of these quantities are, respectively: s, K, K, $m^2/s$, m, W/(mK), m, $W/(m^2K)$, which are expressed in terms of four independent units (s, K, m, W). It is thus possible to specify the problem in terms of $n - k = 4$ independent dimensionless ratios.

We can obviously find a large number of different solutions. We should use physical arguments to guide the first choices, and deduce the others from these.

- $\Pi_1 = r/2R = r^+$: dimensionless distance (we could equally well use $r/R$).
- $\Pi_2 = at/4R^2 = Fo$: the Fourier number representing, in this case, a dimensionless time $t/(4R^2/a)$.
- $\Pi_3 = T/T_0$: as noted above, the ratio of two temperature differences.

- The group $\Pi_4$ remains to be found. It must include $\lambda$ and $h_2$. The simplest is $\Pi_4 = 2h_2R/\lambda = Bi$, the Biot number already encountered in Chapter 2. The particular physical interpretation of $Bi$ in (5.15) will be discussed in Chapter 8.

Without having made any calculations at all, we now know that the solution of (5.15) can be written:

$$f\left(\frac{T}{T_0}, Bi, Fo, \frac{r}{2R}\right) = 0 \tag{5.17}$$

When the system is solved (see Appendix 3 to Part II), it turns out that we could as well write $T/T_0 = g(Bi, Fo, r/2R)$.

We should notice at this point that in this type of finite geometry problem, the **spatial and temporal variables $r/2R$ and $Fo$ play distinctly separate roles.**

**Example 2**

Consider a semi-infinite ($x > 0$) wall (initially at temperature $T_0$) whose temperature distribution depends only on $x$ (Figure 5.3). At time $t = 0$, the temperature of the plane $x = 0$ is forced to a temperature $T_e$. All the relevant physical parameters ($\lambda, a, \rho, c$) are constants.

The temperature $T(x, t)$ evidently satisfies the system of equations:

$$\frac{\partial T}{\partial t} = a\frac{\partial^2 T}{\partial x^2} \tag{5.18.1}$$

$$T(x = 0, t) = T_e \tag{5.18.2}$$

$$T(x = \infty, t) = T_0 \tag{5.18.3}$$

$$T(x, t = 0) = T_0 \tag{5.18.4}$$

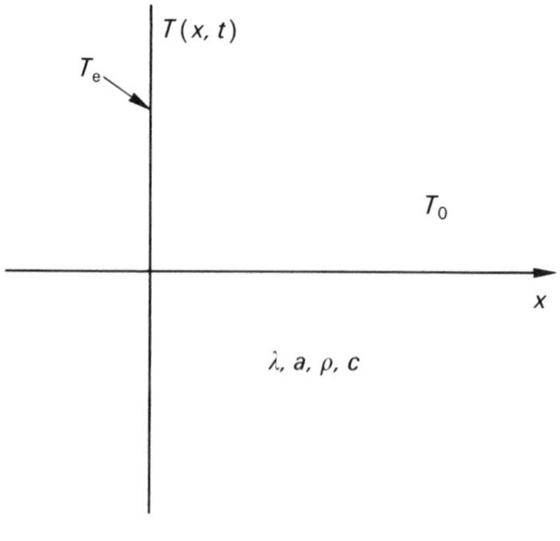


**Figure 5.3**

This system depends on five physical quantities: $(T - T_e, T_0 - T_e, t, a, x)$, whose respective units (K, K, s, $m^2/s$, m) depend on $k = 3$ independent units.

We can thus find the following two dimensionless groups:

- $\Pi_1 = (T - T_e)/(T_0 - T_e) = T^+$ : the dimensionless temperature.
- $\Pi_2$ must involve $t, a, x$. We obviously use $\Pi_2 = at/x^2 = Fo$, the Fourier number.

The system thus has a solution of the type $f((T - T_e)/(T_0 - T_e), Fo) = 0$, which can also be put in the form $(T - T_e)/(T_0 - T_e) = g(Fo)$.

It is interesting to notice that in this second example, involving a semi-infinite geometry, **the variables $x$ and $t$ are no longer independent**, as in the last example, **but appear associated in the group $at/x^2$**. The Fourier number thus has the following two distinct meanings:

- At a given $x$, $Fo$ is a dimensionless time: **the characteristic heat diffusion (from 0 to $x$) is thus $x^2/a$.**
- At a given $t$, we can use $Fo^{-1/2} = x/\sqrt{(at)}$ as a dimensionless space variable; **$\sqrt{(at)}$ is thus a characteristic diffusion length at time $t$.** (Time $t = 0$ is the time of the perturbation.)

It also appears from the group $at/x^2$ that it is equivalent to allow $x$ to tend to infinite (the initial assumption was of a semi-infinite wall) or time $t$ to tend to 0. It can thus be seen that the semi-infinite wall solution can be used to model the response of a system in the period immediately following a perturbation (short-term response).

The rest of this study will be split into the following two parts:

- the study of physical aspects related to semi-infinite geometries (or 'short-term' responses): Chapter 6;
- the study of finite geometries (response at an arbitrary time): Chapter 7.

# 6

# Semi-Infinite Geometry. Response after a Short Time Interval

In this chapter, we will consider some characteristic, but not exhaustive, examples relating to semi-infinite geometry, from a physical point of view:

- the response of a system a brief interval of time after a perturbation (Sections 6.1 and 6.2.1);
- the problem of bringing into contact two bodies with initially different temperatures (Section 6.3).

**Some powerful mathematical tools (e.g. Green's functions, Laplace Transforms), which are presented in Appendices 3–5 of Part II, allow the cases treated in this chapter to be generalized to very much more complex cases.**

## *6.1 Example of the response of a system after a short time interval*

We will look again at the example from Section 5.4 – the initially isothermal semi-infinite medium on which we impose a wall temperature $T_e$ at time $t = 0$. Using dimensionless variables:

$$T^+(u) = \frac{T - T_e}{T_0 - T_e} \tag{6.1}$$

and for mathematical convenience, not $at/x^2$, but a derivative group:

$$u = \frac{x}{2\sqrt{at}} \tag{6.2}$$

the partial differential equation (5.18.1) becomes an ordinary differential equation* in $u$:

$$\frac{d^2T^+}{du^2} + 2u\frac{dT^+}{du} = 0 \tag{6.3.1}$$

The other equations of (5.18) become:

$$T^+(0) = 0 \tag{6.3.2}$$

$$T^+(u \to \infty) = 1 \tag{6.3.3}$$

We should note that the initial condition (5.18.4) and the condition at infinity (5.18.3) have merged: conditions at infinity are not, after a short time, affected by a perturbation at $x = 0$. The solution of (6.3.1) is straightforward, if we substitute:

$$v = \frac{dT^+}{du} \tag{6.4}$$

we obtain, after integration:

$$v = \frac{dT^+}{du} = A\,e^{-u^2} \tag{6.5}$$

The factor of 2 introduced arbitrarily into (6.2) allows this particularly simple form to be obtained.

Taking (6.3.2) and (6.3.3) into account, we obtain:

$$T^+ = \frac{\int_0^u e^{-u'^2}\,du'}{\int_0^\infty e^{-u'^2}\,du'} = \operatorname{erf}(u) = \frac{2}{\sqrt{\pi}}\int_0^u e^{-u'^2}\,du' \tag{6.6}$$

This function is called the error function or erf$(u)$. Some properties and references relating to this function are given in Appendix 2 of this part. The solution to the current problem is shown in Figure 6.1. The rôle of the group $x/2\sqrt{(at)}$, mentioned at the end of Section (5.4), appears clearly in Figure 6.1:

- At a given $x$, $\tau = x^2/a$ is the 'time constant' for the development of the system from its initial temperature to its final temperature. In fact, for $x/2\sqrt{(at)} = 0.5$, the temperature change imposed at $x = 0$ has produced half its effect at $x$ (see Figure 6.1).
- At a given $t$, $\sqrt{(at)}$ appears as the 'diffusion depth' of the heat.

* In fact, for $t \neq 0$, $\partial^2 T/\partial x^2 = (1/4at)\,d^2T/du^2$ and $\partial T/\partial t = -(u/2t)\,dT/du$.

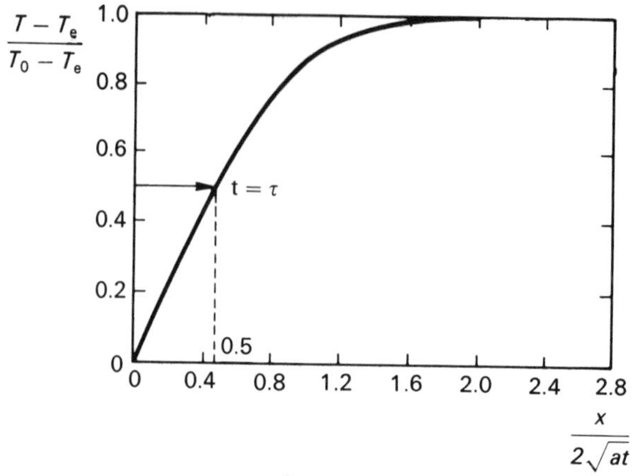


**Figure 6.1**

It is obvious that **the semi-infinite wall model only makes sense when the length of the wall in the** $x$ **direction,** $L$**, satisfies**:

$$L \gg \sqrt{at} \qquad \text{so that:} \quad Fo = \frac{at}{L^2} \ll 1 \tag{6.7}$$

In dimensional variables, the solution to our particular problem is:

$$\frac{T - T_e}{T_0 - T_e} = \frac{2}{\sqrt{\pi}} \int_0^{x/(2\sqrt{at})} e^{-u'^2} \, du' \quad \text{with} \quad \frac{at}{L^2} \ll 1 \tag{6.8}$$

The flux density $\varphi(x, t)$ is obtained using the relation:

$$\varphi = -\lambda \frac{\partial T}{\partial x} = -\frac{1}{\sqrt{\pi t}} \sqrt{\lambda \rho c} \exp\left(-\frac{x^2}{4at}\right) \quad \text{with} \quad \frac{at}{L^2} \ll 1 \tag{6.9}$$

In (6.9), we see the group:

$$b = \sqrt{\lambda \rho c} \tag{6.10}$$

appear. This is called the **effusivity** of a material, and **physically characterizes the response of a system in the period immediately following a perturbation**. The usefulness of this quantity will appear more explicitly in Section 6.3.

Note that the flux falls as $t^{-1/2}$ at the end of the wall ($x = 0$).

## *6.2 Response of a system to a periodic boundary condition. Comparison between diffusion and propagation*

A system can be subjected to periodic thermal conditions (day–night alternation, which is important for solar technology or buildings; chopped beam techniques used in metrology, etc.). In the first case, a sinusoidal variation is a first approximation

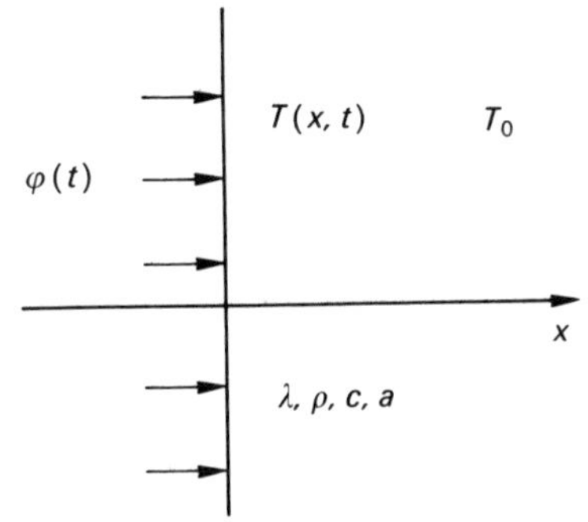


**Figure 6.2**

to the forcing function. In the case of metrology, signal processing techniques allow us to restrict ourselves, in full rigour, to a sinusoidal thermal régime.

Let us consider a semi-infinite wall ($x > 0$), initially at temperature $T_0$, subjected at $t = 0$ to a flux $\varphi = \varphi_0(1 + \cos \omega t)$: that is, modulated sinusoidally about a mean value $\varphi_0$ (Figure 6.2).

The system to be solved is:

$$\frac{\partial T}{\partial t} = a \frac{\partial^2 T}{\partial x^2} \tag{6.11.1}$$

$$-\lambda \frac{\partial T}{\partial t}(0, t) = \varphi_0(1 + \cos \omega t) \tag{6.11.2}$$

$$T(x \to \infty, t) = T_0 \tag{6.11.3}$$

$$T(x, t = 0) = T_0 \tag{6.11.4}$$

We make the assumption that the perturbation will not extend to infinity, so that we remain in the domain of short-term response (equation (6.11.3)). This will be discussed below.

We will use the principle of superposition, and look for a solution of the form:

$$T = T_1(x, t) + T_2(x, t) \tag{6.12}$$

such that $T_1(x, t)$ and $T_2(x, t)$ satisfy the two sets of equations (6.13) and (6.14) respectively:

$$\frac{\partial T_1}{\partial t} = a \frac{\partial^2 T_1}{\partial x^2} \tag{6.13.1}$$

$$-\lambda \frac{\partial T_1}{\partial x}(0, t) = \varphi_0 \tag{6.13.2}$$

$$T_1(x \to \infty, t) = T_0 \tag{6.13.3}$$

$$T_1(x, t = 0) = T_0 \tag{6.13.4}$$

This sub-system represents a problem analogous to that of Section 6.1, but with a flux imposed at the boundary; it will be dealt with in Section 6.2.2.

The other sub-system is then:

$$\frac{\partial T_2}{\partial t} = a\frac{\partial^2 T_2}{\partial x^2} \tag{6.14.1}$$

$$-\lambda\frac{\partial T_2}{\partial x}(0, t) = \varphi_0 \cos \omega t = \varphi_0 \operatorname{Re}(e^{j\omega t}) \tag{6.14.2}$$

(The notation $\operatorname{Re}(e^{j\omega t})$ means the real part of $e^{j\omega t}$.)

$$T_2(x \to \infty, t) = 0 \tag{6.14.3}$$

$$T_2(x, t = 0) = 0 \tag{6.14.4}$$

Using the $\Pi$ theorem (Section 5.4), we can show that the solution of (6.14) depends on three dimensionless groups $\omega t$, $\lambda T_2/\varphi_0 x$ and $x^2\omega/a$. The general solution is obtained using the Green's function method (Appendix 5).

Let us demonstrate the existence of an asymptotic solution for the second sub-system, after a sufficiently long time (which will be defined more precisely below).

### 6.2.1 *Solution for the forced case: degeneration of the diffusion process to a propagation process*

We look for a solution of the form:

$$T_2^{\infty}(x, t) = \operatorname{Re}[\theta_2(x)e^{j\omega t}] \tag{6.15}$$

after an 'infinite' time. In this, $\theta_2(x)$ clearly satisfies:

$$j\omega\theta_2 = a\frac{d^2\theta_2}{dx^2} \tag{6.16.1}$$

$$-\lambda\frac{d\theta_2}{dx}(0) = \varphi_0 \tag{6.16.2}$$

$$\theta_2(\infty) = 0 \tag{6.16.3}$$

The initial condition is no longer relevant.

The characteristic equation of (6.16.1) is:

$$\rho^2 - \frac{j\omega}{a} = 0 \tag{6.17}$$

and the general form of $\theta_2(x)$ is:

$$\theta_2(x) = A\exp\left[-\sqrt{\frac{\omega}{2a}}(1 + j)x\right] + B\exp\left[\sqrt{\frac{\omega}{2a}}(1 + j)x\right] \tag{6.18}$$

From condition (6.16.3), coefficient $B$ is zero. Condition (6.16.2) gives:

$$A = \frac{\varphi_0}{\lambda}\sqrt{\frac{a}{2\omega}}(1-j) \tag{6.19}$$

From (6.15) and (6.19), the asymptotic solution $T_2^\infty(x, t)$ is thus:

$$T_2^\infty(x,t) = \frac{\varphi_0}{\lambda}\sqrt{\frac{a}{\omega}}\exp\left[-\sqrt{\frac{\omega}{2a}}\,x\right]\cos\left(\omega t - \sqrt{\frac{\omega}{2a}}\,x - \frac{\pi}{4}\right) \tag{6.20}$$

This asymptotic solution does not satisfy the initial condition (6.14.4): the system has lost the memory of its initial state. We will not consider here the transient régime which transforms the initial uniformity of temperature at $T_2 = 0$ to this asymptotic sinusoidal régime. It may be formally obtained by the method in Appendix 5, for example.

Let us look at the following important properties of the forced response at circular frequency $\omega$:

1. **At a fixed circular frequency $\omega$, the diffusion phenomenon has degenerated to a propagation phenomenon**: this propagation is in the positive $x$ direction with a celerity $C$ given by:

$$\omega\left(t - \frac{x}{C}\right) = \omega t - \sqrt{\frac{\omega}{2a}}\,x \qquad \text{so that: } C(\omega) = \sqrt{2a\omega} \tag{6.21}$$

This celerity is affected by dispersion, since it depends strongly on $\omega$, as seen in (6.21). Spectral analysis (frequency analysis) of diffusion shows that:

- Infinite frequency signals are propagated infinitely fast – there is no delay between a perturbation and its propagation. This phenomenon, at the physical limit, arises from Fourier's hypothesis (see the discussion in Section 1.1 and Appendix 1 to this part).
- Very low frequency signals are practically not propagated.
- The spectral composition of a signal varies with $x$.
- A signal of circular frequency $\omega$ is strongly attenuated during propagation due to the factor $\exp[-x\sqrt{\omega/2a}]$ in (6.20). The attenuation rises with $\omega$. In other words, the high frequency components of a thermal signal propagate very fast, but are rapidly attenuated. We define the depth of penetration $e_p$ at frequency $\omega$ by:

$$e_p = \sqrt{\frac{2a}{\omega}} \tag{6.22}$$

It is of note that the Fourier number, given by:

$$Fo = \frac{a}{x^2\omega} \tag{6.23}$$

is directly related to this depth of penetration, since $\omega$ is the inverse of time:

$$Fo = \frac{1}{2}\left(\frac{e_p}{x}\right)^2 \tag{6.24}$$

2. We note that there is a phase lag of $\pi/4$ between the excitation flux $\varphi_0 \cos \omega t$ and the temperature at $x = 0$.

All these effects imply the existence of a skin effect for the propagation of a thermal wave of circular frequency $\omega$. We should recall that it is correct to speak of the propagation of a thermal wave at a fixed frequency $\omega$, but if the phenomenon is viewed as a whole, we must consider diffusion characterized by dispersion of the celerity.

*Note: physical discussion of diffusion*
A thermal perturbation, maybe time dependent, imposed on the face $x = 0$ of a semi-infinite wall may be decomposed by spectral analysis in the frequency domain into elementary components of circular frequency $\omega$. Each of these components propagates with celerity $C(\omega)$ and attenuation $e_p(\omega)$. The global signal corresponding to the total of components representing all circular frequencies $\omega$ is thus doubly distorted during its progress through the medium: this is diffusion.

A powerful mathematical technique for solving this type of problem for an arbitrary thermal perturbation is to apply a Laplace transform (or a Fourier transform, which is mathematically equivalent) to the system. The partial differential equation representing the energy equation then becomes an ordinary differential equation, if there is only one space variable involved (see Appendix 4 to this part). The physical basis of this method is clear: the Laplace transformation corresponds to spectral analysis of the given thermal signal. The fact that the energy equation degenerates (at a given frequency) to an ordinary differential equation corresponds to the degeneration of the diffusion process to a propagation process, which is easier to analyse. The inverse transformation allows the diffusion process to be reconstructed by integration of the spectrum.

### 6.2.2 Response to a constant external flux

We wish to determine the solution $T_1(x, t)$ of sub-system (6.13). When $a$ is constant (the condition for (6.13.1)), the energy equation can be written in terms of flux. Putting:

$$\varphi_1(x, t) = -\lambda \frac{\partial T_1}{\partial x} \tag{6.25}$$

we obtain equations in terms of $\varphi_1(x, t)$:

$$\frac{\partial \varphi_1}{\partial t} = a \frac{\partial^2 \varphi_1}{\partial x^2} \tag{6.26.1}$$

$$\varphi_1(0, t) = \varphi_0 \tag{6.26.2}$$

The other two conditions obviously become:

$$\varphi_1(x \to \infty, t) = 0 \tag{6.26.3}$$

and

$$\varphi_1(x, 0) = 0 \tag{6.26.4}$$

System (6.26) is formally of the same type as (5.18), whose solution is given in Section 6.1. We find the solution $\varphi_1(x, t)$:

$$\varphi_1(x, t) = \varphi_0\left[1 - \text{erf}\left(\frac{x}{2\sqrt{at}}\right)\right] \qquad t \neq 0$$

$$= \varphi_0 \,\text{erfc}\left(\frac{x}{2\sqrt{at}}\right) \tag{6.27}$$

where $\text{erfc}(u)$ is the complementary error function (see Appendix 2 to this part).

The solution in terms of temperature follows immediately from (6.27), using conditions (6.13.3) and (6.13.4) (which are identical in the variable $x/(2\sqrt{at})$ and Fourier's law:

$$T_1(x, t) = \frac{2\varphi_0\sqrt{at}}{\lambda} \int_{x/2\sqrt{at}}^{\infty} \text{erfc}(u')\, \text{d}u' + T_0 \qquad \text{with } \frac{at}{L^2} \ll 1 \tag{6.28}$$

$$T_1(x, t) = \frac{2\varphi_0\sqrt{t}}{\sqrt{\lambda\rho c}} \int_{x/2\sqrt{at}}^{\infty} \text{erfc}(u')\, du' + T_0 \qquad \text{with } \frac{at}{L^2} \ll 1 \tag{6.29}$$

The standard integral appearing in (6.29) is tabulated in Appendix 2 to this part.

Once again, the effusivity $b = \sqrt{\lambda\rho c}$ appears in the equation. This characterizes the response a few moments after a perturbation. We note that the temperature of a system with high effusivity will be less perturbed than one with low effusivity.

*Notes*

1. At the wall, at $x = 0$, the temperature rises according to $\sqrt{t}$, as does the interior if the two conditions $x/(2\sqrt{at}) \ll 1$ and $at/L^2 \ll 1$ are satisfied.
2. Dimensional analysis shows that system (6.13) depends on six independent variables $((T_1 - T_0), x, t, a, \lambda, \varphi_0)$, expressible in terms of four independent units (m, s, W, K). The solution thus depends on two independent groups: $at/x^2$ and $\varphi_0 x/[\lambda(T_1 - T_0)]$. In this case, we actually use the groupings $x/(2\sqrt{at})$ and $\varphi_0\sqrt{at}/[\lambda(T_1 - T_0)]$, which appear explicitly in (6.28).

3. The integral function which appears in (6.29) has a magnitude at mid-height given by $x/(2\sqrt{at}) = \frac{1}{2}$; the time constant of this phenomenon is always $\tau = x^2/a$ and the diffusion length $\sqrt{at}$.

## 6.3 Bringing two bodies into thermal contact

A very common problem consists of the study of the thermal development of a system consisting of two bodies (1 and 2), initially isothermal at different temperatures $T_1^0$ and $T_2^0$, and brought suddenly into thermal contact at the instant $t = 0$ (Figure 6.3). We make the following assumptions:

- The thermophysical quantities characterizing the two bodies ($\lambda_i$, $\rho_i$, $c_i$, $a_i$, $b_i$; $i = 1, 2$) are constant.
- The contact is assumed perfect.
- We are not interested in the behaviour except immediately after they are brought into contact; the model of two semi-infinite walls is therefore adopted.

The system to be solved is:

$$\frac{\partial T_1}{\partial t} = a_1 \frac{\partial^2 T_1}{\partial x^2} \tag{6.30.1}$$

$$\frac{\partial T_2}{\partial t} = a_2 \frac{\partial^2 T_2}{\partial x^2} \tag{6.30.2}$$

$$-\lambda \frac{\partial T_1}{\partial x}(0, t) = -\lambda_2 \frac{\partial T_2}{\partial x}(0, t) \tag{6.30.3}$$

$$T_1(0, t) = T_2(0, t) \tag{6.30.4}$$

$$T_1(-\infty, t) = T_1^0 \tag{6.30.5}$$

$$T_2(+\infty, t) = T_2^0 \tag{6.30.6}$$

$$T_1(x, 0) = T_1^0 \tag{6.30.7}$$

$$T_2(x, 0) = T_2^0 \tag{6.30.8}$$

The system depends on nine independent parameters ($T_2^0 - T_1^0$, $T_1 - T_1^0$, $T_2 - T_2^0$, $\lambda_1$, $\lambda_2$, $x$, $t$, $a_1$, $a_2$) which in turn depend on four independent units (K, W, m, s). The solution will be expressed in terms of five independent dimensionless groups. We obtain, initially, $(T_1 - T_1^0)/(T_2^0 - T_1^0)$, $(T_2 - T_2^0)/(T_1^0 - T_2^0)$, $\lambda_1/\lambda_2$, $a_1 t/x^2$, $a_2 t/x^2$. In obtaining these groups, we have endeavoured to preserve the symmetry of the system. $a_1/a_2$ is obtained as the ratio of the last two groups.

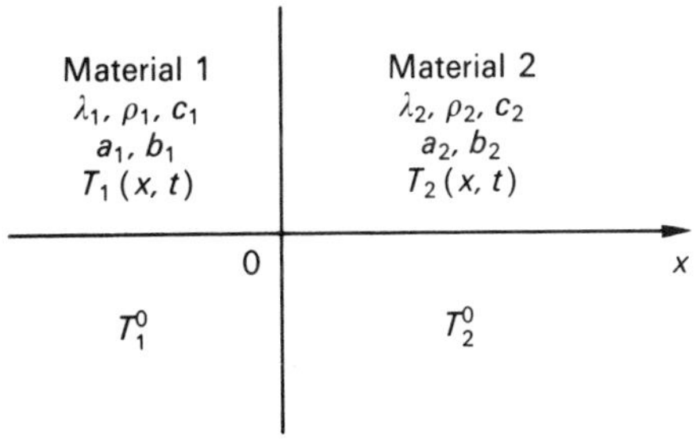


**Figure 6.3**

The mathematical solution of this system, which we will not detail here (see Appendix 3 to this part) is:

$$\frac{T_1(x,t) - T_1^0}{T_2^0 - T_1^0} = \frac{b_2}{b_1 + b_2} \operatorname{erfc}\left(\frac{-x}{2\sqrt{a_1 t}}\right) \qquad x < 0 \tag{6.31}$$

$$\frac{T_2(x,t) - T_2^0}{T_1^0 - T_2^0} = \frac{b_1}{b_1 + b_2} \operatorname{erfc}\left(\frac{x}{2\sqrt{a_2 t}}\right) \qquad x > 0 \tag{6.32}$$

In fact, the ratio of the effusivities $b_1/b_2$ appears in place of $\lambda_1/\lambda_2$ in the simplest formulation. This group can obviously be expressed in terms of those above:

$$\frac{b_1}{b_2} = \sqrt{\frac{\lambda_1 \rho_1 c_1}{\lambda_2 \rho_2 c_2}} = \left(\frac{\lambda_1}{\lambda_2}\right)\left(\frac{a_2 t}{x^2}\right)^{1/2}\left(\frac{a_1 t}{x^2}\right)^{-1/2} \tag{6.33}$$

Let us look at the variation of temperature at the point of contact ($x = 0$) in the moments following contact. It is clear from the definitions of the functions erf and erfc (Appendix 2) that erfc(0) is 1. It follows that:

$$T_1(0,t) = T_2(0,t) = \frac{b_1 T_1^0 + b_2 T_2^0}{b_1 + b_2} \tag{6.34}$$

This result, which is valid immediately after the contact ($t \to 0$), shows that a temperature intermediate between $T_1^0$ and $T_2^0$ is established at the point of contact, and that this is apparently maintained for an indefinite time. In fact it is only maintained for as long as the semi-finite wall model (appropriate to the moments following the perturbation) is valid (that is, $a_1 t/L_1^2 \ll 1$, $a_2 t/L_2^2 \ll 1$).

A significant result is that **the body with the higher effusivity tends to impose its temperature** – if $b_1 \gg b_2$, we obtain:

$$T_1(0,t) = T_2(0,t) \simeq T_1^0 \tag{6.35}$$

The conditions in Section 6.1 are thus attainable with a very high effusivity.

This is the explanation of a common but paradoxical phenomenon: walking barefoot (at 37°C) on a metal plate at 60°C or a plank of wood at the same temperature do not give the same 'feeling of temperature'. The first is unbearable, while the second is perfectly tolerable. The answer is simple:

- The effusivity of feet is similar to that of water: $b_f \simeq 1600$ (SI).
- That of wood: $b_w \simeq 11$ (SI).
- That of steel: $b_s \simeq 13\,000$ (SI).

Following contact between foot and steel, the contact temperature establishes itself immediately (6.34) at $T_{fs} = 57.5$°C. Following contact between foot and wood, we obtain merely $T_{fw} = 37$°C.

If, on the other hand, you persist in standing on the wooden plank, the effect very quickly becomes the same as for the steel plate. This is a facetious, but easily tested, demonstration that the semi-infinite wall model is only valid for a short time. Fire walkers know well that you must keep moving on burning coals!

*Note*

For the three cases in (6.1), (6.2.2) and (6.3), **the effusivity** $\sqrt{\lambda\rho c}$ represents a value characterizing the unsteady response of a system in the moments following a perturbation (so long as the medium can be considered as semi-infinite). Physically, it represents the capacity of a material to 'resist' an abrupt change in the temperature of its surroundings.

# 7

# Finite Geometry. Response at an Arbitrary Time

The response of a finite system to a perturbation has already been discussed (Section 6.3). Here we will consider another case in order to bring out the physical particularities of this type of problem. The mathematical tools outlined in Appendices 3, 4 and 5 of this part allow more complex problems to be tackled, as do standard numerical methods (finite difference, finite element, etc.).

## *7.1 The case of an instantaneous perturbation*

Consider a wall of thickness $L$ in the $x$ direction, insulated on the face at $x = 0$, initially at temperature $T_0$, and exchanging energy with its surroundings at temperature $T_e$ by conducto-convection, with heat transfer coefficient $h$, through the face at $x = L$ (Figure 7.1).

The equations to be solved are:

$$\frac{\partial T}{\partial t} = a \frac{\partial^2 T}{\partial x^2} \tag{7.1.1}$$

$$-\lambda \frac{\partial T}{\partial x}(L, t) = h(T(L, t) - T_e) \tag{7.1.2}$$

$$-\lambda \frac{\partial T}{\partial x}(0, t) = 0 \tag{7.1.3}$$

$$T(x, 0) = T_0 \tag{7.1.4}$$

They depend on eight independent quantities ($T - T_e$, $T_0 - T_e$, $x$, $t$, $a$, $\lambda$, $h$, $L$), which are expressed in terms of four independent units (K, W, m, s); the solution

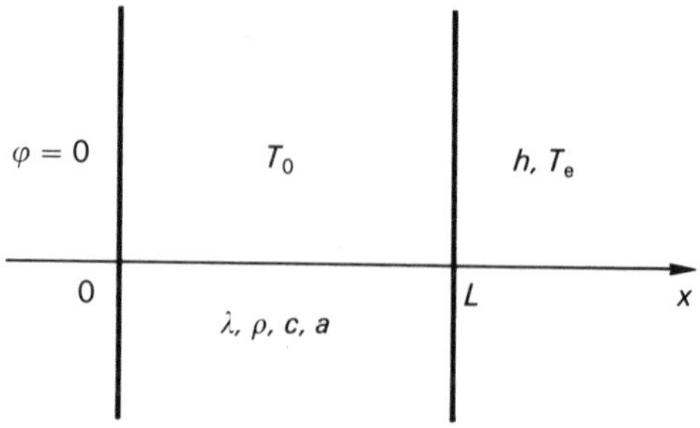


**Figure 7.1**

depends on four dimensionless groups: $T^+ = (T - T_e)/(T_0 - T_e)$, $x^+ = x/L$, $t^+ = Fo = at/L^2$, $Bi = hL/\lambda$. We notice that the Fourier number ($Fo$) and the Biot number ($Bi$), whose physical significance will be defined in Chapter 8, appear.

The system can be re-written in non-dimensional form:

$$\frac{\partial T^+}{\partial t^+} = \frac{\partial^2 T^+}{\partial x^{+2}} \tag{7.2.1}$$

$$\frac{\partial T^+}{\partial x^+}(1, t^+) + BiT^+(1, t^+) = 0 \tag{7.2.2}$$

$$\frac{\partial T^+}{\partial x^+}(0, t^+) = 0 \tag{7.2.3}$$

$$T^+(x^+, 0) = 1 \tag{7.2.4}$$

This linear system consists of just one non-homogeneous equation (7.2.4). It can be solved in this case by various techniques, of which the most common are the Laplace transform method (see Appendix 3) or the method of separation of variables (see Appendix 4).

The solution, detailed in Appendix 4, is:

$$T^+(x^+, t^+) = \sum_{n=0}^{\infty} \left( \frac{2\sin(k_n)}{k_n + \sin(k_n)\cos(k_n)} \right) \cos(k_n x^+) \exp[-k_n^2 t^+] \tag{7.3}$$

where $k_n$ form a set of discrete values which are the solutions of:

$$Bi = k_n \tan(k_n) \tag{7.4}$$

In dimensional form, the solution becomes:

$$\frac{T(x,t) - T_e}{T_0 - T_e} = \sum_{n=0}^{\infty} \left( \frac{2\sin(k_n)}{k_n + \sin(k_n)\cos(k_n)} \right) \cos\left( k_n \frac{x}{L} \right) \exp\left[ -k_n^2 \frac{at}{L^2} \right] \tag{7.5}$$

This example is representative of the solutions obtained for any finite geometry. When it is possible to obtain an analytic solution, it usually appears in the form of an infinite series as in (7.5). We could also consider the solution in Section 2 of Appendix 3. Such a solution is valid for any $t$. It is obvious that the characteristic time $\tau_{cd} = L^2/a$ which appears in the dimensionless time $t^+$ (or Fourier number) represents a 'system time constant':

- When $at/L^2 \gg 1$, we can just use one term; the long-term response of the system is exponential in form.
- When $at/L^2 \ll 1$, that is, in the brief period after a perturbation, all the terms must be taken into account. In this case, the medium may be considered to be semi-infinite, since the boundary at $x = L$ is practically unaffected by diffusion. We could obtain the same solution by adopting a model analogous to that of Section 6.1. The result obtained with this last model is easier to use in these circumstances than (7.5).
- In intermediate cases, it is generally sufficient to take two or three terms in the series.

## *7.2 Response to a forced input*

The analysis in Section 6.2 may be repeated without difficulty for the case of a finite geometry. The result is, for **a given circular frequency** $\omega$, found to be 'thermal waves' propagating simultaneously in the positive and negative $x$ directions, giving rise to standing waves.

## *7.3 Generalization to two- and three-dimensional problems*

In some cases, an analytic solution can be found to a multi-dimensional problem by means of knowledge of the solutions to related one-dimensional problems. Let us look at an example.

Consider a parallelepiped of sides $2c_1, 2c_2, 2c_3$, initially at temperature $T_0$, which exchanges energy from the time $t = 0$ with its surroundings at temperature $T_e$. The transfers are characterized by coefficient $h_1, h_2, h_3$, identical on opposite faces (Figure 7.2). There are some obvious symmetries in the problem. We can associate three one-dimensional problems with it (infinite walls between $x_j = 0$ and $x_j = c_j$), for which we solve three systems of equations ($j = 1, 2, 3$) for $T_j^+$, in the dimensionless variables:

$$x_j^+ = \frac{x_j}{c_j}, \qquad t_j^+ = Fo_j = \frac{at}{c_j^2}, \qquad Bi_j = \frac{h_j c_j}{\lambda}, \qquad T_j^+ = \frac{T_j}{T_0} \tag{7.6.1}$$

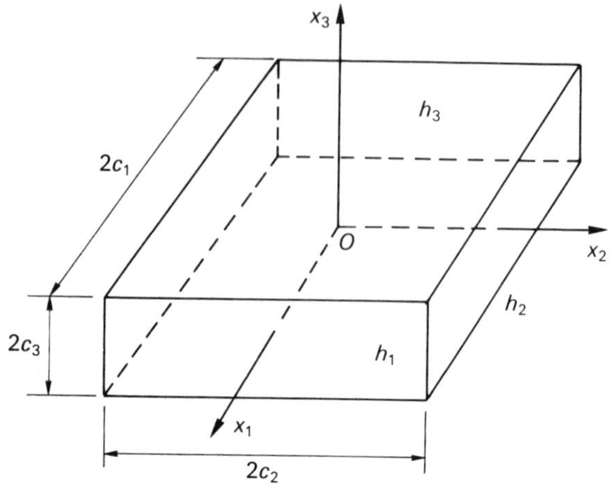


**Figure 7.2**

which are:

$$\frac{\partial T_j^+}{\partial t_j^+} = \frac{\partial^2 T_j^+}{\partial x_j^{+2}} \tag{7.6.2}$$

$$\frac{\partial T_j^+}{\partial x_j^+}(0, t_j^+) = 0 \tag{7.6.3}$$

$$\frac{\partial T_j^+}{\partial x_j^+}(1, t_j^+) + Bi_j T_j^+(1, t_j^+) = 0 \tag{7.6.4}$$

$$T_j^+(x_j^+, 0) = 1 \tag{7.6.5}$$

We will show that the general solution to the current problem is equal to:

$$T^+ = T_1^+ T_2^+ T_3^+ \tag{7.7}$$

The system of equations associated with the three-dimensional problem is:

$$\frac{\partial T^+}{\partial t^+} = \frac{\partial^2 T^+}{\partial X_1^{+2}} + \frac{\partial^2 T^+}{\partial X_2^{+2}} + \frac{\partial^2 T^+}{\partial^2 X_3^{+2}} \tag{7.8.1}$$

$$\left.\frac{\partial T^+}{\partial X_j^+}\right|_{x_j^+ = 0} = 0 \qquad j = 1, 2, 3 \tag{7.8.2}$$

$$\left.\frac{\partial T^+}{\partial X_j^+}\right|_{x_j^+ = 1} + Bi_j\left(\frac{c_1}{c_j}\right) T^+|_{x_j^+ = 1} = 0 \tag{7.8.3}$$

$$T^+|_{t^+ = 0} = 1 \tag{7.8.4}$$

in which the length and time scales are, respectively, $c_1$ and $c_1^2/a$:

$$t^+ = \frac{at}{c_1^2}, \qquad X_j^+ = \frac{x_j}{c_1} = x_j^+\left(\frac{c_j}{c_1}\right) \tag{7.9}$$

If the solution is in the form of (7.7), the two sides of (7.8.1) become:

$$\frac{\partial T^+}{\partial t^+} = T_2^+ T_3^+ \frac{\partial T_1^+}{\partial t_1^+} + T_1^+ T_3^+ \left(\frac{c_1}{c_2}\right)^2 \frac{\partial T_2^+}{\partial t_2^+} + T_1^+ T_2^+ \left(\frac{c_1}{c_3}\right)^2 \frac{\partial T_3^+}{\partial t_3^+} \tag{7.10}$$

$$\frac{\partial^2 T^+}{\partial X_1^{+2}} + \frac{\partial^2 T^+}{\partial X_2^{+2}} + \frac{\partial^2 T^+}{\partial X_3^{+2}} = T_2^+ T_3^+ \frac{\partial^2 T_1^+}{\partial x_1^{+2}} + T_1^+ T_3^+ \left(\frac{c_1}{c_2}\right)^2 \frac{\partial^2 T_2^+}{\partial x_2^{+2}} + T_1^+ T_2^+ \left(\frac{c_1}{c_3}\right)^2 \frac{\partial^2 T_3^+}{\partial x_3^{+2}} \tag{7.11}$$

(7.7) obviously satisfies (7.8.1) and is also the solution of (7.8.2), (7.8.3) and (7.8.4). The solution is thus deduced from equation (7.5); there is no point in writing it out. Suffice it to note that at the limit of a long time period the behaviour of the system with time is as: $\exp\{-[(k_0^1)^2 a/c_1^2 + (k_0^2)a/c_2^2 + (k_0^3)a/c_3^2]t\}$. This corresponds to a characteristic decay time. In the case of a cube such that $c_1 = c_2 = c_3$ and $h_1 = h_2 = h_3$, the characteristic time becomes

$$\tau = \frac{c^2}{3a} \tag{7.12}$$

*Notes*

1. The above technique can be generalized without difficulty to cases where the initial condition is a product of three functions of $x_1$, $x_2$ and $x_3$: $T_0(x_1, x_2, x_3) = \theta_0^1(x_1)\theta_0^2(x_2)\theta_0^3(x_3)$. We need only solve the elementary systems with the initial conditions $\theta_0^j(x_j)$.
2. The same method can be used for other simple geometries: for example, a truncated cylinder of height $H$.

# 8

# Time and Length Scales

## *8.1 Characteristic times*

A thermal system is often characterized by particular characteristic times (periodic excitation, impulse duration, time typical of the movement of a fusion front, time necessary for heat treatment of a material, etc.). These times must be compared among themselves and with others which are appropriate to heat transfer by conduction, convection and radiation as part of the initial analysis of the system. It is then often possible to achieve a considerable simplification of the system model, since certain times may be negligible in comparison to others.

1. **The characteristic time for conduction** $\tau_{cd}$ in a particular direction for which a typical length of the system is $L$ is given by the Fourier number for that direction (for example, the Fourier number for a cylinder obtained in Section 5.4, and, more generally, that for a three-dimensional case in Section 7.3):

$$Fo = \frac{t}{\tau_{cd}} = \frac{at}{L^2} \tag{8.1}$$

whence:

$$\tau_{cd} = \frac{L^2}{a} \tag{8.2}$$

A system often has several characteristic times. A cuboid of dimensions $L_1$, $L_2$, $L_3$ has three characteristic Fourier numbers $at/L_1^2$, $at/L_2^2$ and $at/L_3^2$, with which we can associate three characteristic times $L_1^2/a$, $L_2^2/a$ and $L_3^2/a$ (see Section 7.3).

2. **A characteristic time for conducto-convection** $\tau_{cc}$ can easily be found. Let us suppose that the characteristic time for conduction normal to the surface is zero ('infinite' material thermal conductivity – discussed further below). Consider a bar of radius $R$, initially at temperature $T_0$, which exchanges heat only with a bath at temperature $T_e$ by conducto-convective transfer with coefficient $h$. At any instant the bar will,

on these assumptions, be isothermal at temperature $T(t)$. The temperature evolution $T(t)$ of an element of volume $\mathrm{d}V$, of surface area $\mathrm{d}S$, of the bar is governed by:

$$\rho c \frac{\mathrm{d}T}{\mathrm{d}t} \mathrm{d}V = -h(T - T_e)\,\mathrm{d}S \tag{8.3}$$

$$T(0) = T_0 \tag{8.4}$$

whose solution is:

$$\frac{T - T_e}{T_0 - T_e} = \exp\left[-\frac{h\,\mathrm{d}S}{\rho c\,\mathrm{d}V} t\right] = \exp\left(-\frac{t}{\tau_{cc}}\right)$$

In the absence of delay associated with conduction (the bar is, at any instant, isothermal), the characteristic time associated with conducto-convective heat transfer at the surface is:

$$\tau_{cc} = \frac{\rho c\,\mathrm{d}V}{h\ \mathrm{d}S} = \frac{\rho c R}{2h} \tag{8.5}$$

*Notes:*

- We can use the hypothesis of instantaneous isothermicity on time scales which are large compared to $\tau_{cd}$, when:

$$\frac{\tau_{cd}}{\tau_{cc}} \ll 1 \qquad \text{or} \qquad \frac{8hR}{\lambda} \ll 1$$

  is satisfied. In the opposite case when $\tau_{cc}/\tau_{cd} \ll 1$, we can consider the medium to be semi-infinite since $at/L^2 \ll 1$.

  On time scales small compared to $\tau_{cc}$, we can neglect the effects of conducto-convective transfers.

- **The Biot number** appears here, with a multiplier, as the ratio of the characteristic time for conduction normal to the wall to that for conducto-convective transfer:

$$Bi = \frac{2hR}{\lambda} = \frac{1}{4}\frac{\tau_{cd}}{\tau_{cc}} \tag{8.6}$$

- A very important consequence of (8.2), (8.5) and (8.6) is in regard to the response time of a thermocouple in unsteady conditions. If the junction of the thermocouple is locally like a cylinder of radius $R$, it is apparent that **the response time of a thermocouple falls as $R$ is reduced**. It is therefore of interest to construct a thermocouple whose outside diameter, after drawing through a die, is as small as possible (a few hundredths of a millimetre) so that the response time shall be as short as possible.

3. It is also possible to define a **characteristic time for radiation** at the boundaries of a system. In a simple standard case, (8.3) becomes:

$$\rho c \frac{\mathrm{d}T}{\mathrm{d}t} \mathrm{d}V = -\varepsilon\sigma (T^4 - T_e^4) \, \mathrm{d}S \tag{8.7.1}$$

$$T(0) = T_0 \tag{8.7.2}$$

whose solution is:

$$t - t_0 = \frac{\rho c \, dV}{\varepsilon\sigma \, \mathrm{d}S} \int_{T_0}^{T} \frac{\mathrm{d}T'}{T'^4 - T_e^4} \tag{8.8}$$

Notice that the time is related to the radiative flux, not to the propagation of the photons which is in practice instantaneous (because it depends on the speed of light over finite distances).

## *8.2 Fourier number*

Various interpretations of the Fourier number have already been presented in this work; let us briefly review them:

1. For a finite system of dimension $L$ (or $R$), it is a dimensionless time:

$$Fo = t^+ = \frac{t}{\tau_{cd}} = \frac{at}{L^2} \tag{8.9}$$

   where $\tau_{cd}$ is the characteristic time for conduction in the direction of the dimension $L$ (see Section 8.1).
2. For a system with one semi-infinite dimension (that is, considering only the moments immediately following a perturbation), the Fourier number allows us to:
   - obtain the characteristic distance for heat diffusion ($\sqrt{at}$) at a given time $t$;
   - obtain the characteristic time for diffusion ($\tau = x^2/a$) at a given point $M$ on the $x$ (or $r$) axis.
3. For a sinusoidal forcing system with circular frequency $\omega$, the Fourier number allows the depth of penetration of the wave ($e_p = \sqrt{a/\omega}$) to be obtained (Section 6.2.1).

# Limit of Validity of Fourier's Law (Special Cases) [1–3]

Fourier's law implies an instantaneous response to a perturbation, so that, in one dimension (the $x$ direction):

$$\varphi^{\mathrm{cd}} = -\lambda \frac{\partial T}{\partial x} \tag{1}$$

If we consider the elementary carriers responsible for conduction, it is obvious that the energy transfers leading to the movement of heat through the medium are characterized by a very short time scale, of the order of $10^{-9}$–$10^{-12}$ s. We can take this into Fourier's law by writing:

$$-\lambda \frac{\partial T}{\partial x} = \varphi^{\mathrm{cd}} + \tau \frac{\partial \varphi^{\mathrm{cd}}}{\partial t} \tag{2}$$

in which appears the relaxation time for the carriers, $\tau$ of order $10^{-10}$ s. We have introduced a 'delay' term $\tau\, \partial\varphi^{\mathrm{cd}}/\partial t$.

The energy equation, which is of the diffusion type:

$$\rho c_{\mathrm{p}} \frac{\partial T}{\partial t} = -\frac{\partial}{\partial x}(\varphi^{\mathrm{cd}}) \tag{3}$$

becomes, after substituting for $\varphi^{\mathrm{cd}}$ from (2) and $\partial\varphi^{\mathrm{cd}}/\partial x$ from (3):

$$\rho c_{\mathrm{p}} \frac{\partial T}{\partial t} = -\frac{\partial}{\partial x}\left(-\lambda \frac{\partial T}{\partial x} - \tau \frac{\partial \varphi^{\mathrm{cd}}}{\partial t}\right) = \lambda \frac{\partial^2 T}{\partial x^2} - \rho c_{\mathrm{p}} \tau \frac{\partial^2 T}{\partial t^2} \tag{4}$$

This equation is well known under the name 'telegraphist's equation'. It is hyperbolic, and leads to the propagation of a thermal signal to circular frequency $\omega$ of the form: $T(t) = \theta(t)\mathrm{e}^{\mathrm{j}}\omega^{t}$, with a celerity (phase velocity) which tends to a finite value as

$\omega \rightarrow \infty$:

$$C_\infty = \sqrt{a/\tau} \tag{5}$$

There is always dispersion of the celerity, but infinite frequency signals no longer propagate at infinite speed. We thus resolve the paradox contained in Fourier's law since there is now a limiting speed of propagation of thermal signals.

These effects must be taken into account in certain very uncommon applications:

- Concerning the 'immediate' response of systems to very short impulses. (The pulses of certain lasers last a few times $10^{-12}$ s, but deliver a non-negligible energy, which implies an instantaneous flux density of $10^{12}$ W m$^{-2}$!) But, very often, we are faced with problems which are not in local thermodynamic equilibrium, as in plasmas.
- In rarefied gases.

APPENDIX 2

# Error Functions (erf, erfc)

### *Definitions*

- Error function (Figure 1)

$$\operatorname{erf}(u) = \frac{2}{\sqrt{\pi}} \int_0^u e^{-u'^2}\, du'$$

$$= \left(\int_0^u e^{-u'^2}\, du'\right) \Big/ \left(\int_0^\infty e^{-u'^2}\, du'\right)$$

$$\operatorname{erf}(\infty) = 1 \qquad \operatorname{erf}(0) = 0$$

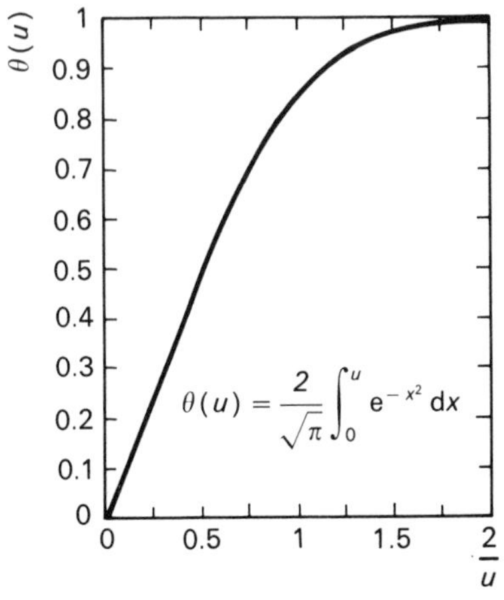


**Figure 1**

- Complementary error function

$$\operatorname{erfc}(u) = 1 - \operatorname{erf}(u) = \frac{2}{\sqrt{\pi}} \int_u^\infty e^{-u'^2}\, du'$$

## *Derivatives and integrals*

$$\frac{d}{du} \operatorname{erf}(u) = \frac{2}{\sqrt{\pi}} e^{-u^2}$$

$$\int_u^\infty \operatorname{erfc}(u')\, du' = \frac{1}{\sqrt{\pi}} e^{-u^2} - u \operatorname{erfc}(u)$$

$\int_u^\infty \operatorname{erfc}(u')\, du'$ is sometimes denoted by $\operatorname{ierfc}(u)$.

**Some numerical values**

| | | | | | | | | | | |
|---|---|---|---|---|---|---|---|---|---|---|
| $u$ | 0 | 0.05 | 0.1 | 0.2 | 0.3 | 0.4 | 0.5 | 0.6 | 0.7 | 0.8 |
| $\operatorname{erf}(u)$ | 0 | 0.056 | 0.112 | 0.223 | 0.329 | 0.428 | 0.520 | 0.604 | 0.678 | 0.742 |
| $\operatorname{erfc}(u)$ | 1 | 0.944 | 0.888 | 0.777 | 0.671 | 0.572 | 0.480 | 0.396 | 0.322 | 0.258 |
| $u$ | 0.9 | 1.0 | 1.1 | 1.2 | 1.3 | 1.4 | 1.5 | 1.7 | 1.8 | 2.0 |
| $\operatorname{erf}(u)$ | 0.797 | 0.843 | 0.880 | 0.910 | 0.934 | 0.952 | 0.966 | 0.984 | 0.989 | 0.995 |
| $\operatorname{erfc}(u)$ | 0.203 | 0.157 | 0.120 | 0.090 | 0.066 | 0.048 | 0.034 | 0.016 | 0.011 | 0.005 |
| $u$ | 0 | 0.05 | 0.1 | 0.2 | 0.3 | 0.4 | 0.5 | 0.6 | 0.7 | 0.8 |
| $2\operatorname{ierfc}(u)$ | 1.128 | 1.031 | 0.940 | 0.773 | 0.628 | 0.540 | 0.399 | 0.312 | 0.240 | 0.182 |
| $2/\sqrt{\pi}\, e^{-u^2}$ | 1.128 | 1.126 | 1.117 | 1.084 | 1.031 | 0.962 | 0.880 | 0.787 | 0.691 | 0.595 |
| $u$ | 0.9 | 1.0 | 1.1 | 1.2 | 1.3 | 1.4 | 1.5 | 1.6 | 1.7 | 2.0 |
| $2\operatorname{ierfc}(u)$ | 0.136 | 0.101 | 0.073 | 0.052 | 0.036 | 0.025 | 0.017 | 0.011 | 0.007 | 0.002 |
| $2/\sqrt{\pi}\, e^{-u^2}$ | 0.502 | 0.415 | 0.336 | 0.267 | 0.208 | 0.159 | 0.119 | 0.087 | 0.063 | 0.021 |

# Example of the Use of the Laplace transform (for a case with two variables x and t)

For a physical interpretation, see the comments in Section 6.2.1.

## 1. Brief summary of the Laplace transformation

### 1.1 Definition

We define the Laplace transform $f(p)$ of a function $F(t)$ as:

$$p \in C: \qquad f(p) = \int_0^{\infty} e^{-pt} F(t)\, dt \tag{1}$$

This transformation is linear.

### 1.2 Some properties

| | |
|---|---|
| $f(p)$ | $F(t)$ |
| $f(p+k)$ | $e^{-kt}F(t)$ |
| $f(p)e^{-kp}$ | $F(t-k) \quad k>0$ |
| $pf(p) - F(0^+)$ | $F'(t)$ |
| $\partial f/\partial x(x,p)$ | $\partial F/\partial x(x,t)$ |
| $\int_{x_1}^{x_2} f(x,p)\, dx$ | $\int_{x_1}^{x_2} F(x,t)\, dx$ |
| $f(p)g(p)$ | $F(t) * G(t)$ (*: convolution operator) |

## 1.3 Table of Laplace transforms of some functions

$$\theta(p) = \int_0^{\infty} F(t)\mathrm{e}^{-pt}\,\mathrm{d}t \qquad \text{with} \qquad q^2 = \frac{p}{a}; \qquad a = \frac{\lambda}{\rho c} \quad \text{(diffusivity)}$$

| $\theta(p)$ | $F(t)$ |
|---|---|
| $1/p$ | $1$ |
| $1/(p^{\nu} + 1)$, $\nu > -1$ | $t^{\nu}/\Gamma(\nu + 1)$ |
| $1/(p + b)$ | $\mathrm{e}^{-bt}$ |
| $\omega/(p^2 + \omega^2)$ | $\sin \omega t$ |
| $p/(p^2 + \omega^2)$ | $\cos \omega t$ |
| $\exp(-qx)$ | $(x/2)(\pi a t^3)^{-1/2} \exp(-x^2/4at)$ |
| $\exp(-qx)/q$ | $(a/\pi t)^{1/2} \exp(-x^2/4at)$ |
| $\exp(-qx)/p$ | $\mathrm{erfc}[x/(2\sqrt{(at)})]$ |
| $\exp(-qx)/pq$ | $2\left(\dfrac{at}{\pi}\right)^{1/2} \mathrm{e}^{-x^2/4at} - x\,\mathrm{erfc}[x/(2\sqrt{at})]$ |
| $\exp(-qx)/p^2$ | $\left(t + \dfrac{x^2}{2a}\right) \mathrm{erfc}[x/(2\sqrt{at})] - x(t/\pi a)^{1/2}\,\mathrm{e}^{-x^2/4at}$ |

Other transforms may be found in references 4 and 5.

## 1.4 The method

The method consists of:

1. Applying the Laplace transformation to the complete system of equations consisting of the energy equation (which thus becomes an ordinary differential equation in $x$), the boundary conditions and the initial condition.
2. Solving the new system thus obtained in the transformed space.
3. Returning to $(x, t)$ space by an inverse transformation. Two methods are possible:
   - in appropriate cases, use look-up tables (see above)
   - otherwise use Mellin's inversion formula (see next paragraph).

## 1.5 The inverse Laplace transform

Mellin's formula is:

$$F(t) = \lim_{X \to 0} \frac{1}{2\pi i} \int_{\gamma - iX}^{\gamma + iX} \mathrm{e}^{pt} f(p)\,\mathrm{d}p \tag{2}$$

where i is the square root of $-1$. This method often makes use of the calculus of residues. On a contour of integration enclosing simple or higher order poles of a function $f(z)$ we calculate:

- the residue of a simple pole at $z = a$ by:

$$Res(a) = \lim_{z \to a} (z - a) f(z) \tag{3}$$

- the residue of a pole of order $k$ at $z = b$ by:

$$Res(b) = \frac{1}{(k-1)!} \lim_{z \to b} \left( \frac{\mathrm{d}^{k-1}}{\mathrm{d}z^{k-1}} (z - b)^k f(z) \right) \tag{4}$$

We will proceed to a complete problem as an illustration in the next section. We have deliberately chosen a cylindrical geometry to demonstrate that the use of Bessel functions (perhaps modified) presents no conceptual difficulties, even if the subsequent calculations may turn out to be laborious.

### 1.6 Limitations

The Laplace transform is a powerful and successful tool for one-dimensional (plane, cylindrical or spherical) unsteady problems. It can be used for the solution of two- and three-dimensional problems, but its interest is considerably less since the energy equation to be solved after transformation is still a partial differential equation in the spatial variables.

Note that for one-dimensional problems, it is applicable when the initial conditions are non-uniform in space.

## 2. Application to the problem of a bar of radius R

Consider a cylindrical bar of radius $R$, initially at temperature $T_0$, which at time $t = 0$ is plunged into a bath at temperature $T_e = 0$, with which the heat exchange coefficient is $h$ (the same problem as in Section 5.3).

### 2.1 Equations in (r, t) space:

$$\frac{\partial T}{\partial t} = \frac{a}{r} \frac{\partial}{\partial r} \left( r \frac{\partial T}{\partial r} \right) \tag{5a}$$

$$\lambda \frac{\partial T}{\partial r}(R, t) + hT(R, t) = 0 \tag{5b}$$

$$\lambda \frac{\partial T}{\partial r}(0, t) = 0 \tag{5c}$$

$$T(r, 0) = T_0 \tag{5d}$$

Putting:

$$T^+ = \frac{T}{T_0}, \qquad Fo = t^+ = \frac{at}{R^2}, \qquad r^+ = \frac{r}{R}, \qquad Bi = \frac{hR}{\lambda} \tag{6}$$

these become:

$$\frac{\partial T^+}{\partial t^+} = \frac{1}{r^+}\frac{\partial}{\partial r^+}\left(r^+\frac{\partial T^+}{\partial r^+}\right) \tag{7a}$$

$$\frac{\partial T^+}{\partial r^+}(1, t^+) + BiT^+(1, t^+) = 0 \tag{7b}$$

$$\frac{\partial T^+}{\partial r^+}(0, t^+) = 0 \tag{7c}$$

$$T^+(r^+, 0) = 1 \tag{7d}$$

## 2.2 Laplace transformation of equations (5)

We will drop the +, to simplify the notation:

$$p\theta - 1 = \frac{1}{r}\frac{\mathrm{d}}{\mathrm{d}r}\left(r\frac{\mathrm{d}\theta}{\mathrm{d}r}\right) \tag{8a}$$

$$\frac{\mathrm{d}\theta}{\mathrm{d}r}(1, p) + Bi\theta(1, p) = 0 \tag{8b}$$

$$\frac{\mathrm{d}\theta}{\mathrm{d}r}(0, p) = 0 \tag{8c}$$

where $\theta(r, p)$ is the Laplace transform of $T^+(r^+, t^+)$. The initial condition has disappeared, being incorporated into equation (8a).

## 2.3 Solution in the transformed space

If we write $p = q^2$ and:

$$u = rp^{1/2} = rq \tag{9}$$

equation (8a) (without its right-hand side) becomes:

$$u^2\frac{\mathrm{d}^2\theta}{\mathrm{d}u^2} + u\frac{\mathrm{d}\theta}{\mathrm{d}u} - u^2\theta = 0 \tag{10}$$

This is of the 'modified Bessel' type [4], that is, it is in the form:

$$u^2 \frac{d^2\omega}{du^2} + u \frac{d\omega}{du} - (u^2 + v^2)\omega = 0 \tag{11}$$

with $v = 0$.

Its general solution is written [4] in terms of the modified Bessel functions $I_0$ and $K_0$:

$$\theta(r, p) = AI_0(qr) + BK_0(qr) \tag{12}$$

An obvious particular integral of the whole equation is:

$$\theta_{\text{part}} = \frac{1}{p} \tag{13}$$

so that the general solution of the whole equation is:

$$\theta(r, p) = AI_0(qr) + BK_0(qr) + \frac{1}{p} \tag{14}$$

The solution to the physical problem is finite everywhere. Since $K(qr) \to \infty$ as $r \to 0$, coefficient $B$ must be zero. This is compatible with (8c), since $I_0(qr)$ is everywhere finite.

Coefficient $A$ is found from (8b):

$$AqI'_0(q) + Bi\left[AI_0(q) + \frac{1}{q}\right] = 0 \tag{15}$$

The solution to the problem in transformed space is thus:

$$\theta(r, p) = -\frac{BiI_0(qr)}{p[qI'_0(q) + BiI_0(q)]} + \frac{1}{q} \tag{16}$$

with:

$$q = p^{1/2} \tag{17}$$

## 2.4 Finding the poles of the solution

The poles are at:

- $p = 0$: a simple pole
- the roots in terms of $p$ of:

$$qI'_0(q) + BiI_0(q) = 0 \tag{18}$$

Formulae given in reference [4] allow this expression to be transformed: considering non-modified Bessel functions, we have:

$$I_0(z) = J_0(iz) \tag{19}$$

$$I'_0(z) = \mathrm{i}J_0(\mathrm{i}z) = -\mathrm{i}J_1(\mathrm{i}z) \tag{20}$$

substituting $k = \mathrm{i}q$, the equation to be solved becomes:

$$-kJ_1(k) + BiJ_0(k) = 0 \tag{21}$$

This equation has an infinite number of simple real roots:

$$k = k_1, \ldots, k_n, \ldots \tag{22}$$

in terms of $p$:

$$p = -k^2 \tag{23}$$

these correspond to an infinite sequence of negative roots:

$$p = -k_1^2, -k_2^2, \ldots, -k_n^2, \ldots \tag{24}$$

We will assume that the roots are known and all the poles are simple.

### 2.5 Contour of integration and evaluation residues (Figure 1)

The integral:

$$J = \frac{1}{2\pi\mathrm{i}} \lim_{X \to 0} \int_{\gamma - \mathrm{i}X}^{\gamma + \mathrm{i}X} \frac{I_0(qr)\mathrm{e}^{pt}\,\mathrm{d}p}{p[qI'_0(q) + BiI_0(q)]} \tag{25}$$

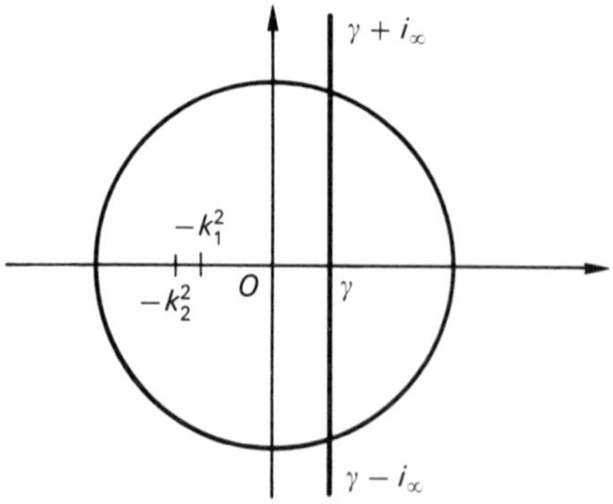


**Figure 1**

is equal to:

$$J = Res(0) + \sum_{n=1}^{\infty} Res(-k_n^2) \tag{26}$$

- At $z = 0$:

$$Res(0) = \frac{1}{Bi} \tag{27}$$

- At $z = -k_n^2$, let:

$$u = -kJ_1(k) + BiJ_0(k) \tag{28}$$

so that:

$$\frac{\mathrm{d}u}{\mathrm{d}p} = \frac{\mathrm{d}u}{\mathrm{d}k}\frac{\mathrm{d}k}{\mathrm{d}p} = +\frac{1}{2k}[J_1(k) + kJ_1'(k) - BiJ_0'(k)] \tag{29}$$

and:

$$\left(p\frac{\mathrm{d}u}{\mathrm{d}p}\right)\bigg|_{k_n} = \frac{1}{2}J_0(k_n)(k_n^2 + Bi) \tag{30}$$

Making use of:

$$J_0'(k) = -J_1(k) = -\frac{1}{k}J_0(k) \tag{31}$$

$$kJ_1'(k) = -J_1(k) + kJ_0(k) \tag{32}$$

we obtain:

$$Res(-k_n^2) = \frac{2J_0(k_n r)\mathrm{e}^{-k_n^2 t}}{(Bi + k_n^2)J_0(k_n)} \tag{33}$$

Whence:

$$J = \frac{1}{Bi} + 2\sum_{n=1}^{\infty} \frac{J_0(k_n r)\mathrm{e}^{-k_n^2 t}}{(Bi + k_n^2)J_0(k_n)} \tag{34}$$

The solution in dimensionless variables, in $(r^+, t^+)$ space, is:

$$T^+(r^+, t^+) = 1 - BiJ$$

$$= 2\sum_{n=1}^{\infty} \frac{J_0(k_n r)\mathrm{e}^{-k_n^2 t^+}}{(Bi + k_n^2)J_0(k_n)} \tag{35}$$

and in dimensional variables:

$$T(r, t) = 2T_0\sum_{n=1}^{\infty} \frac{J_0(k_n r/R)\exp(-k_n^2 at/R^2)}{(hR/\lambda + k_n^2)J_0(k_n)} \tag{36}$$

where $k_n$ is the series of real roots of equation (21).

## 3. Tabulation of solutions

This type of solution has been tabulated in the form $T^+ = f(Fo, Bi, r/R)$, the Biot number $Bi$ and the ratio $r/R$ being parameters [7]. Numerous geometries are included in this reference. If the charts appear to lose much of their interest with the current accessibility of computers, they do nevertheless allow us to obtain rapidly an accurate estimate of the unsteady behaviour of a system with simple geometry.

Let us look at an example related to the problem in Section 1.3, concerning a rod of fissile material (equations (1.17), identical to (5)).

- Take: $R = 10^{-2}$ m.
- The allowable temperature rise is: $\delta T = 100°C$.
- The physical properties (in SI units) are:

$$\lambda_1 = \lambda_2 = 1 \text{ W m}^{-1}\text{ K}^{-1}$$

$$a_1 = a_2 = 10^{-6} \text{ m}^2\text{ s}^{-1}$$

$$h_1 = 3 \times 10^3 \text{ W m}^{-2}\text{ K}^{-1}$$

$$h_2 = 2 \times 10^3 \text{ W m}^{-2}\text{ K}^{-1}$$

- The temperature change of the heat transfer fluid is: $\theta_2 - \theta_1 = 20°C$.
- The power dissipated per unit volume is: $P = 2 \times 10^8 \text{ W m}^{-3}$.
- We then obtain, using the notation of equations (5.17):

$$T_0 = (\theta_1 - \theta_2) + \frac{PR}{2}\left(\frac{1}{h_1} - \frac{1}{h_2}\right) = -16\,700°C!$$

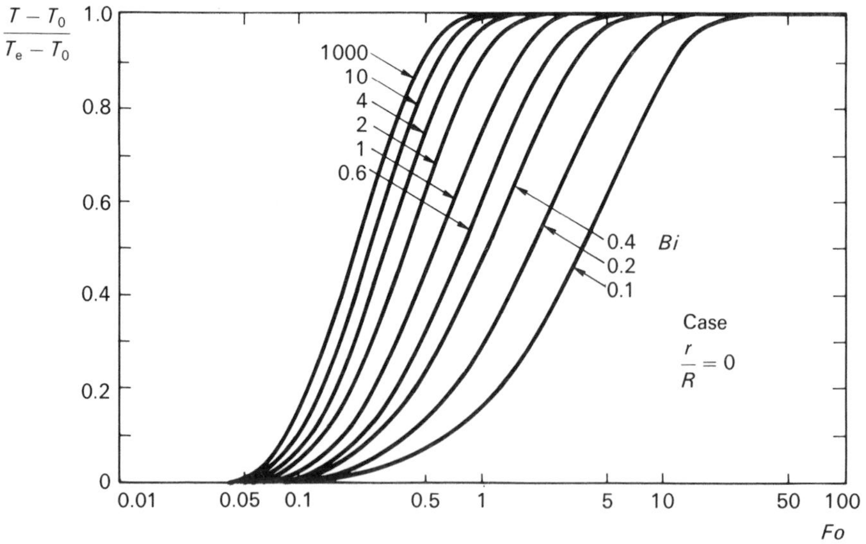


**Figure 2**

(Obviously the steady state $T_2^S$ will never be reached, and there is some urgency in lowering the moderator rods.)

In dimensionless variables, we then obtain:

- $Bi = h_2 R/\lambda = 20$.
- The chart uses: $\theta = (T - T_0)/(0 - T_0) = -\delta T/T_0 = 6 \times 10^{-3}$.
- We consider the temperature at $r = 0$, or $r/R = 0$.

In Figure 2, which is taken from reference [7], we find on the $x$ axis that: $Fo = t^+ = at/R^2 \simeq 5 \times 10^{-2}$, whence $t \simeq 5$ s.

## APPENDIX 4

# Example of the use of the Method of Separation of Variables

As an illustration, we will use the example of Section 3.1. The equations to be solved (in which we have omitted the + for convenience) are:

$$\frac{\partial T}{\partial t} = \frac{\partial^2 T}{\partial x^2} \tag{1a}$$

$$\frac{\partial T}{\partial x}(1, t) + BiT(1, t) = 0 \tag{1b}$$

$$\frac{\partial T}{\partial x}(0, t) = 0 \tag{1c}$$

$$T(x, 0) = 1 \tag{1d}$$

This system is linear and homogeneous, except for the initial condition (1d), which is not homogeneous.

### *1. Principle of the method*

We will look for particular solutions $\theta$ of (1a), in the form:

$$\theta(x, t) = f(x)g(t) \tag{2}$$

which is arbitrarily imposed on them to satisfy the homogeneous conditions, and then seek a linear combination of all these solutions which satisfies the heterogeneous condition. If this exists it must, by the uniqueness theorem, be the solution.

*Important note*
The method is equally applicable to multi-dimensional (steady or unsteady) problems.

## 2. Solution

We seek $g(t)$ and $f(x)$ such that:

$$f(x)g'(t) = g(t)f''(x) \tag{3}$$

It clearly follows that:

$$\frac{g'(t)}{g(t)} = \frac{f''(x)}{f(x)} = \text{const} = -k^2 \tag{4}$$

From this we obtain solutions of the form:

$$g(t) = A\,e^{-k^2 t} \tag{5}$$

$$f(x) = B\sin(kx) + C\cos(kx) \tag{6}$$

so that:

$$\theta(x, t) = e^{-k^2 t}(D\sin(kx) + E\cos(kx)) \tag{7}$$

The three parameters $(k, D, E)$ on which these particular solutions depend are chosen to satisfy the homogeneous conditions (1.b) and (1c):

$$(1c) \Rightarrow D = 0 \tag{8}$$

$$(1b) \Rightarrow -k\sin(k) = Bi\cos(k) = 0 \tag{9}$$

This relation puts a constraint on the values of $k$; the solutions are the infinite sequence of separate positive values $k_n$: $k_0, k_1, k_2, \ldots, k_n, \ldots$ satisfying:

$$k_n\, tg(k_n) = Bi \tag{10}$$

We will therefore seek a solution of the form:

$$T(x, t) = \sum_{n=0}^{\infty} A_n \cos(k_n x)\exp(-k_n^2 t) \tag{11}$$

which certainly satisfies (1a), (1b) and (1c), and which we will force to satisfy (1d). Let:

$$\sum_{n=0}^{\infty} A_n \cos(k_n x) = 1 \tag{12}$$

To find the coefficients $A_n$, we will use the properties of the set of functions $\cos(k_n x)$ which satisfy (10), relating to the scalar product. It can be shown* that:

$$\int_0^1 \cos(k_n x)\cos(k_m x)\,\mathrm{d}x = \delta_{nm}\left(\frac{1}{2} + \frac{\sin(2k_n)}{4k_n}\right) \tag{13}$$

which implies that the set of functions $\cos(k_n x)$ is orthogonal with respect to the scalar product. Multiplying (12) by $\cos(k_m x)$ term by term, and integrating from 0 to 1, we obtain:

$$A_n\left[\frac{1}{2} + \frac{\sin(2k^n)}{4k_n}\right] = \frac{\sin(k_n)}{k_n} \qquad \forall\, n \tag{14}$$

We have thus found the solution to the equations by this method. It is (re-instating the + signs):

$$T^+(x^+, t^+) = \sum_{n=0}^{\infty}\left(\frac{2\sin(k_n)}{k_n + \sin(k_n)\cos(k_n)}\right)\exp(-k_n^2 t^+)\cos(k_n x^+) \tag{15}$$

*Note on the method*

The example solved by the Laplace transform in Appendix 3 could equally well have been done by separation of variables. The converse is also true for the example solved here.

* If $n \neq m$:

$$\int_0^1 \cos(k_n x)\cos(k_m x)\,\mathrm{d}x = \frac{1}{2}\left[\frac{\sin(k_n + k_m)}{k_n + k_m} + \frac{\sin(k_n - k_m)}{k_n - k_m}\right] = X$$

further, equation (10) becomes:

$$\frac{k_n}{k_m} = \frac{tg(k_m)}{tg(k_n)} = \frac{\sin(k_m)\cos(k_n)}{\sin(k_n)\cos(k_m)} \Rightarrow X = 0$$

# Use of Green's Function in Unsteady Conduction [5,8]

## 1. Definition of Green's function

The general problem to be solved in linear unsteady conduction (with $\lambda$ and $h$ constant and radiation excluded or linearized) has a structure of the form:

$$\frac{\partial T}{\partial t} = a\nabla^2 T + \frac{P(\mathbf{r}, t)}{\rho c_p} \tag{1a}$$

which is an energy equation in which $T(\mathbf{r}, t)$ represents the temperature field at the point $\mathbf{r}$ at time $t$,

$$T(\mathbf{r}, t = 0) = T_0(\mathbf{r}) \tag{1b}$$

which is the initial condition, and boundary conditions of the type:

$$T(\mathbf{r}, t) = T_1(\mathbf{r}, t) \qquad \text{over} \qquad S_1 \tag{1c}$$

$$-\lambda\nabla T\cdot\mathbf{n}_{ext} = h[T(\mathbf{r}, t) - T_2(\mathbf{r}, t)] + \varphi_0(\mathbf{r}, t) \qquad \text{over} \qquad S_2 \tag{1d}$$

where $S_1$ and $S_2$ represent two parts of the lateral surface $S$ ($S = S_1 \cup S_2$) and $\mathbf{n}_{ext}$ is the outwards-directed unit vector normal to $S$.

The function $G(\mathbf{r}, \mathbf{r}', t - t')$ is called the Green's function associated with such a problem, and is the solution of the homogeneous system of equations associated with (1a)–(1d):

$$\frac{\partial G}{\partial t} = a\nabla^2 G + \delta(\mathbf{r} - \mathbf{r}')\delta(t - t') \tag{2a}$$

$$G(\mathbf{r}, \mathbf{r}', t - t') = 0 \qquad \text{if} \qquad t < t' \tag{2b}$$

$$G(\mathbf{r}, \mathbf{r}', t - t') = 0 \qquad \text{over} \qquad S_1 \tag{2c}$$

$$-\lambda\nabla G\cdot\mathbf{n}_{ext} = hG \qquad \text{over} \qquad S_2 \tag{2d}$$

The Green's function thus represents the temperature response of the system (2a)–(2d) made up of the homogeneous parts of system (1a)–(1d) to a unit step input applied at point $\mathbf{r}'$ at time $t'$. This response is called **the percussive response of the system**. The physical interpretation we have just given for the solution $G$ of system (2a)–(2d) will be useful to us in the following for the practical determination of Green's functions for some particular systems.

It can be demonstrated [5,8] that the solution system (1a)–(1d) can be obtained from the solution $G$ of system (2a)–(2d) and the non-homogeneous terms of (1a)–(1d) (i.e. $T_0(\mathbf{r})$, $T_1(\mathbf{r}, t)$, $T_2(\mathbf{r}, t)$, $\varphi_0(\mathbf{r}, t)$) by the explicit relationship:

$$T(\mathbf{r}, t) = \int_V G(\mathbf{r}, \mathbf{r}', t) T_0(\mathbf{r}')\, \mathrm{d}V'$$

$$+ a \int_0^t \int_{S_1} T_1(\mathbf{r}', t')(-\mathbf{n}_{\text{ext}}) \cdot \nabla_{\mathbf{r}'} G(\mathbf{r}, \mathbf{r}', t - t')\, \mathrm{d}S_1'\, \mathrm{d}t'$$

$$+ a \int_0^t \int_{S_2} \left[ \frac{h}{\lambda} T_2(\mathbf{r}', t') - \frac{\varphi_0}{\lambda}(\mathbf{r}', t') \right] G(\mathbf{r}, \mathbf{r}', t - t')\, \mathrm{d}S_2'\, \mathrm{d}t'$$

$$+ \int_0^t \int_V G(\mathbf{r}, \mathbf{r}', t - t') \frac{P}{\rho c_{\mathrm{p}}}(\mathbf{r}', t')\, \mathrm{d}V'\, \mathrm{d}t' \tag{3}$$

The notation $\nabla_{\mathbf{r}'}$ means that the gradient is taken relative to the components of the vector $\mathbf{r}'$.

*Note*

- If we compare the first and fourth terms on the right-hand side, we can see that the initial condition can be interpreted as a power $P$ dissipated at the instant $t' = 0$ only.

In practice the problem has been simplified because the main difficulty resides in the solution of equations (2a)–(2d) which may be undertaken by common methods (Laplace, etc.).

## 2. Reference example: Green's function associated with an infinite medium

The following derivation will be inductive, limited to a proof based on physical considerations. The function:

$$G(\mathbf{r}, \mathbf{r}', t) = \frac{1}{8[\pi a(t - t')]^{3/2}} \exp\left[ -\frac{(\mathbf{r} - \mathbf{r}')^2}{4a(t - t')} \right] \tag{4}$$

is the solution of the energy equation:

$$\frac{\partial T}{\partial t} = a\nabla_{\mathbf{r}}^2 T \qquad \text{for} \qquad t > t' \tag{5}$$

Further, as $t$ approaches $t'$, this function tends to 0 at all points except the point $\mathbf{r}'$, where it becomes infinite (corresponding to an impulse imposed at time $t = t'$ at point $\mathbf{r} = \mathbf{r}'$ to a medium characterized by $G = 0$ everywhere).

If we assume that $\rho c_p$ is constant, the total enthalpy gain of the system at any moment $t - t'$ is:

$$\int_V \rho c_p G(\mathbf{r}, \mathbf{r}', t - t')\, \mathrm{d}V = \rho c_p \tag{6}$$

which comes down to saying that the impulse at time $t'$ and point $\mathbf{r}'$ is of unit magnitude (or, alternatively, that the function $G$ is normalized). The function $G(\mathbf{r}, \mathbf{r}', t - t')$ satisfies conditions (2a) and (2b) and can be considered to be the Green's function associated with an infinite medium (obviously conditions (2c) and (2d) are not relevant to such a medium).

**Application**

If we consider an infinite medium of initial temperature $T_0(\mathbf{r})$, the temperature distribution at a later time is:

$$T(\mathbf{r}, t) = \frac{1}{8(\pi a t)^{3/2}} \int_V T_0(\mathbf{r}') \exp\left[ -\frac{(\mathbf{r} - \mathbf{r}')^2}{4at} \right] \mathrm{d}V' \tag{7}$$

## 3. Green's function associated with a semi-infinite medium

We will use Cartesian coordinates as shown in Figure 1.

### 3.1 Flux density $\varphi_0(y, z, t)$ imposed on the face $x = 0$

The Green's function associated with the actual system corresponds to the impulse response at a point $(x', y', z')$ at time $t'$ with an insulated face at $x = 0$.

This is equivalent to considering condition (1d) of the actual system, made homogeneous, with the extra assumption $h = 0$. The most elegant solution is to associate the problem with the mirror image problem: an infinite medium in which all the distributions of physical quantities are symmetrical about the plane $x = 0$. The zero flux condition is always satisfied at $x = 0$ and the situation is identical to the original system for positive $x$.

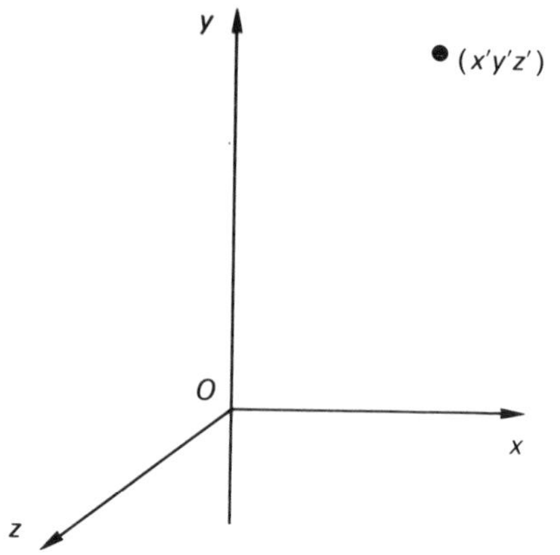


**Figure 1**

We thus consider also an impulse identical to the original impulse, located at the point $(-x', y', z')$, symmetrical to $(x', y', z')$ relative to $x = 0$. The Green's function associated with the original problem is thus that resulting from the two image impulses:

$$G = \frac{1}{8[\pi a(t-t')]^{3/2}} (\exp\{-[(x-x')^2 + (y-y')^2 + (z-z')^2]/4a(t-t')\}$$
$$+ \exp\{-[(x+x')^2 + (y-y')^2 + (z-z')^2]/4a(t-t')\}) \quad (8)$$

The temperature distribution in a system initially at a uniform temperature $T_0$ to which a flux $\varphi_0(y, z, t)$ is applied at point $x = 0$ is thus, using (3):

$$T(\mathbf{r}, t) = T_0 + \frac{a}{\lambda} \int_0^t \int_S \varphi_0(y', z', t') G(x, y, z, x' = 0, y', z', t - t')\, dy'\, dz'\, dt' \quad (9)$$

in which expression $S$ is the surface of the plane $x' = 0$. The first term $T_0$ on the right-hand side of (9) comes from the first integral of the right-hand side of (3) and takes this simple form because $G$ is normalized. Let us consider the solution in three particularly simple cases:

- $\varphi_0(y, z, t) = \varphi_0$:

$$T(x, t) = T_0 + \frac{a\varphi_0}{\lambda} \int_0^t [\pi a(t-t')]^{-1/2} \exp[-x^2/4a(t-t')]\, dt' \quad (10)$$

- $\varphi_0(y, z, t) = \varphi_0 \cos \omega t$:

$$T(x, t) = T_0 + \frac{a\varphi_0}{\lambda} \int_0^t \cos \omega t' [\pi a(t-t')]^{-1/2} \exp[-x^2/4a(t-t')]\, dt' \quad (11)$$

- $\varphi_0(y, z, t) = \varphi_0 \exp[-(y^2 + z^2)/\omega^2]$ (this case corresponds to the absorption of radiation from a Gaussian laser beam operating in mode TEM 00, centred at $O(0, 0, 0)$):

$$T(x, y, z, t) = T_0 + \frac{\varphi_0 \omega}{\lambda\sqrt{\pi}} \int_0^{\sqrt{4at/\omega^2}} (1 + u^2)^{-1} \exp[-R^2/(1 + u^2)] \exp(-X^2/u)\,\mathrm{d}u \tag{12}$$

with: $R^2 = (y^2 + z^2)/\omega^2$ and $X = x/\omega$.

*Note*

For one-dimensional problems (the first and second cases above, for example), we only need to introduce a one-dimensional Green's function:

$$G(x, x', t - t') = \int_{-\infty}^{+\infty} \int_{-\infty}^{+\infty} G(\mathbf{r}, \mathbf{r}', t - t')\,\mathrm{d}y'\,\mathrm{d}z' \tag{13}$$

which, for a semi-infinite medium with a flux condition at the surface, becomes:

$$G(x, x', t - t') = \tfrac{1}{2}[\pi a(t - t')]^{-1/2}\{\exp[-(x - x')^2/4a(t - t')] + \exp[-(x + x')^2/4a(t - t')]\} \tag{14}$$

## 3.2 *Temperature $T_1(y, z, t)$ imposed at $x = 0$*

The problem associated with the actual problem corresponds to a condition of zero temperature imposed at $x = 0$ (see condition (2b)). We can do this by considering an odd Green's function over the whole of space: as well as the unit impulse at $(x', y', z')$ in the positive half-space $x' > 0$, we impose an impulse of opposite sign at $(-x', y', z')$. The normalized Green's function is:

$$G(\mathbf{r}, \mathbf{r}', t - t') = \frac{1}{8}[\pi a(t - t')]^{-3/2}(\exp\{-[(x - x')^2 + (y - y')^2 + (z - z')^2]/4a(t - t')\} - \exp\{-[(x + x')^2 + (y - y')^2 + (z - z')^2]/4a(t - t')\}) \tag{15}$$

Applying (3), we obtain the temperature redistribution in a system initially at temperature $T_0(\mathbf{r})$ with a temperature $T_1(y, z, t)$ imposed on its wall:

$$T(x, y, z, t) = \int_V G'(\mathbf{r}, \mathbf{r}', t) T_0(\mathbf{r}')\,\mathrm{d}V' + a \int_0^t \int_S T_1(y, z, t) \frac{\partial G}{\partial x'}(x, y, z, x' = 0, y', z', t - t')\,\mathrm{d}y'\,\mathrm{d}z'\,\mathrm{d}t' \tag{16}$$

where the surface $S$ is the plane $x = 0$ and:

$$\frac{\partial G}{\partial x'}(x, y, z, x' = 0, y', z', t - t')$$

$$= \frac{x}{8\pi^{3/2}}[a(t - t')]^{-5/2} \exp\{-[x^2 + (y - y')^2 + (z - z')^2]/4a(t - t')\} \quad (17)$$

Some simple special cases are:

- When $T_1(t) = T_1$ and $T_0(x) = T_0$

$$T(x, t) = T_0 \int_0^{\infty} G(x, x', t)\,\mathrm{d}x' + a \int_0^t T_1 \frac{\partial G}{\partial x'}(x, x' = 0, t - t')\,\mathrm{d}t' \quad (18)$$

where, for simplicity, we used the one-dimensional Green's function:

$$G(x, x', t - t') = \tfrac{1}{2}[\pi a(t - t')]^{-1/2}\{\exp[-(x - x')^2/4a(t - t')] - \exp[-(x + x')^2/4a(t - t')]\} \quad (19)$$

$$\frac{\partial G}{\partial x'}(x, x' = 0, t - t') = \frac{x}{2\sqrt{\pi}a^{3/2}(t - t')^{3/2}}\{\exp[-x^2/4a(t - t')]\} \quad (20)$$

So that:

$$T(x, t) = T_0\left[1 - \frac{2}{\sqrt{\pi}}\int_u^{\infty} \mathrm{e}^{-u'^2}\,\mathrm{d}u'\right] + \frac{2T_1}{\sqrt{\pi}}\int_u^{\infty} \mathrm{e}^{-u'^2}\,\mathrm{d}u' \quad (21)$$

with $u = x/(2\sqrt{(at)})$.

With a change of variable, we can recover the result of Section 6.1:

$$\frac{T - T_0}{T_1 - T_0} = \frac{2}{\sqrt{\pi}}\int_u^{\infty} \mathrm{e}^{-u'^2}\,\mathrm{d}u' \quad (22)$$

- When $T_1(\mathbf{r}, t) = T_1(t)$ and $T_0(x) = T_0$ we find:

$$T(x, t) - T_0 = \frac{2}{\sqrt{\pi}}\int_u^{\infty}\left[T_1\left(t - \frac{x^2}{4au'^2}\right) - T_0\right]\mathrm{e}^{-u'^2}\,\mathrm{d}u' \quad (23)$$

### 4. *Green's function associated with a finite medium*

In this section we will limit ourselves to the example of a finite wall of thickness $e$, initially at $T_0$, insulated at $x = e$, to which a flux $\varphi_0$ is applied at $x = 0$ (Figure 2).

The Green's function $G(x, x', t - t')$ corresponds to the response to a unit impulse at position $x'$ and time $t'$, with the two faces of the wall being insulated. The equivalent system is thus an infinite system with identical impulses at $x' + 2ne$ and $-x' + 2ne$, the integer $n$ varying from $-\infty$ to $+\infty$ (Figure 3).

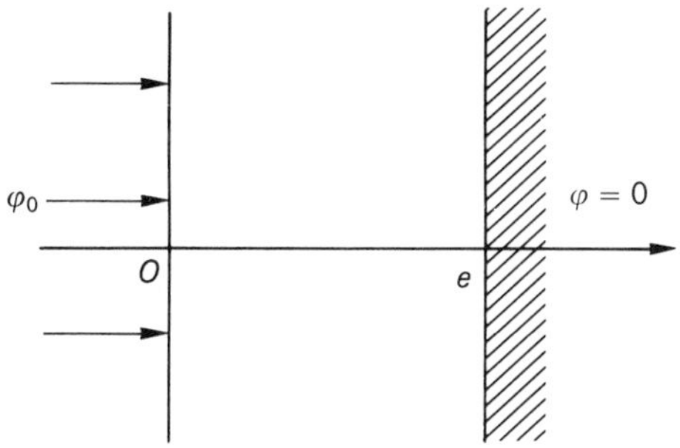


**Figure 2**

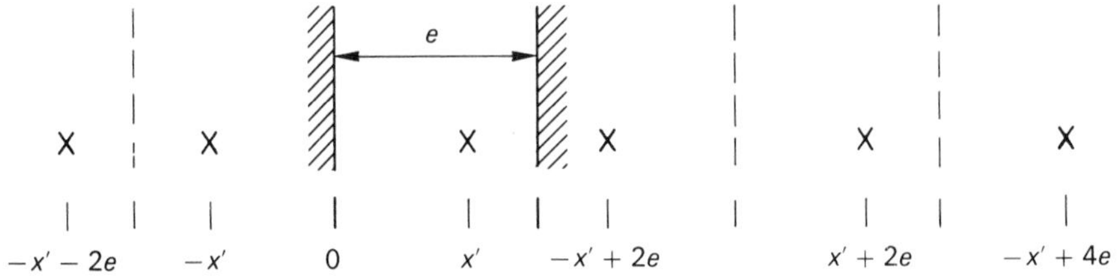


**Figure 3**

The Green's function (one-dimensional for simplicity) is then:

$$G(x, x', t - t') = \frac{1}{2\sqrt{\pi a(t - t')}}\left(\sum_{n=-\infty}^{+\infty} \exp\left[-\frac{(x - x' - 2ne)^2}{4a(t - t')}\right] + \sum_{n=-\infty}^{+\infty} \exp\left[-\frac{(x + x' - 2ne)^2}{4a(t - t')}\right]\right) \quad (24)$$

This Green's function becomes more amenable if put in the form of a Fourier series:

$$G(x, x', t - t') = \frac{1}{e}\left[1 + 2\sum_{n=1}^{+\infty} \exp[-an^2\pi^2(t - t')/e^2] \cos\left(\frac{n\pi x}{e}\right)\cos\left(\frac{n\pi x'}{e}\right)\right] \quad (25)$$

The solution to the problem is then:

$$T(x, t) = T_0 + \frac{a\varphi_0}{\lambda e} t + \frac{2e\varphi_0}{\lambda\pi^2}\sum_{n=1}^{\infty} \frac{1}{n^2}\left[1 - \exp\left(\frac{-an^2\pi^2 t}{e^2}\right)\right]\cos\left(\frac{n\pi x}{e}\right) \quad (26)$$

Green's functions appropriate to other cases may be found in references [5] and [8].

# Solution of the Thermal Contact Problem (Solution of equations (6.30) from Section 6.3)

In dimensionless variables (avoiding breaking the symmetry of the system):

$$T_1^+ = \frac{T_1 - T_1^0}{T_2^0 - T_1^0}, \qquad T_2^+ = \frac{T_2 - T_2^0}{T_1^0 - T_2^0}, \qquad u_1 = \frac{x}{2\sqrt{a_1 t}}, \qquad u_2 = \frac{x}{2\sqrt{a_2 t}}$$

equations (6.30) become:

$$\frac{d^2 T_1^+}{du_1^2} + 2u_1 \frac{dT_1^+}{du_1} = 0 \tag{1}$$

$$\frac{d^2 T_2^+}{du_2^2} + 2u_2 \frac{dT_2^+}{du_2} = 0 \tag{2}$$

$$-\frac{\lambda_1}{2\sqrt{a_1 t}} \frac{dT_1^+}{du_1}(0) = +\frac{\lambda_2}{2\sqrt{a_2 t}} \frac{dT_2^+}{du_2}(0)$$

or:

$$b_1 \frac{dT_1^+}{du_1}(0) + b_2 \frac{dT_2^+}{du_2}(0) = 0 \tag{3}$$

$$T_1(0, t) = T_2(0, t)$$

or:

$$(T_2^0 - T_1^0) T_1^+(0) + T_1^0 = -(T_2^0 - T_1^0) T_2^+(0) + T_2^0$$

$$T_1^+(0) + T_2^+(0) = 1 \tag{4}$$

$$T_1^+(-\infty) = 0 \tag{5}$$

$$T_2^+(+\infty) = 0 \tag{6}$$

The solutions of (1) and (2) satisfy (see Section 6.1):

$$\frac{dT_1^+}{du_1} = A_1 e^{-u_1^2}, \qquad \frac{dT_2^+}{du_2} = A_2 e^{-u_2^2} \tag{7}$$

Equation (3) becomes:

$$b_1 A_1 + b_2 A_2 = 0 \tag{8}$$

Equations (5) and (6) lead to:

$$T_2^+(u_2) = -A_2 \int_{u_2}^{\infty} \frac{e^{-u'^2}\, du'}{\int_0^{\infty} e^{-u'^2}\, du'} = -A_2 \operatorname{erfc}(u_2) \tag{9}$$

$$T_1^+(u_1) = A_1 \int_{-\infty}^{u_1} \frac{e^{-u'^2}\, du'}{\int_{-\infty}^{0} e^{-u'^2}\, du'} = +A_1 \operatorname{erfc}(-u_1) \tag{10}$$

Equation (4) gives:

$$(A_1 - A_2) = 1 \tag{11}$$

We thus obtain:

$$T_1^+ = \frac{b_2}{b_1 + b_2} \operatorname{erfc}\left(\frac{-x}{2\sqrt{a_1 t}}\right) \tag{12}$$

$$T_2^+ = \frac{b_1}{b_1 + b_2} \operatorname{erfc}\left(\frac{+x}{2\sqrt{a_2 t}}\right) \tag{13}$$

These results are discussed in Section 6.3.

PART III

# Radiation

The main difficulty in the calculation of the radiative heat transfer between opaque bodies lies in the accurate knowledge of the relevant radiative properties (emissivities, absorptivities, reflectivities) of the bodies concerned. So the first chapter in this part (Chapter 9) is dedicated to the physical study of the quantities which influence these properties both in ideal conditions (pure substances, smooth surfaces) and in realistic industrial conditions.

Another difficulty in the study of radiative transfer arises from the multiple reflections between $N$ opaque bodies at the system boundaries. This difficulty has been avoided in Chapter 4, which is limited to two important special cases: an opaque body surrounded by a black body, and an opaque body immersed in equilibrium radiation. In Chapter 10 we develop a classical model, based on realistic assumptions which arise from the conclusions of Chapter 9, to tackle a more general problem: this is the method of incident and emergent fluxes, which is valid when the medium separating the opaque bodies is transparent (vacuum and, in some conditions, air).

A completely different approach is used when the medium is semi-transparent, possibly scattering, or has a fine structure (partial or complete absorption in some spectral lines, which is the case for gases such as $CO_2$, $H_2O$, CO, etc.). These ideas are covered in Chapter 11, which concludes with a bibliography of reference articles and books covering the most recent developments in this field.

# 9

# Radiative Properties of Opaque Bodies

The objective of this chapter is to determine the behaviour of the radiative properties ($\varepsilon'_\lambda$, $\alpha'_\lambda$, etc.) of common opaque bodies as functions of the direction of the incident radiation and wavelength; we will look both at conductors and insulators, and at both laboratory and industrial conditions. We will restrict ourselves to relatively weak incident fluxes, that is, we will keep within LTE. The aim of this chapter is not to develop a complete physical model to describe the radiative properties of these opaque bodies; such a model is in the domain of solid state physics, and considerable research effort is being put into this direction. The reader who is interested in the basic physics of such a study should consult references [1] and [2], which offer an excellent introduction.

In the first phase (Section 9.1) we will consider only ideal opaque bodies: pure homogeneous bodies of perfectly defined chemical composition, non-diffusing and with 'optically smooth' surfaces. We will look at the consequences of electromagnetic theory for these materials, in particular Fresnel's law and an elementary theory of absorption. The case of real materials whose radiative properties depend on surface finish and material surface structure is considered thereafter (Section 9.2). These considerations end up in general in a simplified, but none the less realistic, model for radiative properties, which serves as the basis for the method developed in Chapter 10.

## 9.1 The opaque body in laboratory conditions

Let us define what is understood by 'laboratory conditions':

- The material is homogeneous and isotropic, and of perfectly defined chemical composition.
- The bounding surface with the external medium is 'optically smooth', that is, the asperities on the surface do not exceed a fraction of the wavelength $\lambda$ concerned,

typically $\lambda/10$. This is the case, for example, for windows and lenses made of fluoride ($CaF_2$) or sapphire ($Al_2O_3$) or KCl, and for the mirrors used in infra-red or visible optics, which are good thermal conductors covered with a gold film.

- The bulk medium is non-diffusing (which follows from the homogeneity).

Such conditions are only encountered exceptionally by the engineer; they are of interest, nevertheless, in enabling the influences of the various physical quantities on the radiative properties to be identified.

### 9.1.1 *The consequences of electromagnetic theory (Fresnel's laws)*

In ideal conditions, described above, we can consider that a plane incident wave of frequency $\nu$ propagates in **continuous, linear, isotropic and non-diffusing media** (which have a discontinuity of properties at their interfaces), and use Maxwell's equations (see, for example, references [3]–[5]).

#### 9.1.1.1 PHYSICAL QUANTITIES CHARACTERIZING THE MATERIAL

Every medium is characterized by a complex index $\hat{n}$ which depends strongly on the wave frequency $\nu$:

$$\hat{n} = n + j\chi \tag{9.1}$$

where $n$ and $\chi$ are the real and extinction indices of the material, respectively. The complex index $\hat{n}$ is related to the complex relative dielectric permittivity $\hat{\varepsilon}$ and the relative magnetic permeability $\mu$ by:

$$\hat{\varepsilon} = \varepsilon_1 + j\varepsilon_2 = \hat{n}/\mu \tag{9.2}$$

from which we can obtain expressions for the real and imaginary parts of the permittivity:

$$\varepsilon_1 = (n^2 - \chi^2)/\mu \qquad \text{and} \qquad \varepsilon_2 = 2n\chi/\mu \tag{9.3}$$

The latter quantity, $\varepsilon_2$, is directly related [1] to the electrical conductivity $\sigma$ (for materials which are conductors) by:

$$\varepsilon_2 = \frac{\sigma}{2\pi\nu} \tag{9.4}$$

A plane wave travelling through an infinte medium is characterized by an electric field perpendicular to the unit vector $\mathbf{u}$ of the direction of propagation, whose amplitude $\mathbf{E}$ is given by:

$$\mathbf{E} = \mathbf{E}_0 \exp[-2\pi j\nu(t - \hat{n}\mathbf{u}\cdot\mathbf{r}/c_0)] \tag{9.5}$$

In this, $c_0$ designates the speed of light in a vacuum. Then, in another form:

$$\mathbf{E} = \mathbf{E}_0 \exp(-2\pi\nu\chi\mathbf{u}\cdot\mathbf{r}/c_0)\exp[-j(2\pi\nu t - \mathbf{k}'\cdot\mathbf{r})] \tag{9.6}$$

where $\mathbf{k}'$ is the real part of the associated wave vector. The flux density of energy associated with this wave, given by the Poynting vector model, is proportional to the square of the modulus of the electric field:

$$P \sim \mathbf{E}_0^2 \exp(-4\pi\nu\chi\mathbf{u}\cdot\mathbf{r}/c_0) \tag{9.7}$$

The phenomenon of absorption by the opaque body appears in this expression. In the study of semi-transparent media (see Chapter 11), we introduce a monochromatic absorption coefficient $\kappa_\nu$ by means of the relation:

$$\mathrm{d}I'_\nu = -\kappa_\nu I'_\nu(s)\,\mathrm{d}s \tag{9.8}$$

which leads to an exponential law $\exp(-\kappa_\nu s)$ analogous to (9.7). Comparing (9.7) and (9.8), and identifying $s$ with $\mathbf{u}\cdot\mathbf{r}$:

$$\kappa_\nu = 4\pi\nu\chi(\nu)/c_0 \tag{9.9}$$

This expression is the reason that $\chi$ is called the extinction coefficient. We note that the depth of penetration $\eta_\nu$ of the wave into the opaque body is given by:

$$\eta_\nu = \kappa_\nu^{-1} = c_0/(4\pi\nu\chi(\nu)) \tag{9.10}$$

For copper, with a wavelength in a vacuum ($\lambda = c_0/\nu$) of 1 $\mu$m, this comes out at 63 Å. At first sight one might expect that a more absorbent medium (higher absorptivity $\alpha'_\lambda$ or $\alpha'_\nu$) would have a higher coefficient of extinction $\chi_\nu$. But, as we will see, this is, paradoxically, not so. It is, moreover, well known that copper casseroles for good cooking are highly reflecting!

#### 9.1.1.2 FRESNEL'S LAWS

In the following we will, in order to simplify matters, consider radiation which is initially propagating in a medium with complex refractive index $\hat{n}$ equal to 1 (strictly, in a vacuum*), and incident on an opaque material with a complex index $\hat{n}$, but not magnetic ($\mu = 1$), as is nearly always the case. This is shown in Figure 9.1. Since the reference medium has an index equal to 1, we refer all quantities relevant to an opaque body to the wavelength in a vacuum $\lambda$, following convention, rather than to the frequency $\nu$.

Starting from Maxwell's and the field continuity equations, we can derive Descartes' laws of reflection. We can also calculate $\rho'_\lambda$, the specular energy reflectivity of the material, for incident radiation making an angle $\theta_1$ with the normal. For a more detailed analysis, we must distinguish radiation whose electric field vector $\mathbf{E}_\parallel$ is parallel to the plane of incidence $(\mathbf{u}_1, \mathbf{n})$, with which is associated a reflectivity $\rho'_{\lambda\parallel}$, and radiation whose electric field vector $\mathbf{E}_\perp$ is perpendicular to the plane of incidence, with a reflectivity $\rho'_{\lambda\perp}$. Thermal radiation is generally considered, as a first approximation, to be non-polarized. (Statistically, the two planes of polarization are

* Air has a complex refractive index close to 1 (within a few times $10^{-4}$), except for the absorption bands of the molecules which make it up ($H_2O$, $CO_2$ in the infra-red). In these bands, $\chi$ is clearly non-zero, but $n$ remains very close to 1.

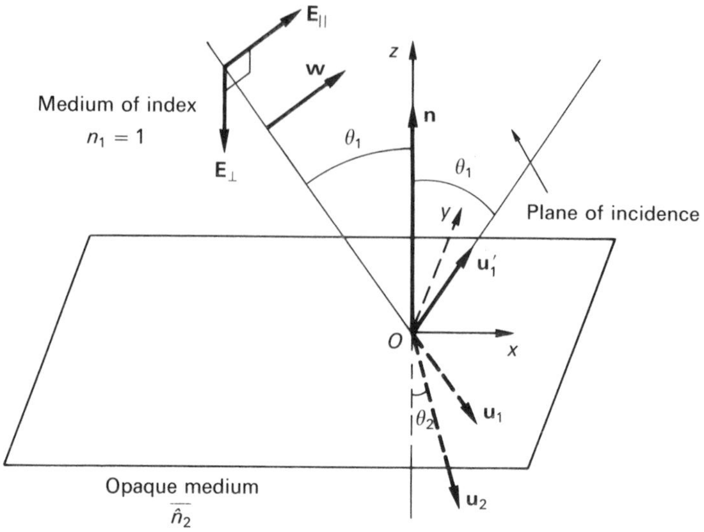


**Figure 9.1**

equally represented in equilibrium radiation, for example.) In such cases, the global energy reflectivity, for specular reflection, $\rho'_\lambda$ is defined as:

$$\rho'_\lambda = (\rho'_{\lambda\|} + \rho'_{\lambda\perp})/2 \tag{9.11}$$

If we start with Maxwell's equations and substitute [6]:

$$\rho'_{\lambda\|} = |r'_{\lambda\|}|^2 \qquad \text{and} \qquad \rho'_{\lambda\perp} = |r'_{\lambda\perp}|^2 \tag{9.12}$$

where $r'_{\lambda\|}$ and $r'_{\lambda\perp}$ represent the reflectivities of the components of the electric field **E** which are polarized parallel and perpendicular to the plane of incidence, respectively (positive in the **w** direction in the former case – Figure 9.1), we obtain [6]:

$$r'_{\lambda\perp} = \frac{k_{1z} - k_{2z}}{k_{1z} + k_{2z}} \tag{9.13}$$

$$r'_{\lambda\|} = \frac{\hat{n}_2^2 k_{1z} - k_{2z}}{\hat{n}_2^2 k_{1z} + k_{2z}} \tag{9.14}$$

where $k_{1z}$ and $k_{2z}$ are the projections of the complex wave vector $\mathbf{k} = \mathbf{k}' + j\mathbf{k}''$ on to the $z$ axis in media 1 and 2:

$$k_{jz} = \frac{\omega}{c}[\hat{n}_j^2 - \sin^2\theta_1]^{1/2} \quad (j = 1, 2) \qquad \text{with} \qquad \text{Im}(k_j z) > 0 \tag{9.15}$$

where $\theta_1$ is the angle of incidence.

The energy transmissivities $\tau'_{\lambda\|}$ and $\tau'_{\lambda\perp}$ may be deduced from the above expressions using:

$$\tau'_{\lambda\|} = 1 - \rho'_{\lambda\|} \qquad \tau'_{\lambda\perp} = 1 - \rho'_{\lambda\perp} \tag{9.16}$$

We consider radiation passing from 1 to 2 across the surface of separation of the two media. In the event that 2 is an opaque body, the radiation is completely absorbed by medium 2 in the layers immediately adjacent to the surface.

### 9.1.1.3 DIRECTIONAL BEHAVIOUR OF RADIATIVE PROPERTIES [7–9]

We see from (9.16) that reflection modifies the polarization of incident radiation. In particular, radiation which is initially unpolarized is, in general, partly polarized after reflection. In two extreme cases, however, the polarization is unchanged:

- For grazing incidence ($\theta_1 = \pi/2$) we obtain the trivial result:

$$\rho'_{\lambda\|}\left(\frac{\pi}{2}\right) = \rho'_{\lambda\perp}\left(\frac{\pi}{2}\right) = 1 \tag{9.17}$$

  The wave does not 'see' the opaque body – it is totally reflected.
- For zero incidence ($\theta_1 = 0$), the plane of polarization degenerates into the normal: polarization is no longer meaningful. The same limit is clearly obtained for the two polarizations:

$$\rho'_{\lambda\|}(0) = \rho'_{\lambda\perp}(0) = \frac{(n-1)^2 + \chi^2}{(n+1)^2 + \chi^2} \tag{9.18}$$

  Notice that the functions $\rho'_{\lambda\|}$ and $\rho'_{\lambda\perp}$ vary very little as functions of $\theta_1$ in the neighbourhood of zero incidence (Figures 9.2 and 9.3).

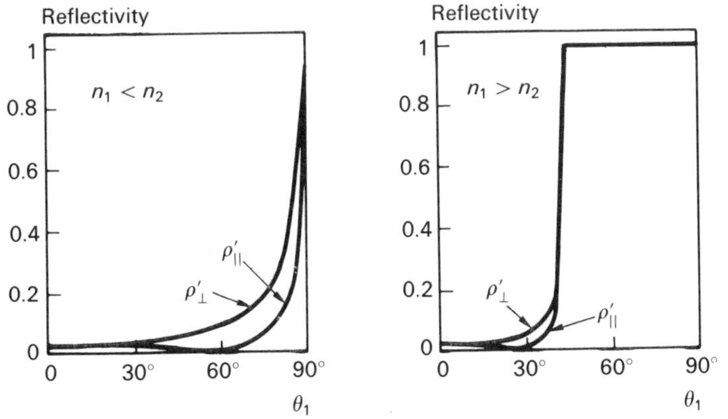


**Figure 9.2** Polarized reflectivities of a dielectric as a function of angle of incidence for external ($n_1 < n_2$) and internal ($n_1 > n_2$) reflection (after reference [7]; notation as in Figure 9.1).

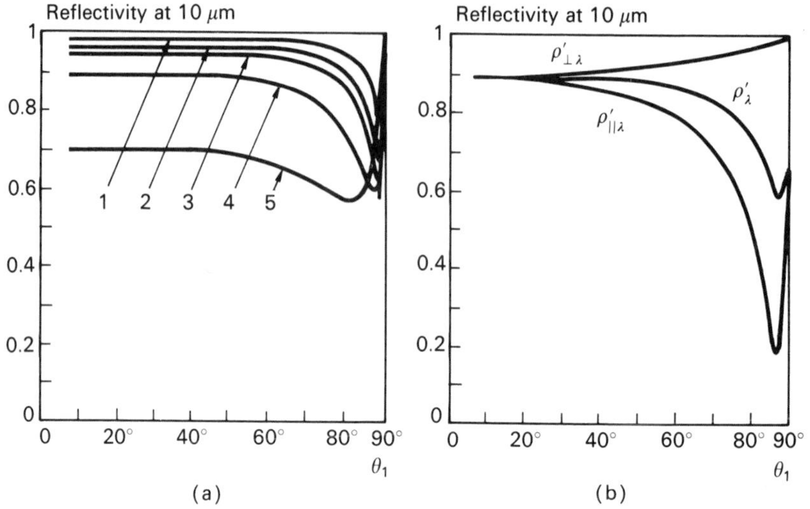


**Figure 9.3** (a) Reflectivity ($\rho'_\lambda$) of polished metals at 10 $\mu$m and 300 K (from reference [7]) 1. gold; 2. platinum; 3. lead; 4. nickel–chrome; 5. graphite (at 2300 K). (b) Typical reflectivity (from reference [7]).

The function $\rho'_{\lambda\perp}$ is a monotonic function of $\theta_1$; $\rho'_{\lambda\parallel}$ passes through a minimum at an angle $\theta_B$, known as the Brewster incidence. This minimum corresponds to total annihilation of the reflectivity $\rho'_{\lambda\parallel}$ for a glassy material (such that $\kappa \to 0$, $\chi \ll n$) (Figure 9.2), but does not occur for other materials (Figure 9.3). Clearly, thermal radiation can be strongly polarized by reflection!

In all cases, the non-polarized reflectivity is very uniform from 0° to 45°, perhaps from 0° to 60°, or even from 0° to 80°. The absorptivity $\alpha'_\lambda$ and the emissivity $\varepsilon'_\lambda$ of an opaque body (see Chapter 4) can be obtained from:

$$\varepsilon'_\lambda = \alpha'_\lambda = 1 - \rho'_\lambda = 1 - (\rho'_{\lambda\parallel} + \rho'_{\lambda\perp})/2 \tag{9.19}$$

We may also assume that $\varepsilon'_\lambda$ and $\alpha'_\lambda$ are, for most opaque bodies in laboratory conditions, constant and largely independent of direction between 0° and 45°, perhaps 0° and 60° (see Figure 9.4).

### 9.1.1.4 THE PARADOX OF ABSORPTION

The non-polarized reflectivity $\rho'_\lambda$ is generally an increasing function of $\theta_1$; the emissivity $\varepsilon'_\lambda$ and absorptivity $\alpha'_\lambda$ are then maximum at zero incidence, which from (9.18) gives the constant value:

$$\alpha'_\lambda(0) = \varepsilon'_\lambda(0) = \frac{4n}{[(n+1)^2 + \chi^2]} \tag{9.20}$$

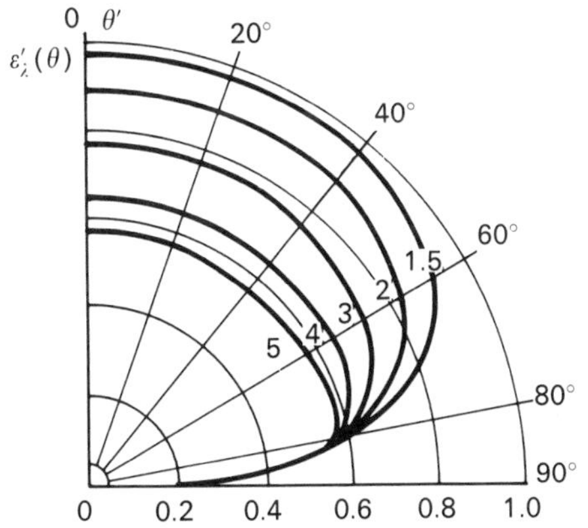


**Figure 9.4** Emissivity $\varepsilon'_\lambda(\theta)$ as a function of the ratio of the moduli of the indices $n_2/n_1$ (after reference [7], notation as on Figure 9.1).

We discover, in these circumstances, the paradox that, for a given $n$, the absorptivity $\alpha'_\lambda$ of an opaque body is large when $\chi$ (or the absorption coefficient $\kappa$ defined in (9.8)) is small. Furthermore, when $\chi$ (or $\kappa$) is large, the absorbtivity is small.

We notice that copper, with a depth of penetration of 300 Å at 5 $\mu$m, which corresponds to an absorption coefficient $\kappa$ of $3 \times 10^7\ \text{m}^{-1}$, is an excellent mirror when polished! A sheet of glass with a depth of penetration of the order of 1 mm at 5 $\mu$m (corresponding to an absorption coefficient of $10^3\ \text{m}^{-1}$) is, on the contrary, almost completely absorbent ($\alpha'_\lambda(0) \simeq 0.95$)!

The interpretation of these apparently paradoxical observations is straightforward: for radiation to be absorbed, it must penetrate deep into the interior of the material; however, for large values of $\chi$ the medium is repulsive: the depth of penetration $\eta$ is then very small. Absorption is a phenomenon which takes place in the interior of the medium.

### *9.1.2 Physical concepts of absorption in solids*

In this section we will determine what physical model is necessary to determine $n$ and $\chi$ (or, equivalently, $\hat{\varepsilon}$). These models are in the domain of solid state physics [1,2], and vary according to the nature of the opaque body concerned (conductor, semi-conductor or insulator). We will exclude gases, which are semi-transparent in bands and give a fine line structure – these will be considered in Chapter 11. Similarly, we will exclude liquids.

The material concerned is made up of groups of atoms whose density and arrangement depend on its physical state (about $10^{23}$ atoms $\text{cm}^{-3}$). In crystalline solids, the atoms form a lattice whose vibrations (of very low amplitudes) are characterized by quanta called phonons. The number and energy of the phonons depend on the temperature of the lattice [2]. The most mobile charges are the free electrons, or even the electrons bound to the atoms in the lattice. The energies of the charge carriers (electrons or holes) are well represented by the valence and conduction band model.

The interactions between the radiation and the material correspond to **photon–electron** or **photon–phonon** interactions. The former are the most rapid (duration less than $10^{-17}$ s at 1 $\mu$m), and are in fact the most probable by a long way. Electrons absorbing a photon by this method acquire an immense energy (14 000 K or 1 eV for a 1 $\mu$m photon). They lose this energy to their environment by various processes: **electron–electron**, **electron–phonon or electron–defect** interactions:

- The **electron–electron** (e–e) interaction has an effective cross-section $\sigma$ of the order of $10^{-15}$ cm$^2$ (which at 300 K for an absorbed photon of 1 $\mu$m corresponds to a mean free path for the electron in the material of about 10 mm). The relaxation time $\tau_{ee}$ for the e–e interaction is $10^{-12}$–$10^{-13}$ s. After the interaction, the energy is divided between the two electrons, and the process repeats in a chain reaction.
- The **electron–phonon** (e–p) interactions is a much more efficient process than the above. The electrons lose their energy to the phonons in the course of their 'collisions' with the lattice. The process is complex; the number of impacts of this type is very high [10]: the relaxation time $\tau_{ep}$ is of the order $10^{-14}$–$10^{-15}$ s, but at each 'impact' only a small amount of energy is transferred ($\simeq \frac{1}{100}$ eV). The energy is transferred to the crystal lattice by the cumulative effect of these 'impacts' for a given electron (one hundred impacts are needed for an initial energy of 1 eV).
- **Electron–impurity** (e–i) and **electron–defect** interactions are equally important in semi-conductors. They have a relaxation time $\tau_{ei}$ [1,2].

All the interactions together (e–e, e–p, e–i) give a global relaxation time $\tau$ (or a related 'global speed constant' $\Gamma$) for the electrons to return to equilibrium after absorbing a photon, given by:

$$\Gamma = \frac{1}{\tau} = \frac{1}{\tau_{ee}} + \frac{1}{\tau_{ep}} + \frac{1}{\tau_{ei}} \tag{9.21}$$

$\Gamma$ and $\tau$ are functions of temperature $T$ and frequency $\nu$.

#### 9.1.2.1 INSULATORS

The electrons in a crystalline lattice are able to oscillate at very high frequency and have characteristic or resonant frequencies $\nu_0$. They behave like harmonic oscillators with a damping or dissipation term linked to the relaxation phenomena discussed above; this term is defined by $\Gamma$. A model originally developed by Lorentz [5] gives a formula for the complex relative permittivity (see equation (9.2)):

$$\mu = 1; \qquad \hat{\varepsilon}^{\text{bound}} = \varepsilon_1 + \mathrm{j}\varepsilon_2 \tag{9.22}$$

with:

$$\varepsilon_1^{\text{bound}} = (n^2 - \chi^2) = 1 + \frac{\omega_p^2(\omega_0^2 - \omega^2)}{(\omega_0^2 - \omega^2)^2 + \Gamma^2\omega^2}$$

and (9.23)

$$\varepsilon_2^{\text{bound}} = 2n\chi = \omega_p^2 \frac{\Gamma\omega}{(\omega_0^2 - \omega^2) + \Gamma^2\omega^2}$$

where we have written:

$$\omega_p^2 = 4\pi^2\nu_p^2 = \frac{4\pi Ne^2}{m} \quad \text{and} \quad \omega = 2\pi\nu \tag{9.24}$$

The quantity $\nu_p$ is called the 'plasma frequency' of the material; it depends on the charge $e$ and the mass $m$ (effective mass in a solid) of an electron.

A summation has been carried out over all the electrons bound to the element of material, which appears as $N$, the number of electrons per unit volume in (9.24).

The behaviour of the functions $\varepsilon_1$ and $\varepsilon_2$ in the neighbourhood of $\omega_0$ is typical of a resonance phenomenon (Figure 9.5), which is also found for the absorption of a gas. We may note that the dispersion curve representing $\varepsilon_1$ spreads out much more than the absorption curve representing $\varepsilon_2$. For a gas, $n \simeq 1$ and $\chi \ll n$, justifying the names of the two curves: the first represents in effect the variation of $n$, the second that of $2\chi$.

The phenomenon corresponding to equations (9.23) is called **selective absorption of radiation**; the quantities $n$ and $\chi$ show significant variations in the neighbourhood of the electron resonance fequency $\nu_0$. In particular, $\chi$ has higher values around this value $\nu_0$. As a result of the absorption paradox discussed in Section 9.1.1.4, this generally makes the material more reflecting than at frequencies far from $\nu_0$ (Figure 9.6).

For these distant frequencies the absorption coefficient $\kappa$ (related to $\chi$) actually has a very small but non-zero value, depending on the denominator of (9.23); the

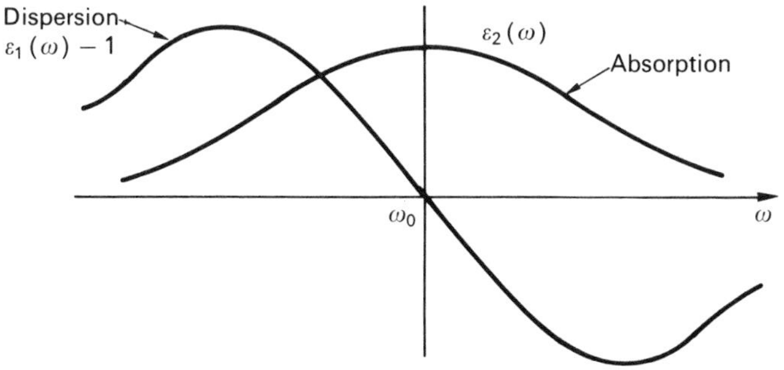


**Figure 9.5** Behaviour of $\varepsilon_1$ and $\varepsilon_2$ in the neighbourhood of resonance (selective absorption).

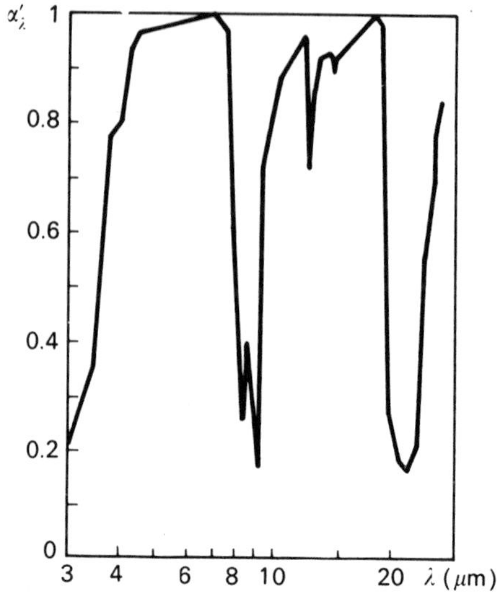


**Figure 9.6** Selective absorption of $SiO_2$ (silica) between 3.9 and 23 $\mu$m (from reference [8]) (insulator).

absorption can then be almost total. Insulating materials only show the type of absorption predicted by (9.23), though perhaps with multiple resonances $\omega_0^1$, $\omega_0^2$, etc. The selective absorption ranges may either overlap or may, on the contrary, be well separated (Figure 9.6).

9.1.2.2 CONDUCTORS

In conductors, as well as photon absorption by bound electrons (or selective absorption), there is absorption of photons by free electrons. This latter is usually the more important. The model describing it is similar to the above, but without resonances. Thus, putting $\nu_0 = 0$ in equation (9.23), we obtain:

$$\varepsilon_1^{\text{free}} = n^2 - \chi^2 = 1 - \frac{\omega_p^2 \tau^2}{1 + \omega^2 \tau^2} \quad \text{and} \quad \varepsilon_2^{\text{free}} = 2n\chi = \frac{\omega_p^2 \tau^2}{1 + \omega^2 \tau^2} \tag{9.25}$$

where $\tau$ is the global relaxation time, defined by (9.21).

The complex permittivity of a metal actually arises from the combination of the effects of free and bound electrons. We obtain $\varepsilon_1$ and $\varepsilon_2$ by relations such as:

$$\varepsilon_1 = a\varepsilon_1^{\text{bound}} + b\varepsilon_1^{\text{free}} \quad \text{and} \quad \varepsilon_2 = c\varepsilon_2^{\text{bound}} + b\varepsilon_2^{\text{free}} \tag{9.26}$$

The coefficients $a$, $b$, $c$, $d$ are normalized.

Outside the spectral regions where selective absorption is important, only the phenomena of (9.25) occur. In particular, in the far infra-red and microwave regions (when $\nu$ and $\omega$ become small and $\lambda$ large), the quantity $n\chi$ is proportional to the

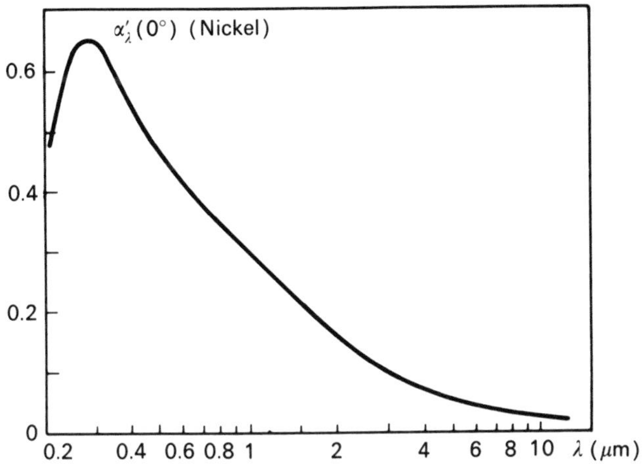


**Figure 9.7** Monochromatic absorptivity of nickel (from reference [7]).

wavelength, since $n^2 - \chi^2$ becomes constant as $\lambda \to \infty$:

$$n \simeq \chi \simeq \frac{\omega_p \tau}{\sqrt{2\omega}} \simeq \frac{\omega_p \tau}{\sqrt{4\pi c_0}} \sqrt{\lambda} \tag{9.27}$$

The absorption coefficient thus increases as $\lambda^{1/2}$: the body becomes more reflective as $\lambda$ increases; this is a well-known result for pure polished metals which are selective materials – relatively absorbent in the visible and highly reflecting in the infra-red (see Figure 9.7, for example).

We should note that the behaviour predicted by (9.27) is in practice a limiting case, outside selective absorption zones for large $\lambda$. Using (9.20), with $n$ and $\chi$ large compared to 1, we can predict the behaviour of the normal emissivity for these conditions:

$$\varepsilon'_\lambda(0) = A\lambda^{-1/2} \tag{9.28}$$

This is commonly known as Drude's law.

## 9.2 Common opaque bodies

Opaque bodies encountered in industrial conditions are generally far from satisfying the various conditions of Section 9.1. It quite often happens that they satisfy none of them. A common opaque body often has the following characteristics:

- a very heterogeneous and time dependent composition;
- scattering properties associated with the heterogeneous material;
- a rough and variable surface finish.

### 9.2.1 *Parameters affecting the radiative properties of opaque bodies*

#### 9.2.1.1 MATERIAL SURFACE COMPOSITION

The depth of penetration of radiation (related to the extinction index $\chi$ of the material) is small on the macroscopic scale (a few Å to a few 100 s of $\mu$m, according to the frequency of the radiation concerned). Only the layers closest to the surface need be considered in determining the radiative properties. These layers are precisely those which are most sensitive to environmental effects: oxidation, corrosion, dirt, surface defects, etc. In fact, **we are not concerned with the intrinsic physical properties of the material, but with those of the surface of the material in a particular environment and taking account of its history.** A given material may, for example, over time become covered with a layer of a mixture of oxides of variable constitution and thickness.

A striking case is the change in properties of a so-called non-oxidizing material (for example, a special Inconel steel) when covered with a very thin layer of oxide (Figure 9.8). The term 'non-oxidizing' means that the oxide layer is stable and protects the lower layers of the material; the mechanical properties are unaffected. This is manifestly not the case for the radiative properties. There is no point in proliferating examples: the reader can readily imagine what happens with an 'oxidizing' material.

#### 9.2.1.2 MATERIAL SURFACE FINISH

Under common conditions of use, the surface of a material is often rough, granular or porous (through oxidation); it may also have undergone surface treatment (shot blasting, sand blasting, shot peening, fluting, etc. – see Figure 9.8 for sandblasting) which will considerably change the radiative properties.

We recall at this stage that an optically smooth surface (one of the conditions required in Section 9.1) has asperities of heights less than $\lambda/10$, that is, less than 1 $\mu$m for a source at 300 K and 0.2 $\mu$m for a source at 1500 K. The influence of surface finish degradation (from polished metal to rough metal) is clear from Figure 9.9 (for copper). The reflection becomes progressively less and less specular as the toughness increases (Figure 9.10). Rays which are not reflected in the direction predicted by Descartes' law emerge in all other directions, which has the effect of increasing the absorption of the material by 'cavity effect'. We may see that it is possible to change continuously, with roughness, from specular (for optically smooth) to quasi-isotropic (where the surface roughness elements are random) reflection, as shown on Figure 9.10.

The surface roughness of a material may also be artificially controlled in order to obtain a particular desired effect: increase of absorptivity by cavity effect, change of diffusion index by fluting, sand blasting operations, shot peening, etc. A systematic study of natural and controlled roughness is presented in reference [11].

#### 9.2.1.3 MATERIAL DIFFUSION CHARACTERISTICS

Diffusion comes about by directional dispersion within a heterogeneous element of volume $d\tau$ with an incident light pencil in an elementary solid angle $d\Omega$ centred on

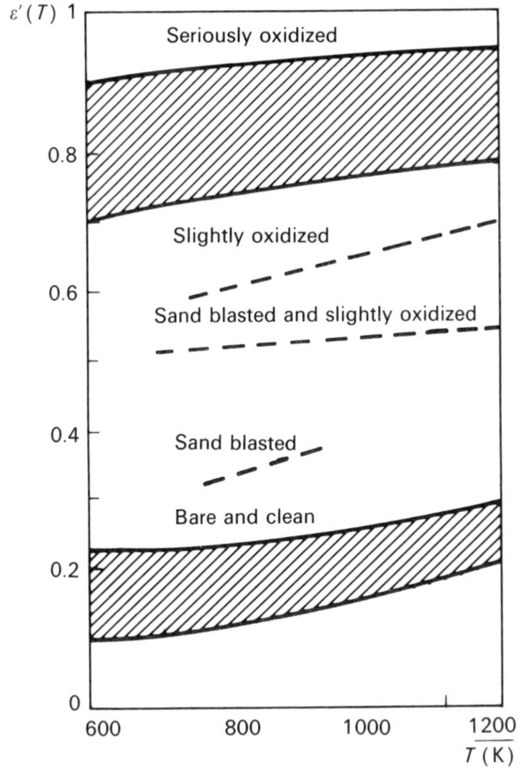


**Figure 9.8** Total emissivity of Inconel as a function of surface treatment and/or state of oxidation (from reference [7]).

$$\varepsilon' = \frac{\int_0^\infty \pi \varepsilon'_\lambda I^0_\lambda(T)\, \mathrm{d}\lambda}{\sigma T^4}$$

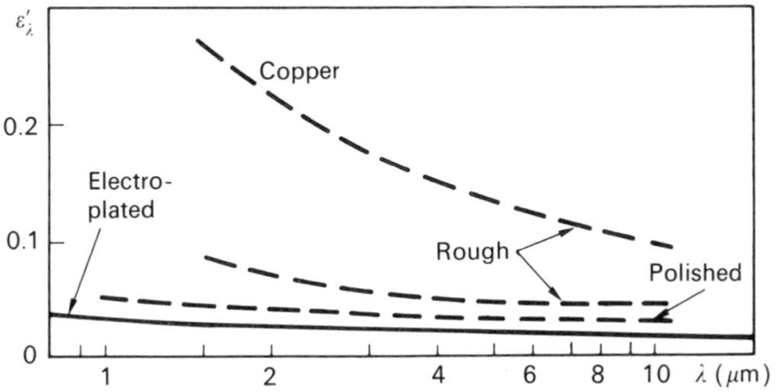


**Figure 9.9** Emissivity of copper as a function of surface finish.

**Figure 9.10** Continuous change from specular reflection to quasi-isotropic reflection, depending on surface finish.

a direction Δ. This is, actually, in the field of semi-transparent media, and is discussed in Chapter 11.

### 9.2.2 Relastic radiative properties

For materials in common situations, it would be vain to try to calculate radiative properties using a theoretical model because the number of local physical parameters needed (the nature and proportions of various oxides, roughness, diffusion, etc.) makes this impossible. We must, then, make some realistic approximations.

1. The first approximation relates to the directional character of radiative properties. Except in the particular case of a point source, it is usually adequate to assume that radiative properties (emissivity, absorptivity, reflectivity) are isotropic, for the calculation of energy transfer. There are two justifications for this:
   - Even for optically smooth materials, $\alpha'_\lambda$ and $\varepsilon'_\lambda$ are largely independent of direction (Section 9.1.1.3). Further, if the intensity of the incident radiation is isotropic, the intensity of specularly reflected radiation is almost so. On this assumption, no significant overall difference appears between specular and isotropic reflection.
   - When we consider real bodies, the reflectivity tends to become at least partially isotropic, partly because of the surface finish (Figure 9.10) and partly because of the diffusing properties of the material (Section 9.2.1.3).

   Under these conditions, we generally consider all the radiative properties of opaque bodies to be isotropic when considering energy exchange with common bodies, since:

   $$\varepsilon_\lambda = \alpha_\lambda = 1 - \rho_\lambda$$

   Note that reflection is now considered independent of angle of incidence and angle of emergence.

   *Note:* An important exception is for vitreous materials, particularly in one-dimensional plane geometries. For these, it is often necessary to make use of Fresnel's laws (see pp. 173–5).
2. On the other hand, the radiative properties of real bodies depend strongly on wavelength. There are two methods for determining this dependence:
   - Estimate the properties compared to reference conditions, taking account of any surface treatment and the material history (e.g. heating for $N$ hours at a

specified temperature), making use of the wealth of information in, for example, references [7–9]. In most cases for common bodies, it is possible to predict the monochromatic radiative properties to within 10% by this method.

- Carry out measurements of the properties on samples of material at various temperatures and wavelengths in specialist laboratories, or even in some cases on site.

Finally, it is often possible to assume, outside the regions where selective absorption is important for an insulator or the continuous absorption of a metal dominates, that the body behaves like a grey body in a certain spectral range (see the modelling of a glass window in Chapter 4). An adequate model often involves considering the body to be grey in discrete bands. The number of bands must be chosen in the light of the accuracy required and the acceptable amount of calculation. Note, for example, that metals (even stainless steels) protected by an adequate oxide film may often be considered as grey in the infra-red, with an emissivity around 0.8, to within about 20%.

# 10

# Radiative Transfer between Opaque Bodies

The objective of this chapter is to develop a realistic calculation method for radiative transfers between opaque bodies of arbitrary shape, through transparent media. This method rests on a model for the radiative properties of materials which follows from the conclusions of Chapter 9.

## *10.1 The general radiative transfer problem*

### *10.1.1 Radiative flux equation*

Let us recall some of the results obtained in Chapter 4. The radiative flux $\varphi^{R}(\mathbf{r})$, considered as positive in the direction of the normal to the surface of an opaque body, is obtained at any point $M(\mathbf{r})$ by integration of the monochromatic flux density $d\varphi_{\lambda}^{R}(\mathbf{r})$ relevant to the interval $[\lambda, \lambda + d\lambda]$ over the whole spectrum of wavelengths $\lambda$:

$$\varphi^{R}(\mathbf{r}) = \int_{\lambda=0}^{\lambda=\infty} d\varphi_{\lambda}^{R}(\mathbf{r}) \qquad (10.1)$$

The monochromatic flux can be written (with an arbitrary choice of the direction of the axes in Figure 10.1) in either of two ways:

$$d\varphi_{\lambda}^{R}(\mathbf{r}) = d\varphi_{\lambda}^{e}(\mathbf{r}) - d\varphi_{\lambda}^{a}(\mathbf{r}) \qquad (10.2)$$

$$d\varphi_{\lambda}^{R}(\mathbf{r}) = d\varphi_{\lambda}^{l}(\mathbf{r}) - d\varphi_{\lambda}^{i}(\mathbf{r}) \qquad (10.3)$$

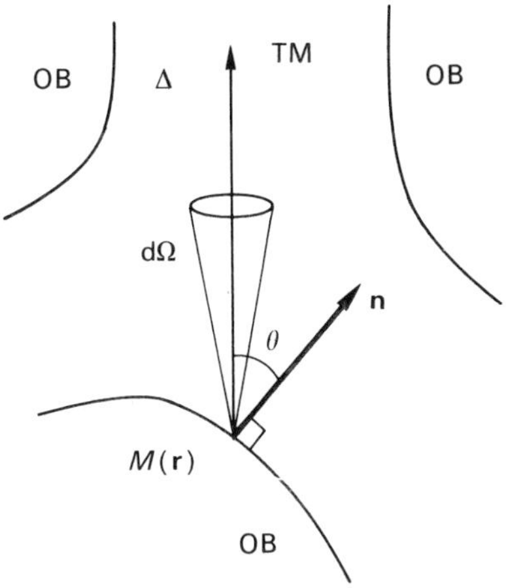


**Figure 10.1**

The indices a, e, i, l refer, respectively, to the quantities absorbed by, emitted by, incident on and emergent from[1] the surface. Remember that the various fluxes on the right-hand sides are arithmetic.

Any monochromatic flux density (emitted, absorbed, emergent, incident) can be expressed by the relation:

$$d\varphi_\lambda(\mathbf{r}) = d\lambda \int_0^{2\pi} I'_\lambda(\mathbf{r}, \mathbf{\Delta}) \cos\theta \, d\Omega \tag{10.4}$$

as a function of the corresponding monochromatic intensity in the solid angle $d\Omega$ around the direction $\mathbf{\Delta}$. Note that if we make the assumption that the monochromatic intensity is isotropic (independent of $\mathbf{\Delta}$), and denoted by $I_\lambda(\mathbf{r})$, equation (10.4) reduces to:

$$d\varphi_\lambda(\mathbf{r}) = \pi I_\lambda(\mathbf{r})\, d\lambda \tag{10.5}$$

The intensity of the radiation emitted from an element of the opaque body around $M(\mathbf{r})$ is expressed, in general, as a function of the directional monochromatic emissivity $\varepsilon'_\lambda(\mathbf{r},\mathbf{\Delta})$ of the body concerned and the intensity of the equilibrium radiation at the local temperature of the point $M$, denoted by $I^0_\lambda(T(\mathbf{r}))$:

$$I'^{e}_\lambda(\mathbf{r}, \mathbf{\Delta}) = \varepsilon'_\lambda(\mathbf{r}, \mathbf{\Delta}) I^0_\lambda(T(\mathbf{r})) \tag{10.6}$$

The monochromatic intensity which will be absorbed by that element of the opaque body in the direction $\mathbf{\Delta}$ is:

$$I'^{a}_\lambda(\mathbf{r}, \mathbf{\Delta}) = \varepsilon'_\lambda(\mathbf{r}, \mathbf{\Delta}) I'^{i}_\lambda(\mathbf{r}, \mathbf{\Delta}) \tag{10.7}$$

[1] Or leaving the surface.

where $I'^{i}_{\lambda}$ is the monochromatic intensity of the incident radiation in the direction $\Delta$. Clearly, (10.7) assumes equality of the directional monochromatic emissivity and absorptivity:

$$\alpha'_{\lambda}(\Delta, \mathbf{r}) = \varepsilon'_{\lambda}(\Delta, \mathbf{r}) \tag{10.8}$$

Then, making use of (10.1), (10.2), (10.4), (10.6) and (10.7), the local radiative flux at $M$ is given by:

$$\varphi^{R} = \varphi^{e} - \varphi^{a} = \int_{0}^{\infty} d\lambda \int_{0}^{2\pi} \varepsilon'_{\lambda}(\mathbf{r}, \Delta)[I^{0}_{\lambda}(T(\mathbf{r})) - I'^{i}_{\lambda}(\mathbf{r}, \Delta)] \cos\theta \, d\Omega \tag{10.9}$$

In a similar manner, using equation (10.3) (instead of (10.2)), the flux may also be written:

$$\varphi^{R} = \varphi^{l} - \varphi^{i} = \int_{0}^{\infty} d\lambda \int_{0}^{2\pi} [I'^{l}_{\lambda}(\Delta, \mathbf{r}) - I'^{i}_{\lambda}(\Delta, \mathbf{r})] \cos\theta \, d\Omega \tag{10.10}$$

where $I'^{l}_{\lambda}$ and $I'^{i}_{\lambda}$ designate the intensities of the radiations incident at, and emerging from, $M$ in $d\Omega$ around the direction $\Delta$, respectively.

The complexity of the problem is large if we consider (10.9):

- In principle, it is necessary to know $\varepsilon_{\lambda}(\mathbf{r}, \Delta)$ at every point. This is a considerable amount of information, since $\Delta$ depends on two angles ($\theta$ and the azimuth $\eta$). In practice, it is unrealistic to take account of the angular dependency of emissivity for opaque, corroded, oxidized and dirty materials in industrial conditions.
- Again the determination of the field of incident intensity $I'^{i}_{\lambda}$ (for all directions $\Delta$ and for every point on every surface) is unrealistic because of multiple reflections. Figure 10.2, showing a case where radiation is initially emitted at just one point $M$, should convince the reader of this.

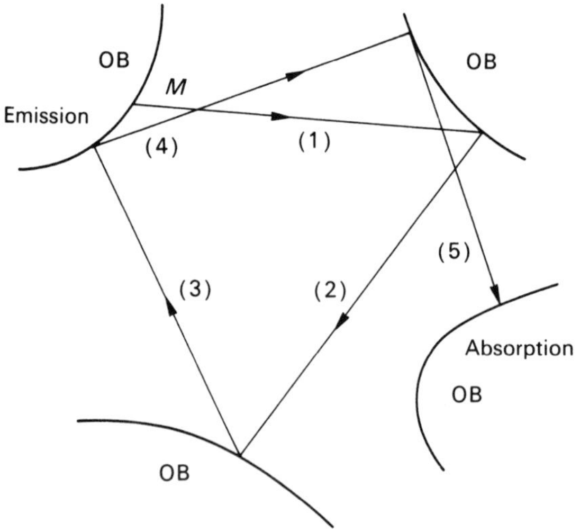


Figure 10.2

The numerical technique known as the Monte Carlo method allows (in principle, if one is not short of computing time) this kind of problem to be solved for relatively simple geometries [12]. This is not a very good solution. In general, we must have recourse to realistic approximations: this is the object of Section 10.2. Nevertheless, in some very special cases with very simple geometry, direct calculation of radiative flux allowing for multiple reflections is possible.

### *10.1.2 Example of direct calculation: radiation shields*

Let us consider two infinite, plane, opaque bodies with isotropic emissivities $\varepsilon_{1\lambda}$ and $\varepsilon_{2\lambda}$, and at temperatures $T_1$ and $T_2$, respectively, separated by a transparent medium of refractive index 1. Using the conventions for directions on Figure 10.3, the radiative fluxes $\varphi_1^{\mathrm{R}}$ and $\varphi_2^{\mathrm{R}}$ are:

$$\varphi_1^{\mathrm{R}} = +\varphi_2^{\mathrm{R}} = \int_0^\infty \varepsilon_{1\lambda}\pi(I_\lambda^0(T_1) - I_{1\lambda}^{\mathrm{i}})\,\mathrm{d}\lambda \tag{10.11}$$

where $I_{1\lambda}^{\mathrm{i}}$ is the intensity of the radiation (isotropic in this particular case) incident on 1 (indeed, $\varepsilon_{1\lambda}$, $\varepsilon_{2\lambda}$, $\rho_{1\lambda}$ and $\rho_{2\lambda}$ are isotropic and the geometry is planar). We need to obtain $I_{1\lambda}^{\mathrm{i}}$. The **incident radiation** on 1 consists of one part originally **emitted by 1** and undergoing an odd number of reflections[2] from 2 and 1, and one part originally **emitted by 2**. The fraction of the intensity of the radiation incident on 1 originally emitted by 2 is made up of the following components:

- $\varepsilon_{2\lambda} I_\lambda^0(T_2)$ (direct from 1 to 2)
- $\varepsilon_{2\lambda}\rho_{1\lambda}\rho_{2\lambda} I_\lambda^0(T_2)$ (+1 return journey)
- ...
- $\varepsilon_{2\lambda}(\rho_{1\lambda}\rho_{2\lambda})^n I_\lambda^0(T_2)$ (+$n$ return journeys)
- ...

The sum of these terms is: $\varepsilon_{2\lambda} I_\lambda^0(T_2)(1-\rho_{1\lambda}\rho_{2\lambda})^{-1}$. By a similar argument, the fraction of the intensity of radiation incident on 1 due to emission by 1 is $I_\lambda^0(T_1)(1-\rho_{1\lambda}\rho_{2\lambda})^{-1}\rho_{2\lambda}\varepsilon_{1\lambda}$.

The intensity of the isotropic radiation incident on 1 is then:

$$I_{1\lambda}^{\mathrm{i}} = [\varepsilon_{2\lambda} I_\lambda^0(T_2) + \varepsilon_{1\lambda}\rho_{2\lambda} I_\lambda^0(T_1)](1-\rho_{1\lambda}\rho_{2\lambda})^{-1} \tag{10.12}$$

Then using (10.11) and (10.12), the flux density is:

$$\varphi_1^{\mathrm{R}} = +\varphi_2^{\mathrm{R}} = \int_0^\infty \frac{\varepsilon_{1\lambda}\varepsilon_{2\lambda}\pi[I_\lambda^0(T_1) - I_\lambda^0(T_2)]}{1-(1-\varepsilon_{1\lambda})(1-\varepsilon_{2\lambda})}\,\mathrm{d}\lambda \tag{10.13}$$

[2] On Figure 10.3, only one ray is shown, for clarity, and the argument is applied here to isotropic reflection.

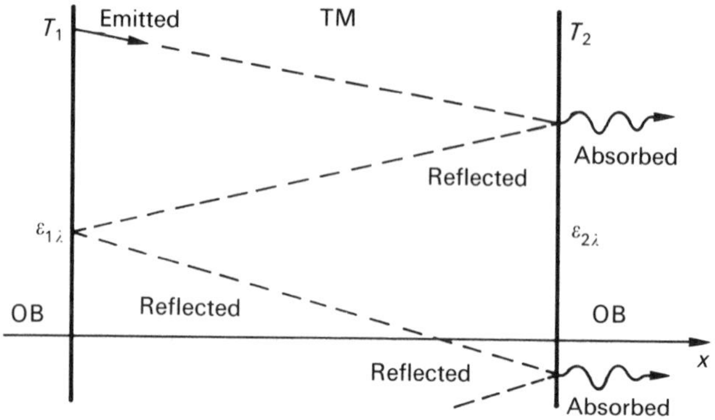


**Figure 10.3** Emission from 1.

in which, obviously, if $T_1 = T_2$ the flux cancels out. In the particular case of two grey bodies, (10.13) becomes:

$$\varphi_1 = +\varphi_2 = \frac{\varepsilon_1 \varepsilon_2 \sigma (T_1^4 - T_2^4)}{1 - (1 - \varepsilon_1)(1 - \varepsilon_2)} \tag{10.14}$$

Equation (10.14) allows a rapid estimate of the performance of a system of $N$ radiations shields between two media at $T_C$ and $T_H$ (Figure 10.4) to be made. In such an array under steady conditions, the flux exchanged is constant. The temperature differences on the hot side are very small (thus minimizing the risk of natural convection), but are large on the cold side. In practice, hexagonal structures (honeycombs) or stacks of tubes ('straws' in cryogenics) are used.

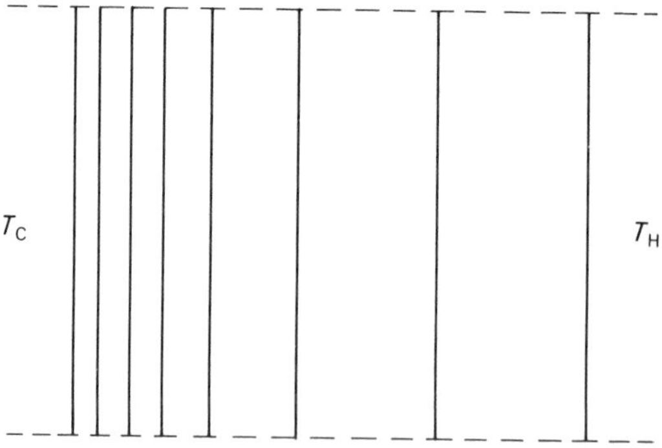


**Figure 10.4**

## 10.2 The method of incident and emergent fluxes

### 10.2.1 General hypotheses

(H1) Following the general discussion of Chapter 9, **the radiative properties of opaque bodies are isotropic** (independent of direction):

$$\alpha_\lambda = \varepsilon_\lambda = 1 - \rho_\lambda \tag{10.15}$$

In particular, the reflectivity is doubly isotropic – it depends neither on the angle of incidence nor on the angle of reflection.

(H2) **The system is divided into $N$ surfaces, denoted by $S_j$, ($j = 1, \ldots, N$), assumed isothermal at $T_j$ and with uniform radiative properties**. The validity of this hypothesis obviously depends on the fineness of the discretization ($N$) adopted, for which an optimum between precision and calculation time must be found.

As a first consequence of H1 and H2, the intensity $I^{\mathrm{e}}_{j\lambda}$ of the radiation emitted by surface $S_j$ is isotropic and uniform (independent of the point on the surface under consideration):

$$I^{\mathrm{e}}_{j\lambda} = \varepsilon_{j\lambda} I^0_\lambda(T_j) \tag{10.16}$$

In this model, it is desirable that the intensity of the emergent radiation, and therefore that of the reflected radiation, should have the same properties of isotropy and uniformity. The isotropy condition follows from hypothesis H1, as applied to $\rho_\lambda$. To obtain the uniformity condition, we need only make a supplementary hypothesis:

(H3) **The monochromatic flux density incident on a surface** $S_j$, defined by:

$$\mathrm{d}\varphi^{\mathrm{i}}_{j\lambda}(\mathbf{r}) = \mathrm{d}\lambda \int_{2\pi} I'^{\mathrm{i}}_{j\lambda}(\mathbf{r}) \cos\theta \, \mathrm{d}\Omega \tag{10.17}$$

(in which $I'^{\mathrm{i}}_{j\lambda}$ is the intensity, dependent on $\theta$ and the azimuth $\eta$, of the incident radiation) **is uniform** (independent of $\mathbf{r}$). In other words, this flux can be characterized by an **equivalent isotropic incident intensity** $I^{\mathrm{i}}_{j\lambda}$ defined by:

$$\mathrm{d}\varphi^{\mathrm{i}}_{j\lambda} = \pi I^{\mathrm{i}}_{j\lambda} \, \mathrm{d}\lambda \tag{10.18}$$

which is independent of the point $\mathbf{r}$. We should note well that the intensity of the incident radiation generally depends on direction. In fact, $I^{\mathrm{i}}_{j\lambda}$ is given by:

$$I^{\mathrm{i}}_{j\lambda} = \frac{1}{\pi} \int_{2\pi} I'^{\mathrm{i}}_{j\lambda}(\theta, \eta) \cos\theta \, \mathrm{d}\Omega \tag{10.19}$$

### 10.2.2 Expression for the radiative flux

With hypotheses H1, H2 and H3, the expression for the radiative flux (10.11) simplifies to:

$$\varphi_j^{\mathrm{R}} = \int_0^{\infty} \varepsilon_{j\lambda}\pi(I_\lambda^0(T_j) - I_{j\lambda}^{\mathrm{i}})\,\mathrm{d}\lambda \tag{10.20}$$

while equation (10.10) becomes:

$$\varphi_j^{\mathrm{R}} = \int_0^{\infty} \pi(I_{j\lambda}^{\mathrm{l}} - I_{j\lambda}^{\mathrm{i}})\,\mathrm{d}\lambda \tag{10.21}$$

The isotropic intensity of the emergent radiation $I_{j\lambda}^{\mathrm{l}}$ is:

$$I_{j\lambda}^{\mathrm{l}} = \varepsilon_{j\lambda} I_\lambda^0(T_j) + (1 - \varepsilon_{j\lambda}) I_{j\lambda}^{\mathrm{i}} \tag{10.22}$$

The first term in equation (10.22) represents the intensity of the emitted radiation, the second that of the reflected radiation, as we will show.

Remember that, by definition (see Chapter 4), the intensity of the radiation incident in dΩ and reflected in a half-space is written:

$$I_{j\lambda}^{\mathrm{r}}(\mathrm{d}\Omega \to 2\pi) = \rho_{j\lambda}^{\prime\bullet} I_{j\lambda}^{\prime\mathrm{i}} = (1 - \alpha_{j\lambda}) I_{j\lambda}^{\prime\mathrm{i}} = (1 - \varepsilon_{j\lambda}) I_{j\lambda}^{\prime\mathrm{i}} \tag{10.23}$$

The corresponding reflected flux is:

$$\mathrm{d}\varphi_{j\lambda}^{\mathrm{r}}(\mathrm{d}\Omega \to 2\pi) = I_{j\lambda}^{\mathrm{r}}(\mathrm{d}\Omega \to 2\pi)\cos\theta\,\mathrm{d}\Omega\,\mathrm{d}\lambda = (1 - \varepsilon_{j\lambda}) I_{j\lambda}^{\prime\mathrm{i}}\cos\theta\,\mathrm{d}\Omega\,\mathrm{d}\lambda \tag{10.24.1}$$

and the global reflected flux:

$$\mathrm{d}\varphi_{j\lambda}^{\mathrm{r}}(2\pi \to 2\pi) = (1 - \varepsilon_{j\lambda})\,\mathrm{d}\lambda \int_{\Omega=0}^{2\pi} I_{j\lambda}^{\prime\mathrm{i}}\cos\theta\,\mathrm{d}\Omega = (1 - \varepsilon_{j\lambda})\pi I_{j\lambda}^{\mathrm{i}} \tag{10.24.2}$$

The isotropic intensity of the radiation reflected by a surface $S_j$ is thus $(1 - \varepsilon_{j\lambda}) I_{j\lambda}^{\mathrm{i}}$, which is the second term in equation (10.22).

It remains to calculate the equivalent isotropic incident intensity $I_{j\lambda}^{\mathrm{i}}$. To do this, we calculate the monochromatic flux $\mathrm{d}\Phi_{j\lambda}^{\mathrm{i}}$ incident on the set of surfaces $S_j$:

$$\mathrm{d}\Phi_{j\lambda}^{\mathrm{i}} = \pi I_{j\lambda}^{\mathrm{i}} S_j\,\mathrm{d}\lambda \tag{10.25}$$

where we have made use of the uniformity property of $I_{j\lambda}^{\mathrm{i}}$ and of (10.18). The flux $\mathrm{d}\Phi_{j\lambda}^{\mathrm{i}}$ originates from the $N$ opaque surfaces making up the system (including the surface $S_j$ itself):

$$\mathrm{d}\Phi_{j\lambda}^{\mathrm{i}} = \sum_{k=1}^{N} \mathrm{d}\Phi_{k\to j,\lambda}^{\mathrm{l}} \tag{10.26}$$

where $\mathrm{d}\Phi_{k\to j,\lambda}^{\mathrm{l}}$ represents the **monochromatic flux leaving $k$ and arriving directly on $j$** (without intermediate reflection). Referring to Figure 10.5, if $M$ and $N$ are the current

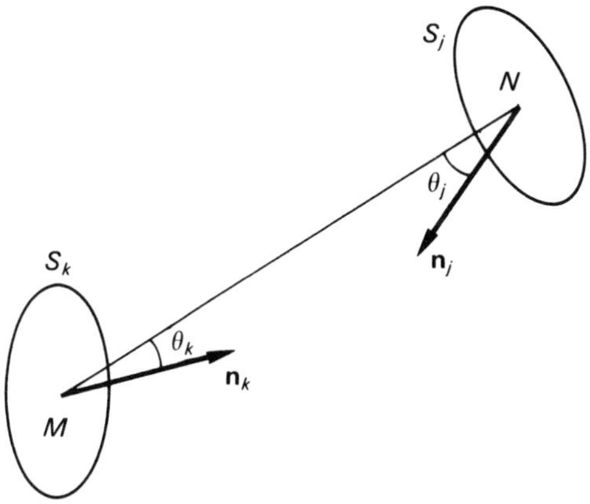


**Figure 10.5**

points of $S_k$ and $S_j$ respectively, we have:

$$d\Phi^1_{k\to j,\lambda} = I^1_{k\lambda}\, d\lambda \int_{S_k} dS_k \int_{S_j} \frac{\cos\theta_j \cos\theta_k}{MN^2}\, dS_j \tag{10.27}$$

in which we have used the isotropy and uniformity properties of $I^1_{k\lambda}$. Taking the global monochromatic flux leaving $k$ (denoted by $d\Phi^1_{k\lambda}$) as a reference:

$$d\Phi^1_{k\lambda} = \pi I^1_{k\lambda}\, d\lambda S_k \tag{10.28}$$

**the purely geometric factor $f_{kj}$ is called the shape factor** for $k$ to $j$, and is defined by the relation:

$$d\Phi^1_{k\to j,\lambda} = f_{kj}\, d\Phi^1_{k\lambda} \tag{10.29}$$

so that:

$$S_k f_{kj} = \frac{1}{\pi}\int_{S_k} dS_k \int_{S_j} dS_j \frac{\cos\theta_j \cos\theta_k}{MN^2} \tag{10.30}$$

Three properties of the shape factor are immediately apparent:

- From (10.30):

$$S_k f_{kj} = S_j f_{jk} \tag{10.31}$$

  (the reciprocity relation);
- From (10.29), if $N$ surfaces, including $k$, make a closed volume, **conservation of energy** implies that:

$$d\Phi^1_{k\lambda} = \sum_{j=1}^{N} d\Phi^1_{k\to j,\lambda} \tag{10.32}$$

or:

$$\sum_{j=1}^{N} f_{kj} = 1 \tag{10.33}$$

- For a convex body $k$, it is clear from (10.29) that:

$$k \text{ convex} \Rightarrow f_{kk} = 0 \tag{10.34}$$

and in particular, if a body $j$ completely surrounds a convex body $k$:

$$f_{kj} = 1 \tag{10.35}$$

Under these conditions, equation (10.26) becomes:

$$\mathrm{d}\Phi^{\mathrm{i}}_{j\lambda} = \pi I^{\mathrm{i}}_{j\lambda} S_j \, \mathrm{d}\lambda = \sum_{k=1}^{N} f_{kj} S_k \pi I^{\mathrm{l}}_{k\lambda} \, \mathrm{d}\lambda \tag{10.36}$$

which leads to a relationship between intensities:

$$S_j I^{\mathrm{i}}_{j\lambda} = \sum_{k=1}^{N} f_{kj} S_k I^{\mathrm{l}}_{k\lambda} \tag{10.37}$$

and, taking account of (10.31) gives:

$$I^{\mathrm{i}}_{j\lambda} = \sum_{k=1}^{N} f_{jk} I^{\mathrm{l}}_{k\lambda} \tag{10.38}$$

Equations (10.22) and (10.38) constitute a **linear system of 2$N$ equations in 2$N$ unknowns** (the intensities of the incident and emergent radiations for the $N$ surfaces $S_j$) from whose solution we can calculate the flux from (10.21). Thus:

$$\varphi^{\mathrm{R}}_j = \int_0^{\infty} \pi (I^{\mathrm{l}}_{j\lambda} - I^{\mathrm{i}}_{j\lambda}) \, \mathrm{d}\lambda \tag{10.21}$$

$$I^{\mathrm{l}}_{j\lambda} = \varepsilon_{j\lambda} I^{0}_{\lambda}(T_j) + (1 - \varepsilon_{j\lambda}) I^{\mathrm{i}}_{j\lambda} \tag{10.22}$$

$$I^{\mathrm{i}}_{j\lambda} = \sum_{k=1}^{N} f_{jk} I^{\mathrm{l}}_{k\lambda} \tag{10.38}$$

### 10.2.3 The special case where all the surfaces are grey

If, in addition, we are able to make the assumption that the $N$ surfaces $S_j$ are grey, or:

$$\forall j: \quad \varepsilon_j = \alpha_j = 1 - \rho_j \tag{10.39}$$

we can obtain equations in terms of flux by integrating (10.21), (10.22) and (10.38) over all wavelengths and a half-space:

$$\varphi^{\mathrm{R}}_j = \varphi^{\mathrm{l}}_j - \varphi^{\mathrm{i}}_j \tag{10.40}$$

$$\varphi_j^{\mathrm{l}} = \varepsilon_j \sigma T_j^4 + (1 - \varepsilon_j)\varphi_j^{\mathrm{i}} \tag{10.41}$$

$$\varphi_j^{\mathrm{i}} = \sum_{k=1}^{N} f_{jk}\varphi_k^{\mathrm{l}} \tag{10.42}$$

which simplifies the theory considerably!

### *10.2.4 Example: Intensity standard – return to equilibrium radiation*

Let us consider a hollow sphere of radius $R$, opaque, with isotropic emissivity $\varepsilon_\lambda$ and isothermal at temperature $T$ (Figure 10.6). We are interested in the radiation which leaves it through a circular orifice of radius $r$. We will use the following numerical values:

$$R = 0.1\ \mathrm{m};\ r = 0.001\ \mathrm{m}$$

$$\varepsilon_\lambda = 0.10,\ \text{so that}\ \rho_\lambda = 0.90$$

We will also make the supplementary assumption that the incident monochromatic flux is uniform in the interior of the sphere (this uniformity is only very slightly affected by the effect of the orifice on the flux). Conditions H1, H2 and H3 are thus satisfied. The incident and emergent intensities $I_\lambda^{\mathrm{i}}$ and $I_\lambda^{\mathrm{l}}$ at every point on the sphere (except the orifice) are then, from (10.22) and (10.38):

$$I_\lambda^{\mathrm{l}} = \varepsilon_\lambda I_\lambda^0(T) + (1 - \varepsilon_\lambda)I_\lambda^{\mathrm{i}} \tag{10.43}$$

$$I_\lambda^{\mathrm{i}} = f_{\Sigma\Sigma} I_\lambda^{\mathrm{l}} + f_{\Sigma S} I_{S\lambda}^{\mathrm{l}} \tag{10.44}$$

where $\Sigma$ and $S$ are the areas, respectively, of the sphere (without the orifice) and the orifice, $f_{S\Sigma}$ and $f_{\Sigma\Sigma}$ the corresponding shape factors, and $I_{S\lambda}^{\mathrm{l}}$ is the intensity of the radiation entering the sphere through the orifice. We take, for example:

$$I_{S\lambda}^{\mathrm{l}} = I_\lambda^0(T_{\mathrm{e}}) \tag{10.45}$$

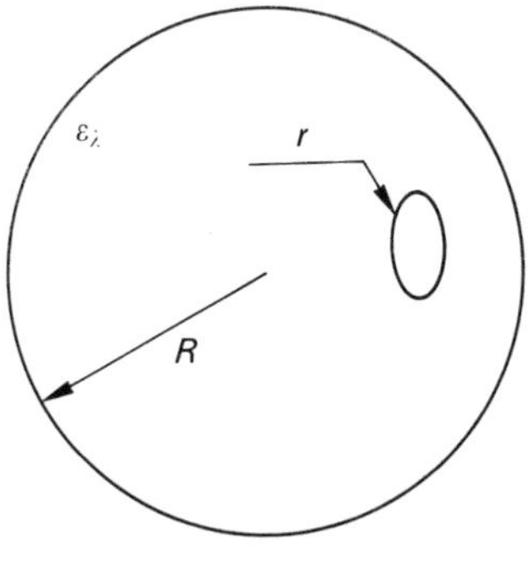


**Figure 10.6**

where $T_e$ is the equilibrium temperature of the room housing the sphere. Furthermore, arising from (10.35), (10.31) and (10.33):

$$f_{S\Sigma} = 1; \qquad f_{\Sigma S} = f_{S\Sigma}\left(\frac{S}{\Sigma}\right) = \frac{S}{\Sigma}; \qquad f_{\Sigma\Sigma} = 1 - \frac{S}{\Sigma} \tag{10.46}$$

After manipulation, we obtain expressions for the incident and emergent intensities at every point on the sphere (except the orifice):

$$I_\lambda^1 = I_\lambda^0(T) + \frac{S(1-\varepsilon_\lambda)}{\varepsilon_\lambda\Sigma + S(1-\varepsilon_\lambda)}[I_\lambda^0(T_e) - I_\lambda^0(T)] \tag{10.47}$$

$$I_\lambda^i = I_\lambda^0(T) + \left[\left(1 - \frac{S}{\Sigma}\frac{S(1-\varepsilon_\lambda)}{\varepsilon_\lambda\Sigma + S(1-\varepsilon_\lambda)}\right) + \frac{S}{\Sigma}\right][I_\lambda^0(T_e) - I_\lambda^0(T)] \tag{10.48}$$

It is clear from close examination of (10.47) and (10.48) that with the chosen values of $S/\Sigma$ and $\varepsilon_\lambda$, and whatever the value of $T_e$ (which will usually be much lower than $T$), we have as a first approximation:

$$I_\lambda^1 \simeq I_\lambda^i \simeq I_\lambda^0(T) \tag{10.49}$$

The incident and emergent radiations inside the sphere are in practice characterized by the intensity of equilibrium radiation at temperature $T$ (i.e. $I_\lambda^0(T)$), even when the surface of the sphere (approximated here by an almost perfect mirror with $\rho_\lambda = 0.9$) is very far from being a black body (for which $\alpha_\lambda = \varepsilon_\lambda = 1$). In this case: $S/\Sigma = 10^{-4}$; $\varepsilon = 10^{-1}$; $S/(\varepsilon_\lambda\Sigma) = 10^{-3}$.

The intensities $I_\lambda^1$ and $I_\lambda^i$ differ from $I_\lambda^0$ by less than $10^{-3}$ in these conditions:

$$\left[\frac{1}{I_\lambda^0(T)}\frac{S(1-\varepsilon_\lambda)[I_\lambda^0(T) - I_\lambda^0(T_e)]}{\varepsilon_\lambda\Sigma + S(1-\varepsilon_\lambda)}\right] < \frac{S(1-\varepsilon_\lambda)}{\Sigma\varepsilon_\lambda} < 10^{-3}$$

This result is of both fundamental interest and has many practical uses.

1. At a fundamental level, so long as a system is, to a first approximation, in thermal equilibrium, and thus in radiative equilibrium, it is evident that one of the laws of thermal radiation can be deduced[3]: **the intensity of an isotropic thermal radiation in a transparent medium of refractive index 1, in equilibrium at temperature $T$, is given for all directions by the function $I_\lambda^0(T)$ (Planck's formula)**. The difference calculated here allows the difference between the system and one in perfect thermal equilibrium to be calculated. The difference is clearly proportional to $[I_\lambda^0(T) - I_\lambda^0(T_e)]$ (see (10.47) and (10.48)), and also increases with the ratio $S/\Sigma$. The system cannot strictly be in thermal equilibrium since it loses energy to its surroundings which are at a lower temperature.
2. At the practical level, we will consider two applications:
   - Suppose that in metrology we wished to have a **standard of intensity at temperature** $T$, that is, $I_\lambda^0(T)$, we can see that we could use a perfectly insulated

[3] Proved, we recall, in statistical physics (Bose–Einstein statistics for indeterminate numbers of bosons).

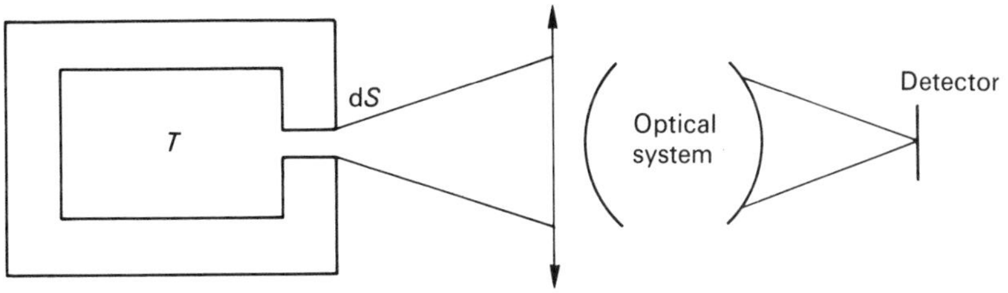


**Figure 10.7**

isothermal cavity (Figure 10.7), which means making a constant temperature oven (which is by no means easy), and providing a small orifice d$S$ in this cavity. It is obviously useful to use somewhat absorbent walls. The radiation is, more or less precisely, depending on the quality of the design, characterized by intensity $I_\lambda^0(T)$.

Pragmatically, the user of the system needs only to consider the radiation leaving the fictitious (non-material) surface d$S$ as behaving as if emitted by a black body at temperature $T$. (Actually, the surface d$S$ is the object point of the optical system used.) The orifice d$S$ (and by misuse of language, the whole system) is also called a standard black body. In fact, we have created an equilibrium radiation. Notice that any radiation entering through the orifice d$S$ has very little chance of coming out again; it is absorbed after multiple reflections: $\alpha_\lambda = 1$.

- Conditions of radiative equilibrium are often achieved even when conditions of thermal equilibrium are not totally achieved. This happens typically when we wish to use a thermocouple to measure the temperature of a transparent, highly anisothermal flowing fluid. The thermocouple temperature is affected not only by the conducto-convective heat transfer, which tends to impose the temperature of the nearby fluid layers, but also (because the fluid is transparent) by radiative transfers over large distances with the walls, which are usually much colder and tend to reduce the thermocouple temperature.

  The classical remedy for this perturbation is to protect the thermocouple with a radiation shield, but the correction so obtained is not generally totally satisfactory [13]. (See also Application 5.3 in Part VI.)

## *10.3 Properties of shape factors*

### *10.3.1 Principal properties and counting*

We recall that the method developed in Section 10.2 depends on the (purely geometric) properties of the shape factors; the radiative flux on a surface $S_j$ is given by (10.21):

$$\varphi_j^{\mathrm{R}} = \int_0^\infty \pi(I_{j\lambda}^{\mathrm{l}} - I_{j\lambda}^{\mathrm{i}})\,\mathrm{d}\lambda \qquad (10.50)$$

with:

$$I^{\mathrm{l}}_{j\lambda} = \varepsilon_{j\lambda}\pi I^{0}_{\lambda}(T_j) + (1 - \varepsilon_{j\lambda})I^{\mathrm{i}}_{j\lambda} \tag{10.51}$$

and:

$$I^{\mathrm{i}}_{j\lambda} = \sum_{k=1}^{N} f_{jk} I^{\mathrm{l}}_{j\lambda} \tag{10.52}$$

In this formula, the shape factors $f_{jk}$ satisfy the conditions given above:

1.

$$\sum_{k=1}^{N} f_{jk} = 1 \tag{10.53}$$

   where the summation is over all the surfaces forming a closed volume (including $S_j$).

2.

$$S_j f_{jk} = S_k f_{kj} \tag{10.54}$$

3. If $S_j$ is convex:

$$f_{jj} = 0 \tag{10.55}$$

It is also true that all the shape factors, from the definition, are between 0 and 1.

In a closed region, delimited by $N$ surfaces, we therefore have:

- $N^2$ shape factors (from $f_{11}$ to $f_{NN}$);
- $N$ energy conservation equations (10.53);
- $C_N^2$ reciprocity relations (10.54);
- $p$ null shape factors ($f_{jj}$) if there are $p$ convex surfaces among the $N$.

To obtain the set of shape factors in the problem, the number of unknown shape factors to be determined is:

$$n = N^2 - N - N(N+1)/2 - p$$

In practice, the region considered will usually have a number of symmetries, which will reduce the number of shape factors, but also the number of relations between them.

**Example 1: Direct calculation of all the factors**

Consider radiative exchanges between two concentric spheres (Figure 10.8). We have:

$$S_1 = 4\pi R_1^2, \qquad S_2 = 4\pi R_2^2 \tag{10.56}$$

$$f_{11} = 0 \quad \text{(convex surface)} \tag{10.57}$$

$$f_{12} = 1 - f_{11} = 1 \tag{10.58}$$

$$f_{21} = \frac{S_1}{S_2} f_{12} = \frac{S_1}{S_2} \tag{10.59}$$

$$f_{22} = 1 - f_{21} = 1 - \frac{S_1}{S_2} \tag{10.60}$$

All the shape factors have been determined directly.

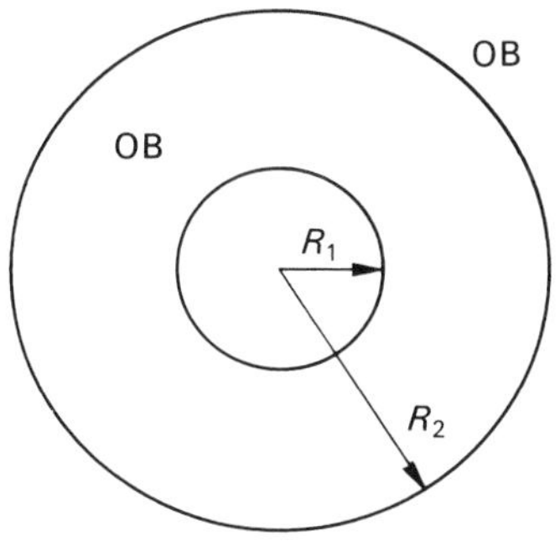


**Figure 10.8**

**Example 2: A cube**

Consider a cube whose six internal faces (1–6) have different but uniform properties. There are in principle (Figure 10.9):

- 36 shape factors;
- $6 + 15 + 6 = 27$ relationships, and thus nine independent shape factors to be calculated directly.

Actually, because of the symmetries of the problem, there are only two distinct shape factors:

- $f$ for adjacent faces;
- $f'$ for opposite faces.

There is only one non-trivial relationship expressing energy conservation:

$$f' + 4f = 1 \tag{10.61}$$

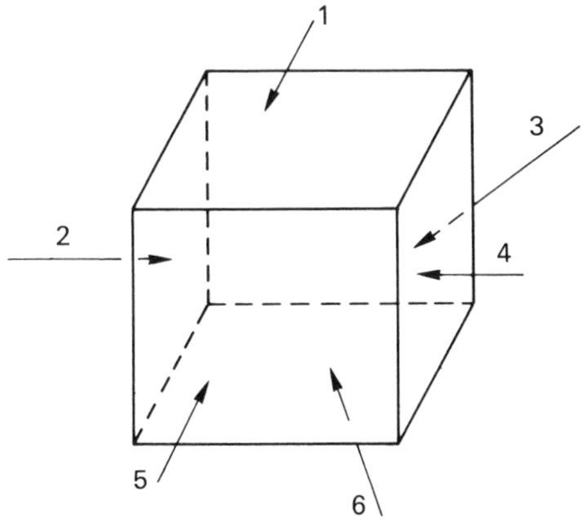


**Figure 10.9**

We therefore need to calculate one factor directly using equation (10.30), or look up the appropriate shape factor in a table (see Part V: 'Basic Data').

### 10.3.2 The fictitious surface technique

It is sometimes useful when calculating a shape factor to introduce an imaginary surface (usually convex, to introduce another relation). As a simple example, we will calculate the shape factors between two half-cylinders of radius $R$ (Figure 10.10(a)). Initially, we need to calculate $f_{12}$ and $f_{11}$. In fact, we only have available one non-trivial equation: $f_{12} + f_{11} = 1$, and we still have one unknown. Introduce a fictitious surface $O$ (Figure 10.10(b)). We now have shape factors $f_{11}$ and $f_{10}$ to be calculated, related by:

$$f_{11} + f_{10} = 1 \tag{10.62}$$

$$f_{01} = 1 \tag{10.63}$$

$$f_{10} = \frac{S_0}{S_1} f_{01} = \frac{S_0}{S_1} = \frac{2}{\pi} \tag{10.64}$$

so that:

$$f_{11} = 1 - \frac{2}{\pi} \tag{10.65}$$

and correspondingly:

$$f_{12} = f_{10} = \frac{2}{\pi} \tag{10.66}$$

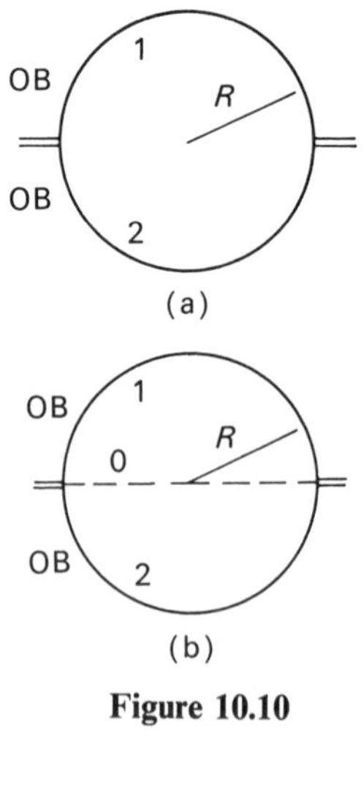


Figure 10.10

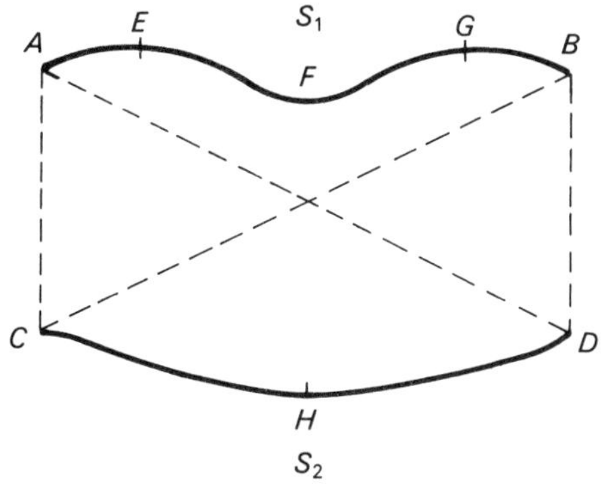


**Figure 10.11** Cross-section in $x$–$y$ plane.

The method can be generalized to much more complex cases: an example is Hottel's intersecting chords method [14]. We consider the shape factors between two cylindrical surfaces $S_1$ and $S_2$, whose generators are parallel to the $z$ axis, but whose cross-sections perpendicular to the $z$ axis are arbitrary (Figure 10.11).

It can be shown by using various fictitious surfaces (left as an exercise for the reader), and introducing chord lengths $AD$, $BC$, $AC$ and $BD$, that:

$$L_1 f_{12} = L_2 f_{21} = \frac{AD + BC - (AC + BD)}{2} \tag{10.67}$$

where $L_1$ and $L_2$ are the lengths of the arcs $AEFGB$ and $CHD$.

### 10.3.3 Differential shape factors – calculation technique

Consider the system made up of $S_1$ and $S_2$ facing each other, and the differential elements $dS_1$ and $dS_2$ (Figure 10.12). We can define various shape factors:

- from $dS_1$ to $dS_2$, denoted by $df_{dS_1 dS_2}$:

$$dS_1\, df_{dS_1 dS_2} \pi I_{1\lambda}\, d\lambda = I_{1\lambda} \frac{dS_1 \cos\theta_1\, dS_2 \cos\theta_2}{M_1 M_2^2}\, d\lambda \tag{10.68}$$

thus:

$$df_{dS_1 dS_2} = \frac{1}{\pi} \cos\theta_1\, d\Omega_1$$

with (10.69)

$$d\Omega_1 = \frac{dS_2 \cos\theta_2}{M_1 M_2^2}$$

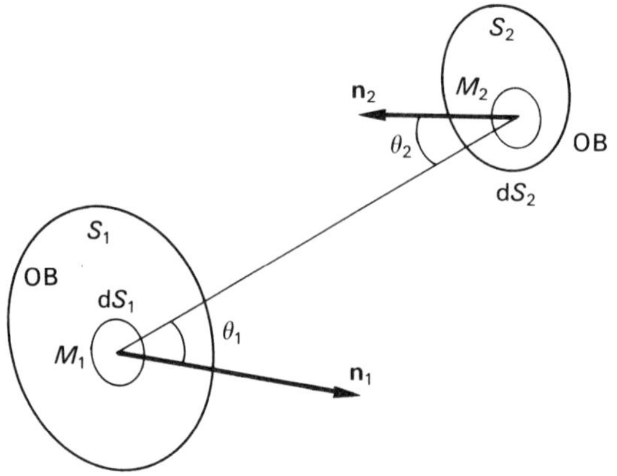


**Figure 10.12**

- from $dS_1$ to $S_2$, denoted by $f_{dS_1S_2}$:

$$f_{dS_1S_2} = \frac{1}{\pi} \int_{S_2} \frac{dS_2 \cos\theta_2 \cos\theta_1}{M_1M_2^2} \tag{10.70}$$

- from $S_1$ to $dS_2$, denoted by $df_{S_1dS_2}$:

$$df_{S_1dS_2} = \frac{dS_2}{\pi S_1} \int_{S_1} \frac{\cos\theta_1 \cos\theta_2 \, dS_1}{M_1M_2^2} \tag{10.71}$$

- and, of course, the global shape factor $f_{S_1S_2}$:

$$f_{S_1S_2} = \frac{1}{\pi S_1} \int_{S_1} dS_1 \int_{S_2} \frac{\cos\theta_1 \cos\theta_2}{M_1M_2^2} dS_2 \tag{10.72}$$

The differential shape factors can often be easily deduced from the global shape factors. As an example, we wish to calculate the shape factor $f_{dxdx'}$ between two elementary cylindrical segments of the inside of a tube of lengths $dx$ and $dx'$ (Figure 10.13). Suppose we know the shape factor $f_{xx'}$ between the two disks of radius $R$,

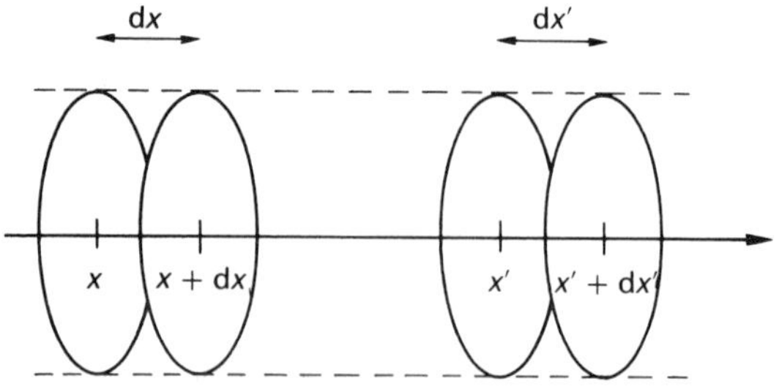


**Figure 10.13**

centred at $x$ and $x'$ respectively. In a straightforward manner:

$$df_{x\,dx'} = f_{xx'} - f_{xx'+dx'}$$

$$df_{x\,dx'} = \frac{\partial}{\partial x'} f_{xx'}\, dx' \qquad (10.73)$$

thus, using the reciprocity relation:

$$f_{dx'x} = \frac{R}{2}\frac{\partial}{\partial x'} f_{xx'} \qquad (10.74)$$

By an analogous argument we find:

$$df_{dx'dx} = \frac{R}{2}\frac{\partial}{\partial x'}\frac{\partial}{\partial x} f_{xx'}\, dx \qquad (10.75)$$

$$df_{dx\,dx'} = \frac{R}{2}\frac{\partial}{\partial x'}\frac{\partial}{\partial x} f_{x'x}\, dx' \qquad (10.76)$$

the shape factors likely to arise in heat transfer in a tube are thus readily deduced from the shape factor $f_{xx'}$ between two disks (see Part V: 'Basic Data').

## *10.4 Simple example of the radiative transfer method: cryogenic insulating structure*

In an evacuated region ($10^{-2}$–$10^{-6}$ torr), we often use a structure consisting of a stack of straws (cylindrical tubes) to limit the radiative transfer between the outside walls and the region which is to be maintained at a very low temperature. In the absence of conducto-convective heat transfer (vacuum) and considering the weakness of the conductive heat transfer (low conductivity of the 'straws' and poor thermal contact between the generators of adjacent straws), radiative heat exchange is important even at the very low temperatures considered.

We will consider a regular stack of straws as shown in Figure 10.14. The straws have an emissivity $\varepsilon$, independent of wavelength and direction. We will assume for simplicity that all the elements in half-layer $2n$ (made up of half-cylinders) are isothermal at $T_{2n}$, and satisfy hypotheses H1, H2 and H3. We will consider the emergent and incident flux densities $\varphi^{1'}_{2n}$, $\varphi^{i'}_{2n}$ and $\varphi^{1''}_{2n}$, $\varphi^{i''}_{2n}$, respectively, according to whether we are considering the interior or the exterior of the cylinders. We may then write:

1. Inside one cylinder (Figure 10.15):

$$\varphi^{1'}_{2n} = \varepsilon\sigma T^4_{2n} + (1-\varepsilon)\varphi^{i'}_{2n} \qquad (10.77)$$

$$\varphi^{i'}_{2n} = f\varphi^{1'}_{2n} + f'\varphi^{1'}_{2n+1} \qquad (10.78)$$

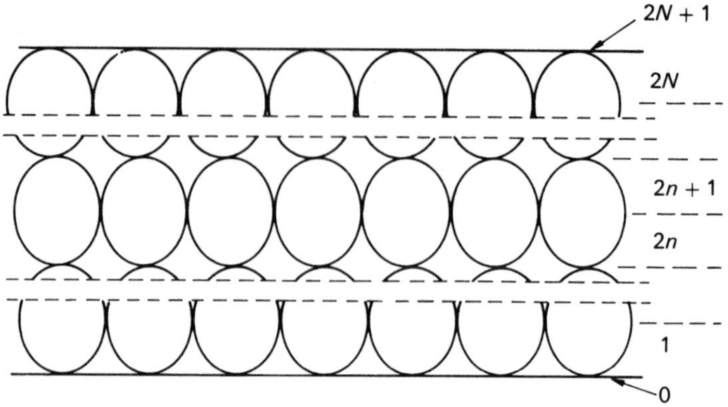


Figure 10.14

where $f = f_{2n2n} = 1 - 2/\pi$ from (10.65), and $f' = f_{2n2n+1} = 2/\pi$ from (10.64), and:

$$\varphi^{l'}_{2n+1} = \varepsilon\sigma T^4_{2n+1} + (1-\varepsilon)\varphi^{i'}_{2n+1} \tag{10.79}$$

$$\varphi^{i'}_{2n+1} = f\varphi^{l'}_{2n+1} + f'\varphi^{l'}_{2n} \tag{10.80}$$

the radiative flux density, positive in the $z$ direction, is:

$$\varphi^{R'}_{2n} = \varphi^{l'}_{2n} - \varphi^{i'}_{2n} = \frac{(1-f')\varepsilon\sigma(T^4_{2n} - T^4_{2n+1})}{1-(1-\varepsilon)(2f-1)} \tag{10.81}$$

2. Outside four quarter cylinders (Figure 10.16).
   The shape factors needed can be calculated using the fictitious surface method:

$$f_{01} = 1; \qquad f_{10} = \frac{S_0}{S_1} = \frac{2R}{\pi R} = \frac{2}{\pi} \tag{10.82}$$

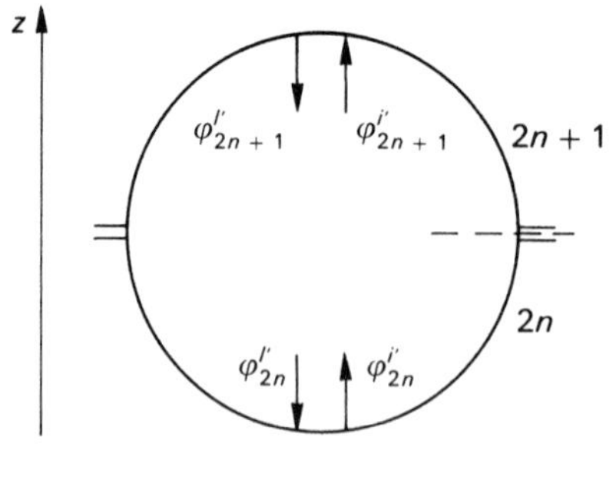


Figure 10.15

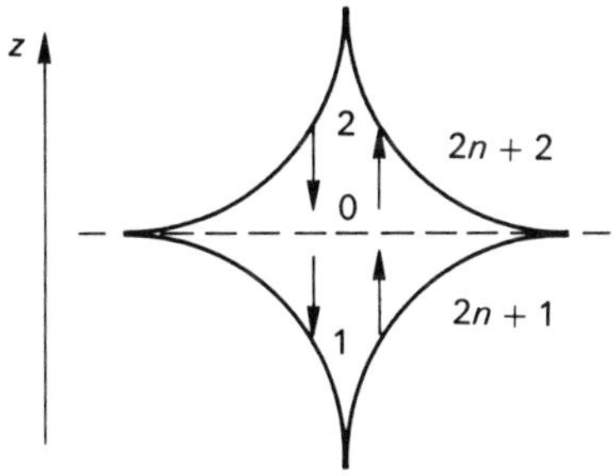


**Figure 10.16**

whence:

$$f_{2n+2\,2n+1} = f_{10} = \frac{2}{\pi} = f' \tag{10.83}$$

$$f_{2n+2\,2n+2} = 1 - f_{2n+2\,2n+1} = 1 - \frac{2}{\pi} = f \tag{10.84}$$

The relations between the flux densities $\varphi^{i''}_{2n+1}$, $\varphi^{1''}_{2n+1}$, etc., are analogous to equations (10.77)–(10.80), giving:

$$\varphi^{R''}_{2n+1} = \varphi^{1''}_{2n+1} - \varphi^{i''}_{2n+1} = \frac{(1-f')\varepsilon\sigma(T^4_{2n+1} - T^4_{2n+2})}{1-(1-\varepsilon)(2f-1)} \tag{10.85}$$

Under steady conditions, the radiative fluxes per unit surface area $\varphi^{R'}$ and $\varphi^{R''}$ are constants from $n = 1$ to $n = 2N$ if transfers other than radiation are neglected, and satisfy:

$$\frac{2}{\pi}\varphi^R_0 = \varphi^{R'}_2 = \varphi^{R''}_3 = \varphi^{R'}_4 = \varphi^{R''}_5 = \cdots = \varphi^{R'}_{2n} = \varphi^{R''}_{2n+1} = \cdots = \varphi^R_{2N+1}\left(\frac{2}{\pi}\right) \tag{10.86}$$

which leads to the required relationships between the temperatures $T_0$, $T_1$, . . ., $T_{2N+1}$.

## *10.5 Generalization of the method*

### *10.5.1 Generalization for partially transparent walls*

10.5.1.1 SOME CONCEPTS FOR SEMI-TRANSPARENT MEDIA

We will not treat the general case of semi-transparent walls in this section (see Chapter 11). We will restrict ourselves to partially transparent walls which are non-absorbent or absorbent but isothermal, **with isotropic radiative properties.**

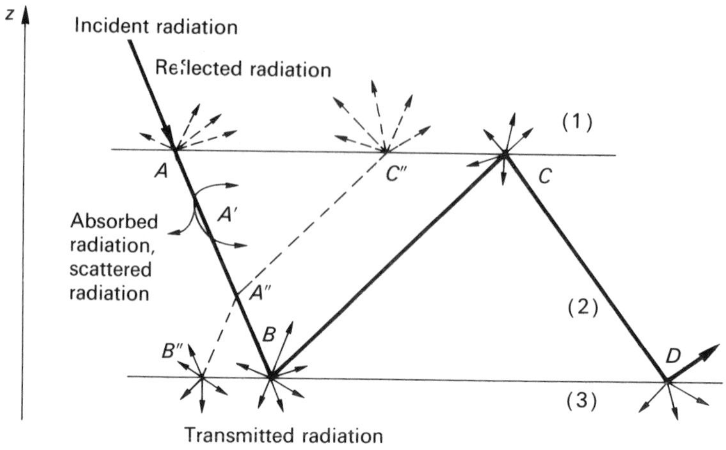


**Figure 10.17**

We will define, for the first time, exactly what we mean by transmitted, absorbed or reflected radiation for such a material, using the example shown in Figure 10.17.

Radiation incident on the surface at $A$ is partly reflected into medium (1) with a complex index, the remainder is propagated through medium (2) (also with a complex index). We will just follow the principal path from $A$ to $B$; between $A$ and $B$, the radiation may be absorbed at any point, such as the single point $A'$ in Figure 10.17. It may also be scattered in other directions at any point, as at point $A''$. Some scattered rays reach the upper face of the wall ($C''$) and others the lower face ($B''$). The part which is neither absorbed nor scattered arrives at $B$, where it is partly internally reflected into medium (2) and partly transmitted into medium (3). If we follow the radiation reflected from $B$ towards $C$, we find the same phenomena as for $A$ to $B$: absorption, scattering, internal reflection (towards $D$) or transmission (in medium (1) this time), and so on.

Reflected radiation is defined as all the radiation originating from the ray incident at $A$ which finally leaves in the positive $z$ direction in medium (1) through the interface 1–2 (whatever path was followed by these rays). We define the **global reflection factor, or reflectance,** $R_\lambda$ as the ratio of the total reflected flux $d\Phi_\lambda^r$ to the incident flux $d\Phi_\lambda^i$:

$$R_\lambda = \frac{d\Phi_\lambda^r}{d\Phi_\lambda^i} \tag{10.87}$$

Similarly, if we introduce the global transmitted flux $d\Phi_\lambda^t$ in (3) in the negative $z$ direction and the global absorbed flux $d\Phi_\lambda^a$, we can define the global transmission factor or **transmittance** $T_\lambda$:

$$T_\lambda = \frac{d\Phi_\lambda^t}{d\Phi_\lambda^i} \tag{10.88}$$

and the global absorption factor or **absorbence** $A_\lambda$:

$$A_\lambda = \frac{d\Phi_\lambda^a}{d\Phi_\lambda^i} \tag{10.89}$$

Conservation of energy obviously requires: $d\Phi_\lambda^i = d\Phi_\lambda^r + d\Phi_\lambda^t + d\Phi_\lambda^a$, or:

$$R_\lambda + T_\lambda + A_\lambda = 1 \tag{10.90}$$

Here we are only considering isotropic radiative properties, for simplicity. $T_\lambda$ and $A_\lambda$ are independent of direction and $R_\lambda$ is independent of both the direction of incidence and that of emergence.

The global radiation emitted by the plate is the sum of the radiation emitted from all points on the surface (medium (2)) which propagate in medium (1) (in the positive $z$ direction) or in medium (3) (in the negative direction), which may have been subject to transmission, diffusion or multiple reflection inside medium (2). In general, medium (2) is not isothermal, and the fluxes emitted into (1) and (2) are different. Because we have no temperature reference, we cannot define a global emission factor (analogous to $A_\lambda$). The calculation of the emitted flux is complex; there is coupling between the temperature field and the intensity field inside the material (see the discussion in the section on unsteady conduction in Chapter 5).

If we limit ourselves to the particular case of an **isothermal** plate (for example, in unsteady conditions, such that $Bi \ll 1$), the problem is simplified. It is possible to define a **global emission factor** $E_\lambda$ by reference to the plate temperature $T$, such that the monochromatic flux density emitted (both towards $z > 0$ and $z < 0$) is:

$$d\varphi_\lambda^e = E_\lambda \pi I_\lambda^0(T)\, d\lambda \tag{10.91}$$

We always consider the simplifying case of isotropic radiative properties. We will show (from equilibrium considerations) that in this particular case the global absorption and emission factors are related:

$$A_\lambda = E_\lambda \tag{10.92}$$

10.5.1.2 CALCULATION OF RADIATIVE FLUX

If in a closed region one of the walls is partially transparent and at temperature $T_0$ (wall 0 in Figure 10.18), it is possible to generalize the calculation method of this chapter. (Hypotheses H1, H2 and H3 are satisfied – see Section 10.2.1.)

We distinguish between the lower and upper faces of the wall (even if it is infinitely thin), and consider **two sub-systems**:

- **Sub-system I** is the inside of the region; the radiation at the wall has intensities $I_{0\lambda}^l$ and $I_{0\lambda}^i$.

$$I_{0\lambda}^i = \sum_k f_{0k} I_{k\lambda}^l; \qquad k = 0, 1, \ldots, N \tag{10.93}$$

$$I_{0\lambda}^l = \underset{\text{emitted}}{E_\lambda I_\lambda^0(T_0)} + \underset{\text{reflected}}{R_\lambda I_{0\lambda}^i} + \underset{\text{transmitted}}{I_{0\lambda}^t} \tag{10.94}$$

where the intensity $I_{0\lambda}^t$ is the radiation issued from II and transmitted through I.

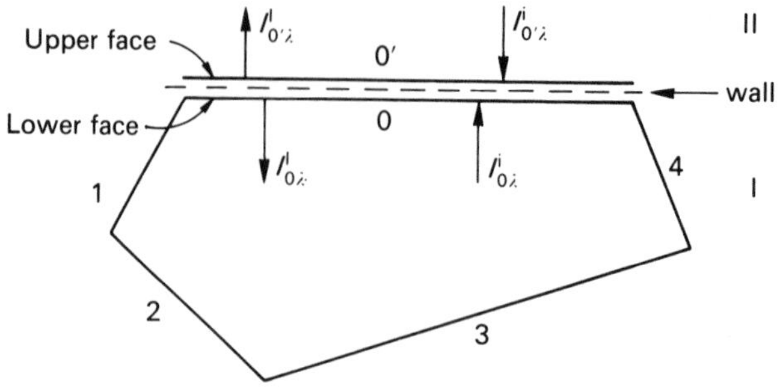


Figure 10.18

- **Sub-system II** comprises the upper face of the wall and the surroundings. The radiation above the wall is characterized by the two intensities $I^{l}_{0'\lambda}$ and $I^{i}_{0'\lambda}$; the intensity $I^{i}_{0'\lambda}$ depends only on the surroundings, while the intensity $I^{l}_{0'\lambda}$ is given by:

$$I^{l}_{0'\lambda} = \underset{\text{emitted}}{E_\lambda I^0_\lambda(T_0)} + \underset{\text{reflected}}{R_\lambda I^{i}_{0'\lambda}} + \underset{\text{transmitted}}{I^{t}_{0'\lambda}} \tag{10.95}$$

The coupling between the two sub-systems is through the equations of the transmitted radiation:

$$I^{t}_{0\lambda} = T_\lambda I^{i}_{0'\lambda} \tag{10.96}$$

$$I^{t}_{0'\lambda} = T_\lambda I^{i}_{0\lambda} \tag{10.97}$$

10.5.1.3 EXAMPLE: OVEN WINDOW

Let us consider a window, made of safety glass, installed opposite a liquid metal surface. For simplicity, we will consider the geometry to be two infinite parallel planes (Figure 10.19).

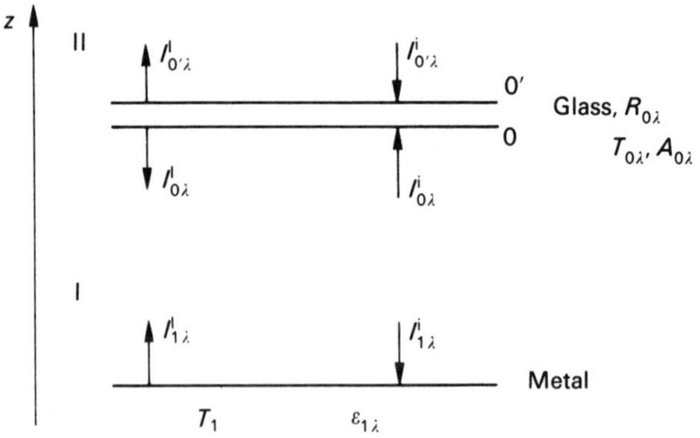


Figure 10.19

We have the following formulae:

- **Sub-system II**:

$$I^{i}_{0'\lambda} = I^{0}_{\lambda}(T_e) \qquad \text{(ambient radiation)} \tag{10.98}$$

$$I^{1}_{0'\lambda} = A_{0\lambda}I^{0}_{\lambda}(T_0) + R_{0\lambda}I^{i}_{0'\lambda} + T_{0\lambda}I^{i}_{0\lambda} \tag{10.99}$$

- **Sub-system I**:

$$I^{1}_{1\lambda} = \varepsilon_{1\lambda}I^{0}_{\lambda}(T_1) + (1 - \varepsilon_1\lambda)I^{i}_{1\lambda} \tag{10.100}$$

$$I^{i}_{1\lambda} = I^{1}_{0\lambda} \tag{10.101}$$

$$I^{i}_{0\lambda} = I^{1}_{1\lambda} \tag{10.102}$$

$$I^{1}_{0\lambda} = A_{0\lambda}I^{0}_{\lambda}(T_0) + R_{0\lambda}I^{i}_{0\lambda} + T_{0\lambda}I^{i}_{0'\lambda} \tag{10.103}$$

This is a linear system of six equations in six unknowns, whose solution is the radiative flux (in the steady state and in the absence of other types of heat transfer):

$$\varphi = \int_0^\infty \pi(I^{1}_{1\lambda} - I^{i}_{1\lambda})\,d\lambda = \int_0^\infty \pi(I^{i}_{0\lambda} - I^{1}_{0\lambda})\,d\lambda = \int_0^\infty \pi(I^{p}_{0'\lambda} - I^{1}_{0'\lambda})\,d\lambda \tag{10.104}$$

so that for example:

$$I^{1}_{1\lambda} - I^{i}_{1\lambda} = \frac{\varepsilon_{1\lambda}[(1 - R_{0\lambda})I^{0}_{\lambda}(T_1) - A_{0\lambda}I^{0}_{\lambda}(T_0) - T_{0\lambda}I^{0}_{\lambda}(T_e)]}{1 - R_{0\lambda}(1 - \varepsilon_{1\lambda})} \tag{10.105}$$

Clearly, in thermal equilibrium ($T_0 = T_1 = T_e$) the numerator disappears ($A_{0\lambda} + R_{0\lambda} + T_{0\lambda} = 1$). Further, if medium 0 is opaque ($T_{0\lambda} = 0$, $A_{0\lambda} = \varepsilon_{0\lambda} = 1 - R_{0\lambda} = 1 - \rho_{0\lambda}$) the expression becomes:

$$I^{1}_{1\lambda} - I^{i}_{1\lambda} = \frac{\varepsilon_{1\lambda}\varepsilon_{0\lambda}[I^{0}_{\lambda}(T_1) - I^{0}_{\lambda}(T_0)]}{1 - (1 - \varepsilon_{0\lambda})(1 - \varepsilon_{1\lambda})} \tag{10.106}$$

thus coming back to result (10.13).

*Note:*
We notice that for glass (averaging over $\theta$):

$\lambda \leqslant 3$ to $4\ \mu\text{m}$: $A_{0\lambda} \simeq 0$; $R_{0\lambda} \simeq 0.05$; $T_{0\lambda} \simeq 0.95$ ('transparent' body)

$\lambda > 3$ to $4\ \mu\text{m}$: $A_{0\lambda} \simeq 0.95$; $R_{0\lambda} \simeq 0.05$; $T_{0\lambda} \simeq 0$ (opaque body)

### *10.5.2 Generalization to directional incident radiation(s)*

#### 10.5.2.1 THE METHOD

In certain applications, directional radiation sources are involved (for example, the portion of solar radiation which is not scattered by the atmosphere). For this type of source, the flux received by the surface cannot be calculated by the shape factor method; the calculation must be carried out directly in terms of geometrical considerations, perhaps having to take account of shadows.

**The assumption of isotropy of the radiative properties of opaque bodies** ($\varepsilon_\lambda$, $\rho_\lambda$, $\alpha_\lambda$) **can always be made under these conditions**, so that after the first reflection from a surface, the flux leaving that surface has an isotropic intensity, and the model of Chapter 9 can be applied to its contributions. The incident radiation on a surface $S_j$ has the following components:

1. Radiation leaving the surfaces $S_k$, (which have isotropic radiative properties) which surround $S_j$. As in Section 10.2, we will introduce an equivalent isotropic intensity $I^{\text{i}}_{j\lambda}$ to describe this radiation:

$$I^{\text{i}}_{j\lambda} = \sum_k f_{jk} I^{\text{l}}_{k\lambda} \tag{10.107}$$

   where $I^{\text{l}}_{k\lambda}$ is the isotropic intensity of the radiation leaving $S_k$.
2. Directional radiation, which, for a single point source, appears only in an elementary solid angle d$\Omega$. This radiation has a intensity $I'^{\text{idir}}_{j\lambda}$. The only difficulty with this type of problem is to evaluate $I'^{\text{idir}}_{j\lambda}$ and the fraction of the flux associated with this intensity which is reflected by the surface $S_j$. We will illustrate the method with an example.

### 10.5.2.2 EXAMPLE: RADIATIVE FLUX IN A BUILDING

**Problem definition**

We will consider a building, very long in the $x$ direction, whose cross-section in the $yz$ plane is shown in Figure 10.20.

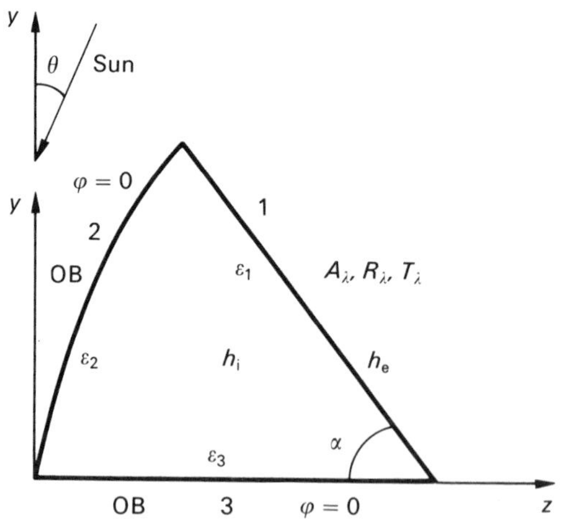


**Figure 10.20**

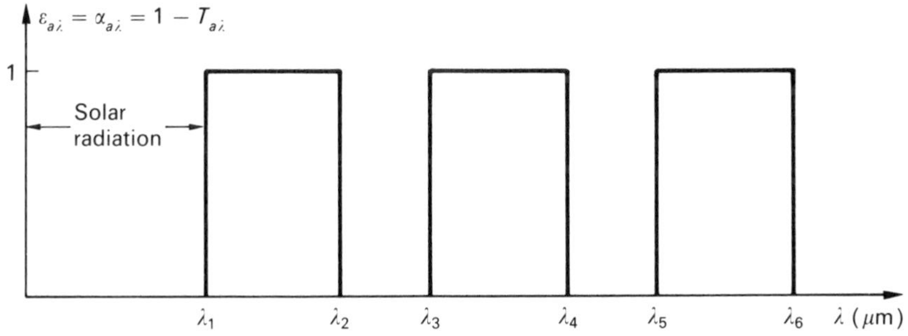


**Figure 10.21**

The glazed surface 1 is flat (inclined at an angle $\alpha$ to the horizontal), and receives directional solar radiation with intensity $I_\lambda'^S$ at an angle $\theta$ with the vertical. To describe the glass (which is assumed to be non-absorbent for the solar spectrum) we will introduce a directional transmittance $T_\lambda'$ for the direction of incidence $(\alpha - \theta)$, and a transmittance $T_\lambda$, a reflectance $R_\lambda$ and an absorbence $A_\lambda$ for non-directional radiation, related by:

$$A_\lambda + R_\lambda + T_\lambda = 1$$

In practice, we will set:

$$\lambda \leqslant \lambda_1: \quad A_\lambda = 0; \quad T_\lambda = T^S; \quad R_\lambda = 1 - T^S$$

$$\lambda \geqslant \lambda_1: \quad A_\lambda = \varepsilon_1; \quad R_\lambda = 1 - \varepsilon_1$$

The internal faces 2 and 3 are grey and are characterized by isotropic emissivities $\varepsilon_2$ and $\varepsilon_3$. The internal air[4], which is humid, and atmosphere are approximated by the radiative properties shown in Figure 10.21. All the solar radiation received is assumed to be transmitted between 0 and $\lambda_1$; the convective heat transfer coefficient between the internal air at $T_i$ and the walls is $h_i$, and that between the external air at $T_e$ and wall 1 by forced convection is $h_e$. Walls 2 and 3 are assumed to be, to a first approximation, thermally insulated. Derive the thermal balance for the system.

**Modelling the radiative transfer**

The simplest model for the radiative transfer consists of **separating the internal faces of the building** into four surfaces $S_1$, $S_2$, $S_{3'}$ and $S_{3''}$, as shown on Figure 10.22.

We split 3 into two parts to allow for the shadow – the directional radiation transmitted by 1 does not fall on 3″.

Using the usual notation, and splitting the spectrum into two zones:

$$\lambda \leqslant \lambda_1; \qquad \lambda > \lambda_1$$

[4] This model requires that the thicknesses of air concerned exceed a few metres, and that the air is rich in $CO_2$ and $H_2O$.

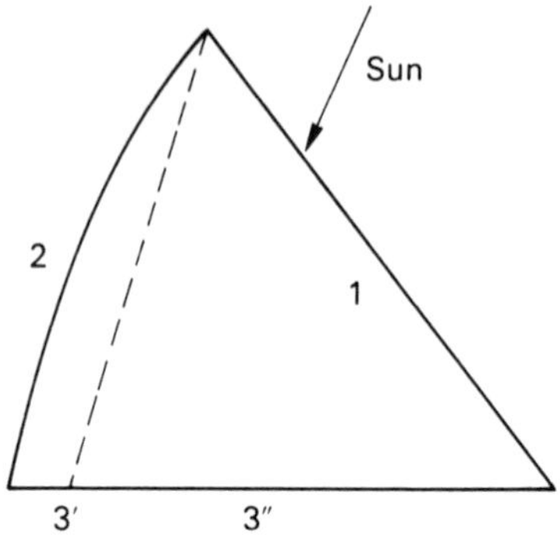


**Figure 10.22**

We can write down the following relationships:

1. Solar spectrum ($\lambda \leqslant \lambda_1$)
   - Surface 1 (internal face):

$$I'^{\text{ldir}}_{1\lambda} = T'_\lambda I'^{\text{S}}_\lambda \qquad \text{(directional portion)} \tag{10.108}$$

$$I^{\text{l}}_{1\lambda} = (1 - T^{\text{S}}_\lambda) I^{\text{i}}_{1\lambda} \qquad \text{(isotropic portion)} \tag{10.109}$$

$$I_{1\lambda} = \sum_k f_{1k} I^{\text{l}}_{k\lambda} \qquad k = 2, 3', 3'' \qquad \text{(isotropic portion)} \tag{10.110}$$

   - Surface 3″ (internal face):

$$I'^{\text{idir}}_{3''\lambda} = I'^{\text{ldir}}_{1\lambda} = T'_\lambda I'^{\text{S}}_\lambda \qquad \text{(directional)} \tag{10.111}$$

Notice that for this directional radiation, shape factors do not appear in equation (10.111). The use of intensities assures conservation of energy.

$$I^{\text{i}}_{3''\lambda} = \sum_k f_{3''k} I^{\text{l}}_{k\lambda} \qquad k = 1, 2 \text{ (isotropic)} \tag{10.112}$$

$$I^{\text{l}}_{3''\lambda} = (1 - \varepsilon_3)(I^{\text{i}}_{3''\lambda} + I'^{\text{idir}}_{3''\lambda}\zeta) \quad \text{(isotropic)} \quad \text{with: } \zeta = \frac{\cos\theta \, d\Omega}{\pi} \tag{10.113}$$

If the incident radiation on 3″ is partly directional ($I'^{\text{idir}}_{3''\lambda}$) and partly diffuse ($I^{\text{i}}_{3''\lambda}$), **all the radiation leaving 3 is isotropic**.

The part of the radiation emitted by 3 (as for the other surfaces) in the solar range is clearly negligible.

   - Surfaces 2 and 3′ (internal faces):

$$I^{\text{i}}_{2\lambda} = \sum_k f_{2k} I^{\text{l}}_{k\lambda} \tag{10.114}$$

$$I^{\text{l}}_{2\lambda} = (1 - \varepsilon_2) I^{\text{i}}_{2\lambda} \tag{10.115}$$

$$I^{i}_{3'\lambda} = \sum_{k} f_{3'k} I^{l}_{k\lambda} \tag{10.116}$$

$$I^{l}_{3'\lambda} = (1 - \varepsilon_3) I^{i}_{3'\lambda} \tag{10.117}$$

We notice that the radiation emitted by the walls in the range $[0, \lambda_1]$ is negligible.

The radiative fluxes in the interval $[0, \lambda_1]$ on the surfaces 2, 3′, 3″ are:

$$k = 2, 3': \quad \varphi_k^{[0,\lambda_1]} = \int_0^{\lambda_1} \pi (I^{l}_{k\lambda} - I^{i}_{k\lambda})\, d\lambda \tag{10.118}$$

$$k = 3'': \quad \varphi_{3''}^{[0,\lambda_1]} = \int_0^{\lambda_1} \pi (I^{l}_{3''\lambda} - I^{i}_{3''\lambda} - I'^{i\,\mathrm{dir}}_{3''\lambda} \zeta)\, d\lambda \tag{10.119}$$

The radiation does not interact with the glazing

2. Infra-red region ($\lambda > \lambda_1$)
   In this region there is no longer any directional radiation; the glass is now an opaque body and the air is transparent or black in bands.
   (2a) Zones where the air is transparent. Here we have the relations:

$$k = 1, 2, 3', 3'': \quad I^{l}_{k\lambda} = \varepsilon_k I^0_\lambda(T_k) + (1 - \varepsilon_k) I^{i}_{k\lambda} \tag{10.120}$$

$$k = 1, 2, 3', 3'': \quad I^{i}_{k\lambda} = \sum_{k'} f_{kk'} I^{l}_{k'\lambda} \tag{10.121}$$

$$k' = 1, 2, 3', 3'': \quad \varphi_k^{(2a)} = \int_{\textit{transparent bands}} \pi (I^{l}_{k\lambda} - I^{i}_{k\lambda})\, d\lambda \tag{10.122}$$

   Notice that in this region, wall 1 emits radiation to its surroundings (without receiving any radiation in exchange – this is a special case).
   (2b) Zones where the air is absorbent. Each of the internal surfaces sees a black body ($\varepsilon_{a\lambda} = 1$) at temperature $T_i$, and similarly the outside of face 1 sees a black body at temperature $T_e$.
   We have:

$$k = 1, 2, 3', 3'': \quad \varphi_k^{(2b)} = \int_{\textit{absorption bands}} \varepsilon_k \pi (I^0_\lambda(T_k) - I^0_\lambda(T_i))\, d\lambda \tag{10.123}$$

   The exchange between wall 1 and the atmosphere is given by:

$$\varphi^{(2b)}_{1-\mathrm{atm}} = \int_{\textit{absorption bands}} \varepsilon_1 \pi (I^0_\lambda(T_1) - I^0_\lambda(T_e))\, d\lambda \tag{10.124}$$

**Thermal balance**

1. Balance for the walls:

- Walls 2, 3′ and 3″:

$$k = 2, 3' \text{ or } 3'': \quad \varphi_k^{[0,\lambda_1]} + \varphi_k^{(2a)} + \varphi_k^{(2b)} + h_1 (T_k - T_i) = 0 \tag{10.125}$$

These three expressions follow from surfaces $S_2$, $S_{3'}$ and $S_{3''}$ being thermally insulated.

- Wall 1:

$$\varphi_1^{(2a)} + \varphi_1^{(2b)} + \varphi_{1\,atm}^{(2b)} + h_i(T_1 - T_i) + h_e(T_1 - T_e) = 0 \qquad (10.126)$$

2. Balance for the inside air:

$$\sum_k S_k[h_i(T_k - T_i) + \varphi_k^{(2b)}] = 0 \qquad (10.127)$$

Temperatures $T_1$, $T_2$, $T_{3'}$, $T_{3''}$ and $T_i$ can be obtained by solution of the five balance equations ((10.125)–(10.128)) by an iterative numerical technique.

# 11

# Semi-transparent Media

## *11.1 General*

Any (semi-transparent) medium is characterized optically by a complex refractive index $\hat{n}$ given by $n - j\chi$. Here, $n$ is the real index and $\chi$ the extinction index of the medium (see Section 9.1 for a discussion of the relative magnitudes of $n$ and $\chi$). The index $\hat{n}$ varies, in semi-transparent media, with position **r** and, in general, with direction **Δ**. In fact, it was demonstrated in Section 9.1.1.4 that, **for semi-transparent media,** $\chi$ **is generally small in comparison to** $n$ (for air, glass, etc.). We will use this characteristic in the following, and only consider the real index $n$ when determining the geometric properties of rays. It is, however, obvious that the bulk absorption of the medium which gives it its semi-transparent character is closely linked to $\chi$, and more precisely, to the complex part of the permittivity $\varepsilon_2$. Because of this, since the frequency of radiation propagating in a medium is invariant in some reference system, its wavelength will be a function of **r** and **Δ**:

$$\lambda(\mathbf{r}, \mathbf{\Delta}) = \frac{c(\mathbf{r}, \mathbf{\Delta})}{v} = \frac{c_0}{vn(\mathbf{r}, \mathbf{\Delta})} \qquad \text{with} \qquad n = \frac{c_0}{c} \tag{11.1}$$

where $c$ and $c_0$ represent the celerities of radiation in the medium and in a vacuum respectively (as a first approximation, $n$ is 1 to a few times $10^{-4}$ for air and more generally for all gases). It is therefore no longer possible to use $\lambda$ to characterize radiation in an arbitrary semi-transparent medium; instead we use either frequency $v$, measured in Hz and its multiples (which is the usage in the so-called Hertzian domain of microwaves) or the wave number in a vacuum, $\sigma$, given by $v/c_0$, measured usually in $\text{cm}^{-1}$ or in kaisers (1 kaiser $= 10^{-3}\ \text{cm}^{-1}$) (which is the convention in the infra-red domain).

The intensity of equilibrium radiation depends on both **r** and **Δ**, and is given by [15].

$$I_v^0(\mathbf{\Delta}, \mathbf{r}, T) = \frac{2hv^3}{c^2(\mathbf{r}, \mathbf{\Delta})}\left[\exp\left(\frac{hv}{kT}\right) - 1\right]^{-1} = n^2(\mathbf{r}, \mathbf{\Delta})I_v^0(T) \tag{11.2}$$

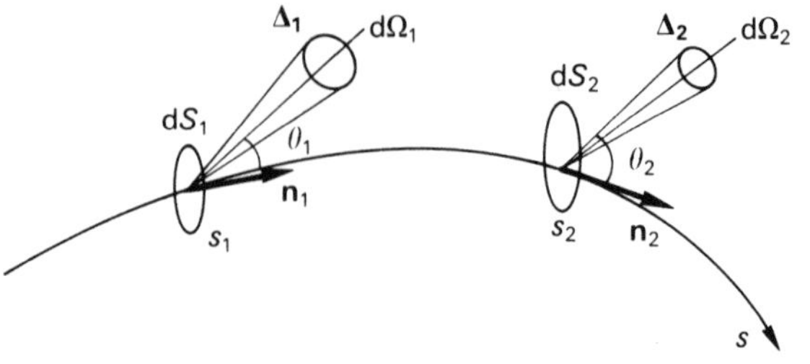


**Figure 11.1**

where $I_\nu^0(T)$ is the isotropic intensity of equilibrium radiation in a vacuum (or in a material with index $n$ equal to 1, such as a gas). The equilibrium intensity given by (11.2) is no longer isotropic except in so far as the index $n$ of the material is isotropic, which it generally is not (e.g. bi-refringent materials).

Result (11.2) can be generalized to any intensities in a transparent material. Consider surface elements $dS_1$ and $dS_2$ and elementary solid angles $d\Omega_1$ and $d\Omega_2$, with principal directions $\boldsymbol{\Delta}_1$ and $\boldsymbol{\Delta}_2$ making angles $\theta_1$ and $\theta_2$ with the normals to the surfaces; all these elements are optically related along a curved optical trajectory with curvilinear coordinate $s$ (Figure 11.1).

Clausius' relationship for the conservation of the optical quantity $n^2(\mathbf{r}, \boldsymbol{\Delta})\, dS \cos\theta\, d\Omega$ (where $n$ is the real index of the material) along a ray pencil [16] is written:

$$n_1^2\, dS_1 \cos\theta_1\, d\Omega_1 = n_2^2\, dS_2 \cos\theta_2\, d\Omega_2 \tag{11.3}$$

The real index is likely to vary between points $s_1$ and $s_2$. In addition, in the same material, conservation of monochromatic energy flux in the spectral interval $d\nu$ gives (introducing intensities $I'_{1\nu}$ and $I'_{2\nu}$ at $s_1$ and $s_2$):

$$d^5\Phi'_\nu = I'_{1\nu}\, dS_1 \cos\theta_1\, d\Omega_1\, d\nu = I'_{2\nu}\, dS_2 \cos\theta_2\, d\Omega_2\, d\nu \tag{11.4}$$

From these two, it follows that for a transparent medium:

$$\frac{I'_{1\nu}}{n_1^2} = \frac{I'_{2\nu}}{n_2^2} \tag{11.5}$$

This gives a result analogous to (11.2): the intensity in a transparent medium is proportional to the square of the local real index of the material.

## 11.2 Absorption, emission and scattering

We are concerned with any medium likely to absorb, emit or scatter radiation in any element of volume.

### 11.2.1 Absorption

Consider the propagation along a curve $s$. The flux $d^5\Phi'_\nu(s)$ which propagates in a solid angle $d\Omega$ centred on $Os$ through an element of area $dS$, normal to the optical path, is partly absorbed. The (arithmetic) flux absorbed is given by:

$$-d^6\Phi'^{a}_\nu = d^6\Phi'_\nu(s) = -\kappa_\nu\, d^5\Phi'_\nu(s)\, ds \tag{11.6}$$

where $\kappa_\nu$ is the monochromatic absorption coefficient, with dimensions of inverse length, and generally expressed in $cm^{-1}$.

*Note: particular case of an elementary column*

Consider a straight **homogeneous and isothermal column**, of length $l$, and of a semi-transparent material (Figure 11.2). The monochromatic absorption coefficient $\kappa_\nu$ will be constant.

Considering absorption alone between $s = 0$ and $s = l$, the flux transmitted at $s = l$ can be written in terms of the incident flux in the $Os$ direction at $s = 0$ by integration of (11.6):

$$d^5\Phi'_\nu(l) = \exp(-\kappa_\nu l)\, d^5\Phi'_\nu(0) \tag{11.7}$$

The transmission and absorption factors (transmissivity and absorbtivity) of the column are obtained from this:

$$\tau'_\nu = 1 - \alpha'_\nu = \exp(-\kappa_\nu l) \tag{11.8}$$

This is known as Beer's law. This configuration is of interest, because any calculation in a heterogeneous, non-isothermal material can be reduced to it by numerical discretization.

### 11.2.2 Emission

Any scattering or **non-scattering** medium which can absorb radiation can also emit it. The emitted flux can be written quite generally:

$$d^6\Phi'^{e}_\nu = \eta_\nu(s)\, dV\, d\Omega\, d\nu \tag{11.9}$$

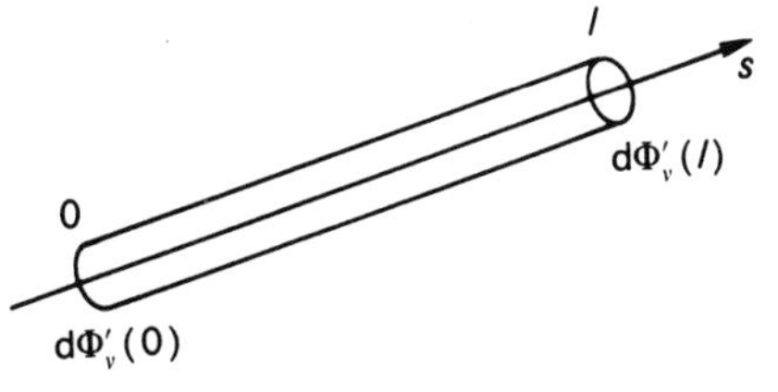


**Figure 11.2**

It is proportional to the element of volume $dV$, the elementary emission solid angle $d\Omega$, and to $d\nu$; $\eta_\nu(s)$ is called the monochromatic emission coefficient at $s$. If we assume thermal equilibrium in the material, there must be an equilibrium radiation with intensity $n^2 I_\nu^0(T)$ everywhere. Furthermore, in thermal equilibrium, the flux absorbed must be equal to the flux emitted:

$$d^6\Phi_\nu'^{a} = \kappa_\nu n^2 I_\nu^0(T)\, dS\, d\Omega\, ds\, d\nu = d^6\Phi_\nu'^{e} = \eta_\nu(s)\, dV\, d\Omega\, d\nu \tag{11.10}$$

Since we are looking at a solid angle $d\Omega$ whose principal direction is normal to $dS$, $dV$ is equal to $dS\, ds$, and it follows, in equilibrium:

$$\eta_\nu(s) = \kappa_\nu n^2 I_\nu^0(T) \tag{11.11}$$

This relationship can be generalized in mild dis-equilibrium (LTE – see the discussion of this idea in Chapter 1). In this case, the flux emitted by the element of volume $dS\, ds$ in the solid angle $d\Omega$, normally to $dS$, is given by:

$$d^6\Phi_\nu'^{e} = \kappa_\nu n^2 I_\nu^0(T)\, dS\, ds\, d\Omega\, d\nu \tag{11.12}$$

*Note: particular case of an elementary column*

Consider again a straight **homogeneous and isothermal column**, of length $l$, made of a material characterized by a constant absorption coefficient $\kappa_\nu$ (Figure 11.2).

**In thermal equilibrium, at temperature** $T$, the incident flux (at $s = 0$) and the emergent flux (at $s = l$) are characterized by the intensity $n^2 I_\nu^0(T)$. If we take $dS$ normal to $Os$, then:

$$d^5\Phi_\nu'(0) = d^5\Phi_\nu'(l) = n^2 I_\nu^0(T)\, dS\, d\Omega\, d\nu \tag{11.13}$$

Since $d^5\Phi_\nu'(l)$ is made up of one flux emitted and one flux transmitted by the column:

$$d^5\Phi_\nu'(l) = d^5\Phi_\nu'^{e}(l) + d^5\Phi_\nu'^{t}(l) \tag{11.14}$$

where:

$$d^5\Phi_\nu'^{t}(l) = \tau_\nu' d^5\Phi_\nu'(0) = \exp(-\kappa_\nu l) n^2 I_\nu^0(T)\, dS\, d\Omega\, d\nu \tag{11.15}$$

we can introduce the monochromatic emissivity of the **isothermal homogeneous** column of material between 0 and $l$ with the relation:

$$d^5\Phi_\nu'^{e} = \varepsilon_\nu' n^2 I_\nu^0(T)\, dS\, d\Omega\, d\nu \tag{11.16}$$

It therefore follows from (11.13) and (11.15):

$$\varepsilon_\nu' = 1 - \tau_\nu' = \alpha_\nu' = 1 - \exp(-\kappa_\nu l) \tag{11.17}$$

Equations (11.16) and (11.17) represent the global flux emitted and the global emissivity of the column of length $l$. They take account of all the auto-absorption effects (that is, of radiation emitted by an element $ds$ between $s$ and $s + ds$ and absorbed by any element $ds'$ between $s'$ and $s' + ds'$ (Figure 11.3)).

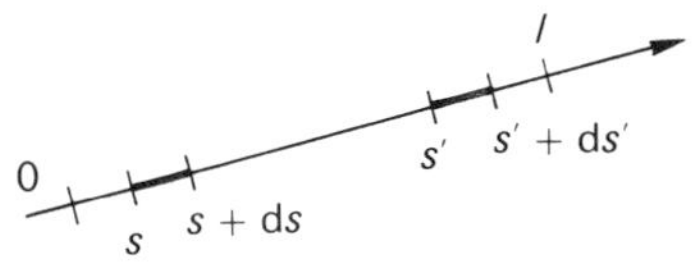


**Figure 11.3**

Equations (11.15) and (11.16) can be generalized to situations of mild dis-equilibrium (LTE assumption); the general expression for flux at $s$ is (in LTE) given by the intensity $I'_\nu(s)$:

$$I'_\nu(s) = \tau'_\nu I'_\nu(0) + (1 - \tau'_\nu) n^2 I^0_\nu(T) \tag{11.18}$$

The first term represents the intensity of the radiation transmitted by the column, the second that of the radiation emitted by the whole of it.

### 11.2.3 Scattering

Scattering of radiation is a volumetric phenomena which has two effects, considering only the flux $d^5\Phi'_\nu(s)$ propagating normal to $dS$ in the solid angle $d\Omega$:

- An attenuation effect: a fraction of the flux propagates at $s + ds$ in directions such as $\mathbf{\Delta}'$ which do not lie within the solid angle $d\Omega$ (Figure 11.4(a)).
- A constructive effect: some radiation which is not within $d\Omega$ at $s$ comes within that solid angle at $s + ds$ (Figure 11.4(b)).

The first of these is analogous to absorption from the energy point of view. The (arithmetic) flux reduction by scattering is given by:

$$d^6\Phi'^{s-}_\nu = \sigma_\nu \, d^5\Phi'_\nu(s) \, ds \tag{11.19}$$

where $\sigma_\nu$ is the scattering coefficient (dimension of inverse length, generally expressed in $cm^{-1}$). Taking account of (11.6) and (11.19), we can define generally an attenuation coefficient $\beta_\nu$ such that:

$$\beta_\nu = \kappa_\nu + \sigma_\nu \tag{11.20}$$

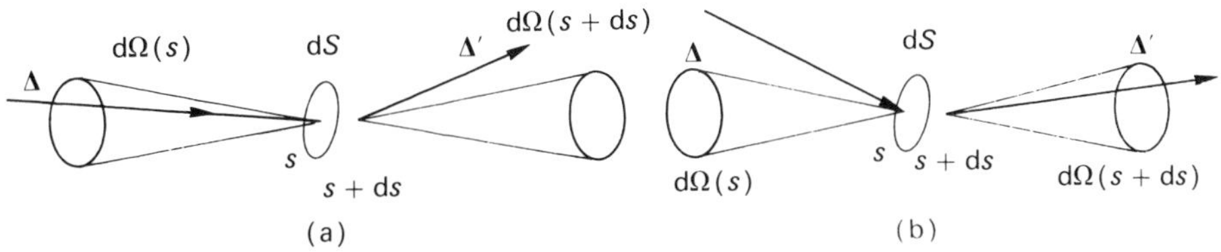


**Figure 11.4**

satisfying:

$$d^6\Phi_\nu'^- = \beta_\nu d^5\Phi_\nu'(s)\,ds \tag{11.21}$$

where $d^6\Phi_\nu'^-$ represents the arithmetic flux loss between $s$ and $s + ds$.

*Note: particular case of an elementary column*
The scattering coefficient $\sigma_\nu$ plays exactly the same role as the absorption coefficient $\kappa_\nu$. If we consider a homogeneous and isothermal column of length $l$, with scattering coefficient $\sigma_\nu$, the transmissivity associated with scattering is:

$$\tau_\nu' = \exp(-\sigma_\nu l) \tag{11.22}$$

and the transmissivity associated with the total attenuation is:

$$\tau_\nu' = \exp[-(\kappa_\nu + \sigma_\nu)l] \tag{11.23}$$

It is more difficult to take account of the constructive effects of scattering shown in Figure 11.4(b). With an incident flux at $s$ in the solid angle $d\Omega'$, centred in direction $\mathbf{\Delta}'$, there is associated an intensity $I_\nu'(\mathbf{\Delta}', s)$; this flux has a probability $ds\sigma_\nu p(\mathbf{\Delta}' \to \mathbf{\Delta}, s)\,d\Omega/4\pi$ of being scattered into the elementary solid angle $d\Omega$ centred in direction $\mathbf{\Delta}$; $p(\mathbf{\Delta}' \to \mathbf{\Delta}, s)$ is called the phase scattering function. The flux coming from an arbitrary point in space to point $s$ and scattered at $s + ds$ in the solid angle $d\Omega$, whose principal direction is normal to $dS$, is thus given by:

$$d^6\Phi_\nu'^{s+} = dS\,ds\,d\nu\,\frac{\sigma_\nu}{4\pi}\,d\Omega \int_0^{4\pi} p(\mathbf{\Delta}' \to \mathbf{\Delta}) I_\nu'(\mathbf{\Delta}', s) d\Omega' \tag{11.24}$$

*Note*
In thermal equilibrium (uniform temperature $T$) the constructive scattering term $d^6\Phi_\nu'^{d+}$ clearly compensates exactly for the attenuation scattering term $d^6\Phi_\nu'^{d-}$. Therefore, the phase function is normalized by the relation:

$$\frac{1}{4\pi}\int_0^{4\pi} n^2(\mathbf{\Delta}', s) p(\mathbf{\Delta}' \to \mathbf{\Delta}, s)\,d\Omega' = n^2(\mathbf{\Delta}, s) \tag{11.25}$$

Also, in equilibrium the incident intensity in direction $\mathbf{\Delta}'$ is $n^2(\mathbf{\Delta}', s) I_\nu^0(T)$, whence

$$d^6\Phi_\nu'^{s+} = d^6\Phi_\nu'^{s-} = n^2(\mathbf{\Delta}, s) I_\nu^0(T)\,dS\,ds\,d\Omega\,d\nu\sigma_\nu \tag{11.26}$$

## 11.3 Radiation transfer equation

Reviewing all the results so far, starting from (11.6), (11.9), (11.19) and (11.24) we obtain a law for the development of the flux incident in the solid angle $d\Omega$ with principal direction $\mathbf{\Delta}$, normal to $dS$:

$$d^6\Phi_\nu' = -d^6\Phi_\nu'^a - d^6\Phi_\nu'^{s-} + d^6\Phi_\nu'^e + d^6\Phi_\nu'^{s+} \tag{11.27}$$

which in local thermodynamic equilibrium (LTE) leads to an expression in terms of the intensity $I'_\nu(s, \mathbf{\Delta})$:

$$\frac{d^6\Phi'_\nu}{dS\, ds\, d\Omega\, d\nu} = -(\kappa_\nu + \sigma_\nu)I'_\nu(\mathbf{\Delta}, s) + \kappa_\nu n^2(\mathbf{\Delta}, s)I^0_\nu(T)$$
$$+ \frac{\sigma_\nu}{4\pi}\int_0^{4\pi} p(\mathbf{\Delta}' \to \mathbf{\Delta}, s)I'_\nu(\mathbf{\Delta}', s)\, d\Omega' \tag{11.28}$$

We can calculate the change in flux between $s$ and $s + ds$, using the conservation of $n^2\, dS \cos\theta\, d\Omega$ (equation (11.3), with $\cos\theta = 0$):

$$d^6\Phi'_\nu = d^5\Phi'_\nu(s + ds) - d^5\Phi'_\nu(s)$$
$$= dS(s)\, d\Omega(s)\, d\nu n^2(s)\left[\frac{I'_\nu(s + ds)}{n^2(s + ds)} - \frac{I'_\nu(s)}{n^2(s)}\right]$$
$$= dS(s)\, d\Omega(s)\, d\nu n^2(s)\frac{d}{ds}\left(\frac{I'_\nu}{n^2}\right)ds \tag{11.29}$$

Using (11.28) and (11.29), we obtain:

$$n^2(\mathbf{\Delta}, s)\frac{d}{ds}\left(\frac{I'_\nu}{n^2}(s, \mathbf{\Delta})\right) = -(\kappa_\nu + \sigma_\nu)I'_\nu(s, \mathbf{\Delta}) + \kappa_\nu n^2(\mathbf{\Delta}, s)I^0_\nu(T)$$
$$+ \frac{\sigma_\nu}{4\pi}\int_0^{4\pi} p(\mathbf{\Delta}' \to \mathbf{\Delta}, s)I'_\nu(\mathbf{\Delta}', s)\, d\Omega' \tag{11.30}$$

This equation for the variation of flux along a ray path is called the **radiation transfer equation**.

*Notes*

1. For a transparent ($\kappa_\nu = 0$) and non-diffusing ($\sigma_\nu = 0$) material, this gives the conservation of $I'_\nu/n^2$, equation (11.5).
2. In most applications, whether for gases, liquids or solids, we usually make the approximation of rectilinear propagation.
3. The transfer equation is coupled to the general energy equation, in its local form:

$$\rho c_p\left(\frac{\partial T}{\partial t} + \mathbf{v}\cdot\nabla T\right) = \nabla\cdot(\lambda\nabla T) + P + P^R \tag{11.31}$$

There are many interactions:

- The intensity $I'_\nu(\mathbf{\Delta}, \mathbf{r})$ at a given point $\mathbf{r}$ depends (through (11.30)) on the temperature field $T(\mathbf{r}, t)$ at every point in the system (both within the semi-transparent medium and on its opaque boundaries): this is due to the instantaneous influences over large distances which occur with radiation.

- The temperature field $T(\mathbf{r}, t)$ depends on the intensity field at point $\mathbf{r}$ in the semi-transparent medium, via the radiative power dissipation term $P^{R}(\mathbf{r}, t)$ (see equations 14.25 and 5.2):

$$P^{R}(\mathbf{r}, t) = -\int_0^{\infty} \mathrm{d}\nu \int_{4\pi} \frac{\partial}{\partial s} I'_{\nu}(\mathbf{r}, \mathbf{\Delta})\mathbf{u}\, \mathrm{d}\Omega \tag{11.32}$$

- A final coupling is provided by the thermal boundary conditions, among which, for example, the radiative flux at an opaque boundary appears (see Section 4.2.3):

$$\varphi_{R}(\mathbf{r}, t) = \int_0^{\infty} \mathrm{d}\nu \int_{4\pi} I'_{\nu}(\mathbf{r}, \mathbf{\Delta})\mathbf{n}\cdot\mathbf{u}\, \mathrm{d}\Omega \tag{11.33}$$

  where $\mathbf{n}$ and $\mathbf{u}$ are the unit vectors normal to the wall and in the $\mathbf{\Delta}$ direction respectively.

An alternative integral form of the transfer equation (11.30) is often used. To introduce it, we will restrict ourselves to the special case of a **non-scattering medium** ($\sigma_{\nu} = 0$), in which (11.30) simplifies to:

$$n^2 \frac{\mathrm{d}}{\mathrm{d}s}\left(\frac{I'_{\nu}}{n^2}\right) + n^2\kappa_{\nu}\left(\frac{I'_{\nu}}{n^2}\right) = \kappa_{\nu} n^2 I^0_{\nu}(T) \tag{11.34}$$

Without the right-hand side, the solution of this equation is:

$$\frac{I'_{\nu}}{n^2} = C \exp\left[-\int_0^{s} \kappa_{\nu}(s')\, \mathrm{d}s'\right] \tag{11.35.1}$$

In its complete form, obtained by the method of variation of the constant, its solution is:

$$\frac{I'_{\nu}(\mathbf{\Delta}, s)}{n^2(\mathbf{\Delta}, s)} = \frac{I'_{\nu}(\mathbf{\Delta}_0, 0)}{n^2(\mathbf{\Delta}_0, 0)} \exp\left[-\int_0^{s} \kappa_{\nu}(s')\, \mathrm{d}s'\right] + \int_0^{s} \kappa_{\nu}(s') I^0_{\nu}(s') \exp\left[-\int_{s'}^{s} \kappa_{\nu}(s'')\, \mathrm{d}s''\right] \mathrm{d}s' \tag{11.35.2}$$

The first term in (11.35.2) represents the radiation which is incident at 0 which is transmitted to $s$; the second term the summed contributions of the radiation emitted in all the segments $[s', s' + \mathrm{d}s']$ (whence the factor $\kappa_{\nu}(s')\mathrm{d}s' I^0_{\nu}(s')$) and transmitted from $s'$ to $s$: this term represents the total emission from the semi-transparent medium from 0 to $s$, taking account of auto-absorption by the medium. If we introduce the monochromatic transmissivity from $s'$ to $s$:

$$\tau'_{\nu s's} = \exp\left(-\int_{s'}^{s} \kappa_{\nu}(s'')\, \mathrm{d}s''\right) \tag{11.36}$$

equation (11.32) becomes:

$$\frac{I'_{\nu}(\mathbf{\Delta}, s)}{n^2(\mathbf{\Delta}, s)} = \frac{I'_{\nu}(\mathbf{\Delta}_0, 0)}{n^2(\mathbf{\Delta}_0, 0)} \tau'_{\nu O s} + \int_O^{s} I^0_{\nu}(T(s')) \frac{\mathrm{d}\tau'_{\nu s's}}{\mathrm{d}s'} \mathrm{d}s' \tag{11.37.1}$$

an integral expression on which numerous methods of radiative transfer depend. The main interest of this formulation is that it gets rid of the absorption coefficient in favour of the **monochromatic transmissivity** of a possibly heterogeneous and non-isothermal path element, $\tau'_{\nu_{s's}}$. In this approach, we need to determine all the transmissivities $\tau'_{\nu_{s's}}$ for all parts of the trajectories followed by the radiative transfer. Usually, a heterogeneous and isothermal column is discretized into $N_c$ homogeneous and isothermal elements, indicated by $j$ ($j = 1, \ldots, N_c$), of lengths $l_j$ and with absorption coefficients $\kappa_{j\nu}$. In discretized form, equation (11.37.1) becomes (Figure 11.5(b)):

$$\frac{I'_\nu(\mathbf{\Delta}, N_c)}{n^2(\mathbf{\Delta}, N_c)} = \frac{I'_\nu(\mathbf{\Delta}_0, 0)}{n^2(\mathbf{\Delta}_0, 0)}\tau'_{\nu 0 N_c} + \sum_{j=1}^{N_c} I^0_\nu(T_j)(\tau'_{\nu j N_c} - \tau'_{\nu j-1 N_c})$$

with:

$$\tau'_{\nu j N_c} = \exp\left[-\sum_{j'=j+1}^{N_c} \kappa_{j'\nu} l_{j'}\right] \qquad (11.37.2)$$

All radiation problems consist of the solution of the radiation transfer equation subject to **radiation boundary conditions**, perhaps coupled at every point in the medium with the local energy equation, which itself is governed by boundary conditions which include radiative contributions. We will now consider several classical boundary conditions for the transfer equation.

An **opaque wall** can be characterized with complete generality by knowledge of the monochromatic bidirectional reflectivity $\rho''_{\nu P_1 P_2}(\mathbf{\Delta}, \mathbf{\Delta}')$ at every point. This information, which is not usually obtainable, depends on the following:

- the incidence direction $\mathbf{\Delta}$ (in a solid angle $d\Omega$);
- the emergence direction $\mathbf{\Delta}'$ (in fact in $d\Omega'$);
- the polarization $P_1$ of the radiation relative to the incidence plane ($\parallel$ or $\perp$ – see Chapter 9);
- the polarization $P_2$ relative to the emergence plane;
- the frequency.

It obviously depends on the complex refractive indices of the wall and the medium and also, strictly, on the surface finish, the material surface composition, its history, etc. These parameters are uncontrollable except in laboratory conditions. In the

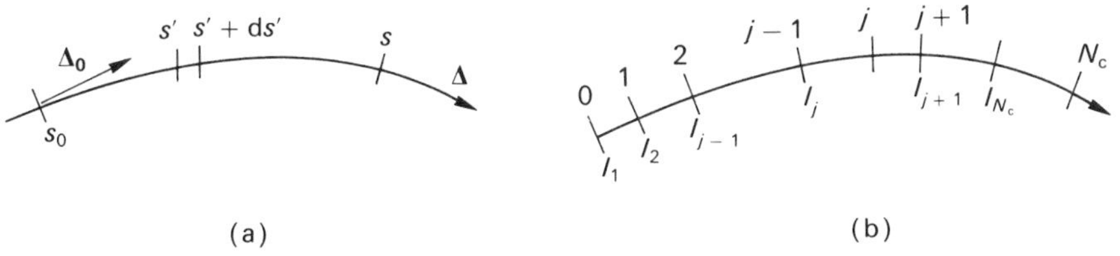


**Figure 11.5**

present case, the boundary condition is found by calculating the intensity $I_\nu'^{l}(\mathbf{\Delta}')$ of the radiation leaving the opaque surface in the direction $\mathbf{\Delta}'$:

$$I_\nu'^{l}(\mathbf{\Delta}') = \varepsilon_\nu'(\mathbf{\Delta}')I_\nu^0(T) + \int_{\Omega=0}^{\Omega=2\pi} \frac{\rho_\nu''}{\pi}(\mathbf{\Delta}, \mathbf{\Delta}')I_\nu'^{i}(\mathbf{\Delta}) \cos\theta \, d\Omega \qquad (11.38.1)$$

where, obviously:

$$\varepsilon_\nu'(\mathbf{\Delta}') = 1 - \rho_\nu^{\bullet\prime}(\mathbf{\Delta}') = 1 - \int_{\Omega=0}^{\Omega=2\pi} \frac{\rho_\nu''}{\pi}(\mathbf{\Delta}, \mathbf{\Delta}') \cos\theta \, d\Omega \qquad (11.38.2)$$

There are two standard assumptions which can be made for reflection (see the discussion in Chapter 9):

- For an optically smooth surface (for infra-red optics, for example), we use Fresnel's law of specular reflection:

$$I_\nu'^{l}(\mathbf{\Delta}') = \varepsilon_\nu'(\mathbf{\Delta}')I_\nu^0(T) + [1 - \varepsilon_\nu'(\mathbf{\Delta}')]I_\nu'^{i}(\mathbf{\Delta}'^*) \qquad (11.39)$$

  where $\mathbf{\Delta}'^*$ designates the direction symmetrical to $\mathbf{\Delta}$ with respect to the local normal.
- For a wall whose surface finish is coarser relative to the wavelength concerned, we assume the emergent radiation to be isotropic ($\varepsilon_\nu$ isotropic and $\rho_\nu$ independent of both the incident and emergent directions):

$$I_\nu^{l} = \varepsilon_\nu I_\nu^0(T) + \frac{(1 - \varepsilon_\nu)}{\pi} \int_0^{2\pi} I_\nu'^{i}(\mathbf{\Delta}) \cos\theta \, d\Omega \qquad (11.40)$$

For a **partly transparent interface**, the boundary conditions are more complex. For example, let us look at the case of the interface between a bath of molten glass and a mixture of combustion gases. The two media are both semi-transparent, but of very different natures. In the ideal case where the surface of separation is optically smooth and stable, we can use Fresnel's[1] laws of specular reflection and refraction; we will restrict ourselves to the common situation where the imaginery parts of the indices are very small compared to the real parts $n_i$, for both materials.

Referring to Figure 11.6, an incident ray in medium 1 (whose index $n_1$ is larger than that, $n_2$, of medium 2) in a solid angle $d\Omega$ around $\mathbf{\Delta}$ is totally reflected if the angle of incidence $\theta$ is greater than the limiting angle $i_L$, defined by:

$$\sin i_L = \frac{n_2}{n_1} \qquad (11.41)$$

Otherwise, it will be partly reflected and partly transmitted. The corresponding intensities obey the energy conservation equation, valid for either polarization $P(\parallel$ or $\perp)$:

$$\frac{I_{\nu P}'^{i}(\mathbf{\Delta})}{n_1^2} = \rho_{\nu P}'(\mathbf{\Delta})\frac{I_{\nu P}'^{i}(\mathbf{\Delta})}{n_1^2} + \frac{I_{\nu P}'^{t}(\mathbf{\Delta}')}{n_2^2} \qquad (11.42)$$

[1] In this case, $\rho_\nu^{\square\prime}(\mathbf{\Delta}')$ is more simply written $\rho_\nu'(\mathbf{\Delta}')$.

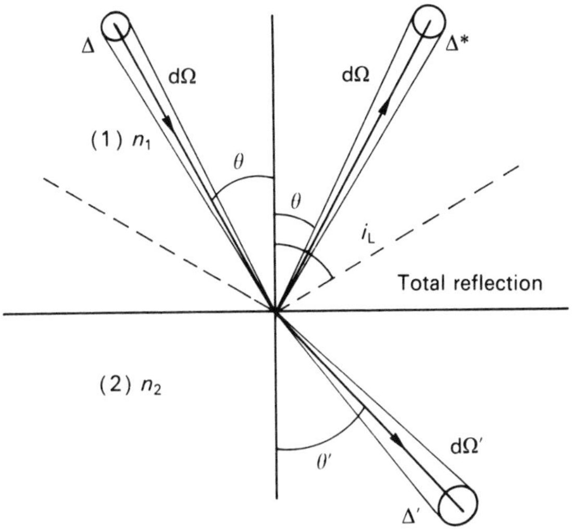


**Figure 11.6** Interface between two semi-transparent media.

where $I'^{\mathrm{t}}(\mathbf{\Delta}')$ is the intensity of the radiation transmitted into medium 2 in a solid angle $\mathrm{d}\Omega'$, different from $\mathrm{d}\Omega$, around a direction $\mathbf{\Delta}'$, different from $\mathbf{\Delta}$, and satisfying:

$$n_1 \sin\theta = n_2 \sin\theta' \tag{11.43}$$

and Clausius's theorem, which here takes the form:

$$n_1^2 \cos\theta \,\mathrm{d}\Omega = n_2^2 \cos\theta' \,\mathrm{d}\Omega' \tag{11.44}$$

The reflectivity in equation (11.42) is given by Fresnel's laws ((11.12)–(11.14)). In the limit where there is no reflection, this obviously reduces to equation (11.5). The preceding equations clearly also apply under time reversal (transmission of radiation from 2 to 1), but total reflection does not then occur.

If the surface separating the two media is of significant roughness in comparison to the wavelength, and the local roughness elements are pseudo-randomly distributed, the total reflection phenomenon disappears. We postulate an isotropic reflection law at the interface in this case:

$$\frac{I_\nu^{\mathrm{i}}}{n_1^2} = \frac{R_\nu}{n_1^2} I_\nu^{\mathrm{i}} + \frac{I_\nu^{\mathrm{t}}}{n_2^2} \tag{11.45.1}$$

where $R_\nu$ is the global reflection factor at the surface and $I_\nu^{\mathrm{i}}$ is the equivalent isotropic intensity

$$I_\nu^{\mathrm{i}} = \frac{1}{\pi}\int_0^{2\pi} I_\nu'^{\mathrm{i}}(\mathbf{\Delta}) \cos\theta \,\mathrm{d}\Omega \tag{11.45.2}$$

The methods used to solve the radiative transfer equations and their boundary conditions in radiative transfer depend strongly on the nature of the media and their radiative properties.

## *11.4 Basics of radiative transfers (without scattering)*

The calculation of radiative heat transfer for realistic geometries in a dense semi-transparent material or in a gas containing suspended particles is difficult for various reasons:

1. The local key quantity, $I'_\nu$ the directional monochromatic intensity of the radiation, depends on the following:
   - The spectral interval concerned, which requires a **discretization of the spectrum**. For a gas, the necessity of using finite spectral intervals $\Delta\nu$ means that **spectral correlations** between emission, transmission and absorption for a given gaseous species at various temperatures and pressures must be taken into account (see Section 11.5).
   - Position in the material $M(\mathbf{r})$, which requires **spatial discretization**.
   - Direction $\mathbf{u}$, which requires a **directional discretization**, which is just as difficult as the spatial discretization, and which must be done for each spectral interval and each mesh point. (This idea of directional discretization is often disconcerting for fluid mechanicists.)
2. Radiative transfer is a long-range effect with, on common scales, instantaneous propagation of radiation. The intensity field at any point depends explicitly on those at all points in and on the boundaries of the material. By virtue of the emitted intensity field, the intensity at a point depends explicitly on the fields of temperature, pressure and concentration at all points in the material. We are thus far from the local closure laws dear to the fluid mechanicist. There is another difficulty related to this long-range action, particularly for gases: the range of the latter (in other words the optical thickness) is strongly dependent on the spectral interval concerned[2]: there is interdependance between the spatial and spectral discretisations (see Figure 11.7).
3. This calculation method, which is already complex, uses the principle of rectilinear propagation of radiation. It gets worse if the medium is locally scattering, so that at each point in space there is strong coupling between the intensity leaving in direction $\mathbf{u}'$ and the incident intensities in all directions $\mathbf{u}$. The directional calculation locally becomes completely implicit. In the absence of bulk scattering it is, we may note, made implicit by another form of scattering: reflections from boundaries.

[2] This may also be compared with spectrum analysis (in wave number $K$), which may be applied to turbulent kinetic energy $k$ or to $\theta = \overline{T'^2}$. The ranges of energy transfers vary as $1/K$ (see Chapter 16).

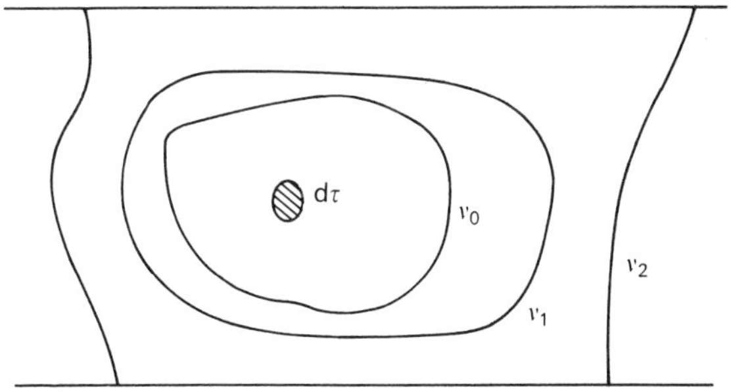


**Figure 11.7** An element of volume $d\tau$ interacts radiatively with various regions according to the frequency concerned ($\nu_1, \nu_2, \nu_3$). This interaction is complicated if opaque, or semi-transparent walls (causing reflection) are involved.

We will not attempt a fully general treatment of radiative transfer here; the field is so involved physically, mathematically and numerically that it will provide an active research field for several generations to come. We will only consider transfers without scattering, using as an example the case of a plane geometry. A greatly simplified approach to two-dimensional[3] and three-dimensional geometries will then allow rapid evaluation of simple cases.

### *11.4.1 The case of a plane wall*

We will consider a semi-transparent medium between two parallel planes, assumed of infinite extent, and characterized by one-dimensional fields of temperature, concentration, pressure, etc., varying in the $x$ direction, normal to the planes concerned. The boundary conditions are also uniform on the plane surface, distance $L$ apart, which bound the medium (Figure 11.8).

The semi-transparent medium absorbs and emits radiation, but does not scatter it. We will calculate the following:

- the radiative flux densities $\varphi_1^R$ and $\varphi_2^R$ on walls 1 and 2;
- the radiative flux at every point $x$ between the walls;
- the power per unit volume $P_R$ dissipated at every point $x$.

Any direction is characterized by a unit vector $\mathbf{u}(\theta, \varphi)$. Because of the cylindrical symmetry of the system, the intensity field only depends on $\cos\theta$; we will write (Figure 11.8):

$$\mu = \cos\theta \qquad \theta \in [0, \pi] \tag{11.46}$$

[3] It is obvious that all radiative transfers are inherently three-dimensional. A problem is described as one-dimensional, two-dimensional or three-dimensional according to whether the medium and its boundary conditions are one-dimensional, two-dimensional or three-dimensional respectively.

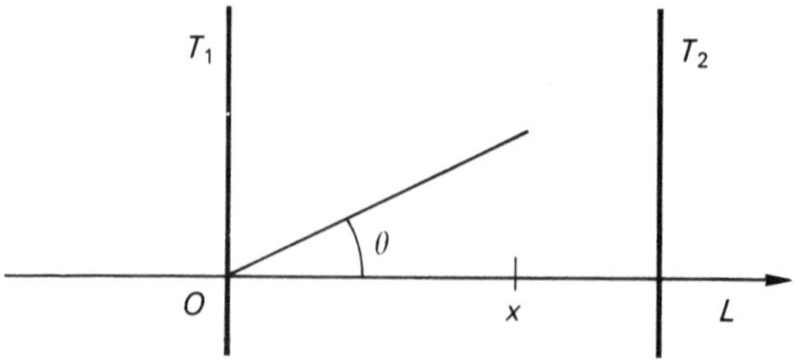


**Figure 11.8**

- The radiation intensity propagating in the **u** direction in the positive $x$ direction ($x^+$ so that $\mu > 0$) is (from (11.35.2)):

$$\left(\frac{I_\nu'^+}{n^2}\right)(x) = \tau'_{\nu 0x}\left(\frac{I_\nu'^+}{n^2}\right)(0) + \int_0^x \kappa_\nu(x')I_\nu^0(x')\tau_{\nu x'x}\frac{dx'}{\mu} \tag{11.47}$$

In this equation the intensity $I_\nu^0$ depends on $x'$, via the local temperature $T(x')$ and concentration; and $\tau'_{\nu x'x}$, is called the monochromatic transmissivity in the **u** direction from $x'$ to $x$:

$$\tau_{\nu x'x}(\mu) = \exp\left[-\frac{e_\nu(x', x)}{\mu}\right] \tag{11.48}$$

where the absolute value of $e_\nu(x', x)$ is the **optical thickness** of a column normal to the walls between $x'$ and $x$, and is a fundamental dimensionless quantity for radiative heat transfer:

$$e_\nu(x', x) = \int_{x'}^x \kappa_\nu \, dx'' \tag{11.49}$$

An alternative formulation, more powerful than (11.47), is to make use of (11.37.1):

$$\frac{\partial}{\partial x'}\tau'_{\nu x'x} = \frac{\kappa_\nu(x')}{\mu}\tau_{\nu x'x} \tag{11.50}$$

Then in terms of transmissivities rather than absorption coefficients:

$$\left(\frac{I_\nu'^+}{n^2}\right)(x) = \tau'_{\nu 0x}\left(\frac{I_\nu'^+}{n^2}\right)(0) + \int_0^x I_\nu^0(x')\frac{\partial}{\partial x'}\tau'_{\nu x'x}\,dx' \tag{11.51}$$

- If we consider the propagation of radiation in the negative $x$ direction ($x; \mu < 0$), equation (11.49) for the optical thickness becomes:

$$x' > x: \qquad e_\nu(x', x) = \int_{x'}^x \kappa_\nu \, dx'' > 0 \tag{11.52}$$

Equation (11.48) can be generalized to $\tau'_{\nu x'x}$, but also:

$$\frac{\partial}{\partial x'}\tau'_{\nu x'x} = +\frac{\kappa_\nu}{\mu}\tau'_{\nu x'x} < 0 \tag{11.53}$$

It therefore follows, by analogy with (11.51):

$$x' > x: \qquad \left(\frac{I_\nu'^-}{n^2}\right)(x) = \tau'_{\nu lx}\left(\frac{I_\nu'^-}{n^2}\right)(L) + \int_L^x I_\nu^0(x')\frac{\partial}{\partial x'}\tau'_{\nu x'x}\,\mathrm{d}x' \qquad (11.54)$$

The integral on the right of (11.54) is always positive, since if $(\partial/\partial x')\tau'_{\nu x'x}$ is negative, the integration limits are exchanged.

- The radiative flux density at any point $x$ is given by:

$$\varphi_{\mathrm{R}}(x) = \int_0^\infty \mathrm{d}\nu \int_0^\pi I_\nu'(x)\cos\theta 2\pi\sin\theta\,\mathrm{d}\theta = \varphi^{\mathrm{R}+}(x) - \varphi^{\mathrm{R}-}(x) \qquad (11.55)$$

where:

$$\varphi^{\mathrm{R}+}(x) = 2\pi\int_0^\infty \mathrm{d}\nu \int_0^1 I_\nu'^+(x)\mu\,\mathrm{d}\mu \qquad (11.56)$$

$$\mu' = -\mu \qquad \varphi^{\mathrm{R}-}(x) = 2\pi\int_0^\infty \mathrm{d}\nu \int_0^1 I_\nu'^-(x)\mu'\,\mathrm{d}\mu' \qquad (11.57)$$

This approach is equivalent to considering angles $\theta$ and $\theta'$, lying between 0 and $\pi/2$, and distinguishing between the directions of propagation. The auxiliary variable $\mu'$ will be called $\mu$ hereafter.

From equations (11.32) and (11.33), the power dissipated per unit volume at a point $x$ in the medium is:

$$P^{\mathrm{R}}(x) = -\frac{\mathrm{d}\varphi^{\mathrm{R}}}{\mathrm{d}x}(x) \qquad (11.58)$$

At this stage it is convenient to distinguish on one hand the general treatment of dense media (and gases using the *ck* or line by line models) based on an absorption coefficient, and on the other hand the special treatment of gases using the narrow band statistical model or the *ckfg* method, based on transmissivities. References concerning the various approaches applicable to gases are given in Section 11.5.

#### 11.4.1.1 METHODS BASED ON THE ABSORPTION COEFFICIENT

These methods are applicable to dense materials (glasses, refractories, etc.) and also to gases (high resolution line by line or *ck* approaches).

### General formulation

We will make the assumption of an isotropic distribution of radiation intensity leaving the walls (whether emitted, reflected or transmitted). This is valid for opaque walls (see the discussion in Chapter 9). We then write:

$$\left(\frac{I_\nu'^+}{n^2}\right)(0) = \left(\frac{I_\nu^+}{n^2}\right)(0); \qquad \left(\frac{I_\nu'^-}{n^2}\right)(L) = \left(\frac{I_\nu^-}{n^2}\right)(L) \qquad (11.59)$$

Then, from (11.48), (11.51), (11.53) and (11.56):

$$\left(\frac{\varphi^{\mathrm{R}+}}{n^2}\right)(x) = 2\pi\int_0^\infty \mathrm{d}\nu\left[\left(\frac{I_\nu^+}{n^2}\right)(0)E_3(e_\nu(0,x)) + \int_0^x I_\nu^0(x')\frac{\partial}{\partial x'}E_3(e_\nu(x',x))\,\mathrm{d}x'\right] \tag{11.60.1}$$

where the third-order integro-exponential function $E_3$ appears. This is introduced in summary in Appendix 1 to Part III:

$$E_3(\xi) = \int_0^1 \mathrm{e}^{-\xi/\mu}\mu\,\mathrm{d}\mu \tag{11.60.2}$$

In practice, this function is tabulated in various ways. After integration by parts, equation (11.60) becomes:

$$\left(\frac{\varphi^{\mathrm{R}+}}{n^2}\right)(x) = \int_0^\infty \left\{\pi\left[\left(\frac{I_\nu^+}{n^2}\right)(0) - I_\nu^0(0)\right]2E_3e_\nu(0,x)) + \pi I_\nu^0(x)\right.$$
$$\left. - \int_0^x 2E_3(e_\nu(x',x))\frac{\mathrm{d}}{\mathrm{d}T}(\pi I_\nu^0(T))\frac{\partial T(x')}{\partial x'}\mathrm{d}x'\right\}\mathrm{d}\nu \tag{11.61}$$

Similarly, we obtain:

$$\left(\frac{\varphi^{\mathrm{R}-}}{n^2}\right)(x) = \int_0^\infty \left\{\pi\left[\left(\frac{I_\nu^-}{n^2}\right)(L) - I_\nu^0(L)\right]2E_3(e_\nu(x,L)) + \pi I_\nu^0(x)\right.$$

$x' > x$

$$\left. - \int_L^x 2E_3(e_\nu(x,x'))\frac{\mathrm{d}}{\mathrm{d}T}(\pi I_\nu^0(x'))\frac{\partial T(x')}{\partial x'}\mathrm{d}x'\right\}\mathrm{d}\nu \tag{11.62}$$

whence we obtain $P^{\mathrm{R}}(x)$ using (11.56), (11.57), (11.58), (11.61) and (11.62).

**The special case of an isothermal medium**

For an isothermal[4] medium at temperature $T$ (distinct from that of the boundaries at $x = 0$ and $x = L$) equation (11.61) becomes:

$$\left(\frac{\varphi^{\mathrm{R}+}}{n^2}\right)(x) = \int_0^\infty \pi\left(\frac{I_\nu^+}{n^2}\right)(0)2E_3(e_\nu(0,x))\mathrm{d}\nu + \int_0^\infty \pi I_\nu^0(T)(1 - 2E_3(e_\nu(0,x))\,\mathrm{d}\nu \tag{11.63}$$

This equation may be readily interpreted by substituting:

$$\begin{aligned}\tau_\nu(0,x) &= 2E_3[e_\nu(0,x)]\\ \varepsilon_\nu(0,x) &= 1 - 2E_3[e_\nu(0,x)] = 1 - \tau_\nu(0,x)\end{aligned} \tag{11.64}$$

[4] Notice that we have not assumed the medium to be homogeneous. The real index may vary from one point to another, due to variations in the composition of the medium (for example, a glass melt may be essentially isothermal, but have a spatially varying composition).

Assuming the intensity of the radiation leaving $x = 0$ is isotropic, the first term in (11.63) is the flux density (per unit area of the plane $x = 0$) transmitted by the material from 0 to $x$, and $\tau_\nu$ is the corresponding transmission factor (integrated over all directions). Clearly (see Appendix):

$$2E_3(0) = 1; \qquad E_3(\infty) = 0 \tag{11.65}$$

The second term in (11.63) represents the total flux emitted by the isothermal material between the planes $x' = 0$ and $x$ towards plane $x$ (per unit area of that plane). The phenomenon of **autoabsorption** of radiation by the medium is taken into account in this expression: $\varepsilon_\nu$ is a **global** or apparent emission factor of the isothermal layer towards the plane. This concept loses validity if the medium is not isothermal. In other words, we have here:

$$\left(\frac{\varphi^{R+}}{n^2}\right)(x) = \int_{\nu=0}^{\nu=\infty} \tau_\nu(0, x)\left(\frac{d\varphi_\nu^+}{n^2}\right)(0) + \int_{\nu=0}^{\nu=\infty} \varepsilon_\nu(0, x)\left(\frac{d\varphi_\nu^e}{n^2}\right)$$

with: (11.66)

$$\left(\frac{d\varphi_\nu^+}{n^2}\right)(0) = \pi\left(\frac{I_\nu^+}{n^2}\right)(0)\, d\nu \qquad \text{and} \qquad \left(\frac{d\varphi_\nu^e}{n^2}\right) = \pi I_\nu^0(T)\, d\nu$$

*Notes*

1. From (11.63), we can see that at equilibrium between the walls and the semi-transparent medium characterized by:

$$\left(\frac{I_\nu^+}{n^2}\right)(0) = I_\nu^0(T)$$

the radiative flux density $(\varphi^{R+}/n^2)(x)$ reduces to $\sigma T^4$ (equilibrium radiation), and we see the obvious point that the radiation which is absorbed by the layèr, characterized by an absorption coefficient:

$$\alpha_\nu = 1 - \tau_\nu \tag{11.67}$$

exactly compensates for the radiation emitted, characterized by: $\varepsilon_\nu = 1 - \tau_\nu$.
2. An analysis along the same lines may be carried out for $\varphi^{R-}$.

**Optical thickness[5] and limiting cases**

Considering the above example of a layer of semi-transparent isothermal material, there are two limiting cases to consider according to the optical thickness $|e_\nu|$ (a dimensionless quantity defined by (11.49)) of the layer, for the frequency concerned:

- If $|e_\nu|$ is greater than 3, we can in practice consider that no radiation is transmitted from one wall to the other: $2E_3(3) = 0.024$. Such a medium is said to be **optically thick**.

[5] Optical thickness is the key parameter for all analyses in semi-transparent media.

- If on the contrary, $e_\nu$ is small compared to 1, the medium is **optically thin**. In this case, to a first approximation, all the radiation is transmitted from one wall to the other, and, furthermore, auto-absorption within the medium is negligible.

The most difficult cases to handle, semi-transparent *par excellence*, have values of $e_\nu$ between 0.3 and 2. On the other hand, the two extreme cases described above generally permit considerably simplified treatments, even in complex geometries.

*Note*

**The concept of optical thickness depends on frequency**; the same medium may be optically thin, thick or semi-transparent for spectral intervals which are very close together. This is, for example, the case for a gas, and does not simplify the treatment.

1. Limiting case of an optically thick medium. Consider a medium such that over a distance $\eta_\nu$ (which is frequency dependent) which satisfies:

$$\eta_\nu > 3/\kappa_\nu$$

the quantities $\kappa_\nu$, $T$ and $\partial T/\partial x$ can be considered uniform to a very good approximation. The medium is thus locally optically thick (over a distance of the order of $\eta_\nu$) and the radiative flux takes a very simple form. We have:

$$\forall\, x > \eta \qquad E_3[e_\nu(0, x)] \simeq 0$$

If we write:

$$u = e_\nu(x'x) \qquad \text{and} \qquad du = -\kappa_\nu(x')\,dx' \tag{11.68}$$

we obtain (see Appendix 1):

$$\int_0^x E_3[e_\nu(x', x)]\,dx' = +\frac{1}{\kappa_\nu(x)}\int_0^U E_3(u)\,du = -\left[\frac{E_4(U) - E_4(0)}{\kappa_\nu(x)}\right] = \frac{E_4(0)}{\kappa_\nu(x)}$$
$$= \frac{1}{3\kappa_\nu(x)}$$

Equations (11.61) and (11.62) become:

$$\left(\frac{\varphi^{R+}}{n^2}\right)(x) = \int_0^\infty \pi I_\nu^0(x)\,d\nu - \frac{2\pi}{3}\left[\int_0^\infty \frac{1}{\kappa_\nu}\frac{d}{dT} I_\nu^0(T)\,d\nu\right]\frac{\partial}{\partial x}T$$
$$\left(\frac{\varphi^{R-}}{n^2}\right)(x) = \int_0^\infty \pi I_\nu^0(x)\,d\nu + \frac{2\pi}{3}\left[\int_0^\infty \frac{1}{\kappa_\nu}\frac{d}{dT} I_\nu^0(T)\,d\nu \nabla \frac{\partial}{\partial x}T \tag{11.69}$$

Whence:

$$\left(\frac{\varphi^{R}}{n^2}\right)(x) = \left(\frac{\varphi^{R+}}{n^2}\right) - \left(\frac{\varphi^{R-}}{n^2}\right) = -\lambda_R(T)\frac{\partial T}{\partial x} \tag{11.70.1}$$

This is a remarkable expression: at this limit **for optically thick materials the radiative transfer phenomenon becomes diffusive in character**; $\lambda_R(T)$ is an equivalent radiative conductivity, which is very strongly dependent on local temperature:

$$\lambda_R(T) = \frac{4\pi}{3}\int_0^\infty \frac{1}{\kappa_\nu}\frac{d}{dT}I_\nu^0(T(x))\,d\nu \tag{11.70.2}$$

At high temperatures for materials such as glasses and some refractories, the **equivalent radiative conductivity** becomes much more significant than the molecular conductivity.

It is important to stress that (11.70) does not arise completely accidentally. There is a very close analogy between kinetic energy transport by gas molecules, which through elastic collisions gives rise to thermal conductivity and the transport of energy quanta $h\nu$ by photons, which through inelastic absorption–emission processes gives rise to radiative conductivity (see the comparison between conductive and radiative flux in Chapter 4). In an optically thick medium, conditions are close to those of local thermodynamic equilibrium. That is, the medium only differs from a state of radiative equilibrium (characterized by intensity $I_\nu^0(T(x))$ at a point $M(x)$) by a small perturbation, of order one. It is precisely in the neighbourhood of LTE, and only in this case, that the conductive flux may be closed, thanks to Fourier's hypothesis (see, for example, references [17] and [18]).

Equation (11.70) can be generalized to three dimensions in the form of a radiative Fourier equation:

$$\varphi^R(x) = -\lambda_R(T)\nabla T \tag{11.71}$$

provided the local optical index of the material is isotropic. Otherwise, we obtain a conductivity tensor, as in thermal conductivity. Notice that the very strong dependence of $\lambda_R(T)$ on $T$ makes radiative conduction problems very strongly non-linear in $T$. We must also watch that the condition $\eta_\nu > 3/\kappa_\nu$ is valid at all points in the system and in all spectral regions which play an appreciable role thermally.

Near the walls, in the boundary layers of a semi-transparent fluid such as a glass being processed in a bath, this condition is rarely satisfied, even if the motion is very small. It is then necessary to use an exact treatment based on (11.61) and (11.62). A locally one-dimensional model is usually appropriate, and poses no particular difficulty.

*Note.* The idea of radiative conductivity can readily be introduced by very different methods, for example an introduction based on the transfer equation in reference [19].

2. Limiting case of an optically thin isothermal medium. The radiative flux $\varphi^{R+}(x)$ is given by (11.61) and (11.62) in this case, with the simplifications:

$$\tau_\nu(0, x) = 2E_3[e_\nu(0, x)] \simeq 1 \quad \text{(to order } e_\nu) \tag{11.72}$$

$$\varepsilon_\nu(0, x) = 1 - \tau_\nu(0, x) = 2e_\nu(0, x)$$

since for an optically thin medium, $e_\nu(0, x)$ is small compared to 1.

11.4.1.2 APPROACHES BASED ON TRANSMISSIVITY

The above approaches cannot be used for gases if a narrow band statistical model or the *ckfg* method is used (for these methods, see the references given in Section 11.5). It is therefore necessary to develop transfer models based on the transmissivities of all the heterogeneous and non-isothermal columns appearing in the system. This is not within the scope of this basic course. However, examples will be found in references [20] and [21].

### 11.4.2 Pragmatic models for semi-transparent isothermal and heterogeneous materials

In this section, a **pragmatic approach** is developed to calculate the **transfer between opaque walls and a homogeneous isothermal semi-transparent material** for relatively general geometries, for dense materials or gases using, in the latter case, the line by line or *ck* methods.[6] This crude model starts from exact considerations which are valid at the limit of an optically thin medium, and then generalizes them pragmatically to a general medium.

11.4.2.1 PARTICULAR CASE OF A HEMISPHERE (EXACT CALCULATION)

We consider a semi-transparent homogeneous and isothermal medium, with a uniform real index $n$, contained within a hemisphere of radius $R$, which emits radiation towards the element of area $\mathrm{d}S$ at the centre of the base of the hemisphere (Figure 11.9).

The flux emitted by the hemisphere to the element of area $\mathrm{d}S$, taking account of autoabsorption of radiation by the medium, is:

$$\mathrm{d}\Phi_\nu^{\mathrm{e}} = \int_0^{\pi/2} (1 - \exp(-\kappa_\nu R))n^2 I_\nu^0(T) \cos\theta 2\pi \sin\theta \, \mathrm{d}\theta \, \mathrm{d}\nu \, \mathrm{d}S \qquad (11.73.1)$$

$$\mathrm{d}\Phi_\nu^{\mathrm{e}} = \varepsilon_{\nu\Sigma O} \pi n^2 I_\nu^0(T) \, \mathrm{d}\nu \, \mathrm{d}S \qquad (11.73.2)$$

in which expression the emissivity of the medium towards a surface centred at $O$ appears:

$$\varepsilon_{\nu\Sigma O} = 1 - \exp(-\kappa_\nu R) \qquad (11.74)$$

This expression is structurally identical to that for the emissivity of a homogeneous isothermal column of gas of length $R$. With the obvious notation:

$$\alpha_{\nu\Sigma O} = 1 - \tau_{\nu\Sigma O} = \varepsilon_{\nu\Sigma O} \qquad (11.75)$$

*Note*

In the particular case of an optically thin medium ($\kappa_\nu R \ll 1$), the emissivity of a hemisphere of homogeneous and isothermal material is:

$$\varepsilon_{\nu\Sigma O} = \kappa_\nu R \qquad (11.76)$$

[6] See the appropriate references in Section 11.5.

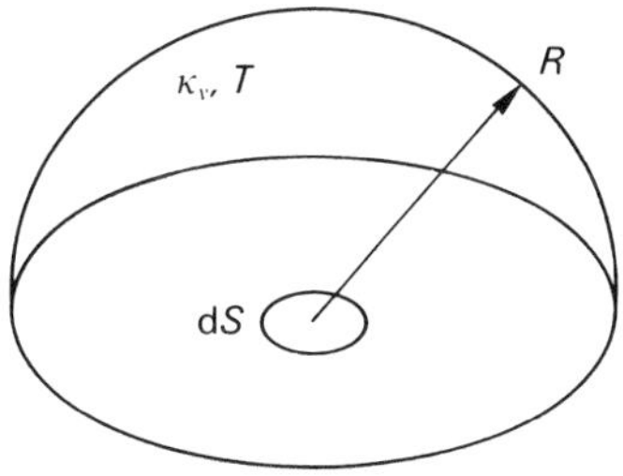


**Figure 11.9**

11.4.2.2 LIMITING CASE OF AN OPTICALLY THIN MEDIUM OF ARBITRARY GEOMETRY

Consider an arbitrary volume $V$ of a homogeneous and isothermal semi-transparent medium, bounded by a surface $S$ (Figure 11.10).

We assume that:

$$e_\nu = \kappa_\nu D_c \ll 1 \tag{11.77}$$

where $D_c$ is a characteristic dimension of the system.

Under these conditions, the flux emitted by the medium is totally transmitted to the boundaries of the surface $S$, without autoabsorption. The total monochromatic flux emitted is, by reference to Section 11.1:

$$\mathrm{d}\Phi_\nu^e = \mathrm{d}\nu \int_{\Omega=0}^{4\pi} \mathrm{d}\Omega \int_0^V \kappa_\nu n^2 I_\nu^0(T)\,\mathrm{d}V \tag{11.78}$$

Because of the assumption that the medium is homogeneous and isothermal:

$$\mathrm{d}\Phi_\nu^e = 4\pi V \kappa_\nu n^2 I_\nu^0(T)\,\mathrm{d}\nu \tag{11.79}$$

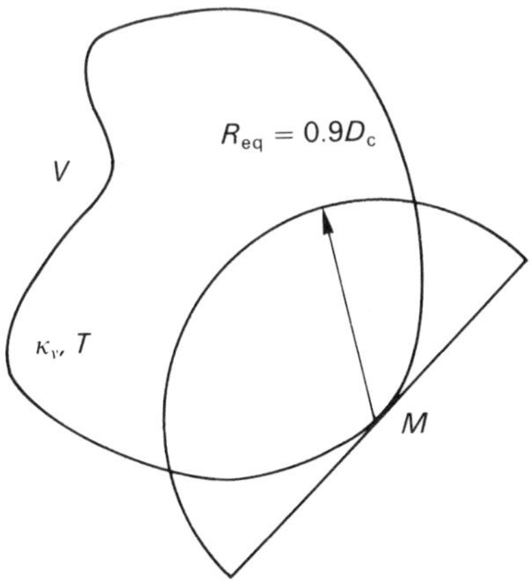


**Figure 11.10**

If we introduce an average flux density on the boundary surface $S$, denoted $\langle d\varphi_\nu^e \rangle_S$, we obtain:

$$\langle d\varphi_\nu^e \rangle_S = \frac{d\Phi_\nu^e}{S} = \langle \varepsilon_\nu \rangle_S \pi n^2 I_\nu^0(T)\, d\nu \qquad (11.80)$$

writing:

$$\langle \varepsilon_\nu \rangle_S = \frac{4V}{S} \kappa_\nu \qquad (11.81)$$

$\langle \varepsilon_\nu \rangle_S$ is the average emissivity of the isothermal and homogeneous medium towards an arbitrary elementary area of $S$ in the spectral interval $[\nu, \nu + d\nu]$. This remarkably simple result allows us to define the characteristic length of the system required for equation (11.77):

$$D_c = \frac{4V}{S} \qquad (11.82)$$

We can put a simple physical explanation on to equation (11.81) by comparing it with equation (11.76), which relates to a homogeneous isothermal hemisphere of semi-transparent material, similarly for an optically thin material. $D_c$ can be considered as the radius of an equivalent hemisphere which, from the viewpoint of an 'average' point on $S$, emits the same radiation as the volume $V$ of the medium concerned (see Figure 11.10). Note that everything up to this stage has been rigorous, within the assumptions made.

#### 11.4.2.3 (SWEEPING) GENERALIZATION TO ANY HOMOGENEOUS AND ISOTHERMAL MEDIUM

The very pragmatic argument which follows is originally due to Hottel and Sarofim [14]. The results of Section 11.4.2.2 are rigorous when applied to optically thin media. But if the equivalent hemisphere of radius $R = D_c$ has physical meaning for an optically thin material, it is tempting to hope that it might still be credible for a medium of any optical thickness. This is obviously the case for a thick homogeneous and isothermal material; in this other limiting case the mean emissivity $\langle \varepsilon_\nu \rangle$ becomes 1. On this assumption, the average emissivity of a homogeneous isothermal medium towards an 'average' element of its surface is (from (11.74) and (11.82)) given by:

$$\langle \varepsilon_\nu \rangle_S = 1 - \exp\left( -\kappa_\nu \frac{4V}{S} \right) \qquad (11.83)$$

which equation is arbitrarily (and audaciously) assumed value for $4\kappa_\nu V/S$ of order 1.

Comparisons with exact calculations for a certain number of geometries reveal that this crude approach introduces errors of the order of 10–15%. An empirical coefficient $C$ usually leads to improved results [14]:

$$\langle \varepsilon_\nu \rangle_S = 1 - \exp[-\kappa_\nu C D_c] \qquad C \simeq 0.9 \qquad (11.84)$$

The physical interpretation of $\langle \varepsilon_\nu \rangle_S$ follows from its definition (11.80). We will now show that we can similarly introduce an overall monochromatic absorptivity and transmissivity for a semi-transparent medium, relative to radiation leaving the surface of area $S$ uniformly and with isotropic intensity. Let $\mathrm{d}\varphi_\nu^{l}$ be the uniform flux density leaving $S$. We will define the absorptivity $\langle \alpha_\nu \rangle_S$ and transmissivity $\langle \tau_\nu \rangle_S$ by:

$$\mathrm{d}\Phi_\nu^{a} = \langle \alpha_\nu \rangle_S S \,\mathrm{d}\varphi_\nu^{l}$$

$$\mathrm{d}\Phi_\nu^{t} = S\,\mathrm{d}\varphi_\nu^{l} - \mathrm{d}\Phi_\nu^{a} = [1 - \langle \alpha_\nu \rangle_S] S\,\mathrm{d}\varphi_\nu^{l} \tag{11.85}$$

where $\mathrm{d}\Phi_\nu^{a}$ and $\mathrm{d}\Phi_\nu^{t}$ are the fluxes absorbed and transmitted by the semi-transparent medium. Note here that $\mathrm{d}\Phi_\nu^{a}$ is the total flux absorbed in so far as $S\,\mathrm{d}\varphi_\nu^{l}$ takes account of all the successive reflections from the boundary of $S$. (It follows that if the wall is opaque, $\mathrm{d}\Phi_\nu^{t}$ is completely absorbed by $S$.)

If we consider the situation where the semi-transparent medium and the wall are in thermal equilibrium at temperature $T$, the total fluxes emitted (allowing for autoabsorption) and absorbed by the gas are equal:

$$\mathrm{d}\Phi_\nu^{e} = \mathrm{d}\Phi_\nu^{a} \tag{11.86}$$

Since in equilibrium, the radiation leaving surface $S$ is:

$$\mathrm{d}\Phi_\nu^{l} = S\pi n^2 I_\nu^0(T)\,\mathrm{d}\nu \tag{11.87}$$

it follows, using (11.85)–(11.87):

$$\langle \alpha_\nu \rangle_S = \langle \varepsilon_\nu \rangle_S \tag{11.88}$$

This result can clearly be generalized to the non-equilibrium (but still isothermal) case of a uniform flux with an isotropic intensity leaving the surface. Following (11.85) and (11.88), we obtain:

$$\langle \tau_\nu \rangle_S = 1 - \langle \alpha_\nu \rangle_S = 1 - \langle \varepsilon_\nu \rangle_S \tag{11.89}$$

where $\langle \tau_\nu \rangle_S$ represents the global transmissivity of the total flux transmitted from surface $S$ to surface $S$ through the semi-transparent medium, which here is uniform and isothermal.

#### 11.4.2.4 PRACTICAL EXAMPLE

**Problem**

Calculate the radiative flux on the internal wall of an infinite cylindrical tube of square cross-section of side $a$. A homogeneous and isothermal semi-transparent material at temperature $T$ (as a first approximation) and of constant index $n$ flows through the tube. The opaque walls are uniformly at temperature $T_w$ and have an isotropic emissivity of $\varepsilon_{w\nu}$.

**Solution**

The required flux per unit surface area is:

$$\varphi^{R} = \varphi^{i} - \varphi^{l} \tag{11.90}$$

where $\varphi^{\mathrm{i}}$ and $\varphi^{\mathrm{l}}$ are the incident and leaving flux densities on the wall, assumed to be uniform to a first approximation. The average flux density leaving a wall is:

$$\varphi^{\mathrm{l}} = \int_0^\infty \pi I_\nu^{\mathrm{l}}\,\mathrm{d}\nu = \int_0^\infty [\varepsilon_{\mathrm{w}\nu}\pi n^2 I_\nu^0(T_{\mathrm{w}}) + (1-\varepsilon_{\mathrm{w}\nu})\pi I_\nu^{\mathrm{i}}]\,\mathrm{d}\nu \tag{11.91}$$

where $I_\nu^{\mathrm{i}}$ and $I_\nu^{\mathrm{l}}$ are the monochromatic intensities of the incident and leaving radiations. The latter is assumed to be isotropic; the former is the equivalent isotropic intensity. Furthermore, the incident flux density is given by:

$$S\varphi^{\mathrm{i}} = S\int_0^\infty \pi I_\nu^{\mathrm{i}}\,\mathrm{d}\nu = \int_0^\infty \langle\tau_\nu\rangle_S S\pi I_\nu^{\mathrm{l}}\,\mathrm{d}\nu + \int_0^\infty [1-\langle\tau_\nu\rangle_S]S\pi n^2 I_\nu^0(T)\,\mathrm{d}\nu \tag{11.92}$$

The first term in (11.92) is the part of the flux which was incident on the internal surface $S$ of the tube, after transmission through the medium from the whole of the same surface. The second term is the contribution of the semi-transparent medium itself to the flux incident on the wall of area $S$. After all the calculations, the radiative flux density is:

$$\varphi^{\mathrm{R}} = \int_0^\infty \frac{\varepsilon_{\mathrm{w}\nu}\langle\varepsilon_\nu\rangle_S\pi n^2[I_\nu^0(T) - I_\nu^0(T_{\mathrm{w}})]}{1-(1-\langle\varepsilon_\nu\rangle_S)(1-\varepsilon_{\mathrm{w}\nu})}\,\mathrm{d}\nu \tag{11.93}$$

It remains to calculate the total emissivity $\langle\varepsilon_\nu\rangle_S$ of the medium. If $L$ is the length of the tube, the volume of material is $V = a^2L$, and the lateral surface area $S = 4aL$. The radius of the equivalent hemisphere at any point in the medium is, to a first approximation:

$$R_{\mathrm{eq}} = 0.9\frac{4V}{S} = 0.9a \tag{11.94.1}$$

Thus:

$$\langle\varepsilon_\nu\rangle_S = 1 - \exp(-0.9\kappa_\nu a) \tag{11.94.2}$$

This allows the flux to be calculated for a dense medium characterized by an absorption coefficient $\kappa_\nu$.

**Particular case of a gas**

If we consider a gaseous[7] semi-transparent medium (containing $CO_2$, $H_2O$, CO, $N_2O$, etc.), equation (11.93) requires us to take account of fine structural spectral correlations in order to calculate the factor:

$$\frac{1}{\Delta\nu}\int_{\nu-(\Delta\nu/2)}^{\nu+(\Delta\nu/2)} \frac{\langle\varepsilon_\nu\rangle_S}{1-(1-\langle\varepsilon_\nu\rangle_S)(1-\varepsilon_{\mathrm{w}\nu})}\,\mathrm{d}\nu$$

[7] Remember that the above approach is only valid if the emissivities $\langle\varepsilon_\nu\rangle_s$ are considered at high resolution, calculated line by line or using the *ck* method – see references in Section 11.5.

precisely, for an interval $\Delta\nu$ over which the intensities at equilibrium $I_\nu^0$ are constant. A standard route is to have recourse to statistical narrow band models (see Section 11.5). Nevertheless, in certain very special cases the above expression may be considerably simplified provided the following hypotheses are valid:

(H1) **The walls are black.** This 'hypothesis' makes the denominator of (11.93) disappear.

(H2) **The walls are cold relative to the gas.** For example, in absolute temperature, if $T_w/T < 0.5$, the radiation emitted by the walls is negligible compared to that emitted by the gas, and:

$$\varphi^R \simeq \int_0^\infty \langle \varepsilon_\nu(T) \rangle_S \pi I_\nu^0(T)\, d\nu \simeq \varepsilon\sigma T^4 \tag{11.95}$$

Equation (11.95) could be considered as the definition of the total emissivity of the gas. One could then **in this particular case, and this case only, use Hottel's** [14] **charts** for $\varepsilon$ to calculate the radiative flux.

Note that if hypothesis H2 is not satisfied, equation (11.93) becomes:

$$\begin{aligned}\varphi^R &= \int_0^\infty \langle \varepsilon_\nu(T) \rangle_S \pi I_\nu^0(T)\, d\nu - \int_0^\infty \langle \varepsilon_\nu(T_w) \rangle_S \pi I_\nu^0(T)\, d\nu \\ &= \varepsilon(T)\sigma T^4 - \alpha(T, T_w)\sigma T_w^4 \end{aligned} \tag{11.96}$$

where $\alpha(T, T_w)$, defined by equation (11.96), can be considered as a total absorption factor for the gas at temperature $T$ for radiation from black walls at temperature $T_w$. This quantity cannot be calculated from Hottel's charts since information on the spectrum structure has been lost. Experience shows that if $T$ and $T_w$ are appreciably different, $\alpha(T, T_w)$ differs strongly from $\varepsilon(T)$, and Hottel's charts cannot then be usable for transfer calculations.

For calculations of radiative heat transfer in a non-isothermal gas, the use of grey gas models (with $\varepsilon$ and particularly the absorption coefficient $\kappa$ constant) will clearly give incorrect results.[8] The pseudo-quantity $\alpha$ would then depend on the whole distribution of temperatures and concentrations in the medium, and would have no relationship to $\varepsilon$.

## 11.5 Dense materials and gases

The general study of radiative transfers in a semi-transparent medium is beyond the objectives of this course. In practice, the media involved are generally highly non-isothermal and often heterogeneous (particularly gases). The radiative transfers are in most cases associated with conduction (in glasses or ceramics) or with

[8] The same applies to the so-called '$N$ grey gases, 1 clear gas' model.

convective transfers (in gases and glasses). Semi-transparent media can be classified into two categories according to their spectral properties: dense materials and gases.

- **Dense materials** (liquids, glasses, ceramics) are characterized by radiative properties ($\kappa_\nu$, $\sigma_\nu$) which may be considered constant over spectral intervals $\Delta\nu$ such that $I_\nu^0(T)$ does not vary by more than a few percent at the temperature concerned. The typical size of $\Delta\nu$ is of the order of 200 $cm^{-1}$. For such materials, the transfer equation may be applied directly using absorption coefficients as in Section 11.4. The references (books and articles) are very numerous (e.g. references [14], [22–23]). But this happy situation is generally not the case for gases.
- **Gases**, which are absorbent in the infra-red, either pure or with a small proportion of suspended particles, have a fine spectral structure of quantum origin, associated with the existence of absorption lines grouped into absorption bands. The typical width of a line for $CO_2$ is 0.15 $cm^{-1}$ at 300 K and $2.5 \times 10^{-2}$ $cm^{-1}$ at 3000 K, both at atmospheric pressure (see the example in Figure 11.11).

Under these conditions, the variation of $\kappa_\nu$ with frequency is extremely complex, even in an interval $\Delta\nu$ sufficiently small for $I_\nu^0$ to be considered constant. A simplistic idea would be to discretize the spectrum (the ensemble of frequencies) into adjacent intervals $\Delta\nu$ of this order (around 25–200 $cm^{-1}$), and to use as a first attempt absorption coefficients $\bar{\kappa}_\nu$ averaged over $\Delta\nu$. Unfortunately, **an average coefficient of absorption, $\bar{\kappa}_\nu$, has no physical meaning**. Such an idea cannot take account of **spectral correlations** related to the fine line structure of the spectrum. A mathematical method of demonstrating this is to consider the transmissivity $\bar{\tau}_\nu$ of a homogeneous and

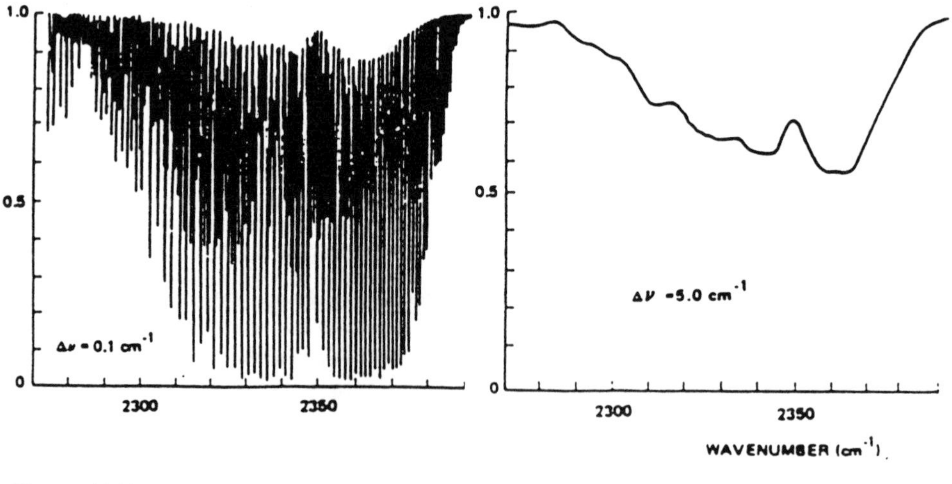


**Figure 11.11** Transmission spectra for $CO_2$–air in the same conditions, with different resolutions $\Delta\nu$: (a) $\Delta\nu = 0.1$ $cm^{-1}$; (b) $\Delta\nu = 5$ $cm^{-1}$ (by courtesy of John Selby).

isothermal rod of length $l$, averaged over $\Delta\nu$:

$$\bar{\tau}_\nu = \frac{1}{\Delta\nu}\int_{\nu_0-\Delta\nu_0/2}^{\nu_0+\Delta\nu_0/2} \exp(-\kappa_\nu l)\,d\nu \qquad (11.97)$$

If we define:

$$\bar{\kappa}_\nu = \frac{1}{\Delta\nu}\int_{\nu_0-\Delta\nu_0/2}^{\nu_0+\Delta\nu_0/2} \kappa_\nu\,d\nu \qquad (11.98)$$

it is clear that $\bar{\tau}_\nu \neq \exp(-\bar{\kappa}_\nu l)$ for a spectrum such as that in Figure 11.11. These conclusions are generalizable to heterogeneous and non-isothermal media. The error involved in using an absorption coefficient $\bar{\kappa}_\nu$ averaged over 10–200 $cm^{-1}$ is uncontrollable, and can reach several hundred % on the transmissivity $\bar{\tau}_\nu$.

The existence of the fine structure has important consequences for energy transfer. Imagine a very simple case where radiation is emitted by an element of column 1, of a homogeneous and isothermal gas of temperature $T_1$, and absorbed by the same gas in adjacent column 2. The identical monochromatic emissivities of the two elements $\varepsilon_{1_\nu}$ and $\varepsilon_{2_\nu}$ are shown diagrammatically by the simple model of Figure 11.12. The 'black' absorption lines, of width $\delta\nu$ (of the order of a few times $10^{-2}$ $cm^{-1}$) are separated by transparent regions of width $\delta\nu'$ (of the order of a fraction of a $cm^{-1}$).

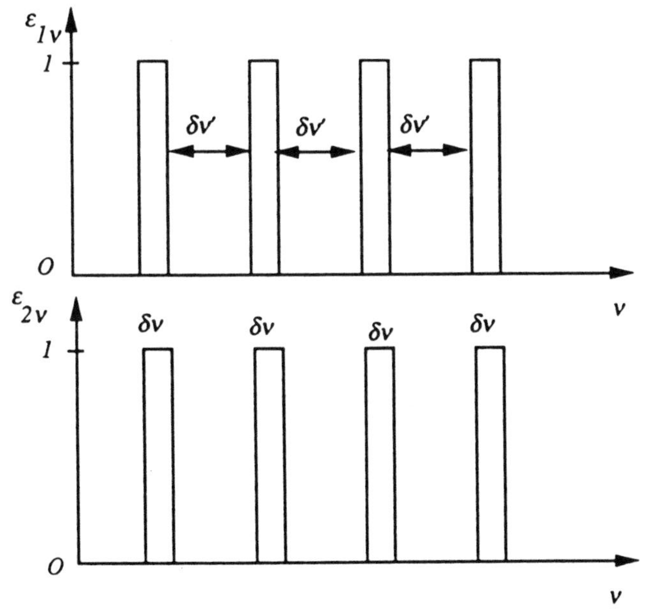


**Figure 11.12** Simplified model of an absorption line structure for two adjacent columns of gas with monochromatic emissivities $\varepsilon_{1\nu}$ and $\varepsilon_{2\nu}$.

The emissivity of element 1 averaged over $\Delta\nu$, so that:

$$\overline{\varepsilon_{1\nu}} = \overline{\varepsilon_{2\nu}} = \overline{\alpha_{2\nu}} = \frac{\delta\nu}{(\delta\nu' + \delta\nu)} \tag{11.99}$$

In the non-correlated calculation, the flux emitted by 1 and absorbed by 2, for an interval $\Delta\nu$, is given by:

$$\Phi^{\mathrm{nc}} = (\overline{\varepsilon_{1\nu} I_\nu^0(T_1)})\overline{\alpha_{2\nu}}\, \mathrm{d}S\, \mathrm{d}\Omega\, \mathrm{d}\nu \tag{11.100}$$

The same flux, calculated exactly (correlated) is:

$$\Phi^{\mathrm{c}} = (\overline{\varepsilon_{1\nu} I_\nu^0(T_1)})\, \mathrm{d}S\, \mathrm{d}\Omega\, \mathrm{d}\nu \tag{11.101}$$

All the radiation emitted by 1 is absorbed by 2, so that:

$$\frac{\Phi^{\mathrm{c}}}{\Phi^{\mathrm{nc}}} = \frac{1}{\overline{\alpha_{2\nu}}} = \frac{\delta\nu'}{\delta\nu} + 1 \tag{11.102}$$

This ratio is typically 10, if the lines are relatively wide spaced. It is essential, therefore, to use a radiation model which takes account, perhaps in a statistical manner, of the effects of correlations between emission, transmission and absorption in a material which may be heterogeneous and isothermal, but contains the same absorbent constituents:

- The most rigorous method of taking account of the spectral correlations in a heterogeneous and non-isothermal medium is clearly the **line by line** calculation. (For examples of the application of line by line to radiative transfer, see references [24–26].
- The **narrow band statistical models** (references [27–37], and for comprehensive bibliographies, references [31], [35] and particularly [37]) are currently the models giving the best representations, with acceptable computation time, of spectral correlations for three-dimensional radiative transfer in strongly anisothermal media. A more basic approach, due to Edwards [30] developed during the 1970s can be used with satisfactory accuracy.
- Another way is to use radiation models based on a random variation of the absorption coefficient in a given spectral interval. The simplest form is the *ck* **model** mainly developed by Goody [31]. This method is very satisfactory for weakly non-isothermal media, but can withstand large concentration and pressure gradients. This is appropriate in particular to the atmosphere, for meterological applications.
- A more complex and accurate model is to consider a real gas as being composed of $N$ fictitious gases, each characterized by a series of lines from the real gas which behave similarly with temperature. The *ck* method is then applied to the $N$ independent fictitious gases; this is the *ckfg* method, developed by the author's group: R. Levi di Léon *et al.* [38], P. Rivière *et al.* [32].

All these models (line by line, narrow band statistical, *ck* and *ckfg*) are compared in reference [32]. It is concluded in this reference that the primary line by line model could be replaced by other models to advantage, according to the type of application concerned, to reduce the computational size and time:

- By the *ck* model for media which may be heterogeneous, but have only weak temperature gradients. A typical case would be in free convection under Boussinesq's hypothesis. Note, further, that such a model is compatible with possible scattering of radiation by particles. It is based, in fact, on calculation of absorption coefficients in the radiation transfer equation.
- By the *ckfg* model, for applications such as long-range detection which require high precision in the calculation of the radiation intensities of highly isothermal and possibly heterogeneous gas columns (e.g. the infra-red signature of rockets, environmental effects of flares, etc.). This model, however, requires the use of a transfer equation in terms of transmissivities (see, for example, [20–21]), and is not very compatible with handling scattering by particles.
- By narrow band statistical models for three-dimensional radiative transfer in complex industrial systems such as furnaces, aircraft engine combustion chambers, rocket motors, etc. As with the preceding model, this approach needs the transfer equation in transmissivities [20–21], and is poorly compatible with simultaneous treatment of scattering from particles.

# Integro-exponential Functions

There are various forms of integro-exponential functions (see, for example, reference [4] to Part II). Here, we will confine ourselves to that which is nearest to the radiative equations:

$x \geqslant 0$

$$E_n(x) = \int_0^1 \mu^{n-2} \exp\left(-\frac{x}{\mu}\right) d\mu \tag{1}$$

Also:

$$E_n(x) = \int_1^\infty t^{-n} \exp(-xt)\, dt \tag{2}$$

From (1), we can write a recurrence formula:

$n \geqslant 2$

$$\frac{d}{dx} E_n(x) = -E_{n-1}(x) \tag{3}$$

$$\frac{d}{dx} E_1(x) = -\frac{1}{x} \exp(-x) \tag{4}$$

and also:

$n \geqslant 1$

$$nE_{n+1}(x) = \exp(-x) - xE_n(x) \tag{5}$$

Below are some useful values. For more details, see reference [4] of Part II:

| $x$ | 0 | 0.05 | 0.1 | 0.2 | 0.3 | 0.4 | 0.5 | 0.6 |
|---|---|---|---|---|---|---|---|---|
| $E_3(x)$ | 0.5000 | 0.4549 | 0.4163 | 0.3159 | 0.3000 | 0.2573 | 0.2216 | 0.1916 |
| $E_4(x)$ | 0.3333 | 0.3095 | 0.2877 | 0.2494 | 0.2169 | 0.1891 | 0.1652 | 0.1446 |

| $x$ | 0.8 | 1 | 1.5 | 2 | 2.5 | 3 | 3.5 | 5 | 10 |
|---|---|---|---|---|---|---|---|---|---|
| $E_3(x)$ | 0.1443 | 0.1097 | 0.0567 | 0.0301 | 0.0163 | 0.0120 | 0.0049 | $8.8 \times 10^{-4}$ | $3.5 \times 10^{-6}$ |
| $E_4(x)$ | 0.1113 | 0.0861 | 0.0460 | 0.0250 | 0.0138 | 0.01303 | 0.0043 | $7.8 \times 10^{-4}$ | $3.3 \times 10^{-6}$ |

## *Asymptotic limits*

- $x \to \infty$
$$E_n(x) \sim \frac{e^{-x}}{x} \to 0 \tag{6}$$

- $$E_n(0) = \frac{1}{n-1} \tag{7}$$

PART IV

# Mechanics of Non-isothermal Fluids. Convection

The object of this part is limited but essential:

- to study the physical mechanisms of convective transfers in the laminar and turbulent régimes;
- to consider the determination under given conditions of the convective heat transfer coefficient $h$ at the wall bounding a flow, in forced or free convection.

The object of Chapter 12 is to use dimensional methods to introduce the characteristic numbers which control forced or free convection, and to define the concept of similarity in thermal convection. At this stage, it is alreay possible to understand the origin of most of the correlations assembled in 'Basic Data' (Part V) which under certain conditions allow a local or average convective heat transfer coefficient to be determined.

Chapter 13 is concerned with the derivation of the instantaneous balance equations appropriate to all convection problems (excluding here local interactions with radiation). We will restrict ourselves in the following to cases where the thermophysical properties of the fluid (conductivity, viscosity, etc.) are assumed independent of temperature and pressure, so as to accentuate the physical aspect of the phenomena in the simplest possible manner. A detailed discussion of the limitations of such a model, for both gases and liquids, is given. It is only possible to take account of varying thermophysical properties by numerical methods, in either the laminar or the turbulent régime.

Chapters 14–17 cover laminar and then turbulent heat transfer, by means of simple but significant examples which allow both the physical aspects of such transfers to be drawn out and the practical implications to be underlined. The methods used are generalizable to other cases and other geometries; note that the results for the most standard geometries are reproduced in 'Basic Data' (Part V).

Chapter 14 is concerned with problems of external forced or natural convection, using a flat plate as an example. Calculation methods based on the momentum and thermal boundary layer models are developed. The importance of heat transfer on leading edges is underlined.

Chapter 15 treats laminar internal flows, for forced convection only. The common special case of conduits of constant cross-section is developed at length; the developments of the mechanical and thermal régimes are studied; they are repeated briefly for turbulent flow in Chapters 16 and 17.

Chapter 16 is an attempt at a first physical approach to turbulent heat transfer. This, which is very complex, is at the present time poorly understood. Nevertheless, models useful to the engineer, based on sometimes doubtful hyopotheses, are used to calculate the development of the temperature and velocity fields at any point in the flow, averaged over a time period long in comparison to that of the turbulent fluctuations. The object of Chapter 16 is to show the basis of these models, which may lead to results which can be used by the heat transfer engineer, particularly in the neighbourhood of a wall. To understand these models, it is necessary to study the structures of the different regions of a turbulent flow and to define scales of length, speed and time for the different types of eddies appearing in the fluid.

The objective of Chapter 17 is more pragmatic. It is to use the dimensional considerations of Chapter 12 and a compilation of experimental results as a basis, and obtain practical correlations for the heat transfer coefficient $h$ in turbulent convection (forced or free, external or internal) for certain simple cases.

Finally, Chapter 18 is a brief introduction to problems which are more complex, but which appear in numerous industrial applications: coupled convective and radiative heat transfer.

# 12

# Dimensional Approach to Convection

Whatever type of convective heat transfer is considered (laminar or turbulent, free or forced), the phenomena which are involved in the expression for conducto-convective[1] flux at a wall within a transparent fluid are as follows:

- an axial **convective momentum transfer** (parallel to the wall);
- a viscous dissipation perpendicular to the direction of motion (**momentum diffusion**);
- an axial **convective enthalpy transfer**;
- a transverse conductive transfer (**enthalpy diffusion**).

This situation is sketched in Figure 12.1; the study of convection consists of modelling the various coupled phenomena.

In this chapter we will study, without writing the balance equations and starting only from considerations of dimensional analysis, the behaviour of the heat transfer coefficient at a wall, $h$, which was introduced in Chapter 1. We will introduce a set of non-dimensional groupings of physical quantities which control thermal convection. The systematic study based on the balance equations will be taken up again in later chapters.

We recall that the convective heat transfer coefficient $h$ allows the conducto-convective flux $\varphi^{cc}$ to be written in the form:

$$\varphi^{cc} = -\lambda \left.\frac{\partial T}{\partial y}\right|_{w} = h(T_w - T_f) \tag{12.1}$$

where $\lambda$ is the thermal conductivity of the fluid and $T_w$ and $T_f$ are the temperature of the wall and a temperature characteristic of the fluid respectively (the temperature field $T(\mathbf{r})$ within a flowing fluid is not necessarily uniform).

[1] The concept of conducto-convective flux was introduced in Chapter 1, Section 1.4.

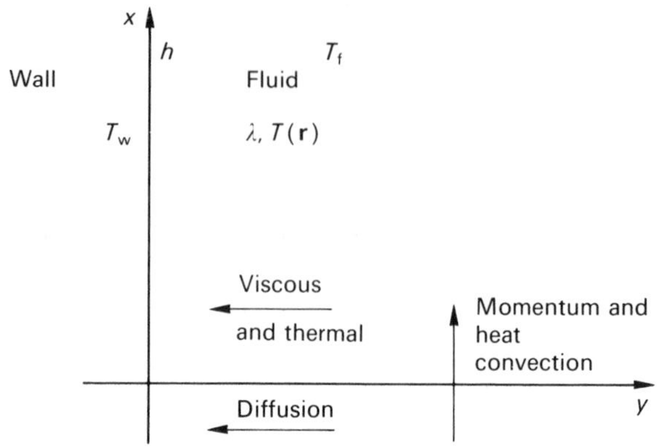


**Figure 12.1** Transfer processes near a wall.

A typical example involving forced convection is treated in Section 12.1 and an example involving free convection is given in Section 12.2.

*Note*
In the case of a semi-transparent fluid (such as combustion products) the situation in Figure 12.1 is complicated by the existence of radiative transfer within the fluid. The heat flux $-\lambda \partial T/\partial y|_w$ near the wall therefore results from the interaction between conduction, convection and radiation. At high temperatures in such conditions, it is very strongly influenced by the radiation. These complex phenomena are not treated systematically in this work; they are introduced, however, in Chapter 18.

## 12.1 Example 1: forced convection

Consider a plane wall at some angle with reference to axes $Oxyz$, of length $L$ in the $x$ direction, maintained at a uniform temperature $T_w$ (Figure 12.2). A fluid of speed $u_0$ and temperature $T_0$ far away from the plane exchanges energy with the plate between the leading edge ($x = 0$) and the trailing edge ($x = L$). We will assume, in this first attempt, that the physical properties of the fluid are constant; more precisely, we will use values of physical properties corresponding to the mean temperature $[T_0 + T_p)/2$:

$\rho$: density ($ML^{-3}$);

$\lambda$: thermal conductivity ($MLT^{-3}\theta^{-1}$);

$\mu$: dynamic viscosity ($ML^{-1}T^{-1}$);

$c_p$: specific heat at constant pressure ($L^2T^{-2}\theta^{-1}$).

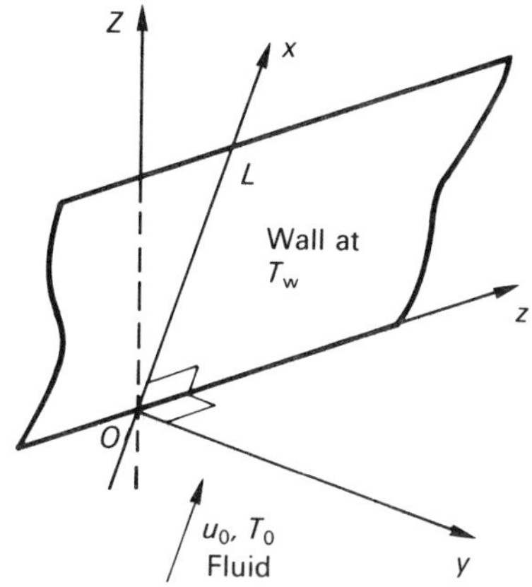


**Figure 12.2**

The dimensions of the properties are given in parentheses: $M$ represents mass, $L$ length, $T$ time and $\theta$ temperature.

We will demonstrate below that the flux density exchanged between the plate and the fluid, and thus the local heat transfer coefficient, varies in general from the leading to the trailing edge. We define the following:

- $h(x)$ the local heat transfer coefficient (at position $x$) associated with the local conducto-convective flux density:

$$\varphi(x) = h(x)(T_w - T_0) = -\lambda \left.\frac{\partial T}{\partial y}\right|_w (x) \tag{12.2}$$

- $\bar{h}$ the average heat transfer coefficient, associated with an average flux density:

$$\bar{\varphi} = \bar{h}(T_w - T_0) = \frac{1}{L}\int_0^L h(x)(T_w - T_0)\,dx \tag{12.3}$$

If we eliminate $\bar{\varphi}$ and $(T_w - T_0)$, using equation (12.3), the thermal problem is completely determined by seven physical quantities: $\bar{h}$, $\rho$, $\mu$, $\lambda$, $c_p$, $L$ and $u_0$, whose respective dimensions are $MT^{-3}\theta^{-1}$, $ML^{-3}$, $ML^{-1}T^{-1}$, $MLT^{-3}\theta^{-1}$, $L$ and $LT^{-1}$, expressed in terms of the four fundamental independent dimensions ($M$, $T$, $L$ and $\theta$). A straightforward application of the $\Pi$ theorem shows that the solution may be expressed as a function of three independent dimensionless groups.

Further, we will make the assumption, well-known in fluid mechanics, that the solution may be written in the form of the product of powers of these groups. This assumption will be valid over a finite range of variation of the parameters. Let us suppose that

$$\bar{h}\,\rho^a\,\mu^b\,\lambda^c\,c_p^d\,u_0^e\,L^f = \text{const}$$

considering of the dimensions of the quantities, this leads to the set of equations:

$$c = -d - 1 \qquad \text{(from } \theta\text{)}$$

$$b = d - a \qquad \text{(from } M\text{)}$$

$$e = a \qquad \text{(from } T\text{)}$$

$$f = a + 1 \qquad \text{(from } L\text{)}$$

solving, with the substitutions

$$\alpha = -a \quad \text{and} \quad \beta = -d$$

we obtain a solution of the form

$$\frac{\bar{h}L}{\lambda} = \text{const}\left(\rho\frac{u_0 L}{\mu}\right)^{\alpha}\left(\frac{\mu c_p}{\lambda}\right)^{\beta} \tag{12.4}$$

which, introducing three dimensionless groups, can also be written

$$Nu_L = \text{const } Re_L^{\alpha} Pr^{\beta} \tag{12.5}$$

| | |
|---|---|
| $Nu_L = \bar{h}L/\lambda$ | Nusselt number based on $L$ |
| $Re_L = \rho u_0 L/\mu$ | Reynolds number based on $L$ |
| $Pr = \mu c_p/\lambda$ | Prandtl number |

Similarly, if we consider the local heat transfer (substitute $x$ and $h(x)$ for $L$ and $\bar{h}$), we obtain

$$\frac{h(x)x}{\lambda} = \text{const}\left(\rho\frac{u_0 x}{\mu}\right)^{\alpha}\left(\frac{\mu c_p}{\lambda}\right)^{\beta} \tag{12.6}$$

or

$$Nu_x(x) = \text{const } (Re_x)^{\alpha} Pr^{\beta} \tag{12.7}$$

| | |
|---|---|
| $Nu_x(x) = h(x)x/\lambda$ | local Nusselt number, at position $x$ and based on $x$ |
| $Re_x = \rho u_0 x/\mu$ | Reynolds number, based on $x$ |

We note that the exponents in equation (12.7) are the same as those in (12.5).

It is possible to give a physical interpretation to the groups which appear in equations (12.5) and (12.7):

1. The Reynolds number $Re_L = \rho u L/\mu$ represents the ratio of the inertia (momentum convection) forces to the viscous forces:

$$Re = 2\left(\frac{\rho u^2}{2L}\right)\left(\frac{\mu u}{L^2}\right)^{-1} \tag{12.8}$$

This number characterizes the flow, in particular the régime (laminar or turbulent, etc.). Clearly, it is the Reynolds number which has most effect on the Nusselt number.

2. The Prandtl number $Pr = \mu c_p/\lambda$ represents the ratio of the momentum diffusivity $\nu$ (or kinematic viscosity) to the thermal diffusivity $a$:

$$Pr = \frac{\mu c_p}{\lambda} = \left(\frac{\mu}{\rho}\right)\left(\frac{\lambda}{\rho c_p}\right)^{-1} = \frac{\nu}{a} \tag{12.9}$$

It thus compares the two transverse dissipative diffusion phenomena, thermal conduction and viscous dissipation.

The Prandtl number depends only on the physical properties of the fluid. We will principally consider three classes of fluid in the following:

- Gases (for which the Prandtl number is close to 1) and common liquids (water, light oils, ...) for which the Prandtl number is of order of magnitude around 1 ($Pr = 7$ for water at 20°C).
- The liquid metals (K, Na, Li, NaK, etc.). These are good conductors (often used as heat transfer fluids), whose Prandtl numbers are very low ($\simeq 10^{-2}$). These fluids have a very different behaviour from those preceding as regards enthalpy transfer.
- Very heavy oils whose Prandtl number can exceed 1000.

3. The local Nusselt number $Nu_x = h(x)x/\lambda$ represents the non-dimensional temperature gradient at the wall. Substituting

$$y^+ = \frac{y}{x}; \qquad T^+ = \frac{T - T_w}{T_0 - T_w} \tag{12.10}$$

equation (12.2) becomes:

$$Nu_x(x) = \left.\frac{\partial T^+}{\partial y^+}\right|_w (x) \tag{12.11}$$

We may compare this expression to that for the friction coefficient $c_F$ in fluid mechanics, defined as the ratio of the viscous stress at the wall to the kinetic energy per unit volume (of the approaching flow in our case) [1]:

$$c_F(x) = \frac{\tau_w}{\frac{1}{2}\rho u_0^2} = \frac{\mu\, \partial u/\partial y|_w(x)}{\frac{1}{2}\rho u_0^2} \tag{12.12}$$

Substituting:

$$y^+ = \frac{y}{x} \quad \text{and} \quad u^+ = \frac{u}{u_0} \tag{12.13}$$

we obtain:

$$c_F(x) = \left.\frac{2}{Re_x}\frac{\partial u^+}{\partial y^+}\right|_w (x) \tag{12.14}$$

4. The Nusselt number $Nu_L = \bar{h}L/\lambda$ is used to calculate the overall heat transfer for the plate. It corresponds to the mean heat transfer coefficient $\bar{h}$. Thus

$$Nu_L = \int_0^L \frac{Nu_x(x)}{x}\,\mathrm{d}x \tag{12.15}$$

Note that the mean obtained is not the mean value of $Nu_x(x)$.

*Notes:*

(a) Other dimensionless groups, relevant to a particular geometry, may arise.

(b) Groupings of the previous groups often appear:
- the Péclet[2] number: $Pe = Re \cdot Pr$;
- the Stanton number: $St = Nu/Pe$ (sometimes also called the Margoulis number), etc.

(c) The reference lengths which appear in the various numbers ($Nu$, $Re$, etc.) in place of $L$ are various and are dependent on the geometry under consideration (cylinder diameter, sphere diameter, ...). It may also be necessary to introduce other characteristic numbers such as the ratio of the diameter of a tube to its length, $D/L$, etc.

(d) In forced convection in a closed geometry (internal convection), we encounter the local Nusselt number at a point $x$ ($x$ measured in the flow direction) associated with a transverse dimension (the tube diameter $D$ for example), denoted by $Nu_D(x)$:

$$Nu_D(x) = \frac{h(x)D}{\lambda} \tag{12.16}$$

In the fully developed régime (a certain distance along the tube) this Nusselt number (and $h$) becomes independent of $x$. Then we write simply:

$$Nu_D = \frac{hD}{\lambda} \tag{12.16.1}$$

Finally, if we consider the Nusselt number appropriate to the average heat transfer coefficient $\bar{h}$ from the tube entry at $x = 0$ to $x = L$, we write:

$$Nu_{D,L} = \frac{\bar{h}D}{\lambda} \tag{12.16.2}$$

## 12.2 Example 2: natural convection

We consider the same geometry as before, but with a vertical plate at a temperature $T_w$ in a fluid of temperature $T_0$ at rest far from the plate (Figure 12.3). The thermophysical properties of the fluid ($\mu$, $\lambda$, $c_p$) are assumed to be constant, with the

[2] Co-founder of the Ecole Centrale Paris.

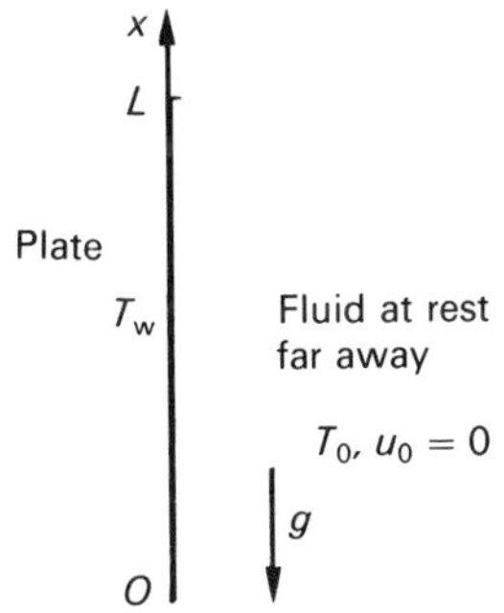


**Figure 12.3** Vertical flat plate at uniform temperature in natural convection.

exception of the density $\rho$ whose variation is given by the formula

$$\rho(T) = \rho_0(T_0)[1 - \beta(T - T_0)] \tag{12.17}$$

where $T_0$ represents, for example, the temperature of the undisturbed fluid.

In fact, it is the change of density which is the origin of the phenomena of natural convection, since it creates a buoyancy (Archimedean) force $\rho_0\beta(T - T_0)g$ which is balanced mainly by viscous friction forces.

We will make the approximations attributed to **Boussinesq**:

- The variations in temperature are very small, so that

$$\beta(T - T_0) \ll 1 \tag{12.18}$$

- The variation of $\rho$ with $T$ is neglected except in the driving (Archimedean) term in the momentum balance equation. It can be shown that this follows from the preceding condition. The discussion of this point is continued in Section 13.1.3.3.

Equations (12.2) and (12.3) represent the local and mean fluxes through the plate, as before, and are associated with the local and mean heat transfer coefficients $h(x)$ and $\bar{h}$. The overall heat transfer problem is completely defined by seven quantities: $\bar{h}$, $\rho_0$, $\mu$, $\lambda$, $c_p$, $L$ and the buoyancy term $\rho_0 g\beta(T_w - T_0)$.

Instead of repeating the method developed for the case of forced convection, we will work by analogy. Having introduced the buoyancy force $\rho_0 g\beta(T_w - T_0)$ as one of the parameters, the choice of a reference velocity $u_r$ (analogous to $u_0$) is not immediately obvious. However, if we find an expression for $u_r$ which depends on $g\beta(T_w - T_0)$ and other parameters, we can use the previous results (equations (12.5) and (12.7)) directly by substituting $u_r$ for $u_0$. To achieve this, let us equate the maximum buoyancy force per unit volume $\rho_0 g\beta(T_w - T_0)$ to a quantity of the order of magnitude of the viscous force per unit volume which opposes it $\mu u_r/L^2$. This argument leads to:

$$u_r = \frac{\rho_0 g\beta(T_w - T_0)L^2}{\mu} \tag{12.19}$$

The problem is now defined by the seven physical quantities $\bar{h}$, $\rho_0$, $\mu$, $\lambda$, $c_p$, $L$ and $u_r$. From (12.5), the Reynolds number has the form:

$$\frac{\rho_0 u_r L}{\mu} = \frac{\rho_0^2 g\beta(T_w - T_0)L^3}{\mu^2} = Gr_L \tag{12.20}$$

and is called the Grashof number, $Gr_L$, based on the length $L$.

The overall Nusselt number is then given by:

$$Nu_L = \text{const } Gr_L^{\alpha} Pr^{\beta} \tag{12.21}$$

similarly, the local Nusselt number at position $x$ is:

$$Nu_x(x) = \text{const } Gr_x^{\alpha} Pr^{\beta} \tag{12.22}$$

in which the local Grashof number, based on $x$, appears

$$Gr_x(x) = \frac{\rho_0^2 g\beta(T_w - T_0)x^3}{\mu^2} \tag{12.23}$$

From this simple dimensional study we can deduce an important result for free convection: although the heat transfer rises with the buoyancy term $\rho_0 g\beta(T_w - T_0)$, it rises as the cube of the characteristic dimension $x$ (or $L$). The viscosity obviously plays a restraining role on the heat transfer – it appears as $1/\mu^2$ in the Grashof number.

*Notes*

1. The results obtained are quite general, but the characteristic length $L$ which appears in the definitions of $Gr_L$ and $Nu_L$ clearly depends on the geometry under consideration, as for forced convection.
2. The Rayleigh number, $Ra$, may also be defined:

$$Ra_x(x) = Gr_x Pr = \frac{g\beta(T_w - T_0)x^3}{\nu a} \tag{12.24}$$

This number may obviously be based on $L$ or $x$. It links the buoyancy term and the two dissipative diffusion phenomena: viscous friction ($\nu$) and thermal conductivity ($a$). Whence:

$$Nu_L = \text{const } Ra_L^{\alpha'} Pr^{\beta'} \tag{12.21.1}$$

$$Nu_x(x) = \text{const } Ra_x^{\alpha'} Pr^{\beta'} \tag{12.22.1}$$

## 12.3 Convective similarity

The very abstract expressions for Nusselt number obtained above (equations (12.5), (12.7), (12.21) and (12.22)) can be generalized to other situations. The problems considered must have:

- A specified geometry, defined at the least by a reference length $H$ (the plate length $L$ in our examples) appearing in the expressions for $Nu_L$, $Gr_L$ and $Re_L$. Other lengths may also appear.
- Specified mechanical and thermal boundary conditions – temperature or heat flux on the plate, upstream or far-distant fluid temperature, zero velocity on the wall, upstream or far-distant fluid velocity, etc.
- A specified type of flow, characterized by the Reynolds number in forced convection or the Rayleigh number in free convection.
- A specified fluid, described by its Prandtl number. We distinguish between common fluids whose Prandtl number is of order 1, liquid metals for which $Pr$ is around $10^{-2}$ and oil with high values of $Pr$.

**Two problems are said to be thermally similar if they are geometrically similar, have identical non-dimensional boundary conditions, the same value of Reynolds (or Rayleigh) number and the same Prandtl number. The Nusselt number must then be the same.**

More generally, the formulae for Nusselt number of the form:

$$Nu = \text{const } Re^{\alpha} Pr^{\beta}, \qquad Nu = \text{const } Ra^{\alpha''} Pr^{\beta''} \tag{12.25}$$

are the same when:

- the geometries are similar;
- the non-dimensional boundary conditions are identical;
- the Reynolds numbers correspond to the same flow régime (for example, laminar or turbulent);
- the fluids belong to the same class, as regards Prandtl number.

The equations used may derive from the following:

1. A theoretical model. The accuracy is then dependent on knowledge of the fluid properties ($\lambda$, $\rho$, $c_p$, $\mu$) and on the accuracy of the analytical and/or numerical methods used. An accuracy of 1 part in $10^3$ can be achieved in laminar flow, much less in turbulent.
2. A semi-theoretical semi-experimental approach, with a certain amount of empiricism necessary due to lack of understanding of turbulent flows. The expected accuracy is 10–30%.

Some well-known expressions applicable to forced and free convection in various conditions will be found in the fifth part of this work, 'Basic Data'.

## 12.4 The laminar to turbulent transition

The aim of this section is to recall various basic results in fluid mechanics concerning the transition between laminar and turbulent flow régimes. We consider simple cases of internal and external flows.

1. In external convection, the fluid flow is only affected by the surface considered (flat plate, exterior of a tube with flow parallel or normal to its axis, engine block, etc.). The reference length is generally the distance $x$ from the leading edge of the plate or structure. For a flat plate, at a point $x$ from the leading edge, we distinguish the cases of forced and free convection:
   - In external forced convection, the régime is laminar in the neighbourhood of the flat plate from the leading edge ($x = 0$) up to a distance $x_c$ from there, such that

     $$Re_{x_c} = \frac{\rho u_0 x_c}{\mu} \simeq 10^5 \qquad (12.26)$$

     Beyond $x_c$, turbulence appears first intermittently, then fully developed as $x$ increases (see Chapter 16). We stress at this point that the limit in equation (12.26) corresponds to the start of the transition zone.
   - In external natural convection, the régime is laminar in the neighbourhood of a flat vertical isothermal plate (under the Boussinesq approximations, see Section 12.2) from $x = 0$ to $x = x_c$ defined by

     $$Gr_{x_c} = \frac{g\beta(T - T_0)x_c^3}{\nu^2} \simeq 10^9 \qquad (12.27)$$

     Near the critical value $x_c$, intermittent turbulence appears as before, and thereafter develops with increasing $x$.
2. In forced internal convection, the fluid flow often occurs inside a tube or structure which can be characterized by a hydraulic diameter (see the Appendix to Part IV). The hydraulic diameter represents the diameter of a circular tube equivalent to the system under consideration. The nature of the flow régime is in this instance governed by a Reynolds number:

   $$Re_{D_h} = \frac{\rho u_0 D_h}{\mu} \qquad (12.28)$$

   Roughly speaking, transition will occur at a value of $Re_{D_h}$ of 2500:

   $$Re_{D_h} < 2500\text{: laminar flow}$$
   $$Re_{D_h} > 2500\text{: turbulent flow} \qquad (12.29)$$

   In fact, turbulence may not appear until values of $Re_{D_h}$ above $10^4$ under very special conditions where the flow is carefully smoothed through successive contractions and the system is isolated from any vibration. The transition values given here correspond to normal conditions. Note also that with strongly non-isothermal fluids, a mixed convection phenomenon may also delay (depending on $D_h$) the onset of turbulence.

Internal free convection is much harder to analyse because of the existence of flows in opposing directions at one cross-section (see Part V, 'Basic Data').

# 13

# Balance equations

## 13.1 Dimensional equations

### 13.1.1 General equations

The general convection equations were established in Chapter 3 (Section 3.3.2). They can be applied even in turbulent flow so long as instantaneous values are considered. They consist of the equation of state of the fluid and the balance equations of the various quantities, which are expressed in tensor and vector notation in this part.

1. **The equation of state** can be written in the form:

$$f(p, \rho, T) = 0 \tag{13.1}$$

where $p$ is the local pressure. Two particular cases are often used: a perfect gas has the equation:

$$p = r\rho T \tag{13.2}$$

where $r$ is the gas constant for unit mass, while an incompressible, undilatable liquid is characterized by:

$$\rho = \text{const} \tag{13.3}$$

2. **The mass balance equation**, often called the continuity equation, is written:

$$\frac{\partial \rho}{\partial t} + \frac{\partial}{\partial x_j}(\rho v_j) = 0 \qquad \frac{\partial \rho}{\partial t} + \boldsymbol{\nabla}\cdot(\rho \mathbf{v}) = 0 \tag{13.4.1}$$

where $\mathbf{v}$ is the velocity of the fluid at $M(\mathbf{r})$. In steady conditions, this reduces to:

$$\frac{\partial}{\partial x_j}(\rho v_j) = 0 \qquad \boldsymbol{\nabla}\cdot(\rho \mathbf{v}) = 0 \tag{13.4.2}$$

If we assume equation of state (13.3), it becomes:

$$\frac{\partial}{\partial x_j} v_j = 0 \qquad \nabla \cdot \mathbf{v} = 0 \tag{13.4.3}$$

which implies that the fluid flows in a constant volume manner.

3. **The momentum balance equation** for a Newtonian fluid (see Chapter 3) is:

$$\rho \frac{\mathrm{d}v_i}{\mathrm{d}t} = \rho\left(\frac{\partial v_i}{\partial t} + v_j \frac{\partial v_i}{\partial x_j}\right)$$
$$= \rho g_i - \frac{\partial}{\partial x_i} p + \frac{\partial}{\partial x_j}\left[\mu\left(\frac{\partial v_i}{\partial x_j} + \frac{\partial v_j}{\partial x_i}\right) + \left(\kappa - \frac{2}{3}\mu\right)\delta_{ij}\frac{\partial v_k}{\partial x_k}\right] \tag{13.5.1}$$

where $g_i$ is the resultant body force per unit mass, or in vector notation:

$$\rho \frac{\mathrm{d}\mathbf{v}}{\mathrm{d}t} = \rho\left(\frac{\partial \mathbf{v}}{\mathrm{d}t} + \mathbf{v}\cdot\nabla\mathbf{v}\right) = \rho g - \nabla\cdot\left[\mu(\nabla\mathbf{v} + \nabla^{\mathrm{T}}\mathbf{v}) + \left(\kappa - \frac{2}{3}\mu\right)\underline{\mathbf{U}}\nabla\cdot\mathbf{v}\right] \tag{13.5.2}$$

where $\nabla^{\mathrm{T}}\mathbf{v}$ is the transpose tensor of $\nabla\mathbf{v}$ and $\underline{\mathbf{U}}$ is the unit tensor. In steady flow, the left-hand side of (13.5) becomes $\rho\mathbf{v}\cdot\nabla\mathbf{v}$, and represents the effect of momentum convection.

4. **The energy equation** is:

$$\rho c_{\mathrm{p}} \frac{\mathrm{d}T}{\mathrm{d}t} = \rho c_{\mathrm{p}}\left(\frac{\partial T}{\partial t} + v_j \frac{\partial}{\partial x_j} T\right) = -\frac{\partial}{\partial x_j}(q_j^{\mathrm{cd}} + q_j^{\mathrm{R}}) + P + T\beta\frac{\mathrm{d}p}{\mathrm{d}t} + \mu\Phi_{\mathrm{D}} \tag{13.6.1}$$

or:

$$\rho c_p \frac{\mathrm{d}T}{\mathrm{d}t} = \rho c_p\left(\frac{\partial T}{\partial t} + \mathbf{v}\cdot\nabla T\right) = -\nabla\cdot(\mathbf{q}^{\mathrm{cd}} + \mathbf{q}^{\mathrm{R}}) + P + T\beta\frac{\mathrm{d}p}{\mathrm{d}t} + \mu\Phi_{\mathrm{D}} \tag{13.6.2}$$

Its terms are explained in Chapter 3:

- $\rho c_{\mathrm{p}}\mathbf{v}\cdot\nabla T$: effect of enthalpy convection;
- $\mathbf{q}^{\mathrm{cd}}, \mathbf{q}^{\mathrm{R}}$: flux density vectors for conduction and radiation;
- $P$: power dissipated per unit volume;
- $T\beta\, \mathrm{d}p/\mathrm{d}t$: effect of reversible adiabatic pressure forces (negligible in incompressible flow);
- $\mu\Phi_{\mathrm{D}}$: irreversible dissipation due to viscous friction (often negligible and usually neglected).

Other balance equations (entropy, vorticity, or for turbulent flow, turbulent kinetic energy per unit mass or its dissipation) may be added to these principal equations.

Associated with the whole of the above system of equations are **initial** and **boundary conditions**:

- **Mechanical**: zero relative velocity at a wall, velocity far away; particular cases with phase change, for example vapour bubbles in a liquid.
- **Thermal**: very variable, associated with conduction and radiation (often coupled) and convection (in open systems), perhaps chemical reaction, latent heat of change of state, etc.

The system of equations (13.1)–(13.6) is extremely complex if we take account of the effect of pressure $p$ and particularly temperature $T$ on the thermophysical properties ($\rho$, $\lambda$, $\mu$, $c_p$) or quantities directly dependent on them (thermal diffusivity $a = \lambda/(\rho c_p)$ and mechanical diffusivity $\nu = \mu/\rho$) as well as the Prandtl number ($Pr = \nu/a$). Note in particular that we can decouple the momentum equation (13.5) and the energy equation (13.6) if we assume that $\rho$ and $\mu$ are, over a wide region, independent of temperature $T$. It is therefore essential to study the behaviour of the thermophysical properties of common fluids as functions of $T$ and $p$.

### *13.1.2 Dependence of thermophysical properties on temperature and pressure*

Comprehensive tables showing the variations of $\lambda$, $\rho$, $c_p$, $\mu$, $\nu$, $a$ and $Pr$ for common fluids are given in the 'Basic Data' section, which the reader should consult while studying this section.

#### 13.1.2.1 DEPENDENCE ON PRESSURE

**1. Gases**

According to the kinetic theory of gases, under normal conditions (scales large compared with the mean free path and pressures not high enough to consider other than binary collisions) the conductivity $\lambda$ and the dynamic viscosity $\mu$ are independent of pressure. For a perfect gas, $c_p$ does not depend on pressure.

On the other hand $\rho$, and consequently $\nu$ and $a$, depend strongly on pressure. Under the perfect gas model, we use equation (13.2).

**2. Liquids**

To an excellent approximation, liquids may be considered incompressible and undilatable (except of course in the driving term for natural convection), so that all processes are at constant volume:

$$\frac{\partial \rho}{\partial p} \simeq 0 \qquad \text{and} \qquad \frac{\partial \rho}{\partial T} \simeq 0 \quad \Rightarrow \rho = \text{const} \tag{13.7}$$

13.1.2.2 DEPENDENCE ON TEMPERATURE

**1. Gases**

The density $\rho$ is increased as $1/T$ for a perfect gas. The quantities $\lambda$ and $\mu$ have a similar dependence on $T$ (in theory as $\sqrt{T}$ for a simple kinetic model; in fact the actual dependence is more like $T^{0.8}$). The specific heat at constant pressure $c_p$ varies little with $T$ (less than 20% for air between 250 and 1300 K at atmospheric pressure – see 'Basic Data').

A consequence of the similar behaviour of $\lambda(T)$ and $\mu(T)$ and the weak dependence of $c_p$ on $T$ is that the Prandtl number ($Pr = \mu c_p/\lambda$) varies very little with temperature for gases. For air it is about 0.7, with variations of less than 3% between 250 and 1300 K.

The Reynolds number varies appreciably with temperature through the dynamic viscosity $\mu(T)$. We should note that in the group $Re = \rho v D/\mu$, the quantity $\rho v$ is constant in a flux tube (for example, a pipe) whatever the temperature gradient.

**2. Liquids**

We generally assume that the density $\rho$ is constant. Variations in the thermal conductivity $\lambda$ with temperature are generally moderate: for water between 0 and 300 °C $\lambda$ rises from 0.54 to 0.68 W m$^{-1}$ K$^{-1}$. On the contrary, the variations in the dynamic viscosity $\mu$ (and consequently those of $\nu$ since $\rho$ is constant) with temperature are large for water and many other liquids. Between 20 and 100 °C, $\mu$ for water varies from $10^{-2}$ to $0.281 \times 10^{-2}$ kg m$^{-1}$ s$^{-1}$, and reaches $10^{-3}$ kg m$^{-1}$ s$^{-1}$ at 280 °C! Because of this, the Prandtl number varies significantly with temperature. For water, $Pr$ goes from 7 to 1.74 between 20 °C and 100 °C. Similarly, the Reynolds number varies strongly with temperature on account of the variation in $\mu(T)$ ($\rho u$ being constant).

### 13.1.3 *Limitation of the problem*

13.1.3.1 VARIATION OF $\rho$

Liquids generally present no difficulty; we use (13.7) except, obviously, in the driving term for free convection.

For gases, while it may be reasonable to assume incompressible flow (relatively small variations in $p$), it is much less so to assume that it is undilatable ($\partial\rho/\partial T = 0$). Therefore, in most cases we have to make the Boussinesq approximation (see Section 12.2). Thus, in this course we will generally set $\rho$ = const with respect to $T$ and $p$, except in the driving term for free convection.

We will not consider the case of a gas subjected to significant temperature or pressure changes in forced or free convection. This kind of problem can only be solved by direct numerical methods. They present no particular physical problems, particularly in the laminar régime, but can be difficult to handle numerically (see, for example, references [2]–[4]).

13.1.3.2 VARIATION OF $\mu$

The second difficulty arises from the variation of dynamic viscosity $\mu$ with $T$, which cannot be neglected either in gases (where it affects $Re$) or in liquids (where it affects both $Re$ and $Pr$) if there are significant temperature changes. Nevertheless, we will assume that **dynamic viscosity is constant**, by using an average temperature as a reference temperature. Various studies have shown that in free or forced internal or external convection with laminar or turbulent flow, the values of $Nu$ obtained may be corrected by a factor $(\mu_m/\mu_w)^{0.14}$, where $\mu_w$ and $\mu_m$ are the viscosities at the wall and mixed fluid temperatures respectively (see Sections 17.2 and 17.3).

13.1.3.3 THE CHOSEN MODEL

We will suppose that, over at least a certain distance, the thermophysical properties ($\lambda$, $\rho$, $\mu$, $c_p$, $\nu$, $a$, $Pr$) are independent of temperature (except in the driving term in the case of $\rho$). If we further assume **steady flow**, the system of equations to be solved is:

1. For forced convection:

$$\rho = \text{const}, \qquad \nabla \cdot \mathbf{v} = 0 \tag{13.8.1}$$

$$\rho \mathbf{v} \cdot \nabla \mathbf{v} = -\nabla p + \mu \nabla^2 \mathbf{v} \tag{13.8.2}$$

$$\rho c_p \mathbf{v} \cdot \nabla T = \lambda \nabla^2 T + P + (\mu \phi_D + T\beta \mathbf{v} \cdot \nabla p) \tag{13.8.3}$$

- + (perhaps) balance of other quantities;
- + **mechanical and thermal boundary conditions**.

We notice that in these conditions the mechanical and thermal problem is de-coupled from the thermal one. The solutions of (13.8.1) and (13.8.2) allow (13.8.3) to be solved.

2. For free convection, we will make the Boussinesq approximation (see Section 12.2), that is, on the one hand $\beta(T - T_0)$ is small compared to 1, and on the other variations of $\rho$ with $T$ are only taken into account by the driving (Archimedean) term $\rho_0 g \beta(T - T_0)$. We use the same equations (13.8), modifying the momentum equation using equation (13.5):

$$\rho_0 \mathbf{v} \cdot \nabla \mathbf{v} = -\rho_0 g[1 - \beta(T - T_0)]\nabla Z - \nabla p + \mu \nabla^2 \mathbf{v} \tag{13.8.2a}$$

$$\rho_0 \mathbf{v} \cdot \nabla \mathbf{v} = -\nabla(p + \rho_0 g Z) + \beta(T - T_0)\rho_0 g \nabla Z + \mu \nabla^2 \mathbf{v} \tag{13.8.2b}$$

In this equation, $Z$ represents the upward vertical distance, whose direction depends on the coordinate system chosen. The equation replaces (13.8.2) for free convection. The piezometric pressure $p + \rho_0 g Z$ appears on the right-hand side. Notice that in the absence of any thermal perturbation (and thus of motion – $\mathbf{v} = 0$), we obtain hydrostatic equilibrium. In particular, if $OZ$ coincides with $Ox$ (Figure 12.3), $\partial p/\partial x = -\rho_0 g$.

*Note*

The Boussinesq approximation can be justified by a systematic analysis of the orders of magnitude of all the thermophyiscal variables (mass, momentum, energy) in all the equations. To use a simple physical argument, in the absence of the driving term (which is of order $\beta(T-T_0)$), no motion occurs, and the static pressure gradient $(\partial p/\partial x)$ is equal to $-\rho_0 g$. Consider the driving term $\rho_0 g\beta(T-T_0)$, the system response which is of order 1 relative to $\beta(T-T_0)$ can only come from contributions of order zero of the other terms in $\rho$. Further, the variations in $\rho$ due to $T$ in a term which is not the driving term will introduce terms of order 2 in $\beta(T-T_0)$, which can therefore be neglected.

## 13.2 Non-dimensional balance equations

We will restrict ourselves to cases where the only body force is weight. Natural convection will develop in such a system, and with forced convection we obtain mixed convection.

We will adopt the dimensionless variables:

$$x^+ = \frac{x}{H}, \qquad y^+ = \frac{y}{H}, \qquad z^+ = \frac{z}{H}, \qquad Z^+ = \frac{Z}{H} \tag{13.9}$$

where $H$ is a characteristic length for the geometry to be defined, and is the same for all three dimensions.

$$\mathbf{v}^+ = \frac{\mathbf{v}}{v_0} \tag{13.10}$$

where $v_0$ is a positive reference velocity, characteristic of the forced convection.

$$p^+ = \frac{p}{\rho_0 v_0^2} \tag{13.11}$$

where the reference pressure is $\rho_0 v_0^2$.

$$T^+ = \frac{T - T_r}{\delta T} \tag{13.12}$$

where $T_r$ is a reference temperature and $\delta T$ is a temperature difference (real or defined in terms of flux density, for example $H\varphi/\lambda$).

Making the assumptions in Section 13.1.3, the momentum equation becomes:

$$\left(\frac{\rho_0 v_0^2}{H}\right)\mathbf{v}^+\cdot\nabla^+\mathbf{v}^+ = -\rho_0 g\nabla^+ Z^+(1-\beta\,\delta T T^+) - \frac{\rho_0 v_0^2}{H}\nabla^+ p^+ + \frac{\mu v_0}{H^2}(\nabla^{+2}\mathbf{v}^+) \tag{13.13}$$

which becomes, after manipulation:

$$\mathbf{v}^+ \cdot \nabla^+ \mathbf{v}^+ = -\nabla^+ Z^+ \left( \frac{1}{Fr_H} - \frac{Gr_H}{Re_H^2} T^+ \right) - \nabla^+ p^+ + \frac{1}{Re_H} \nabla^{+2} \mathbf{v}^+ \tag{13.14}$$

In this expression the following numbers appear:

- Grashof: $Gr_H = \rho_0^2 g \beta (\delta T) H^3 / \mu^2$;
- Reynolds: $Re_H = \rho_0 v_0 H / \mu$;
- Froude: $Fr_H = v_0^2 / Hg$.

(This last number compares the reference kinetic energy to the reference gravitational potential energy.)

If we take account of the viscous dissipation term ($\Phi_D$) and the compressibility term ($-\mathbf{v} \cdot \nabla p$), which are practically negligible, and introduce the friction coefficient $c_F$ defined in equation (12.15), the energy equation becomes:

$$\left( \frac{\rho_0 c_p v_0 \, \delta T}{H} \right) \mathbf{v}^+ \cdot \nabla^+ T^+ = \left( \frac{\lambda}{H^2} \delta T \right) \nabla^{+2} T^+ + \frac{\rho v_0^2}{2} \left( \frac{v_0}{H} \right) (c_F \Phi_D^+ + 2(T\beta) \mathbf{v}^+ \cdot \nabla^+ p^+) \tag{13.15}$$

where we have set:

$$\Phi_D^+ = \frac{2\mu \Phi_D}{\rho v_0^2 c_F (v_0 / H)} \tag{13.16}$$

which becomes after manipulation:

$$\mathbf{v}^+ \cdot \nabla^+ T^+ = \frac{1}{Pe_H} \nabla^{+2} T^+ + Ec(c_F \Phi_D^+ + 2(T\beta) \mathbf{v}^+ \cdot \nabla^+ p^+) \tag{13.17}$$

In practice, $Ec(c_F \Phi_D^+ + 2(T\beta)\mathbf{v}^+ \cdot \nabla^+ p^+)$ is generally negligible. In this expression we see the Péclet number ($Pe = Re\,Pr$) and the Eckert number $Ec$, which compares the kinetic energy per unit volume to the enthalpy transfer per unit volume:

$$Ec = \frac{1}{2} \frac{\rho v_0^2}{\rho c_p \, \delta T} \tag{13.18}$$

Equations (13.14) and (13.17) re-emphasize the controlling role played by the various non-dimensional numbers $Re$, $Gr$, $Pe$, $Pr$ (see Chapter 12).

The term $Gr/(Re^2) T^+$ in equation (13.14), which incorporates the Richardson number $Ri = Gr/Re^2$, allows the domains of forced and free convection to be defined precisely, as follows:

- If $Gr < Re^2$, **forced convection** dominates: the term $Fr^{-1}$ is then often negligible compared to 1 (except in central heating, for example). The non-dimensional fields of velocity and temperature then depend separately on $Pe$ and $Re$ or (if you prefer), on $Pr$ and $Re$.

- If $Gr > Re^2$, **natural convection** dominates: in the limit of $v_0$ tends to 0, it is convenient to abandon the non-dimensional velocities ($v_0$ is no longer meaningful) and the Reynolds number disappears from the expressions.
- If $Gr \simeq Re^2$, we are in the domain of **mixed convection**.

Convection problems can be very difficult to analyse, obviously so in the turbulent régime, but also in laminar flow. In order of increasing difficulty, we find problems of forced convection, free convection and finally mixed convection.

# 14

# External Laminar Convection. Boundary Layers

We are going to study the heat transfer between a fluid and a flat plate, a cylinder and a sphere, under conditions where the flow far away is undisturbed (or, for natural convection, is at rest). This is known as **external convection** or **open geometry**.

This type of problem depends, as in fluid mechanics, on the modelling of boundary layers. Here we will just introduce the two principal boundary layers: the momentum and thermal boundary layers, **in laminar flow**.

## *14.1 Momentum and thermal boundary layers*

The momentum boundary layer is due to diffusion of momentum through viscous friction; the velocity field of the fluid is perturbed by the proximity of the wall, in the immediate neighbourhood of which the velocity reduces to zero, while far away the flow is undisturbed.

- In forced convection, the momentum boundary layer is defined as the region of fluid between the wall and the position of the points (along a normal to the wall) where the velocity is 99% of the velocity of the undisturbed flow (Figure 14.1(a)).
- In free convection, the momentum boundary layer is the region of fluid between the wall and the position of points near the undisturbed region where the velocity is 1% of the maximum velocity reached along the normal under consideration (Figure 14.1(b)).

The thermal boundary layer is due to enthalpy diffusion. The temperature field of the fluid is perturbed by the presence of the wall, which imposes extremely diverse thermal conditions at the wall–fluid interface (constant temperature or flux, etc., always with continuity of energy flux).

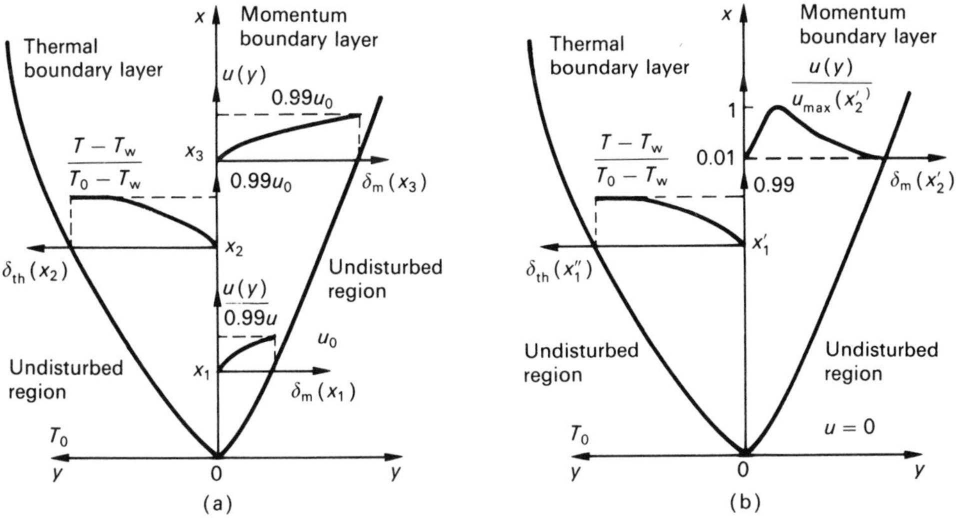


**Figure 14.1** (a) Forced convection. Laminar flow. (b) Free convection. For clarity, the momentum and thermal boundary layers are shown on the right and left of each figure respectively. In reality, of course, they are superimposed. The non-dimensional velocity and temperature profiles are shown as functions of $y$, at fixed positions $x_i$.

- The thermal boundary layer is the region of fluid between the wall and the position of points where the temperature difference with the wall, $T - T_w$, measured along the normal to the wall, is 99% of the temperature difference $T_0 - T_w$ between the undisturbed part of the fluid and the wall (Figures 14.1 (a) and 14.1 (b)).

  The momentum boundary layer thickness at a position $x$, measured normal to the wall, is denoted by $\delta_m(x)$, and the thermal boundary layer thickness by $\delta_{th}(x)$. These thicknesses may be very closely the same or very different according to the Prandtl number ($Pr = \nu/a$), as we will see below.

*Notes*

1. Experience shows that, except in the immediate vicinity of the leading edge, the momentum and thermal boundary layer thicknesses are very small compared to $x$:

$$\frac{\delta_{th}}{x}, \quad \frac{\delta_m}{x} \ll 1$$

2. In fluid mechanics, there are many other physical quantities associated with boundary layers [1]; we will not require them in this work.

## 14.2 Isothermal flat plate in an isothermal flow, with constant velocity far away and in the steady state

Neglecting free convection here, we will look again at the problem in Section 12.1, which is controlled by a number of mechanical and thermal equations with their associated boundary conditions. We will represent the velocity components in the $x$ and $y$ directions by $u$ and $v$, and use $u_0$ for the $x$ component of the upstream velocity $\mathbf{v}_0$. The boundary conditions (Figure 14.2) are:

$$T(x,0) = T_w; \qquad u(x,0) = 0 \tag{14.0.1}$$

$$T(x,\infty) = T_0(x); \qquad u(x,\infty) = u_0(x) \tag{14.0.2}$$

$$T(0,y) = T_0(y); \qquad u(0,y) = u_0(y) \tag{14.0.3}$$

Using the general assumptions of Section 13.1.3.3, the various balance equations are:

- Conservation of mass:

$$\frac{\partial u}{\partial x} + \frac{\partial v}{\partial y} = 0 \tag{14.1.1}$$

- $x$ and $y$ components of the momentum balance equation:

$$u\frac{\partial}{\partial x}u + v\frac{\partial}{\partial y}u = -\frac{\partial p}{\partial x} + \nu\left(\frac{\partial^2 u}{\partial x^2} + \frac{\partial^2 u}{\partial y^2}\right) \tag{14.1.2}$$

$$u\frac{\partial}{\partial x}v + v\frac{\partial}{\partial y}v = -\frac{\partial p}{\partial y} + \nu\left(\frac{\partial^2 v}{\partial x^2} + \frac{\partial^2 v}{\partial y^2}\right) \tag{14.1.3}$$

- Conservation of energy (without source term):

$$u\frac{\partial}{\partial x}T + v\frac{\partial}{\partial y}T = a\left(\frac{\partial^2 T}{\partial x^2} + \frac{\partial^2 T}{\partial y^2}\right) \tag{14.1.4}$$

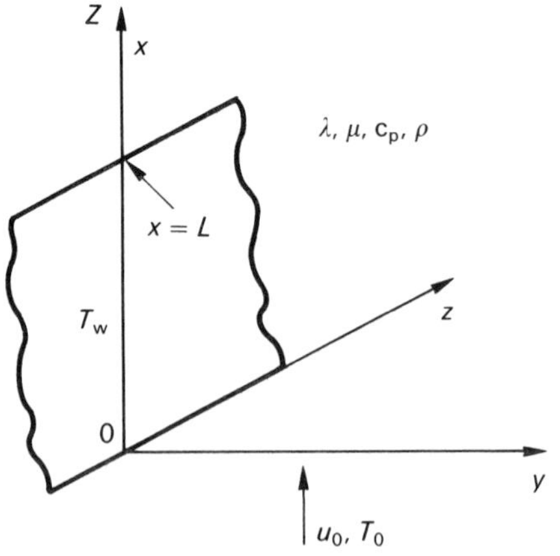


**Figure 14.2**

In a boundary layer the gradients in the $x$ direction are inversely proportional, in order of magnitude, to the distance $L$:

$$\frac{\partial}{\partial x} \simeq \frac{1}{L}, \qquad \frac{\partial^2}{\partial x^2} \simeq \frac{1}{L^2} \tag{14.2}$$

The gradients in the $y$ direction are inversely proportional, in order of magnitude, to the boundary layer thickness, $\delta_{\mathrm{m}}$ or $\delta_{\mathrm{th}}$ as appropriate:

$$\frac{\partial}{\partial y} \simeq \frac{1}{\delta_{\mathrm{m}}} \text{ or } \frac{1}{\delta_{\mathrm{th}}} \qquad \frac{\partial^2}{\partial y^2} \simeq \frac{1}{\delta_{\mathrm{m}}^2} \text{ or } \frac{1}{\delta_{\mathrm{th}}^2} \tag{14.3}$$

We will only retain the dominant terms in (14.1). Equation (14.1.1) becomes:

$$v \simeq \frac{\delta_{\mathrm{m}}}{L} u \tag{14.4}$$

This implies that the $v$ velocity component is of the infinitely small order of $\delta_{\mathrm{m}}/L$ compared to $u$. From this we see that:

$$u \frac{\partial}{\partial x} \simeq v \frac{\partial}{\partial y} \simeq \frac{u}{L} \tag{14.5.1}$$

The terms on the left-hand sides of (14.1.2), (14.1.3) and (14.1.4) are the same order of magnitude (within one equation). Furthermore:

$$\left(\frac{\partial^2}{\partial x^2} \simeq \frac{1}{L^2}\right) \ll \left(\frac{\partial^2}{\partial y^2} \simeq \frac{1}{\delta_{\mathrm{m}}^2} \text{ or } \frac{1}{\delta_{\mathrm{th}}^2}\right) \tag{14.5.2}$$

The terms in $\partial^2/\partial x^2$ are negligible in all the equations in comparison with the corresponding terms in $\partial^2/\partial y^2$, being second order relative to $(\delta/L)$.

Let us compare the terms in equations (14.1.2) and (14.1.3) among themselves: obviously $u\,\partial v/\partial x$ is negligible compared to $u\,\partial u/\partial x$, from (14.4); the same goes for $v\,\partial v/\partial y$ in comparison to $v\,\partial u/\partial y$ and for $v\,\partial^2 v/\partial y^2$ in comparison to $v\,\partial^2 u/\partial y^2$. As a consequence, equation (14.1.3) reduces (to first order in $\delta_{\mathrm{m}}/L$) to:

$$\left(\frac{\rho_0 u^2}{L}\right)^{-1} \frac{\partial p}{\partial y} = O\left(\frac{\delta}{L}\right) \tag{14.5.3}$$

This expression, which is valid for all values of $x$ and $y$, suggests that $p$ is a function of $x$ only, both within the boundary layer and in the undisturbed region (where $y$ tends to infinity). An important consequence of this is that the pressure gradient $\partial p/\partial x$ in the boundary layer is the same as in the undisturbed region. If the undisturbed flow far away has a constant velocity $u_0$, then $\partial p/\partial x$ is zero (or at least completely negligible):

$$\frac{\partial p}{\partial x} \simeq 0 \tag{14.5.4}$$

Consequently, by making use of this analysis of orders of magnitude, we obtain a simplified system of non-dimensional equations by substituting:

$$x^+ = \frac{x}{L}, \qquad y^+ = \frac{y}{L}, \qquad u^+ = \frac{u}{u_0}, \qquad u^+ = \frac{v}{u_0}, \qquad T^+ = \frac{T - T_w}{T_0 - T_w} \tag{14.6.0}$$

$$\frac{\partial u^+}{\partial x^+} + \frac{\partial v^+}{\partial y^+} = 0 \tag{14.6.1}$$

$$u^+ \frac{\partial u^+}{\partial x^+} + v^+ \frac{\partial u^+}{\partial y^+} = \frac{1}{Re_L} \frac{\partial^2 u^+}{\partial y^{+2}}; \qquad Re_L = \frac{u_0 L}{\nu} \tag{14.6.2}$$

$$\frac{\partial p^+}{\partial y^+} = 0 \tag{14.6.3}$$

$$u^+ \frac{\partial T^+}{\partial x^+} + v^+ \frac{\partial T^+}{\partial y^+} = \frac{1}{Re_L Pr} \frac{\partial^2 T^+}{\partial y^{+2}} = \frac{1}{Pe_L} \frac{\partial^2 T^+}{\partial y^{+2}}; \qquad Pe_L = \frac{u_0 L}{a} \tag{14.6.4}$$

$$T^+(x^+, 0) = 0, \qquad u^+(x^+, 0) = 0 \tag{14.6.5}$$

$$T^+(x^+, \infty) = 1, \qquad u^+(x^+, \infty) = 1 \tag{14.6.6}$$

$$T^+(0, y^+) = 1, \qquad u^+(0, y^+) = 1 \tag{14.6.7}$$

*Important note*

There is a singular zone of very limited extent in the immediate vicinity of the leading edge in which the approximations made above are not justified either for the mechanical or the thermal equations. Indeed, in this zone the temperature and velocity variations in the $x$ and $y$ directions are comparable, and $v$ is not small compared to $u$. It is strictly necessary to solve the two-dimensional problem using the general equations (14.1). The two models developed in Sections 14.2.1 and 14.2.2 are based on the simplifying assumptions introduced above ((14.3)–(14.5.4)): their results do not therefore apply in the immediate vicinity of the leading edge. The infinite value of heat transfer coefficient that they predict at $x = 0$ is not realistic. Nevertheless, the results of the first model (Section 14.2.1) are valid everywhere else and are satisfactory in most applications.

### 14.2.1 Exact solution of equations (14.6)

The exact solution is classically obtained by Blasius's method for the velocity field [1]. In terms of the solution $f(\eta)$, it is given by Blasius's equation:

$$f'''(\eta) + \frac{1}{2} f(\eta) f''(\eta) = 0 \tag{14.7}$$

with boundary conditions:

$$f(0) = f'(0); \qquad f'(\infty) = 1 \tag{14.8}$$

where:

$$\eta = y^+ \sqrt{\frac{Re_L}{x^+}} \tag{14.9}$$

It follows that:

$$u^+ = f'(\eta)$$

$$v^+ = \frac{1}{2} Re_x^{-1/2} (\eta f' - f)$$

The friction coefficient $c_F$ defined by (12.3) is given by:

$$c_F = \frac{f''(0)}{\sqrt{Re_x}} = \frac{0.664}{\sqrt{Re_x}} \tag{14.10.1}$$

and the thickness $\delta_m$ of the momentum boundary layer:

$$\frac{\delta_m(x)}{x} = \frac{4.9}{\sqrt{Re_x}} \tag{14.10.2}$$

We are primarily interested in the solution to the thermal problem. We will first look at the case where $Pr = 1$, then the general case of arbitrary $Pr$.

14.2.1.1 SPECIAL CASE OF $Pr = 1$

In this case, equations (14.6.4) and (14.6.2) are identical if $T^+$ is substituted for $u^+$, and the same applies to the boundary conditions. Consequently, the local Nusselt number $Nu_x$ defined by:

$$Nu_x(x) = \frac{\partial T^+}{\partial y^+}(x^+, 0) \tag{14.11}$$

can be deduced from the friction coefficient $c_F$ given by (14.10):

$$c_F(x) = \frac{2}{Re_x} \frac{\partial u^+}{\partial y^+}(x^+, 0) = \frac{0.664}{\sqrt{Re_x}} \tag{14.12}$$

Whence:

$$Pr = 1; \qquad Nu_x(x) = 0.332\sqrt{Re_x} \tag{14.13}$$

We note that in this case the momentum and thermal boundary layers are coincident since the diffusivities $a$ and $\nu$ are equal ($Pr = \nu/a = 1$).

14.2.1.2 GENERAL CASE

Normally, the momentum and thermal boundary layers are distinct. By analogy with the method of solving for $u^+$, we can make the substitution:

$$T^+(x^+, y^+) = T^+(\eta) = T^+\left(y^+ \sqrt{\frac{Re_L}{x^+}}\right) \tag{14.14}$$

in equation (14.6.4), which gives, after making substitutions for $u^+$ and $v^+$:

$$\frac{\mathrm{d}^2 T^+}{\mathrm{d}\eta^2} + \frac{1}{2} Pr f(\eta) \frac{\mathrm{d}T^+}{\mathrm{d}\eta} = 0 \tag{14.15.1}$$

$$T^+(0) = 0 \tag{14.15.2}$$

$$T^+(\infty) = 1 \tag{14.15.3}$$

whose solution is:

$$T^+(\eta) = \frac{\displaystyle\int_0^{\eta} \exp\left[-Pr/2 \int_0^{\eta'} f(\eta'')\,\mathrm{d}\eta''\right] \mathrm{d}\eta'}{\displaystyle\int_0^{\infty} \exp\left[-Pr/2 \int_0^{\eta'} f(\eta'')\,\mathrm{d}\eta''\right] \mathrm{d}\eta'} \tag{14.16}$$

The local Nusselt number $Nu_x(x)$ is then given by:

$$Nu_x(x) = \frac{\partial T^+}{\partial y^+} = \frac{\partial T^+}{\partial \eta} \frac{\mathrm{d}\eta}{\mathrm{d}y^+} = \sqrt{\frac{Re_L}{x^+}} \frac{\partial T^+}{\partial \eta} \tag{14.17}$$

Schlichting [6] gives an excellent approximation to this result for common fluids:

$$0.6 \leqslant Pr \leqslant 10 \qquad Nu_x(x) = 0.332 Re_x^{1/2} Pr^{1/3} \tag{14.18}$$

The Nusselt number based on the overall length $L$ of the plate, $Nu_L$, can be deduced from $Nu_x$ by:

$$0.6 \leqslant Pr \leqslant 10 \qquad Nu_L = \int_0^L \frac{Nu_x(x)}{x} \mathrm{d}x = 0.664 Re_L^{1/2} Pr^{1/3} \tag{14.19}$$

### *14.2.2 Approximate solution (integral method)*

The advantage of the approximate solution is to enhance the links between the various physical quantities ($Pr$, $\delta_{\mathrm{m}}$, $\delta_{\mathrm{th}}$, $c_{\mathrm{F}}(x)$, $Nu_x(x)$, etc.). It also leads to simple analytic solutions which are usually sufficiently accurate (except in the neighbourhood of the leading edge and around the edge of the boundary layer).

The approximations made are as follows:

- Assume that the undisturbed stream conditions ($T_0$, $u_0$) are satisfied exactly at the edge of the boundary layer (which does not precisely agree with the definition of the latter), so that:

$$T(x, \delta_{\mathrm{th}}) = T_0 \tag{14.20}$$

$$u(x, \delta_{\mathrm{m}}) = u_0 \tag{14.21}$$

and that the functions $u$, $v$ and $T$ and their derivatives are continuous.

- Assume that the ratio $\delta_{\mathrm{th}}/\delta_{\mathrm{m}}$ is independent of $x$ (a result which is justified by exact calculation in the present case).

The principle of the method is to look for velocity and temperature fields in the form of polynomials in $y/\delta$, using forms of the momentum and energy equations integrated over the respective boundary layer thicknesses.

#### 14.2.2.1 MOMENTUM BOUNDARY LAYER

Let us set:

$$u(x, 0) = 0 \quad (1)$$

$$u(x, \delta_{\mathrm{m}}) = u_0 \quad (2) \quad \Rightarrow \frac{\partial u}{\partial x}(x, \delta_{\mathrm{m}}) = 0 \quad (3) \tag{14.22}$$

$$\frac{\partial u}{\partial y}(x, \delta_{\mathrm{m}}) = 0 \tag{14.23}$$

Equations (14.22.3), (14.23) and (14.6.2) lead to:

$$\frac{\partial^2 u}{\partial y^2}(x, \delta_{\mathrm{m}}) = 0 \tag{14.24}$$

If we look for a solution in the form of a third-order polynomial:

$$u = u_0 + u_1\left(\frac{y}{\delta_{\mathrm{m}}}\right) + u_2\left(\frac{y}{\delta_{\mathrm{m}}}\right)^2 + u_3\left(\frac{y}{\delta_{\mathrm{m}}}\right)^3 \tag{14.25}$$

equations (14.22.1), (14.22.2), (14.23) and (14.24) allow us to evaluate the four coefficients:

$$u(x, y) = \frac{u_0}{2}\left[3\left(\frac{y}{\delta_{\mathrm{m}}(x)}\right) - \left(\frac{y}{\delta_{\mathrm{m}}(x)}\right)^3\right] \tag{14.26}$$

while $v(x, y)$ can be obtained by integration of the continuity equation (14.1.1):

$$v(x, y) = v(x, 0) - \int_0^y \frac{\partial u}{\partial x}\,\mathrm{d}y \tag{14.27.1}$$

so that:

$$v(x, y) = \frac{3u_0}{4}\frac{\mathrm{d}\delta_{\mathrm{m}}(x)}{\mathrm{d}x}\left[\left(\frac{y}{\delta_{\mathrm{m}}(x)}\right)^2 - \frac{1}{2}\left(\frac{y}{\delta_{\mathrm{m}}(x)}\right)^4\right] \tag{14.27.2}$$

The remaining quantity $\delta_m(x)$ may be found by integrating the momentum equation (14.1.2), making use of (14.5.2):

$$\int_0^{\delta_m}\left(u\frac{\partial u}{\partial x}+v\frac{\partial u}{\partial y}\right)\mathrm{d}y=\int_0^{\delta_m}\nu\frac{\partial^2 u}{\partial y^2}\,\mathrm{d}y=-\nu\frac{\partial u}{\partial y}(x,0) \tag{14.28.1}$$

which leads to:

$$\frac{\delta_m(x)}{x}=\frac{4.64}{\sqrt{Re_x}} \tag{14.28.2}$$

This result only differs from the exact result by a few % (see equation (14.10)).

14.2.2.2 THERMAL BOUNDARY LAYER

In the same manner, we set:

$$T(x,0)-T_p=0 \tag{14.29.1}$$

$$T(x,\delta_{th})-T_p=T_0-T_p \tag{14.29.2}$$

$$\frac{\partial T}{\partial x}(x,\delta_{th})=0 \tag{14.29.3}$$

$$\frac{\partial T}{\partial y}(x,\delta_{th})=0 \tag{14.29.4}$$

The energy equation (14.6.4) combined with these equations leads to:

$$\frac{\partial^2 T}{\partial y^2}(x,\delta_{th})=0 \tag{14.29.5}$$

If we assume a solution for $T$ in the form of a third-order polynomial in $(y/\delta_{th})$, we obtain from (14.29):

$$T(x,y)-T_w=\left(\frac{T_0-T_w}{2}\right)\left[3\left(\frac{y}{\delta_{th}}\right)-\left(\frac{y}{\delta_{th}}\right)^3\right] \tag{14.30}$$

which is essentially the same expression as for $u(x,y)$.

The ratio $\delta_{th}/\delta_m$, assumed independent of $x$ (which is also the case for the ratios of the derivatives $\delta'_{th}/\delta'_m(x)$), is obtained by integrating the energy equation over the thermal boundary layer thickness:

$$\int_0^{\delta_{th}(x)}\left(u\frac{\partial T}{\partial x}+v\frac{\partial T}{\partial y}\right)\mathrm{d}y=a\int_0^{\delta_{th}(x)}\frac{\partial^2 T}{\partial y^2}\,\mathrm{d}y=-a\frac{\partial T}{\partial y}(x,0)$$

in which we substitute the previously obtained expressions for $u(x,y)$, $v(x,y)$ and $T(x,y)$. After all calculations, we obtain the implicit relationship:

$$\left(\frac{\delta_{th}}{\delta_m}\right)^3-\frac{1}{14}\left(\frac{\delta_{th}}{\delta_m}\right)^5=\frac{0.93}{Pr} \tag{14.31}$$

When $Pr=1\,(a=\nu)$ this again gives $\delta_{th}=\delta_m$, and the boundary layers are coincident.

For common fluids, a good approximation to (14.31) is given by:

$$Pr \geqslant 0.7 \qquad \frac{\delta_{th}}{\delta_m} = Pr^{-1/3} \tag{14.32}$$

a valuable result which establishes the relationship between the boundary layer thicknesses and the Prandtl number, or ratio of the momentum and thermal diffusivities $\nu/a$:

- For common gases, the Prandtl number is around 0.7. The boundary layers are then essentially coincident ($\delta_{th}/\delta_m \simeq 1.13$).
- For heavy oils, *Pr* is very high. The thermal boundary layer is much thinner than the momentum boundary layer. This is also true for water between 0 °C and 100 °C (*Pr* between 1.7 and 7).
- For liquid metals ($Pr \simeq$ a few times $10^{-2}$), the opposite is true. The thermal boundary layer is thicker, in theory, than the momentum boundary layer.

Let us calculate the local Nusselt number $Nu_x$ for a common fluid:

$$Nu_x(x) = \frac{h(x)x}{\lambda} = \frac{x}{\lambda}\frac{[-\lambda\,\partial T/\partial y(x,0)]}{(T_0 - T_w)} = \frac{3}{2}\frac{x}{\delta_{th}(x)} = \frac{3}{2}\frac{xPr^{1/3}}{\delta_m(x)}$$

so that:

$$Pr \geqslant 0.7 \qquad Nu_x(x) = 0.323 Re_x^{0.5} Pr^{1/3} \tag{14.33.1}$$

This value only differs by a few percent from the exact value (equation (14.18): $Nu_x(x) = 0.332 Re_x^{0,5} Pr^{1/3}$).

Note particularly the relationship between $Nu_x$, $h(x)$ and the thermal boundary layer thickness $\delta_{th}$ in this model:

$$Nu_x(x) = \frac{3}{2}\frac{x}{\delta_{th}(x)} \qquad h(x) = \frac{3}{2}\frac{\lambda}{\delta_{th}(x)} \tag{14.33.2}$$

If we recall the simplistic model for $h(x)$ developed in Section 1.4.2, we see that the system is equivalent to a conducting film of thickness $2/3\,\delta_{th}(x)$, whose two faces are held at temperatures $T_w$ and $T_0$.

### *14.2.3 Discussion of results – applications*

The dependence of the heat transfer coefficient on a flat plate on $x$, the distance from the leading edge, is very important in the study and control of heat transfer in many applications, particularly in heat exchangers. From the exact equation, we obtain the local heat transfer coefficient:

$$0.6 \leqslant Pr \leqslant 10 \qquad h(x) = \frac{Nu_x(x)\lambda}{x} = 0.332 Pr^{1/3}\left(\frac{u_0}{\nu}\right)^{1/2}\lambda x^{-1/2} \tag{14.34}$$

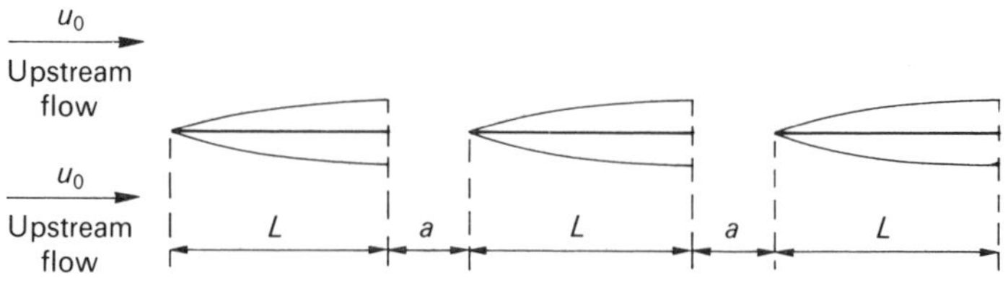


**Figure 14.3**

which shows a theoretical variation as $x^{-1/2}$ in the region of the leading edge ($x = 0$); the local heat transfer coefficient falls with $x$ in laminar flow. This result is also true in turbulent flow, but with a much weaker dependence, as $x^{-n}$ (see Chapter 17).

*Notes*

1. At $x = 0$, the assumption $\partial^2 T/\partial x^2 \ll \partial^2 T/\partial y^2$ is not strictly true, but it becomes valid very rapidly for small values of $x$.
2. A practical outcome of the variation of $h(x)$ with $x$ is that it is useful, if possible, to have multiple leading edges in industrial designs (see Figure 14.3):
3. The distance $a$ between the plates and the plate length $L$ must be optimized: $a$ must be sufficiently short that the loss of heat exchange area involved in this configuration is small compared to the gain resulting from the increase in $h(x)$; conversely, for the leading edge effect to be important, upstream conditions (particularly temperature) must be substantially re-established at the beginning of the next plate.
4. Practical designs based on this principle are very diverse (boundary layer separation by inclination of the trailing edge, large scale roughness, spikes, etc.), but are outside the scope of this work.

## 14.3 Vertical isothermal flat plate in an isothermal fluid at rest far away (free convection)

We will use the same conditions as in Section 14.2, using the Boussinesq approximation:

1. The term $\beta(T - T_0)$ is much less than 1.
2. Variations of $\rho$ with $T$ are neglected except in the driving term, in which:

$$\rho = \rho_0[1 - \beta(T - T_0)] \tag{14.35}$$

The system of equations to be solved, in dimensional variables, follows from (13.8.1), (13.8.4) and (13.8.3):

- The continuity equation:

$$\frac{\partial u}{\partial x} + \frac{\partial v}{\partial y} = 0 \tag{14.36}$$

- The momentum equations (assuming the direction $OZ$ coincides with the $x$ axis):

$$u\frac{\partial u}{\partial x}+v\frac{\partial u}{\partial y}=-\frac{1}{\rho_0}\frac{\partial}{\partial x}(p+\rho_0 gx)+\nu\left(\frac{\partial^2 u}{\partial y^2}+\frac{\partial^2 u}{\partial x^2}\right)+\beta(T-T_0)g \tag{14.37.1}$$

$$u\frac{\partial v}{\partial x}+v\frac{\partial v}{\partial y}=-\frac{1}{\rho_0}\frac{\partial p}{\partial y}+\nu\left(\frac{\partial^2 v}{\partial x^2}+\frac{\partial^2 v}{\partial y^2}\right) \tag{14.37.2}$$

- The energy equation:

$$u\frac{\partial T}{\partial x}+v\frac{\partial T}{\partial y}=a\left(\frac{\partial^2 T}{\partial x^2}+\frac{\partial^2 T}{\partial y^2}\right) \tag{14.38}$$

An analysis of orders of magnitude along the lines of that undertaken in (14.2) leads to important simplification of this system. We will introduce the plate length $L$ and the boundary layer thicknesses $\delta_m$ and $\delta_{th}$ as characteristic dimensions.

Equation (14.36) becomes:

$$v\simeq\frac{\delta_m}{L}u \tag{14.39}$$

In equations (14.37), the following terms appear:

$$\frac{u^2}{L},\ \frac{u^2}{L},\ \frac{1}{\rho_0}\frac{\partial}{\partial x}(p+\rho_0 gx),\ \nu\frac{u}{\delta_m^2},\ \nu\frac{u}{L^2},\ g\beta(T-T_0)$$

and (14.40)

$$\frac{u^2}{L}\frac{\delta_m}{L},\ \left(u\frac{\delta_m}{L}\right)\frac{u}{L},\ \frac{1}{\rho_0}\frac{\partial p}{\partial y},\ \frac{\nu}{L^2}u\frac{\delta_m}{L},\ \frac{\nu}{\delta_m}\frac{u}{L}$$

With the exception of the pressure terms, all the terms in the $y$ equation are negligible compared with the corresponding terms in the $x$ equation, which means that:

$$\left(\frac{\rho_0 u^2}{L}\right)^{-1}\frac{\partial p}{\partial y}=O\left(\frac{\delta_m}{L}\right) \tag{14.41}$$

Therefore the pressure $p$ only depends on $x$, to first order in $\delta_m/L$; $\partial p/\partial x$ in the boundary layer is thus equal to the (static) pressure gradient in the undisturbed flow, or:

$$\rho_0 g=-\frac{\partial p}{\partial x} \tag{14.42}$$

Finally, neglecting the $\partial^2/\partial x^2$ term in comparison with $\partial^2/\partial y^2$:

$$u\frac{\partial u}{\partial x}+v\frac{\partial u}{\partial y}=\nu\frac{\partial^2 u}{\partial y^2}+g\beta(T-T_0) \tag{14.43}$$

If we introduce a reference velocity $v_r$ and dimensionless variables:

$$x^+ = \frac{x}{L}, \qquad y^+ = \frac{y}{L}, \qquad u^+ = \frac{u}{v_r}, \qquad v^+ = \frac{v}{v_r}, \qquad T^+ = \frac{T - T_0}{T_w - T_0} \tag{14.45}$$

the system of equations becomes:

$$\frac{\partial u^+}{\partial x^+} + \frac{\partial v^+}{\partial y^+} = 0 \tag{14.46}$$

$$u^+ \frac{\partial u^+}{\partial x^+} + v^+ \frac{\partial u^+}{\partial y^+} = \frac{\nu}{L v_r} \frac{\partial^2 u^+}{\partial y^{+2}} + \frac{g\beta(T_w - T_0)L}{v_r^2} T^+ \tag{14.47}$$

In this case we will choose a different reference velocity from that used in Section 12.2, putting:

$$v_r = \frac{\nu}{L} \tag{14.48}$$

which gives:

$$u^+ \frac{\partial u^+}{\partial x^+} + v^+ \frac{\partial u^+}{\partial y^+} = \frac{\partial^2 u^+}{\partial y^{+2}} + Gr_L T^+ \tag{14.49}$$

where, as in Section 12.2, we have put:

$$Gr_L = \frac{g\beta(T_w - T_0)L^3}{\nu^2} \tag{14.50}$$

In the same way, the energy equation becomes:

$$u^+ \frac{\partial T^+}{\partial x^+} + v^+ \frac{\partial T^+}{\partial y^+} = \frac{1}{Pr} \frac{\partial^2 T^+}{\partial y^{+2}} \tag{14.51}$$

The exact solution of this problem can be found by Pollhausen's method, which generalizes this in detail to handle forced convection. The reader interested in the details of the solution may refer to Knudsen and Katz [5]. We finally obtain (from Pollhausen's solution):

$$Nu_x = 0.39(Gr_x Pr)^{1/4} \tag{14.52}$$

which leads to a Nusselt number for the whole plate:

$$Nu_L = \frac{1}{L}\int_0^L \frac{Nu_x}{x}\,\mathrm{d}x = \frac{4}{3} \times 0.39(Gr_L Pr)^{1/4} = 0.52(Gr_L Pr)^{1/4}$$

or:

$$Nu_L = 0.52 Ra_L^{1/4} \tag{14.53}$$

Here we find an analogous result, with a weaker dependence on $x^{-n}$, to that found in forced convection for heat transfer at the leading edge:

$$h(x) = \lambda Nu_x/x = \text{const } x^{-1/4}$$

The thicknesses of the momentum and thermal boundary layers (assumed coincident) calculated by the integral method are:

$$\delta_m(x) \sim \delta_{th}(x) = 3.93 Pr^{-0.5}\left[\frac{20}{21} + Pr\right]^{1/4}\left[\frac{g\beta(T_w - T_0)}{\nu^2}\right]^{-1/4} x^{1/4} \quad (14.54)$$

## 14.4 Other geometries

Refer to Part V of this book ('Basic Data').

# 15

# Internal Laminar Convection. Development of Flow Conditions in Forced Convection

In this chapter, we will be concerned solely with forced convection in laminar flow. Further, we use the term **internal convection** to refer to heat transfer inside a conduit (flows between parallel plates, in rectangular, hexagonal or circular tubes, etc.); these are also described as closed geometries. We will restrict ourselves to **geometries of constant cross-section**.

For flow over a flat plate, we saw in Chapter 14 that at a certain distance away from the plate in the normal direction, undisturbed flow conditions (velocity and temperature profiles) are recovered. This is not the case for the velocity distribution in the entry of a pipe for a closed geometry. Indeed, the discharge is constant at any cross-section of the tube: as the fluid velocity falls near the walls, due to viscous friction, it must rise near the axis. In the region near the axis, viscous effects are weak, and the velocity profile retains a value $u_{max}(x)$ independent of distance across the flow, but strongly dependent on $x$ – locally there is plug flow. The value of $u_{max}(x)$ increases with $x$ going away from the entrance (Figure 15.1(a)).

On the other hand, the temperature near the axis retains its upstream value for a certain distance from the entrance (for example, a uniform value $T_0$). We will see in the remainder of this chapter that temperature itself is of little direct interest in this type of flow. We will work in terms of the temperature difference with the wall, and particularly the dimensionless temperature difference $T^+$. We should note that if the flux at the wall is constant, the wall temperature $T_w(x)$ varies with $x$ from the entry, and, even in the region of the axis, the difference between local and wall temperature varies from the entry in a similar manner to the speed. Because of this, boundary layers in the entry regions of closed geometries have different definitions from those in Chapter 14 (concerned with external convection).

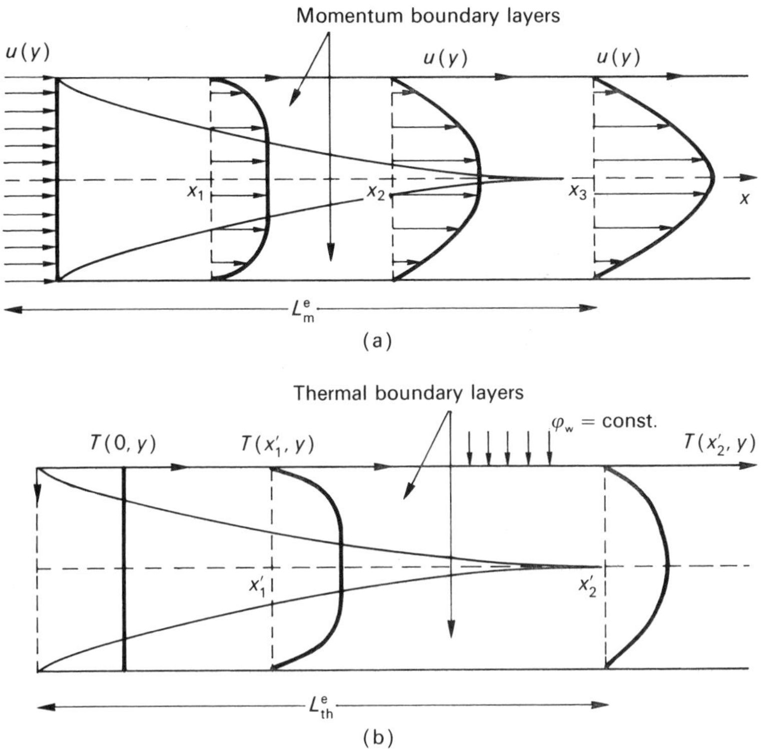


**Figure 15.1** (a) Momentum boundary layer (special case of an isothermal flow with uniform, i.e. independent of $y$, velocity at entry). (b) Thermal boundary layer (uniform fluid temperature $T_0$ at entry, constant flux, variable wall temperature).

The momentum and thermal boundary layers develop from the entry of the particular geometry (leading edges) in a manner depending on the operating conditions chosen. The momentum boundary layer is clearly no longer defined with reference to a flow velocity far away (as in Chapter 14), but relative to the maximum velocity at any cross-section (usually on the axis, where there is an axis of symmetry for the geometry and boundary conditions).

**The momentum boundary layer** lies between the wall and the point where the velocity is 0.99 times the maximum velocity at a given cross-section (Figure 15.1(a)).

**The thermal boundary layer** lies between the wall and the point where the difference between the local and wall temperature is 0.99 times that between the most extreme temperature and the wall temperature (Figure 15.1(b)).

The thicknesses of the momentum and thermal boundary layers will be denoted by $\delta_m(x)$ and $\delta_{th}(x)$.

We postulate that after a certain length $L^e_m$, called the velocity entry length, the momentum boundary layers merge ($x_3$ on Figure 15.1(a)). Similarly, after a thermal entry length $L^e_{th}$, the thermal boundary layers merge ($x'_2$ in Figure 15.1(b)).

*Important notes*

1. In certain situations, a tube with a large diameter $D$ but a short length $L$ can be considered as an **open geometry**; the exit boundary layer thicknesses $\delta_m(L)$ and $\delta_{th}(L)$ may be much smaller than $D$, and the flat plate model (Chapter 14) may be used so long as $\delta_m$ and $\delta_{th}$ remain small compared with the radius of curvature and the characteristic transverse dimension of the geometry concerned.
2. In a closed circuit without sudden changes of section, there will be no entry region.
3. When there is an entry region, the thermophysical properties of the fluid in general change significantly with temperature. The calculation of the coupled velocity and temperature fields then has only numerical solutions. Soufiani [4] analyses such a case and gives a comprehensive bibliography for laminar flow cases.

If we make the following two restrictive hypotheses:

(H1) **the cross-sections normal to the flow direction (the $x$ direction) are constant,**
(H2) **the thermophysical properties** ($\rho$, $\lambda$, $\mu$, $c_p$, etc) **of the fluid are independent of temperature,**

we can see that we may reach **developed flow conditions** at a certain distance from the entry, such that the profiles of certain quantities (to be specified) are independent of $x$. We will look first at development of velocity (Section 15.1) then at thermal development (Section 15.2).

## 15.1 Development of the velocity régime

A developed velocity régime is characterized by (using the previous notation):

$$\frac{\partial u}{\partial z}(y,z) = 0 \qquad \text{or} \qquad \frac{\partial u}{\partial x}(r,\theta) = 0 \tag{15.1.1}$$

If we consider a tube of radius $R$, the continuity equation gives an expression for the radial velocity $v$:

$$\frac{1}{r}\left(\frac{\partial(rv)}{\partial r}\right) = 0 \tag{15.1.2}$$

this means, since the velocity at the wall ($r = R$) is zero, that at all $r$:

$$v(r) = 0 \tag{15.1.3}$$

The velocity in the developed régime is thus parallel to the generators, and to the $x$ axis.

The governing equations for a tube of radius $R$ are, in the general case:

$$u\frac{\partial u}{\partial x} = -\frac{\partial p}{\partial x} + \nu\left(\frac{\partial^2 u}{\partial r^2} + \frac{1}{r}\frac{\partial u}{\partial r}\right) \tag{15.2.1}$$

$$u(x,R) = 0 \tag{15.2.3}$$

$$\frac{\partial u}{\partial r}(x, 0) = 0 \tag{15.2.4}$$

We can see from the first of these that after a certain development length the fluid will have lost memory of its entry conditions $u_e(r)$ and that we can find a solution $u_\infty(r)$ satisfying (15.2.1) and compatible with the boundary conditions (15.2.3) and (15.2.4). The term $\partial p/\partial x$ can be eliminated by using conservation of mass flux $\dot{m}$ across a cross-section $\Sigma$:

$$\frac{\dot{m}}{\rho} = \int_\Sigma u_\infty(r)\,\mathrm{d}\Sigma = \int_\Sigma u_e(r)\,\mathrm{d}\Sigma = \Sigma u_m \tag{15.3}$$

where $u_m$ is the discharge velocity, equal to $u_0$ for the particular case of plug flow at velocity $u_0$ at entry.

- For the case of the tube:

$$\frac{u_\infty(r)}{u_m} = 2\left[1 - \left(\frac{r}{R}\right)^2\right] \tag{15.4}$$

  and the maximum velocity is thus $2u_m$.
- For two parallel plates a distance $2e$ apart (in the $y$ direction):

$$\frac{u_\infty(y)}{u_m} = \frac{3}{2}\left[1 - \left(\frac{y}{e}\right)^2\right] \tag{15.5}$$

  The maximum velocity is now only $1.5u_m$.

  Various authors have calculated the velocity development length $L_m$, defined as the length at the end of which the axial velocity differs by only 1% from the asymptotic velocity; this is obviously the same thing as the entry length [5].
- For a tube of diameter $D = 2R$:

$$\frac{L_m}{D} = 0.0575 Re_D \qquad \text{with} \qquad Re_D = \frac{u_m D}{\nu} \tag{15.6}$$

- For two parallel plates, a distance $e$ apart:

$$\frac{L_m}{D_h} = 0.006\,48 Re_{D_h} \qquad \text{with} \qquad Re_{D_h} = \frac{u_m D_h}{\nu} \tag{15.7}$$

  where:

$$D_h = \frac{4V}{\Sigma} = \frac{4e\Sigma}{2\Sigma} = 2e \tag{15.8}$$

$D_h$ is the hydraulic diameter, whose definition appears in the appendix to this part. We see that equations (15.6) and (15.7) are similar in form, but that developed flow is much more quickly attained between two plates than in a tube. This last

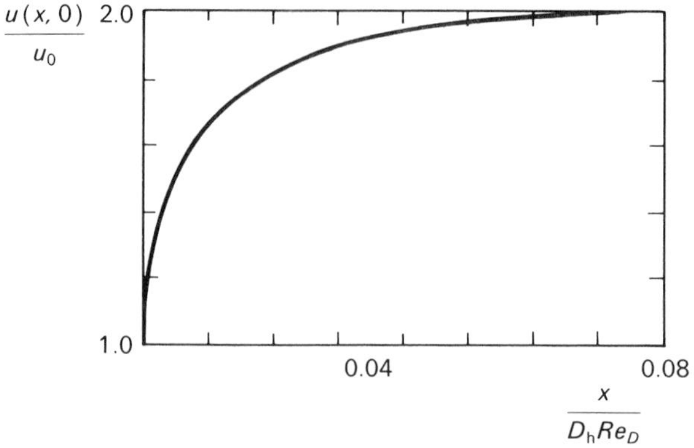


**Figure 15.2** Axial velocity $u(x, 0)$ in a circular tube as a function of distance from the leading edge, with plug flow at entry (from reference [6]).

point is explained by the fact that the maximum velocities given by (15.4) and (15.5) are different ($2u_m$ and $1.5u_m$). It is striking to notice that on Figure 15.2 for the case of a tube, the velocity $1.5u_m$ is reached after a distance in agreement with equation (15.7).

## *15.2 Development of the thermal régime*

A developed thermal régime can be defined by extension of (15.1.1). However, the choice of quantity to which to apply the relationship analogous to (15.1.1) is difficult.

It is obvious, on the one hand, that only a temperature difference can be analogous to velocity, an extensive quantity. It is also obvious, on the other, that if the wall temperature $T_w$ is fixed, we cannot reduce the quantity $\partial/\partial x(T - T_w)$ without reaching isothermal conditions, which is a trivial solution. We will actually define a local dimensionless temperature $T^+$ by taking the difference between the wall temperature $T_w(x)$ and the mixed fluid temperature $T_m(x)$ (introduced in Section 3.4.2.1 and defined by formula:

$$\dot{m}T_m(x) = \rho \int_\Sigma u(x, r)T(x, r)\,d\Sigma \tag{15.9}$$

if $c_p$ is assumed constant) as reference. We then define:

$$T^+(x, r) = \frac{T(x, r) - T_w(x)}{T_m(x) - T_w(x)} \tag{15.10}$$

We will show in this section that for large values of $x$ we can obtain a developed thermal régime satisfying:

$$\frac{\partial T^+}{\partial x} = 0 \tag{15.11}$$

and that we can do this for various thermal boundary conditions. Equation (15.11) is analogous to (15.1.1) since the latter implies that:

$$\frac{\partial u^+}{\partial x} = 0 \qquad \text{with} \qquad u^+ = \frac{u}{u_m} \tag{15.12}$$

We will look at the temperature field $T^+$ in various situations depending (within the limits of hypotheses H1 and H2) on various conditions:

1. Entry conditions:
   - Thermal conditions are imposed from the leading edge of the particular geometry concerned (tube, parallel plates, etc.). In this event the velocity development and the thermal development are interdependent (for example, if at entry ($x = 0$) we have a plug flow ($u$ constant in the $y$ direction) and a fixed temperature).
   - The duct is adiabatic (or isothermal at the constant temperature of the fluid) over a length longer than $L_m$, with thermal conditions only imposed thereafter (the so-called Graetz problem).

   First of all the velocity régime develops itself by the end of the distance $L_m$, then the thermal régime develops itself, starting at $l$, over a length $L_{th}$ (Figure 15.3).
2. Thermal conditions at the walls, for example:
   - constant temperature $T_w$;
   - constant flux $\varphi_w$;
   - linear temperature profile;
   - etc.

Detailed studies for various geometries and thermal conditions are given in Knudsen and Katz's work [5], for example.

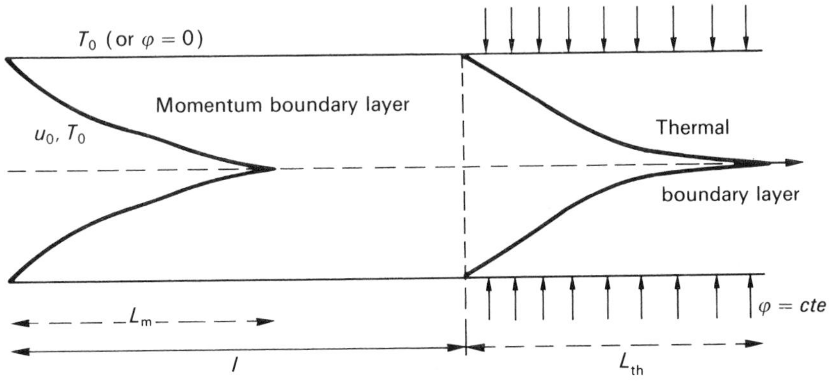


**Figure 15.3** The Graetz problem.

### 15.2.1 *Case 1: flow régime already developed*

This class of problem was originally solved by Graetz for a tube, then generalized to other geometries. Let us consider the equations to be solved for the thermal development in a tube of radius $R$, using the same notation as in the last section:

$$u\frac{\partial T}{\partial x} = a\left(\frac{\partial^2 T}{\partial r^2} + \frac{1}{r}\frac{\partial T}{\partial r}\right) \tag{15.3.1}$$

$$T(0, r) = T_i(r) \tag{15.13.2}$$

$$T(x, R) = T_w \quad \text{or} \quad -\lambda\frac{\partial T}{\partial r}(x, R) = \varphi_w, \text{ etc.} \tag{15.13.3}$$

$$\frac{\partial T}{\partial r}(x, 0) = 0 \tag{15.13.4}$$

Using the dimensionless variable $T^+$, equation (15.13.1) becomes:

$$u_\infty\left[\frac{T^+}{T_m(x) - T_w(x)}\frac{d}{dx}(T_m - T_w) + \frac{\partial T^+}{\partial x} + \frac{1}{T_m - T_w}\frac{dT_w}{dx}\right] = a\left(\frac{\partial^2 T^+}{\partial r^2} + \frac{1}{r}\frac{\partial T^+}{\partial r}\right) \tag{15.14}$$

15.2.1.1 CONSTANT WALL FLUX $\varphi_w$

For a tube of radius $R$, the boundary conditions associated with (15.14) are (changing to cylindrical coordinates):

$$T^+(0, r) = T_i^+(r) \ (2); \quad -\lambda\frac{\partial T^+}{\partial r}(x, R) = \frac{\varphi_w}{T_m - T_w} \ (3); \quad \frac{\partial T^+}{\partial r}(x, 0) = 0 \ (4) \tag{15.14.1}$$

Direct solution of equations (15.14(1–4)) shows that the local Nusselt number defined by:

$$Nu_D(x) = -\frac{\partial T^+}{\partial r^+}(x, 1) = \frac{h(x)D}{\lambda} = \frac{2Rh(x)}{\lambda} \tag{15.15}$$

where $h(x)$ is the local heat transfer coefficient at position $x$, tends asymptotically to a value $Nu_D$ given by (Figure 15.4):

$$Nu_D = 48/11 \simeq 4.364; \qquad Gz^{-1} \geqslant 0.25 \tag{15.16}$$

where $Gz$ is the Graetz number, defined by:

$$Gz = \frac{RPe_D}{x} = \frac{D^2 u_m}{2xa} \tag{15.17}$$

It follows from (15.15) and (15.16) that $h$ and consequently $(T_m - T_w)$ are constants. From this:

$$\frac{dT_w}{dx} = \frac{dT_m}{dx} \tag{15.18.1}$$

A heat balance of the tube gives:

$$\frac{dT_w}{dx} = \frac{dT_m}{dx} = \frac{h\pi D}{\dot{m}c_p}(T_w - T_m) = \text{const} \tag{15.18.2}$$

The wall and mixed fluid temperatures rise linearly, with a constant difference. Putting all these results into (15.14), we confirm that we obtain a solution $T^+(r)$ in developed conditions satisfying:

$$\frac{\partial T^+}{\partial x} = 0 \tag{15.11}$$

and the energy equation:

$$a\left(\frac{\partial^2 T^+}{\partial r^2} + \frac{1}{r}\frac{\partial T^+}{\partial r}\right) = \frac{u_\infty(r)}{T_m - T_w}\frac{dT_w}{dx} = -\frac{4Nu_D u_\infty(r)}{Pe_D D} \tag{15.19.1}$$

where $u_\infty(r)$ is given by equation (15.4), and with boundary conditions:

$$-\frac{\partial T^+}{\partial r^+}(x, 1) = 4.364 = Nu_D \tag{15.19.2}$$

$$\frac{\partial T^+}{\partial r^+}(x, 0) = 0 \tag{15.19.4}$$

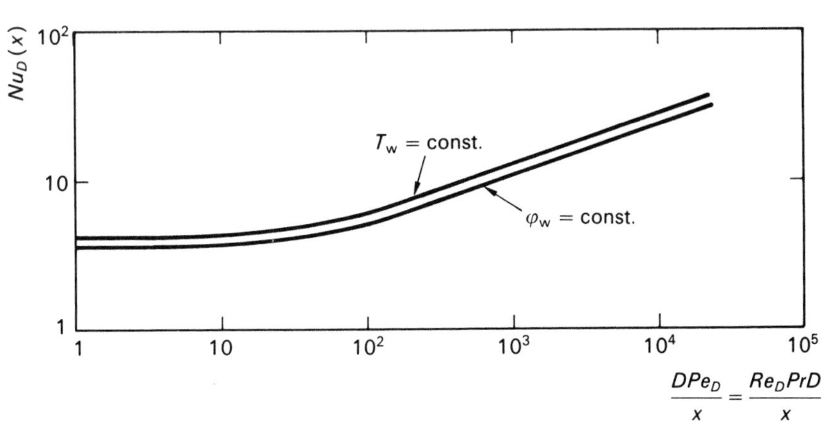


**Figure 15.4** Variation of local Nusselt number $Nu_D(x)$.

The solution of equations (15.19) is valid for $Gz^{-1} \geqslant 0.25$, from (15.16). The thermal development length is then, from (15.16):

$$\frac{L_{\mathrm{th}}}{2R} = \frac{L_{\mathrm{th}}}{D} = 0.125 Pe_D = 0.125 Re_D Pr \tag{15.20}$$

This expression can be compared with the preceding velocity development length given by (15.6). It is obvious that $L_{\mathrm{th}}$ is the thermal entry length.

*Note*

It is very important to note that the local Nusselt number has a large value at the entry to the thermally perturbed region, and falls steadily to the asymptotic value $Nu_D$, which is reached at the end of the development length (Figure 15.4). The situation near the entry of the duct is comparable to that in an open geometry. Note, however, that far from the entry the two situations are totally different – for an open geometry no asymptotic limit is reached, but there is a transition to turbulence for $Re_x \geqslant 10^5$.

15.2.1.2 CONSTANT WALL TEMPERATURE

For a tube, the boundary conditions associated with (15.14) are, in cylindrical coordinates:

$$T^+(0, r) = T_i^+(r) \ (2); \quad T^+(x, R) = 0 \ (3); \quad \frac{\partial T^+}{\partial r}(x, 0) = 0 \ (4) \tag{15.21}$$

The solution leads as before to an asymptotic value $Nu_D$ of local Nusselt number $Nu_D(x)$ (Figure 15.4):

$$Nu_D = 3.656 \qquad \text{if} \qquad Gz^{-1} \geqslant 0.25 \tag{15.22}$$

What consequences has this for the structure of the energy equation (15.14)? An overall balance for the tube, for $Gz^{-1} \geqslant 0.25$, leads to:

$$\dot{m} c_{\mathrm{p}} \frac{\mathrm{d}T_{\mathrm{m}}}{\mathrm{d}x} = h\pi D (T_{\mathrm{w}} - T_{\mathrm{m}}) \tag{15.23}$$

$$\frac{1}{T_{\mathrm{m}} - T_{\mathrm{w}}} \frac{\mathrm{d}}{\mathrm{d}x}(T_{\mathrm{m}} - T_{\mathrm{w}}) = -\frac{h\pi D}{\dot{m} c_{\mathrm{p}}} = -\frac{4 Nu_D}{Pe_D D} \tag{15.24}$$

The energy equation (15.14) thus simplifies, if we assume $\partial T^+/\partial x$ to be zero, to:

$$-\frac{4 Nu_D}{Pe_D} u_\infty(r) T^+(r) = a\left(\frac{\partial^2 T^+}{\partial r^2} + \frac{1}{r}\frac{\partial T^+}{\partial r}\right) \tag{15.25}$$

Associated with boundary conditions (15.21), this leads to a unique solution independent of $x$ when the thermal régime is developed in the sense that:

$$\frac{\partial T^+}{\partial x} = 0 \tag{15.11}$$

The development length for a tube is, as before, given by (15.20).

The same comments about the variation of $h(x)$ and $Nu_D(x)$ between the duct entry and the developed position can be made as for the constant flux case.

*Notes*

Approximate expressions for $Nu_D(x)$ and $Nu_{D,L}$ are given in Part V: 'Basic Data', for various geometries.

### 15.2.2 Case 2: Flow conditions not developed at entry

This case applies to the majority of industrial applications.

We have to take account of the variation of velocity profile with $x$ in the energy equation, which is written:

$$\rho c_p \left( u(x,r)\frac{\partial T}{\partial x} + v(x,r)\frac{\partial T}{\partial r} \right) = \lambda \nabla^2 T(x,r) \tag{15.26}$$

Clearly, for large values of $x$ the same asymptotic solutions for local Nusselt number $Nu_D(x)$ will be found.

This type of problem was first approached by Kays. For a tube, he gives a number of relationships for various thermal boundary conditions (see Knudsen and Katz [5]). See also 'Basic Data'.

# 16

# Turbulent Forced Convection

Present understanding of turbulence is fragmentary, often empirical. Various approaches are used: statistical, spectral, etc., not forgetting direct numerical simulation. They do not allow the effects of turbulence on energy transfer to be predicted in all practical situations.

Conversely, turbulence is a phenomenon present in nearly all heat transfer situations of industrial interest; we must attempt to understand it. The aim of the next chapter, 17, is pragmatic: to study the behaviour of the convective heat transfer coefficient $h$ in turbulent flow. The aim of the present chapter is to present the fundamentals of the most widely used methods of calculating turbulent heat transfer. To do this, we must identify various aspects of the phenomenon – generation, dissipation, flow structures, etc. In particular, we will concentrate on scales of length, velocity, time and frequency in the different regions of the flow.

In this chapter we will limit ourselves to the following:

- Flows steady in the mean – that is to say that at any point in space, the values of physical quantities averaged over a time $\tau$, long in comparison to that of the turbulent fluctuations, are independent of time.
- Forced convection.
- Problems involving boundaries; one of the major objectives in the study of heat transfer is to determine the conducto-convective flux exchanged between a fluid and a wall. This (very important) type of turbulence problem is particularly intractable, and the literature is comparatively poor in this area. We will consider primarily the cases of the flat plate and the tube; the results may be generalized to other geometries.

## 16.1 Basics of turbulence

### 16.1.1 Flat plate

Considering a fluid flowing at velocity $u_0$ near a flat plate: we saw in Chapter 14 that two laminar boundary layers, one of momentum and one thermal, develop from the leading edge. These undergo change far from the leading edge. At a critical Reynolds number $Re_x^c$, based on distance $x$ from the leading edge:

$$Re_x^c = \frac{u_0 x}{\nu} \simeq 5 \times 10^5 \tag{16.1}$$

experience shows that **intermittent turbulent plumes** appear, that is, eddies are likely to appear, but after a while the flow (in this region) tends to return to laminar until the next eddies appear. This is the **transition** zone.

As we move further from the leading edge ($Re_x$ rising), we arrive at a region of fully developed turbulence. Here we can define **statistically** a turbulent boundary layer and turbulent boundary layer thickness (Figure 16.1). This artificial boundary is crossed by large eddies (typically of the order of magnitude of the boundary layer thickness).

We postulate, partly on experimental grounds, that in immediate contact with the wall there will be a very thin sub-layer of almost constant thickness (which can be measured by strioscopy) which can be considered **viscous**: while turbulent effects are not totally negligible, they are small compared with viscous effects and the flow resembles a laminar flow. This zone plays a crucial rôle in heat transfer, because it is here that the conducto-convective exchange with the wall takes place. Corresponding to this viscous sub-layer there is a conductive sub-layer in which

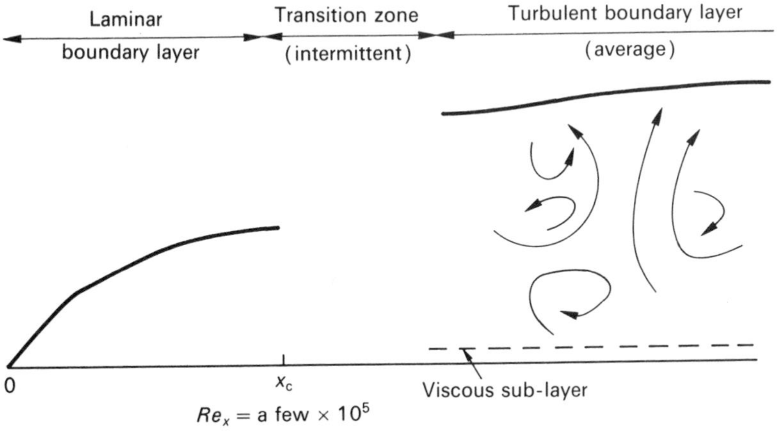


**Figure 16.1** Flat plate: structure of the momentum boundary layer as a function of $x$.

turbulent thermal effects (i.e. enthalpy transport by turbulent eddies) are negligible compared with conductive effects near the wall.

In the remainder of the boundary layer we can identify complex structures which we will return to in Section 16.3; in particular, the turbulence production (and dissipation) is greatest in the zone of fluid immediately adjacent to the viscous sub-layer. From there, large eddies of size comparable to the turbulent boundary layer thickness $\delta$ are generated. These eddies decay thereafter to smaller and smaller eddies before disappearing through viscous dissipation (this is the energy cascade, discussed in Section 16.2.3).

### 16.1.2 Tube of uniform section

We will next consider the flow of a fluid starting from the entry of a tube of diameter $D$, in which the Reynolds number based on diameter, $Re_D$, is above the critical value 2400:

$$D = 2R, \qquad Re_D^c = \frac{u_m D}{\nu} \simeq 2400 \tag{16.2}$$

Here, $u_m$ is the discharge velocity defined by equation (15.3). Initially, a laminar boundary layer, thinner as $Re_D$ is higher, will develop at the tube entry – this is the development region. But when another Reynolds number $Re_x = u_m x/\nu$, based this time on the distance $x$ measured from the tube entry, reaches a value of $5 \times 10^5$, the boundary layers rapidly thicken and, as before, intermittent turbulence appears – this is the transition region. As a result of this, the momentum and thermal régimes are developed much more rapidly than in laminar flow, in a length of 10 or 20 diameters (see Figure 16.2(a)). In industrial applications, the development length is nearly always ignored in turbulent flow, which reduces the heat transfer slightly in this region (see 'Basic Data').

It is found experimentally that once the flow is developed, a flow structure similar to that on the plate exists; in particular, it has a viscous sub-layer near the wall, maximum turbulent kinetic energy production immediately adjacent to the viscous sub-layer, and large eddies filling the whole tube (see Figure 16.2(b)).

*Note*

When $Re_x$ reaches around $3 \times 10^5$, another Reynolds number $Re_\delta = u_m \delta/\nu$, based on the boundary layer thickness $\delta$, will be round about 2400. The criterion for the appearance of turbulence in a tube ($Re_D > 2400$) corresponds to that for a plate ($Re_x > 3$ to $5 \times 10^5$) if $\delta$ is of the same order as $D$.

From a large number of experimental results for flow in pipes, various authors have proposed universal velocity profiles $\bar{u}$; a classic result is that of Schlichting [7]:

$$\frac{\bar{u}(r)}{u_0} = \left(\frac{R - r}{R}\right)^{1/n} \quad \text{with} \quad n = 3Re_D^{0.08} \tag{16.3.1}$$

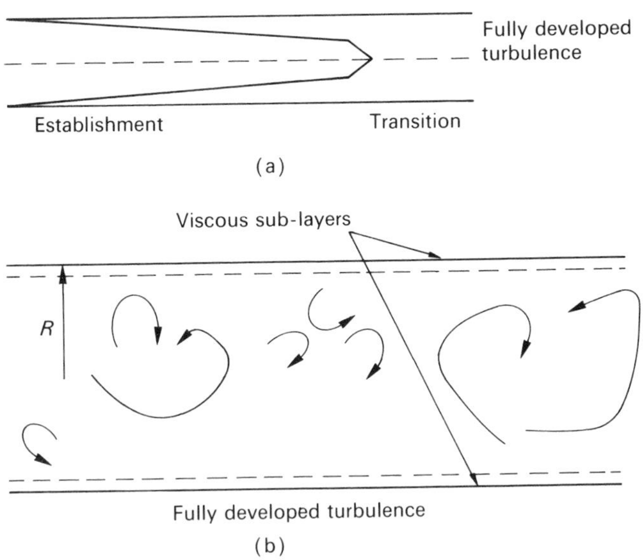


**Figure 16.2**

($n$ has the commonly used value of 7 when $Re_D \simeq 10^5$). In this equation $u_0$ is the maximum velocity reached at the centre of the tube. This profile is in excellent agreement with experimental results for values of $Re_D$ ranging from about $10^3$ to about $10^6$, in all regions of the flow except in the immediate neighbourhood of the wall where the derivative $\partial\bar{u}/\partial r$ is infinite (which is clearly nonsense physically).

This profile is shown in Figure 16.3 for a value of $n$ equal to 7. Note that to start off a numerical calculation, we often use the following purely empirical profile,

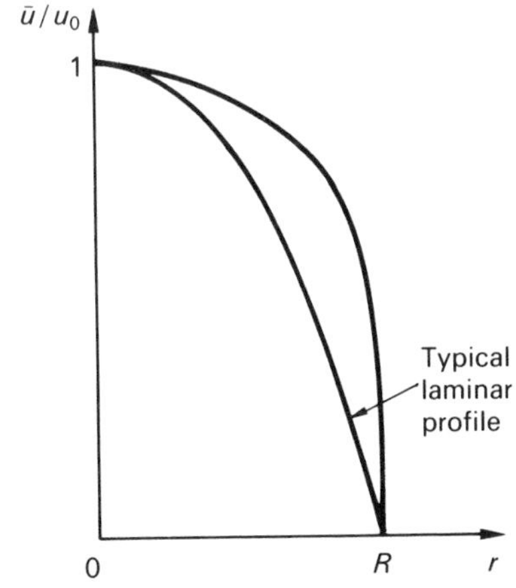


**Figure 16.3** Schlichting's universal profile ($n = 7$).

which has the merit of being much more realistic at the wall (for a two-dimensional case):

$$\bar{u}(r) = u_0(1 - 0.3293Y^2 - 0.6707Y^{32})$$

with (16.3.2)

$$Y = y/e$$

where $y$ and $e$ are the transverse coordinate and the half-width of the channel, respectively.

## 16.2 Balance equations

### 16.2.1 Turbulence quantities

We will assume that the thermophysical properties of the fluid ($\rho$, $\lambda$, $c_p$, etc.) are independent of temperature, which will simplify the theory in this introductory treatment considerably (see the discussion of this point in Section 13.1.2). This assumption, which is seriously contravened for gases which are obviously dilatable (even though in many applications they can be considered incompressible), or for liquids whose viscosity varies strongly with $T$, can be removed by Favre's method [8].

Some authors have successfully modelled turbulence directly, for very simple cases in isothermal fluids, by very extensive calculations from the unsteady equations of motion (of the Navier–Stokes type). This approach is currently (1988) impracticable for heat transfer, particularly in realistic situations, because of limitations in available computer power.

An extra assumption which is generally made is that all the instantaneous physical quantities $A$ may be split into a mean part $\bar{A}$, averaged over a period which is long compared with that associated with the turbulent fluctuations, and a fluctuating part $A'$, which satisfy:

$$A = \bar{A} + A' \qquad \text{with} \qquad \overline{A'} = 0 \tag{16.4}$$

This is called Reynolds decomposition.

For example:

$$T = \bar{T} + T' \tag{16.5.1}$$

$$\mathbf{v} = \bar{\mathbf{v}} + \mathbf{v}' \begin{cases} x & u = \bar{u} + u' \\ y \text{ (or } r) & v = \bar{v} + v' \\ z \text{ (or } \theta) & w = \bar{w} + w' \end{cases} \tag{16.5.2}$$

$$p = \bar{p} + p' \tag{16.6}$$

or, for the product of two quantities such as $vT$:

$$vT = (\overline{vT}) + (vT)' = [\bar{v}\bar{T} + \overline{v'T'}] + [v'\bar{T} + \bar{v}T' + v'T' - \overline{v'T'}] \qquad (16.7)$$

**Notice that the product of the fluctuating terms $v'$ and $T'$ can be split into two parts: one part, $\overline{v'T'}$ whose time mean is non-zero (over a period which is long in comparison to the turbulent fluctuations) because of correlations between the fluctuations in $v$ and $T$, and one part which is zero in the mean, written as $v'T' - \overline{v'T'}$.**

Many authors have measured turbulence related velocity and thermal fluctuations under various conditions. They have discovered important correlations between the fluctuations in various quantities. For example:

- Laufer [9], studying flow in a pipe, measured the quantities $\sqrt{\overline{u'^2}}/u^*$ and $\sqrt{\overline{v'^2}}/u^*$ (where $u^*$ is the wall friction velocity which will be defined below, equation (16.42)) in the near wall region and the quantities $\sqrt{\overline{u'^2}}/u_0$ and $\sqrt{\overline{v'^2}}/u_0$ (where $u_0$ is the mean velocity on the axis) in the centre of the flow (Figure 16.4). It appears that the root-mean-square fluctuating velocity $\sqrt{\overline{u'^2}}$ and $\sqrt{\overline{v'^2}}$ are the same order of magnitude, which suggests that $u'$ and $v'$ are themselves the same order of magnitude ($w'$ does not come into it because of the axi-symmetry of the system). The same worker measured the correlation coefficient $C = -\overline{u'v'}/(\sqrt{\overline{u'^2}}\sqrt{\overline{v'^2}})$, which is of order 1 throughout the pipe, which demonstrates a strong correlation between $u'$ and $v'$ (Figures 16.4 and 16.5).
- Various authors (references [11]–[13]), using a heated tube with constant surface flux density $\varphi_p$, have measured $\sqrt{\overline{T'^2}}/T^*$ (where $T^*$ is the friction temperature $\varphi_p(\rho c_p u^*)$ defined in Section 16.3) and the correlation coefficient $C' = -\overline{v'T'}/\sqrt{\overline{v'^2}}\sqrt{\overline{T'^2}}$ (Figures 16.6 and 16.7). Similar conclusions to those above may be drawn.

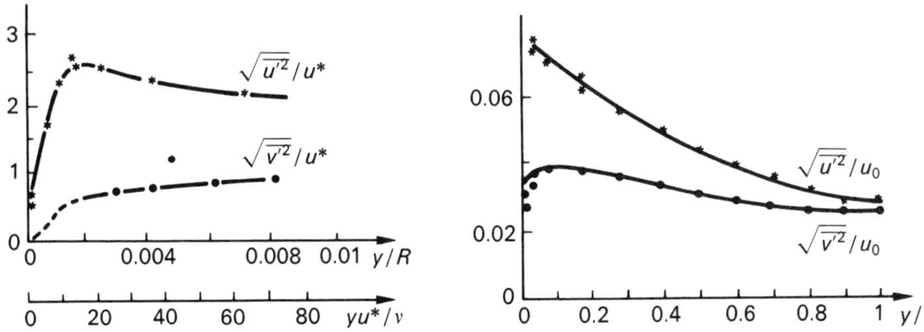


**Figure 16.4** Velocity fluctuations measured by Laufer [9] (from Pimont [10]) for $Re_R = 2.5 \times 10^5$ and $u_0/u^* = 0.035$, fully developed pipe flow (putting $y = R - r$).

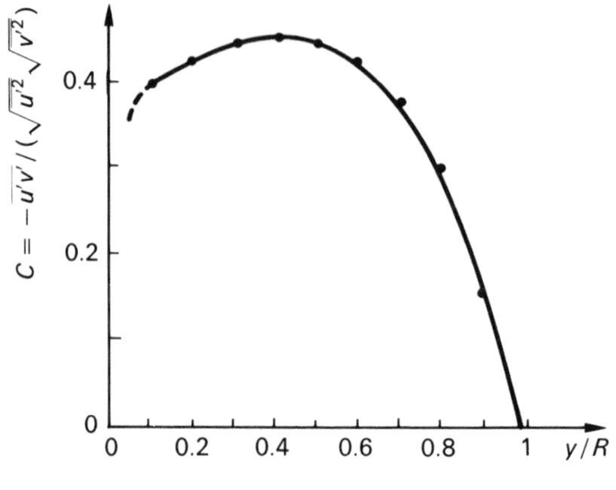


**Figure 16.5** (Same conditions as Figure 16.4.)

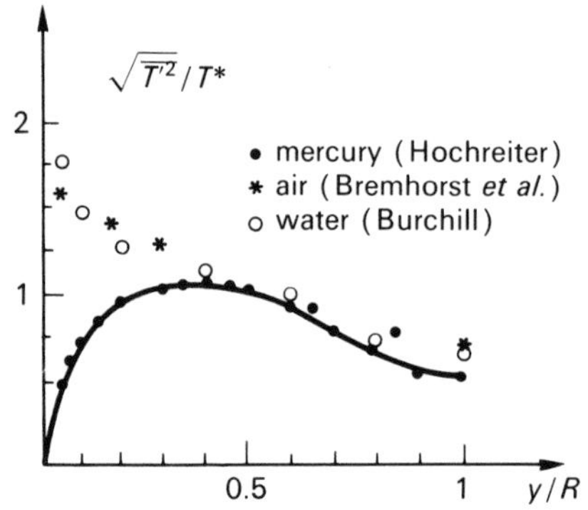


**Figure 16.6** (References 10–13.)

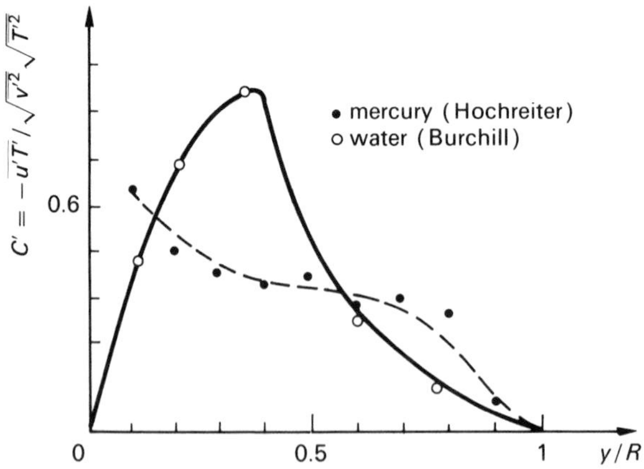


**Figure 16.7** (Same conditions as Figure 16.6.)

### 16.2.2 Mean balance equations

The instantaneous balance equations are time dependent. At a point $M(\mathbf{r})$ in some reference, the velocity is given by $\mathbf{v}(\mathbf{r}, t)$ and the temperature by $T(\mathbf{r}, t)$.

The energy equation for **a fluid with constant thermophysical properties** is given in Cartesian coordinates (with no energy source term) by:

$$\rho c_p\left(\frac{\partial T}{\partial t} + u\frac{\partial T}{\partial x} + v\frac{\partial T}{\partial y} + w\frac{\partial T}{\partial z}\right) = \lambda\left(\frac{\partial^2 T}{\partial x^2} + \frac{\partial^2 T}{\partial y^2} + \frac{\partial^2 T}{\partial z^2}\right) \tag{16.8}$$

We can distinguish two types of time dependence in this equation: let us suppose that we average equation (16.8) over a period $\Delta t$ (which might, for example, be the temporal resolution of a measuring technique), between the instants $t$ and $t + \Delta t$. There are two limiting cases when we compare $\Delta t$ to $\tau$, the typical period of the turbulent fluctuations:

1. $\Delta t < \tau$: observation shows $\bar{T}$ and other quantities averaged over $\Delta t$ depending explicitly on time at a fixed point $M$. This is turbulent unsteadiness.
2. $\Delta t \gg \tau$: we cannot directly observe the effect of turbulent fluctuations. Nevertheless, the system can show two different types of behaviour, as in laminar flow:
   - Temperature averaged over a period long compared with that of the turbulent fluctuations depends explicitly on time at a point $M$ ($\partial\bar{T}/\partial t \neq 0$): **the flow is said to be unsteady in the mean**. Obviously, the same goes for the other quantities ($\bar{u}$, $\bar{v}$, $\bar{w}$, etc.).
   - Temperature averaged over $\Delta t$ does not depend explicitly on time at a point $M$ ($\partial\bar{T}/\partial t = 0$): **the flow is said to be steady in the mean.**

The most rigorous method (solving the unsteady equations in real time) is not practicable at present, although some initial experiments have borne fruit for isothermal flows in very simple geometries. Most turbulence models involve averaging the instantaneous equations over a time period which is long in comparison to $\tau$. The **mean balance equations** given below are then obtained. Here we use Cartesian coordinates, but the method can be adapted to any other system.

1. Mass balance (with $\rho$ constant in this case):

$$\frac{\partial\bar{u}}{\partial x} + \frac{\partial\bar{v}}{\partial y} + \frac{\partial\bar{w}}{\partial z} = 0 \tag{16.9}$$

2. Momentum balance ($x$-component). To obtain this equation, we make use of (16.9) to replace the quantity $\bar{u}\,\partial\bar{u}/\partial x + \bar{v}\partial\bar{u}/\partial y + \bar{w}\,\partial\bar{u}/\partial z$ by $\partial\bar{u}^2/\partial x + \partial\bar{u}\bar{v}/\partial y + \partial\bar{u}\bar{w}/\partial z$; we then use relationships analogous to (16.7) averaged over $\Delta t$. As a result, terms of the form $(\bar{u})^2$, $\bar{u}\bar{w}$, etc., appear on the

left-hand side, and terms such as $-\overline{u'^2} - \overline{u'v'}$, etc., appear on the right-hand side.

$$\rho\left(\frac{\partial \bar{u}}{\partial t} + \frac{\partial}{\partial x}(\bar{u})^2 + \frac{\partial}{\partial y}\bar{u}\bar{v} + \frac{\partial}{\partial z}\bar{u}\bar{w}\right) = -\frac{\partial}{\partial x}\bar{p} + \mu\left(\frac{\partial^2}{\partial x^2}\bar{u} + \frac{\partial^2}{\partial y^2}\bar{u} + \frac{\partial^2}{\partial z^2}\bar{u}^2\right)$$
$$-\rho\left(\frac{\partial}{\partial x}\overline{u'^2} + \frac{\partial}{\partial y}\overline{u'v'} + \frac{\partial}{\partial z}\overline{u'w'}\right) \tag{16.10.1}$$

3. Momentum balance ($y$-component):

$$\rho\left(\frac{\partial \bar{v}}{\partial t} + \frac{\partial}{\partial x}\bar{u}\bar{v} + \frac{\partial}{\partial y}(\bar{v})^2 + \frac{\partial}{\partial z}\bar{v}\bar{w}\right) = -\frac{\partial}{\partial y}\bar{p} + \mu\left(\frac{\partial^2}{\partial x^2}\bar{v} + \frac{\partial^2}{\partial y^2}\bar{v} + \frac{\partial^2}{\partial z^2}\bar{v}\right)$$
$$-\rho\left(\frac{\partial}{\partial x}\overline{u'v'} + \frac{\partial}{\partial y}\overline{v'^2} + \frac{\partial}{\partial z}\overline{v'w'}\right) \tag{16.10.2}$$

4. Momentum balance ($z$-component):

$$\rho\left(\frac{\partial \bar{w}}{\partial t} + \frac{\partial}{\partial x}\bar{u}\bar{w} + \frac{\partial}{\partial y}\bar{v}\bar{w} + \frac{\partial}{\partial z}(\bar{w})^2\right) = -\frac{\partial}{\partial z}\bar{p} + \mu\left(\frac{\partial^2}{\partial x^2}\bar{w} + \frac{\partial^2}{\partial y^2}\bar{w} + \frac{\partial^2}{\partial z^2}\bar{w}\right)$$
$$-\rho\left(\frac{\partial}{\partial x}\overline{u'w'} + \frac{\partial}{\partial y}\overline{v'w'} + \frac{\partial}{\partial z}\overline{w'^2}\right) \tag{16.10.3}$$

5. Energy balance:

$$\rho c_p\left(\frac{\partial \bar{T}}{\partial t} + \bar{u}\frac{\partial T}{\partial x} + \bar{v}\frac{\partial \bar{T}}{\partial y} + \bar{w}\frac{\partial \bar{T}}{\partial z}\right) = \lambda\left(\frac{\partial^2 \bar{T}}{\partial x^2} + \frac{\partial^2 \bar{T}}{\partial y^2} + \frac{\partial^2 \bar{T}}{\partial z^2}\right)$$
$$-\rho c_p\left(\frac{\partial}{\partial x}\overline{u'T'} + \frac{\partial}{\partial y}\overline{v'T'} + \frac{\partial}{\partial z}\overline{w'T'}\right) \tag{16.11.1}$$

or:

$$\rho c_p\left(\frac{\partial \bar{T}}{\partial t} + \bar{u}\frac{\partial \bar{T}}{\partial x} + \bar{v}\frac{\partial \bar{T}}{\partial y} + \bar{w}\frac{\partial \bar{T}}{\partial z}\right) = -\frac{\partial}{\partial x}\left(-\lambda\frac{\partial \bar{T}}{\partial x} + \rho c_p\overline{u'T'}\right)$$
$$-\frac{\partial}{\partial y}\left(-\lambda\frac{\partial \bar{T}}{\partial y} + \rho c_p\overline{v'T'}\right)$$
$$-\frac{\partial}{\partial z}\left(-\lambda\frac{\partial \bar{T}}{\partial z} + \rho c_p\overline{w'T'}\right) \tag{16.11.2}$$

The last terms of (16.10) and (16.11) are analogous: they represent the transfer of momentum and enthalpy by the correlated effects of turbulent fluctuations. The vector $(\rho c_p \overline{u'T'}, \rho c_p \overline{v'T'}, \rho c_p \overline{w'T'})$ represents the enthalpy transfer due to turbulent fluctuations.

The fundamental problem in turbulence is to calculate the terms $\partial \overline{u'^2}/\partial x$, $\partial \overline{u'v'}/\partial y$, etc., and $\partial \overline{v'T'}/\partial y$, etc., i.e. to find closure equations for the system. This will be considered in Section 16.3.

6. Other mean balance equations are often used – for example of average turbulent kinetic energy per unit mass $k$, and its dissipation rate $\varepsilon$. Turbulent kinetic energy per unit mass is defined as:

$$k = \frac{1}{2}(\overline{u'^2} + \overline{v'^2} + \overline{w'^2}) \tag{16.12}$$

It is associated with the average of the instantaneous kinetic energy per unit mass over a time $\Delta t$. Its mean equation is obtained from the instantaneous equation after some manipulation [18]. We will accept the result, and for convenience use tensor notation (see conventions in Appendix 1 of Part I):

$$\rho\left(\frac{\partial k}{\partial t} + \bar{v}_j \frac{\partial k}{\partial x_j}\right) = -\frac{\partial}{\partial x_j}\left(\overline{p'v'_j} + \frac{1}{2}\rho\overline{v'_i v'_i v'_j} - \mu\frac{\partial k}{\partial x_j}\right) - \rho\overline{v'_i v'_j}\frac{\partial}{\partial x_j}\bar{v}_i - \mu\overline{\frac{\partial v'_i}{\partial x_j}\frac{\partial v'_i}{\partial x_j}} \tag{16.13.1}$$

In this equation, the indices $i$, $j$, $k$ refer to the coordinate axes (thus $v_1 = u$, $v_2 = v$, $v_3 = w$, etc.), and $d'_{ij}$ is the fluctuating part of the local rate of strain tensor defined in Appendix 2 of Part I:

$$d'_{ij} = \frac{1}{2}\left(\frac{\partial v'_i}{\partial x_j} + \frac{\partial v'_j}{\partial x_i}\right) \tag{16.13.2}$$

Three types of term appear on the right-hand side of equation (16.13.1), in a similar way to the other mean balance equations:

- A term of the form $-\partial/\partial x_j(\ )$, which in most cases can be thought of as a **diffusion** term.
- A **production** term (of turbulent kinetic energy, in this case), which is always positive:

$$\Pi = -\overline{v'_i v'_j}\frac{\partial}{\partial x_j}\bar{v}_i > 0 \tag{16.14}$$

If, for example, we consider developed pipe flow, the only term of this type is $-\overline{u'v'}\,\partial\bar{u}/\partial r$. We deduce from this that the turbulent kinetic energy production increases as the shear $\partial\bar{v}_i/\partial x_j$ increases. It follows that the turbulence production $\Pi$ is greatest in the vicinity of the wall (Figures 16.8(a) and (b)). The eddies created in this way fill the whole pipe, as we will see in Section 16.2.3, using simple dimensional arguments.

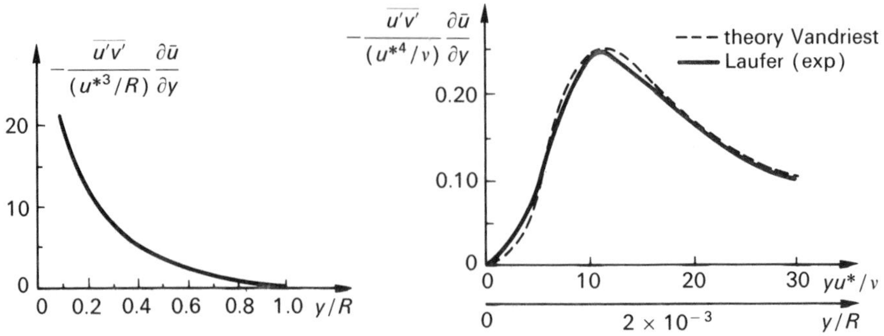


**Figure 16.8** Kinetic energy production (from reference [9]), under the same conditions as Figure 16.4 (putting $y = R - r$).

- A **dissipation** term (of turbulent kinetic energy per unit mass in this case):

$$-\varepsilon = -\frac{\mu}{\rho}\overline{\frac{\partial v_i'}{\partial x_j}}\,\overline{\frac{\partial v_i'}{\partial x_j}} < 0 \tag{16.15}$$

This term is always negative; it corresponds to a loss of turbulent kinetic energy. This loss is due to purely molecular (viscous) effects. However, it is important to note that the characteristic scale of this dissipation is very different from that of turbulent kinetic energy production, as will be shown in the next section.

### *16.2.3 Turbulence production and dissipation length scales; the energy cascade*

A production term $\Pi$ and a dissipation term $\varepsilon$ appear in the mean balance equation for $k$, as we have just seen. It is interesting to look at these terms and the corresponding scales (of velocity, time, etc.), to gain a better understanding of the turbulence phenomenon.

Certain eddies produced by the shear stress $\partial \bar{v}_i/\partial x_j$ may occupy practically the whole cross-section of the pipe (or the whole boundary layer thickness on a flat plate): these eddies have a length scale $R$ for a pipe (or $\delta$ for a boundary layer). These eddies play an important role, since they mix the fluid and tend to make the velocities and temperatures averaged over a period $\Delta t$ more uniform within the flow.

We can see from Figure 16.8(a) that, over most of the section of the pipe ($0 \leqslant r \leqslant 8/10R$, approximately), the quantity $\Pi$ defined by (16.14) is of the order:

$$\Pi = -\overline{u'v'}\frac{\partial}{\partial y}\bar{u} \simeq \frac{u^{*3}}{R} \tag{16.16}$$

where $u^*$ is the friction velocity at the wall, which will be defined in (16.42). In view of (16.16), it is natural to adopt (for pipe flow):

- $R$ as the length scale of the large eddies produced by shear.
- $u^*$ as a velocity scale, in place of the mean velocity $\bar{u}$, for the velocity fluctuations $u'$ and $v'$ (assuming that $\sqrt{\overline{u'v'}}$, $\sqrt{\overline{u'^2}}$, $\sqrt{\overline{v'^2}}$ are of the same order of magnitude as $u'$ and $v'$): see Figure 16.5. The scales of time and frequency can be deduced from these by simple dimensional considerations.

In summary, the scales for production of large eddies, indicated by the subscript p, are:

- length: $l_p = R$
- mean velocity: $\bar{v}_p = u^*$
- velocity fluctuations: $v'_p = u^*$ (16.17)
- time: $t_p = R/u^*$
- frequency: $\omega_p = u^*/R$

It is found experimentally that the eddies produced will, after a certain time, degenerate into smaller and smaller eddies. These last will disappear through viscous (molecular) dissipation, characterized by $\nu$, which introduces an irreversible term like (16.15) into the balance equation for $k$ (15.13).

We can find the reference scales for the small eddies, which are dissipated by viscosity, by simple dimensional analysis based on (16.15): $\varepsilon \to L^2T^{-3}$, $\nu \to L^2T^{-1}$.

The length scale of the small eddies which are dissipated by viscosity is thus given by $(\nu^3/\varepsilon)^{1/4}$, the time scale by $(\nu/\varepsilon)^{1/2}$, the velocity scale by $(\nu\varepsilon)^{1/4}$ and the frequency scale by $(\varepsilon/\nu)^{1/2}$.

We next assume that the production and dissipation terms $\Pi$ and $\varepsilon$ are locally of the same order of magnitude:

$$\varepsilon \simeq \Pi \simeq \frac{u^{*3}}{R} \tag{16.18}$$

This assumption, which is called **Kolmogorov's equilibrium hypothesis**, corresponds to a kind of micro-reversibility ('detailed balance'): enough turbulence is produced locally to replace that disappearing (although the scales of production and dissipation are very different). The dissipation scales for small eddies, indicated by the subscript d, are thus:

- length: $l_d = R(\nu/Ru^*)^{3/4} = R/Re^{*3/4}$
- velocity (mean and fluctuating): $\bar{v}_d = v'_d = u^*(\nu/Ru^*)^{1/4} = u^*/Re^{*1/4}$
- time: $t_d = (R/u^*)/Re^{*1/2}$
- frequency: $\omega_d = (u^*/R)R^{*1/2}$

where:

$$Re^* = u^*R/\nu$$

Since the Reynolds number based on the friction velocity is, in turbulent flow, large in comparison to 1, it follows that:

$$\begin{gathered} l_d \ll l_p \\ v'_d,\ \bar{v}_d \ll v'_p,\ \bar{v}_p \\ t_d \ll t_p \\ \omega_d \gg \omega_p \end{gathered} \tag{16.19}$$

The small eddies which are dissipated by turbulence have much higher angular velocities and much smaller ranges than those of the large eddies which are created. Their lifetimes are, moreover, much shorter than those of the large eddies.

**The whole physical problem of turbulence consists of correctly modelling the energy cascade which starts from the large generated eddies, passes through intermediate eddies and terminates with the small eddies which are dissipated by viscosity.**

*Important note*

Notice that the dissipation phenomenon is not linear with respect to the viscosity $\nu$; all other things being equal, **a doubling of viscosity has no appreciable effect on the dissipation term**. Indeed, the terms $\partial v'_i/\partial x_j \partial v'_i/\partial x_j$ which appear in the equation for $\varepsilon$ are very complex; their calculation depends on an accurate modelling of all turbulent phenomena, at all scales, which is, we repeat, the basic problem of turbulence.

## 16.3 The closure problem

It is necessary, as we saw in the preceding section, to evaluate terms of the form $\partial \overline{u'^2}/\partial x$, $\partial \overline{u'v'}/\partial y$, $\partial \overline{v'T'}/\partial z$, etc., in the mean balance equations.

### 16.3.1 Closure by supplementary balance equations

An initial route, the most physical, is to **write statistical balance equations for the relevant quantities** ($\overline{u'^2}$, $\overline{u'v'}$, $\overline{v'T'}$, etc.). One can easily see that the problem is merely displaced – the new equations will require closure, and so on. The physical ideal underlying this is that the higher the order at which the closure is made, the less risk there is that an approximate model will affect the result. This approach has been used notably by Vandromme *et al.* [15]. Such models are currently limited by the performance of computing systems as soon as the geometry becomes at all complex. They have, however, the advantage that they avoid extra assumptions whose foundations are disputable.

Here, we will restrict ourselves to the most common closure models, which replace the terms $\partial \overline{u'v'}/\partial y$, $\partial \overline{v'T'}/\partial z$, etc., with diffusion terms.

### 16.3.2 General ideas on turbulent diffusion

Almost all practicable solutions depend on the concepts of:

- Turbulent viscosity ($\mu_t$ or $\nu_t$), obtained by writing:

$$-\rho\overline{v_i'v_j'} = \mu_t\left(\frac{\partial \bar{v}_i}{\partial x_j} + \frac{\partial \bar{v}_j}{\partial x_i}\right) - \frac{2}{3}\rho k\delta_{ij} \tag{16.20}$$

(which is compatible with the definition of $k$, when $\rho$ is constant), or in the simplest case of a shear flow:

$$-\rho\overline{u'v'} = \mu_t\frac{\partial \bar{u}}{\partial y} \tag{16.21}$$

In an obvious manner, we can introduce the turbulent momentum diffusivity:

$$\nu_t = \frac{\mu_t}{\rho} \tag{16.22}$$

- Turbulent thermal conductivity and diffusivity ($\lambda_t$ and $a_t$) are obtained by analogy with (16.20):

$$-\rho c_p\overline{v_j'T'} = \lambda_t\frac{\partial \bar{T}}{\partial x_j} \tag{16.23}$$

or:

$$-\overline{v_j'T'} = a_t\frac{\partial \bar{T}}{\partial x_j} \qquad \text{with:} \qquad a_t = \frac{\lambda_t}{\rho c_p} \tag{16.24}$$

The analogy between the two processes has led a number of authors to introduce a 'turbulent Prandtl number' $Pr_t$, defined by:

$$Pr_t = \frac{\nu_t}{a_t} = \frac{\mu_t c_p}{\lambda_t} \tag{16.25}$$

The turbulence quantities thus introduced by analogy with the molecular transport mechanisms (viscous and conductive) are clearly not intrinsic properties of the fluid concerned. Introducing $\lambda_t$ in (16.23) has no physical foundation other than making the tubulent convection term in the energy equation (16.11), $-\rho c_p\partial(\overline{v'T'})/\partial y$, look like the molecular diffusion term $\partial(\lambda\,\partial\bar{T}/\partial y)/\partial y$ by transforming it to $\partial(\lambda_t\,\partial\bar{T}/\partial y)/\partial y$. Although such methods are physically contestable, they are used in the majority of models which produce results that can be used by the engineer. (Realistically, the results of these models should be compared with experience as often as possible.)

The analogy between equations (16.20) and (16.24), and thus between the diffusivities $\nu_t$ and $a_t$, extends the analogy already developed in laminar flow. The correlations $\overline{v_i'v_j'}$ and $\overline{T'v_j'}$ will appear together, within the same eddy. Nevertheless, there are important differences, because in one case we are transporting enthalpy (a scalar quantity) (related to $T'$) and, in the other, momentum (a vector quantity).

The closure methods based on the concepts of turbulent viscosity, conductivity and diffusivity comprise the following:

1. Finding direct expressions for the quantities $\mu_t$ (or $\nu_t$) and $\lambda_t$ (or $a_t$) which appear in equations (16.20)–(16.24). This is the approach of the Prandtl model and its generalizations (see Section 16.3.3).
2. Using supplementary statistical balance equations to find $\nu_t$ and $a_t$. The most common are models using:
   - $k$ (the Prandtl–Kolmogorov model),
   - both $k$ and $\varepsilon$ (the $k$–$\varepsilon$ model)

   which are discussed in Section 16.6.

The latter approach re-presents the closure problem in the new equations introduced, and needs numerous approximations.

The energy equation leads to the introduction of terms in $\overline{v_j'T'}$, which are generally expressed in terms of $\mu_t$, from which one obtains an expression from the turbulent Prandtl number at any point in the flow (based, for example, on van Driest's or Cebeci's models, discussed in Section 16.4). Then, whether method 1 or some other method 2 is used, the energy equation is closed through knowledge of $Pr_t$.

### 16.3.3 Prandtl's closure model

This model is based on an analogy with molecular diffusion. Consider a flow with mean velocity $\bar{u}$ in the $x$ direction, dependent only on $y$. A packet of fluid which is transported across the flow from $y$ (where the mean velocity is $\bar{u}(y)$) to $y + dy$ (where it is $\bar{u}(y + dy)$) is likely to have a fluctuation velocity $u'$ of the order of $|\bar{u}(y + dy) - \bar{u}(y)|$ (Figure 16.9).

We assume that an eddy retains its identity (as regards velocity) only over a finite distance $l_m$, known as the momentum mixing length. Then:

$$|u'| \propto l_m \frac{\partial \bar{u}}{\partial y} \tag{16.26}$$

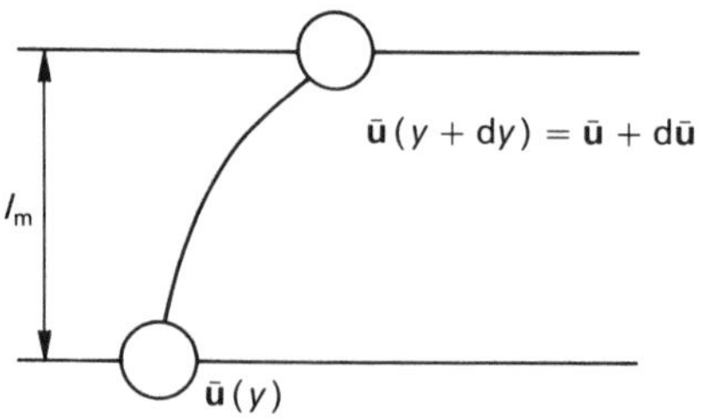


**Figure 16.9**

This length has no obvious relationship to the size of the eddy. As we saw above (Figures 16.4 and 16.5), since:

- $|u'|$ and $|v'|$ are the same order of magnitude,
- $\overline{u'v'}$ is of the order of $\sqrt{\overline{u'^2}}\sqrt{\overline{v'^2}}$,

it is natural to write:

$$-\overline{u'v'} \propto l_m^2\left(\frac{\partial \bar{u}}{\partial y}\right)^2 \tag{16.27}$$

Since, further, from equation (16.21) we have:

$$-\rho\overline{u'v'} = \mu_t \frac{\partial \bar{u}}{\partial y} \tag{16.28}$$

it follows that:

$$\mu_t = \rho l_m^2 \frac{\partial \bar{u}}{\partial y} \tag{16.29}$$

This very simple expression gives the turbulent viscosity as a function of known macroscopic quantities and the mixing length $l_m$. We will see that van Driest has suggested a precise formula for $l_m$ in the inner momentum region, that is, the layer of the flow adjacent to the wall (Section 16.4.3.3).

By similar arguments, we can obtain:

$$|T'| \propto l_{th} \frac{\partial \bar{T}}{\partial y} \tag{16.30}$$

where $l_{th}$ is the thermal mixing length, the distance over which an eddy retains its thermal identity, assumed to be different from $l_m$. Since $\overline{v'T'}$ is of the order of magnitude of $\sqrt{\overline{v'^2}}\sqrt{\overline{T'^2}}$, and thus of $|v'|\,|T'|$, we can deduce that:

$$-\overline{v'T'} = l_m l_{th} \frac{\partial \bar{T}}{\partial y}\frac{\partial \bar{u}}{\partial y} \tag{16.31}$$

Similarly, by writing:

$$-\overline{v'T'} = a_t \frac{\partial \bar{T}}{\partial y} \tag{16.32}$$

we obtain:

$$a_t = l_m l_{th} \frac{\partial \bar{u}}{\partial y} \tag{16.33}$$

The turbulent Prandtl number defined by:

$$Pr_t = \frac{\mu_t}{\rho a_t} = \frac{\nu_t}{a_t} \tag{16.34}$$

turns out, on this very simple model, to be the ratio of the momentum and thermal mixing lengths:

$$Pr_t = \frac{l_m}{l_{th}} \tag{16.35}$$

Expressions for the momentum and thermal mixing lengths can be obtained by a detailed analysis of the various processes occurring in the layers of fluid near the wall; this is the object of Section 16.4. Various methods of setting up the $k$–$\varepsilon$ model in the neighbourhood of the wall develop from this analysis

*Note*

Note that **the analogy with molecular processes which is the basis of this model is arguable**; it is, however, consistent with assumptions (16.28) and (16.32) which associate turbulence transport phenomena with diffusion.

## 16.4 Turbulent flow structures in a pipe. Van Driest and Cebeci models[1]

By way of example, we will consider the **fully developed** flow in a pipe, on whose walls there is a constant heat flux density $\varphi_w$. The idea of fully developed laminar flow discussed in Chapter 15 is readily extended to turbulent flow.

Most physical quantities $A$, when averaged over a time period long in comparison to that of the turbulent fluctuations, are invariant in the flow ($x$) direction, i.e.:

$$\frac{\partial}{\partial x}\bar{A} = 0 \tag{16.36.1}$$

This property applies to:

- the mean velocity, the dimensionless mean velocity $u^+ = \bar{u}/u_m$, where $u_m$ is the discharge velocity, and the mean dimensionless temperature defined as in Chapter 15 by $T^+ = (T - T_w)/(T_m - T_w)$, where $T_w$ and $T_m$ are the wall and mixed fluid temperatures at a cross-section, so:

$$\frac{\partial \bar{u}}{\partial x} = 0; \qquad \frac{\partial \bar{u}^+}{\partial x} = 0; \qquad \frac{\partial \bar{T}^+}{\partial x} = 0 \tag{16.36.2}$$

- the correlations $\overline{u'v'}$, $\overline{u'T'}$, $\overline{v'T'}$, $\overline{u'^2}$, $\overline{T'^2}$, etc., so:

$$\frac{\partial}{\partial x}\overline{u'v'} = 0; \qquad \frac{\partial}{\partial x}\overline{v'T'} = 0; \qquad \text{etc.} \tag{16.36.3}$$

[1] This section is in part inspired by a thesis compiled at C.E.A. (CENG) by Vincent Pimont [10], under the supervision of Jean-Marc Delhaye. (Figures in Chapter 16 deriving from this work are reproduced with permission from J-M.D.)

They clearly do not apply to temperature $\bar{T}$ or $\bar{T}(x) - T_w(x)$, nor to the pressure, whose gradient provides the force maintaining the motion.

Starting from expressions for the total shear stresses and the total radial heat flux, we will identify various zones and sub-layers in the flow and various scales of length, velocity and temperature (both mean and fluctuating) to be considered, and eventually obtain semi-theoretical and semi-experimental expressions for the mean profiles of velocity and temperature and also for the turbulent quantities $\overline{v'T'}$ and $\overline{u'v'}$.

### 16.4.1 Total shear stress $\bar{\tau}_{\text{Tot}}$

We will use cylindrical coordinates $(x, r, \theta)$, with instantaneous velocity components $u$, $v$, $w$ in the axial, radial and circumferential directions, and corresponding velocity fluctuations $u'$, $v'$, $w'$. As the problem is axially symmetric, all magnitudes are independent of $\theta$: $\bar{w}$ and the correlation terms $\overline{u'w'}$, $\overline{v'w'}$ and $\overline{T'w'}$ are zero.

The mean continuity equation is (since $\partial\bar{u}/\partial x$ is zero):

$$\frac{1}{r}\frac{\partial}{\partial r}(r\bar{v}) = 0 \qquad \text{with} \qquad \bar{v}(R) = 0 \tag{16.37}$$

which implies:

$$\bar{v}(r) = 0 \qquad \forall\, r \tag{16.38}$$

The $x$-component of the mean momentum equation is in this case:

$$-\frac{\partial\bar{p}}{\partial x} + \frac{1}{r}\frac{\mathrm{d}}{\mathrm{d}r}\left(r\mu\frac{\mathrm{d}\bar{u}}{\mathrm{d}r} - r\rho\overline{u'v'}\right) = 0 \tag{16.39}$$

The $y$-component gives, after differentiation with respect to $x$:

$$\frac{\partial}{\partial r}\left(\frac{\partial\bar{p}}{\partial x}\right) = 0 \tag{16.40}$$

Since $\partial\bar{p}/\partial x$ is independent of $r$, integrating the last but one equation from 0 to $r$ gives:

$$\bar{\tau}_{\text{Tot}} = -\mu\frac{\mathrm{d}\bar{u}}{\mathrm{d}r} + \rho\overline{u'v'} = -\frac{r}{2}\frac{\mathrm{d}\bar{p}}{\mathrm{d}x} = \rho u^{*2}\frac{r}{R} \tag{16.41}$$

in which appears the wall friction velocity $u^*$, obtained by integrating (16.39) from 0 to $R$ (neglecting $\rho\overline{u'v'}$ at the wall):

$$u^* = \sqrt{\frac{\tau_w}{\rho}} = \sqrt{-\nu\frac{\mathrm{d}\bar{u}}{\mathrm{d}r}(R)} = \sqrt{-\frac{R}{2\rho}\frac{\mathrm{d}\bar{p}}{\mathrm{d}x}} \tag{16.42}$$

where $\tau_w$ is the viscous shear stress at the wall. Since we are interested in the near-wall region, we write:

$$y = R - r \qquad \mathrm{d}y = -\mathrm{d}r \tag{16.43}$$

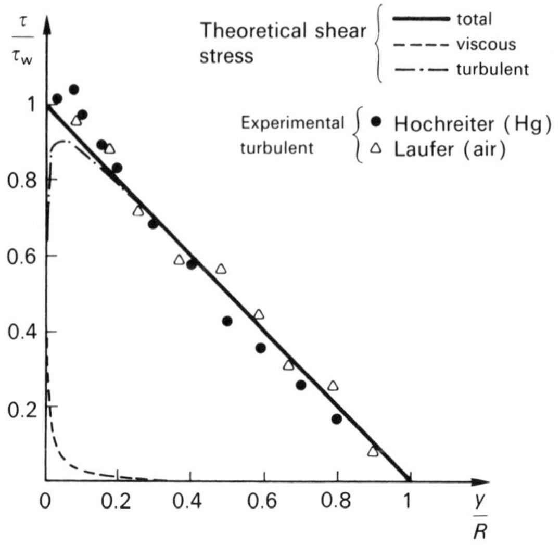


**Figure 16.10** Shear stress in a pipe in turbulent flow (from references [9]–[11]).

which gives a new expression for $\bar{\tau}_{\text{Tot}}$, in terms of $(x, y)$:[2]

$$\bar{\tau}_{\text{Tot}} = \mu \frac{d\bar{u}}{dy} - \rho \overline{u'v'} = \rho u^{*2}\left(1 - \frac{y}{R}\right) \tag{16.44}$$

The modulus of the total shear stress falls linearly from $\tau_w$ at the wall to 0 on the flow axis. Experimental measurements of the turbulent stress $\rho\overline{u'v'}$ show that it dominates in the centre of the flow and falls very sharply near the wall (Figure 16.10).

## *16.4.2 Radial thermal flux density*

The mean energy balance equation is:

$$\rho c_p \bar{u} \frac{\partial \bar{T}}{\partial x} = \frac{1}{r}\frac{\partial}{\partial r}\left(r\lambda \frac{\partial \bar{T}}{\partial r} - r\rho c_p \overline{v'T'}\right) \tag{16.45}$$

The total radial flux density (conductive and turbulent) is defined by:

$$\bar{\varphi}_{\text{Tot}}(r) = -\frac{\partial \bar{T}}{\partial r} + \rho c_p \overline{v'T'} \tag{16.46}$$

[2] From equation (16.44) to the end, $v$ and $v'$ will be taken as positive in the positive $y$ direction. This leads to a change of sign in some terms; this new convention is compatible with that used for plane geometry.

It depends only on $r$ (assuming developed thermal conditions). Indeed, as we have seen, in developed conditions $\overline{v'T'}$ is independent of $x$ and, further, it can be shown by a derivation identical to that in Section 15.2.1.1 (which concerns laminar flow) that $\partial \bar{T}/\partial r$ depends only on $r$.

If we adopt the universal mean velocity profile $\bar{u}(r)$ (16.3.1[7]):

$$\frac{\bar{u}}{u_0} = \left(\frac{y}{R}\right)^{1/n} \quad \text{with} \quad n = 3Re_D^{0.08} \tag{16.47}$$

(where $u_0$ is the maximum velocity (at $r = 0$) and $y$ the distance $(R - r)$ from the wall) we can integrate the energy equation from 0 to $y$ (using the sign conventions of (16.44)), and obtain:

$$\bar{\varphi}_{\text{Tot}}(y) = +\lambda \frac{\partial \bar{T}}{\partial y} - \rho c_p \overline{v'T'} = \varphi_w \left(1 - \frac{y}{R}\right) f\left(\frac{y}{R}, n\right) \tag{16.48}$$

where:

$$f\left(\frac{y}{R}, n\right) = \left(\frac{R}{R - y}\right)^2 \left(1 - \frac{(2n + 1)Ry^{(n+1)/n} - (n + 1)y^{(2n+1)/n}}{nR^{(2n+1)/n}}\right) \tag{16.49}$$

The function $f$ depends only on $y/R$ and the Reynolds number (through $n$). In the limit where $Re_D$ tends to infinity, $f$ tends to 1 and the radial heat flux is then:

$$Re_D \to \infty: \quad \bar{\varphi}_{\text{Tot}}(y) = +\lambda \frac{\partial \bar{T}}{\partial y} - \rho c_p \overline{v'T'} = \varphi_w \left(1 - \frac{y}{R}\right) \tag{16.50}$$

The expression thus obtained for the total radial heat flux $\bar{\varphi}_{\text{Tot}}(r)$ is analogous for that for the total shear stress $\bar{\tau}_{\text{Tot}}(y)$, whose derivation, however, did not need any assumption regarding $Re_D$. The deviations from linearity of (16.48) are fairly weak – about 20% for $Re$ around 5000, and decreasing as $Re$ rises. In the following, we will use (16.50). The magnitude of the radial heat flux then increases linearly from 0 at the centre of the flow to $\varphi_w$ at the wall. The turbulent contribution is small at the wall, but dominates in the core of the flow; the converse is true for the conductive contribution.

### *16.4.3 Flow structure and reference scales*

Traditionally, two types of region are distinguished in fully developed turbulent pipe flow, defined in terms of the turbulent shear stress terms $\overline{u'v'}$ and $\overline{v'T'}$:

- **Inner regions**, close to the wall, in which the contributions of molecular effects (viscosity and conductivity) are not negligible. These effects dominate at the wall, but become very weak at the other limit of the inner region.
- **Outer regions**, in which only turbulent effects are present: $\overline{u'v'}$ and $\overline{v'T'}$. These regions, which are geometrically the larger, constitute the heart or core of the turbulent flow.

Clearly, the boundary between the inner and outer regions is somewhat fuzzy; the zones are differentiated essentially by the scales of reference for the mean and turbulent quantities (temperatures, speeds, lengths, etc.) which are adopted. We speak of the inner momentum region (IMR) and the outer momentum region (OMR) when referring to the turbulent stress $\overline{u'v'}$, and the inner thermal region (ITR) and outer thermal region (OTR) for $\overline{v'T'}$.

16.4.3.1 REFERENCE SCALES IN THE OUTER REGIONS

Since the outer regions occupy most of the pipe, the length $l_o$ will be used for both the OMR and OTR:

$$l_o = R \tag{16.51}$$

In the outer momentum region, equation (16.44) becomes, in the absence of viscous dissipation:

$$-\frac{\overline{u'v'}}{u^{*2}} = 1 - \frac{y}{R} \tag{16.52}$$

The velocity fluctuation scale is thus:

$$v'_o = u'_o = u^* \tag{16.53}$$

The turbulent kinetic energy production due to shear $-\overline{u'v'}\,\partial\bar{u}/\partial y$ is of the order of $u^{*3}/R$ (Figure 16.8(a)), so as a scale of mean velocity we adopt:

$$u_o = u^* \tag{16.54}$$

In the outer thermal region, we similarly adopt the scales (16.51), (16.53) and (16.54). The equation giving the total flux (16.50) becomes:

$$-\overline{v'T'} = \frac{\varphi_w}{\rho c_p}\left(1 - \frac{y}{R}\right) \tag{16.55}$$

Since, from the experimental results (Figure 16.7), $\overline{v'T'}$ appears to be of the order of magnitude of $\sqrt{\overline{v'^2}}\sqrt{\overline{T'^2}}$, we are led to adopt the friction temperature $T^*$, defined by:

$$T'_o = \frac{\varphi_w}{\rho c_p u^*} = T^* \tag{16.56}$$

as the scale for temperature fluctuations.

In dimensionless variables, we substitute:

$$y_o^+ = \frac{y}{R}, \qquad \bar{u}^+ = \frac{\bar{u}}{u^*}, \qquad v'^+ = \frac{v'}{u^*}, \qquad T'^+ = \frac{T'}{T^*}, \qquad \bar{T}^+ = \frac{\overline{\delta T}}{\delta T_o} \tag{16.57}$$

where $\overline{\delta T}$ is the mean temperature difference between the temperature of the fluid and that of the wall, and $\delta T_o$ a reference temperature difference. The total radial heat flux, given by (16.50), then becomes:

$$\frac{\bar{\varphi}_{Tot}}{\rho c_p u^* T^*} = \left(\frac{a}{u^* R}\right)\left(\frac{\delta T_o}{T^*}\right)\frac{\partial}{\partial y_o^+}\bar{T}^+ - \overline{v'^+ T'^+} = 1 - y_o^+ \tag{16.58}$$

in which the Péclet number:

$$Pe_{u^*,R} = \frac{u^* R}{a} \tag{16.59}$$

appears. Since in turbulent flow, $Pe_{u^*,R}$ is large compared to 1, equation (16.58) becomes simply:

$$-\overline{v'^+ T'^+} = 1 - y_o^+ \tag{16.60}$$

if we adopt:

$$\delta T_o = T^* \tag{16.61}$$

as a reference scale for temperature difference in the outer thermal region.

#### 16.4.3.2 REFERENCE SCALES IN THE INNER REGIONS

In the inner momentum and thermal regions (near to the wall) it is logical to take the wall friction velocity $u^*$ as the reference scale for the mean velocity $u_i$ and the fluctuating velocities $u_i'$ and $v_i'$:

$$u_i = v_i' = u_i' = u^* \tag{16.62}$$

Putting the viscous and turbulent contributions to the shear stress as the same order of magnitude in equation (16.44), we obtain:

$$\frac{\nu u^*}{l_i^m}\frac{d\bar{u}^+}{dy_i^+} = -\overline{u'^+ v'^+}\, u^{*2} \tag{16.63}$$

where we have put:

$$y_i^+ = \frac{y}{l_i^m}, \qquad \bar{u}^+ = \frac{\bar{u}}{u^*}, \qquad u'^+ = \frac{u'}{u^*}, \qquad v'^+ = \frac{v'}{u^*} \tag{16.64}$$

The reference length in the inner momentum zone, $l_i^m$, is thus (from (16.63)):

$$l_i^m = \frac{\nu}{u^*} \tag{16.65}$$

which leads to an expression for $y_i^+$:

$$y_i^+ = \frac{u^* y}{\nu} = Re_{u^*,y} \tag{16.66}$$

If $y/R$ is negligible compared to 1, equation (16.44) for the total shear stress becomes:

$$\frac{\bar{\tau}_{\mathrm{Tot}}}{\rho u^{*2}} = \frac{\mathrm{d}\bar{u}^*}{\mathrm{d}y_i^+} - \overline{u'^+ v'^+} \simeq 1 \tag{16.67}$$

**The inner momentum region is a region of constant shear stress.** In the inner thermal region, close to the wall, we will adopt the wall friction temperature $T^*$ as a reference for the mean and fluctuating temperature differences $\delta T_i$ and $T_i'$, by analogy with (16.62):

$$\delta T_i = T_i' = T^* \tag{16.68}$$

which is confirmed by experimental observations.

In the equation for the radial heat flux (16.50), the contributions of conduction and turbulence are of the same order of magnitude, so that:

$$\frac{a}{l_i^{\mathrm{th}}} T^* \frac{\partial \bar{T}}{\partial y_i^{++}} \sim -\overline{v'^+ T'^+} u^* T^* \tag{16.69}$$

where we have obviously put:

$$\overline{\delta T^+} = \frac{\overline{\delta T}}{T^*} \qquad T'^+ = \frac{T'}{T^*} \tag{16.70}$$

By analogy with (16.65), we are led to take the reference length for the inner thermal layer as:

$$l_i^{\mathrm{th}} = \frac{a}{u^*} \tag{16.71}$$

with the substitution:

$$y_i^{++} = \frac{u^* y}{a} = Pe_{u^*,y} \tag{16.72}$$

**The inner thermal region has a constant radial heat flux:**

$$\frac{\overline{\varphi_{\mathrm{Tot}}}}{\rho c_p u^* T^*} = \frac{\mathrm{d}\bar{T}^+}{\mathrm{d}y_i^{++}} - \overline{v'^+ T'^+} \simeq 1 \tag{16.73}$$

which is analogous to (16.67).

In summary, for whichever region we consider:

- The velocity scale is $u^*$.
- The temperature or temperature difference scale is $T^*$.
- The length scales are as follows:

| | OMR | OTR | IMR | ITR |
|---|---|---|---|---|
| $l$ (length<br>$y^+$ | $R$<br>$y/R$ | $R$<br>$y/R$ | $\nu/u^*$<br>$y_i^+ = Re_{u^*,y}$ | $a/u^*$<br>$y_i^{++} = Pe_{u^*,y}$ |

16.4.3.3 STRUCTURE OF THE INNER REGIONS (CLOSE TO THE WALLS)

By comparison of the orders of magnitude of the molecular and turbulent effects we can distinguish the following three sub-layers in each of the inner regions:

1. **In the part of the inner region which is immediately adjacent to the wall, the turbulent stresses are negligible** (although observation suggests that they do exist). From (16.67) and (16.73) we obtain:

$$0 \leqslant y_i^+ \leqslant 3 \text{ to } 5 \qquad \frac{d\bar{u}^+}{dy_i^+} = 1 \Rightarrow \bar{u}^+ = y_i^+ \tag{16.74}$$

This sub-layer where viscous effects are dominant is called the **viscous sub-layer**:

$$0 \leqslant y_i^{++} \leqslant 3 \text{ to } 5 \qquad \frac{d\bar{T}^+}{dy_i^{++}} = 1 \Rightarrow \overline{\delta T^+} = y_i^{++} \tag{16.75}$$

This sub-layer, where conductive effects are dominant, although turbulent effects are not completely absent, is, by analogy with the above, the **conductive sub-layer.**

2. At the other extreme, **in the part of the inner region which is next to the outer region, molecular effects are very small compared to viscous effects.** This part of the inner region is little different from the adjacent part of the outer region. The (somewhat arbitrary) distinction between these regions is the reference length used (see the preceding section). We can thus write the mean velocity gradient in two different ways for the inner momentum region at the boundary between the two regions:

$$\frac{d\bar{u}}{dy} = \frac{u^{*2}}{\nu}\frac{d\bar{u}^+}{dy_i^+} \qquad \text{and} \qquad \frac{d\bar{u}}{dy} = \frac{u^*}{R}\frac{d\bar{u}^+}{dy_o^+} \tag{16.76}$$

Equating these two and multiplying by $y/u^*$, we obtain:

$$\frac{yu^*}{\nu}\frac{d\bar{u}^+}{dy_i^+} = y_i^+\frac{d\bar{u}^+}{dy_i^+} = y_o^+\frac{d\bar{u}^+}{dy_o^+} = \frac{y}{R}\frac{d\bar{u}^+}{dy_o^+} = \frac{1}{k} \tag{16.77}$$

Equality of two functions of different variables implies that they must be constant, here written as $1/k$. This leads to logarithmic velocity profiles, both in the outer region (see Section 16.4.3.4) and the inner region:

$$\bar{u}^+ = \frac{1}{k}\log\left(\frac{yu^*}{\nu}\right) + C \tag{16.78}$$

The constant $k$, called **Von Karman's constant**, is usually taken as 0.4. Tennekes [14] proposes a dependency of $k$ on $Re_{u^*,R}$, based on a large amount of experimental data:

$$\frac{1}{k} = 3 - \frac{5}{Re_{u^*,R}^{1/3}} \tag{16.79}$$

This equation gives values of $k$ close to 0.4 over a wide range of $Re_{u^*,R}$ in turbulent flow. Nikuradze proposed a profile of the form of (16.78) on the basis of many experimental results, and used values $k = 0.4$ and $C = 5.5$. The sub-layer of the inner region in which (16.78) is valid is often called the **inertial sub-layer** of the inner momentum region.

In the same way, we can introduce an **inertial sub-layer** in the inner thermal region (near the outer region) by looking at mean temperature gradients in the inner and outer regions respectively:

$$\frac{\partial \bar{T}}{\partial y} = \frac{T^* u^*}{a} \frac{\partial \bar{T}^*}{\partial y_i^{++}} \quad \text{and} \quad \frac{\partial \bar{T}}{\partial y} = \frac{T^*}{R} \frac{\partial \bar{T}^*}{\partial y_o^+} \tag{16.80}$$

By analogy with (16.78), we obtain a logarithmic temperature profile in both the outer region near the inner region and in the inner region near the outer region; this latter is given by:

$$\delta \bar{T}^+ = \frac{\bar{T} - T_w}{T^*} = \frac{1}{k'} \log\left(\frac{yu^*}{a}\right) + C' \tag{16.81}$$

Pimont [10] proposes a correlation analogous to (16.79), based on various experimental results:

$$\frac{1}{k'} \sim 2 + \frac{65}{Pe_{u^*,R}} \tag{16.82}$$

The values of $k'$ obtained are around 0.44, and are thus similar to Von Karman's $k$. The two boundaries of the momentum and inertial sub-layers are imprecise because:

- the logarithmic profiles are continuous at the boundary further away from the wall (where $y_i^+$ = a few $\times 10^2$ and $y_o^+ = 0.1$ to $0.2$).
- molecular phenomena (conduction and viscosity) become progressively more important in the neighbourhood of the wall. The inner boundaries of the inertial sub-layers are generally taken at $y_i^+ = 40$ and $y_i^{++} = 40$.

The Prandtl mixing lengths $l_m$ and $l_{th}$ within the inertial sub-layers can be deduced from equations (16.77) and (16.80). We use:

$$\frac{\mathrm{d}\bar{u}}{\mathrm{d}y} = \frac{u^*}{ky}, \qquad \frac{\mathrm{d}\bar{T}}{\mathrm{d}y} = \frac{T^*}{k'y} \tag{16.83}$$

In the inertial sub-layers of the inner regions turbulence effects are dominant compared to molecular effects. Equations (16.67) and (16.73) reduce to:

$$\frac{y}{R} \ll 1; \qquad -\overline{u'v'} \simeq u^{*2}; \qquad -\overline{v'T'} \simeq u^* T^* \tag{16.84}$$

Whence, using (16.20), (16.24), (16.29), (16.33) and (16.84):

$$\nu_t \frac{\partial \bar{u}}{\partial y} = l_m^2 \left(\frac{\partial \bar{u}}{\partial y}\right)^2 = u^{*2}; \qquad a_t \frac{\partial \bar{T}}{\partial y} = l_m l_{th} \frac{\partial \bar{u}}{\partial y} \frac{\partial \bar{T}}{\partial y} = u^* T^* \tag{16.85}$$

Equations (16.83) and (16.85) lead to:

$$l_{\mathrm{m}} = ky \qquad l_{\mathrm{th}} = k'y \tag{16.86}$$

which gives the **momentum and thermal Prandtl mixing lengths** in the inertial sub-layers.

3. Between the conductive and viscous sub-layers on one side and the inertial sub-layers on the other, there are **blending regions** whose description is important because it is in them that the **turbulent kinetic energy production** $\Pi$ and its dissipation $\varepsilon$ are greatest (see Figure 16.8(b), for example).

   In the momentum blending region, Van Driest has approximately modelled the attenuation of the velocity fluctuations due to the wall. By introducing an attenuation term, he allows the expression for the turbulent shear stress in this region to be generalized:

$$-\overline{u'v'} = (ky)^2 \left(\frac{\mathrm{d}\bar{u}}{\mathrm{d}y}\right)^2 \left[1 - \exp\left(-\frac{y_{\mathrm{i}}^+}{A}\right)\right]^2 \quad \text{where} \quad A = 26 \tag{16.87}$$

This model has a doubtful theoretical basis (see, for example, reference [16]), but is in agreement with some experimental results. In terms of the momentum mixing length of Prandtl's model, Van Driest's equation becomes:

$$\nu_{\mathrm{t}} = l_{\mathrm{m}}^2 \frac{\partial \bar{u}}{\partial y} = \left\{ky\left[1 - \exp\left(-\frac{y_{\mathrm{i}}^+}{A}\right)\right]\right\}^2 \frac{\partial \bar{u}}{\partial y} \tag{16.88}$$

so that:

$$l_{\mathrm{m}} = ky\left[1 - \exp\left(-\frac{y_{\mathrm{i}}^+}{A}\right)\right] \tag{16.89}$$

which is a generalization of the value in the inner logarithmic (inertial) region. The kinetic energy production $k$ calculated in this way has a maximum at $y_{\mathrm{i}}^+ \simeq 12$, in agreement with experimental results (Figure 16.8(b)).

*Notes*

1. For $y_{\mathrm{i}}^+/A > 3$, this reduces to the expression for the Prandtl mixing length in the logarithmic zone.
2. Near the wall, $l_{\mathrm{m}}$ decays as $y^2$, and thus $\nu_{\mathrm{t}}$ as $y^4\, \partial\bar{u}/\partial y$, which is consistent with the viscous sub-layer model, but does not seem to be confirmed by recent results, which tend to show a dependence as $y^3$.

   Cebeci [17] has generalized Van Driest's approach for temperature fluctuations $T'$; it produces an expression for $\overline{v'T'}$ similar to that for $\overline{u'v'}$:

$$-\frac{\overline{v'T'}}{u^* T^*} = kk' y_{\mathrm{i}}^+ y_{\mathrm{i}}^{++} \left[1 - \exp\left(-\frac{y_{\mathrm{i}}^+}{A}\right)\right]\left[1 - \exp\left(-\frac{y_{\mathrm{i}}^{++}}{B(Pr)}\right)\right] \frac{\partial \bar{u}^+}{\partial y_{\mathrm{i}}^+} \frac{\partial \bar{T}^+}{\partial y_{\mathrm{i}}^{++}} \tag{16.90}$$

where: $y_{\mathrm{i}}^+ = y_{\mathrm{i}}^{++}/Pr$; $A = 26$; $k = 0.4$; $k' = 0.44$.

$B$ has been determined from experimental data as a function of Prandtl number by Na and Habib [18]:

$$B = Pr^{1/2}(34.96 + 28.79\zeta + 33.95\zeta^2 + 6.33\zeta^3 - 1.186\zeta^4) \quad \text{where } \zeta = \log_{10} Pr \tag{16.91}$$

This function grows monotonically, with value 35 for $Pr = 1$, 26 for $Pr = 0.7$ (which is also the value of $A$) and 6 for $Pr = 0.02$. For gases, the constants $A$ and $B$ are very similar.

In terms of momentum and thermal mixing lengths, Cebeci's model gives:

$$a_t = l_m l_{th} \frac{\partial \bar{u}}{\partial y} = kk'y^2 \left[1 - \exp\left(-\frac{y_i^+}{A}\right)\right]\left[1 - \exp\left(-\frac{y_i^{++}}{B}\right)\right]\frac{\partial \bar{u}}{\partial y} \tag{16.92}$$

(from equations (16.24) and (16.90)), so that we obtain an analogous expression to that for $l_m$:

$$l_{th} = k'y\left[1 - \exp\left(-\frac{y_i^{++}}{B(Pr)}\right)\right] \tag{16.93}$$

Using van Driest's and Cebeci's models we can obtain the turbulent Prandtl number:

$$Pr_t = \frac{v_t}{a_t} = \frac{k}{k'} \frac{[1 - \exp(-y_i^+/A)]}{[1 - \exp(-y_i^+ Pr/B(Pr))]} \tag{16.94}$$

For a gas, the turbulent Prandtl number has a value around $k/k'$, or about 0.9, since the expressions in square brackets are similar. For water and (particularly) liquid

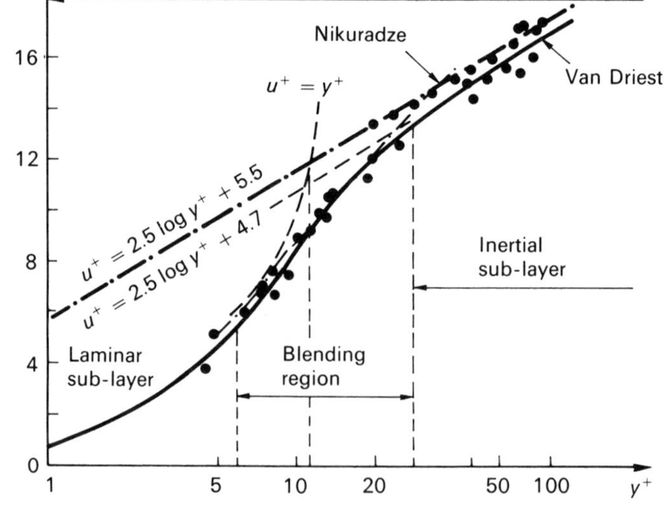


**Figure 16.11** The inner momentum region.

metals, $Pr_t$ differs from 1, and depends more or less strongly on $y_i^+$ according to Na and Habib's formula for $B$.

**Recapitulation: the structure of the inner zones (see Figure 16.11) in the generalized Prandtl model:**

| Inner momentum region | Inner thermal region |
|---|---|
| $y_i^+ = yu^*/\nu$ | $y_i^{++} = yu^*/a = y_i^+ Pr$ |
| ($\alpha$) *Viscous sub-layer* | ($\alpha'$) *Conductive sub-layer* |
| $0 < y_i^+ < 3$ to 5 | $0 < y_i^{++} < 3$ to 5 |
| $\bar{u}^+ = y_i^+$ | $\bar{T}^+ = y_i^{++}$ |
| $(l_m \sim kyy_i^+/A)$ | $(l_{th} \sim k'yy_i^{++}/\mathrm{B})$ |
| ($\beta$) *Blending region* | ($\beta'$) *Blending region* |
| $5 < y_i^+ < 40$ | $5 < y_i^{++} < 40$ |
| $\nu_t = l_m^2\, \partial\bar{u}/\partial y$ | $a_t = l_m l_{th}\, \partial\bar{u}/\partial y$ |
| $l_m = ky[1 - \exp(-y_i^+/A)]$ | $l_{th} = k'y[1 - \exp(-y_i^{++}/B)]$ |
| (Van Driest) $A = 26$ | (Cebeci, Na and Habib) $\Rightarrow B$ |
| (Von Karman) $k = 0.4$ | $k' = 0.44$ |
| ($\gamma$) *Inertial sub-layer* | ($\gamma'$) *Inertial sub-layer* |
| (logarithmic) | (logarithmic) |
| $40 < y_i^+ <$ a few $\times 10^2$ | $40 < y_i^{++} <$ a few $\times 10^2$ |
| $y_o^+ < 0.1$ or 0.2 | $y_o^{++} < 0.1$ or 0.2 |
| $l_m = ky$ | $l_{th} = k'y$ |
| $\bar{u}^+ = 1/k \log y_i^+ + C$ | $\bar{T}^+ = (\bar{T} - T_w)/T^* = 1/k' \log y_i^{++} + C'$ |
| $y_i^+ > 20$ | |
| $k = 0.4; C = 5.5$ (Nikuradze) | |

*Notes*

1. The analyses of the various scales of reference and the structure of the inner region carried out in Section 16.4.3.3 are quite general and are based on sound physical principles. However, the mixing length model, which has the virtue of taking a number of phenomena into account, at least approximately, with a simple model, leads to a certain number of predictions which are contrary to experience (notably with regard to the behaviour of turbulence in the vicinity of the walls). It is also unsuitable for modelling the processes taking place in the outer region, in which, for want of a better model, the mixing lengths at the boundary of the inner zone are often assumed to apply. Finally, this model is unsuitable for handling physical properties which vary with temperature, and can, only with difficulty, handle the interaction with radiation which occurs within the fluid in semi-transparent media.
2. $k$–$\varepsilon$ models, or at least those variants which allow calculations right up to the vicinity of the walls, give a better description of turbulent heat transfer (see Section 16.6).

3. Turbulent stress ($\overline{\rho u'v'}$, $\overline{\rho v'T'}$, etc.) transport models show promising prospects (see reference [14], for example).

16.4.3.4 STRUCTURE OF THE OUTER REGIONS (CENTRAL PART OF THE FLOW)
In the outer region, but adjacent to the inner layer, equation (16.77) can be integrated to give the logarithmic profile:

$$\frac{\bar{u} - \bar{u}_0}{u^*} = \frac{1}{k} \log y_o^+ + D \tag{16.95}$$

which is called the 'velocity defect law', since it is referred to the flow velocity $u_0$ of the flow at the centre of the pipe. This expression is valid over most of the tube, but obviously not near the axis. The velocity should be constant at $y_o^+ = 1$ (or $y = R$) – see Figure 16.3, for example.

We can similarly obtain a logarithmic law in the outer thermal region near the inner region:

$$\bar{T}^+ = \frac{\overline{\delta T}}{T^*} = \frac{1}{k} \log y_e^+ + E \tag{16.96}$$

## 16.5 Turbulent flow structure in other geometries

The results derived above for a circular tube can easily be generalized to other geometries (for internal convection in a closed channel of some constant cross-section). All that is required is to substitute the half-hydraulic-diameter $D_h/2$ of the section for the radius $R$ in the theory in Section 16.3. (The idea of a hydraulic diameter is summarized in the Appendix to this part.)

For flow over a flat plate, or more generally an arbitrary profile (an open geometry), we again find the same structure:

- No change in the structure of the inner region; the viscous sub-layer is of constant thickness.
- The reference length for the outer region is now the turbulent boundary layer thickness, $\delta$, rather than the pipe radius $R$.

## 16.6 Introduction to models with one supplementary equation (Prandtl–Kolmogorov) and two supplementary equations ($k$–$\varepsilon$)

The difficulty in turbulence, the closure problem, is to find expressions for the correlation terms such as $\overline{u'^2}$, $\overline{u'v'}$, $\overline{v'T'}$, etc., in the balance equations (16.9)–(16.11).

In this section, we will use a closure method based on the concepts of turbulent viscosity $\mu_t$ and $\nu_t$ (introduced in (16.20)) and turbulent thermal diffusivity $a_t$ and conductivity (introduced in (16.24)). We have to find equations for these various quantities. One solution, called direct closure (or the zero supplementary equation model), is to describe $\nu_t$ and $a_t$ by the analytic relationships of Prandtl's model, perhaps generalized (van Driest and Cebeci – see Section 16.4.3.2 and equations (16.88), (16.89), (16.92) and (16.93)). The weakness of this model is that it is necessary to know the mixing lengths $l_m$ and $l_{th}$ at every point in the flow, in particular the central region, to use in the equations giving $\nu_t$ and $a_t$.

Another solution is to calculate the mean turbulent kinetic energy per unit mass $k$ using the balance equation introduced in Section 16.2.2 (equation (16.13)). Then $\nu_t$ is related to $k$ by simple dimensional arguments:

$$\nu_t \doteq \frac{\mu_t}{\rho} \rightarrow L^2T^{-1}$$

$$k \rightarrow L^2T^{-2}$$

Introducing an arbitrary length $l$, we write:

$$\mu_t = \rho\nu_t = \rho[k(\mathbf{r})]^{0.5}l(\mathbf{r}) \tag{16.97}$$

$k$ is obtained as the solution of the coupled equations (16.9), (16.10), (16.11) and (16.13). The terms of the form $\overline{u'v'}$ in the momentum equations (16.10) are expressed using equations (16.20).

Terms such as $\overline{v'T'}$ in the energy equation (16.11) are given by (16.24), so long as the turbulent Prandtl number $Pr_t$ is known. This closure model is called 'one supplementary equation' (or Kolmogorov's model). It is related to Prandtl's mixing length model in that a reference length $l$ must be known at all points in the flow. A discussion of this point is given in reference [1], for example.

A more complex solution, but one which avoids the need for an arbitrary reference length, is to calculate not only $k$ at every point, but also its rate of dissipation $\varepsilon$ (defined by (16.15)), which appears as the dissipation term in the equation for $k$ (equation (16.13)):

$$\varepsilon \rightarrow L^2T^{-3}$$

Considering the dimensions of $\mu_t$, $k$ and $\varepsilon$, we write:

$$\mu_t = \rho c_\mu[k(\mathbf{r})]^2/\varepsilon(\mathbf{r}) \tag{16.98}$$

where $c_\mu$ is a constant, generally taken as 0.09.

The complete system of equations is now, with the restrictions of Section 16.2.3:

- The equation of state.
- The continuity equation (16.9).
- The momentum equations (16.10.1)–(16.10.3).
- The energy equation (16.11).

- The $k$ balance equation (16.13), usually written in the form [18]:

$$\rho \frac{\partial k}{\partial t} + \rho \bar{v}_j \frac{\partial k}{\partial x_j} = -\frac{\partial}{\partial x_j}\left(\overline{p' v'_j} + \frac{1}{2}\rho\overline{v'_i v'_i v'_j}\right) - \rho\overline{v'_i v'_j}\frac{\partial \bar{v}_i}{\partial x_j} + 2\mu\left[\frac{\partial}{\partial x_j}(\overline{v'_i \mathrm{d}'_{ij}}) - \overline{\mathrm{d}'_{ij}\mathrm{d}'_{ij}}\right] \tag{16.99}$$

where $\mathrm{d}'_{ij}$ is the fluctuating part of the stress tensor $\mathrm{d}_{ij}$. After some manipulation, we obtain:

$$+2\mu\left[\frac{\partial}{\partial x_j}(\overline{v'_i \mathrm{d}'_{ij}}) - \overline{\mathrm{d}'_{ij}\mathrm{d}'_{ij}}\right] = +\mu\frac{\partial^2}{\partial x_j^2}k - \varepsilon \tag{16.100.1}$$

where $\varepsilon$ is the rate of dissipation of $k$, defined by (16.15), which leads to equation (16.13).

The difficult term to evaluate in (16.99) is the first term on the right-hand side. By analogy with the closure equations for $-\overline{v'_i v'_j}$ and $\overline{v'_j T'}$, we write (assuming the dominant effect is the turbulent transport of $k$, which exceeds the $p'/\rho$ term):

$$-\rho\overline{v'_j\left(\frac{p'}{\rho} + \frac{v_i'^2}{2}\right)} = \frac{\mu_t}{\sigma_k}\frac{\partial k}{\partial x_j} \tag{16.100.2}$$

where $\sigma_k$ is a constant, generally taken as 1.

After closure, the $k$ equation becomes:

$$\rho \frac{\partial k}{\partial t} + \rho \bar{v}_j \frac{\partial k}{\partial x_j} = \frac{\partial}{\partial x_j}\left(\left[\mu + \frac{\mu_t}{\sigma_k}\right]\frac{\partial k}{\partial x_j}\right) + \mu_t\left(\frac{\partial \bar{v}_i}{\partial x_j} + \frac{\partial \bar{v}_j}{\partial x_i}\right)\frac{\partial \bar{v}_i}{\partial x_j} - \rho\varepsilon$$

⇑ CONVECTION ⇑ DIFFUSION ⇑ PRODUCTION ⇑ DISSIPATION

(16.101)

- The $\varepsilon$ balance equation (which is laborious to derive), which we will take as:

$$\rho \frac{\partial \varepsilon}{\partial t} + \rho \bar{v}_j \frac{\partial \varepsilon}{\partial x_j} = \frac{\partial}{\partial x_j}\left(\left[\mu + \frac{\mu_t}{\sigma_\varepsilon}\right]\frac{\partial \varepsilon}{\partial x_j}\right) + c_{1\varepsilon}\frac{\varepsilon}{k}\left[\mu_t\left(\frac{\partial \bar{v}_i}{\partial x_j} + \frac{\partial \bar{v}_j}{\partial x_i}\right)\frac{\partial \bar{v}_i}{\partial x_j}\right] - \rho c_{2\varepsilon}\frac{\varepsilon^2}{k}$$

⇑ CONVECTION ⇑ DIFFUSION ⇑ PRODUCTION ⇑ DISSIPATION

(16.102)

where $c_{1\varepsilon}$ and $c_{2\varepsilon}$ are constant parameters. The generally adopted values are:

$$c_\mu \simeq 0.09; \quad \sigma_k \simeq 1; \quad c_{1\varepsilon} \simeq 1.4; \quad c_{2\varepsilon} \simeq 1.92; \quad \sigma_\varepsilon \simeq 1.3$$

We notice that this system of equations has similar structures (despite the various modelling assumptions), which generally include the following:

- a convective transport term on the left-hand side, of the form $\bar{v}_j \partial\bar{A}/\partial x_j$ (to which is added an unsteady contribution, over a period long in comparison to the turbulent fluctuations, of the form $\partial\bar{A}/\partial t$);

- a diffusion term of the form $\partial^2 \bar{A}/\partial x_j x_j$;
- a production term for the quantity concerned (always positive);
- a dissipation term (always negative) for the same quantity.

The $k$–$\varepsilon$ method, however, gives **acute convergence problems in numerical methods near the walls**. Although this problem is not too important for the study of free jets and mixing regions, which are the simplest cases to handle, it is crucial for applications involving energy flux transfers between fluids and walls. Various solutions to this major inconvenience are in current use.

One joining method is to use experience to choose universal profiles of velocity and temperature near the walls (as defined above at the end of Section 16.3), and carry out a $k$–$\varepsilon$ calculation in the central part of the flow. The matching is carried out at a point $M$ on a normal to the wall in the inner region (either the inertial or the blending sub-layer). At $M$, $\bar{T}$ and $\bar{\mathbf{v}}$ are made continuous in the two parts of the flow. Values of $k$ and $\varepsilon$ are chosen for the $k$–$\varepsilon$ method at the joining point $M(\mathbf{r})$ using the following relationships (for the logarithmic region):

- Use (16.16), (16.18), (16.83) and (16.84) to find $\varepsilon$:

$$\varepsilon(\mathbf{r}) = (-\overline{u'v'})\frac{\partial \bar{u}}{\partial y} = u^{*2}\frac{u^*}{ky} \tag{16.103}$$

- Find the turbulent kinetic energy per unit mass at $M$ using (16.98):

$$k(\mathbf{r}) = \left(\frac{\nu_t \varepsilon}{c_\mu}\right)^{1/2} = \frac{u^{*2}}{\sqrt{c_\mu}} \tag{16.104}$$

Matching in other zones can be done using the same approach.

A variant of the above, which is much more computationally intensive, is to treat the fluid near the wall as a pseudo-fluid with turbulent diffusivities $a_t$ and $\nu_t$ given by the Prandtl, Van Driest and Cebeci models, and use $k$–$\varepsilon$ in the central part of the flow. The matching is in terms of continuity of $T$ and $\mathbf{v}$, and is obtained (see (16.2.3)) from:

$$\varepsilon(\mathbf{r}) = (-\overline{u'v'})\frac{\partial \bar{u}}{\partial y} = \frac{\mu_t}{\rho}\left(\frac{\partial \bar{u}}{\partial y}\right)^2; \qquad k(\mathbf{r}) = \left(\frac{\nu_t \varepsilon}{c_\mu}\right)^{1/2} \tag{16.105}$$

Finally, a method principally developed by Lam and Bremhorst [19] consists in essence of using a $k$–$\varepsilon$ model throughout the fluid, but introducing attenuation functions for $\mu_t$ and $\varepsilon$ near the walls. This allows for turbulence anisotropy near the walls, but also allows the calculation code to converge in the vicinity of the wall. A detailed study of these models has been made by Patel, Rodi and Scheurer [20].

These functions, denoted by $f_\mu$, $f_1$ and $f_2$, are applied to $c_\mu$ (in the expression for $\mu_t$), $c_{1\varepsilon}$ and $c_{2\varepsilon}$ (in the $\varepsilon$ transport equation) respectively. The expressions used by Lam and Bremhorst, for example, are:

$$\begin{aligned} f_\mu &= (1 - \exp[0.0165 Re_y])^2(1 + 20.5/Re_T) \\ f_1 &= 1 + (0.05/f_\mu)^3 \\ f_2 &= 1 - \exp[-Re_T^2] \end{aligned} \tag{16.106}$$

where:

$$Re_y = \frac{\sqrt{k}\,y}{\nu} \tag{16.107}$$

$$Re_T = \frac{\sqrt{k}}{\nu}\left(\frac{k^{3/2}}{\varepsilon}\right) = \frac{k^2}{\nu\varepsilon} \tag{16.108}$$

where $\sqrt{k}$ is a characteristic local scale for velocity fluctuations, $y$ the distance from the wall, and $k^{3/2}/\varepsilon$ is a local length scale associated with the turbulence.

*Note*

The affinity between the attenuation function in $f_\mu$ and that in Van Driest's model (equation (16.87), where $y^+$ is a Reynolds number based on $u^*$ and $y$) is obvious. The extra factor in (16.106) is important for low values of local Reynolds number.

# 17

# Empirical Correlations in Turbulent Flow

For common geometries and flows, it is not necessary to calculate the whole flow field to obtain the global flux transferred between the fluid and the wall in **developed** turbulent flow. Instead we use correlations of experimental origin, in line with the conclusions of Chapter 12, of the form:

$$Nu = \text{const } Re^n Pr^m \tag{17.1}$$

A large number of correlations of this type are collected in Part V 'Basic Data' in this book; here by way of example we will consider the classic cases of fully developed flow in a circular pipe and of a flat plate.

The expressions for Nusselt number depend on the following:

- The **geometry** concerned (geometrical similarity).
- The **flow régime** – turbulent in this chapter. Criteria for the laminar to turbulent transition in some common situations are given in Section 12.4.
- The **nature of the fluid** concerned. The common fluids (air, gases, water, etc.) whose Prandtl numbers are around 1 ($0.5 < Pr < 7$) are described by correlations very different from those applicable to liquid metals ($Pr < 2 \times 10^{-2}$) or heavy oils ($Pr > 100$).
- The **Reynolds number range** involved. For some geometries in turbulent flow, the correlations can vary with $Re$.
- The **thermal conditions at the wall** – constant flux, constant temperature, etc.

## *17.1 Correlations for a flat plate (in forced or free convection)*

### 17.1.1 Forced convection

Let us recall that in laminar flow, such that the distance $x$ from the leading edge satisfies the inequality:

$$Re_x = \frac{u_0 x}{\nu} < 10^5 \tag{17.2}$$

the local or global Nusselt numbers for an isothermal flat plate are given by:

$$Nu_x(x) = \frac{h(x)x}{\lambda} = 0.332 Re_x^{0.5} Pr^{1/3} \tag{17.3.1}$$

$$Nu_L = \frac{\bar{h}L}{\lambda} = 0.644 Re_L^{0.5} Pr^{1/3} \tag{17.3.2}$$

After a transition region (which starts at $Re_x \simeq 10^5$), and once turbulence has fully developed, we obtain an experimental correlation of the form:

$$\left\{\begin{array}{lll} Re_x > 5 \times 10^5 & Nu_x(x) = \dfrac{h(x)x}{\lambda} = 0.0288 Re_x^{0.8} Pr^{1/3} & (17.4.1) \\ Pr \geqslant 0.5 & Nu_L = \dfrac{\bar{h}L}{\lambda} = 0.035 Re_L^{0.8} Pr^{1/3} & (17.4.2) \end{array}\right.$$

We note that the average of $h(x)$, which gives $\bar{h}$ and appears in the expression for $Nu_L$ is taken from the leading edge of the plate. It therefore takes account of the laminar and transitional parts of the boundary layer. The experimental correlations (17.4.1) and (17.4.2) give a precision of the order of 20%, while the corresponding expressions in laminar flow are exact.

If we compare (17.3) and (17.4), we notice the following:

- that the heat transfer coefficient $h(x)$ falls as $x^{-1/2}$ in laminar flow, as $x^{-0.2}$ in turbulent flow;
- that the Nusselt numbers, and thus $h$, increase noticeably in the laminar to turbulent transition zone (see Figure 17.1).

A problem in practice is to determine whether increased energy transfer will be obtained by creating turbulent flow over the plate, or alternatively, by increasing the leading-edge effects which occur in laminar flow at $x = 0$ ($h_x \to \infty$). The answer depends on the particular situation. The variation of the heat transfer coefficient $h(x)$ from the leading edge is shown in Figure 17.1.

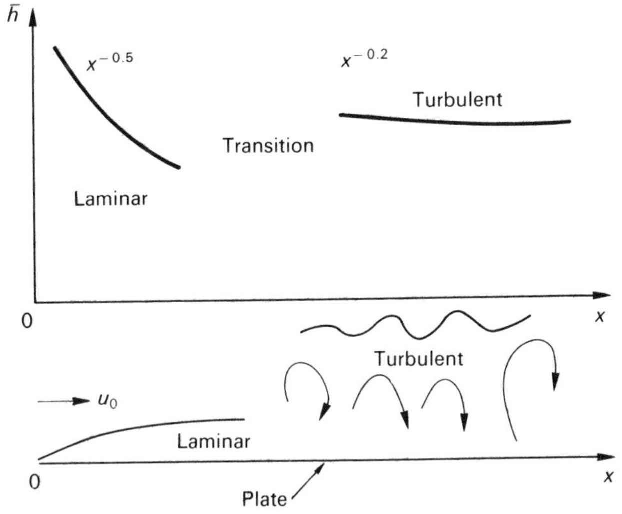


**Figure 17.1**

### *17.1.2 Natural convection*

We have obtained expressions for the local and global Nusselt numbers for laminar flow past a vertical isothermal flat plate, in an isothermal medium which is at rest far away:

$$10^4 < Ra < 10^9 \qquad Nu_x(x) = \frac{h(x)x}{\lambda} = 0.39 Ra_x^{1/4} \tag{17.5.1}$$

$$Nu_L = \frac{\bar{h}L}{\lambda} = 0.59 Ra_L^{1/4} \tag{17.5.2}$$

In turbulent flow under the same conditions we use the well-known correlation due to MacAdams:

$$10^9 < Ra < 10^{13} \qquad Nu_x(x) = \frac{h(x)x}{\lambda} = 0.12 Ra_x^{1/3} \tag{17.6.1}$$

$$Nu_L = \frac{\bar{h}L}{\lambda} = 0.13 Ra_L^{1/3} \tag{17.6.2}$$

Notice that **in turbulent flow in these conditions, $h(x)$ is independent of $x$ (and similarly $\bar{h}$ of $L$)**, since $Ra_x$ is proportional to $x^3$:

$$Ra_x = \frac{g\beta(T_w - T_0)x^3}{a\nu} \tag{17.7}$$

Further correlations for various other conditions are given in 'Basic Data'.

## 17.2 Correlations for developed pipe flow (forced convection)

For common fluids with constant physical properties in fully developed turbulent flow in a pipe (defined in Section 16.4), there are various equivalent correlations in the literature. The accuracy of these correlations is of the order of 20%. The best known is that of Colburn, which leads to the Stanton number (see Chapter 13), which can be written in the form:

$$Nu_{D_h} = 0.023 Re_{D_h}^{0.8} Pr^{0.33} \tag{17.8}$$

where:

$$Nu_{D_h} = \frac{hD_h}{\lambda} \qquad Re_{D_h} = \frac{\rho u_m D_h}{\mu} \tag{17.9}$$

where $u_m$ is the discharge velocity in the section.

The quantity $\rho u_m$ is independent of temperature; $\lambda$, $\mu$ and $c_p$ are taken at a temperature midway between the mixed fluid temperature $T_m$ and the wall temperature $T_w$: $T = (T_m + T_w)/2$.

The region of validity for this formula, given by the author, is:

$$Re > 10^4; \qquad 0.7 < Pr < 160; \qquad \frac{L}{D} > 60$$

The last condition corresponds to that of a thermally developed régime (see Section 16.4).

*Notes*

1. In developed turbulent flow (as in laminar) the local Nusselt number and $h(x)$ are constants, denoted by $Nu_{D_h}$ and $h$ respectively. **In contrast to laminar flow, these quantities depend very strongly on Reynolds number in turbulent flow.**
2. The thermal conditions at the wall (constant flux, constant temperature, etc.) have little effect in practice on the expression for $Nu_{D_h}$ turbulent flow.
3. As for a flat plate, the Nusselt number in a pipe increases as $Re^{0.8}$ in turbulent flow, a similarity which does not apply in laminar flow. This arises from the fact that the turbulence structures in the flows are very similar, particularly in the inner region (the viscous/conductive sub-layer, blending and inertial layers: see Chapter 16).

## 17.3 Empirical treatment of temperature-dependent viscosity

We have seen that although (for water, for example) $\rho$ may vary little with $T$, the viscosity varies considerably for a variation of $T$ from around 20°C to 300°C. Sieder and Tate [21] have suggested a correlation based on experiment which takes account of the dependence of $\mu$ on temperature. Colburn's formula becomes:

$$Nu_{D_h} = 0.027 Re_{D_h}^{0.8} Pr^{0.33} \left(\frac{\mu_m}{\mu_w}\right)^{0.14} \tag{17.10}$$

where $\mu_m$ and $\mu_w$ are the viscosities at the mixed fluid and wall temperatures respectively. Whitaker [22] recommends the use of the same correction factor for many other geometries in turbulent flow.

## 17.4 Turbulent flow development length

There has been very little work on the development of the turbulent flow régime in closed conduits. We may cite the work of Aladyev [23] for water and Deissler [24] for air. The conclusions they draw are as follows (referring to Figure 17.2):

- The Nusselt number $Nu_{LD_h}$, equal to $\bar{h}L/\lambda$ and associated with $Re_{D_h}$, reaches an asymptotic value which depends on $Re_{D_h}$, after a distance of approximately $40D_h$, independent of Reynolds number.
- The ratio between the average heat transfer coefficient and its asymptotic limit, at a given distance from the leading edge, falls with increasing $Re_{D_h}$ (note that no results for $x < 0.4D_h$ are shown in Figure 17.2).

Identical conclusions may be drawn for a gas. Entry effects may be taken into account using a correlation attributed to Nusselt to replace that of Colburn:

$$Nu_{D_h} = 0.036 Re_{D_h}^{0.8} Pr^{1/3} \left( \frac{L}{D_h} \right)^{-0.054} \quad \text{for} \quad 10D_h < L < 400D_h \quad (17.11)$$

This predicts a slow decline of the heat transfer coefficient $\bar{h}$ with tube length $L$. Colburn's formula is recovered approximately for a tube of length $400D_h$.

Knudsen and Katz [5] quote Boetler [25], giving several expressions for $\bar{h}$ for ducts with very diverse entry conditions. They are also given in 'Basic Data'.

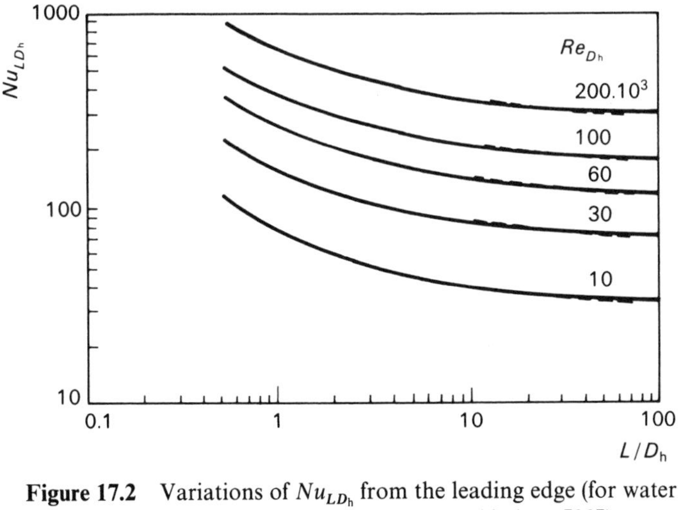


**Figure 17.2** Variations of $Nu_{LD_h}$ from the leading edge (for water in a pipe with fixed wall temperature (Aladyev [23]).

# 18

# Coupling between Convection and Radiation

This chapter is concerned with coupling of phenomena in real materials whose thermophysical properties are strongly dependent on temperature; such coupling takes place at the scale of elementary volumes of the system concerned. Fluid media have the distinction of having convective transfers, mainly turbulent, simultaneously with conductive and radiative transfers. Among fluids, gases need the most attention since, in contrast to liquids, they are strongly dilatable (which introduces many problems in modelling convection processes) and also they show a fine spectral structure, of quantum origin, which makes modelling of radiative transfer difficult. Coupling between conduction and radiation and between convection and radiation in dense semi-transparent media thus appears in particularly simple cases with gases, which are considered here.

## *18.1 Examples*

In most of the cases we will consider, we are concerned with coupled transfers within a flowing (usually turbulent) semi-transparent fluid, and wish to determine the complex energy fluxes exchanged between the fluid and its containment. In certain cases, our objective is to modify the thermal conditions in order to minimize the relevant fluxes; for example:

- the effects of emergency flares on offshore platforms on their environment;
- heat transfer to walls in fires, and remote ignition problems;
- wall heat transfer in combustion chambers (perhaps with ablation); aircraft engines with or without afterburners, flame tubes (solid propellent combustion), cryogenic rocket motors, etc.;
- atmospheric re-entry phenomena – semi-transparent windows, heat shields, etc.;

- problems related to the safety of pressurized water reactors in conditions of partial loss of coolant in the reactor core;
- heat transfer within media not in LTE – heat treatment by power lasers, controlled and uncontrolled fusion plasmas, etc.

In other situations, the objective is to maximize the flux exchanged between the fluid and the structure; for example:

- the turbulent heat transfer in metal fusion processes using intense magnetic fields, and interactions with the fusion front;
- the maximization of the flux on the charge in industrial combustion furnaces (combustion in air, in oxygen, at high pressure, etc.);
- etc.

## 18.2 Physical nature of the coupling

Within a moving fluid which is semi-transparent to radiation, the instantaneous balance equations (possibly time dependent) are as follows, in tensor notation:

- The equation of state.
- Continuity of mass:

$$\frac{\partial \rho}{\partial t} + \frac{\partial}{\partial x_j}(\rho v_j) = 0 \tag{18.1}$$

- Momentum balance:

$$\rho\left(\frac{\partial v_i}{\partial t} + v_j \frac{\partial v_i}{\partial x_j}\right) = -\frac{\partial p}{\partial x_i} - \rho \frac{\partial U}{\partial x_i} + \mu \frac{\partial^2 v_i}{\partial x_j^2} + \frac{1}{3}\frac{\partial}{\partial x_j}\left(\mu \frac{\partial}{\partial x_k} v_k\right) + \cdots \tag{18.2}$$

- The energy equation:

$$\rho c_{\mathrm{p}}\left(\frac{\partial T}{\partial t} + v_j \frac{\partial}{\partial x_j} T\right) = \frac{\partial}{\partial x_j}\left(\lambda \frac{\partial}{\partial x_j} T\right) + P^{\mathrm{R}} + P + \cdots \tag{18.3}$$

where $T$, $p$ and $v_i$ are the temperature, pressure and velocity vector; $\rho$, $\lambda$, $\mu$ and $c_{\mathrm{p}}$ are the density, thermal conductivity, dynamic viscosity and specific heat at constant pressure; these quantities depend strongly on temperature; $U$ is the potential which generates body forces (gravity, etc.).

Additional to equations (18.1)–(18.3) is the radiation transfer equation applicable at every point in the medium, in the spectral interval $[\nu, \nu + \mathrm{d}\nu]$ and in the elementary solid angle $\mathrm{d}\Omega$, whose axis is characterized by the unit vector $u_i$ and a curvilinear coordinate $s$. Assuming LTE, we write (see Chapter 11):

$$\frac{1}{c}\frac{\partial I'_\nu}{\mathrm{d}t} + n^2 \frac{\partial(I'_\nu/n^2)}{\partial s} = \kappa_\nu n^2 I^0_\nu(T) - (\kappa_\nu + \sigma_\nu)I'_\nu + \frac{\sigma_\nu}{4\pi}\int_0^{4\pi} I'_\nu(s, u'_j)p_\nu(u'_j \to u_i)\,\mathrm{d}\Omega' \tag{18.4}$$

where $I'_\nu(s, u_i)$ is the directional monochromatic intensity of the radiation in the medium, $\kappa_\nu$ and $\sigma_\nu$ the coefficients of monochromatic absorption and diffusion and $p_\nu$ the diffusion phase function. The power dissipation per unit volume of the radiation, $P^R$, which appears in equation (18.3), is given by:

$$P^R = -\frac{\partial}{\partial x_j} q_j^R = -\frac{\partial}{\partial x_j}\left[\int_0^\infty \mathrm{d}\nu \int_0^{4\pi} I'_\nu u_j \, \mathrm{d}\Omega\right] \tag{18.5}$$

where $q_j^R$ is the radiative flux density vector, defined at every point, including on the boundaries.

To specify the problem completely, we must add the flow, thermal and radiative boundary conditions to equations (18.1)–(18.5). The **interactions** between the various processes are complex:

- The first interaction arises from the fact that the **thermophysical properties of fluids** ($\lambda$, $\rho$, $c_p$ and $\mu$, and also $\kappa_\nu$ and $\sigma_\nu$) **are strongly dependent on the temperature** $T$. (For a gas, which is inherently dilatable, we often find variations of temperature of 3000 K over 1 mm, which corresponds to a change in $\rho$ of a factor of 10; $\mu$ varies by a factor of 5 for liquid water between 10 and 100 °C.) The temperature field arising from (18.3) strongly influences the velocity field in (18.1) and (18.2), via $\rho$ and $\mu$. Conversely, the latter influences the temperature field through the convection term in (18.3). Models with variable thermophysical properties are often unavoidable [26,27]; they are particularly difficult in turbulent flow [3,28]. Note that dimensionless groups such as the Nusselt number are now meaningless since there is no longer a reference conductivity (the concept of film temperature is now completely inadequate).
- The second interaction, however, is between conductive and convective transfers on one hand and radiative transfer on the other. The energy equation (18.3) depends on the radiative source term $P^R$, defined by (18.5), in any elementary volume; this latter depends on the intensity field $L'_\nu$ for all directions and frequencies, which is a very complex function, very sensitive to the temperature field at all points within and on the boundaries of the system. The effects due to the radiation field produce interactions at very large distances and quasi-instantaneously. (The time dependent term in (18.4) is generally negligible on the time scales considered, except when we are concerned with ultra-fast pulsed lasers or in applications concerned with controlled or uncontrolled nuclear fusion.) It follows that all the usual ideas of similarity and non-dimensionalization which are used in the absence of radiation and with constant physical properties lose their physical significance [29]. They are, unfortunately however, widely used in the literature.

The thermal engineer will generally consider the energy flux exchanged with a moving semi-transparent fluid at a system boundary in a pragmatic manner. In a simple case where the wall normal to the $y$ axis is an opaque solid with a temperature field $T_s$ and conductivity $\lambda_s$ (while within the fluid the temperature and conductivity are $T$ and $\lambda$), the flux continuity equation at the interface is:

$$\varphi = -\lambda_s \frac{\partial T_s}{\partial y}\bigg|_w = -\lambda \frac{\partial T}{\partial y}\bigg|_w + \varphi_w^R \tag{18.6}$$

where $\varphi_w^R$ is the radiative flux. This introduces an additional interaction (through the boundary conditions) between the heat transfer modes, an interaction which will even occur in transparent media. A detailed analysis shows the following for semi-transparent media (Figure 18.1):

- The conductive near-wall flux in the fluid is strongly coupled to the radiation and convection through the temperature field which they produce within the fluid. The idea of a (convective heat transfer coefficient $h'$) is no longer meaningful in this situation. We ought to speak of conducto-convecto-radiative flux! We will simply call it conductive flux.
- The radiative wall flux (the difference between the fluxes absorbed and emitted by the wall) is closely coupled to the conductive and (particularly) convective fluxes through the temperature field which determines the intensity field.

Some examples of results are given in Section 18.4.

*Note*
Many investigations of coupled heat transfer have been carried out outside local thermodynamic equilibrium (LTE): cold and hot plasmas (gas heating, atmospheric re-entry, electric arcs, gas lasers, fusion, etc.). In such conditions the flow equations (18.2)–(18.3) generally remain valid (there is partial equilibrium of internal translational and rotational degrees of freedom), but with modified thermophysical properties. The radiation equation (18.4) is extended by introducing a spontaneous emission term and an absorption coefficient which is a function of the distribution of the populations over the energy levels of the system. These populations are obtained from a kinetic model based on the effective total collision cross-sections. The science of heat transfer in non-LTE systems is expanding rapidly.

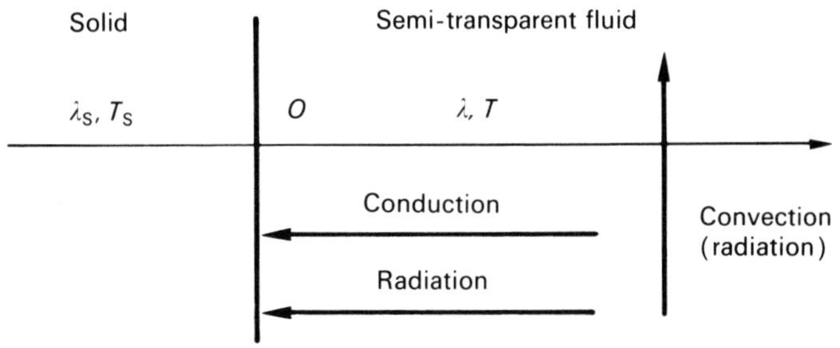


**Figure 18.1**

## 18.3 Modelling the transfer

The model based on equations (18.1)–(18.5) is valid if we adopt a fully unsteady viewpoint from the flow point of view (reference [30], by way of example) and if we work from the most monochromatic scale possible for the radiation, in the case of a gas. Actually, it is not applicable in practice in this form, when the capabilities of calculating systems are considered. There are two major limitations: modelling of heat transfer taking account of turbulent flow, and taking account of semi-transparent fluids with fine structure (notably gases). Various practical approaches are briefly introduced in this section, which builds on the conclusions of Chapter 11 and 16.

### 18.3.1 Modelling of radiation

Dense materials (solids and liquids) have absorption spectra without fine structure – $\kappa_\nu$ and $\sigma_\nu$ are practically constant over a spectral interval $\Delta\nu$ such that the intensity $I_\nu^0(T)$ at a given temperature varies by less than 1% with $\nu$ ($\Delta\nu \simeq 10$ to $50\ \text{cm}^{-1}$). The radiation transfer equation (18.4) can then be used directly (see Chapter 11).

Gases show a high resolution spectral structure in the infra-red (again, see Chapter 11). This fine structure generates large **spectral correlation** phenomena between emission, transmission and absorption. Because of this, the low resolution absorption coefficient $\bar{\kappa}$, averaged over an interval $\Delta\nu$ from 10 to $50\ \text{cm}^{-1}$ (that is, deduced from a degraded spectrum), makes no physical sense. If we wish to use equation (18.4), it must be done at high resolution (that is, using about $10^6$ points in the discretization of the spectrum) when calculating $P^R$ (using (18.5)) or $\varphi^R$, at each node of the calculation.

Other approaches are thus necessary. They consist of a model for the radiative properties of a gas, generalizable to highly heterogeneous and non-isothermal columns, and a transfer model. These models must in addition be compatible with the presence of particles (emitting, absorbing and scattering) and with a convective heat transfer model. The most rigorous analysis is to use a code which calculates synthetic spectra for heterogeneous and non-isothermal materials on a line by line basis [31], starting from data on the intensities (for example, reference [31]) and shapes (widths, for example [33,34]) of the lines. The parameters of approximate models, which take account of spectral correlations and have the advantage of being much more economical in calculation time than the line by line methods, can be deduced from the results of the line by line models. We may mention in particular the regular models [35], narrow band statistical models [36,37], wide band statistical models [38], fictitious gas model [39], etc. The correlated models which prove most useful in the range 300–3000 K for $H_2O$, $CO_2$ and CO at high temperatures are the statistical narrow band models with inverse exponential law associated with the Curtis–Godson [40] or Lindquist–Simmons [41] approximations for heterogeneous or non-isothermal media. They show improvement over the wide band statistical models of the Edwards [38] type in that they correctly take account of the significant movement of the spectral bands with temperature [42]. The narrow band statistical models require knowledge of about 200 different parameter values for the calculation of the radiative properties of an arbitrary mixture of $CO_2$, $H_2O$, CO, $N_2$ and $O_2$ over the whole spectrum (in comparison to about $10^6$ values required for a line-by-line calculation). We should note here that the radiant properties of a medium depend very strongly on the proportions of an absorbent constituent and on the gas pressure. For example, the necessary models for allowing for water vapour will be very different if the molar fraction of $H_2O$ is a few percent (e.g. products of combustion with air) or if it is mainly water vapour (combustion with $O_2$). Similarly, from $10^{-2}$ to 1 or 100 atm, the models to be used will be very different (due to different line profiles or non-additive effects of isolated lines at very high pressures).

The only model for calculating global radiative heat transfer (finding $\varphi^R$ and $P^R$ at every point), allowing for spectral correlations and compatible with a convection model, is to discretize the three-dimensional system into $N_1$ spatial nodes, into $N_2$ elementary solid angles (about 100) at each spatial node (so that the whole $4\pi$ steradians is covered), and into $N_3$ (about 100) spectral intervals $\Delta\nu$ covering the whole spectrum [27,43–44]. The intensities at each node and for each direction are calculated in a correlated manner by a direct method [27] or by successive approximation [43]. The calculation may be speeded up by recourse to a correlation index [43].

Very diverse methods for energy transfer by radiation scattering have been developed for the simplest cases of dense media (for example references [45], [46]). Here, we may mention briefly scattering by particles carried in the fluid (droplets, soot, wall ablation or thermo-degradation products, etc.). The long-established Mie model [47], which is valid for spherical particles, has been generalized to particles of any shape [48]. We recall that the nature of the scattering depends on the size parameter $\alpha = 2\pi d/\lambda$, where $d$ and $\lambda$ are characteristic dimensions of the particle and the wavelength. For very low values of $\alpha$ ($\alpha \ll 1$), Rayleigh scattering is generally negligible if we are interested in global transfer – only absorption and emission by the particles need be considered. For $\alpha$ around 1, scattering is important, and is generally far from isotropic. For large $\alpha$ ($\alpha \gg 1$), the phase function is sharply peaked (with very pronounced forward scattering), while the particles are radiatively grey ($\kappa$ and $\sigma$ independent of $\nu$). This is the case for radiative transfer with very large particles, as are found in the combustion gases from solid propellants or in fires. In the latter case, the radiation from the particles generally dominates over that from the gas; at very high temperatures (above 3000 K) the spectral ranges of these two types of radiation do not overlap much (the regions of the strongly absorbing bands are 0.5–3 $\mu$m for particles, 2.7–15 $\mu$m for the gas).

### 18.3.2 Modelling turbulent heat transfer

Direct calculation of turbulent heat transfer by solution of the coupled instantaneous equations ((18.1)–(18.3)) (direct simulation, e.g. reference [30]) is not at present feasible for complex geometries and realistic fluids. It is usual to use Reynolds' formulation (or that of Favre [49] for dilatable fluids), which involves statistical equations averaged over a time $\Delta t$ which is long in comparison to the period of turbulent fluctuations. Time averaged quantities such as $\overline{v_i'v_j'}$ or $\overline{v_i'T'}$, which represent correlations between the fluctuating velocity components or between the fluctuations in one velocity component and those in temperature, appear in these equations. The key problem is to calculate these quantities – the closure of the statistical equations. Attempts at direct closure using the balance equations of $\overline{v_i'v_j'}$ or $\overline{v_i'T'}$ themselves (higher-order moment methods) are currently in progress [14,50], but it is not yet possible to say whether these methods will lead to definite progress.

A supplementary hypothesis, without a strong physical basis, is made in most practical models: turbulent mechanical and thermal diffusivity terms $\nu_t$ and $a_t$ are introduced – closure of the correlation terms is by diffusion terms, although $\nu_t$ and $a_t$ depend on position, local velocity field, turbulent kinetic energy field, etc. In the best-known model, $\nu_t$ depends on the average kinetic energy $k$ associated with the turbulent fluctuations, and on its dissipation $\varepsilon$ (the $k$–$\varepsilon$ model, originally due to Johns and Launder [51]); $a_t$ is generally found from $\nu_t$ through a turbulent Prandtl number. This model is, however, difficult to use close to a wall (in the viscous and blending sub-layers of the inner region of a turbulent flow). Most codes use phenomenological matching methods at the wall, discussed at the end of Chapter 16, using universal velocity and temperature profiles in the molecular and blending sub-layers. These methods are discussed critically in reference [52]. Such an approach is unsatisfactory with the two assumptions used here:

- A fluid whose thermophysical properties vary strongly with temperature is likely to considerably modify the profiles in the near wall region, where there are large variations in temperature.
- Significant coupling between conduction, convection and radiation is also likely to modify the velocity and temperature profiles, particularly near the wall.

Various modifications of the $k$–$\varepsilon$ method introduce attenuation functions of $\mu_t$ and $\varepsilon$, which allows the codes to converge near the wall [19,20]; this approach, due primarily to Lam and Bremhorst, was introduced at the end of Chapter 16. These methods for isothermal flows have been applied to strongly non-isothermal flows by allowing for variable thermophysical properties in the energy equation [28].

*Note*

Transport models with four supplementary equations ($k$, $\varepsilon$ and also, in a similar manner, $\overline{T'^2}$ and its dissipation $\varepsilon_\theta$) have recently been developed [53,54]. These avoid the need for a turbulent Prandtl number.

### 18.3.3 Interaction between radiation and turbulence

Coupling between the statistical equations representing the mean turbulence quantities and radiation in semi-transparent media is discussed in Section 18.4. Here,

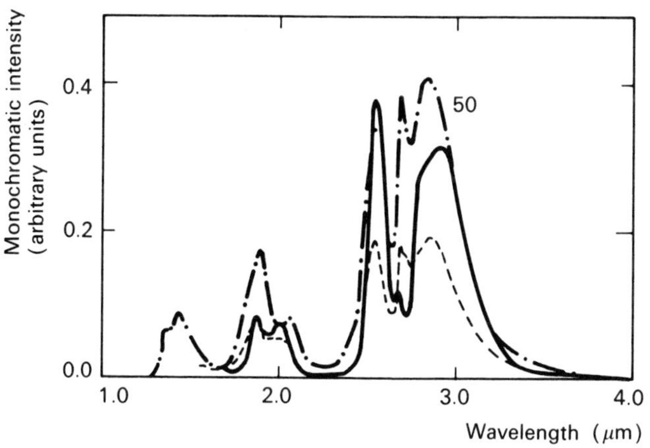


**Figure 18.2** Comparison of the intensities of a column of a mixture of combustion gases, in turbulent flow (from reference [55]): (——) experiment; (– – – –) statistical model using mean flow data; (–·–·–) statistical model with data from a stochastic turbulence model.

we are interested in smaller scale effects: the effect of turbulent fluctuations on the radiation field and of the radiation on these same turbulent fluctuations.

Radiation phenomena are strongly non-linear as regards temperature. For this reason, fluctuations in temperature or density, which are zero in the mean, produce effects on the radiation fields $\varphi^R$ and $P^R$ which are non-zero in the mean. This phenomenon has been investigated experimentally and theoretically, using a narrow band statistical model and a stochastic turbulence model, in turbulent flames [55] (see Figure 18.2). It has also been investigated for channel flow [28]. The size of the effect is very variable over the energy range considered – from 5 to 50% – according to conditions.

The influence of radiation on the turbulence spectrum has been widely studied in the fields of meteorology [56] and astrophysics. Radiation has an important (destructive) effect on the turbulence structures whose sizes correspond to an optical thickness of order 1. It appears necessary to carry out similar work for the field of heat transfer in high temperature gases.

## 18.4 Examples of coupled transfers

In this section we will look at a few important examples of the modelling of coupled transfers. We will basically consider fundamental studies.

For molten metals, the radiation source term can arise from an intense electromagnetic field, which causes the fluid flow, or amplifies the movement. The resulting turbulent effects tend to shorten the solidification process, permitting it to be better controlled [57].

The test case of coupling between conduction and radiation in air (assumed to be at rest) containing traces of $CO_2$ has been studied by Soufiani *et al.* [58]. The temperature field and, particularly, the conductive boundary flux resulting from the coupling are found to be extremely sensitive to the radiation model chosen (Figure 8.3). If the results deriving from a line by line approach and narrow band statistical models are identical, Edwards' model gives an error in the combined flux of 30% in the conditions considered. Other models (grey, grey in bands) should be rejected. (These are, alas, the most widely used methods!)

Coupling between forced laminar convection and radiation in non-developed channel flow, for a mixture of various proportions of $H_2O$ and $CO_2$, has been analysed using a narrow band statistical model and strongly temperature dependent thermophysical properties [27,42]. The agreement with experimental results [61] is very satisfactory. This study showed that results obtained by wide band statistical models have large errors compared with those from narrow and statistical models for medium optical thicknesses (channel width $E = 2.5$ cm) [42]. For example, the radiative power dissipated by the gas is overestimated by 35% by the wide band model compared to the narrow band statistical model (Figure 18.4).

The effect of the coupling on the temperature field and the conductive flux at the wall are shown clearly in Figure 18.5.

The coupling between free laminar convection and radiation has recently been studied by Yamada [60] in a vertical channel and Lauriat [61] in pure $CO_2$. The former author has also demonstrated the insufficiency of grey gas models. A more

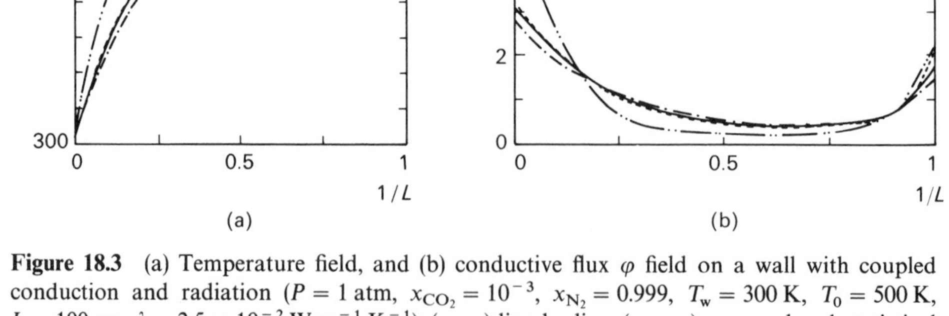

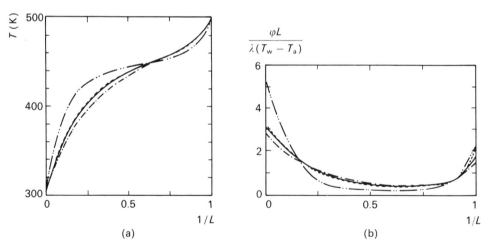

**Figure 18.3** (a) Temperature field, and (b) conductive flux $\varphi$ field on a wall with coupled conduction and radiation ($P = 1$ atm, $x_{CO_2} = 10^{-3}$, $x_{N_2} = 0.999$, $T_w = 300$ K, $T_0 = 500$ K, $L = 100$ cm, $\lambda = 2.5 \times 10^{-2}$ W m$^{-1}$ K$^{-1}$): (——) line by line; (– – – –) narrow band statistical model; (–·–·–) Edwards' model; (–··–··–) grey in bands (from reference [58]).

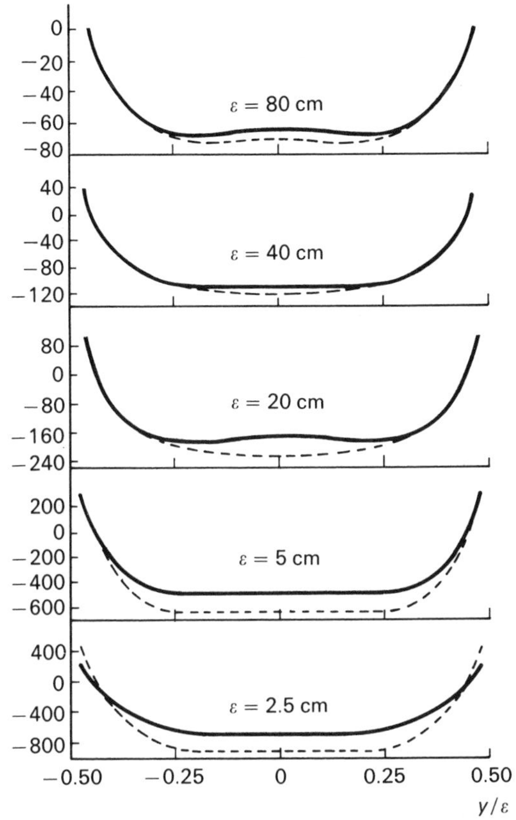


**Figure 18.4** Power per unit volume $P^R$ dissipated at a channel cross-section: (——) narrow band statistical model; (– – – –) Edwards' model; for various channel widths $E$ and pure $H_2O$ [42]. The walls are black, as required by Edwards' model.

general investigation by Zhang *et al.* [44] has used the narrow band statistical model [61] to look at the coupling between mixed laminar convection and radiation on a vertical flat plate. The fields of the velocity component parallel to the plate, the temperature and the wall flux are very strongly affected by the coupling (Figure 18.6).

Coupling between forced turbulent convection and radiation in a channel has been studied with strongly temperature dependent thermophysical properties ($\lambda$, $\rho$, $c_p$) under conditions close to industrial situations [28]. The turbulence model (of the $k$–$\varepsilon$ type) generalizes the approach of Lam and Bremhorst recommended by reference [19] to strongly non-isothermal media. The radiation model is the same as that cited above [42]. Profiles of temperature, radiative power dissipation $P^R$ and wall heat

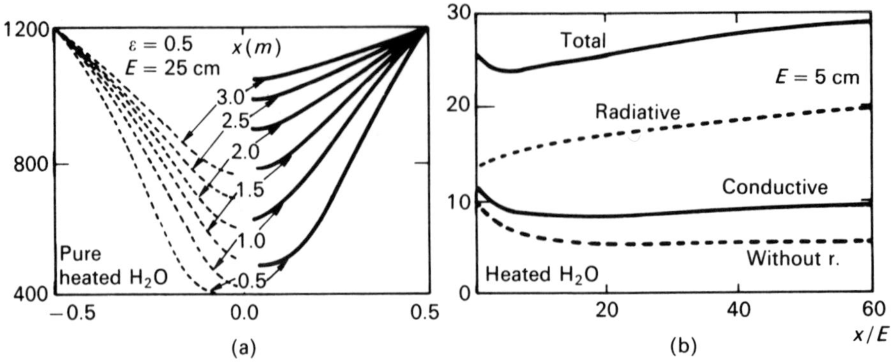


**Figure 18.5** From reference [42] (a) Temperature field with (——) and without (- - - -) coupling between convection and radiation in a channel at various distances $x$ from the entry. Heating of water vapour ($T_w = 1200$ K, $T_i$ = (fluid) = 400 K, $\varepsilon_w = 0.5$). (b) Conductive heat transfer coefficient ($\varphi/(T_w - T_0)$ with radiation coupling, without coupling, and total as a function of the ratio $x/E$ [42].

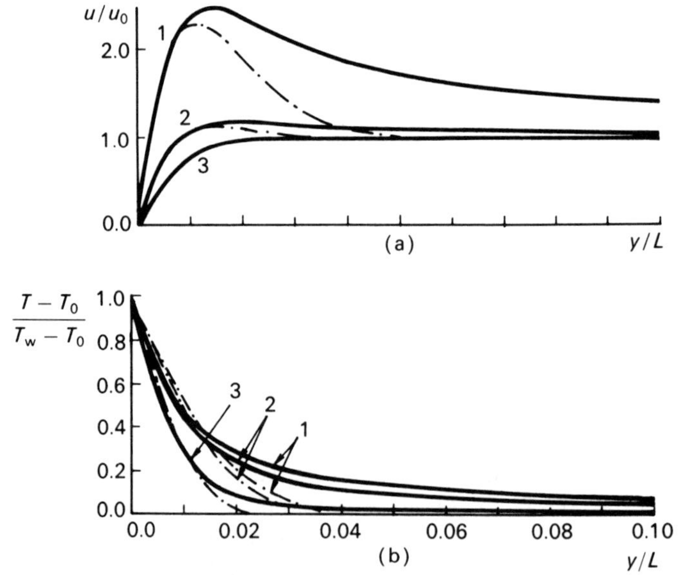


**Figure 18.6** From reference [42], velocity and temperature fields at the top of a plate in pure mixed convection (–·–·–) or coupled with radiation (——), for various values of external velocity: (1) $u_0 = 0.2$ m s$^{-1}$; (2) $u_0 = 0.5$ m s$^{-1}$; (3) $u_0 = 2$ m s$^{-1}$ in water vapour with $T_w = 650$ K, $T_0 = 600$ K, $\varepsilon_w = 0.5$. The corresponding Reynolds numbers are $2.49 \times 10^3$, $6.22 \times 10^3$, $2.49 \times 10^4$.

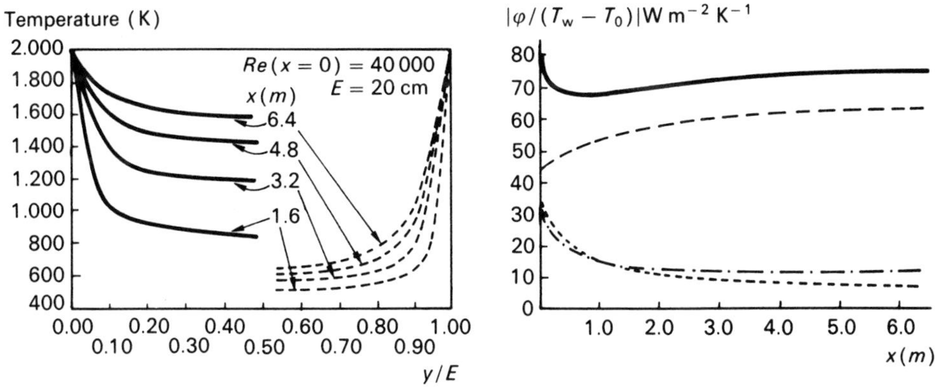


**Figure 18.7** (From reference [28].) (a) Calculated temperature profiles at different $x$ positions taking account of radiation (——), and not taking account of radiation (– – – –); $T_0 = 400$ K, $T_w = 2000$ K. (b) Heat transfer coefficients: (–·–·–) conductive; (– – –) radiative; (——) total; and (·····) conductive, for the same fluid, assumed transparent.

transfer coefficient $\varphi/(T_w - T_m)$ for conduction (coupled to radiation and convection), radiation (coupled to conduction and convection) and conduction (coupled to convection) for the same fluid, assumed transparent, are shown in Figure 18.7. The effect of radiation on the wall conduction profile appears very nicely in this example.

# Hydraulic Diameter

## *Idea of hydraulic diameter*

We postulate that various results (entry length, development length, friction factor, Nusselt number, etc.), which we have obtained for one particular closed geometry with a collection of (thermal and momentum) boundary conditions, can be generalized with reasonable accuracy to other geometries, not too remote, so long as the boundary conditions are of the same type. We will discuss this approach in more detail for laminar and turbulent flows at the end of this Appendix.

The basic idea is to take a tube of circular section with diameter $D$ as reference, and for another geometry, to define an equivalent tube with an equivalent diameter, traditionally called the **hydraulic diameter**, $D_h$. Consider, for example, cylindrical tubes whose cross-sections have the following geometries (Figure 1).

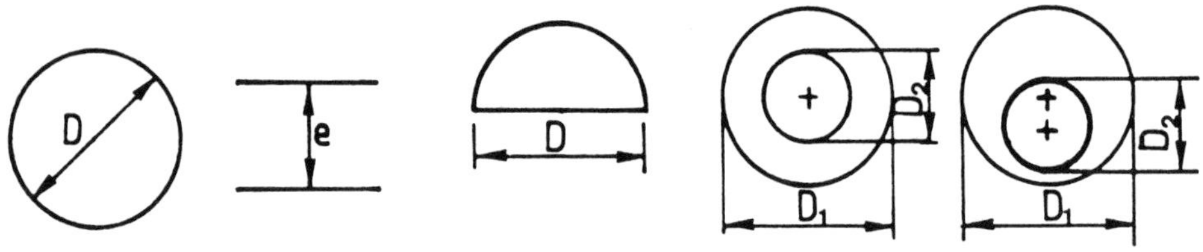


**Figure 1**

We can characterize them by two quantities:

- the perimeter which is wetted by the fluid, $\mathscr{P}$ ($\pi[D_1 + D_2]$ in the last two cases);
- the area of the passage, $\Sigma$, which is open to the fluid ($\pi/4[D_2^2 - D_1^2]$ in the last two cases).

We propose that the diameter of an equivalent tube is a characteristic length proportional to $\Sigma/\mathscr{P}$. By convention, the **hydraulic diameter** is defined as $D_h = 4\Sigma/\mathscr{P}$, which gives the diameter $D$ for a circular tube (whence the coefficient 4).

For example, for two infinite parallel plates a distance $e$ apart: $\Sigma = ed$ and $\mathscr{P} = 2d$, so that $D_h = 2e$. For an annular space with diameters $D_2$ and $D_1$: $D_h = D_2 - D_1$.

## *Limitations on the use of $D_h$*

### *Geometric limitations*

It is obvious that this approach is only of interest for compact shapes without points which are too singular (see the examples above). We notice (see Section 15.1) that the distance from the entry of a tube to where the maximum velocity on the pipe centre-line is $1.5u_0$ is much the same for a circular tube or two parallel plates with the same hydraulic diameter (although the development lengths are different).

### *Limitations related to the flow régime*

The idea of hydraulic diameter is of particular interest in turbulent flow. As we saw in Chapter 16, in this type of flow the variations in mean velocity $\bar{u}$ and mean temperature $\bar{T}$ take place essentially in the inner regions (momentum and thermal), which are narrow and close to the wall – typically of thickness $\eta = R/10$ for a circular section tube.

Also the wetted perimeter is directly associated with the extent of these inner regions, and could be characterized by flow cross-section of $\eta\mathscr{P}$. The section $\Sigma$ available to the fluid is mainly occupied by the outer region (the core of a turbulent flow). We can see, therefore, that the Nusselt number, for example, will depend primarily on the hydraulic diameter, and relatively little on the detailed geometry.

In laminar flow, however, temperature and velocity variations are appreciable throughout the section of the tube, and not just near the walls. Therefore the results depend much more on the particular geometry, for a given hydraulic diameter.

PART V

# Basic Data

# Section 1

# Forced Internal Convection

## *1.1 Circular pipe* [1–10]

### *1.1.1 Laminar flow*

$$\frac{L_m}{D_h} = 0.59 + 0.056 Re_{D_h}$$

$$\frac{L_m}{D_h} = 0.0575 Re_{D_h}$$

1.1.1.1. $V$ AND $T$ PROFILES FULLY DEVELOPED [11,12]

(a) $T_w = \text{const}$:

$$Pe_{D_h}\frac{D_h}{L} \leqslant 12 \qquad Nu_{D_h} = \frac{hD_h}{\lambda} = 3.66$$

$$Pe_{D_h}\frac{D_h}{L} > 12 \qquad Nu_{D_h} = 1.61\left(\frac{Pe_{D_h}D_h}{L}\right)^{1/3}$$

$$L\text{: pipe length, } \frac{Pe_{D_h}D_h}{L} = Gz_{D_h} = \text{the Graetz number}$$

(b) $\varphi_w = \text{const}$:

$$Nu_{D_h} = \frac{48}{11} = 4.36$$

1.1.1.2 $V$ PROFILE FULLY DEVELOPED, $T$ PROFILE NOT FULLY DEVELOPED

(a) $T_w = \text{const}$ (Graetz's problem [13]):

Local Nusselt number:

$$Nu_{D_h}(x) = \frac{h(x)D_h}{\lambda} = \frac{\sum_{n=0}^{\infty} G_n \exp[-\lambda_n^2 x^+]}{2\sum_{n=0}^{\infty} \frac{G_n}{\lambda_n^2} \exp[-\lambda_n^2 x^+]}$$

$$x^+ = \frac{x}{RPe_{D_h}}$$

for $n > 2$ $\quad \lambda_n = 4n + 8/3 \quad G_n = 1.012\,76\lambda_n^{-1/3}$

| $n$ | $\lambda_n^2$ | $G_n$ |
|---|---|---|
| 0 | 7.312 | 0.749 |
| 1 | 44.62 | 0.544 |
| 2 | 113.8 | 0.463 |
| 3 | 215.2 | 0.414 |
| 4 | 348.5 | 0.382 |

Approximate expressions [14] ($\pm 0.5\%$):

$x^+ \leqslant 0.02 \quad Nu_{D_h}(x) = 1.357(x^+)^{-1/3} - 0.7$

$x^+ > 0.02 \quad Nu_{D_h}(x) = 3.657 + 9.641(10^3 x^+)^{-0.488} e^{-28.6x^+}$

- Average Nusselt number between 0 and $x^+$:

$$Nu_{D_h,x} = \frac{\bar{h}D_h}{\lambda} = \frac{1}{2x^+} \ln\left\{\frac{1}{8\sum_{n=0}^{\infty} \frac{G_n}{\lambda_n^2} \exp[-\lambda_n^2 x^+]}\right\}$$

| $x^+$ | $Nu_{D_h}(x)$ | $Nu_{D_h,x}$ |
|---|---|---|
| 0 | $\infty$ | $\infty$ |
| 0.001 | 12.86 | 22.96 |
| 0.004 | 7.91 | 12.59 |
| 0.01 | 5.99 | 8.99 |
| 0.04 | 4.18 | 5.87 |
| 0.08 | 3.79 | 4.89 |
| 0.10 | 3.71 | 4.66 |
| 0.20 | 3.66 | 4.16 |
| $\infty$ | 3.66 | 3.66 |

Approximate expressions [14] ($\pm 0.3\%$):

$$x^+ \leqslant 0.01 \qquad Nu_{D_h,x} = 2.035(x^+)^{-1/3} - 0.7$$

$$0.01 < x^+ < 0.06 \qquad Nu_{D_h,x} = 2.035(x^+)^{-1/3} - 0.2$$

$$x^+ \geqslant 0.06 \qquad Nu_{D_h,x} = 3.657 + 0.0998/x^+$$

- Thermal development length [14]:

$$\frac{L_{th}}{D_h Pe_{D_h}} = 0.0335$$

(b) $\varphi_w = \text{const}$:

- Local Nusselt number:

$$Nu_{D_h}(x) = \frac{h(x)D_h}{\lambda} = \left[\frac{11}{48} - \frac{1}{2}\sum_{m=1}^{\infty}\frac{\exp[-\gamma_m^2 x^+]}{A_m \gamma_m^4}\right]^{-1}$$

$$x^+ = \frac{x}{RPe_{D_h}}$$

for $m > 5$ $\qquad \gamma_m = 4m + 4/3, \qquad A_m = 0.358\gamma_m^{-2.32}$

| $m$ | $\gamma_m^2$ | $A_m$ |
|---|---|---|
| 1 | 25.68 | $7.63 \times 10^{-3}$ |
| 2 | 83.86 | $2.058 \times 10^{-3}$ |
| 3 | 174.2 | $0.901 \times 10^{-3}$ |
| 4 | 296.5 | $0.487 \times 10^{-3}$ |
| 5 | 450.9 | $0.297 \times 10^{-3}$ |

approximate expressions [14,15] ($\pm 1\%$):

$$x^+ \leqslant 0.0001 \qquad Nu_{D_h}(x) = 1.640(x^+)^{-1/3} - 1$$

$$0.0001 < x^+ \leqslant 0.003 \qquad Nu_{D_h}(x) = 1.640(x^+)^{-1/3} - 0.5$$

$$x^+ > 0.003 \qquad Nu_{D_h}(x) = 4.364 + 12.327(10^3 x^+)^{-0.506} e^{-20.5x^+}$$

- Average Nusselt number between 0 and $x^+$ [14] ($\pm 3\%$):

$$x^+ \leqslant 0.06 \qquad Nu_{D_h,x} = \frac{\bar{h}D_h}{\lambda} = 2.461(x^+)^{-1/3}$$

$$x^+ > 0.06 \qquad Nu_{D_h,x} = 4.364 + 0.1444/x^+$$

- Thermal development length [14]:

$$\frac{L_{th}}{D_h Pe_{D_h}} = 0.0431$$

1.1.1.3 $V$ AND $T$ PROFILES NOT FULLY DEVELOPED

(a) $T_w = \text{const}$:

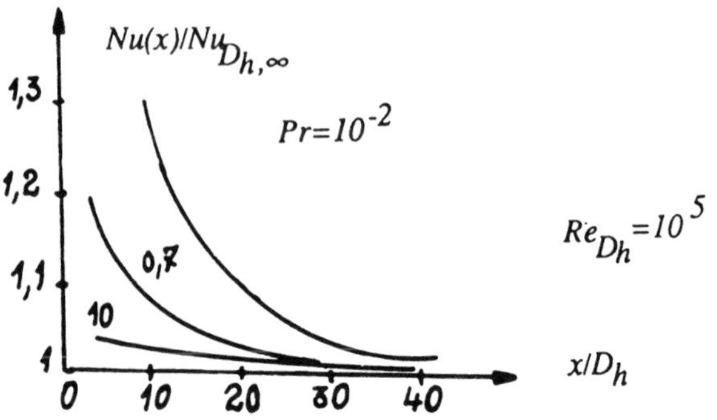


$$\frac{L_{\text{th}}}{D_{\text{h}} Pe_{D_{\text{h}}}} = 0.037 \qquad \text{for} \qquad Pr = 0.7$$

$$\frac{L_{\text{th}}}{D_h Pe_{D_{\text{h}}}} = 0.033 \qquad \text{for} \qquad Pr = \infty$$

| $x^+$ | $Nu_{D_h}(x)$ $Pr = 0.7$ | $Pr = 2$ | $Pr = 5$ | $Nu_{D_h,x}$ $Pr = 0.7$ | $Pr = 2$ | $Pr = 5$ |
|---|---|---|---|---|---|---|
| $4 \times 10^{-4}$ | 24.8 | 21.2 | 18.9 | 44.1 | 36.4 | 30.3 |
| 8 | 18.6 | 16.2 | 14.7 | 33.6 | 27.7 | 24.0 |
| $1.2 \times 10^{-3}$ | 15.5 | 13.8 | 12.7 | 28.1 | 23.4 | 20.7 |
| 1.6 | 13.7 | 12.4 | 11.5 | 24.6 | 20.8 | 18.4 |
| 2 | 12.6 | 11.4 | 10.6 | 22.2 | 19.1 | 16.9 |
| 4 | 9.6 | 8.8 | 8.2 | 16.7 | 14.4 | 12.8 |
| 8 | 7.3 | 6.8 | 6.5 | 12.4 | 11.1 | 9.9 |
| $1.2 \times 10^{-2}$ | 6.25 | 5.8 | 5.7 | 10.6 | 9.5 | 8.6 |
| 1.6 | 5.6 | 5.3 | 5.27 | 9.5 | 8.5 | 7.7 |
| 2 | 5.25 | 4.93 | 4.92 | 8.7 | 7.8 | 7.2 |
| 3 | 4.6 | 4.44 | 4.44 | 7.5 | 6.8 | 6.3 |
| 4 | 4.286 | 4.17 | 4.17 | 6.8 | 6.2 | 5.8 |
| $\infty$ | 3.66 | 3.66 | 3.66 | 3.66 | 3.66 | 3.66 |

(From Hornbeck [16].)

(b) $\varphi_w = \text{const}$:

| $x^+$ | $Nu_{D_h}(x)$ $Pr = 0.01$ | $Pr = 0.7$ | $Pr = 10$ |
|---|---|---|---|
| 0.0002 | | 51.9 | 39.1 |
| 0.002 | 24.2 | 17.8 | 14.3 |
| 0.01 | 12.0 | 9.12 | 7.87 |
| 0.02 | 9.10 | 7.14 | 6.32 |
| 0.10 | 6.08 | 4.72 | 4.51 |
| 0.20 | 5.73 | 4.41 | 4.38 |
| $\infty$ | 4.36 | 4.36 | 4.36 |

(From Heaton, Reynolds and Kays [17].)

$$\frac{L_{th}}{D_h Pe_{D_h}} = 0.053 \quad \text{for} \quad Pr = 0.7$$

$$\frac{L_{th}}{D_h Pe_{D_h}} = 0.043 \quad \text{for} \quad Pr = \infty$$

## 1.1.2 Turbulent flow

### 1.1.2.1 *V* AND *T* PROFILES FULLY DEVELOPED

$$Nu(\varphi_w) \sim Nu(T_w) \quad \text{for} \quad Pr \geqslant 0.7$$

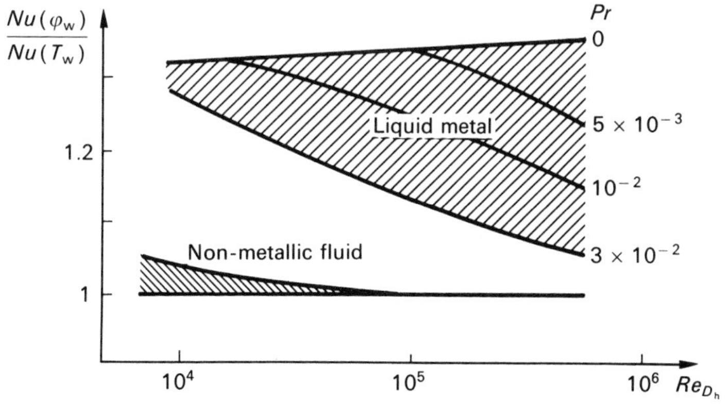

The thermal entry length is much less than in laminar flow – the asymptotic value of Nusselt number is often used from the pipe entry.

- $\dfrac{L}{D_h} \geqslant 60$:

  The Dittus–Boelter equation [18]:

$$Nu_{D_h} = 0.023 Re_{D_h}^{0.8} Pr^{0.4}$$

$$0.7 \leqslant Pr \leqslant 120$$

$$10^4 \leqslant Re_{D_h} \leqslant 1.2 \times 10^5$$

  The Colburn equation [19]:

$$Nu_{D_h} = 0.023 Re_{D_h}^{0.8} Pr^{1/3} \qquad \text{(non-metallic fluid)}$$

  For gases ($0.5 \leqslant Pr \leqslant 1$) in the region where $Nu(\varphi) \neq Nu(T)$, use:

$$\varphi_w = \text{const} \rightarrow \qquad Nu_{D_h} = 0.022 Re_{D_h}^{0.8} Pr^{0.6}$$

$$T_w = \text{const} \rightarrow \qquad Nu_{D_h} = 0.021 Re_{D_h}^{0.8} Pr^{0.6}$$

- $20 \leqslant \dfrac{L}{D_h} < 60$

$$Nu_{D_h,L} = Nu_{D_h}\left(1 + 6\frac{D_h}{L}\right)$$

- $2 \leqslant \dfrac{L}{D_h} < 20$:

$$Nu_{D_h,L} = Nu_{D_h}\left[1 + \left(\frac{D_h}{L}\right)^{0.7}\right]$$

In the two preceding formulae, $Nu_{D_h}$ is the value for a long tube $\left(\dfrac{L}{D_h} > 60\right)$.

For liquid metals (with very low $Pr$):

$\varphi_w = \text{const}$: the Lyon equation [20]:

$$Nu_{D_h} = 7 + 0.025 Pe_{D_h}^{0.8}$$

$T_w = \text{const}$: the Seban–Shimazaki equation [21]:

$$Nu_{D_h} = 4.8 + 0.025 Pe_{D_h}^{0.8}$$

1.1.2.2 $V$ PROFILE FULLY DEVELOPED, $T$ PROFILE NOT FULLY DEVELOPED

Solutions are obtained in the same way as for laminar flow:

(a) $T_w = \text{const}$ [22]:

$$Nu_{D_h}(x) = \frac{h(x)D_h}{\lambda} = \frac{\sum_{n=0}^{\infty} G_n \exp[-\lambda_n^2 x^+]}{2\sum_{n=0}^{\infty} \frac{G_n}{\lambda_n^2}\exp[-\lambda_n^2 x^+]}$$

$$Nu_{D_h,x} = \frac{\bar{h}D_h}{\lambda} = \frac{1}{2x^+}\ln\left[\frac{1}{8\sum_{n=0}^{\infty} \frac{G_n}{\lambda_n^2}\exp[-\lambda_n^2 x^+]}\right]$$

$x^+ \to \infty$

$$Nu = \frac{\lambda_2^0}{2} = \text{Established Nusselt number}$$

(b) $\varphi_w = \text{const}$ [35]:

$$Nu_{D_h}(x) = \frac{1}{\dfrac{1}{Nu_{D_h,\infty}} - \dfrac{1}{2}\sum_{m=1}^{\infty}\dfrac{\exp[-\gamma_m^2 x^+]}{A_m \gamma_m^4}}$$

$Nu_{D_h,\infty}$ is the value obtained for fully developed $V$ and $T$ by the preceding formulae, or from:

$$Nu_{D_h,\infty} = \left[16\sum_{n=0}^{\infty}\frac{G_n}{\lambda_n^4}\right]^{-1}$$

- Development length for $\varphi_w = \text{const}$. For $Pr = 0.7$, this is practically independent of $Re_{D_h}$ in the range $5 \times 10^4 < Re_{D_h} < 2 \times 10^5$.

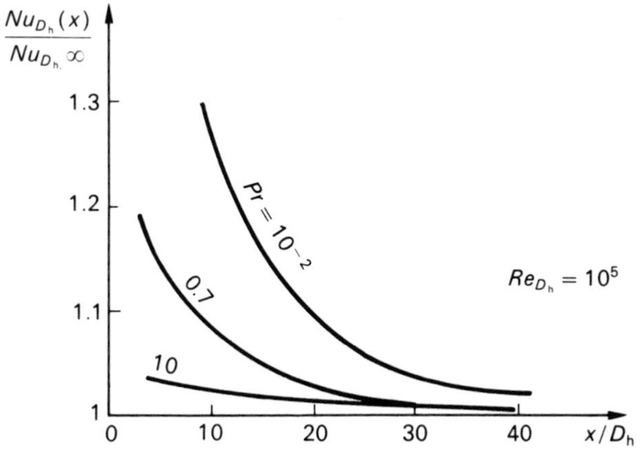

| $Re_{D_h}$ | $n$ / $m$ | $Pr = 10^{-2}$: $\lambda_n^2$ | $G_n$ | $\lambda_m^2$ | $A_m \times 10^3$ | $Pr = 0.7$: $\lambda_n^2$ | $G_n$ | $\gamma_m^2$ | $A_m \times 10^5$ | $Pr = 10$: $\lambda_n^2$ | $G_n$ | $\gamma_m^2 \times 10^{+4}$ | $A_m \times 10^6$ |
|---|---|---|---|---|---|---|---|---|---|---|---|---|---|
| $5 \times 10^4$ | 0 | 11.7 | 1.11 | | | 235 | 28.6 | | | | | | |
| | 1 | 65 | 0.95 | 34.5 | 6.45 | 2640 | 5.51 | 1947 | 7.51 | | | 2.736 | 5.021 |
| | 2 | 163 | 0.88 | 113 | 1.81 | 7400 | 3.62 | 5230 | 1.97 | | | 7.316 | 1.21 |
| | 3 | 305 | 0.835 | 237 | 0.828 | | | 9875 | 0.8 | | | 13.73 | 0.436 |
| | 4 | 491 | 0.802 | 406 | 0.468 | | | 15 900 | 0.402 | | | 21.96 | 0.187 |
| | 5 | 722 | 0.777 | 621 | 0.299 | | | 23 270 | 0.228 | | | 31.98 | 0.0879 |
| $10^5$ | 0 | 13.2 | 1.3 | | | 400 | 49 | | | | | | |
| | 1 | 74 | 1.04 | 40.9 | 5.39 | 4430 | 9.12 | 3557 | 4.12 | | | 5.04 | 2.78 |
| | 2 | 190 | 0.935 | 135 | 1.40 | 12 800 | 5.66 | 9530 | 1.08 | | | 13.46 | 0.697 |
| | 3 | 360 | 0.869 | 284 | 0.625 | | | 17 970 | 0.443 | | | 25.28 | 0.268 |
| | 4 | 583 | 0.823 | 488 | 0.326 | | | 28 840 | 0.226 | | | 40.46 | 0.126 |
| | 5 | 860 | 0.788 | 746 | 0.204 | | | 42 230 | 0.130 | | | 59.04 | 0.0656 |
| $2 \times 10^5$ | 0 | 16.9 | 1.7 | | | | | | | | | | |
| | 1 | 96 | 1.24 | 55.5 | 3.86 | | | | | | | | |
| | 2 | 247 | 1.07 | 181 | 0.989 | | | | | | | | |
| | 3 | 467 | 0.978 | 376 | 0.449 | | | | | | | | |
| | 4 | 757 | 0.901 | 642 | 0.252 | | | | | | | | |
| | 5 | 1116 | 0.847 | 980 | 0.160 | | | | | | | | |

## *1.2 Rectangular channel* [1–10]

$$\zeta = \frac{b}{a}$$

$$D_h = \frac{2ab}{a+b} = \frac{2b}{1+\zeta}$$

$\zeta \to 0$: parallel plates

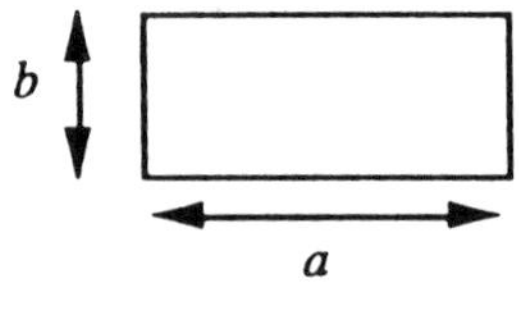


### *1.2.1. Laminar flow* [24,25]

1.2.1.1 PROFILES OF $V$ AND $T$ FULLY DEVELOPED

(a) $T_w = \text{const}$ [26]:

| $\zeta$ | $Nu_{D_h}$  |  |  | 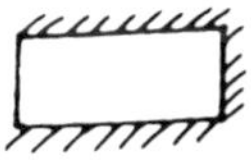 | 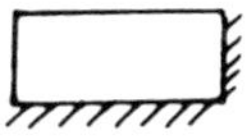 |
|---|---|---|---|---|---|
| 0 | 7.541 | 7.541 | 7.541 | 0 | 4.861 |
| 0.1 | 5.858 | 6.095 | 6.399 | 0.457 | 3.823 |
| 0.2 | 4.803 | 5.195 | 5.703 | 0.833 | 3.330 |
| 0.3 | 4.114 | 4.579 | 5.224 | 1.148 | 2.996 |
| 0.4 | 3.670 | 4.153 | 4.884 | 1.416 | 2.768 |
| 0.5 | 3.383 | 3.842 | 4.619 | 1.647 | 2.613 |
| 0.6 | 3.198 | | | | 2.509 |
| 0.7 | 3.083 | 3.408 | 4.192 | 2.023 | 2.442 |
| 0.8 | 3.014 | | | | 2.401 |
| 0.9 | 2.98 | | | | 2.381 |
| 1 | 2.970 | 3.018 | 3.703 | 2.437 | 2.375 |
| 2 | 3.383 | 3.842 | 4.619 | 1.647 | 2.613 |
| 5 | 4.803 | 5.195 | 5.703 | 0.833 | 3.330 |
| 10 | 5.858 | 6.095 | 6.399 | 0.457 | 3.823 |
| $\infty$ | 7.541 | 7.541 | 7.541 | 0 | 4.861 |

Adiabatic wall

Approximate expression for the case where all four walls are at $T_w$ ($\pm 0.1\%$):

$$Nu_{D_h} = 7.541[1 - 2.610\zeta + 4.970\zeta^2 - 5.119\zeta^3 + 2.702\zeta^4 - 0.548\zeta^5]$$

(b) $\varphi_w = \text{const}$ [26]:

| $\zeta$ | $Nu_{D_h}$  |  |  | 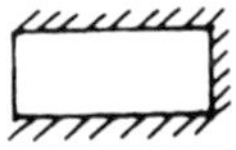 |  |
|---|---|---|---|---|---|
| 0 | 8.235 | 8.235 | 8.235 | 0 | 5.385 |
| 0.1 | 6.7 | 6.939 | 7.248 | 0.538 | 4.410 |
| 0.2 | 5.704 | 6.072 | 6.561 | 0.964 | 3.914 |
| 0.3 | 4.969 | 5.393 | 5.997 | 1.312 | 3.538 |
| 0.4 | 4.457 | 4.885 | 5.555 | 1.604 | 3.279 |
| 0.5 | 4.111 | 4.505 | 5.203 | 1.854 | 3.104 |
| 0.6 | 3.884 | | | | 2.987 |
| 0.7 | 3.740 | 3.991 | 4.662 | 2.263 | 2.911 |
| 0.8 | 3.655 | | | | 2.866 |
| 0.9 | 3.612 | | | | 2.843 |
| 1 | 3.599 | 3.556 | 4.094 | 2.712 | 2.836 |
| 2 | 4.111 | 4.505 | 5.203 | 1.854 | 3.104 |
| 5 | 5.704 | 6.072 | 6.561 | 0.964 | 3.914 |
| 10 | 6.78 | 6.939 | 7.248 | 0.538 | 4.410 |
| $\infty$ | 8.235 | 8.235 | 8.235 | 0 | 5.385 |

Adiabatic wall

Approximate expression for the case where all four walls are subject to a flux $\varphi_w$ ($\pm 0.03\%$):

$$Nu_{D_h} = 8.235[1 - 2.0421\zeta + 3.0853\zeta^2 - 2.4765\zeta^3 + 1.0578\zeta^4 - 0.1861\zeta^5]$$

1.2.1.2 *V* PROFILE FULLY DEVELOPED, *T* PROFILE NOT FULLY DEVELOPED

(a) $T_w$ = const on all four walls [27]:

| | | | $Nu_{D_h}(x)$ | | | |
|---|---|---|---|---|---|---|
| $1/x^*$ | $\zeta = 1$ | $\zeta = 0.5$ | $\zeta = 1/3$ | $\zeta = 0.25$ | $\zeta = 0.2$ | $\zeta = 1/6$ |
| 0 | 2.65 | 3.39 | 3.96 | 4.51 | 4.92 | 5.22 |
| 10 | 2.86 | 3.43 | 4.02 | 4.53 | 4.94 | 5.24 |
| 20 | 3.08 | 3.54 | 4.17 | 4.65 | 5.04 | 5.34 |
| 30 | 3.24 | 3.70 | 4.29 | 4.76 | 5.31 | 5.41 |
| 40 | 3.43 | 3.85 | 4.42 | 4.87 | 5.22 | 5.48 |
| 60 | 3.78 | 4.16 | 4.67 | 5.08 | 5.40 | 5.64 |
| 80 | 4.10 | 4.46 | 4.94 | 5.32 | 5.62 | 5.86 |
| 100 | 4.35 | 4.72 | 5.17 | 5.55 | 5.83 | 6.07 |
| 120 | 4.62 | 4.93 | 5.42 | 5.77 | 6.06 | 6.27 |
| 140 | 4.85 | 5.15 | 5.62 | 5.98 | 6.26 | 6.47 |
| 160 | 5.03 | 5.34 | 5.80 | 6.18 | 6.45 | 6.66 |
| 180 | 5.24 | 5.54 | 5.99 | 6.37 | 6.63 | 6.86 |
| 200 | 5.41 | 5.72 | 6.18 | 6.57 | 6.80 | 7.02 |

| | | | $Nu_{D_h,x}$ | | | |
|---|---|---|---|---|---|---|
| $1/x^*$ | $\zeta = 1$ | $\zeta = 0.5$ | $\zeta = 1/3$ | $\zeta = 0.25$ | $\zeta = 0.2$ | $\zeta = 1/6$ |
| 0 | 2.65 | 3.39 | 3.96 | 4.51 | 4.92 | 5.22 |
| 10 | 3.50 | 3.95 | 4.54 | 5.00 | 5.36 | 5.66 |
| 20 | 4.03 | 4.46 | 5 | 5.44 | 5.77 | 6.04 |
| 30 | 4.47 | 4.86 | 5.39 | 5.81 | 6.13 | 6.37 |
| 40 | 4.85 | 5.24 | 5.74 | 6.16 | 6.45 | 6.70 |
| 60 | 5.50 | 5.85 | 6.35 | 6.73 | 7.03 | 7.26 |
| 80 | 6.03 | 6.37 | 6.89 | 7.24 | 7.53 | 7.77 |
| 100 | 6.46 | 6.84 | 7.33 | 7.71 | 7.99 | 8.17 |
| 120 | 6.86 | 7.24 | 7.74 | 8.13 | 8.39 | 8.63 |
| 140 | 7.22 | 7.62 | 8.11 | 8.50 | 8.77 | 9 |
| 160 | 7.56 | 7.97 | 8.45 | 8.86 | 9.14 | 9.35 |
| 180 | 7.87 | 8.29 | 8.77 | 9.17 | 9.46 | 9.67 |
| 200 | 8.15 | 8.58 | 9.07 | 9.47 | 9.79 | 10.01 |

(b) $\varphi_w$ = const on all four walls [27]:

| $1/x^*$ | $Nu_{D_h}(x)$ $\zeta = 1$ | $\zeta = 0.5$ | $\zeta = 1/3$ | $\zeta = 0.25$ |
|---|---|---|---|---|
| 0 | 3.60 | 4.11 | 4.77 | 5.35 |
| 10 | 3.71 | 4.22 | 4.85 | 5.45 |
| 20 | 3.91 | 4.38 | 5.00 | 5.62 |
| 30 | 4.18 | 4.61 | 5.17 | 5.77 |
| 40 | 4.45 | 4.84 | 5.39 | 5.87 |
| 60 | 4.91 | 5.28 | 5.82 | 6.26 |
| 80 | 5.33 | 5.70 | 6.21 | 6.63 |
| 100 | 5.69 | 6.05 | 6.57 | 7.00 |
| 120 | 6.02 | 6.37 | 6.92 | 7.32 |
| 140 | 6.32 | 6.68 | 7.22 | 7.63 |
| 160 | 6.60 | 6.96 | 7.50 | 7.92 |
| 180 | 6.86 | 7.23 | 7.76 | 8.18 |
| 200 | 7.10 | 7.46 | 8.02 | 8.44 |

| $1/x^*$ | $Nu_{D_{h,x}}$ $\zeta = 1$ | $\zeta = 0.5$ | $\zeta = 1/3$ | $\zeta = 0.25$ |
|---|---|---|---|---|
| 0 | 3.60 | 4.11 | 4.77 | 5.35 |
| 10 | 4.48 | 4.94 | 5.45 | 6.03 |
| 20 | 5.19 | 5.60 | 6.06 | 6.57 |
| 30 | 5.76 | 6.15 | 6.60 | 7.07 |
| 40 | 6.24 | 6.64 | 7.09 | 7.51 |
| 60 | 7.02 | 7.45 | 7.85 | 8.25 |
| 80 | 7.66 | 8.10 | 8.48 | 8.87 |
| 100 | 8.22 | 8.66 | 9.02 | 9.39 |
| 120 | 8.69 | 9.13 | 9.52 | 9.83 |
| 140 | 9.09 | 9.57 | 9.93 | 10.24 |
| 160 | 9.50 | 9.96 | 10.31 | 10.61 |
| 180 | 9.85 | 10.31 | 10.67 | 10.92 |
| 200 | 10.18 | 10.64 | 10.97 | 11.23 |

(c) Thermal development length [27]:

| $\zeta$ | $T_w$ = const. $L^*_{th}$ | $\varphi_w$ = const. $L^*_{th}$ |
|---|---|---|
| 0 | $8 \times 10^{-3}$ | $1.15 \times 10^{-2}$ |
| 0.25 | $5.4 \times 10^{-2}$ | $4.2 \times 10^{-2}$ |
| $\frac{1}{3}$ | | $4.8 \times 10^{-2}$ |
| 0.5 | $4.9 \times 10^{-2}$ | $5.7 \times 10^{-2}$ |
| 1 | $4.1 \times 10^{-2}$ | $6.6 \times 10^{-2}$ |

$$L^*_{th} = \frac{L_{th}}{D_h Pe_{D_h}}$$

1.2.1.3 PROFILES OF $V$ AND $T$ NOT FULLY DEVELOPED

(a) $T_w$ = const on all four walls [27]:

| $Pr = 0.72$ | | | $Nu_{D_h,x}$ | | |
|---|---|---|---|---|---|
| $x^*$ | $\zeta = 1$ | $\zeta = 0.5$ | $\zeta = 1/3$ | $\zeta = 0.25$ | $\zeta = 1/6$ |
| $4.55 \times 10^{-3}$ | 9.70 | 10.00 | 10.30 | 10.58 | 10.90 |
| 5 | 9.30 | 9.60 | 9.91 | 10.18 | 10.51 |
| 5.56 | 8.91 | 9.20 | 9.50 | 9.77 | 10.12 |
| 6.25 | 8.50 | 8.80 | 9.10 | 9.36 | 9.72 |
| 7.14 | 8.06 | 8.37 | 8.66 | 8.93 | 9.28 |
| 8.33 | 7.61 | 7.91 | 8.18 | 8.48 | 8.85 |
| $1 \times 10^{-2}$ | 7.10 | 7.42 | 7.70 | 7.98 | 8.38 |
| 1.25 | 6.57 | 6.88 | 7.17 | 7.47 | 7.90 |
| 1.67 | 5.95 | 6.27 | 6.60 | 6.90 | 7.35 |
| 2 | 5.63 | 5.95 | 6.28 | 6.61 | 7.07 |
| 2.5 | 5.27 | 5.61 | 5.96 | 6.27 | 6.78 |
| 3.33 | 4.88 | 5.23 | 5.60 | 5.93 | 6.47 |
| 5 | 4.39 | 4.79 | 5.17 | 5.56 | 6.13 |
| $1 \times 10^{-1}$ | 3.75 | 4.20 | 4.67 | 5.11 | 5.72 |

(b) $\varphi_w$ = const on all four walls [27]:

| $Pr = 0.72$ | | | $Nu_{D_h}(x)$ | |
|---|---|---|---|---|
| $x^*$ | $\zeta = 1$ | $\zeta = 0.5$ | $\zeta = 1/3$ | $\zeta = 0.25$ |
| $5 \times 10^{-3}$ | 9.69 | 9.88 | 10.06 | 10.24 |
| 5.56 | 9.28 | 9.47 | 9.70 | 9.87 |
| 6.25 | 8.84 | 9.05 | 9.38 | 9.59 |
| 7.14 | 8.38 | 8.61 | 8.84 | 9.05 |
| 8.33 | 7.90 | 8.11 | 8.37 | 8.58 |
| $1 \times 10^{-2}$ | 7.38 | 7.59 | 7.86 | 8.08 |
| 1.25 | 6.80 | 7.02 | 7.32 | 7.55 |
| 1.67 | 6.14 | 6.42 | 6.74 | 7.00 |
| 2 | 5.83 | 6.09 | 6.44 | 6.70 |
| 2.5 | 5.47 | 5.75 | 6.13 | 6.43 |
| 3.33 | 5.07 | 5.40 | 5.82 | 6.17 |
| 5 | 4.66 | 5.01 | 5.50 | 5.92 |
| $1 \times 10^{-1}$ | 4.18 | 4.60 | 5.18 | 5.66 |

| $Pr = 0.72$ | $Nu_{D_h,x}$ | | | |
|---|---|---|---|---|
| $x^*$ | $\zeta = 1$ | $\zeta = 0.5$ | $\zeta = 1/3$ | $\zeta = 0.25$ |
| $4.55 \times 10^{-3}$ | 15.03 | 15.36 | 15.83 | 16.02 |
| 5 | 14.55 | 14.88 | 15.21 | 15.49 |
| 5.56 | 14.05 | 14.35 | 14.70 | 14.95 |
| 6.25 | 13.50 | 13.79 | 14.10 | 14.48 |
| 7.14 | 12.87 | 13.15 | 13.47 | 13.73 |
| 8.33 | 12.19 | 12.48 | 12.78 | 13.03 |
| $1 \times 10^{-2}$ | 11.43 | 11.70 | 12.00 | 12.23 |
| 1.25 | 10.53 | 10.83 | 11.13 | 11.35 |
| 1.67 | 9.49 | 9.77 | 10.07 | 10.32 |
| 2 | 8.90 | 9.17 | 9.48 | 9.70 |
| 2.5 | 8.25 | 8.54 | 8.85 | 9.07 |
| 3.33 | 7.52 | 7.83 | 8.13 | 8.37 |
| 5 | 6.60 | 6.94 | 7.31 | 7.58 |
| $1 \times 10^{-1}$ | 5.43 | 5.77 | 6.27 | 6.65 |
| $2 \times 10^{-1}$ | 4.60 | 5.00 | 5.57 | 6.06 |

| $\zeta = 0.5$ | $Nu_{D_h,x}$ | | | | |
|---|---|---|---|---|---|
| $x^*$ | $Pr = \infty$ | 10 | 0.72 | 0.1 | 0 |
| $2.5 \times 10^{-3}$ | 13.00 | 15.40 | 18.50 | 19.90 | 20.65 |
| 2.86 | 12.55 | 14.75 | 17.75 | 19.10 | 19.80 |
| 3.33 | 12.00 | 14.05 | 17.00 | 18.30 | 18.90 |
| 3.85 | 11.50 | 13.45 | 16.25 | 17.60 | 18.10 |
| 4.55 | 10.95 | 12.75 | 15.35 | 16.70 | 17.20 |
| 5.56 | 10.31 | 11.95 | 14.35 | 15.65 | 16.15 |
| 7.14 | 9.57 | 11.05 | 13.15 | 14.50 | 14.95 |
| $1 \times 10^{-2}$ | 8.66 | 9.90 | 11.70 | 13.05 | 13.50 |
| 1.25 | 8.10 | 9.20 | 10.83 | 12.15 | 12.65 |
| 1.67 | 7.45 | 8.40 | 9.77 | 11.10 | 11.65 |
| 2.5 | 6.64 | 7.50 | 8.54 | 9.75 | 10.40 |
| 5 | 5.60 | 6.15 | 6.94 | 7.90 | 8.65 |

(c) Thermal development length [26]: $Pr = 0.72$ and $\varphi_w = \text{const}$:

$$\frac{L_{th}}{D_h Pe_{D_h}} = 0.017 \quad \text{for} \quad \zeta = 0$$
$$= 0.136 \qquad \zeta = 0.25$$
$$= 0.17 \qquad \zeta = 1/3$$
$$= 0.23 \qquad \zeta = 0.5$$
$$= 0.34 \qquad \zeta = 1$$

### 1.2.2 *Turbulent flow* [1–3]

The Sieder–Tate equation:

$$Nu_b = 0.027 Re_b^{0.8} Pr^{1/3} \left(\frac{\mu_m}{\mu_w}\right)^{0.14}$$

$\mu_m$ and $\mu_w$ are the viscosities taken at the mean and wall temperatures, respectively.

For liquid metal, the Hartnett–Irvine relation:

$$Nu_{D_h} = \frac{2}{3} Nu_s + 0.015 Pe_{D_h}^{0.8}$$

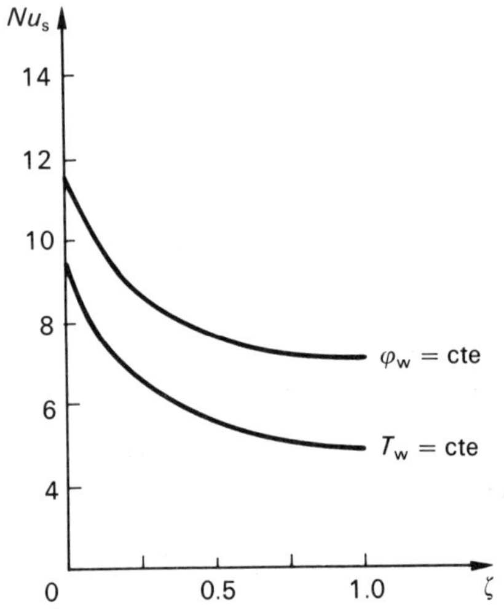


## *1.3 Concentric annulus* [1–10]

$$D_h = d_{ext} - d_{int}$$

$$Y = \frac{d_{int}}{d_{ext}}$$

$Y = 1 \rightarrow$ parallel plates

## 1.3.1 *Laminar flow* [28–30]

| $\frac{L_m}{D_h Re_{D_k}}$ | $Y$ |
|---|---|
| — | 0 |
| 0.0375 | $10^{-3}$ |
| 0.0303 | $10^{-2}$ |
| 0.0241 | $5 \times 10^{-2}$ |
| 0.0210 | $10^{-1}$ |
| 0.0171 | $2 \times 10^{-1}$ |
| 0.0131 | $4 \times 10^{-1}$ |
| 0.0118 | $8 \times 10^{-1}$ |

(Sparrow and Lin [28].)

1.3.1.1 PROFILES OF $V$ AND $T$ FULLY DEVELOPED [26,31]
(a) $T_{wi} = \text{const}$, $\varphi_{we} = 0$ or $\varphi_{wi} = 0$, $T_{we} = \text{const}$:

| $Y$ | $Nu_{ii}^{(T)}$ | $Nu_{ee}^{(T)}$ |
|---|---|---|
| 0 | $\infty$ | 3.66 |
| 0.02 | 32.34 | 3.99 |
| 0.05 | 17.46 | 4.06 |
| 0.10 | 11.56 | 4.11 |
| 0.25 | 7.37 | 4.23 |
| 0.5 | 5.74 | 4.43 |
| 1.00 | 4.86 | 4.86 |

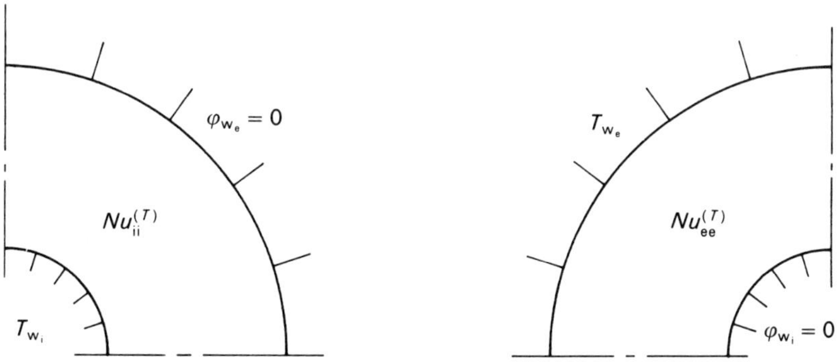

(b) $T_{wi} \neq T_{we}$ or $T_{wi} = T_{we} = T$:

| $Y$ | $Nu_i^{T_{wi}}$ | $Nu_e^{T_{we}}$ | $Nu_i^{(T)}$ | $Nu_e^{(T)}$ |
|---|---|---|---|---|
| 0 | $\infty$ | 2.67 | $\infty$ | 3.66 |
| $1 \times 10^{-4}$ | 2355.3 | 2.78 | | |
| $1 \times 10^{-3}$ | 322.3 | 2.82 | | |
| $1 \times 10^{-2}$ | 50.5 | 2.91 | | |
| $2 \times 10^{-2}$ | 30.18 | 2.95 | 58.08 | 4.59 |
| $4 \times 10^{-2}$ | 18.61 | 3 | | |
| $5 \times 10^{-2}$ | 16.06 | 3.02 | 31.23 | 4.84 |
| $6 \times 10^{-2}$ | 14.28 | 3.04 | | |
| $8 \times 10^{-2}$ | 11.94 | 3.07 | | |
| $1 \times 10^{-1}$ | 10.46 | 3.1 | 20.4 | 5.13 |
| $1.5 \times 10^{-1}$ | 8.34 | 3.16 | | |
| $2 \times 10^{-1}$ | 7.2 | 3.21 | | |
| $2.5 \times 10^{-1}$ | 6.47 | 3.27 | 12.63 | 5.7 |
| $3 \times 10^{-1}$ | 5.97 | 3.32 | | |
| $4 \times 10^{-1}$ | 5.31 | 3.42 | | |
| $5 \times 10^{-1}$ | 4.89 | 3.52 | 9.44 | 6.40 |
| $6 \times 10^{-1}$ | 4.60 | 3.62 | | |
| $7 \times 10^{-1}$ | 4.39 | 3.72 | | |
| $8 \times 10^{-1}$ | 4.23 | 3.81 | | |
| $9 \times 10^{-1}$ | 4.10 | 3.91 | | |
| 1.00 | 4.00 | 4.00 | 7.54 | 7.54 |

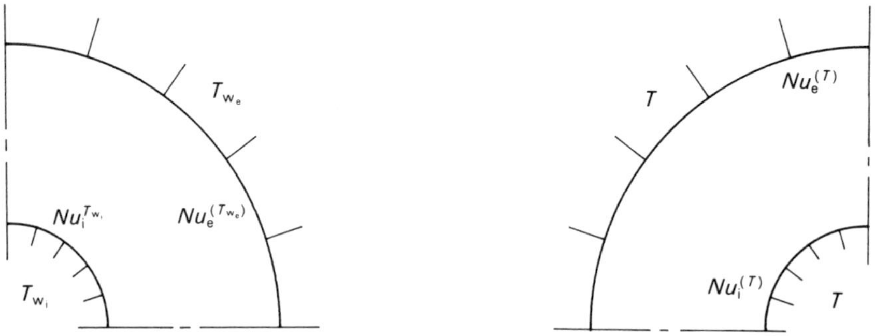

(c) $\varphi_{wi} = \text{const}$, $\varphi_{we} = 0$, or $\varphi_{wi} = 0$, $\varphi_{we} = \text{const}$:

| $Y$ | $Nu^{\varphi}_{ii}$ | $Nu^{\varphi}_{ee}$ | $\theta^*_i$ | $\theta^*_e$ |
|---|---|---|---|---|
| 0 | $\infty$ | 4.36 | $\infty$ | 0 |
| $1 \times 10^{-3}$ | 337.04 | 4.59 | | |
| $1 \times 10^{-2}$ | 54.02 | 4.69 | | |
| $2 \times 10^{-2}$ | 32.71 | 4.73 | | |
| $4 \times 10^{-2}$ | 20.51 | 4.78 | | |
| $5 \times 10^{-2}$ | 17.81 | 4.79 | 2.18 | 0.0294 |
| $6 \times 10^{-2}$ | 15.93 | 4.80 | | |
| $8 \times 10^{-2}$ | 13.47 | 4.80 | | |
| $1 \times 10^{-1}$ | 11.91 | 4.83 | 1.383 | 0.0562 |
| $1.5 \times 10^{-1}$ | 9.69 | 4.86 | | |
| $2 \times 10^{-1}$ | 8.5 | 4.88 | 0.905 | 0.1041 |
| $2.5 \times 10^{-1}$ | 7.75 | 4.90 | | |
| $3 \times 10^{-1}$ | 7.24 | 4.93 | | |
| $4 \times 10^{-1}$ | 6.58 | 4.98 | 0.603 | 0.1823 |
| $5 \times 10^{-1}$ | 6.18 | 5.04 | | |
| $6 \times 10^{-1}$ | 5.91 | 5.1 | 0.473 | 0.2455 |
| $7 \times 10^{-1}$ | 5.72 | 5.17 | | |
| $8 \times 10^{-1}$ | 5.58 | 5.24 | 0.401 | 0.299 |
| $9 \times 10^{-1}$ | 5.47 | 5.31 | | |
| 1.00 | 5.385 | 5.385 | 0.346 | 0.346 |

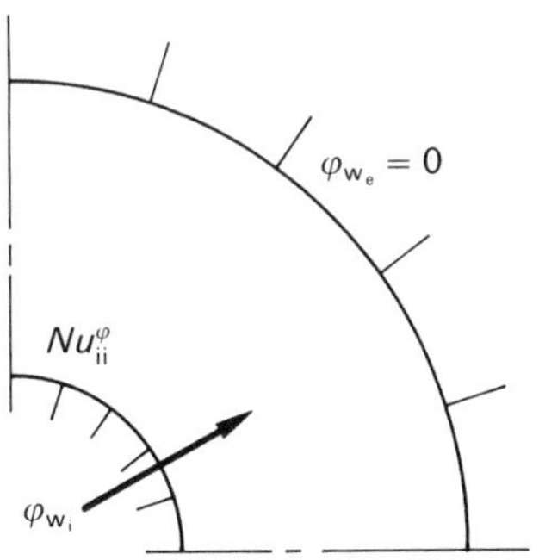


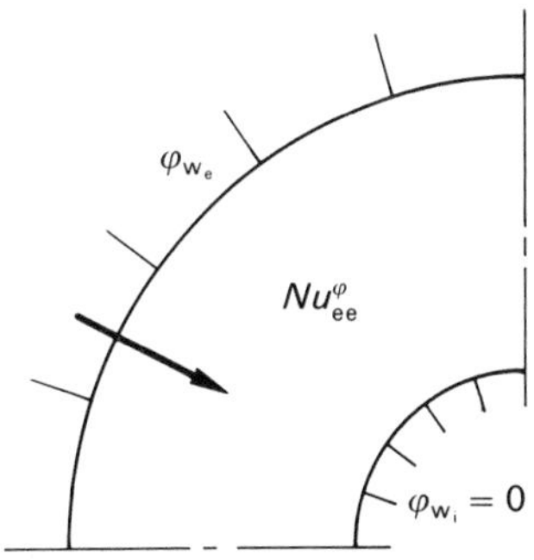

(d) $\varphi_{wi} \neq \varphi_{we}$:

$$Nu_i^\varphi = \frac{h_i D_h}{\lambda} \qquad Nu_e^\varphi = \frac{h_e D_h}{\lambda}$$

$$Nu_i^\varphi = \frac{Nu_{ii}^\varphi}{1 - \left(\dfrac{\varphi_{we}}{\varphi_{wi}}\right)\theta_i^*}$$

$$Nu_e^\mu = \frac{Nu_{ee}^\varphi}{1 - \left(\dfrac{\varphi_{wi}}{\varphi_{we}}\right)\theta_e^*}$$

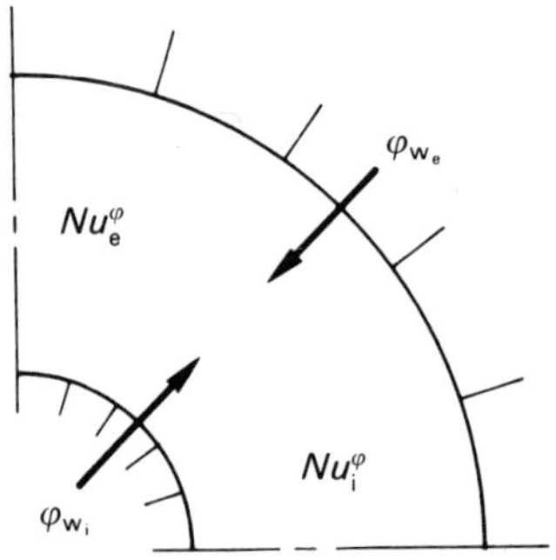


1.3.1.2 *V* PROFILE FULLY DEVELOPED, *T* PROFILE NOT FULLY DEVELOPED [26,31]
(a) Case I: $T_{wi} = \text{const}$, $T_{we} = T_e$ or $T_{wi} = T_e$, $T_{we} = \text{const}$:

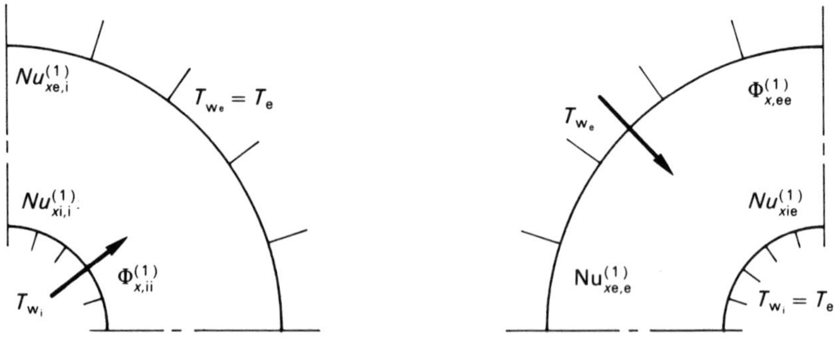


$T_e$ = fluid temperature at heater entry (fluid assumed isothermal).

| $x^*$ | $\varphi^{(1)}_{x,ii}$ | $\theta^{(1)}_{x,mi}$ | $Nu^{(1)}_{x,ii}$ | $Nu^{(1)}_{x,ei}$ | $\varphi^{(1)}_{x,ee}$ | $\theta^{(1)}_{x,me}$ | $Nu^{(1)}_{x,ee}$ | $Nu^{(1)}_{x,ie}$ |
|---|---|---|---|---|---|---|---|---|
| | | | | $Y = 0.02$ | | | | |
| $1 \times 10^{-5}$ | | | | | 51.08 | 0.003 | 51.24 | |
| 5 | | | | | 29.35 | 0.009 | 29.61 | |
| $1 \times 10^{-4}$ | 78.5 | 0.001 | 78.5 | | 23.03 | 0.014 | 23.36 | |
| 5 | 57.5 | 0.003 | 57.7 | | 12.93 | 0.039 | 13.46 | |
| $1 \times 10^{-3}$ | 50.87 | 0.005 | 51.14 | | 9.99 | 0.061 | 10.65 | |
| 5 | 39.28 | 0.019 | 40.03 | | 5.27 | 0.168 | 6.34 | |
| $1 \times 10^{-2}$ | 35.48 | 0.033 | 36.7 | 0.04 | 3.88 | 0.257 | 5.22 | 0.567 |
| 5 | 28.38 | 0.113 | 31.99 | 2.13 | 1.54 | 0.606 | 3.9 | 19.96 |
| $1 \times 10^{-1}$ | 26.12 | 0.151 | 30.79 | 2.75 | 0.835 | 0.757 | 3.44 | 27.5 |
| 5 | 25.05 | 0.17 | 30.18 | 2.95 | 0.501 | 0.830 | 2.95 | 30.18 |
| $\infty$ | 25.05 | 0.17 | 30.18 | 2.95 | 0.501 | 0.830 | 2.95 | 30.18 |
| | | | | $Y = 0.05$ | | | | |
| $1 \times 10^{-5}$ | | | | | 51.63 | 0.003 | 51.78 | |
| 5 | | | | | 29.68 | 0.009 | 29.93 | |
| $1 \times 10^{-4}$ | 52.0 | 0.001 | 52.1 | | 23.3 | 0.014 | 23.62 | |
| 5 | 35.4 | 0.005 | 35.6 | | 13.1 | 0.039 | 13.62 | |
| $1 \times 10^{-3}$ | 30.43 | 0.008 | 30.67 | | 10.13 | 0.060 | 10.77 | |
| 5 | 22.03 | 0.027 | 22.63 | | 5.36 | 0.166 | 6.42 | |
| $1 \times 10^{-2}$ | 19.4 | 0.046 | 20.33 | 0.05 | 3.95 | 0.252 | 5.29 | 0.17 |
| 5 | 14.67 | 0.147 | 17.2 | 2.24 | 1.59 | 0.593 | 3.91 | 11.09 |
| $1 \times 10^{-1}$ | 13.27 | 0.191 | 16.41 | 2.84 | 0.92 | 0.732 | 3.41 | 14.86 |
| 5 | 12.69 | 0.210 | 16.06 | 3.02 | 0.634 | 0.79 | 3.02 | 16.06 |
| $\infty$ | 12.69 | 0.210 | 16.06 | 3.02 | 0.634 | 0.79 | 3.02 | 16.06 |

| $x^*$ | $\varphi^{(1)}_{x,ii}$ | $\theta^{(1)}_{x,mi}$ | $Nu^{(1)}_{x,ii}$ | $Nu^{(1)}_{x,ei}$ | $\varphi^{(1)}_{x,ee}$ | $\theta^{(1)}_{x,me}$ | $Nu^{(1)}_{x,ee}$ | $Nu^{(1)}_{x,ie}$ |
|---|---|---|---|---|---|---|---|---|
| | | | | $Y = 0.1$ | | | | |
| $1\times10^{-5}$ | 80.29 | 0.0004 | 80.32 | | 52.19 | 0.003 | 52.34 | |
| 5 | 49.63 | 0.0013 | 49.7 | | 30.02 | 0.008 | 30.27 | |
| $1\times10^{-4}$ | 40.68 | 0.002 | 40.77 | | 23.58 | 0.013 | 23.89 | |
| 5 | 26.25 | 0.007 | 26.42 | | 13.28 | 0.037 | 13.79 | |
| $1\times10^{-3}$ | 21.95 | 0.011 | 22.19 | | 10.28 | 0.058 | 10.91 | |
| 5 | 13.83 | 0.035 | 14.34 | | 5.46 | 0.161 | 6.51 | |
| $1\times10^{-2}$ | 12.92 | 0.061 | 13.76 | 0.06 | 4.04 | 0.245 | 5.36 | 0.16 |
| 5 | 9.23 | 0.184 | 11.31 | 2.34 | 1.67 | 0.575 | 3.93 | 7.49 |
| $1\times10^{-1}$ | 8.2 | 0.234 | 10.70 | 2.93 | 1.02 | 0.700 | 3.41 | 9.79 |
| 5 | 7.82 | 0.253 | 10.46 | 3.1 | 0.782 | 0.747 | 3.1 | 10.5 |
| $\infty$ | 7.82 | 0.253 | 10.46 | 3.1 | 0.782 | 0.747 | 3.1 | 10.5 |
| | | | | $Y = 0.25$ | | | | |
| $1\times10^{-5}$ | 66.50 | 0.0008 | 66.56 | | 53.28 | 0.0026 | 53.41 | |
| 5 | 39.73 | 0.0023 | 39.83 | | 30.71 | 0.0075 | 30.94 | |
| $1\times10^{-4}$ | 31.95 | 0.0038 | 32.07 | | 24.15 | 0.0118 | 24.44 | |
| 5 | 19.48 | 0.0113 | 19.70 | | 13.67 | 0.0337 | 14.14 | |
| $1\times10^{-3}$ | 15.84 | 0.0183 | 16.14 | | 10.61 | 0.0527 | 11.20 | |
| 5 | 9.98 | 0.0564 | 10.57 | | 5.72 | 0.147 | 6.7 | |
| $1\times10^{-2}$ | 8.24 | 0.0923 | 9.07 | 0.08 | 4.28 | 0.225 | 5.52 | 0.13 |
| 5 | 5.32 | 0.253 | 7.12 | 2.53 | 1.89 | 0.528 | 4 | 4.84 |
| $1\times10^{-1}$ | 4.57 | 0.312 | 6.64 | 3.12 | 1.28 | 0.635 | 3.49 | 6.14 |
| 5 | 4.33 | 0.331 | 6.47 | 3.27 | 1.08 | 0.669 | 3.27 | 6.47 |
| $\infty$ | 4.33 | 0.331 | 6.47 | 3.27 | 1.08 | 0.669 | 3.27 | 6.47 |
| | | | | $Y = 0.5$ | | | | |
| $1\times10^{-5}$ | 60.47 | 0.0012 | 60.54 | | 54.61 | 0.0022 | 54.73 | |
| 5 | 35.54 | 0.0035 | 35.67 | | 31.58 | 0.0064 | 31.79 | |
| $1\times10^{-4}$ | 28.3 | 0.0056 | 28.46 | | 24.89 | 0.0101 | 25.14 | |
| 5 | 16.71 | 0.0166 | 16.99 | | 14.19 | 0.03 | 14.61 | |
| $1\times10^{-3}$ | 13.34 | 0.0264 | 13.70 | | 11.08 | 0.045 | 11.61 | |
| 5 | 7.93 | 0.0782 | 8.60 | | 6.09 | 0.128 | 6.98 | |
| $1\times10^{-2}$ | 6.34 | 0.125 | 7.25 | 0.09 | 4.62 | 0.198 | 5.76 | 0.12 |
| 5 | 3.71 | 0.322 | 5.48 | 2.74 | 2.20 | 0.472 | 4.17 | 3.75 |
| $1\times10^{-1}$ | 3.07 | 0.391 | 5.04 | 3.37 | 1.62 | 0.563 | 3.7 | 4.67 |
| 5 | 2.89 | 0.41 | 4.89 | 3.52 | 1.44 | 0.59 | 3.52 | 4.89 |
| $\infty$ | 2.89 | 0.41 | 4.89 | 3.52 | 1.44 | 0.59 | 3.52 | 4.89 |
| | | | | $Y = 1$ | | | | |
| $1\times10^{-5}$ | 56.80 | 0.0017 | 56.90 | | 56.80 | 0.0017 | 56.90 | |
| 5 | 33.05 | 0.005 | 33.21 | | 33.05 | 0.005 | 33.21 | |
| $1\times10^{-4}$ | 26.14 | 0.008 | 26.35 | | 26.14 | 0.008 | 26.35 | |
| 5 | 15.11 | 0.023 | 15.46 | | 15.11 | 0.023 | 15.46 | |
| $1\times10^{-3}$ | 11.9 | 0.036 | 12.34 | | 11.9 | 0.036 | 12.34 | |
| 5 | 6.75 | 0.104 | 7.53 | | 6.75 | 0.104 | 7.53 | |
| $1\times10^{-2}$ | 5.24 | 0.163 | 6.25 | 0.06 | 5.24 | 0.163 | 6.25 | 0.06 |
| 5 | 2.76 | 0.4 | 4.6 | 3.11 | 2.76 | 0.4 | 4.6 | 3.11 |
| $1\times10^{-1}$ | 2.17 | 0.48 | 4.15 | 3.84 | 2.17 | 0.48 | 4.15 | 3.84 |
| 5 | 2 | 0.5 | 4 | 4 | 2 | 0.5 | 4 | 4 |
| $\infty$ | 2 | 0.5 | 4 | 4 | 2 | 0.5 | 4 | 4 |

$$\theta^{(1)}_{x,\mathrm{mi}} = \frac{T_\mathrm{m}(x) - T_\mathrm{wi}}{T_\mathrm{e} - T_\mathrm{wi}}$$

$$\theta^{(1)}_{x,\mathrm{me}} = \frac{T_\mathrm{m}(x) - T_\mathrm{we}}{T_\mathrm{e} - T_\mathrm{we}}$$

$T_\mathrm{m}(x)$ = average temperature in the cross-section at position $x$.

(b) Case I′: $T_\mathrm{wi} = \mathrm{const}$, $T_\mathrm{we} = \mathrm{const}$, $T_\mathrm{wi} \neq T_\mathrm{we} \neq T_\mathrm{e}$ [31]:

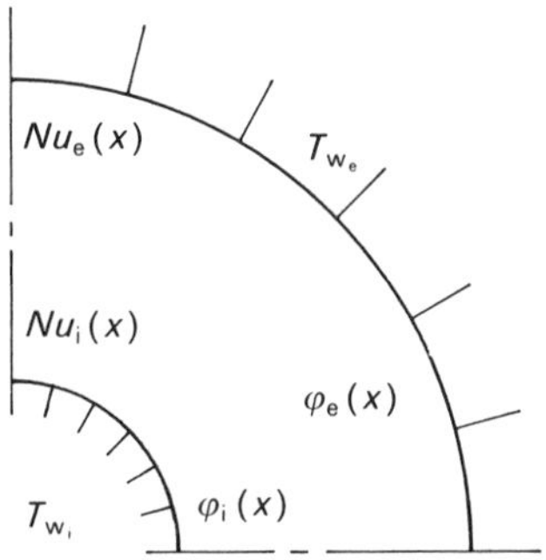


$$T_\mathrm{m}(x) = (T_\mathrm{wi} - T_\mathrm{e})\theta^{(1)}_{x,\mathrm{mi}} + (T_\mathrm{we} - T_\mathrm{e})\theta^{(1)}_{x,\mathrm{me}} + T_\mathrm{e}$$

$$\varphi_\mathrm{i}(x) = \frac{\lambda}{D_\mathrm{h}}[(T_\mathrm{wi} - T_\mathrm{e})\varphi^{(1)}_{x,\mathrm{ii}} + (T_\mathrm{we} - T_\mathrm{e})\varphi^{(1)}_{x,\mathrm{ie}}]$$

$$\varphi_\mathrm{e}(x) = \frac{\lambda}{D_\mathrm{h}}[(T_\mathrm{wi} - T_\mathrm{e})\varphi^{(1)}_{x,\mathrm{ei}} + (T_\mathrm{we} - T_\mathrm{e})\varphi^{(1)}_{x,\mathrm{ee}}]$$

$$Nu_\mathrm{i}(x) = \frac{(T_\mathrm{wi} - T_\mathrm{we})\varphi^{(1)}_{x,\mathrm{ii}} + (T_\mathrm{we} - T_\mathrm{e})\varphi^{(1)}_{x,\mathrm{ie}}}{(T_\mathrm{wi} - T_\mathrm{e})(1 - \theta^{(1)}_{x,\mathrm{mi}}) - (T_\mathrm{we} - T_\mathrm{e})\theta^{(1)}_{x,\mathrm{me}}}$$

$$Nu_\mathrm{e}(x) = \frac{(T_\mathrm{wi} - T_\mathrm{e})\varphi^{(1)}_{x,\mathrm{ei}} + (T_\mathrm{we} - T_\mathrm{e})\varphi^{(1)}_{x,\mathrm{ee}}}{(T_\mathrm{we} - T_\mathrm{e})(1 - \theta^{(1)}_{x,\mathrm{me}} - (T_\mathrm{we} - T_\mathrm{e})\theta^{(1)}_{x,\mathrm{mi}}}$$

$\theta_{x,\mathrm{mi}}$, $\theta_{x,\mathrm{me}}$, $\varphi_{x,\mathrm{ii}}$ and $\varphi_{x,\mathrm{ee}}$ are given in the above tables.

$$\varphi_{x,\mathrm{ei}} = -\theta_{x,\mathrm{mi}} Nu_{x,\mathrm{ei}}$$

$$\varphi_{x,\mathrm{ie}} = -\theta_{x,\mathrm{me}} Nu_{x,\mathrm{ie}} \qquad \varphi = -D_\mathrm{h}\frac{\partial \theta}{\partial n}$$

(c) Case II: $\varphi_{\mathrm{wi}} = \mathrm{const}$, $\varphi_{\mathrm{we}} = 0$ or $\varphi_{\mathrm{wi}} = 0$, $\varphi_{\mathrm{we}} = \mathrm{const}$:

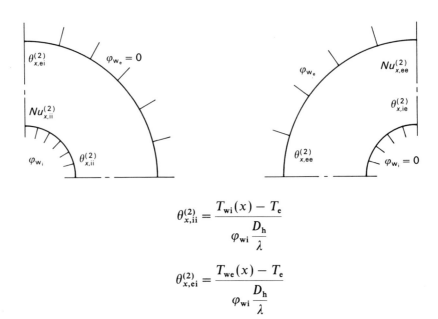

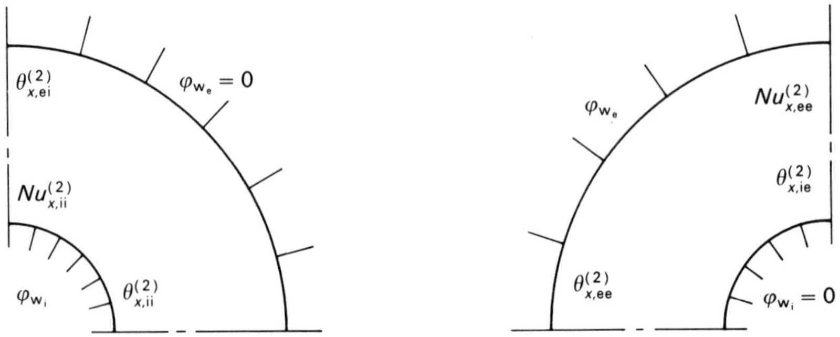


$$\theta^{(2)}_{x,\mathrm{ii}} = \frac{T_{\mathrm{wi}}(x) - T_{\mathrm{e}}}{\varphi_{\mathrm{wi}} \dfrac{D_{\mathrm{h}}}{\lambda}}$$

$$\theta^{(2)}_{x,\mathrm{ei}} = \frac{T_{\mathrm{we}}(x) - T_{\mathrm{e}}}{\varphi_{\mathrm{wi}} \dfrac{D_{\mathrm{h}}}{\lambda}}$$

$T_{\mathrm{e}}$ = fluid temperature at heater entry (fluid assumed isothermal).

| $x^*$ | $\theta_{x,\text{ii}}^{(2)}$ | $\theta_{x,\text{ei}}^{(2)}$ | $Nu_{x,\text{ii}}^{(2)}$ | $\theta_{x,\text{ee}}^{(2)}$ | $\theta_{x,\text{ie}}^{(2)}$ | $Nu_{x,\text{ee}}^{(2)}$ |
|---|---|---|---|---|---|---|
| | | | $Y = 0.02$ | | | |
| $5 \times 10^{-5}$ | | | | 0.0279 | | 36.16 |
| $1 \times 10^{-4}$ | 0.0115 | | 86.9 | 0.0354 | | 28.58 |
| $5 \times 10^{-4}$ | 0.0164 | | 61.2 | 0.0622 | | 16.59 |
| $1 \times 10^{-3}$ | 0.0186 | | 54.01 | 0.0799 | | 13.16 |
| $5 \times 10^{-3}$ | 0.0243 | | 41.85 | 0.1462 | | 7.9 |
| $1 \times 10^{-2}$ | 0.0270 | $2 \times 10_{\text{w}}^{-6}$ | 38.09 | 0.1930 | $1 \times 10^{-4}$ | 6.50 |
| $5 \times 10^{-2}$ | 0.0341 | $1.57 \times 10^{-3}$ | 33.13 | 0.4019 | 0.0784 | 4.86 |
| $1 \times 10^{-1}$ | 0.0384 | 0.0053 | 32.73 | 0.6031 | 0.2648 | 4.74 |
| 0.5 | 0.0698 | 0.0367 | 32.705 | 2.172 | 1.833 | 4.734 |
| 1 | 0.109 | 0.076 | 32.705 | 4.133 | 3.794 | 4.734 |
| $\infty$ | $\infty$ | $\infty$ | 32.705 | $\infty$ | $\infty$ | 4.734 |
| | | | $Y = 0.05$ | | | |
| $5 \times 10^{-5}$ | | | | 0.0276 | | 36.54 |
| $1 \times 10^{-4}$ | 0.0173 | | 58.0 | 0.035 | | 28.89 |
| $5 \times 10^{-4}$ | 0.0260 | | 38.6 | 0.062 | | 16.77 |
| $1 \times 10^{-3}$ | 0.0303 | | 33.17 | 0.0789 | | 13.31 |
| $5 \times 10^{-2}$ | 0.0423 | | 24.20 | 0.1442 | | 7.99 |
| $1 \times 10^{-2}$ | 0.0484 | $5 \times 10^{-6}$ | 21.52 | 0.1901 | $1.08 \times 10^{-4}$ | 6.58 |
| $5 \times 10^{-1}$ | 0.0648 | 0.0039 | 18.09 | 0.3939 | 0.07743 | 4.92 |
| $1 \times 10^{-1}$ | 0.0741 | 0.0129 | 17.83 | 0.5893 | 0.2589 | 4.8 |
| 0.5 | 0.1514 | 0.0891 | 17.811 | 2.1134 | 1.7822 | 4.792 |
| 1 | 0.2466 | 0.1843 | 17.811 | 4.018 | 3.687 | 4.792 |
| $\infty$ | $\infty$ | $\infty$ | 17.811 | $\infty$ | $\infty$ | 4.792 |
| | | | $Y = 0.1$ | | | |
| $5 \times 10^{-5}$ | 0.0173 | | 58.02 | 0.0273 | | 36.94 |
| $1 \times 10^{--4}$ | 0.0212 | | 47.27 | 0.0346 | | 29.21 |
| $5 \times 10^{-4}$ | 0.0335 | | 30.03 | 0.0608 | | 16:97 |
| $1 \times 10^{-3}$ | 0.0404 | | 24.96 | 0.07779 | | 13.47 |
| $5 \times 10^{-2}$ | 0.0602 | | 17.12 | 0.1419 | | 8.08 |
| $1 \times 10^{--2}$ | 0.0707 | $1.2 \times 10^{-5}$ | 14.90 | 0.1867 | $1.16 \times 10^{-4}$ | 6.65 |
| $5 \times 10^{-1}$ | 0.1006 | 0.0075 | 12.13 | 0.3834 | 0.0746 | 4.96 |
| $1 \times 10^{-1}$ | 0.1203 | 0.0248 | 11.92 | 0.5702 | 0.2479 | 4.84 |
| 0.5 | 0.2658 | 0.1702 | 11.906 | 2.0250 | 1,702 | 4.834 |
| 1 | 0.4476 | 0.3520 | 11.906 | 3.8432 | 3,5201 | 4.834 |
| $\infty$ | $\infty$ | $\infty$ | 11.906 | $\infty$ | $\infty$ | 4.834 |

| $x^*$ | $\theta^{(2)}_{x,\mathrm{ii}}$ | $\theta^{(2)}_{x,\mathrm{ei}}$ | $Nu^{(2)}_{x,\mathrm{ii}}$ | $\theta^{(2)}_{x,\mathrm{ee}}$ | $\theta^{(2)}_{x,\mathrm{ie}}$ | $Nu^{(2)}_{x,\mathrm{ee}}$ |
|---|---|---|---|---|---|---|
| | | | $Y = 0.25$ | | | |
| $5\times10^{-5}$ | 0.0211 | | 47.48 | 0.0267 | | 37.71 |
| $1\times10^{-4}$ | 0.0263 | | 38.1 | 0.0338 | | 29.84 |
| $5\times10^{-4}$ | 0.0436 | | 23.15 | 0.0592 | | 17.36 |
| $1\times10^{-3}$ | 0.0539 | | 18.84 | 0.0757 | | 13.79 |
| $5\times10^{-3}$ | 0.0868 | | 12.08 | 0.1368 | | 8.28 |
| $1\times10^{-2}$ | 0.1059 | $3\times10^{-5}$ | 10.22 | 0.1790 | $1.2\times10^{-4}$ | 6.80 |
| $5\times10^{-2}$ | 0.166 | 0.0165 | 7.94 | 0.3583 | 0.066 | 5.04 |
| $1\times10^{-1}$ | 0.2088 | 0.0546 | 7.76 | 0.5235 | 0.2183 | 4.91 |
| 0.5 | 0.529 | 0.3744 | 7.753 | 1.8039 | 1.4978 | 4.905 |
| 1 | 0.929 | 0.7744 | 7.753 | 3.4039 | 3.0978 | 4.905 |
| $\infty$ | $\infty$ | $\infty$ | 7.753 | $\infty$ | $\infty$ | 4.905 |
| | | | $Y = 0.5$ | | | |
| $5\times10^{-5}$ | 0.0233 | | 42.96 | 0.026 | | 38.67 |
| $1\times10^{-4}$ | 0.0293 | | 34.23 | 0.0329 | | 30.63 |
| $5\times10^{-4}$ | 0.0498 | | 20.35 | 0.0573 | | 17.86 |
| $1\times10^{-3}$ | 0.0625 | | 16.36 | 0.0731 | | 14.21 |
| $5\times10^{-2}$ | 0.1054 | | 10.13 | 0.1303 | | 8.55 |
| $1\times10^{-2}$ | 0.1319 | $5.2\times10^{-5}$ | 8.43 | 0.169 | $1.05\times10^{-4}$ | 7.03 |
| $5\times10^{-1}$ | 0.2241 | 0.0243 | 6.35 | 0.3262 | 0.0549 | 5.19 |
| $1\times10^{-1}$ | 0.2948 | 0.0906 | 6.19 | 0.4649 | 0.1816 | 5.05 |
| 0.5 | 0.8285 | 0.6239 | 6.181 | 1.5319 | 1.2478 | 5.037 |
| 1 | 1,4951 | 1,2906 | 6.181 | 2.8652 | 2.5811 | 5.037 |
| $\infty$ | $\infty$ | $\infty$ | 6.181 | $\infty$ | $\infty$ | 5.037 |
| | | | $Y = 1$ | | | |
| $5\times10^{-5}$ | 0.0249 | | 40.26 | 0.0249 | | 40.26 |
| $1\times10^{-4}$ | 0.0315 | | 31.95 | 0.0315 | | 31.95 |
| $5\times10^{-4}$ | 0.0543 | | 18.75 | 0.0543 | | 18.75 |
| $1\times10^{-3}$ | 0.0688 | | 14.97 | 0.0688 | | 14.97 |
| $5\times10^{-2}$ | 0.1201 | | 9.08 | 0.1201 | | 9.08 |
| $1\times10^{-2}$ | 0.1535 | $8\times10^{-5}$ | 7.49 | 0.1535 | $8\times10^{-4}$ | 7.49 |
| $5\times10^{-1}$ | 0.2803 | 0.0411 | 5.55 | 0.2803 | 0.0411 | 5.55 |
| $1\times10^{-1}$ | 0.3854 | 0.1361 | 5.395 | 0.3854 | 0.1361 | 5.395 |
| 0.5 | 1.1857 | 0.9357 | 5.385 | 1,1857 | 0.9357 | 5.385 |
| $\infty$ | $\infty$ | $\infty$ | 5.385 | $\infty$ | $\infty$ | 5.385 |

$$\theta^{(2)}_{x,\mathrm{mi}} = \frac{T_\mathrm{m}(x) - T_\mathrm{wi}}{T_\mathrm{e} - T_\mathrm{wi}} = \frac{4Y}{1+Y}x^*$$

$$\theta^{(2)}_{x,\mathrm{me}} = \frac{T_\mathrm{m}(x) - T_\mathrm{we}}{T_\mathrm{e} - T_\mathrm{we}} = \frac{Y}{1+Y}x^*$$

(d) Case II′: $\varphi_{wi} \neq \varphi_{we} \neq 0$:

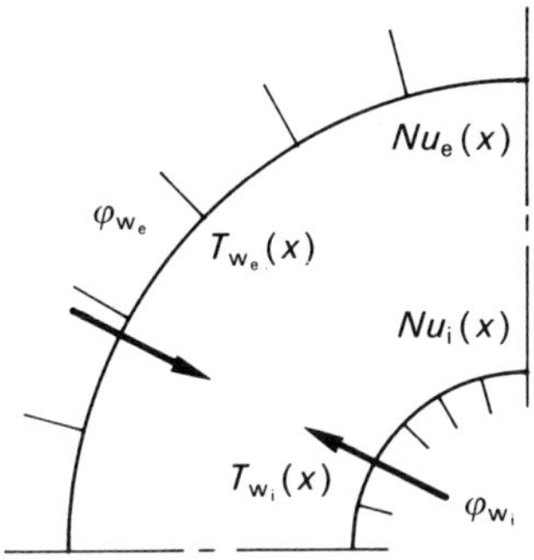


$$T_m(x) = \frac{D_h}{\lambda}[\varphi_{wi}\theta^{(2)}_{x,mi} + \varphi_{we}\theta^{(2)}_{x,me}] + T_e$$

$$T_{wi}(x) = \frac{D_h}{\lambda}[\varphi_{wi}\theta^{(2)}_{x,ii} + \varphi_{we}\theta^{(2)}_{x,ie}] + T_e$$

$$T_{we}(x) = \frac{D_h}{\lambda}[\varphi_{wi}\theta^{(2)}_{x,ei} + \varphi_{we}\theta^{(2)}_{x,ee}] + T_e$$

$$Nu_i(x) = \frac{\varphi_{wi}}{\varphi_{wi}[\theta^{(2)}_{x,ii} - \theta^{(2)}_{x,mi}] - [\theta^{(2)}_{x,me} - \theta^{(2)}_{x,ie}]\varphi_{we}}$$

$$Nu_e(x) = \frac{\varphi_{we}}{\varphi_{we}[\theta^{(2)}_{x,ee} - \theta^{(2)}_{x,me}] - [\theta^{(2)}_{x,mi} - \theta^{(2)}_{x,ei}]\varphi_{wi}}$$

(e) Case III: $T_{wi} = \text{const}$, $\varphi_{we} = 0$ or $\varphi_{wi} = 0$, $T_{we} = \text{const}$ [31]:

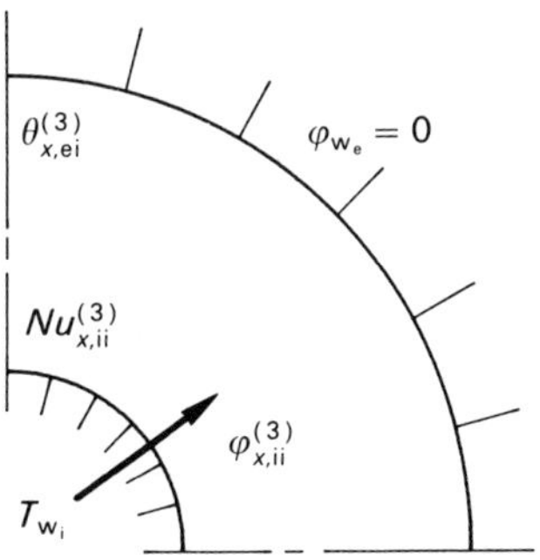


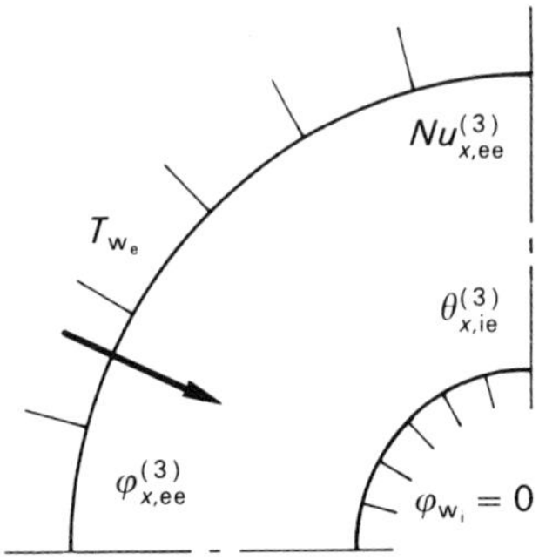


$$\theta^{(3)}_{x,ei} = \frac{T_{we}(x) - T_e}{T_{wi} - T_e} \qquad \theta^{(3)}_{x,ie} = \frac{T_{wi}(x) - T_e}{T_{we} - T_e}$$

| $x^*$ | $\varphi^{(3)}_{x,ii}$ | $\theta^{(3)}_{x,ei}$ | $\theta^{(3)}_{x,mi}$ | $Nu^{(3)}_{x,ii}$ | $\varphi^{(3)}_{x,ee}$ | $\theta^{(3)}_{x,ie}$ | $\theta^{(3)}_{x,me}$ | $Nu^{(3)}_{x,ee}$ |
|---|---|---|---|---|---|---|---|---|
| | | | | $Y = 0.02$ | | | | |
| 0.01 | 35.394 | 0.00012 | 0.0331 | 36.78 | 3.3810 | 0.00118 | 0.2562 | 5.22 |
| 0.05 | 28.207 | 0.05729 | 0.1341 | 32.57 | 1.5241 | 0.34276 | 0.6219 | 4.03 |
| 0.1 | 24.667 | 0.16699 | 0.2374 | 32.34 | 0.6889 | 0.69722 | 0.8275 | 3.994 |
| 0.5 | 8.941 | 0.69793 | 0.7235 | 32.337 | 0.00131 | 0.99942 | 0.9997 | 3.993 |
| 1 | 2.516 | 0.91501 | 0.9222 | 32.337 | 0.0000 | 1.00 | 1.00 | 3.993 |
| ∞ | 0 | 1 | 1 | 32.337 | 0 | 1 | 1 | 3.993 |
| | | | | $Y = 0.05$ | | | | |
| 0.01 | 19.405 | 0.00020 | 0.0456 | 20.33 | 3.9517 | 0.00140 | 0.2525 | 5.29 |
| 0.05 | 14.605 | 0.07801 | 0.1698 | 17.59 | 1.5705 | 0.3431 | 0.6163 | 4.09 |
| 0.1 | 12.273 | 0.21677 | 0.2972 | 17.461 | 0.7177 | 0.6944 | 0.8231 | 4.06 |
| 0.5 | 3.2443 | 0.79290 | 0.81419 | 17.460 | 0.0015 | 0.9994 | 0.9996 | 4.057 |
| 1 | 0.6151 | 0.96074 | 0.96477 | 17.460 | 0.0000 | 1.00 | 1.00 | 4.057 |
| ∞ | 0 | 1 | 1 | 17.460 | 0 | 1 | 1 | 4.057 |
| | | | | $Y = 0.1$ | | | | |
| 0.01 | 12.920 | 0.00032 | 0.0612 | 13.76 | 4.0422 | 0.00153 | 0.2454 | 5.36 |
| 0.05 | 9.160 | 0.10117 | 0.2136 | 11.65 | 1.6427 | 0.33688 | 0.6042 | 4.15 |
| 0.1 | 7.366 | 0.26918 | 0.363 | 11.562 | 0.7697 | 0.6840 | 0.8129 | 4.114 |
| 0.5 | 1.3705 | 0.86398 | 0.8814 | 11.560 | 0.00194 | 0.9992 | 0.9995 | 4.114 |
| 1 | 0.1675 | 0.98337 | 0.9855 | 11.560 | 0.00 | 1.00 | 1.00 | 4.114 |
| ∞ | 0 | 1 | 1 | 11.560 | 0 | 1 | 1 | 4.114 |
| | | | | $Y = 0.25$ | | | | |
| 0.01 | 8.2382 | 0.00058 | 0.0923 | 9.08 | 4.2773 | 0.00159 | 0.2248 | 5.52 |
| 0.05 | 5.2488 | 0.14493 | 0.2935 | 7.43 | 1.8451 | 0.31296 | 0.5679 | 4.27 |
| 0.1 | 3.8763 | 0.36106 | 0.4742 | 7.372 | 0.9280 | 0.6489 | 0.7807 | 4.233 |
| 0.5 | 0.3664 | 0.93959 | 0.9503 | 7.371 | 0.0041 | 0.9984 | 0.9990 | 4.232 |
| 1 | 0.0192 | 0.99683 | 0.9974 | 7.371 | 0.0000 | 1.00 | 1.00 | 4.232 |
| ∞ | 0 | 1 | 1 | 7.371 | 0 | 1 | 1 | 4.232 |
| | | | | $Y = 0.5$ | | | | |
| 0.01 | 6.3404 | 0.00087 | 0.125 | 7.246 | 4.6214 | 0.00147 | 0.1979 | 5.76 |
| 0.05 | 3.6492 | 0.18808 | 0.3692 | 5.79 | 2.1535 | 0.2797 | 0.5180 | 4.47 |
| 0.1 | 2.4675 | 0.44405 | 0.5700 | 5.739 | 1.1814 | 0.5989 | 0.7333 | 4.430 |
| 0.5 | 0.1156 | 0.97394 | 0.9798 | 5.738 | 0.0105 | 0.9964 | 0.9976 | 4.429 |
| 1 | 0.0025 | 0.99943 | 0.9996 | 5.738 | 0.00003 | 0.99999 | 0.999 | 4.429 |
| ∞ | 0 | 1 | 1 | 5.738 | 0 | 1 | 1 | 4.429 |
| | | | | $Y = 1$ | | | | |
| 0.01 | 5.2421 | 0.00119 | 0.163 | 6.26 | 5.2421 | 0.00119 | 0.1625 | 6.26 |
| 0.05 | 2.7028 | 0.23561 | 0.4487 | 4.902 | 2.7028 | 0.23561 | 0.4487 | 4.902 |
| 0.1 | 1.6468 | 0.52780 | 0.6612 | 4.861 | 1.6468 | 0.52780 | 0.6612 | 4.861 |
| 0.5 | 0.0337 | 0.99033 | 0.9931 | 4.861 | 0.0337 | 0.99033 | 0.9931 | 4.861 |
| ∞ | 0 | 1 | 1 | 4.861 | 0 | 1 | 1 | 4.861 |

$$\theta^{(3)}_{x,mi} = \frac{T_m(x) - T_{wi}}{T_e - T_{wi}}$$

$$\theta^{(3)}_{x,me} = \frac{T_m(x) - T_{we}}{T_e - T_{we}}$$

(f) Case IV: $\varphi_{wi} = \text{const}$, $T_{we} = T_e$ or $T_{wi} = T_e$, $\varphi_{we} = \text{const}$ [31]:

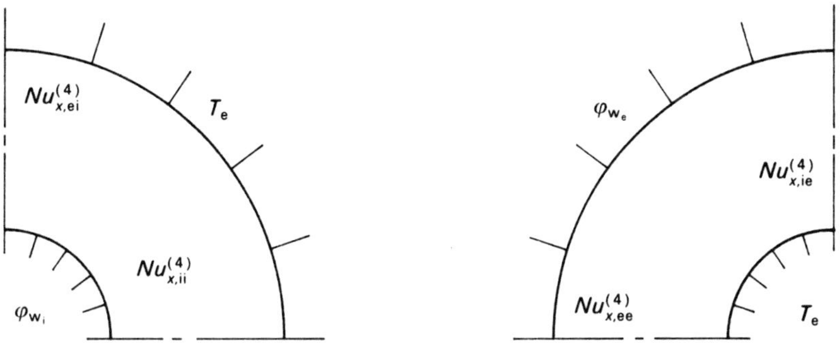


$$\theta^{(4)}_{x,mi} = \frac{T_m(x) - T_e}{\varphi_{wi} D_h/\lambda}$$

$$\theta^{(4)}_{x,me} = \frac{T_m(x) - T_e}{\varphi_{we} D_h/\lambda}$$

| $x^*$ | $\theta^{(4)}_{x,mi}$ | $Nu^{(4)}_{x,mi}$ | $Nu^{(4)}_{x,ei}$ | $\theta^{(4)}_{x,me}$ | $Nu^{(4)}_{x,ee}$ | $Nu^{(4)}_{x,ie}$ |
|---|---|---|---|---|---|---|
| | | | $Y = 0.02$ | | | |
| 0.01 | 0.00078 | 38.09 | 0.03 | 0.03922 | 6.81 | |
| 0.05 | 0.00347 | 32.69 | 1.98 | 0.19608 | 4.94 | 15.13 |
| 0.1 | 0.00527 | 31.26 | 2.65 | 0.37264 | 4.53 | 22.76 |
| 0.5 | 0.00678 | 30.181 | 2.948 | 1.1911 | 3.375 | 29.338 |
| 1 | 0.00678 | 30.179 | 2.948 | 1.5257 | 3.057 | 29.995 |
| $\infty$ | 0.00678 | 30.179 | 2.948 | 1.6568 | 2.948 | 30.179 |
| | | | $Y = 0.05$ | | | |
| 0.01 | 0.00191 | 21.52 | 0.037 | 0.03751 | 6.552 | |
| 0.05 | 0.00842 | 17.78 | 2.038 | 0.18544 | 4.807 | 8.355 |
| 0.1 | 0.0128 | 16.8 | 2.713 | 0.3477 | 4.379 | 12.44 |
| 0.5 | 0.0166 | 16,060 | 3.018 | 1.0081 | 3.289 | 15.728 |
| 1 | 0.0166 | 16.059 | 3.019 | 1.2005 | 3.067 | 16,006 |
| $\infty$ | 0.0166 | 16.058 | 3.019 | 1.2455 | 3.019 | 16.058 |
| | | | $Y = 0.1$ | | | |
| 0.01 | 0.00364 | 14.90 | 0.042 | 0.03625 | 6.646 | 0.034 |
| 0.05 | 0.01609 | 11.86 | 2.093 | 0.17599 | 4.836 | 5.739 |
| 0.1 | 0.02463 | 11.07 | 2.778 | 0.32377 | 4.350 | 8.309 |
| 0.5 | 0.03229 | 10.46 | 3.095 | 0.8385 | 3.271 | 10.30 |
| 1 | 0.03231 | 10.459 | 3.095 | 0.9418 | 3.116 | 10.442 |
| $\infty$ | 0.03231 | 10.459 | 3.095 | 0.9561 | 3,095 | 10.459 |
| | | | $Y = 0.25$ | | | |
| 0.01 | 0.008 | 10.225 | 0.050 | 0.03199 | 6.80 | 0.066 |
| 0.05 | 0.03573 | 7.71 | 2.19 | 0.15253 | 4.88 | 3.80 |
| 0.1 | 0.0558 | 7.04 | 2.91 | 0.27133 | 4.32 | 5.32 |
| 0.5 | 0.07643 | 6.474 | 3.266 | 0.5853 | 3.344 | 6.42 |
| 1 | 0.07652 | 6.471 | 3,267 | 0.61639 | 3.271 | 6.469 |
| $\infty$ | 0.07652 | 6.471 | 3.267 | 0.61810 | 3.267 | 6.471 |
| | | | $Y = 0.50$ | | | |
| 0.01 | 0.013332 | 8.433 | 0.055 | 0.026664 | 7.026 | 0.064 |
| 0.05 | 0.06034 | 6.137 | 2.317 | 0.12508 | 4.998 | 3.007 |
| 0.1 | 0.096747 | 5.503 | 3.095 | 0.21530 | 4.400 | 4.125 |
| 0.5 | 0.14163 | 4.894 | 3.518 | 0.40000 | 3,554 | 4.870 |
| 1 | 0.14203 | 4.889 | 3.520 | 0.40889 | 3.521 | 4.889 |
| $\infty$ | 0.14203 | 4.889 | 3,520 | 0.40908 | 3,520 | 4,889 |
| | | | $Y = 1.0$ | | | |
| 0.01 | 0.02009 | 7.495 | 0.254 | 0.02009 | 7.495 | 0.254 |
| 0.05 | 0.09211 | 5.341 | 2.559 | 0.09211 | 5.341 | 2.559 |
| 0.1 | 0.15285 | 4.723 | 3.453 | 0.15285 | 4.723 | 3,453 |
| 0.5 | 0.24801 | 4,013 | 3.993 | 0.24801 | 4.013 | 3.993 |
| 1 | 0.24998 | 4.000 | 4.000 | 0.24998 | 4.000 | 4.000 |
| $\infty$ | 0.25000 | 4.000 | 4.000 | 0.25000 | 4.000 | 4.000 |

$T_e$ is also the temperature of the fluid entering the annulus.

(g) $\varphi_{\mathrm{we}} = \mathrm{const}$, $T_{\mathrm{wi}} = \mathrm{const}$:

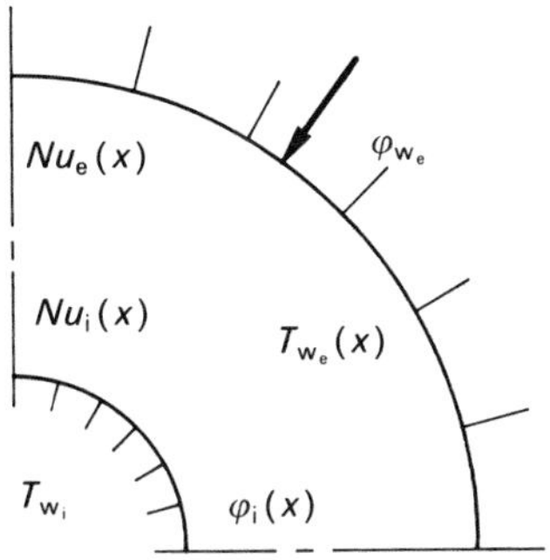


$$T_{\mathrm{we}}(x) = (T_{\mathrm{wi}} - T_{\mathrm{e}})\theta^{(3)}_{x,\mathrm{ei}} + \frac{D_{\mathrm{h}}}{\lambda}\varphi_{\mathrm{we}}\theta^{(4)}_{x,\mathrm{ee}} + T_{\mathrm{e}}$$

$$T_{\mathrm{m}}(x) = (T_{\mathrm{wi}} - T_{\mathrm{e}})\theta^{(3)}_{x,\mathrm{mi}} + \frac{D_{\mathrm{h}}}{\lambda}\varphi_{\mathrm{we}}\theta^{(4)}_{x,\mathrm{me}} + T_{\mathrm{e}}$$

$$\varphi_{\mathrm{wi}}(x) = \frac{\lambda}{D_{\mathrm{h}}}(T_{\mathrm{wi}} - T_{\mathrm{e}})\varphi^{(3)}_{x,\mathrm{ii}} + \varphi_{\mathrm{we}}\varphi^{(4)}_{x,\mathrm{ie}}$$

$$Nu_{\mathrm{i}}(x) = \frac{(T_{\mathrm{wi}} - T_{\mathrm{e}})\varphi^{(3)}_{x,\mathrm{ii}} + \varphi_{\mathrm{we}}\dfrac{D_{\mathrm{h}}}{\lambda}\varphi^{(4)}_{x,\mathrm{ie}}}{(T_{\mathrm{wi}} - T_{\mathrm{e}})(1 - \theta^{(3)}_{x,\mathrm{mi}}) - \varphi_{\mathrm{we}}\dfrac{D_{\mathrm{h}}}{\lambda}\theta^{(4)}_{x,\mathrm{me}}}$$

$$Nu_{\mathrm{e}}(x) = \frac{1}{(\theta^{(4)}_{x,\mathrm{ee}} - \theta^{(4)}_{x,\mathrm{me}}) + \dfrac{(T_{\mathrm{wi}} - T_{\mathrm{e}})\lambda}{\varphi_{\mathrm{we}}D_{\mathrm{h}}}(\theta^{(3)}_{x,\mathrm{ei}} - \theta^{(3)}_{x,\mathrm{mi}})}$$

(h) Thermal development length [32]:

$$L^{*(n)}_{\mathrm{th},j} = \frac{L_{\mathrm{th}}}{D_{\mathrm{h}}Pe_{D_{\mathrm{h}}}}$$

$n$: case under consideration
$j$: wall on which $T_{\mathrm{w}}$ *or* $\varphi_{\mathrm{w}}$ is fixed.

| $Y$ | $L^{*(1)}_{th,i}$ | $L^{*(1)}_{th,e}$ | $L^{*(2)}_{th,i}$ | $L^{*(2)}_{th,e}$ |
|---|---|---|---|---|
| 0.02 | 0.0584 | 0.165 | 0.027 | 0.0390 |
| 0.05 | 0.0649 | 0.146 | 0.0304 | 0.0389 |
| 0.10 | 0.0695 | 0.131 | 0.0333 | 0.0391 |
| 0.25 | 0.0762 | 0.113 | 0.0373 | 0.0401 |
| 0.50 | 0.0824 | 0.100 | 0.0398 | 0.0409 |
| 1.00 | 0.0902 | 0.0902 | 0.0410 | 0.0410 |

| $Y$ | $L^{*(3)}_{th,i}$ | $L^{*(3)}_{th,e}$ | $L^{*(4)}_{th,i}$ | $L^{*(4)}_{th,e}$ |
|---|---|---|---|---|
| 0.02 | 0.0225 | 0.0300 | 0.0796 | 0.0424 |
| 0.05 | 0.0243 | 0.0297 | 0.0949 | 0.664 |
| 0.10 | 0.0256 | 0.0296 | 0.110 | 0.5284 |
| 0.25 | 0.0272 | 0.0296 | 0.131 | 0.377 |
| 0.50 | 0.0283 | 0.0296 | 0.172 | 0.288 |
| 1.00 | 0.0291 | 0.0291 | 0.220 | 0.220 |

1.3.1.3. $V$ AND $T$ PROFILES NOT FULLY DEVELOPED

(a) $T_{wi}$ = const, $T_{we} = T_e$ or $T_{wi} = T_e$, $T_{we}$ = const [33]:

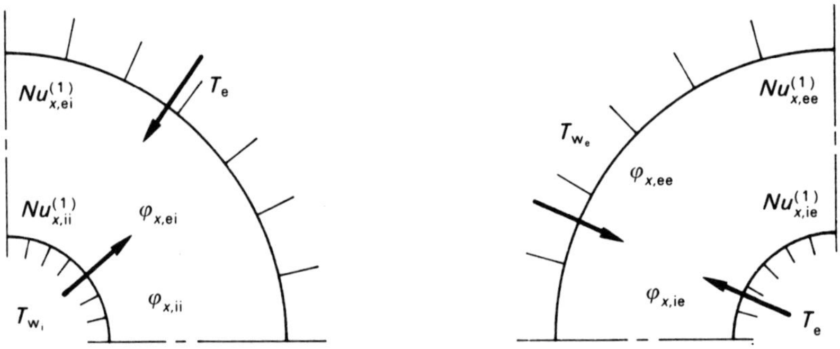


$T_e$: fluid entry temperature

$Pr = 0.72, \qquad Y = 0.25$

| $x^*$ | $\varphi^{(1)}_{x,\mathrm{ii}}$ | $-\varphi^{(1)}_{x,\mathrm{ei}}$ | $\theta^{(1)}_{x,\mathrm{mi}}$ | $Nu^{(1)}_{x,\mathrm{ii}}$ | $Nu^{(1)}_{x,\mathrm{ei}}$ |
|---|---|---|---|---|---|
| 0.0001 | 41.40 | | 0.00615 | 41.65 | |
| 0.0005 | 21.39 | | 0.01482 | 21.71 | |
| 0.001 | 16.48 | | 0.02215 | 16.95 | |
| 0.005 | 9.73 | 0.0001 | 0.05987 | 10.35 | 0.00083 |
| 0.01 | 8.05 | 0.0088 | 0.09475 | 8.89 | 0.0936 |
| 0.05 | 5.31 | 0.641 | 0.253 | 7.11 | 2.53 |
| 0.1 | 4.57 | 0.974 | 0.312 | 6.64 | 3.12 |

| $x^*$ | $\varphi^{(1)}_{x,\mathrm{ee}}$ | $-\varphi^{(1)}_{x,\mathrm{ie}}$ | $\theta^{(1)}_{x,\mathrm{me}}$ | $Nu^{(1)}_{x,\mathrm{ie}}$ | $Nu^{(1)}_{x,\mathrm{ee}}$ |
|---|---|---|---|---|---|
| 0.0001 | 36.86 | | 0.02310 | | 37.73 |
| 0.0005 | 16.91 | | 0.05205 | | 17.84 |
| 0.001 | 12.11 | | 0.0743 | | 13.08 |
| 0.005 | 5.66 | 0.00041 | 0.1718 | 0.0024 | 6.833 |
| 0.01 | 4.12 | 0.0508 | 0.2476 | 0.2050 | 5.474 |
| 0.05 | 1.84 | 2.65 | 0.5350 | 4.947 | 3.965 |
| 0.1 | 1.27 | 3.92 | 0.6362 | 6.154 | 3.485 |

(b) $\varphi_{\mathrm{wi}} = \mathrm{const}$, $\varphi_{\mathrm{we}} = 0$ or $\varphi_{\mathrm{wi}} = 0$, $\varphi_{\mathrm{we}} = \mathrm{const}$. [17]:

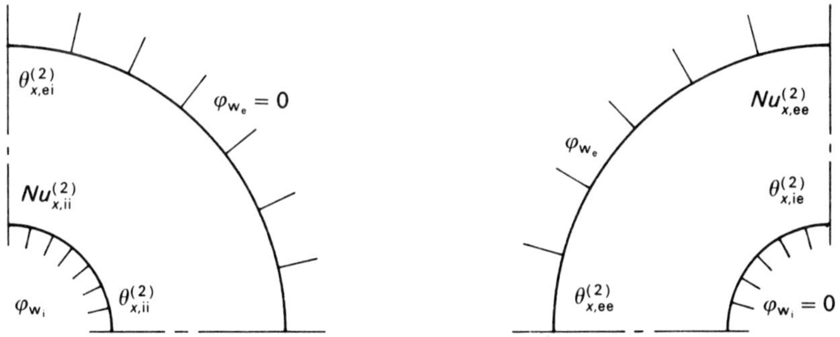

$$Pr = 0.72, \qquad Y = 0.25$$

| $x^*$ | $\theta^{(2)}_{x,ii}$ | $\theta^{(2)}_{x,ei}$ | $Nu^{(2)}_{x,ii}$ | $\theta^{(2)}_{x,ee}$ | $\theta^{(2)}_{x,ie}$ | $Nu^{(2)}_{x,ee}$ |
|---|---|---|---|---|---|---|
| $1 \times 10^{-4}$ | 0.0182 | | 55.4 | 0.0184 | | 52.4 |
| $5 \times 10^{-4}$ | 0.0364 | | 27.8 | 0.0420 | | 24.7 |
| $1 \times 10^{-3}$ | 0.0480 | | 21.2 | 0.0582 | | 18.2 |
| $5 \times 10^{-3}$ | 0.0860 | | 12.20 | 0.1226 | | 9.38 |
| $1 \times 10^{-2}$ | 0.1053 | | 10.28 | 0.1667 | | 7.43 |
| $5 \times 10^{-2}$ | 0.1659 | 0.0165 | 7.94 | 0.3538 | 0.0660 | 5.16 |
| $1 \times 10^{-1}$ | 0.2087 | 0.0546 | 7.77 | 0.5230 | 0.2180 | 4.93 |
| 0.5 | 0.5290 | 0.3745 | 7.75 | 1.804 | 1.498 | 4.91 |
| 1 | 0.9290 | 0.7745 | 7.75 | 3.404 | 3.098 | 4.91 |
| $\infty$ | $\infty$ | $\infty$ | 7.75 | $\infty$ | $\infty$ | 4.91 |

$$\theta^{(2)}_{x,ii} = \frac{T_{wi}(x) - T_e}{\phi_{wi}(D_h/\lambda)}$$

$$\theta^{(2)}_{x,ei} = \frac{T_{we}(x) - T_e}{\phi_{wi}(D_h/\lambda)}$$

(c) $T_{wi} = \text{const}$, $\varphi_{we} = 0$ or $\varphi_{wi} = 0$, $T_{we} = \text{const}$ [39]:

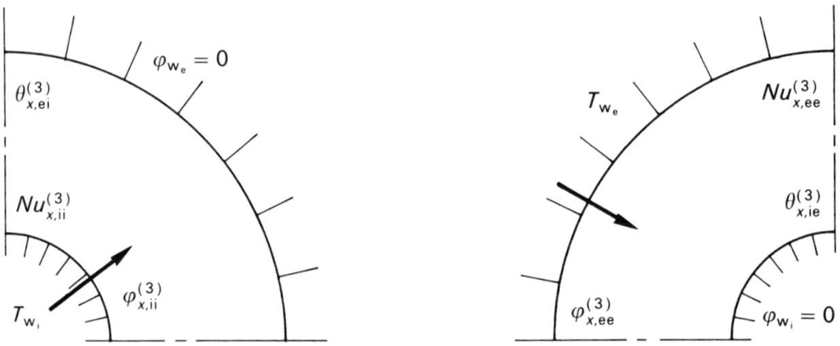

$Pr = 0.72, \qquad Y = 0.25$

| $x^*$ | $\varphi^{(3)}_{x,\mathrm{ii}}$ | $\theta^{(3)}_{x,\mathrm{ei}}$ | $\theta^{(3)}_{x,\mathrm{mi}}$ | $Nu^{(3)}_{x,\mathrm{ii}}$ |
|---|---|---|---|---|
| $1 \times 10^{-4}$ | 41.40 | | 0.00615 | 41.65 |
| $5 \times 10^{-4}$ | 21.39 | | 0.01482 | 21.71 |
| $1 \times 10^{-3}$ | 16.48 | | 0.02215 | 16.85 |
| $5 \times 10^{-3}$ | 9.73 | 0.0001 | 0.05987 | 10.35 |
| $1 \times 10^{-2}$ | 8.05 | 0.00067 | 0.09480 | 8.89 |
| $5 \times 10^{-2}$ | 5.24 | 0.1461 | 0.2941 | 7.424 |
| $1 \times 10^{-1}$ | 3.872 | 0.3615 | 0.4744 | 7.367 |

| $x^*$ | $\varphi^{(3)}_{x,\mathrm{ee}}$ | $\theta^{(3)}_{x,\mathrm{ie}}$ | $\theta^{(3)}_{x,\mathrm{me}}$ | $Nu^{(3)}_{x,\mathrm{ee}}$ |
|---|---|---|---|---|
| $1 \times 10^{-4}$ | 36.86 | | 0.02310 | 37.73 |
| $5 \times 10^{-4}$ | 16.91 | | 0.05205 | 17.84 |
| $1 \times 10^{-3}$ | 12.11 | | 0.07430 | 13.08 |
| $5 \times 10^{-3}$ | 5.663 | 0.00025 | 0.1718 | 6.838 |
| $1 \times 10^{-2}$ | 4.119 | 0.00285 | 0.2477 | 5.474 |
| $5 \times 10^{-2}$ | 1.795 | 0.3294 | 0.5785 | 4.259 |
| $1 \times 10^{-1}$ | 0.9069 | 0.6570 | 0.7857 | 4.231 |

(d) $\varphi_{\mathrm{wi}} = \mathrm{const}$, $T_{\mathrm{we}} = T_{\mathrm{e}}$ or $T_{\mathrm{wi}} = T_{\mathrm{e}}$, $\varphi_{\mathrm{we}} = \mathrm{const}$ [33]:

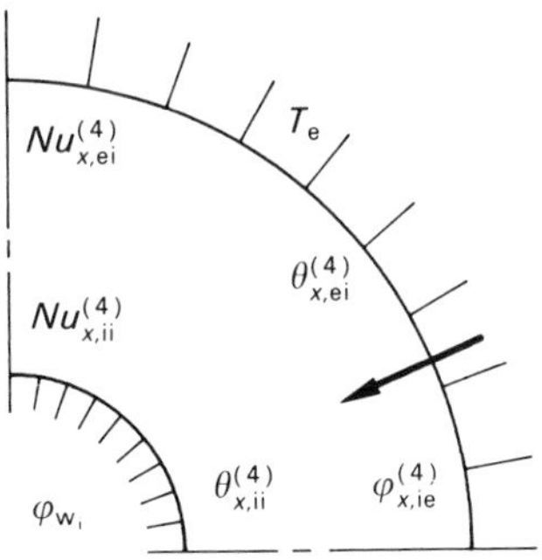


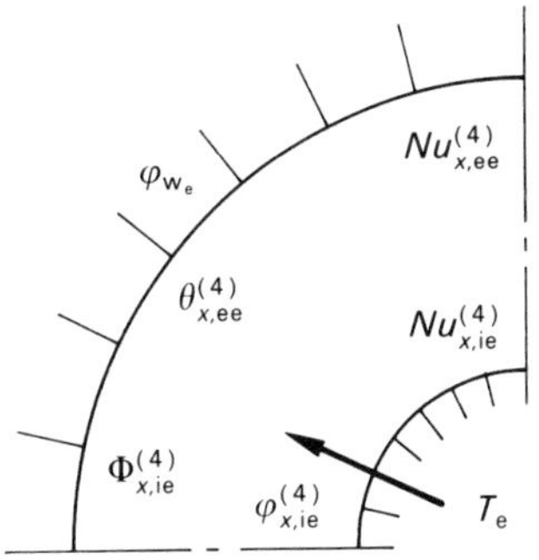

$$Pr = 0.72, \qquad Y = 0.25$$

| $x^*$ | $\theta^{(4)}_{x,\mathrm{ii}}$ | $-\varphi^{(4)}_{x,\mathrm{ei}}$ | $\theta^{(4)}_{x,\mathrm{mi}}$ | $Nu^{(4)}_{x,\mathrm{ii}}$ | $Nu^{(4)}_{x,\mathrm{ei}}$ |
|---|---|---|---|---|---|
| 0.0001 | 0.0181 | | 0.00008 | 55.55 | |
| 0.0005 | 0.0364 | | 0.00040 | 27.76 | |
| 0.01 | 0.0482 | | 0.00080 | 21.12 | |
| 0.005 | 0.0858 | 0.0001 | 0.00399 | 12.23 | 0.001 |
| 0.01 | 0.1061 | 0.00043 | 0.00799 | 10.19 | 0.0537 |
| 0.05 | 0.1655 | 0.07809 | 0.03567 | 7.704 | 2.189 |
| 0.1 | 0.1978 | 0.162 | 0.05570 | 7.038 | 2.908 |

| $x^*$ | $\theta^{(4)}_{x,\mathrm{ee}}$ | $-\varphi^{(4)}_{x,\mathrm{ie}}$ | $\theta^{(4)}_{x,\mathrm{me}}$ | $Nu^{(4)}_{x,\mathrm{ie}}$ | $Nu^{(4)}_{x,\mathrm{ee}}$ |
|---|---|---|---|---|---|
| 0.0001 | 0.0196 | | 0.00032 | | 51.89 |
| 0.0005 | 0.0428 | | 0.00160 | | 24.25 |
| 0.001 | 0.0572 | | 0.0032 | | 17.69 |
| 0.005 | 0.1266 | 0.0001 | 0.01598 | 0.001 | 9.041 |
| 0.01 | 0.1731 | 0.00302 | 0.03196 | 0.0946 | 7.088 |
| 0.05 | 0.3575 | 0.5812 | 0.1525 | 3.811 | 4.879 |
| 0.1 | 0.5027 | 1.443 | 0.2712 | 5.322 | 4.320 |

### *1.3.2 Turbulent flow* [1–3]

$$Nu_{D_\mathrm{h}} = 0.023 Re_{D_\mathrm{h}}^{0.8} Pr^{1/3} \left(\frac{D_\mathrm{e}}{D_\mathrm{i}}\right)^{0.14}$$

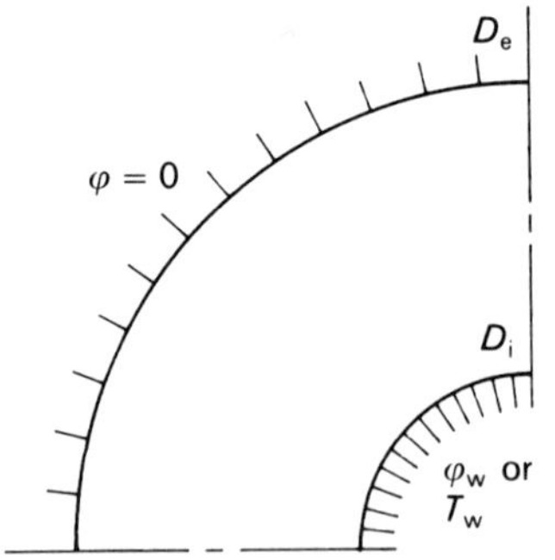

## 1.4 Parallel plates

$$D_h = 2e$$

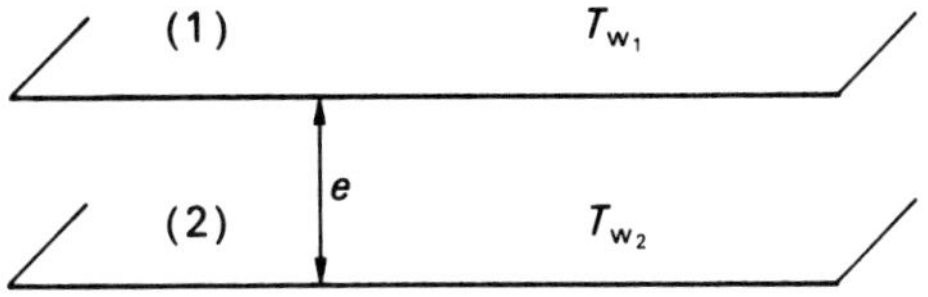


### 1.4.1 Laminar flow [24,25]

$$\frac{L_m}{D_h} = 0.3125 + 0.011 Re_{D_h} \ [34]$$

#### 1.4.1.1 V AND T PROFILES FULLY DEVELOPED

$$T_{w1} \neq T_{w2}$$

$$Nu_{D_h}^{(1)} = Nu_{D_h}^{(2)} = 4$$

$$T_{w1} = T_{w2}$$

$$Nu_{D_h}^{(1)} = Nu_{D_h}^{(2)} = 7.54$$

$$\varphi_{wi} \neq \varphi_{w2}$$

$$Nu_{D_h}^{(1)} = \frac{140}{26 - 9(\varphi_{w2}/\varphi_{w1})}$$

$$Nu_{D_h}^{(2)} = \frac{140}{26 - 9(\varphi_{w1}/\varphi_{w2})}$$

$$\varphi_{w1} = \varphi_{w2} \Rightarrow Nu_{D_h}^{(1)} = Nu_{D_h}^{(2)} = \frac{140}{17} = 8.24$$

$$\varphi_{w1} = 0 \text{ (one wall adiabatic)} \qquad Nu_{D_h}^{(1)} = 0, \qquad Nu_{D_h}^{(2)} = \frac{140}{26} = 5.385$$

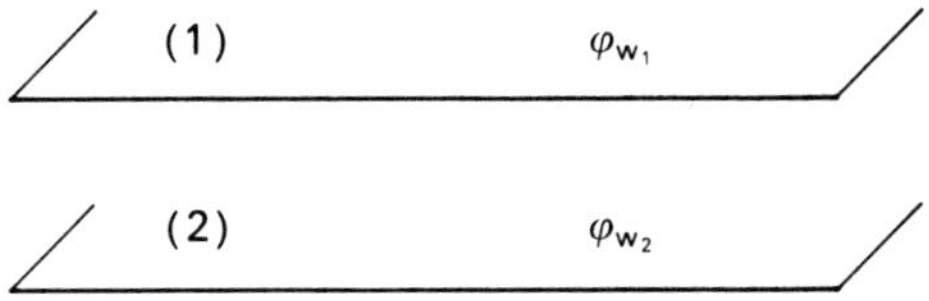

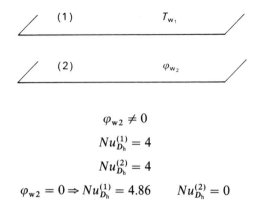

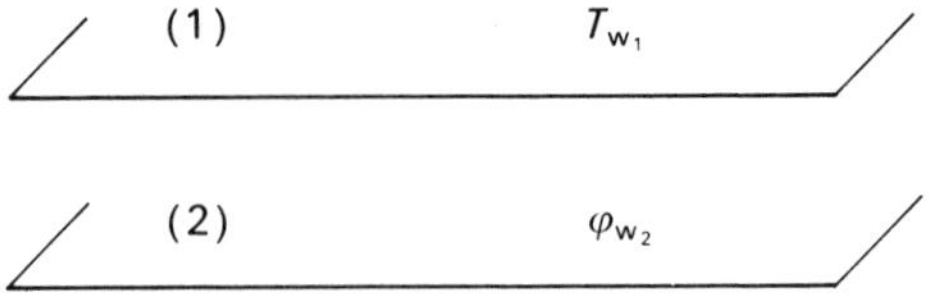


$$\varphi_{w2} \neq 0$$

$$Nu_{D_h}^{(1)} = 4$$

$$Nu_{D_h}^{(2)} = 4$$

$$\varphi_{w2} = 0 \Rightarrow Nu_{D_h}^{(1)} = 4.86 \qquad Nu_{D_h}^{(2)} = 0$$

1.4.1.2 *V* PROFILE FULLY DEVELOPED, *T* PROFILE NOT FULLY DEVELOPED

(a) $T_{w1} = T_{w2} = T_w =$ const: the Graetz–Nusselt problem [1]:

- Local Nusselt number:

$$Nu_{D_h}(x) = \frac{h(x)D_h}{\lambda} = \frac{8}{3}\,\frac{\sum_{n=0}^{\infty} G_n \exp\left[-\frac{32}{3}\lambda_n^2 x^*\right]}{\sum_{n=0}^{\infty} \frac{G_n}{\lambda_n^2} \exp\left[-\frac{32}{3}\lambda_n^2 x^*\right]}$$

$$x^* = \frac{x}{D_h Pe_{D_h}}$$

Approximate expressions ($\pm 1\%$) [14]:

$$Nu_{D_h}(x) = 1.233(x^*)^{-1/3} + 0.4 \qquad \text{for} \quad x^* \leqslant 10^{-3}$$

$$Nu_{D_h}(x) = 7.541 + 6.874(10^3 x^*)^{-0.488}\,\mathrm{e}^{-245x^*} \qquad \text{for} \quad x^* > 10^{-3}$$

- Average between 0 and $x^*$:

$$Nu_{D_h,x} = \frac{\bar{h}D_h}{\lambda} = \frac{1}{4x^*}\ln\left(\frac{1}{3\sum_{n=0}^{\infty} \frac{G_n}{\lambda_n^2}\exp\left[-\frac{32}{3}\lambda_n^2 x^*\right]}\right)$$

Approximate expressions ($\pm 3\%$) [14]:

$$x^* \leqslant 5 \times 10^{-4} \qquad Nu_{D_h,x} = 1.849(x^*)^{-1/3}$$

$$5 \times 10^{-4} < x^* \leqslant 6 \times 10^{-3} \qquad Nu_{D_h,x} = 1.849(x^*)^{-1/3} + 0.6$$

$$x^* > 6 \times 10^{-3} \qquad Nu_{D_h,x} = 7.541 + \frac{0.0235}{x^*}$$

| $n$ | $\lambda_n^2$ | $G_n$ |
|---|---|---|
| 0 | 2.828 | 0.858 |
| 1 | 32.147 | 0.569 |
| 2 | 93.475 | 0.476 |
| 3 | 186.796 | 0.424 |
| 4 | 312.136 | 0.389 |
| 5 | 469.468 | 0.363 |
| 6 | 658.8 | 0.344 |
| 7 | 880.132 | 0.327 |
| 8 | 1133.46 | 0.314 |
| 9 | 1418.8 | 0.302 |

For $n \geqslant 10$,

$$\lambda_n = 4n + \frac{5}{3} \qquad G_n = 1.013\lambda_n^{-1/3}$$

- Thermal development length [14]:

$$\frac{L_{\text{th}}}{D_{\text{h}} Pe_{D_{\text{h}}}} = 8 \times 10^{-3}$$

(b) $\varphi_{\text{w}1} = \varphi_{\text{w}2} = \text{const} = \varphi_{\text{w}}$ [28]:
- Local Nusselt number:

$$Nu_{D_{\text{h}}}(x) = \left[\frac{17}{140} + \frac{1}{4}\sum_{n=1}^{\infty} C_n \exp\left[-\frac{32}{3}\beta_n^2 x^*\right]\right]^{-1}$$

| $n$ | $\beta_n^2$ | $-C_n$ |
|---|---|---|
| 1 | 18.38 | 0.2222 |
| 2 | 68.952 | 0.0725 |
| 3 | 151.551 | 0.0374 |
| 4 | 266.163 | 0.0233 |
| 5 | 412.785 | 0.0161 |
| 6 | 591.409 | 0.0119 |
| 7 | 802.039 | 0.0092 |
| 8 | 1044.67 | 0.0074 |
| 9 | 1319.31 | 0.0061 |
| 10 | 1625.95 | 0.0051 |

for $n > 10$

$$\beta_{\text{n}} = 4n + 1/3 \qquad C_n = -2.401\beta_n^{-5/3}$$

Approximate expressions ($\pm 0.8\%$) [14]:

$x^* \leqslant 2 \times 10^{-4}$ $\quad Nu_{D_h}(x) = 1.490(x^*)^{-1/3}$

$2 \times 10^{-4} < x^* \leqslant 10^{-3}$ $\quad Nu_{D_h}(x) = 1.490(x^*)^{-1/3} - 0.4$

$x^* > 10^{-3}$ $\quad Nu_{D_h}(x) = 8.235 + 8.68(10^3 x^*)^{-0.506} e^{-164x^*}$

- Average between 0 and $x^*$ ($\pm 2.5\%$) [14]:

$x^* \leqslant 10^{-3}$ $\quad Nu_{D_h,x} = \frac{\bar{h}D_h}{\lambda} = 2.236(x^*)^{-1/3}$

$10^{-3} < x^* < 10^{-2}$ $\quad Nu_{D_h,x} = 2.236(x^*)^{-1/3} + 0.9$

$x^* \geqslant 10^{-2}$ $\quad Nu_{D_h,x} = 8.235 + \frac{0.0364}{x^*}$

- Thermal development length [14]:

$$\frac{L_{th}}{D_h Pe_{D_h}} = 0.0115$$

1.4.1.3 $V$ AND $T$ PROFILES NOT FULLY DEVELOPED

(a) $T_{w1} = T_{w2} = T_w = \text{const}$ [35]:

$$Nu_{D_h,x} = \frac{\bar{h}D_h}{\lambda} = 7.55 + \frac{0.024(x^*)^{-1.14}}{1 + 0.0358(x^*)^{0.64} Pr^{0.17}}$$

(b) $\varphi_{w1} = \varphi_{w2} = \varphi_w = \text{const}$ [44]:

| $Pr = 10^{-2}$ | | $Pr = 0.7$ | | $Pr = 10$ | |
|---|---|---|---|---|---|
| $x^*$ | $Nu_\varphi(x)$ | $x^*$ | $Nu_\varphi(x)$ | $x^*$ | $Nu_\varphi(x)$ |
| $10^{-3}$ | 24.5 | $7.14 \times 10^{-4}$ | 21.98 | $5 \times 10^{-5}$ | 50.74 |
| $2 \times 10^{-3}$ | 18.5 | $1.79 \times 10^{-3}$ | 15.11 | $1.25 \times 10^{-4}$ | 34.07 |
| $4 \times 10^{-3}$ | 13.7 | $6.25 \times 10^{-3}$ | 10.03 | $4.38 \times 10^{-4}$ | 20.66 |
| $7 \times 10^{-3}$ | 11.1 | $1.07 \times 10^{-2}$ | 8.90 | $7.5 \times 10^{-4}$ | 17.03 |
| $10^{-2}$ | 10.0 | $2.86 \times 10^{-2}$ | 8.24 | $2 \times 10^{-3}$ | 12.60 |
| $2 \times 10^{-2}$ | 9.0 | $8.93 \times 10^{-2}$ | 8.22 | $6.25 \times 10^{-3}$ | 9.50 |
| $4 \times 10^{-2}$ | 8.5 | 0.143 | 8.22 | $10^{-2}$ | 8.80 |
| $7 \times 10^{-2}$ | 8.3 | | | | |
| $2 \times 10^{-1}$ | 8.23 | | | | |

$$\frac{L_{th}}{D_h Pe_{D_h}} = 0.030 \qquad \text{for } Pr = 10^{-2}$$

$$= 0.017 \qquad \text{for } Pr = 0.7$$

$$= 0.014 \qquad \text{for } Pr = 1$$

$$= 0.012 \qquad \text{for } Pr = 10$$

$$= 0.0115 \qquad \text{for } Pr \to \infty$$

### 1.4.2 *Turbulent flow* [1–3]

- $Pr \geqslant 0.5$ $\quad \varphi_w$ or $T_w$ on both plates:

$$Nu_{D_h} = 0.023 Re_{D_h}^{0.8} Pr^{1/3}$$

- $Pr < 0.02$ $\quad \varphi_w$ on one plate [20], the other insulated:

$$Nu_{D_h} = 5.8 + 0.02 Pe_{D_h}^{0.8}$$

# Section 2

# Forced External Convection

## *2.1 Flat (or small curvature) plate* [1–5]

### *2.1.1 Laminar flow* [37]

$$Nu_x(x) = 0.324 Re_x^{0.5} Pr^{1/3}$$
$$Nu_L = 0.628 Re_L^{0.5} Pr^{1/3}$$

$$Re_x < 3 \times 10^5 \qquad Re_L < 3 \times 10^5$$

$$0.5 \leqslant Pr \leqslant 10$$

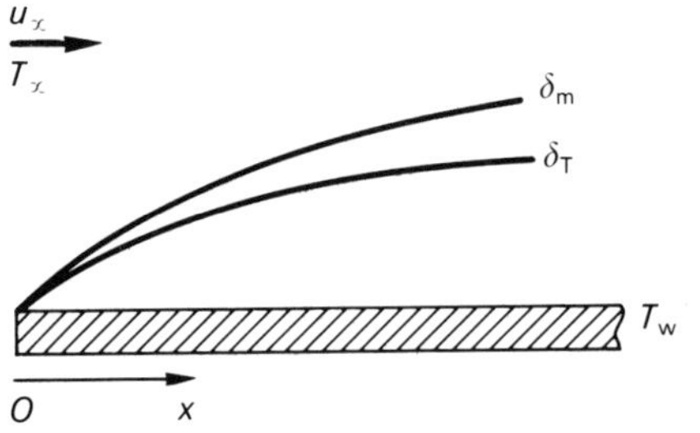


if $Pr > 10$, $\quad Nu_x(x) = 0.339 Re_x^{0.5} Pr^{1/3}$

if $Pr \ll 1$ (liquid metal), $\quad Nu_x(x) = 0.565 Re_x^{0.5} Pr^{1/2}$

$$Nu_L = 1.1\sqrt{(1 - Pr^{1/3}) Pe_L}$$

Physical properties are evaluated at the film temperature: $T_f = (T_w + T_\infty)/2$:

$$Nu_x(x) = \frac{0.324 Re_x^{0.5} Pr^{1/3}}{[1 - (x_0/x)^{3/4}]^{1/3}}$$

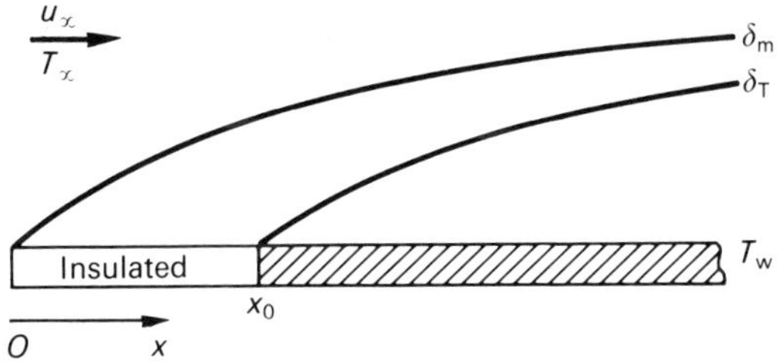


### 2.1.2 Turbulent flow

$$Re_x > 5 \times 10^5$$

$$Nu_x(x) = 0.0288 Re_x^{4/5} Pr^{1/3}$$
$$Nu_L = 0.035 Re_L^{4/5} Pr^{1/3}$$

$$Pr \geqslant 0.5$$

$$\begin{cases} Nu_L = 0.59 Pe_L^{0.61} \\ Pr \ll 1 \text{ (liquid metal)} \end{cases}$$

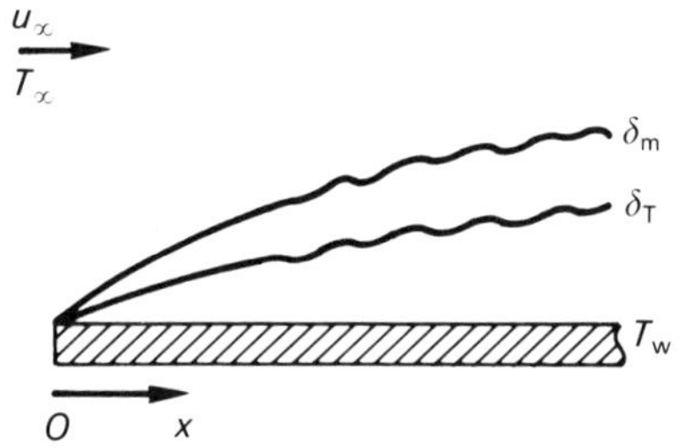


Physical properties are evaluated at the film temperature: $T_f = (T_w + T_\infty)/2$:

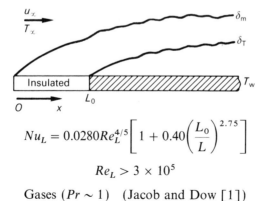

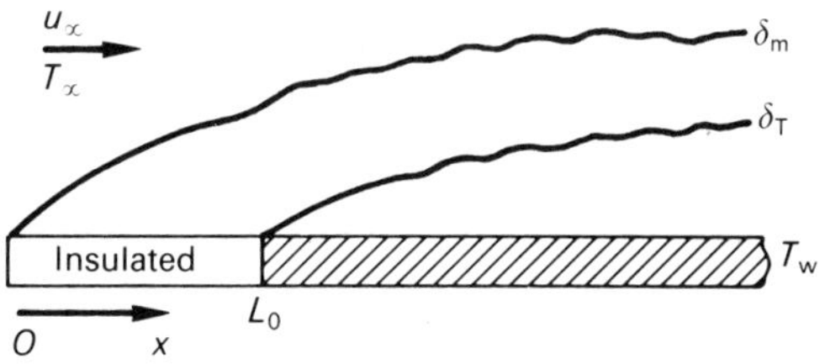


$$Nu_L = 0.0280 Re_L^{4/5}\left[1 + 0.40\left(\frac{L_0}{L}\right)^{2.75}\right]$$

$$Re_L > 3 \times 10^5$$

Gases ($Pr \sim 1$) (Jacob and Dow [1])

For other fluids, $Nu_L$ may be estimated using the same formula, scaling according to $Pr^{1/3}$.

### 2.2 Cylinder with its axis perpendicular to the flow

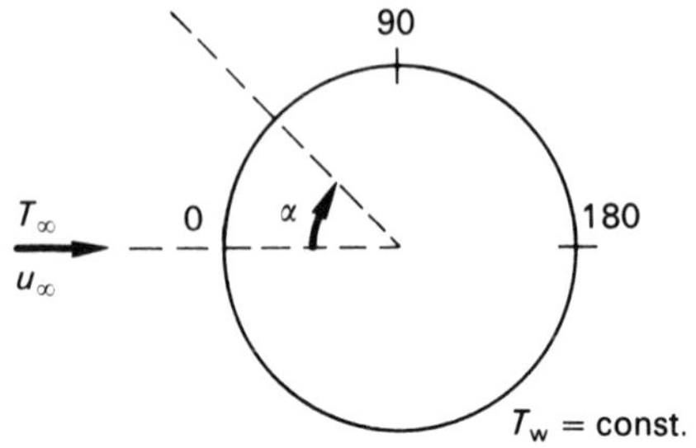


*D*: cylinder diameter

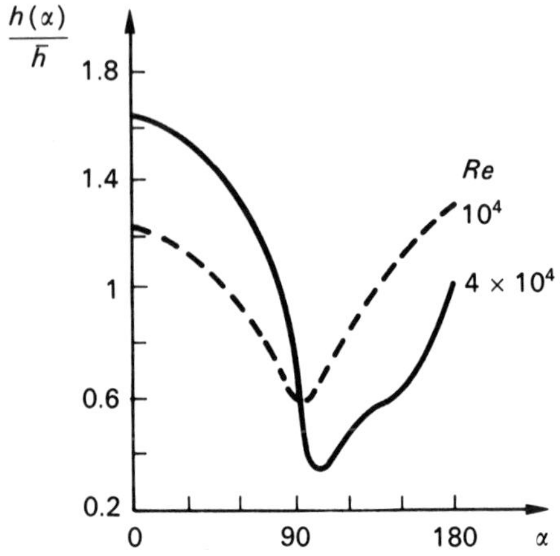


*Pr* = 0.7

Turbulence level below 1%, physical properties evaluated at $(T_w + T_\infty)/2$:

for $0 \leqslant \alpha \leqslant 80°$ [38], $$Nu_D(\alpha) = \frac{h(\alpha)D}{\lambda} = 1.05\left[1 - \left(\frac{\alpha}{90}\right)^3\right] Re_D^{1/2} Pr^{1/3}$$

$$Nu_D = \frac{\bar{h}D}{\lambda} = C Re_D^n$$

| $Re_D$ | $n$ | $C$ Gas | $C$ Liquid |
|---|---|---|---|
| 1–4 | 0.330 | 0.891 | $0.989 Pr^{1/3}$ |
| 4–40 | 0.385 | 0.821 | $0.911 Pr^{1/3}$ |
| 40–4 000 | 0.466 | 0.615 | $0.683 Pr^{1/3}$ |
| 4 000–40 000 | 0.618 | 0.174 | $0.193 Pr^{1/3}$ |
| 40 000–250 000 | 0.805 | 0.0239 | $0.0266 Pr^{1/3}$ |

Cylinders of various cross-section, $T_w = \text{const}$:

$$Nu_d = \frac{\bar{h}d}{\lambda} = CRe_d^m Pr^{0.35}$$

| | $Re_d$ | $C$ | $m$ | Ref. |
|---|---|---|---|---|
| 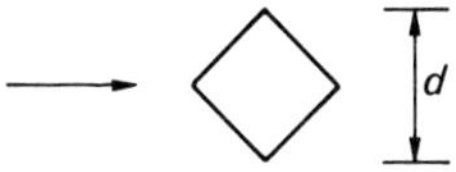  | $5 \times 10^3$ to $10^5$ | 0.25 | 0.588 | 39 |
| 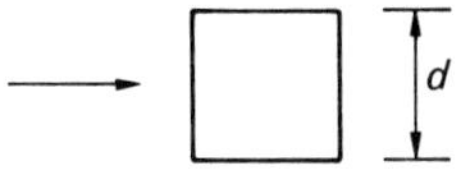  | $2.5 \times 10^3$ to $8 \times 10^3$<br>$5 \times 10^3$ to $10^5$ | 0.180<br>0.104 | 0.699<br>0.675 | 40<br>40 |
|   | $2.5 \times 10^3$ to $1.5 \times 10^4$ | 0.25 | 0.612 | 40 |
| 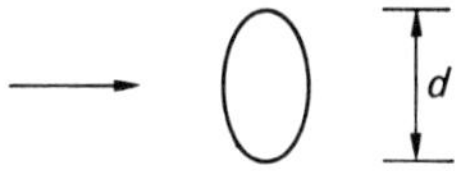  | $3 \times 10^3$ to $1.5 \times 10^4$ | 0.096 | 0.804 | 40 |
| 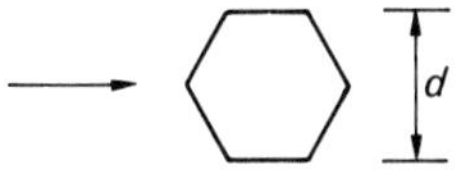  | $5 \times 10^3$ to $10^5$ | 0.156 | 0.638 | 39 |
| 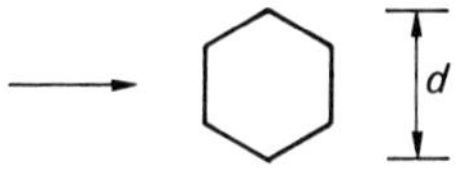  | $5 \times 10^3$ to $1.95 \times 10^4$<br>$1.95 \times 10^4$ to $10^5$ | 0.162<br>0.0395 | 0.638<br>0.782 | 39 |
| 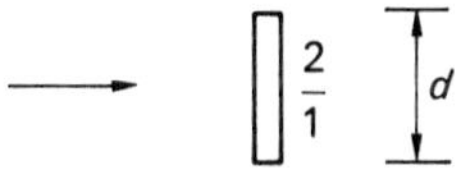  | $3 \times 10^3$ to $2 \times 10^4$ | 0.264 | 0.66 | |
| 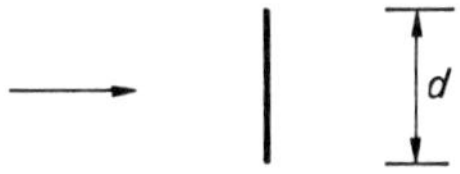  | $4 \times 10^3$ to $1.5 \times 10^4$ | 0.232 | 0.731 | 40 |
| 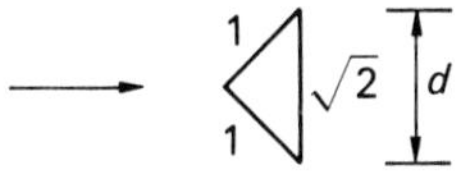  | $3 \times 10^3$ to $2 \times 10^4$ | 0.246 | 0.61 | |

### 2.3 Sphere

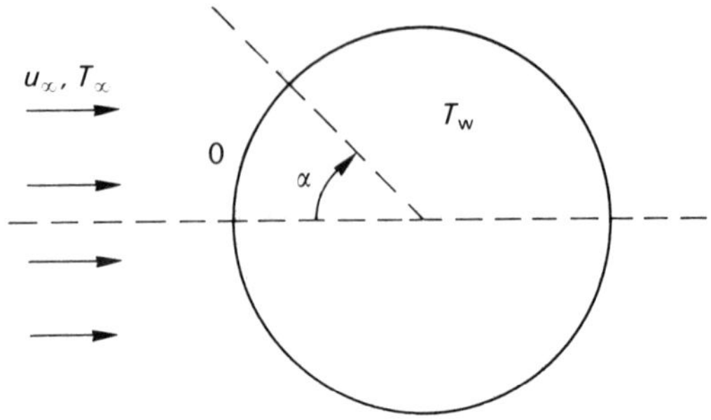


Natural convection may be neglected so long as $Gr_D/Re_D^2 < 1.5$.

- At the stagnation point ($\alpha = 0$):

$$Nu_D(\alpha = 0) = 0.37 Re_D^{0.53}$$

$$4.4 \times 10^4 \leqslant Re_D \leqslant 1.5 \times 10^5$$

- Average Nusselt number:

$$Nu_D = 2.0 + 0.60 Pr^{1/3} Re_D^{1/2}$$

$$1 \leqslant Re_D \leqslant 7 \times 10^4 \qquad \text{Froessling [2]}$$

$$0.6 \leqslant Pr \leqslant 400$$

$$Nu_D = 2.0 + 1.3 Pr^{0.15} + 0.66 Pr^{0.31} Re_D^{1/2}$$

$$10 \leqslant Re_D \leqslant 10^5 \qquad \text{Kramers [41]}$$

$$0.6 \leqslant Pr \leqslant 380$$

$$Nu_D = 2.0 + 0.03 Pr^{0.33} Re_D^{0.54} + 0.35 Pr^{0.36} Re_D^{0.58}$$

for any value of $Re_D$ (Katsnel'son and Timofeyeva [4]).

$$Nu_D = 2.2 Pr + 0.48 Re_D^{1/2} Pr$$

$$1 < Re_D < 25 \qquad \text{Kreith [2]}$$

$$Nu_D = 2 + Pr^{0.4}(0.4 Re_D^{1/2} + 0.06 Re_D^{0.67})(\mu_w/\mu_\infty)^{0.25}$$

$$3.5 \leqslant Re_D \leqslant 7.6 \times 10^4 \qquad \text{Whitaker [3]}$$

$$0.6 \leqslant Pr \leqslant 380$$

$\mu_\infty$ and $\mu_w$ are the viscosities at the temperature far away $T_\infty$ and at the wall temperature $T_w$.

# Section 3

## Free External Convection

### 3.1 Vertical flat plate

#### 3.1.1 Isothermal plate $T_w = const.$

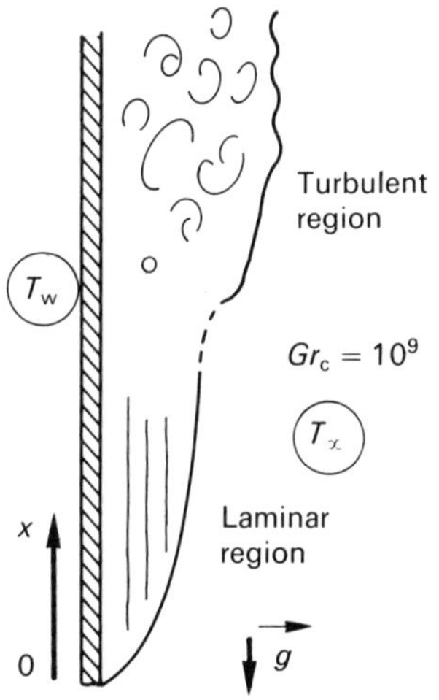


Laminar flow:

$$Nu_x(x) = \frac{h(x)x}{\lambda} = 0.39 Ra_x^{1/4}$$

$$Nu_L = 0.59 Ra_L^{1/4} \qquad \text{McAdams [5]}$$

$$10^4 < Ra_L < 10^9$$

Physical properties taken at:

$$T_f = \frac{T_w + T_\infty}{2}$$

Turbulent flow:

$$Nu_x(x) = 0.12 Ra_x^{1/3}$$

$$Nu_L = 0.13 Ra_L^{1/3}$$

$$10^9 < Ra_L < 10^{12}$$

$$Ra_x = \frac{g\beta \Delta T x^3}{av} \qquad \Delta T = |T_w - T_\infty|$$

$\beta$ = coefficient of thermal expansion

Physical properties evaluated at: $T_f = (T_w + T_\infty)/2$.

### 3.1.2 Plate at $\varphi_w$ = const:

Laminar flow:

$$Nu_L = \frac{0.670 Ra_L^{*1/4}}{[1 + (0.437/Pr)^{9/16}]^{4/9}} \qquad \text{Churchill and Ozoé [40]}$$

Turbulent flow:

$$Nu_L = 0.13 Ra_L^{*1/3}$$

$$10^9 < Ra_L < 10^{12} \qquad \text{McAdams [5]}$$

or for water:

$$Nu_x(x) = 0.658 Ra_x^{*0.22}$$

$$10^{13} < Ra_x^* < 10^{16}$$

$$Ra_x^* = \frac{g\beta\varphi_w x^4}{\lambda a v}$$

## 3.2 Inclined plate

### 13.2.1 $T_w$ = const:

$0 < \alpha < 45°$: the same formulae as for the vertical plate, replacing $Gr_x$ by $Gr_x \cos \alpha$.

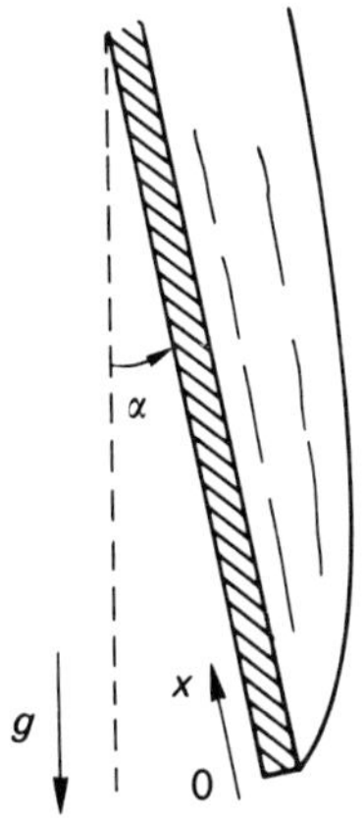


*3.2.2* $\varphi_w = const$ *Vliet and Ross* [42]:

Laminar flow:

$$Nu_x(x) = 0.55(Ra_x^* \cos\alpha)^{0.2}$$

$$Ra_x^* = \frac{g\beta\varphi_w x^4}{\lambda a \nu}$$

Turbulent flow:

- hot face upwards:

$$Nu_x(x) = 0.17 Ra_x^{*0.25}$$

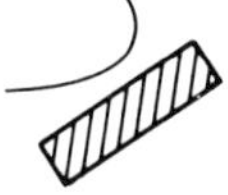

- hot face downwards:

$$Nu_x(x) = 0.17[Ra_x^* \cos^2\alpha]^{0.25}$$

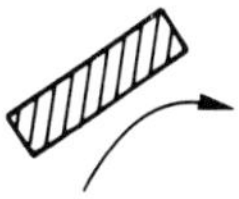

### 3.3 Horizontal plate

- Heated face below:

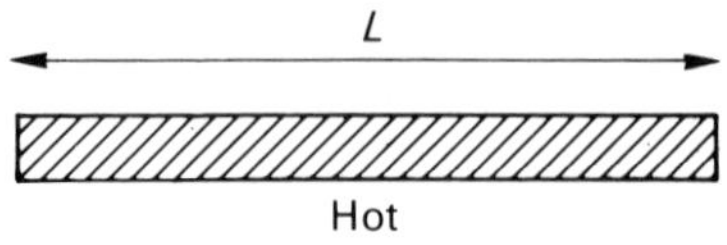


$$Nu_L = 0.27 Ra_L^{1/4} \qquad \text{McAdams [5]}$$
$$3 \times 10^5 \leqslant Ra_L \leqslant 3 \times 10^{10}$$
$$Nu_L = 0.58 Ra_L^{1/5} \qquad \text{Fuji and Imura [42]}$$
$$10^6 \leqslant Ra_L \leqslant 10^{11}$$

- Heated face above:

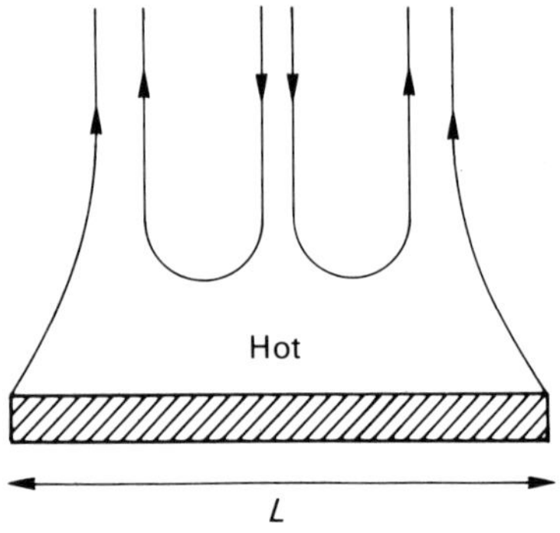


Laminar flow:

$$Nu_L = 0.54 Ra_L^{1/4}$$
$$10^5 < Ra_L < 10^7$$

Turbulent flow Fischenden and Saunders [42]:

$$Nu_L = 0.14 Ra_L^{1/3}$$
$$10^7 < Ra_L < 3 \times 10^{10}$$

### 3.4 Horizontal isothermal cylinder

$$30\ \mu\text{m} < D < 10\ \text{cm}$$

$$0.5 < Pr < 10^3$$

$$Nu_D = 0.53 Ra_D^{1/4} \qquad \text{McAdams [5]}$$

$$10^4 < Ra_D < 10^9$$

$$Nu_D = 0.13 Ra_D^{1/3}$$
$$10^9 < Ra_D < 10^{12}$$

$T_w - T_\infty$ varying by a few degrees at 1500°

$$Nu_D = A Ra_D^n \qquad \text{Morgan [42]}$$

| $A$ | $n$ | $Ra_D$ |
|---|---|---|
| 0.675 | 0.058 | $10^{-10} \leftrightarrow 10^{-2}$ |
| 1.020 | 0.148 | $10^{-2} \leftrightarrow 10^{2}$ |
| 0.850 | 0.188 | $10^{2} \leftrightarrow 10^{4}$ |
| 0.480 | 0.250 | $10^{4} \leftrightarrow 10^{7}$ |
| 0.125 | 0.333 | $10^{7} \leftrightarrow 10^{12}$ |

For $\varphi_w = \text{const}$, $Ra_D$ may be replaced by $Ra_D^* = \dfrac{g\beta\varphi_w D^4}{\lambda a v}$.

### 3.5 Isothermal sphere

$$Nu_D = 2 + 0.43 Ra_D^{1/4} \qquad \text{Yagamata [42]}$$

$$1 < Ra_D < 10^5 \qquad \text{air}$$

$$Nu_D = 2 + 0.56\left[\frac{Ra_D Pr}{0.846 + Pr}\right]^{1/4} \qquad \text{Raithby Hollands [43]}$$

$$Ra_D > 1 \qquad \text{for any } Pr$$

### 3.6 Vertical cylinder

If $D/L \geqslant 35Gr_L^{-1/4}$, the Nusselt number is given by the formulae for a vertical flat plate. Otherwise:

$$Nu_D \exp\left(\frac{-2}{Nu_D}\right) = 0.6\left(\frac{D}{L}\right)^{1/4} Ra_L^{1/4} \qquad \text{Elenbaas [42]}$$

where $L$ is the cylinder length.

# Section 4

## Free Internal Convection

$\lambda_e$: equivalent thermal conductivity defined as:

- rectangular gap: $$\varphi_{FC} = \frac{\lambda_e}{\delta}(T_h - T_c);$$
- cylindrical annular gap: $$\varphi_{FC} = \frac{2\pi\lambda_e}{\ln(d_e/d_i)}(T_e - T_i);$$
- spherical gap: $$\varphi_{FC} = \frac{2\pi\lambda_e(T_e - T_i)}{1/d_i - 1/d_e};$$

$\varphi_{FC}$: flux exchanged between the two walls as a result of free convection.

### *4.1 Horizontal isothermal parallel plates. Horizontal rectangular cell*

#### 4.1.1 Heated from above

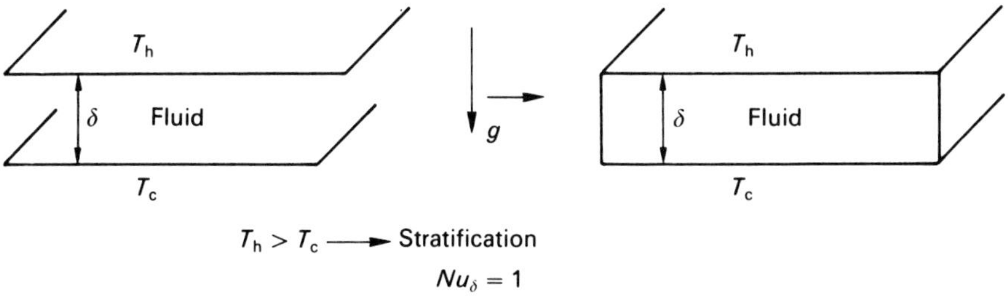

### 4.1.2 Heated from below

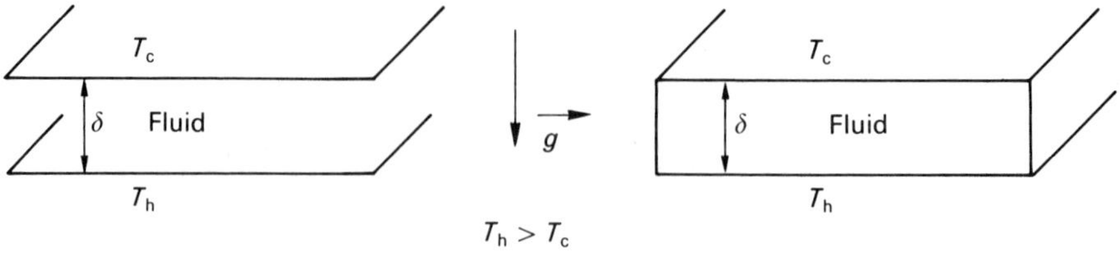


$$Ra_\delta = \frac{g\beta(T_h - T_c)\delta^3}{a\nu}$$

- Gas:

$$0.5 \leqslant Pr \leqslant 2$$

$$Ra_\delta \leqslant 1708 \qquad \frac{\lambda_e}{\lambda} = 1$$

$\lambda_e$: equivalent conductivity

$$1708 < Ra_\delta < 7000 \qquad \frac{\lambda_e}{\lambda} = 0.059 Ra_\delta^{0.4}$$

$$7000 < Ra_\delta < 3.2 \times 10^5 \qquad \frac{\lambda_e}{\lambda} = 0.212 Ra_\delta^{1/4}$$

$$Ra_\delta > 3.2 \times 10^5 \qquad \frac{\lambda_e}{\lambda} = 0.061 Ra_\delta^{1/3}$$

- Liquid:

$$1 \leqslant Pr \leqslant 5000$$

$$1708 < Ra_\delta < 6000 \qquad \frac{\lambda_e}{\lambda} = 0.012 Ra_\delta^{0.6}$$

$$6000 < Ra_\delta < 37\,000 \qquad \frac{\lambda_e}{\lambda} = 0.375 Ra_\delta^{0.2}$$

$$1 \leqslant Pr \leqslant 20$$

$$3.7 \times 10^4 < Ra_\delta < 10^8 \qquad \frac{\lambda_e}{\lambda} = 0.13 Ra_\delta^{0.3}$$

$$Ra_\delta > 10^8 \qquad \frac{\lambda_e}{\lambda} = 0.057 Ra_\delta^{1/3}$$

The physical properties of gases and liquids are evaluated at $(T_h + T_c)/2$.

## 4.2 Vertical rectangular cell

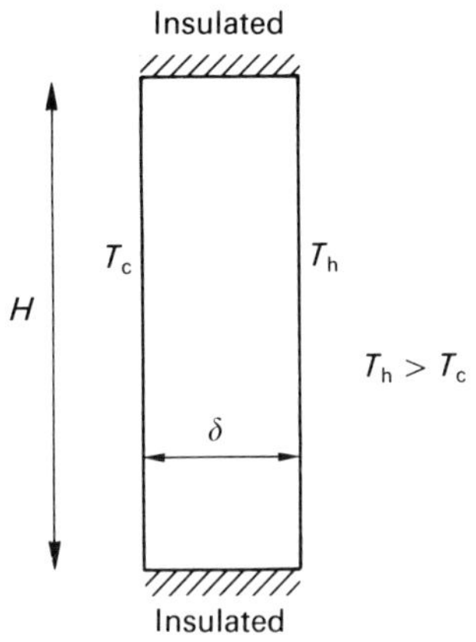


$$\text{Aspect ratio:} \quad Al = \frac{H}{\delta}$$

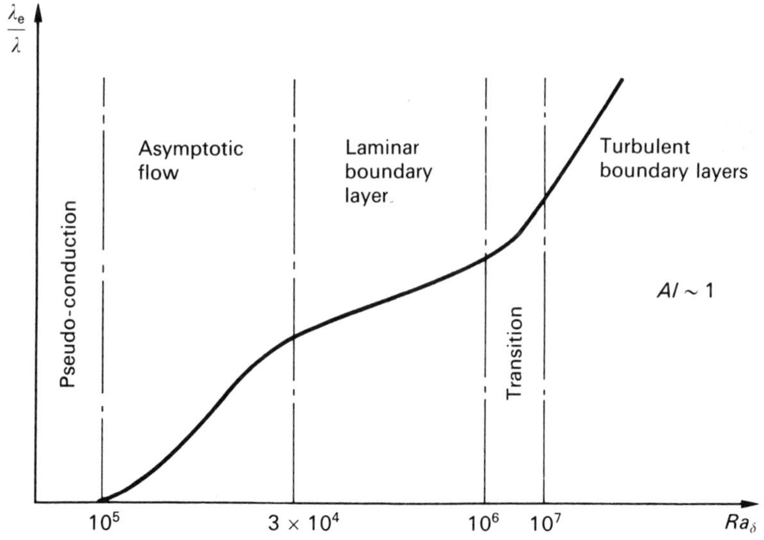


(McGregor and Emery [42])

For any aspect ratio $Al$, the boundary between pseudo-conduction and established free convection is:

$$Ra_\delta = 500Al \qquad \text{Batchelor [42]}$$

Physical properties are evaluated at the average temperature $T_h + T_c/2$. There are many formulae which do not always give consistent values. We may, none the less, mention the following:

- For gases: Mull, Reiher [42]:

$$0.5 \leqslant Pr \leqslant 2 \qquad 3 < Al < 42$$

$$Ra_\delta < 2 \times 10^3 \qquad \rightarrow \frac{\lambda_e}{\lambda} = 1$$

$$6 \times 10^3 < Ra_\delta < 2 \times 10^5 \qquad \frac{\lambda_e}{\lambda} = 0.197 Ra_\delta^{1/4} Al^{-1/9}$$

$$2 \times 10^5 < Ra_\delta < 1.1 \times 10^7 \qquad \frac{\lambda_e}{\lambda} = 0.073 Ra_\delta^{1/3} Al^{-1/9}$$

- For liquids: McGregor *et al.* [42]:

$$1 \leqslant Pr \leqslant 20\,000 \qquad 10 < Al < 40$$

$$10^4 < Ra_\delta < 10^7 \qquad \frac{\lambda_e}{\lambda} = 0.42 Ra_\delta^{1/4} Al^{-0.3} Pr^{0.012}$$

$$1 \leqslant Pr \leqslant 20 \qquad 1 < Al < 40$$

$$10^6 < Ra_\delta < 10^9 \qquad \frac{\lambda_e}{\lambda} = 0.046 Ra_\delta^{1/3}$$

which also apply for $\varphi_w = \text{const.}$

## 4.3 Inclined rectangular cell

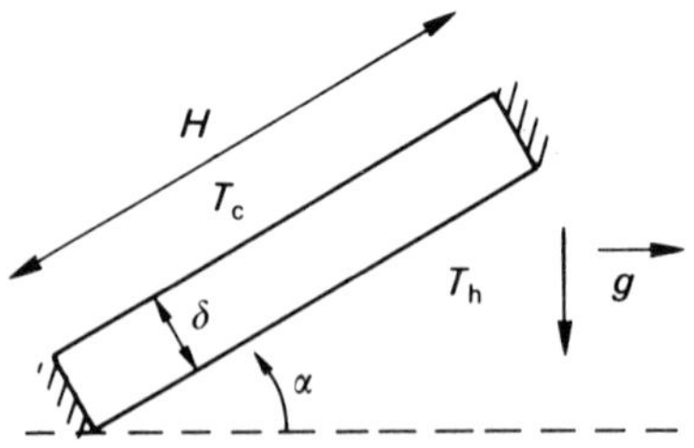


$\alpha^*$: angle for which $\lambda_e/\lambda$ is minimum

$$\text{(Catton [44])} \quad Al = \frac{H}{\delta}$$

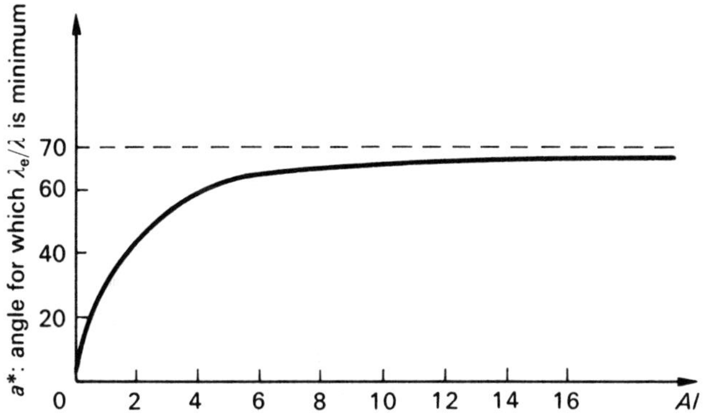


- $0 < \alpha < \alpha^*$ $Al > 10$ Raithby and Hollands [43]:

$$Nu = \frac{\lambda_e}{\lambda} = 1 + 1.44\left[1 - \frac{1708}{Ra_\delta \cos\alpha}\right]^*\left[1 - \frac{(\sin 1.8\alpha)^{1.6}1708}{Ra_\delta \cos\alpha}\right] + \left[\left(\frac{Ra_\delta \cos\alpha}{5830}\right)^{1/3} - 1\right]^*$$

where

$$[X]^* = \frac{|X| + X}{2}$$

- $0 < \alpha < \alpha^*$ $Al < 10$:

$$Nu = \frac{\lambda_e}{\lambda} = Nu(0^\circ)\left[\frac{Nu(90^\circ)}{Nu(0^\circ)}\right]^{\alpha/\alpha^*}(\sin\alpha^*)^{(1/4)(\alpha/\alpha^*)}$$

$Nu(0^\circ)$ and $Nu(90^\circ)$ are the values of Nusselt number with the cell horizontal ($\alpha = 0^\circ$) and vertical ($\alpha = 90^\circ$).

- $\alpha^* < \alpha < 90^\circ$ (Ayyasnamy and Catton [27]), for any $Al$:

$$Nu(\alpha) = \frac{\lambda_e}{\lambda} = Nu(90^\circ)(\sin\alpha)^{1/4}$$

- $90^\circ < \alpha < 180^\circ$ (Arnold *et al.* [42]), for any $Al$:

$$Nu(\alpha) = 1 + [Nu(90^\circ) - 1]\sin\alpha$$

Physical properties are evaluated at $(T_h + T_c)/2$.

### 4.4 Parallel vertical plates

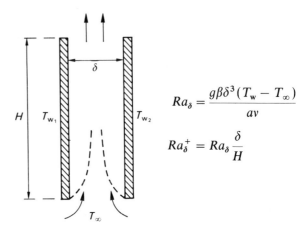

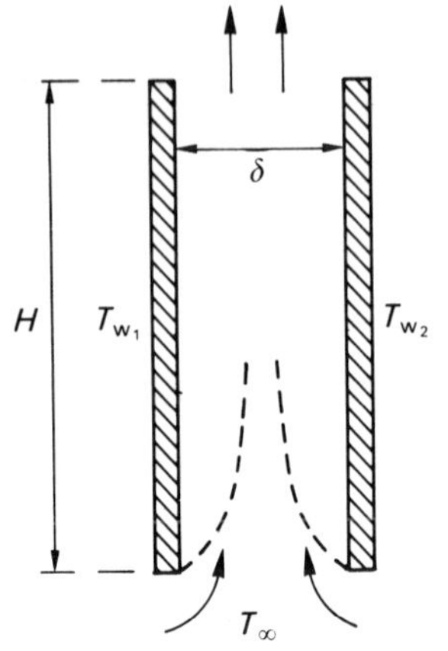


$$Ra_\delta = \frac{g\beta\delta^3(T_w - T_\infty)}{a\nu}$$

$$Ra_\delta^+ = Ra_\delta \frac{\delta}{H}$$

The physical properties of air are evaluated at $T_w$.

- $T_{w1} = T_{w2} = T_w$:

$$Nu_\delta = \frac{\bar{h}\delta}{\lambda} = \frac{Ra_\delta^+}{24}\left[1 - \exp\left(-\frac{35}{Ra_\delta^+}\right)^{3/4}\right] \qquad \text{(Elenbaas [42])}$$

$$\text{for low } Ra_\delta^+ \,(<10), \qquad Nu_\delta = \frac{Ra_\delta^+}{24}$$

$$\text{for high } Ra_\delta^+ \,(>300), \qquad Nu_\delta = \frac{35^{3/4}}{24} Ra_\delta^{1/4} = 0.60 Ra_\delta^{1/4}$$

An alternative expression (Churchill and Usagi [42]):

$$Nu_\delta = \left[\frac{576}{Ra_\delta^{+2}} + \frac{2.873}{\sqrt{Ra_\delta^+}}\right]^{-1/2} \qquad \text{for any } Ra_\delta^+$$

Optimum thickness (for maximum transfer between the plates and air): $\delta_{opt}$ is given by:

$$Ra_{\delta_{opt}}^+ = (2.714)^4 = \frac{g\beta(T_w - T_\infty)\delta_{opt}^4}{a\nu H}$$

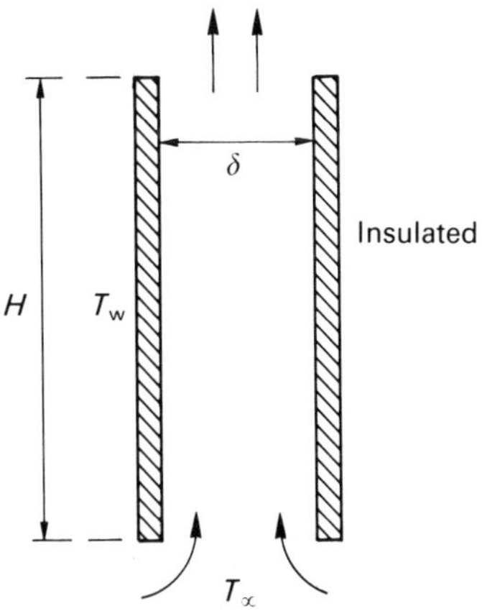


Physical properties evaluated at $T_w$ (Churchill and Usagi [42]):

$$Nu_\delta = \left[ \frac{144}{Ra_\delta^{+2}} + \frac{2.873}{\sqrt{Ra_\delta^+}} \right]^{-1/2} \qquad \text{for any } Ra_\delta^+$$

Optimum thickness: $Ra^+_{\delta_{opt}} = (2.154)^4$

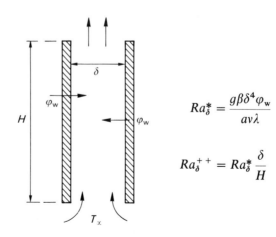

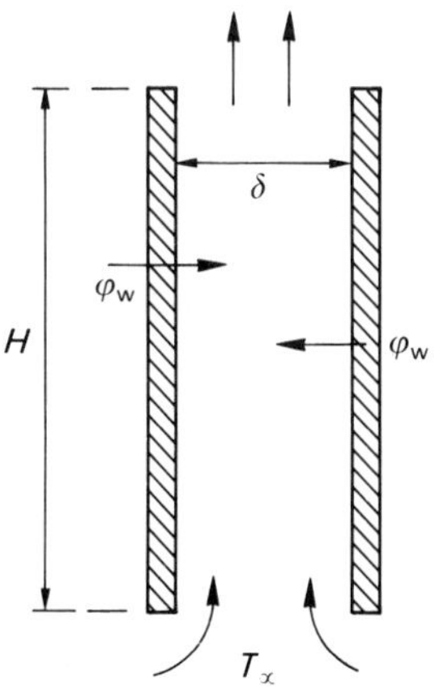


$$Ra_\delta^* = \frac{g\beta\delta^4\varphi_w}{a\nu\lambda}$$

$$Ra_\delta^{++} = Ra_\delta^* \frac{\delta}{H}$$

Physical properties evaluated at $T_\infty$ (Churchill and Usagi [42]):

$$Nu_\delta = \frac{\bar{h}\delta}{\lambda} = \left[\frac{12}{Ra_\delta^{++}} + \frac{1.88}{Ra_\delta^{++2/5}}\right]^{-1/2} \qquad \text{for any } Ra_\delta^{++2/5}$$

Optimum thickness: $Ra_{\delta_{opt}}^{++} = (1.472)^5$

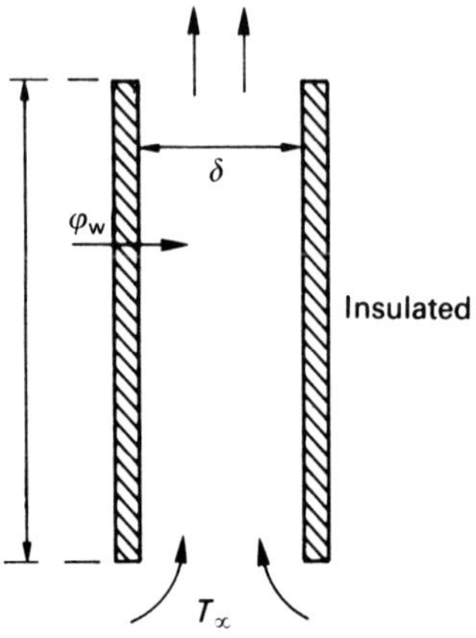


(Churchill and Usagi [42]):

$$Nu_\delta = \left[\frac{6}{Ra_\delta^{++}} + \frac{1.88}{Ra_\delta^{++2/5}}\right]^{-1/2}$$

Optimum thickness: $Ra_{\delta_{opt}}^{++} = (1.169)^5$

## 4.5 Open vertical tubes

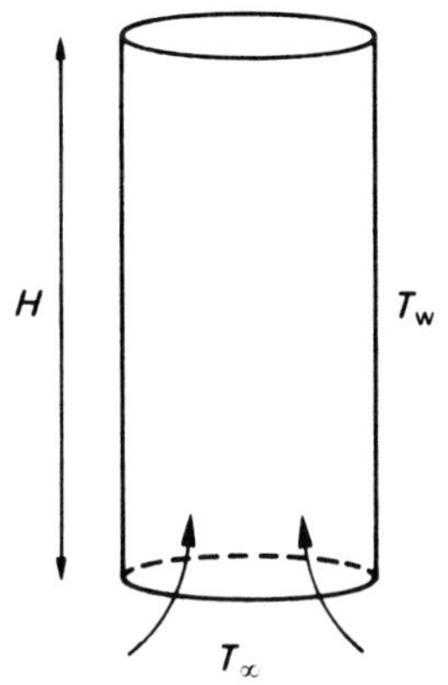


(From Elenbaas [42])

$$Ra_{r_h} = \frac{g\beta(T_w - T_\infty)r_h^3}{a\nu}$$

$r_h = \dfrac{d_h}{2}$ = hydraulic radius of the cross-section

$$Ra_{r_h}^+ = Ra_{r_h} \times \frac{r_h}{H}$$

$$Nu_{r_h} = \frac{\bar{h}r_h}{\lambda} = \frac{Ra_{r_h}^+}{A}\left[1 - e^{-A(1/2Ra_{r_h}^+)^{3/4}}\right]$$

$A = 16$ circular section
$= 14.2$ square section
$= 13^{1/3}$ equilateral triangular section

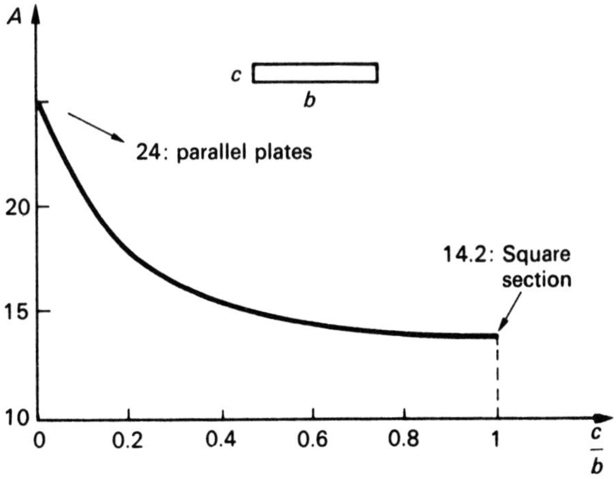

## 4.6 Concentric annular gap

- Horizontal

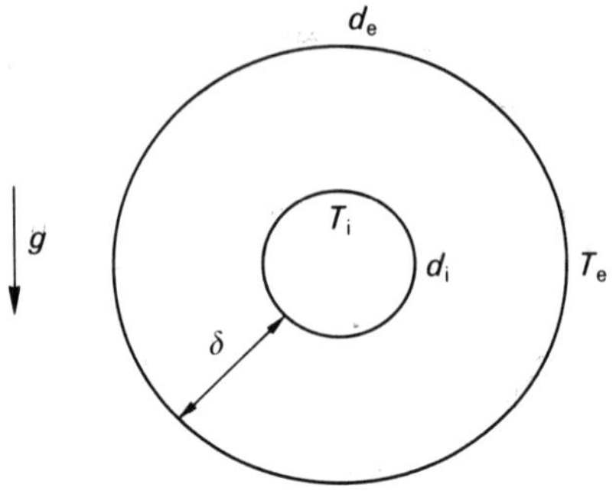


(From Raithby, Hollands [43])

$$Ra_\delta = \frac{g\beta\delta^3 |T_e - T_i|}{a\nu}$$

$$Ra_\delta^+ = \frac{\left(\ln \dfrac{d_e}{d_i}\right)^4}{\delta^3 [d_e^{-3/5} + d_i^{-3/5}]^5} Ra_\delta$$

$$\delta = \frac{d_e - d_i}{2}$$

$$\frac{\lambda_e}{\lambda} = \frac{0.386 Pr^{1/4} \cdot}{[0.861 + Pr]^{1/4}} Ra_\delta^{+\,1/4}$$

$10^2 < Ra_\delta^+ < 10^7$

classical fluid

$$Ra_\delta^+ < 10^2 \rightarrow \frac{\lambda_e}{\lambda} = 1$$

- Vertical

$$Ra_\delta = \frac{g\beta\delta^3 |T_e - T_i|}{a\nu}$$

$$\delta = \frac{d_e - d_i}{2}$$

$$H^+ = \frac{H}{\delta}$$

$$K = \frac{d_e}{d_i}$$

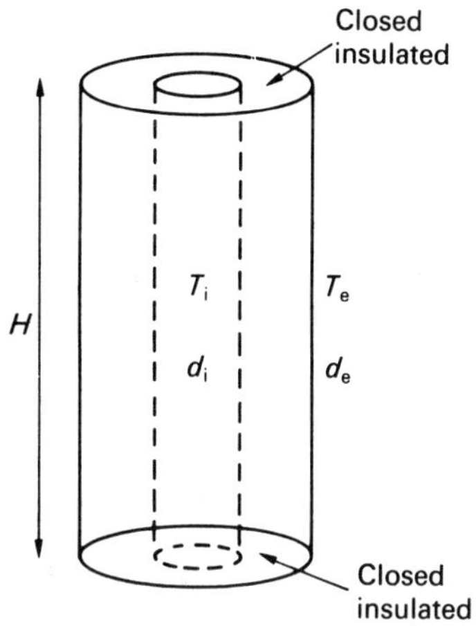


(From de Vahl Davis and Thomas [42])

- Conductive régime

$$\frac{Ra_\delta}{H^+} \leqslant 400 \qquad (H^+ > 5,\ Pr = 1)$$

$$\frac{Ra_\delta}{H^+} \leqslant 10^3 (H^+ = 1)$$

$$Nu_\delta = \frac{\bar{h}\delta}{\lambda} = 0.595 \frac{Ra_\delta^{0.101} K^{0.505} Pr^{0.024}}{(H^+)^{0.052}}$$

- Transition

$$Nu_\delta = 0.202 \frac{Ra_\delta^{0.294} K^{0.423} Pr^{0.097}}{(H^+)^{0.264}}$$

- Boundary layer régime

$$\frac{Ra_\delta}{H^+} \geqslant 3 \times 10^3 \qquad (H^+ \leqslant 5,\ Pr = 1)$$

$$\frac{Ra_\delta}{H^+} \geqslant 8 \times 10^3 \qquad (H^+ = 1)$$

$$Nu_\delta = 0.286 \frac{Ra_\delta^{0.258} K^{0.442} Pr^{0.006}}{(H^+)^{0.238}}$$

These expressions are valid for:

$$1 \leqslant K \leqslant 10$$

$$1 \leqslant H^+ \leqslant 33$$

$$0.5 \leqslant Pr \leqslant 10^4$$

$$Ra_\delta \leqslant 2 \times 10^5$$

Physical properties are evaluated at $\dfrac{T_e + T_i}{2}$.

## 4.7 Geometric spherical gap

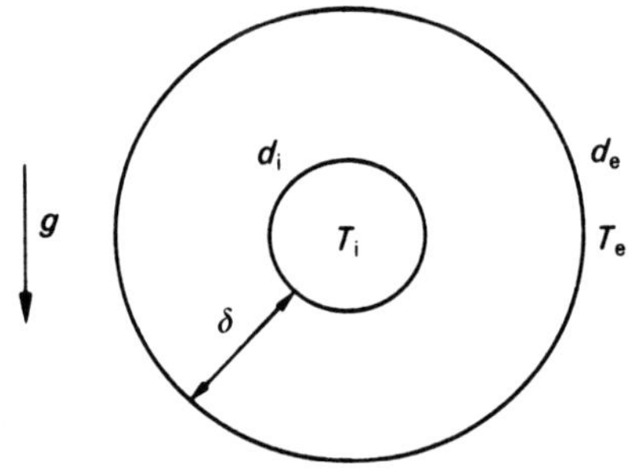


$$Ra_\delta = \frac{g\beta\delta^3 |T_e - T_i|}{a\nu}$$

Physical properties are evaluated at $(T_e + T_i)/2$:

$$0.7 \leqslant Pr \leqslant 4150 \qquad 0.09 \leqslant \frac{2\delta}{d_i} \leqslant 1.81$$

$$1.2 \times 10^2 \leqslant Ra_\delta \frac{2\delta}{d_i} \leqslant 1.1 \times 10^9$$

$$\frac{\lambda_e}{\lambda} = 0.202 Ra_\delta^{0.228} \left(\frac{2\delta}{d_i}\right)^{0.252} Pr^{0.029}$$ (Scanlan *et al.* [42])

$$\frac{\lambda_e}{\lambda} = \frac{0.74}{d_e d_i} \frac{Pr^{1/4}}{(0.861 + Pr)^{1/4}} \frac{Ra_\delta^{1/4}}{[d_i^{-7/5} + d_e^{-7/5}]^{5/4}}$$ (Raithby and Hollands [43])

# Section 5

## Radiation

### *5.1 Shape factors* [45]

1. Elementary strips of finite length $b$ and width $\delta a$, with parallel generators:

$$\delta f_{12} = \frac{\cos\theta}{\pi}\,\delta\theta\,\arctan\frac{b}{r}$$

$$\text{if } b \text{ is infinite:}\quad \delta f_{12} = \frac{\cos\theta}{2}\,\delta\theta = \frac{\delta(\sin\theta)}{2}$$

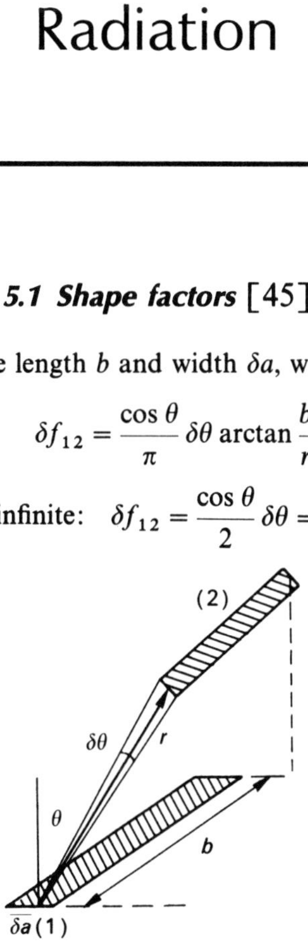


2. Two plane surfaces of infinite length and width $w$, forming a corner of angle $\alpha$:

$$f_{12} = f_{21} = 1 - \sin\frac{\alpha}{2}$$

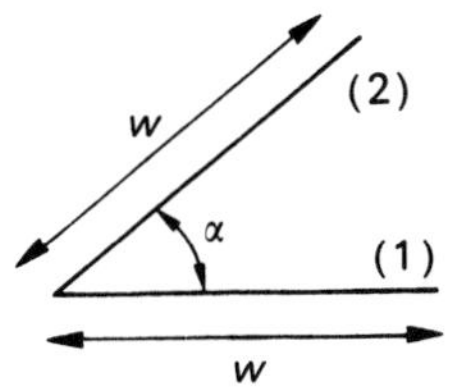

3. Two plane surfaces of infinite length, of widths $h$ and $w$ respectively, forming a right-angled corner:

$$H = \frac{h}{w}$$

$$f_{12} = \frac{1}{2}[1 + H - \sqrt{1 + H^2}]$$

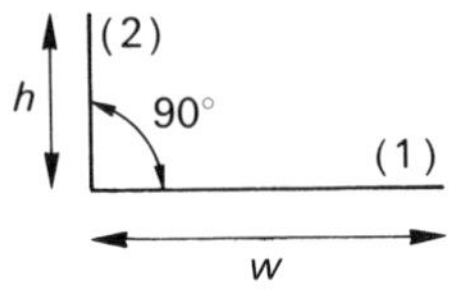


4. Elementary strip parallel to a rectangle (see diagram):

$$X = \frac{a}{c} \qquad Y = \frac{b}{c}$$

$$f_{12} = \frac{1}{\pi Y}\left[\sqrt{1 + Y^2}\arctan\frac{X}{\sqrt{1 + Y^2}} - \arctan X + \frac{XY}{\sqrt{1 + X^2}}\arctan\frac{Y}{\sqrt{1 + X^2}}\right]$$

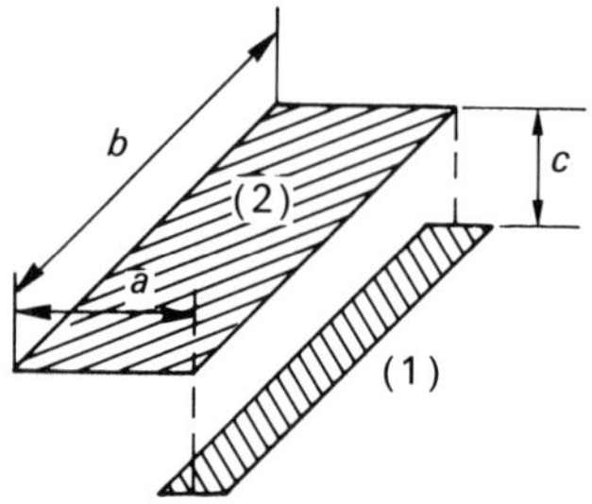


5. Elementary strip in a plane perpendicular to a rectangle (see diagram):

$$X = \frac{a}{b} \qquad Y = \frac{c}{b}$$

$$f_{12} = \frac{1}{\pi}\left\{\arctan\frac{1}{Y} + \frac{Y}{2}\ln\left[\frac{Y^2(X^2 + Y^2 + 1)}{(Y^2 + 1)(X^2 + Y^2)}\right] - \frac{Y}{\sqrt{X^2 + Y^2}}\arctan\left[\frac{1}{\sqrt{X^2 + Y^2}}\right]\right\}$$

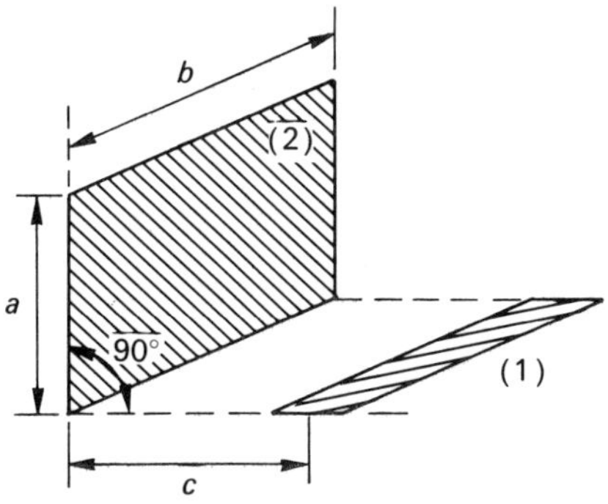


6. Infinite length parallel strips of the same width $w$, opposite each other:

$$H = \frac{h}{w}$$

$$f_{12} = f_{21} = \sqrt{1 + H^2} - H$$

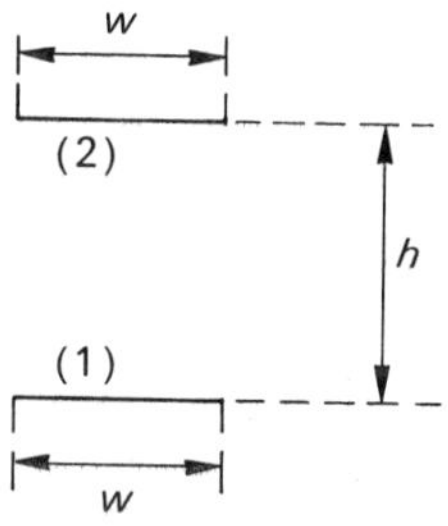


7. Identical parallel rectangles opposite each other:

$$X = \frac{a}{c} \qquad Y = \frac{b}{c}$$

$$f_{12} = \frac{2}{\pi X Y}\left\{\ln\left[\frac{(1 + X^2)(1 + Y^2)}{1 + X^2 + Y^2}\right]^{1/2} + X\sqrt{1 + Y^2}\arctan\frac{X}{\sqrt{1 + Y^2}}\right.$$

$$\left. + Y\sqrt{1 + X^2}\arctan\frac{Y}{\sqrt{1 + X^2}} - X\arctan X - Y\arctan Y\right\}$$

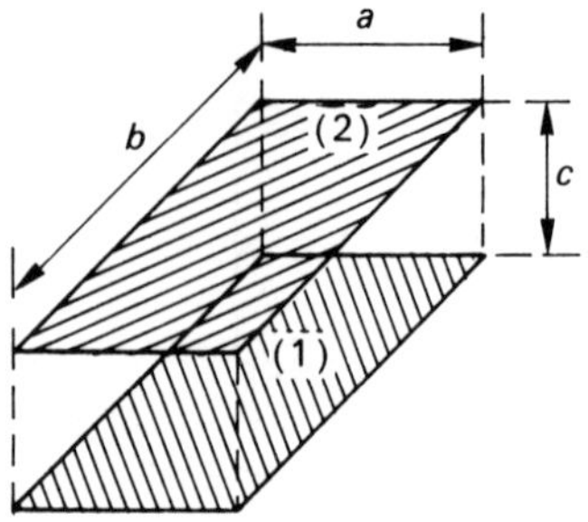


8. Rectangles of the same length $a$ with a common edge and in perpendicular planes:

$$H = \frac{h}{a} \qquad W = \frac{w}{a}$$

$$f_{12} = \frac{1}{\pi W}\left[W \arctan\frac{1}{W} + H \arctan\frac{1}{H} - \sqrt{H^2 + W^2}\arctan\frac{1}{\sqrt{H^2 + W^2}} + \frac{1}{4}\ln\left\{\left[\frac{(1 + W^2)(1 + H^2)}{(1 + W^2 + H^2)}\right]\left[\frac{W^2(1 + W^2 + H^2)}{(1 + W^2)(W^2 + H^2)}\right]^{W^2} \times \left[\frac{H^2(1 + H^2 + W^2)}{(1 + H^2)(H^2 + W^2)}\right]^{H^2}\right\}\right]$$

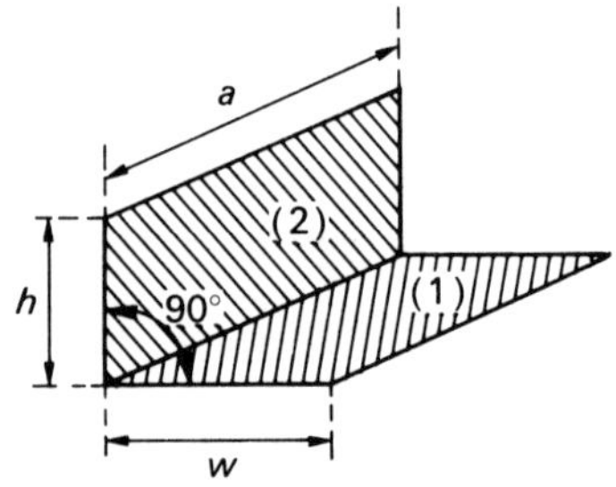


9. Parallel circular dics on the same axis:

$$R_1 = \frac{r_1}{h} \qquad R_2 = \frac{r_2}{h}$$

$$X = 1 + \frac{1 + R_2^2}{R_1^2}$$

$$f_{12} = \frac{1}{2}\left[X - \sqrt{X^2 - 4\left(\frac{R_2}{R_1}\right)^2}\right]$$

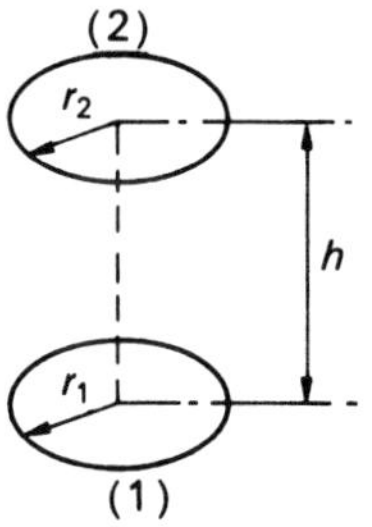


10. Elementary strip of any length, parallel to the axis of an infinitely long cylinder:

$$X = \frac{x}{r} \qquad Y = \frac{y}{r}$$

$$f_{21} = \frac{Y}{X^2 + Y^2}$$

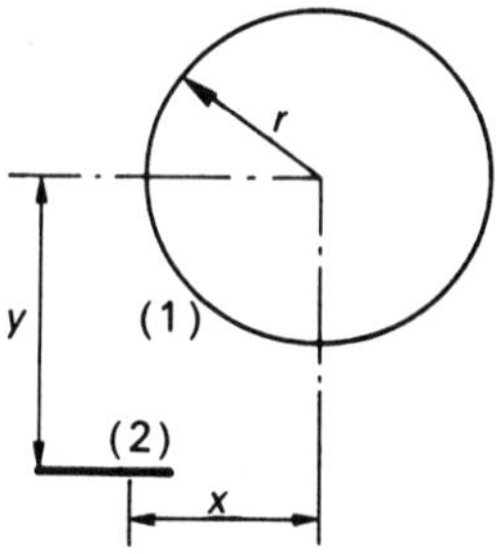


11. Finite width, infinite length, strip parallel to the axis of an infinitely long cylinder:

$$f_{12} = \frac{r}{b - a}\left[\arctan\frac{b}{c} - \arctan\frac{a}{c}\right]$$

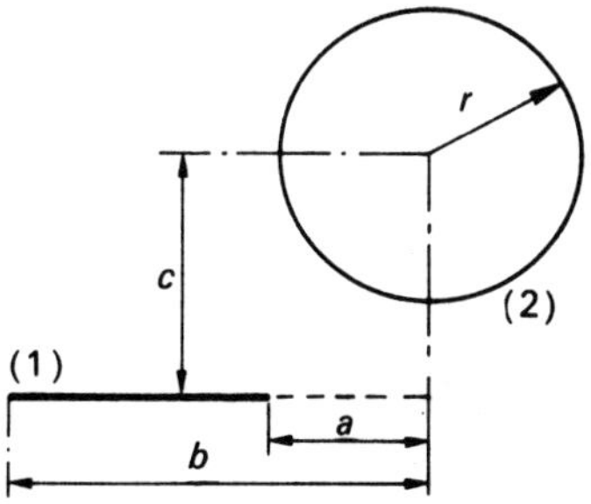

12. Infinitely long cylinders of the same diameter with parallel axes:

$$X = 1 + \frac{s}{2r}$$

$$f_{12} = f_{21} = \frac{1}{\pi}\left[\sqrt{X^2 - 1} + \arcsin\frac{1}{X} - X\right]$$

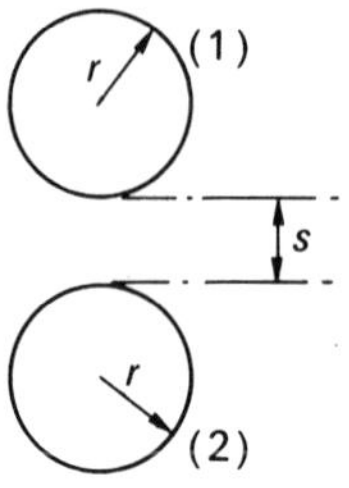


13. Two concentric cylinders with the same length $a$:

$$R = \frac{r_2}{r_1} \qquad L = \frac{a}{r_1}$$

$$A = L^2 + R^2 - 1 \qquad B = L^2 - R^2 + 1$$

$$f_{21} = \frac{1}{R} - \frac{1}{\pi R}\left\{\arccos\frac{B}{A} - \frac{1}{2L}\left[\sqrt{(A+2)^2 - (2R)^2}\arccos\frac{B}{RA} + B\arcsin\frac{1}{R} - \frac{\pi A}{2}\right]\right\}$$

$$f_{22} = 1 - \frac{1}{R} + \frac{2}{\pi R}\arctan\left[\frac{2\sqrt{R^2-1}}{L}\right] - \frac{L}{2\pi R}\left\{\frac{\sqrt{4R^2+L^2}}{L}\arcsin\left[\frac{4(R^2-1) + (L^2/R^2)(R^2-2)}{L^2 + 4(R^2-1)}\right] - \arcsin\frac{R^2-2}{R^2} + \frac{\pi}{2}\left[\frac{\sqrt{4R^2+L^2}}{L} - 1\right]\right\}$$

where for any argument $\zeta$: $-\frac{\pi}{2} \leqslant \arcsin\zeta \leqslant \frac{\pi}{2}$, $0 \leqslant \arccos\zeta \leqslant \pi$.

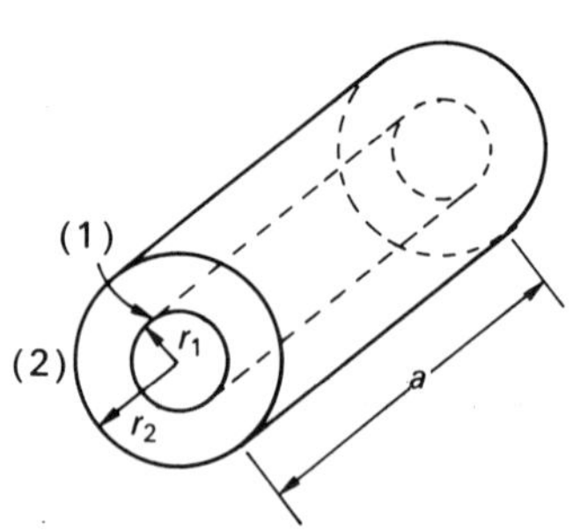

14. Sphere of radius $r_1$ and disk of radius $r_2$, on the same axis:

$$R_2 = \frac{r_2}{h}$$

$$f_{12} = \frac{1}{2}\left[1 - \frac{1}{\sqrt{1 + R_2^2}}\right]$$

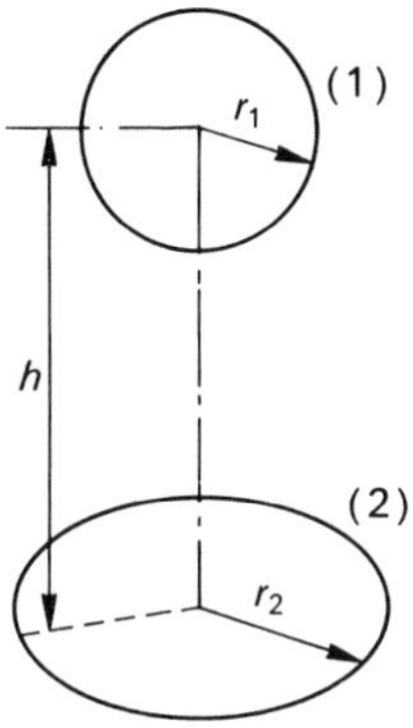


15. Infinite length prisms of triangular cross-section. Sides $AB$, $BC$ and $AC$ may be straight or convex (towards the interior). We find:

$$f_{11} = f_{22} = f_{33} = 0$$

From the geometry and by reciprocity:

$$f_{12} = \frac{L_1 + L_2 - L_3}{2L_1}$$

$f_{13}$ and $f_{23}$ may be deduced from $f_{12}$ by changing the indices.

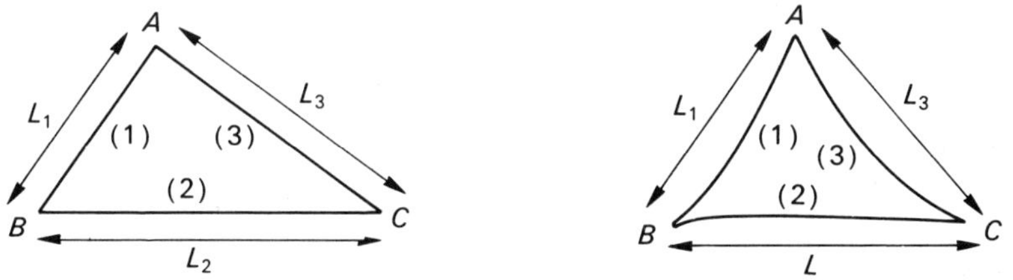


16. Elements (1) and (2), of infinite length in the $z$ direction, and arbitrary cross-section independent of $z$. Use Hottel's crossed chords method.

Consider elements (1), (2), (3) and (4). We wish to determine $f_{12}$ and $f_{21}$. To do this, we consider regions of triangular cross-section with plane or convex faces, ($aghcba$) and ($agheda$). From 15, in region ($aghcba$):

$$f_{agh,abc} = \frac{L_{agh} + L_{abc} + -L_{ch}}{2L_{agh}}$$

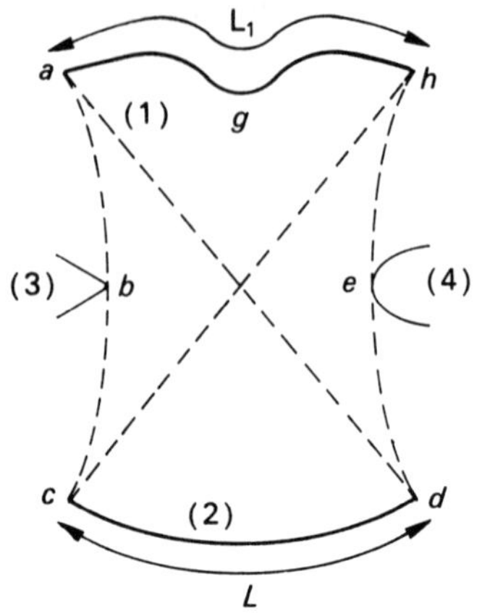


and in region (*agheda*):

$$f_{agh,hed} = \frac{L_{agh} + L_{hed} - L_{ad}}{2L_{agh}}$$

further, for the region (*aghedcba*):

$$f_{agh,abc} + f_{agh,(2)} + f_{agh,hed} = 1$$

The three equations above lead to:

$$L_{agh} f_{agh,(2)} = L_{agh} - \frac{L_{agh} + L_{abc} - L_{ch}}{2} - \frac{L_{agh} + L_{hed} - L_{ad}}{2}$$

whence:

$$L_{agh} f_{agh,(2)} = \frac{L_{ch} + L_{ad} - L_{abc} - L_{hed}}{2}$$

By reciprocity:

$$L_{agh} f_{agh,(2)} = L_2 f_{(2),agh} = L_2 f_{21} = L_1 f_{12}$$

whence:

$$L_2 f_{21} = L_1 f_{12} = \frac{L_{ch} + L_{ad} - L_{abc} - L_{hed}}{2}$$

$$L_2 f_{21} = L_1 f_{12} = \frac{\textit{length of crossed chords} - \textit{length of uncrossed chords}}{2}$$

This method can be used to quickly derive some of the preceding results, such as 6, 11, 12.

## 5.2 *Planck's law* [45]

Table of normalized values of the monochromatic intensity of equilibrium radiation

$$x = \frac{\lambda}{\lambda_m} \qquad y = \frac{I^0_\lambda}{I^0_{\lambda_m}}$$

| $x$ | $y$ | $x$ | $y$ | $x$ | $y$ | $x$ | $y$ | $x$ | $y$ |
|---|---|---|---|---|---|---|---|---|---|
| 0.10 | $4.17\times10^{-15}$ | 0.61 | $49.17\times10^{-2}$ | 1.06 | $99.20\times10^{-2}$ | 1.51 | $70.30\times10^{-2}$ | 2.30 | $28.87\times10^{-2}$ |
| 0.15 | $7.91\times10^{-9}$ | 0.62 | $51.70\times10^{-2}$ | 1.07 | $98.92\times10^{-2}$ | 1.52 | $69.56\times10^{-2}$ | 2.35 | $27.31\times10^{-2}$ |
| 0.20 | $7.37\times10^{-6}$ | 0.63 | $54.20\times10^{-2}$ | 1.08 | $98.60\times10^{-2}$ | 1.53 | $68.83\times10^{-2}$ | 2.40 | $25.85\times10^{-2}$ |
| | | 0.64 | $56.67\times10^{-2}$ | 1.09 | $98.26\times10^{-2}$ | 1.54 | $68.10\times10^{-2}$ | 2.45 | $24.47\times10^{-2}$ |
| | | 0.65 | $59.08\times10^{-2}$ | 1.10 | $97.88\times10^{-2}$ | 1.55 | $67.38\times10^{-2}$ | 2.50 | $23.18\times10^{-2}$ |
| 0.21 | $1.88\times10^{-5}$ | 0.66 | $61.45\times10^{-2}$ | 1.11 | $97.47\times10^{-2}$ | 1.56 | $66.66\times10^{-2}$ | 2.55 | $21.96\times10^{-2}$ |
| 0.22 | $4.37\times10^{-5}$ | 0.67 | $63.78\times10^{-2}$ | 1.12 | $97.04\times10^{-2}$ | 1.57 | $65.94\times10^{-2}$ | 2.60 | $20.83\times10^{-2}$ |
| 0.23 | $9.31\times10^{-5}$ | 0.68 | $66.06\times10^{-2}$ | 1.13 | $96.59\times10^{-2}$ | 1.58 | $65.22\times10^{-2}$ | 2.65 | $19.76\times10^{-2}$ |
| 0.24 | $1.85\times10^{-4}$ | 0.69 | $68.25\times10^{-2}$ | 1.14 | $96.12\times10^{-2}$ | 1.59 | $64.51\times10^{-2}$ | 2.70 | $18.75\times10^{-2}$ |
| 0.25 | $3.45\times10^{-4}$ | 0.70 | $70.42\times10^{-2}$ | 1.15 | $95.63\times10^{-2}$ | 1.60 | $63.80\times10^{-2}$ | 2.75 | $17.79\times10^{-2}$ |
| 0.26 | $6.10\times10^{-4}$ | 0.71 | $72.48\times10^{-2}$ | 1.16 | $95.11\times10^{-2}$ | 1.61 | $63.10\times10^{-2}$ | 2.80 | $16.88\times10^{-2}$ |
| 0.27 | $1.02\times10^{-3}$ | 0.72 | $74.48\times10^{-2}$ | 1.17 | $94.56\times10^{-2}$ | 1.62 | $62.41\times10^{-2}$ | 2.85 | $16.05\times10^{-2}$ |
| 0.28 | $1.62\times10^{-3}$ | 0.73 | $76.42\times10^{-2}$ | 1.18 | $93.99\times10^{-2}$ | 1.63 | $61.73\times10^{-2}$ | 2.90 | $15.28\times10^{-2}$ |
| 0.29 | $2.54\times10^{-3}$ | 0.74 | $78.28\times10^{-2}$ | 1.19 | $93.39\times10^{-2}$ | 1.64 | $61.05\times10^{-2}$ | 2.95 | $14.53\times10^{-2}$ |
| 0.30 | $3.80\times10^{-3}$ | 0.75 | $80.05\times10^{-2}$ | 1.20 | $92.77\times10^{-2}$ | 1.65 | $60.38\times10^{-2}$ | 3.00 | $13.83\times10^{-2}$ |
| 0.31 | $5.50\times10^{-3}$ | 0.76 | $81.74\times10^{-2}$ | 1.21 | $92.14\times10^{-2}$ | 1.66 | $59.72\times10^{-2}$ | 3.10 | $12.57\times10^{-2}$ |
| 0.32 | $7.74\times10^{-3}$ | 0.77 | $83.36\times10^{-2}$ | 1.22 | $91.50\times10^{-2}$ | 1.67 | $59.06\times10^{-2}$ | 3.20 | $11.41\times10^{-2}$ |
| 0.33 | $1.062\times10^{-2}$ | 0.78 | $84.91\times10^{-2}$ | 1.23 | $90.85\times10^{-2}$ | 1.68 | $58.40\times10^{-2}$ | 3.30 | $10.38\times10^{-2}$ |
| 0.34 | $1.425\times10^{-2}$ | 0.79 | $86.36\times10^{-2}$ | 1.24 | $90.19\times10^{-2}$ | 1.69 | $57.75\times10^{-2}$ | 3.40 | $9.47\times10^{-2}$ |
| 0.35 | $1.870\times10^{-2}$ | 0.80 | $87.74\times10^{-2}$ | 1.25 | $89.51\times10^{-2}$ | 1.70 | $57.11\times10^{-2}$ | 3.50 | $8.66\times10^{-2}$ |
| 0.36 | $2.410\times10^{-2}$ | 0.81 | $89.04\times10^{-2}$ | 1.26 | $88.82\times10^{-2}$ | 1.71 | $56.47\times10^{-2}$ | 3.60 | $7.92\times10^{-2}$ |
| 0.37 | $3.051\times10^{-2}$ | 0.82 | $90.26\times10^{-2}$ | 1.27 | $88.12\times10^{-2}$ | 1.72 | $55.84\times10^{-2}$ | 3.70 | $7.26\times10^{-2}$ |
| 0.38 | $3.801\times10^{-2}$ | 0.83 | $91.40\times10^{-2}$ | 1.28 | $87.41\times10^{-2}$ | 1.73 | $55.21\times10^{-2}$ | 3.80 | $6.67\times10^{-2}$ |
| 0.39 | $4.667\times10^{-2}$ | 0.84 | $92.46\times10^{-2}$ | 1.29 | $86.70\times10^{-2}$ | 1.74 | $54.59\times10^{-2}$ | 3.90 | $6.14\times10^{-2}$ |
| 0.40 | $5.648\times10^{-2}$ | 0.85 | $93.45\times10^{-2}$ | 1.30 | $85.98\times10^{-2}$ | 1.75 | $53.98\times10^{-2}$ | 4.00 | $5.65\times10^{-2}$ |
| 0.41 | $6.76\times10^{-2}$ | 0.86 | $94.35\times10^{-2}$ | 1.31 | $85.26\times10^{-2}$ | 1.76 | $53.37\times10^{-2}$ | 4.50 | $3.83\times10^{-2}$ |
| 0.42 | $8.00\times10^{-2}$ | 0.87 | $95.19\times10^{-2}$ | 1.32 | $84.53\times10^{-2}$ | 1.77 | $52.76\times10^{-2}$ | 5.00 | $2.68\times10^{-2}$ |
| 0.43 | $9.36\times10^{-2}$ | 0.88 | $95.95\times10^{-2}$ | 1.33 | $83.79\times10^{-2}$ | 1.78 | $52.16\times10^{-2}$ | 6.00 | $1.421\times10^{-2}$ |
| 0.44 | $10.84\times10^{-2}$ | 0.89 | $96.63\times10^{-2}$ | 1.34 | $83.05\times10^{-2}$ | 1.79 | $51.57\times10^{-2}$ | 7.00 | $8.20\times10^{-3}$ |
| 0.45 | $12.45\times10^{-2}$ | 0.90 | $97.24\times10^{-2}$ | 1.35 | $82.30\times10^{-2}$ | 1.80 | $50.99\times10^{-2}$ | 8.00 | $5.05\times10^{-3}$ |
| 0.46 | $14.18\times10^{-2}$ | 0.91 | $97.78\times10^{-2}$ | 1.36 | $81.55\times10^{-2}$ | 1.82 | $49.84\times10^{-2}$ | 9.00 | $3.27\times10^{-3}$ |
| 0.47 | $16.02\times10^{-2}$ | 0.92 | $98.26\times10^{-2}$ | 1.37 | $80.80\times10^{-2}$ | 1.84 | $48.70\times10^{-2}$ | 10.00 | $2.23\times10^{-3}$ |
| 0.48 | $17.97\times10^{-2}$ | 0.93 | $98.68\times10^{-2}$ | 1.38 | $80.04\times10^{-2}$ | 1.86 | $47.60\times10^{-2}$ | 15.00 | $4.78\times10^{-4}$ |
| 0.49 | $20.03\times10^{-2}$ | 0.94 | $99.04\times10^{-2}$ | 1.39 | $79.29\times10^{-2}$ | 1.88 | $46.52\times10^{-2}$ | 20.00 | $1.58\times10^{-4}$ |
| 0.50 | $22.17\times10^{-2}$ | 0.95 | $99.34\times10^{-2}$ | 1.40 | $78.53\times10^{-2}$ | 1.90 | $45.46\times10^{-2}$ | 25.00 | $6.2\times10^{-5}$ |
| 0.51 | $24.39\times10^{-2}$ | 0.96 | $99.59\times10^{-2}$ | 1.41 | $77.77\times10^{-2}$ | 1.92 | $44.43\times10^{-2}$ | 30.00 | $3.25\times10^{-5}$ |
| 0.52 | $26.70\times10^{-2}$ | 0.97 | $99.78\times10^{-2}$ | 1.42 | $77.01\times10^{-2}$ | 1.94 | $43.42\times10^{-2}$ | 35.00 | $1.7\times10^{-5}$ |
| 0.53 | $29.06\times10^{-2}$ | 0.98 | $99.90\times10^{-2}$ | 1.43 | $76.26\times10^{-2}$ | 1.96 | $42.43\times10^{-2}$ | 40.00 | $1.05\times10^{-5}$ |
| 0.54 | $31.48\times10^{-2}$ | 0.99 | $99.97\times10^{-2}$ | 1.44 | $75.51\times10^{-2}$ | 1.98 | $41.47\times10^{-2}$ | 45.00 | $6.7\times10^{-6}$ |
| 0.55 | $33.95\times10^{-2}$ | 1.00 | $100.00\times10^{-2}$ | 1.45 | $74.76\times10^{-2}$ | 2.00 | $40.54\times10^{-2}$ | 50.00 | $4.36\times10^{-6}$ |
| 0.56 | $36.45\times10^{-2}$ | 1.01 | $99.98\times10^{-2}$ | 1.46 | $74.01\times10^{-2}$ | 2.05 | $38.27\times10^{-2}$ | | |
| 0.57 | $38.98\times10^{-2}$ | 1.02 | $99.91\times10^{-2}$ | 1.47 | $73.27\times10^{-2}$ | 2.10 | $36.14\times10^{-2}$ | | |
| 0.58 | $41.52\times10^{-2}$ | 1.03 | $99.79\times10^{-2}$ | 1.48 | $72.52\times10^{-2}$ | 2.15 | $34.14\times10^{-2}$ | | |
| 0.59 | $44.08\times10^{-2}$ | 1.04 | $99.63\times10^{-2}$ | 1.49 | $71.78\times10^{-2}$ | 2.20 | $32.28\times10^{-2}$ | | |
| 0.60 | $46.63\times10^{-2}$ | 1.05 | $99.44\times10^{-2}$ | 1.50 | $71.04\times10^{-2}$ | 2.25 | $30.53\times10^{-2}$ | | |

### *5.3 Function* $z(0, \lambda_0/\lambda_m(T))$ *for equilibrium radiation* [45]

$$x = \frac{\lambda_0}{\lambda_m(T)}$$

$$z\left(0, \frac{\lambda_0}{\lambda_m(T)}\right) = \frac{\int_0^{\lambda_0} \pi I_\lambda^0(T)\,\mathrm{d}\lambda}{\sigma T^4}$$

We recall that:

1. $T\lambda_m(T) = 2898\ \mu\text{mK}$.

2. 98% of the energy is distributed between $\frac{\lambda_m}{2}$ and $7\lambda_m$.

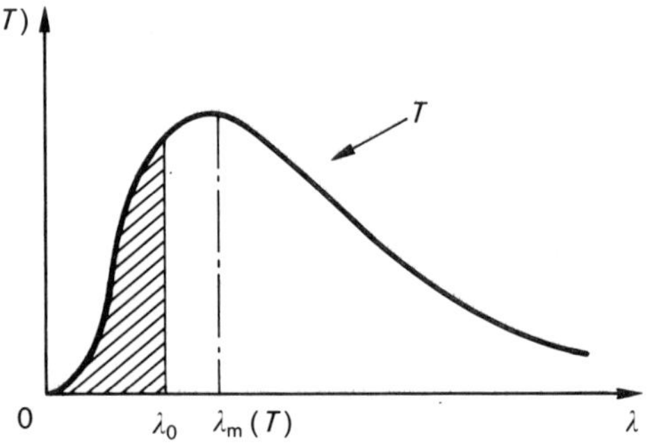

| $x$ | $Z$ | $x$ | $Z$ | $x$ | $Z$ | $x$ | $y$ |
|---|---|---|---|---|---|---|---|
| 0.10 | $5.5 \times 10^{-18}$ | | | | | | |
| 0.20 | $4.10 \times 10^{-8}$ | | | | | | |
| 0.22 | $3.1 \times 10^{-7}$ | 0.92 | $19.78 \times 10^{-2}$ | 1.62 | $59.29 \times 10^{-2}$ | 3.6 | $92.24 \times 10^{-2}$ |
| 0.24 | $1.6 \times 10^{-6}$ | 0.94 | $21.08 \times 10^{-2}$ | 1.64 | $60.10 \times 10^{-2}$ | 3.7 | $92.74 \times 10^{-2}$ |
| 0.26 | $6.4 \times 10^{-6}$ | 0.96 | $22.39 \times 10^{-2}$ | 1.66 | $60.90 \times 10^{-2}$ | 3.8 | $93.20 \times 10^{-2}$ |
| 0.28 | $2.03 \times 10^{-5}$ | 0.98 | $23.69 \times 10^{-2}$ | 1.68 | $61.68 \times 10^{-2}$ | 3.9 | $93.62 \times 10^{-2}$ |
| 0.30 | $5.47 \times 10^{-5}$ | 1.00 | $25.00 \times 10^{-2}$ | 1.70 | $62.43 \times 10^{-2}$ | 4.0 | $94.01 \times 10^{-2}$ |
| 0.32 | $1.28 \times 10^{-4}$ | 1.02 | $26.32 \times 10^{-2}$ | 1.72 | $63.17 \times 10^{-2}$ | 5.0 | $96.61 \times 10^{-2}$ |
| 0.34 | $2.69 \times 10^{-4}$ | 1.04 | $27.63 \times 10^{-2}$ | 1.74 | $63.90 \times 10^{-2}$ | 6.0 | $97.89 \times 10^{-2}$ |
| 0.36 | $5.17 \times 10^{-4}$ | 1.06 | $28.94 \times 10^{-2}$ | 1.76 | $64.61 \times 10^{-2}$ | 7.0 | $98.61 \times 10^{-2}$ |
| 0.38 | $9.21 \times 10^{-4}$ | 1.08 | $30.25 \times 10^{-2}$ | 1.78 | $65.30 \times 10^{-2}$ | 8.0 | $99.03 \times 10^{-2}$ |
| 0.40 | $1.54 \times 10^{-3}$ | 1.10 | $31.55 \times 10^{-2}$ | 1.80 | $65.98 \times 10^{-2}$ | 9.0 | $99.30 \times 10^{-2}$ |
| | | | | | | 10.0 | $99.48 \times 10^{-2}$ |
| 0.42 | $2.43 \times 10^{-3}$ | 1.12 | $32.83 \times 10^{-2}$ | 1.82 | $66.65 \times 10^{-2}$ | 15.0 | $99.84 \times 10^{-2}$ |
| 0.44 | $3.66 \times 10^{-3}$ | 1.14 | $34.09 \times 10^{-2}$ | 1.84 | $67.29 \times 10^{-2}$ | 20.0 | $99.927 \times 10^{-2}$ |
| 0.46 | $5.30 \times 10^{-3}$ | 1.16 | $35.34 \times 10^{-2}$ | 1.86 | $67.92 \times 10^{-2}$ | 30.0 | $99.978 \times 10^{-2}$ |
| 0.48 | $7.41 \times 10^{-3}$ | 1.18 | $36.58 \times 10^{-2}$ | 1.88 | $68.54 \times 10^{-2}$ | 40.0 | $99.991 \times 10^{-2}$ |
| 0.50 | $10.05 \times 10^{-3}$ | 1.20 | $37.81 \times 10^{-2}$ | 1.90 | $69.15 \times 10^{-2}$ | 50.0 | $99.995 \times 10^{-2}$ |
| | | | | | | $\infty$ | $100.0 \times 10^{-2}$ |
| 0.52 | $1.33 \times 10^{-2}$ | 1.22 | $39.02 \times 10^{-2}$ | 1.92 | $69.75 \times 10^{-2}$ | | |
| 0.54 | $1.71 \times 10^{-2}$ | 1.24 | $40.22 \times 10^{-2}$ | 1.94 | $70.33 \times 10^{-2}$ | | |
| 0.56 | $2.16 \times 10^{-2}$ | 1.26 | $41.40 \times 10^{-2}$ | 1.96 | $70.89 \times 10^{-2}$ | | |
| 0.58 | $2.67 \times 10^{-2}$ | 1.28 | $42.56 \times 10^{-2}$ | 1.98 | $71.43 \times 10^{-2}$ | | |
| 0.60 | $3.25 \times 10^{-2}$ | 1.30 | $43.71 \times 10^{-2}$ | 2.00 | $71.96 \times 10^{-2}$ | | |
| 0.62 | $3.90 \times 10^{-2}$ | 1.32 | $44.83 \times 10^{-2}$ | 2.1 | $74.48 \times 10^{-2}$ | | |
| 0.64 | $4.61 \times 10^{-2}$ | 1.34 | $45.93 \times 10^{-2}$ | 2.2 | $76.72 \times 10^{-2}$ | | |
| 0.66 | $5.39 \times 10^{-2}$ | 1.36 | $47.01 \times 10^{-2}$ | 2.3 | $78.73 \times 10^{-2}$ | | |
| 0.68 | $6.22 \times 10^{-2}$ | 1.38 | $48.07 \times 10^{-2}$ | 2.4 | $80.53 \times 10^{-2}$ | | |
| 0.70 | $7.12 \times 10^{-2}$ | 1.40 | $49.11 \times 10^{-2}$ | 2.5 | $82.14 \times 10^{-2}$ | | |
| 0.72 | $8.07 \times 10^{-2}$ | 1.42 | $50.13 \times 10^{-2}$ | 2.6 | $83.58 \times 10^{-2}$ | | |
| 0.74 | $9.08 \times 10^{-2}$ | 1.44 | $51.14 \times 10^{-2}$ | 2.7 | $84.88 \times 10^{-2}$ | | |
| 0.76 | $10.14 \times 10^{-2}$ | 1.46 | $52.12 \times 10^{-2}$ | 2.8 | $86.05 \times 10^{-2}$ | | |
| 0.78 | $11.23 \times 10^{-2}$ | 1.48 | $53.08 \times 10^{-2}$ | 2.9 | $87.11 \times 10^{-2}$ | | |
| 0.80 | $12.37 \times 10^{-2}$ | 1.50 | $54.03 \times 10^{-2}$ | 3.0 | $88.07 \times 10^{-2}$ | | |
| 0.82 | $13.55 \times 10^{-2}$ | 1.52 | $54.95 \times 10^{-2}$ | 3.1 | $88.93 \times 10^{-2}$ | | |
| 0.84 | $14.75 \times 10^{-2}$ | 1.54 | $55.86 \times 10^{-2}$ | 3.2 | $89.72 \times 10^{-2}$ | | |
| 0.86 | $15.98 \times 10^{-2}$ | 1.56 | $56.75 \times 10^{-2}$ | 3.3 | $90.44 \times 10^{-2}$ | | |
| 0.88 | $17.23 \times 10^{-2}$ | 1.58 | $57.61 \times 10^{-2}$ | 3.4 | $91.10 \times 10^{-2}$ | | |
| 0.90 | $18.50 \times 10^{-2}$ | 1.60 | $58.46 \times 10^{-2}$ | 3.5 | $91.70 \times 10^{-2}$ | | |

# Section 6

## Conversion Factors, Temperature Scales

| | | Absolute zero | Melting ice | Boiling water |
|---|---|---|---|---|
| Kelvin | K | 0 | 273.15 | 373.15 |
| Celsius | °C | −273.15 | 0 | 100 |
| Fahrenheit | °F | −459.67 | 32 | 212 |
| Rankine | °R | 0 | 491.67 | 671.67 |

$$T\,(\mathrm{K}) = \theta\,(^\circ\mathrm{C}) + 273.15$$

$$T\,(^\circ\mathrm{R}) = \theta\,(^\circ\mathrm{F}) + 459.67$$

$$\theta\,(^\circ\mathrm{F}) = 32 + \tfrac{9}{5}\theta\,(^\circ\mathrm{C})$$

$$\theta\,(^\circ\mathrm{C}) = \tfrac{5}{9}[\theta\,(^\circ\mathrm{F}) - 32]$$

$$T\,(^\circ\mathrm{R}) = \tfrac{9}{5}T\,(\mathrm{K})$$

$$T\,(\mathrm{K}) = \tfrac{5}{9}T\,(^\circ\mathrm{R})$$

Energy $Q$, $\mathrm{ML^2\,T^{-2}}$

| *Joule (J)* | *BTU* | *kcal* |
|---|---|---|
| 1 | $9.481 \times 10^{-4}$ | $2.389 \times 10^{-4}$ |
| $1.055 \times 10^{3}$ | 1 | 0.252 |
| 4185.5 | 3.968 | 1 |

Flux $\Phi$, power $ML^2 T^{-3}$

| *Watt (W)* | *BTU/h* | *kcal/h* |
|---|---|---|
| 1 | 3.413 | 0.86 |
| 0.293 | 1 | 0.252 |
| 1.163 | 3.97 | 1 |

Flux density $\varphi$ $MT^{-3}$

| *$W/m^2$* | *$BTU/h\ ft^2$* | *$kcal/h\ m^2$* |
|---|---|---|
| 1 | 0.3168 | 0.86 |
| 3.154 | 1 | 2.712 |
| 1.163 | 0.3684 | 1 |

Thermal conductivity $\lambda$ $MLT^{-3}\theta^{-1}$

| *W/m °C* | *BTU/h ft °F* | *kcal/h m °C* |
|---|---|---|
| 1 | 0.578 | 0.86 |
| 1.731 | 1 | 1.488 |
| 1.163 | 0.673 | 1 |

Heat transfer coefficient $h$ $MT^{-3}\theta^{-1}$

| *$W/m^2$ °C* | *$BTU/h\ ft^2$ °F* | *$kcal/h\ m^2$ °C* |
|---|---|---|
| 1 | 0.1761 | 0.86 |
| 5.677 | 1 | 4.878 |
| 1.163 | 0.2048 | 1 |

Specific heat $c$ $L^2 T^{-2}\theta^{-1}$

| *J/kg °C* | *BTU/lb °F* | *kcal/kg °C* |
|---|---|---|
| 1 | $2.389 \times 10^{-4}$ | $2.389 \times 10^{-4}$ |
| 4187 | 1 | 1 |
| 4187 | 1 | 1 |

Density $\rho$ $ML^{-3}$

| *kg/m³* | *lb/ft²* | *kg/m³* |
|---|---|---|
| 1 | 0.0624 | 1 |
| 16.026 | 1 | 16.026 |
| 1 | 0.0624 | 1 |

$\rho c$ $ML^{-1}\ T^{-2}\ \theta^{-1}$

| *J/m³ °C* | *BTU/ft³ °F* | *kcal/h m³ °C* |
|---|---|---|
| 1 | $1.49 \times 10^{-5}$ | $2.389 \times 10^{-4}$ |
| $6.71 \times 10^{4}$ | 1 | 16.026 |
| 4187 | $6.24 \times 10^{-2}$ | 1 |

Diffusivity $v$, $a$, $D$ $L^2\ T^{-1}$

| *m²/s* | *ft²/h* | *m²/h* |
|---|---|---|
| 1 | $3.86 \times 10^{+4}$ | 3600 |
| $2.58 \times 10^{-5}$ | 1 | 0.093 |
| $2.78 \times 10^{-4}$ | 10.74 | 1 |

Dynamic viscosity $\mu$ $ML^{-1}\ T^{-1}$

| *kg/ms* (*Poiseuille*) | *lb/h ft* | *kg/h m* |
|---|---|---|
| 1 | 2419 | 3600 |
| $4.134 \times 10^{-4}$ | 1 | 1.4882 |
| $2.78 \times 10^{-4}$ | 0.672 | 1 |

## Conversion of pressure units

| *Units* | *Pascal* | *Bar Hectopièze (hpz)* | *Barye* | *Technical atmosphere (at) (kgf/cm²)* | *Standard atmosphere (atm)* | *mm Hg* | *inch Hg* | *mm $H_2O$* | *inch $H_2O$* | *psi* |
|---|---|---|---|---|---|---|---|---|---|---|
| Pascal | 1 | $10^{-5}$ | $10^{+1}$ | $1.0197 \times 10^{-5}$ | $0.9869 \times 10^{-5}$ | $0.7501 \times 10^{-2}$ | $0.2953 \times 10^{-3}$ | $1.0197 \times 10^{-1}$ | $0.4014 \times 10^{-2}$ | $14.504 \times 10^{-5}$ |
| Bar Hectopièze (hpz) | $10^5$ | 1 | $10^6$ | 1.0197 | 0.9869 | $7.501 \times 10^2$ | $0.2953 \times 10^2$ | $1.0197 \times 10^4$ | $0.4014 \times 10^3$ | 14.504 |
| Barye | $10^{-1}$ | $10^{-6}$ | 1 | $1.0197 \times 10^{-6}$ | $0.9869 \times 10^{-6}$ | $0.7501 \times 10^{-3}$ | 0.2953 | $1.0197 \times 10^{-2}$ | $0.4014 \times 10^{-3}$ | $14.504 \times 10^{-6}$ |
| Technical atmosphere (at) kgf/cm² | $0.9807 \times 10^5$ | 0.9807 | $0.9807 \times 10^6$ | 1 | 0.9678 | 735.62 | 28.959 | $10^4$ | 393.64 | 14.224 |
| Standard atmosphere (atm) | $1.0133 \times 10^5$ | 1.0133 | $1.0133 \times 10^6$ | 1.0332 | 1 | 760.00 | $0.2992 \times 10^2$ | $10.332 \times 10^3$ | $0.4067 \times 10^{-3}$ | 14.696 |
| mm Hg (Torr) | $1.3332 \times 10^2$ | $1.3332 \times 10^{-3}$ | $1.3332 \times 10^{-3}$ | $1.3595 \times 10^{-3}$ | $1.3158 \times 10^{-3}$ | 1 | 0.0394 | 13.595 | 0.5352 | $19.336 \times 10^{-3}$ |
| inch Hg | $3.3864 \times 10^3$ | $3.3864 \times 10^{-2}$ | $3.3864 \times 10^4$ | 0.0345 | $3.3421 \times 10^{-2}$ | 25.400 | 1 | 345.31 | 13,593 | 0.4912 |
| mm $H_2O$ | 9.807 | $9.807 \times 10^{-5}$ | 98.07 | $10^{-4}$ | $9.6782 \times 10^{-5}$ | 0.0735 | $2.8959 \times 10^{-3}$ | 1 | $3.9364 \times 10^{-2}$ | $142.24 \times 10^{-5}$ |
| inch $H_2O$ | $2.4909 \times 10^2$ | $2.4909 \times 10^{+3}$ | $2.4909 \times 10^{-3}$ | 0.0025 | $2.4587 \times 10^{-3}$ | 1.8686 | 0.7357 | 25.400 | 1 | $36.128 \times 10^{-3}$ |
| Pound per square inch (psi) | $0.0689 \times 10^5$ | 0.0689 | $0.6895 \times 10^6$ | 0.0703 | 0.0680 | 51.717 | 0.0204 | $7.0306 \times 10^2$ | 27.679 | 1 |

# Section 7

## Thermophysical Properties

| | $\theta$ °C | $\lambda$ ($W/m$ °$C$) | $\rho$ ($kg/m^3$) | $c_p$ ($J/kg$ °$C$) | $a \times 10^7$ ($m^2/s$) |
|---|---|---|---|---|---|
| Asphalt | 20–55 | 0.74–0.76 | | | |
| Brick: common | 20 | 0.69 | 1600 | 840 | 5.2 |
| | | 1.32 | 2000 | | |
| carborundum | 600 | 18.5 | | | |
| | 1400 | 11.1 | | | |
| chrome | 200 | 2.32 | 3000 | 840 | 9.2 |
| | 550 | 2.47 | | | 9.8 |
| | 900 | 1.99 | | | 7.9 |
| fireclay | 200 | 0.24 | | | |
| | 870 | 0.31 | | | |
| baked clay (1330 °C) | 500 | 1.04 | 2000 | 960 | 5.4 |
| | 800 | 1.07 | | | |
| | 1100 | 1.09 | | | |
| baked clay (1450 °C) | 500 | 1.28 | 2300 | 960 | 5.8 |
| | 800 | 1.37 | | | |
| | 1100 | 1.40 | | | |
| missouri | 200 | 1.00 | 2600 | 960 | 4.0 |
| | 600 | 1.47 | | | |
| | 1400 | 1.77 | | | |
| magnesite | 200 | 3.81 | | 1130 | |
| | 650 | 2.77 | | | |
| | 1200 | 1.90 | | | |
| Cement: portland | | 0.29 | 1500 | | |
| mortar | 23 | 1.16 | | | |
| Concrete | 23 | 0.76 | | | |
| Rock and concrete | 20 | 1.36 | 1900–3000 | 880 | 6.8–8.2 |
| Glass: window | 20 | ≃0.78 | 2700 | 840 | 3.4 |
| corosilicate | 30–75 | 1.09 | 2200 | | |
| Plaster, gypsum | 20 | 0.48 | 1440 | 840 | 4 |
| Granite | | 1.73–3.98 | 2640 | 820 | 8–18 |
| Limestone | 100–300 | 1.26–1.33 | 2500 | 900 | 5.6–5.9 |
| Marble | | 2.07–2.94 | 2500–2700 | 800 | 10–13.6 |
| Sandstone | 40 | 1.83 | 2160–2300 | 710 | 11.2–11.9 |
| Balsa wood | 30 | 0.055 | 140 | | |
| Cypress wood | 30 | 0.097 | 460 | | |
| Fir | 23 | 0.11 | 420 | 2720 | 0.96 |
| Maple, oak | 30 | 0.166 | 540 | 2400 | 1.28 |
| Pine: white | 23 | 0.147 | 640 | 2800 | 0.82 |
| yellow | 30 | 0.112 | 430 | | |

| | $\theta$ °C | $\lambda$ ($W/m$ °$C$) | $\rho$ ($kg/m^3$) | $c_p$ ($J/kg$ °$C$) | $a \times 10^7$ ($m^2/s$) |
|---|---|---|---|---|---|
| Asbestos, loosely packed | −45 | 0.149 | | | |
| | 0 | 0.154 | 470–570 | 816 | 3.3–4 |
| | 100 | 0.161 | | | |
| Asbestos and cement board | 20 | 0.74 | | | |
| Asbestos sheet | 51 | 0.166 | | | |
| Asbestos felt | | | | | |
| 1600 laminations/m | 38 | 0.057 | | | |
| | 150 | 0.069 | | | |
| | 260 | 0.083 | | | |
| 800 laminations/m | 38 | 0.078 | | | |
| | 150 | 0.095 | | | |
| | 260 | 0.112 | | | |
| Corrugated asbestos | 38 | 0.087 | | | |
| | 93 | 0.100 | | | |
| | 150 | 0.119 | | | |
| Asbestos cement | | 2.08 | | | |
| Corrugated cardboard | | 0.064 | | | |
| Cork | 32 | 0.045 | 45–120 | 1880 | 2–5.3 |
| reconstituted | 30 | 0.043 | 160 | | |
| Felt (hair) | 30 | 0.036 | 130–200 | | |
| Wool | 30 | 0.052 | 330 | | |
| Glass wool | 23 | 0.038 | 24 | 700 | 22.6 |
| Rock wool | 32 | 0.040 | 160 | | |
| loosely packed | 150 | 0.067 | 64 | | |
| | 260 | 0.087 | | | |
| Magnesia, 85% | 38 | 0.067 | 270 | | |
| | 93 | 0.071 | | | |
| | 150 | 0.074 | | | |
| | 204 | 0.080 | | | |
| Wood sawdust | 23 | 0.059 | | | |
| Wood shavings | 23 | 0.059 | | | |
| Kapok | 30 | 0.035 | | | |
| Polyvinylchloride (PVC) | 0 at 60° | 0.15 | 1340 at 1450 | 1000 | |
| Polyvinylidenefluoride (PVDF) | 0 at 100° | 0.17 | $1.78 \times 10^3$ | 200 | |
| Asbestolith (cement reinforced with asbestos fibres) | up to 250° at 400°C | 0.30 at 0.65 | 1600 at 1800 | $1 \times 10^3$ at $1.25 \times 10^3$ | 0.2 at 3 |

| | at 20°C | | | | λ (W/m°C) | | | | | | | | | |
|---|---|---|---|---|---|---|---|---|---|---|---|---|---|---|
| | ρ (kg/m³) | $c_p$ (J/kg°C) | λ (W/m°C) | $a \times 10^5$ (m²/s) | −100 | 0°C | +100 | 200 | 300 | 400 | 600 | 800 | 1000 | 1200 |
| Aluminium, pure | 2707 | 896 | 204 | 8.418 | 215 | 202 | 206 | 215 | 228 | 249 | | | | |
| Duralumin | 2787 | 883 | 164 | 6.676 | 126 | 159 | 182 | 194 | | | | | | |
| Silumin (86.5% Al+1% Cu) | 2659 | 867 | 137 | 5.933 | 119 | 137 | 144 | 152 | 161 | | | | | |
| Alusil (80% Al+20% Si) | 2627 | 854 | 161 | 7.172 | 144 | 157 | 168 | 175 | 178 | | | | | |
| (97% Al, 1% Mg, 1% Si, 1% Mn) | 2707 | 892 | 177 | 7.311 | | 175 | 189 | 204 | | | | | | |
| Carbon steel 0.5% C | 7833 | 465 | 54 | 1.474 | | 55 | 52 | 48 | 45 | 42 | 35 | 31 | 29 | 31 |
| 1.0% C | 7801 | 473 | 43 | 1.172 | | 43 | 43 | 42 | 40 | 36 | 33 | 29 | 28 | 29 |
| 1.5% C | 7753 | 486 | 36 | 0.97 | | 36 | 36 | 36 | 35 | 33 | 31 | 28 | 28 | 29 |
| Nickel steel ≃0% Ni | 7897 | 452 | 73 | 2.046 | | | | | | | | | | |
| 20% Ni | 7933 | 460 | 19 | 0.526 | | | | | | | | | | |
| 40% Ni | 8169 | 460 | 10 | 0.279 | | | | | | | | | | |
| 80% Ni | 8618 | 460 | 35 | 0.872 | | | | | | | | | | |
| Invar 36% Ni | 8137 | 460 | 10.7 | 0.286 | | | | | | | | | | |
| Chrome steel ≃0% | 7897 | 452 | 73 | 2.026 | 87 | 73 | 67 | 62 | 55 | 48 | 40 | 36 | 35 | 36 |
| 1% | 7865 | 460 | 61 | 1.665 | | 62 | 55 | 52 | 47 | 42 | 36 | 33 | 33 | |
| 5% | 7833 | 460 | 40 | 1.11 | | 40 | 38 | 36 | 36 | 33 | 29 | 29 | 29 | |
| 20% | 7689 | 460 | 22 | 0.635 | | 22 | 22 | 22 | 22 | 24 | 24 | 26 | 29 | |
| Chrome–Nickel steel: | | | | | | | | | | | | | | |
| 15% Cr, 10% Ni | 7865 | 460 | 19 | 0.526 | | | | | | | | | | |
| 18% Cr, 8% Ni | 7817 | 460 | 16.3 | 0.444 | | 16.3 | 17 | 17 | 19 | 19 | 22 | 26 | 31 | |
| 20% Cr, 15% Ni | 7833 | 460 | 15.1 | 0.415 | | | | | | | | | | |
| 25% Cr, 20% Ni | 7865 | 460 | 12.8 | 0.361 | | | | | | | | | | |
| Tungsten steel | | | | | | | | | | | | | | |
| ≃0% W | 7897 | 452 | 73 | 2.026 | | | | | | | | | | |
| 1% W | 7913 | 448 | 66 | 1.858 | | | | | | | | | | |
| 5% W | 8073 | 435 | 54 | 1.525 | | | | | | | | | | |
| 10% W | 8314 | 419 | 48 | 1.391 | | | | | | | | | | |
| Iron, pure | 7897 | 452 | 73 | 2.034 | 87 | 73 | 67 | 62 | 55 | 48 | 40 | 36 | 35 | 36 |
| Iron (0.5% C) | 7849 | 460 | 59 | 1.626 | | 59 | 57 | 52 | 49 | 45 | 36 | 33 | 33 | 33 |

| | at 20°C | | | | $\lambda$ ($W/m°C$) | | | | | | | | | |
|---|---|---|---|---|---|---|---|---|---|---|---|---|---|---|
| | $\rho$ ($kg/m^3$) | $c_p$ ($J/kg°C$) | $\lambda$ ($W/m°C$) | $\alpha \times 10^5$ ($m^2/s$) | −100 | 0°C | +100 | 200 | 300 | 400 | 600 | 800 | 1000 | 1200 |
| Copper, pure | 8954 | 383.1 | 386 | 11.234 | 407 | 386 | 379 | 374 | 369 | 363 | 353 | | | |
| Bronze (75% Cu, 25% Sn) | 8666 | 343 | 26 | 0.859 | | | | | | | | | | |
| Aluminium bronze (95% Cu, 5% Al) | 8666 | 410 | 83 | 2.33 | | | | | | | | | | |
| Red brass (85% Cu, 9% Sn, 6% Zn) | 8714 | 385 | 61 | 1.804 | | 59 | 71 | | | | | | | |
| Brass (70% Cu, 30% Zn) | 8522 | 385 | 111 | 3.412 | 88 | | 128 | 144 | 147 | 147 | | | | |
| Nickel silver (62% Cu, 15% Ni, 22% Zn) | 8618 | 394 | 24.9 | 0.733 | 19.2 | | 31 | 40 | 45 | 48 | | | | |
| Constantan (60% Cu, 40% Ni) | 8922 | 410 | 22.7 | 0.612 | 21 | | 22.2 | 26 | | | | | | |
| Magnesium, pure | 1746 | 1013 | 171 | 9.708 | 178 | 171 | 168 | 163 | 157 | | | | | |
| Mg–Al (6–8% Al, 1–2% Zn) | 1810 | 1000 | 66 | 3.605 | | 52 | 62 | 74 | 83 | | | | | |
| Molybdenum | 10220 | 251 | 123 | 4.78 | 138 | 125 | 118 | 114 | 111 | 109 | 106 | 102 | 99 | 92 |
| Nickel, pure (99.9%) | 8906 | 445.9 | 90 | 2.266 | 104 | 93 | 83 | 73 | 64 | 59 | | | | |
| Ni–Cr (90% Ni, 10% Cr) | 8666 | 444 | 17 | 0.444 | | 17.1 | 18.9 | 20.9 | 22.8 | 24.6 | | | | |
| Ni–Cr (80% Ni, 20% Cr) | 8314 | 444 | 12.6 | 0.343 | | 12.3 | 13.8 | 15.6 | 17.1 | 18.0 | 22.5 | | | |
| Lead | 11373 | 130 | 35 | 2.343 | 36.9 | 35.1 | 33.4 | 31.5 | 29.8 | | | | | |
| Silver, pure (99.9%) | 10524 | 234 | 407 | 16.563 | 419 | 410 | 415 | 374 | 362 | 360 | | | | |
| Tin, pure | 7304 | 226.5 | 64 | 3.884 | 74 | 65.9 | 59 | 57 | | | | | | |
| Tungsten | 19350 | 134.4 | 163 | 6.271 | | 166 | 151 | 142 | 133 | 126 | 112 | 76 | | |
| Zinc, pure | 7144 | 384.3 | 112.2 | 4.106 | 114 | 112 | 109 | 106 | 100 | 93 | | | | |

| $T$ (K) | $\rho$ ($kg/m^3$) | $\mu$ ($kg/ms$) | $\nu$ ($m^2/s$) | $c_p$ ($J/kg\ K$) | $\lambda$ ($W/m\ K$) | $a$ ($m^2/s$) | $Pr$ | $\beta$ (1/Kelvin) |
|---|---|---|---|---|---|---|---|---|
| | | AIR | | | | | | |
| 250 | 1.413 | $1.60 \times 10^{-5}$ | $0.949 \times 10^{-5}$ | 1005 | 0.0223 | $1.32 \times 10^{-5}$ | 0.722 | |
| 300 | 1.177 | $1.85 \times 10^{-5}$ | $1.57 \times 10^{-5}$ | 1006 | 0.0262 | $2.22 \times 10^{-5}$ | 0.708 | |
| 350 | 0.998 | $2.08 \times 10^{-5}$ | $2.08 \times 10^{-5}$ | 1009 | 0.0300 | $2.98 \times 10^{-5}$ | 0.697 | |
| 400 | 0.883 | $2.29 \times 10^{-5}$ | $2.59 \times 10^{-5}$ | 1014 | 0.0337 | $3.76 \times 10^{-5}$ | 0.689 | |
| 450 | 0.783 | $2.48 \times 10^{-5}$ | $2.89 \times 10^{-5}$ | 1021 | 0.0371 | $4.12 \times 10^{-5}$ | 0.683 | |
| 500 | 0.705 | $2.67 \times 10^{-5}$ | $3.69 \times 10^{-5}$ | 1030 | 0.0404 | $5.57 \times 10^{-5}$ | 0.680 | |
| 550 | 0.642 | $2.85 \times 10^{-5}$ | $4.43 \times 10^{-5}$ | 1039 | 0.0436 | $6.53 \times 10^{-5}$ | 0.680 | |
| 600 | 0.588 | $3.02 \times 10^{-5}$ | $5.13 \times 10^{-5}$ | 1055 | 0.0466 | $7.51 \times 10^{-5}$ | 0.680 | |
| 650 | 0.543 | $3.18 \times 10^{-5}$ | $5.85 \times 10^{-5}$ | 1063 | 0.0495 | $8.58 \times 10^{-5}$ | 0.682 | $1/T$ |
| 700 | 0.503 | $3.33 \times 10^{-5}$ | $6.63 \times 10^{-5}$ | 1075 | 0.0523 | $9.67 \times 10^{-5}$ | 0.684 | |
| 750 | 0.471 | $3.48 \times 10^{-5}$ | $7.39 \times 10^{-5}$ | 1086 | 0.0551 | $10.8 \times 10^{-5}$ | 0.686 | |
| 800 | 0.441 | $3.63 \times 10^{-5}$ | $8.23 \times 10^{-5}$ | 1098 | 0.0578 | $12.0 \times 10^{-5}$ | 0.689 | |
| 850 | 0.415 | $3.77 \times 10^{-5}$ | $9.07 \times 10^{-5}$ | 1110 | 0.0603 | $13.1 \times 10^{-5}$ | 0.692 | |
| 900 | 0.392 | $3.90 \times 10^{-5}$ | $9.93 \times 10^{-5}$ | 1121 | 0.0628 | $14.3 \times 10^{-5}$ | 0.696 | |
| 950 | 0.372 | $4.02 \times 10^{-5}$ | $10.8 \times 10^{-5}$ | 1132 | 0.0653 | $15.5 \times 10^{-5}$ | 0.699 | |
| 1000 | 0.352 | $4.15 \times 10^{-5}$ | $11.8 \times 10^{-5}$ | 1142 | 0.0675 | $16.8 \times 10^{-5}$ | 0.702 | |
| 1100 | 0.320 | $4.40 \times 10^{-5}$ | $13.7 \times 10^{-5}$ | 1161 | 0.0723 | $19.5 \times 10^{-5}$ | 0.706 | |
| 1200 | 0.295 | $4.63 \times 10^{-5}$ | $15.7 \times 10^{-5}$ | 1179 | 0.0763 | $22.0 \times 10^{-5}$ | 0.714 | |
| 1300 | 0.271 | $4.85 \times 10^{-5}$ | $17.9 \times 10^{-5}$ | 1197 | 0.0803 | $24.8 \times 10^{-5}$ | 0.722 | |
| | | NITROGEN | | | | | | |
| 200 | 1.711 | $12.9 \times 10^{-6}$ | $0.757 \times 10^{-5}$ | 1043 | 0.0182 | $1.02 \times 10^{-5}$ | 0.747 | |
| 300 | 1.142 | $17.8 \times 10^{-6}$ | $1.563 \times 10^{-5}$ | 1041 | 0.0262 | $2.21 \times 10^{-5}$ | 0.713 | |
| 400 | 0.854 | $22.0 \times 10^{-6}$ | $2.574 \times 10^{-5}$ | 1046 | 0.0333 | $3.74 \times 10^{-5}$ | 0.691 | |
| 500 | 0.682 | $25.7 \times 10^{-6}$ | $3.766 \times 10^{-5}$ | 1056 | 0.0398 | $5.53 \times 10^{-5}$ | 0.684 | |
| 600 | 0.569 | $29.1 \times 10^{-6}$ | $5.119 \times 10^{-5}$ | 1076 | 0.0458 | $7.49 \times 10^{-5}$ | 0.686 | |
| 700 | 0.493 | $32.1 \times 10^{-6}$ | $6.512 \times 10^{-5}$ | 1097 | 0.0512 | $9.47 \times 10^{-5}$ | 0.691 | |
| 800 | 0.428 | $34.8 \times 10^{-6}$ | $8.145 \times 10^{-5}$ | 1123 | 0.0561 | $11.7 \times 10^{-5}$ | 0.700 | $1/T$ |
| 900 | 0.380 | $37.5 \times 10^{-6}$ | $9.106 \times 10^{-5}$ | 1146 | 0.0607 | $13.9 \times 10^{-5}$ | 0.711 | |
| 1000 | 0.341 | $40.0 \times 10^{-6}$ | $11.72 \times 10^{-5}$ | 1168 | 0.0648 | $16.3 \times 10^{-5}$ | 0.724 | |
| 1100 | 0.311 | $42.3 \times 10^{-6}$ | $13.60 \times 10^{-5}$ | 1186 | 0.0685 | $18.6 \times 10^{-5}$ | 0.737 | |
| 1200 | 0.285 | $44.5 \times 10^{-6}$ | $15.61 \times 10^{-5}$ | 1204 | 0.0719 | $20.9 \times 10^{-5}$ | 0.748 | |
| | | CARBON DIOXIDE | | | | | | |
| 250 | 2.166 | $12.6 \times 10^{-6}$ | $0.581 \times 10^{-5}$ | 803.9 | 0.0129 | $0.740 \times 10^{-5}$ | 0.793 | |
| 300 | 1.797 | $15.0 \times 10^{-6}$ | $0.832 \times 10^{-5}$ | 870.9 | 0.0166 | $1.06 \times 10^{-5}$ | 0.770 | |
| 350 | 1.536 | $17.2 \times 10^{-6}$ | $1.119 \times 10^{-5}$ | 900.2 | 0.0205 | $1.48 \times 10^{-5}$ | 0.755 | |
| 400 | 1.342 | $19.3 \times 10^{-6}$ | $1.439 \times 10^{-5}$ | 942.0 | 0.0246 | $1.95 \times 10^{-5}$ | 0.738 | $1/T$ |
| 450 | 1.192 | $21.3 \times 10^{-6}$ | $1.790 \times 10^{-5}$ | 979.7 | 0.0290 | $2.48 \times 10^{-5}$ | 0.721 | |
| 500 | 1.073 | $23.3 \times 10^{-6}$ | $2.167 \times 10^{-5}$ | 1013 | 0.0335 | $3.08 \times 10^{-5}$ | 0.702 | |
| 550 | 0.974 | $25.1 \times 10^{-6}$ | $2.574 \times 10^{-5}$ | 1047 | 0.0382 | $3.75 \times 10^{-5}$ | 0.685 | |
| 600 | 0.894 | $26.8 \times 10^{-6}$ | $3.002 \times 10^{-5}$ | 1076 | 0.0431 | $4.48 \times 10^{-5}$ | 0.668 | |

| $T$ (K) | $\rho$ ($kg/m^3$) | $\mu$ ($kg/ms$) | $\nu$ ($m^2/s$) | $c_p$ ($J/kg\ K$) | $\lambda$ ($W/m\ K$) | $a$ ($m^2/s$) | $Pr$ | $\beta$ ($1/Kelvin$) |
|---|---|---|---|---|---|---|---|---|
| | HYDROGEN | | | | | | | |
| 250 | 0.0981 | $7.92 \times 10^{-6}$ | $8.06 \times 10^{-5}$ | 14060 | 0.156 | $11.3 \times 10^{-5}$ | 0.713 | |
| 300 | 0.0819 | $8.96 \times 10^{-6}$ | $10.9 \times 10^{-5}$ | 14320 | 0.182 | $15.5 \times 10^{-5}$ | 0.706 | |
| 350 | 0.0702 | $9.95 \times 10^{-6}$ | $14.2 \times 10^{-5}$ | 14440 | 0.206 | $20.3 \times 10^{-5}$ | 0.697 | |
| 400 | 0.0614 | $10.9 \times 10^{-6}$ | $17.7 \times 10^{-5}$ | 14490 | 0.229 | $25.7 \times 10^{-5}$ | 0.690 | |
| 450 | 0.0546 | $11.8 \times 10^{-6}$ | $21.6 \times 10^{-5}$ | 14500 | 0.251 | $31.6 \times 10^{-5}$ | 0.682 | $1/T$ |
| 500 | 0.0492 | $12.6 \times 10^{-6}$ | $25.7 \times 10^{-5}$ | 14510 | 0.272 | $38.2 \times 10^{-5}$ | 0.675 | |
| 550 | 0.0447 | $13.5 \times 10^{-6}$ | $30.2 \times 10^{-5}$ | 14330 | 0.293 | $45.2 \times 10^{-5}$ | 0.668 | |
| 600 | 0.0408 | $14.3 \times 10^{-6}$ | $35.0 \times 10^{-5}$ | 14540 | 0.315 | $53.1 \times 10^{-5}$ | 0.664 | |
| 650 | 0.0349 | $15.9 \times 10^{-6}$ | $45.5 \times 10^{-5}$ | 14570 | 0.351 | $69.0 \times 10^{-5}$ | 0.659 | |
| 700 | 0.0306 | $17.4 \times 10^{-6}$ | $56.9 \times 10^{-5}$ | 14680 | 0.384 | $85.6 \times 10^{-5}$ | 0.664 | |
| 750 | 0.0272 | $18.8 \times 10^{-6}$ | $69.0 \times 10^{-5}$ | 14820 | 0.412 | $102 \times 10^{-5}$ | 0.676 | |
| 800 | 0.0245 | $20.2 \times 10^{-6}$ | $82.2 \times 10^{-5}$ | 14970 | 0.440 | $120 \times 10^{-5}$ | 0.686 | |
| 850 | 0.0223 | $21.5 \times 10^{-6}$ | $96.5 \times 10^{-5}$ | 15170 | 0.464 | $137 \times 10^{-5}$ | 0.703 | |
| | OXYGEN | | | | | | | |
| 200 | 1.956 | $14.9 \times 10^{-6}$ | $0.795 \times 10^{-5}$ | 913.1 | 0.0182 | $1.02 \times 10^{-5}$ | 0.745 | |
| 250 | 1.562 | $17.9 \times 10^{-6}$ | $1.144 \times 10^{-5}$ | 915.6 | 0.0226 | $1.58 \times 10^{-5}$ | 0.725 | |
| 300 | 1.301 | $20.6 \times 10^{-6}$ | $1.586 \times 10^{-5}$ | 920.3 | 0.0267 | $2.24 \times 10^{-5}$ | 0.709 | |
| 350 | 1.113 | $23.2 \times 10^{-6}$ | $2.080 \times 10^{-5}$ | 929.3 | 0.0307 | $2.97 \times 10^{-5}$ | 0.702 | $1/T$ |
| 400 | 0.976 | $25.5 \times 10^{-6}$ | $2.618 \times 10^{-5}$ | 942.0 | 0.0346 | $3.77 \times 10^{-5}$ | 0.695 | |
| 450 | 0.868 | $27.8 \times 10^{-6}$ | $3.199 \times 10^{-5}$ | 956.7 | 0.0383 | $4.61 \times 10^{-5}$ | 0.694 | |
| 500 | 0.780 | $29.9 \times 10^{-6}$ | $3.834 \times 10^{-5}$ | 972.2 | 0.0417 | $5.50 \times 10^{-5}$ | 0.697 | |
| 550 | 0.710 | $32.0 \times 10^{-6}$ | $4.505 \times 10^{-5}$ | 988.1 | 0.0452 | $6.44 \times 10^{-5}$ | 0.700 | |
| 600 | 0.650 | $33.9 \times 10^{-6}$ | $5.214 \times 10^{-5}$ | 1004 | 0.0483 | $7.40 \times 10^{-5}$ | 0.704 | |

| $T$ (°C) | $\rho$ ($kg/m^3$) | $\mu$ ($kg/ms$) | $\nu$ ($m^2/s$) | $c_p$ ($J/kg\ K$) | $\lambda$ ($W/m\ K$) | $a$ ($m^2/s$) | $Pr$ | $\beta$ ($1/Kelvin$) |
|---|---|---|---|---|---|---|---|---|
| | | | ETHYLENE GLYCOL | | | | | |
| 0 | 1130 | $65 \times 10^{-3}$ | $5.75 \times 10^{-5}$ | 2294 | 0.242 | $9.34 \times 10^{-8}$ | 615 | |
| 20 | 1117 | $21.4 \times 10^{-3}$ | $1.92 \times 10^{-5}$ | 2382 | 0.249 | $9.39 \times 10^{-8}$ | 204 | $0.648 \times 10^{-3}$ |
| 40 | 1101 | $9.6 \times 10^{-3}$ | $0.869 \times 10^{-5}$ | 2474 | 0.256 | $9.39 \times 10^{-8}$ | 93 | |
| 60 | 1008 | $5.2 \times 10^{-3}$ | $0.475 \times 10^{-5}$ | 2562 | 0.260 | $9.31 \times 10^{-8}$ | 51 | |
| 80 | 1078 | $3.2 \times 10^{-3}$ | $0.298 \times 10^{-5}$ | 2650 | 0.261 | $9.21 \times 10^{-8}$ | 32.4 | |
| 100 | 1059 | $2.15 \times 10^{-3}$ | $0.203 \times 10^{-5}$ | 2742 | 0.263 | $9.08 \times 10^{-8}$ | 22.4 | |
| | | | MINERAL LUBRICATING OIL (SAE 50) (UNUSED) | | | | | |
| 0 | 899 | 3.850 | $4.28 \times 10^{-3}$ | 1796 | 0.147 | $9.11 \times 10^{-8}$ | 47100 | |
| 20 | 888 | 0.80 | 0.90 | 1880 | 0.145 | $8.72 \times 10^{-8}$ | 10400 | $0.702 \times 10^{-3}$ |
| 40 | 876 | 0.21 | 0.24 | 1964 | 0.144 | $8.33 \times 10^{-8}$ | 2870 | |
| 60 | 864 | 0.072 | 0.0839 | 2047 | 0.140 | $8.00 \times 10^{-8}$ | 1050 | |
| 80 | 852 | 0.032 | 0.0375 | 2131 | 0.138 | $7.69 \times 10^{-8}$ | 490 | |
| 100 | 840 | 0.0170 | 0.0202 | 2219 | 0.137 | $7.38 \times 10^{-8}$ | 276 | |
| 120 | 829 | 0.0102 | 0.0123 | 2307 | 0.135 | $7.10 \times 10^{-8}$ | 175 | |
| 140 | 817 | 0.0065 | 0.0080 | 2395 | 0.133 | $6.86 \times 10^{-8}$ | 116 | |
| 160 | 806 | 0.0045 | 0.0056 | 2483 | 0.132 | $6.63 \times 10^{-8}$ | 84 | |
| | | | WATER | | | | | |
| 0 | 1002 | $1.78 \times 10^{-3}$ | $0.179 \times 10^{-5}$ | 4218 | 0.552 | $13.1 \times 10^{-8}$ | 13.6 | $0.66 \times 10^{-4}$ |
| 10 | 1001 | $1.30 \times 10^{-3}$ | $0.130 \times 10^{-5}$ | 4192 | 0.586 | $13.7 \times 10^{-8}$ | 9.30 | $0.88 \times 10^{-4}$ |
| 20 | 1001 | $1.00 \times 10^{-3}$ | $0.101 \times 10^{-5}$ | 4182 | 0.597 | $14.3 \times 10^{-8}$ | 7.02 | $2.06 \times 10^{-4}$ |
| 40 | 994.6 | $0.651 \times 10^{-3}$ | $0.0658 \times 10^{-5}$ | 4178 | 0.628 | $15.1 \times 10^{-8}$ | 4.34 | $3.72 \times 10^{-4}$ |
| 60 | 985.4 | $0.469 \times 10^{-3}$ | $0.0477 \times 10^{-5}$ | 4184 | 0.651 | $15.5 \times 10^{-8}$ | 3.02 | $5.15 \times 10^{-4}$ |
| 80 | 974.1 | $0.354 \times 10^{-3}$ | $0.0364 \times 10^{-5}$ | 4196 | 0.668 | $16.4 \times 10^{-8}$ | 2.22 | $6.55 \times 10^{-4}$ |
| 100 | 960.6 | $0.281 \times 10^{-3}$ | $0.0294 \times 10^{-5}$ | 4216 | 0.680 | $16.8 \times 10^{-8}$ | 1.74 | $7.49 \times 10^{-4}$ |
| 120 | 945.3 | $0.234 \times 10^{-3}$ | $0.0247 \times 10^{-5}$ | 4250 | 0.685 | $17.1 \times 10^{-8}$ | 1.446 | $8.92 \times 10^{-4}$ |
| 140 | 928.3 | $0.198 \times 10^{-3}$ | $0.0214 \times 10^{-5}$ | 4283 | 0.684 | $17.2 \times 10^{-8}$ | 1.241 | $10.0 \times 10^{-4}$ |

| | | | | | | | | |
|---|---|---|---|---|---|---|---|---|
| 160 | 909.7 | $0.172\times10^{-3}$ | $0.0189\times10^{-5}$ | 4342 | 0.680 | $17.3\times10^{-8}$ | 1.099 | $10.7\times10^{-4}$ |
| 180 | 889.0 | $0.154\times10^{-3}$ | $0.0173\times10^{-5}$ | 4417 | 0.675 | $17.2\times10^{-8}$ | 1.004 | $11.4\times10^{-4}$ |
| 200 | 866.7 | $0.138\times10^{-3}$ | $0.0160\times10^{-5}$ | 4504 | 0.665 | $17.1\times10^{-8}$ | 0.937 | $14.1\times10^{-4}$ |
| 220 | 842.4 | $0.125\times10^{-3}$ | $0.0149\times10^{-5}$ | 4610 | 0.653 | $16.8\times10^{-8}$ | 0.891 | $15.0\times10^{-4}$ |
| 240 | 815.7 | $0.117\times10^{-3}$ | $0.0143\times10^{-5}$ | 4756 | 0.635 | $16.4\times10^{-8}$ | 0.871 | $18.0\times10^{-4}$ |
| 260 | 785.9 | $0.108\times10^{-3}$ | $0.0137\times10^{-5}$ | 4949 | 0.611 | $15.6\times10^{-8}$ | 0.874 | $21.3\times10^{-4}$ |
| 280 | 752.5 | $0.102\times10^{-3}$ | $0.0135\times10^{-5}$ | 5208 | 0.580 | $14.8\times10^{-8}$ | 0.910 | $26.8\times10^{-4}$ |
| 300 | 714.3 | $0.096\times10^{-3}$ | $0.0135\times10^{-5}$ | 5728 | 0.540 | $13.2\times10^{-8}$ | 1.019 | |
| | | | FREON 12 | | | | | |
| −50 | 1547 | $0.480\times10^{-3}$ | $0.0310\times10^{-5}$ | 875.0 | 0.0675 | $5.01\times10^{-8}$ | 6.2 | $2.63\times10^{-3}$ |
| −40 | 1519 | $0.423\times10^{-3}$ | $0.0279\times10^{-5}$ | 884.7 | 0.0692 | $5.13\times10^{-8}$ | 5.4 | |
| −30 | 1490 | $0.377-10^{-3}$ | $0.0253\times10^{-5}$ | 895.6 | 0.0692 | $5.26\times10^{-8}$ | 4.8 | |
| −20 | 1461 | $0.343\times10^{-3}$ | $0.0235\times10^{-5}$ | 907.3 | 0.0710 | $5.39\times10^{-8}$ | 4.4 | |
| −10 | 1430 | $0.316\times10^{-3}$ | $0.0221\times10^{-5}$ | 920.3 | 0.0727 | $5.50\times10^{-8}$ | 4.0 | |
| 0 | 1397 | $0.229\times10^{-3}$ | $0.0214\times10^{-5}$ | 934.5 | 0.0727 | $5.57\times10^{-8}$ | 3.8 | |
| 10 | 1364 | $0.277\times10^{-3}$ | $0.0203\times10^{-5}$ | 949.6 | 0.0727 | $5.60\times10^{-8}$ | 3.6 | |
| 20 | 1330 | $0.263\times10^{-3}$ | $0.0198\times10^{-5}$ | 965.9 | 0.0727 | $5.60\times10^{-8}$ | 3.5 | |
| 30 | 1295 | $0.251\times10^{-3}$ | $0.0194\times10^{-5}$ | 983.5 | 0.0710 | $5.60\times10^{-8}$ | 3.5 | |
| 40 | 1257 | $0.240\times10^{-3}$ | $0.0191\times10^{-5}$ | 1002 | 0.0692 | $5.55\times10^{-8}$ | 3.5 | |
| 50 | 1216 | $0.229\times10^{-3}$ | $0.0189\times10^{-5}$ | 1022 | 0.0675 | $5.44\times10^{-8}$ | 3.5 | |
| | | | METHYL CHLORIDE $CH_3Cl$ | | | | | |
| −50 | 1053 | $0.337\times10^{-3}$ | $0.032\times10^{-5}$ | 1476 | 0.215 | $13.9\times10^{-8}$ | 2.31 | |
| −40 | 1033 | $0.328\times10^{-3}$ | $0.0318\times10^{-5}$ | 1483 | 0.209 | $13.7\times10^{-8}$ | 2.32 | |
| −30 | 1017 | $0.319\times10^{-3}$ | $0.0314\times10^{-5}$ | 1492 | 0.202 | $13.4\times10^{-8}$ | 2.35 | |
| −20 | 999.4 | $0.309\times10^{-3}$ | $0.0309\times10^{-5}$ | 1504 | 0.196 | $13.0\times10^{-8}$ | 2.38 | |
| −10 | 981.4 | $0.300\times10^{-3}$ | $0.0306\times10^{-5}$ | 1519 | 0.187 | $12.6\times10^{-8}$ | 2.43 | |
| 0 | 962.4 | $0.291\times10^{-3}$ | $0.0302\times10^{-5}$ | 1538 | 0.178 | $12.1\times10^{-8}$ | 2.49 | |
| 10 | 924.4 | $0.279\times10^{-3}$ | $0.0298\times10^{-5}$ | 1560 | 0.171 | $11.7\times10^{-8}$ | 2.55 | |
| 20 | 923.3 | $0.270\times10^{-3}$ | $0.0292\times10^{-5}$ | 1586 | 0.163 | $11.1\times10^{-8}$ | 2.63 | |
| 30 | 903.1 | $0.259\times10^{-3}$ | $0.0287\times10^{-5}$ | 1616 | 0.154 | $10.6\times10^{-8}$ | 2.72 | |
| 40 | 883.1 | $0.248\times10^{-3}$ | $0.0281\times10^{-5}$ | 1650 | 0.144 | $9.96\times10^{-8}$ | 2.83 | |
| 50 | 861.2 | $0.235\times10^{-3}$ | $0.0274\times10^{-5}$ | 1689 | 0.133 | $9.21\times10^{-8}$ | 2.97 | |

| *T* (°C) *Fusion* | *Boiling* | *T* (°C) | $\rho$ $(kg/m^3)$ | *c* $(J/kg\,°C)$ | $\mu \times 10^4$ $(kg/ms)$ | $\nu \times 10^6$ $(m^2/s)$ | $\lambda$ $(W/m\,K)$ | $a \times 10^6$ $(n^2/s)$ | *Pr* |
|---|---|---|---|---|---|---|---|---|---|
| | | BISMUTH | | | | | | | |
| | | 315 | 10001 | 144 | 16.2 | 0.160 | 16.4 | 11.25 | 0.0142 |
| 271 | 1477 | 538 | 9739 | 155 | 11.0 | 0.113 | 15.6 | 10.34 | 0.0110 |
| | | 760 | 9467 | 165 | 7.9 | 0.083 | 15.6 | 9.98 | 0.0083 |
| | | LITHIUM | | | | | | | |
| | | 204.4 | 509.2 | 4365 | 5.416 | 1.1098 | 46.37 | 20.96 | 0.051 |
| 179 | 1317 | 315.6 | 498.8 | 4270 | 4.465 | 0.8982 | 43.08 | 20.32 | 0.0443 |
| | | 426.7 | 489.1 | 4211 | 3.927 | 0.8053 | 38.24 | 18.65 | 0.0432 |
| | | 537.8 | 476.3 | 4171 | 3.473 | 0.7304 | 30.45 | 15.40 | 0.0476 |
| | | MERCURY | | | | | | | |
| −38.9 | 357 | −17.8 | 13707.1 | 141.5 | 18.334 | 0.1342 | 9.76 | 5.038 | 0.0266 |
| | | 93.3 | 13409.4 | 136.5 | 12.224 | 0.0903 | 10.38 | 5.619 | 0.0161 |
| | | 204.4 | 13168.1 | 135.6 | 10.046 | 0.0748 | 12.63 | 7.087 | 0.0108 |
| | | LEAD | | | | | | | |
| 327 | 1737 | 371 | 10540 | 159 | 2.40 | 0.023 | 16.1 | 9.61 | 0.024 |
| | | 704 | 10140 | 155 | 1.37 | 0.014 | 14.9 | 9.48 | 0.0143 |
| | | SODIUM | | | | | | | |
| | | 100 | 926.5 | 1384 | 6.70 | 0.723 | 85.7 | 66.9 | 0.01081 |
| | | 200 | 903.7 | 1339 | 4.46 | 0.493 | 80.9 | 66.8 | 0.00738 |
| 97.6 | 882 | 300 | 880.5 | 1304 | 3.40 | 0.387 | 76 | 66.2 | 0.00584 |
| | | 400 | 857 | 1278 | 2.81 | 0.328 | 71.2 | 65 | 0.00504 |
| | | 500 | 833.1 | 1262 | 2.43 | 0.291 | 66.3 | 63.1 | 0.00462 |
| | | 600 | 808.9 | 1255 | 2.16 | 0.267 | 61.5 | 60.6 | 0.00441 |
| | | 700 | 784.3 | 1257 | 1.97 | 0.251 | 56.6 | 57.4 | 0.00437 |
| | | 800 | 759.4 | 1268 | 1.82 | 0.239 | 51.8 | 53.8 | 0.00445 |
| | | POTASSIUM | | | | | | | |
| | | 100 | 824.1 | 813 | 4.41 | 0.535 | 48 | 71.7 | 0.00746 |
| | | 200 | 796.3 | 791 | 2.99 | 0.376 | 45.1 | 71.6 | 0.00525 |
| 63.2 | 757 | 300 | 769.8 | 775 | 2.32 | 0.301 | 42.3 | 71 | 0.00424 |
| | | 400 | 744.5 | 765 | 1.93 | 0.260 | 39.8 | 69.9 | 0.00372 |
| | | 500 | 720.6 | 762 | 1.69 | 0.234 | 37.5 | 68.3 | 0.00343 |
| | | 600 | 697.9 | 765 | 1.52 | 0.218 | 35.4 | 66.3 | 0.00329 |
| | | 700 | 676.5 | 774 | 1.40 | 0.206 | 33.5 | 64 | 0.00322 |
| | | NaK (56% K, 44% Na) | | | | | | | |
| | | 100 | 879.7 | 1064 | 5.32 | 0.605 | 22.8 | 24.4 | 0.02478 |
| | | 200 | 854.4 | 1032 | 3.57 | 0.418 | 24.8 | 28.1 | 0.01486 |
| 5.7 | 812 | 300 | 829.4 | 1008 | 2.75 | 0.331 | 26.2 | 31.3 | 0.01058 |
| | | 400 | 804.9 | 991 | 2.28 | 0.283 | 26.9 | 33.8 | 0.00838 |
| | | 500 | 780.9 | 982 | 1.98 | 0.253 | 27.1 | 35.4 | 0.00717 |
| | | 600 | 757.3 | 980 | 1.77 | 0.234 | 26.7 | 35.9 | 0.00651 |
| | | 700 | 734.2 | 986 | 1.62 | 0.221 | 25.6 | 35.4 | 0.00623 |
| | | 800 | 711.6 | 1000 | 1.50 | 0.211 | 24 | 33.7 | 0.00628 |

# Section 8

## Convection Equations in Various Coordinate Systems

### *Continuity equation*

Cartesian coordinate $(x, y, z)$:

$$\frac{\partial \rho}{\partial t} + \frac{\partial}{\partial x}(\rho V_x) + \frac{\partial}{\partial y}(\rho V_y) + \frac{\partial}{\partial z}(\rho V_z) = 0$$

Cylindrical coordinates $(r, \theta, z)$:

$$\frac{\partial \rho}{\partial t} + \frac{1}{r}\frac{\partial}{\partial r}(r\rho V_r) + \frac{1}{r}\frac{\partial}{\partial \theta}(\rho V_\theta) + \frac{\partial}{\partial z}(\rho V_z) = 0$$

Spherical coordinates $(r, \theta, \Phi)$:

$$\frac{\partial \rho}{\partial t} + \frac{1}{r^2}\frac{\partial}{\partial r}(\rho r^2 V_r) + \frac{1}{r \sin\theta}\frac{\partial}{\partial \theta}(\rho V_\theta \sin\theta) + \frac{1}{r \sin\theta}\frac{\partial}{\partial \Phi}(\rho V_\Phi) = 0$$

### *Momentum equation for Newtonian fluids with constant $\mu$ and $\rho$ (body force $\rho g$)*

Cartesian coordinates $(x, y, z)$:

$$\rho\left[\frac{\partial V_x}{\partial t} + V_x\frac{\partial V_x}{\partial x} + V_y\frac{\partial V_x}{\partial y} + V_z\frac{\partial V_x}{\partial z}\right] = -\frac{\partial p}{\partial x} + \rho g_x + \mu\left[\frac{\partial^2 V_x}{\partial x^2} + \frac{\partial^2 V_x}{\partial y^2} + \frac{\partial^2 V_x}{\partial z^2}\right]$$

$$\rho\left[\frac{\partial V_y}{\partial t} + V_x\frac{\partial V_y}{\partial x} + V_y\frac{\partial V_y}{\partial y} + V_z\frac{\partial V_y}{\partial z}\right] = -\frac{\partial p}{\partial y} + \rho g_y + \mu\left[\frac{\partial^2 V_y}{\partial x^2} + \frac{\partial^2 V_y}{\partial y^2} + \frac{\partial^2 V_y}{\partial z^2}\right]$$

$$\rho\left[\frac{\partial V_z}{\partial t} + V_x\frac{\partial V_z}{\partial x} + V_y\frac{\partial V_z}{\partial y} + V_z\frac{\partial V_z}{\partial z}\right] = -\frac{\partial p}{\partial z} + \rho g_z + \mu\left[\frac{\partial^2 V_z}{\partial x^2} + \frac{\partial^2 V_z}{\partial y^2} + \frac{\partial^2 V_z}{\partial z^2}\right]$$

Cylindrical coordinates $(r, \theta, z)$:

$$\rho\left\{\frac{\partial V_r}{\partial t} + V_r\frac{\partial V_r}{\partial r} + \frac{V_\theta}{r}\frac{\partial V_r}{\partial \theta} - \frac{V_\theta^2}{r} + V_z\frac{\partial V_r}{\partial z}\right\}$$

$$= -\frac{\partial p}{\partial r} + \rho g_r + \mu\left\{\frac{\partial}{\partial r}\left(\frac{1}{r}\frac{\partial}{\partial r}(rV_r)\right) + \frac{1}{r^2}\frac{\partial^2 V_r}{\partial \theta^2} - \frac{2}{r^2}\frac{\partial V_\theta}{\partial \theta} + \frac{\partial^2 V_r}{\partial z^2}\right\}$$

$$\rho\left\{\frac{\partial V_\theta}{\partial t} + V_r\frac{\partial V_\theta}{\partial r} + \frac{V_\theta}{r}\frac{\partial V_\theta}{\partial \theta} + \frac{V_r V_\theta}{r} + V_z\frac{\partial V_\theta}{\partial z}\right\}$$

$$= -\frac{1}{r}\frac{\partial p}{\partial \theta} + \rho g_\theta + \mu\left\{\frac{\partial}{\partial r}\left(\frac{1}{r}\frac{\partial}{\partial r}(rV_\theta)\right) + \frac{1}{r^2}\frac{\partial^2 V_\theta}{\partial \theta^2} + \frac{2}{r^2}\frac{\partial V_r}{\partial \theta} + \frac{\partial^2 V_\theta}{\partial z^2}\right\}$$

$$\rho\left\{\frac{\partial V_z}{\partial t} + V_r\frac{\partial V_z}{\partial r} + \frac{V_\theta}{r}\frac{\partial V_z}{\partial \theta} + V_z\frac{\partial V_z}{\partial z}\right\}$$

$$= -\frac{\partial p}{\partial z} + \rho g_z + \mu\left\{\frac{1}{r}\frac{\partial}{\partial r}\left(r\frac{\partial V_z}{\partial r}\right) + \frac{1}{r^2}\frac{\partial^2 V_z}{\partial \theta^2} + \frac{\partial^2 V_z}{\partial z^2}\right\}$$

Spherical coordinates $(r, \theta, \Phi)$:

$$\rho\left\{\frac{\partial V_r}{\partial t} + V_r\frac{\partial V_r}{\partial r} + \frac{V_\theta}{r}\frac{\partial V_r}{\partial \theta} + \frac{V_\Phi}{r\sin\theta}\frac{\partial V_r}{\partial \Phi} - \frac{V_\theta^2 + V_\Phi^2}{r}\right\}$$

$$= -\frac{\partial p}{\partial r} + \rho g_r + \mu\left\{\nabla^2 V_r - \frac{2}{r^2}V_r - \frac{2}{r^2}\frac{\partial V_\theta}{\partial \theta} - \frac{2}{r^2}V_\theta\cot\theta - \frac{2}{r^2\sin\theta}\frac{\partial V_\Phi}{\partial \Phi}\right\}$$

$$\rho\left\{\frac{\partial V_\theta}{\partial t} + V_r\frac{\partial V_\theta}{\partial r} + \frac{V_\theta}{r}\frac{\partial V_\theta}{\partial \theta} + \frac{V_\Phi}{r\sin\theta}\frac{\partial V_\theta}{\partial \Phi} + \frac{V_r V_\theta}{r} - \frac{V_\Phi^2\cot\theta}{r}\right\}$$

$$= -\frac{1}{r}\frac{\partial p}{\partial \theta} + \rho g_\theta + \mu\left\{\nabla^2 V_\theta + \frac{2}{r^2}\frac{\partial V_r}{\partial \theta} - \frac{V_\theta}{r^2\sin^2\theta} - \frac{2\cos\theta}{r^2\sin^2\theta}\frac{\partial V_\Phi}{\partial \Phi}\right\}$$

$$\rho\left\{\frac{\partial V_\Phi}{\partial t} + V_r\frac{\partial V_\Phi}{\partial r} + \frac{V_\theta}{r}\frac{\partial V_\Phi}{\partial \theta} + \frac{V_\Phi}{r\sin\theta}\frac{\partial V_\Phi}{\partial \Phi} + \frac{V_\Phi V_r}{r} + \frac{V_\Phi V_\theta\cot\theta}{r}\right\}$$

$$= -\frac{1}{r\sin\theta}\frac{\partial p}{\partial \phi} + \rho g_\Phi + \mu\left\{\nabla^2 V_\Phi - \frac{V_\Phi}{r^2\sin^2\theta} + \frac{2}{r^2\sin\theta}\frac{\partial V_r}{\partial \Phi} + \frac{2\cos\theta}{r^2\sin^2\theta}\frac{\partial V_\theta}{\partial \Phi}\right\}$$

$$\nabla^2 = \frac{1}{r^2}\frac{\partial}{\partial r}\left(r^2\frac{\partial}{\partial r}\right) + \frac{1}{r^2\sin\theta}\frac{\partial}{\partial \theta}\left(\sin\theta\frac{\partial}{\partial \theta}\right) + \frac{1}{r^2\sin^2\theta}\frac{\partial^2}{\partial \Phi^2}$$

## *Energy equation for Newtonian fluids with constant ρ and λ (omitting the dissipation term)*

Cartesian coordinates $(x, y, z)$:

$$\rho c_{\mathrm{p}}\left[\frac{\partial T}{\partial t}+V_x\frac{\partial T}{\partial x}+V_y\frac{\partial T}{\partial y}+V_z\frac{\partial T}{\partial z}\right]=\lambda\left[\frac{\partial^2 T}{\partial x^2}+\frac{\partial^2 T}{\partial y^2}+\frac{\partial^2 T}{\partial z^2}\right]$$

Cylindrical coordinates $(r, \theta, z)$:

$$\rho c_{\mathrm{p}}\left[\frac{\partial T}{\partial t}+V_r\frac{\partial T}{\partial r}+\frac{V_\theta}{r}\frac{\partial T}{\partial \theta}+V_z\frac{\partial T}{\partial z}\right]=\lambda\left[\frac{1}{r}\frac{\partial}{\partial r}\left(r\frac{\partial T}{\partial r}\right)+\frac{1}{r^2}\frac{\partial^2 T}{\partial \theta^2}+\frac{\partial^2 T}{\partial z^2}\right]$$

Spherical coordinates $(r, \theta, \Phi)$:

$$\rho c_{\mathrm{p}}\left[\frac{\partial T}{\partial t}+V_r\frac{\partial T}{\partial r}+\frac{V_\theta}{r}\frac{\partial T}{\partial \theta}+\frac{V_\Phi}{r\sin\theta}\frac{\partial T}{\partial \Phi}\right]$$
$$=\lambda\left[\frac{1}{r^2}\frac{\partial}{\partial r}\left(r^2\frac{\partial T}{\partial r}\right)+\frac{1}{r^2\sin\theta}\frac{\partial}{\partial \theta}\left(\sin\theta\frac{\partial T}{\partial \theta}\right)+\frac{1}{r^2\sin^2\theta}\frac{\partial^2 T}{\partial \Phi^2}\right]$$

Part VI

# Applications

Heat transfer is an engineering science *par excellence*, in that it appears in all fields of activity and in practically all applications. Indeed, thermal phenomena are irreversible parasitic phenomena which appear whenever there is a temperature difference due to chemical, mechanical, radiative, or nuclear processes, or simply due to heating. The field of application therefore ranges from heat treatment by laser, high-quality casting, laser austenitization of sheet steel, to architectural design of houses and factories, taking account of the climatic and solar environment, the operation of a nuclear power station, glass working, passive refrigeration systems, yoghurt production or mastery of the culinary arts. This list is far from being exhaustive.

The objective of this part is to introduce the student to the *physical modelling* of various real systems, such as those given above, and help him to acquire progressively the experience required by the thermal engineer. The sequence followed in many cases is listed below.

1. Analysis of a system, often defined in terms of a specification; that is, to *understand* its operation qualitatively, *identify* the forms of heat transfer physically, and *estimate* the orders of magnitude of the controlling parameters.
2. Definition of an overall strategy for *modelling* the system; that is, to proceed from a specification type of description to a *realistic* physical model. The authors have endeavoured not to cheat real life, that is, to use realistic dimensions for systems and for physical properties of real materials, even though certain approaches might be disputable. A physical model is not the real thing; in the most favourable case, it is a fairly good approximation to it. The alternative approach leads to academic results which are formally rigorous, but cannot be applied.
3. Solution of the problems once they have been physically modelled.

Stages (1) and (2) are the most important. They require knowledge of real situations and the real skill of the engineer, which is not essentially to solve the problems, but to identify them and to pose them clearly. The solution in complex cases is generally only a mathematical or numerical problem, of more or less difficulty, which should not be underestimated, but which is not within our objectives here.

As a result, the specification is generally non-academic and it is not definitive. The method of solution to be followed is not prescribed: that is the central point of the exercise. The numerical values required must often be sought in Part V; conversely, data given in the problem specification may (or may not) be superfluous and may not be needed for the solution.

# Application 1

## Insulation of a Cryogenic Container

### *Required knowledge*

- Steady-state energy balance.
- Conductive and conducto-convective flux (Chapter 1).
- Electrical analogy (Chapters 1 and 2).

### *Features*

- This is a contrived example for a first attempt at a heat transfer problem. In a simple example, the student is confronted with the need to construct and validate a model.
- At the application level, the comparison (in steady state) between the insulating material and insulation by vacuum (or by air without natural convection) is important.

### *The problem*

The container in the sketch, assumed to be spherically symmetric, is required to provide thermal insulation on the outside of a space initially filled with liquid nitrogen. The wall $r = R_0$ is thus held at 77 K. A small vent, whose effects will be neglected, ensures atmospheric pressure within the cavity.

The external face of the first metal container ($R_0 < r < R_1$) and the internal face of the second ($R_2 < r < R_3$) are polished, so that the radiative exchanges are negligible. The intermediate space $R_1 < r < R_2$ is evacuated. Four supports in the form of truncated cones, described below, locate the inner container. These supports are also covered with a perfectly reflecting cladding.

The second metallic container is surrounded by a layer of thermal insulation $R_3 < r < R_4$. The outer surface of the canister at $r = R_4$ is surrounded by ambient air at $T_e$. Only linear heat transfer (free convection of radiation*) is to be considered; a heat constant transfer coefficient $h = 10\ \mathrm{W\,m^{-2}\,K^{-1}}$ may be used.

1. Assuming steady-state conditions and that the thermal problem is rotationally symmetric between $R_0$ and $R_1$, and between $R_2$ and $R_4$, **how long will it take for half of the liquid nitrogen to evaporate?**
2. Comment on the assumptions.

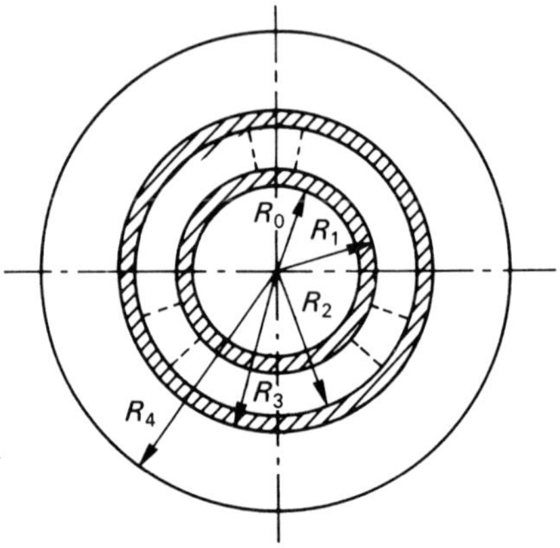


$R_0 = 0.149$ m; $R_1 = 0.150$ m; $R_2 = 0.200$ m; $R_3 = 0.201$ m;
$R_4 = 0.300$ m; $h = 10\ \mathrm{W\,m^{-2}\,K^{-1}}$ $Te = 298$ K

**Truncated conical supports:** perspex, properties from Touloukian.

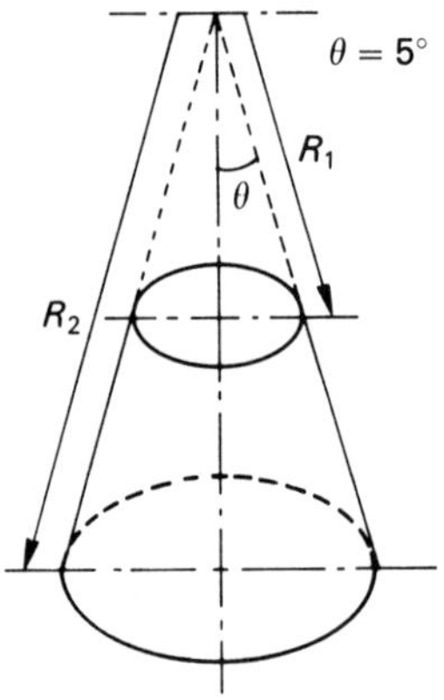


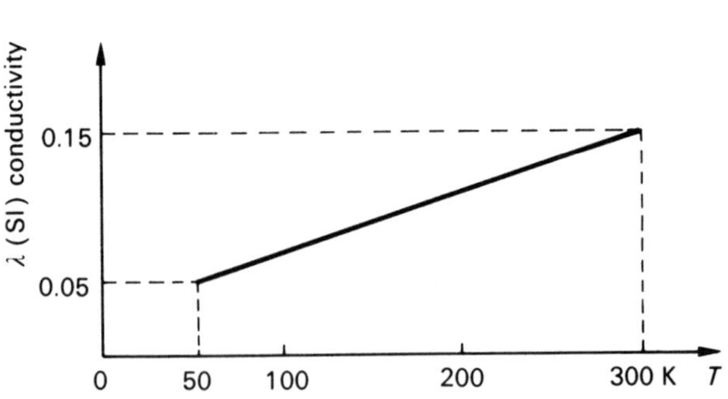


* The linearization of radiation, and its limitations, is discussed in Chapter 4.

**Metal containers**

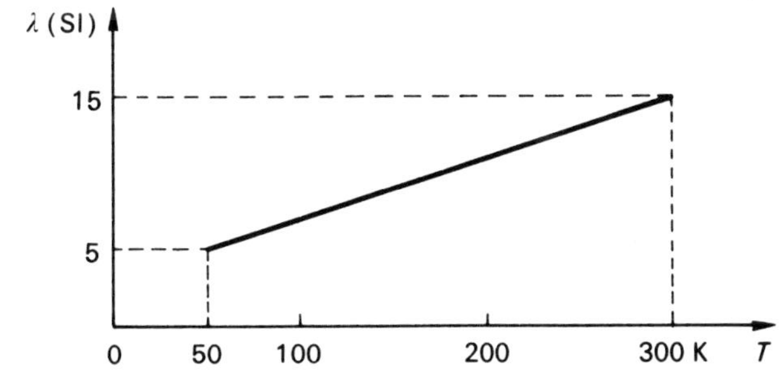


**External insulation:** glass wool, $\lambda = 3 \times 10^{-2}\ \mathrm{W\,m^{-1}\,K^{-1}}$, independent of $T$.

**Liquid nitrogen:** $\rho = 808\ \mathrm{kg\,m^{-3}}$; $L = 2 \times 10^5\ \mathrm{J\,kg^{-1}}$ at 77 K.
$\Lambda$, the molar latent heat of vaporization at atmospheric pressure, is actually given by:

$$\frac{\Lambda}{RT} = \log(82.07T_e) \qquad T_e = 77\ \mathrm{K}$$

# *The solution*

## *1. Method*

Faced with a real problem it is useful to make **quick estimates of the orders of magnitude** of the contributions of different elements to the insulation in the present case. This enables us to build a **realistic model**, which is the simplest possible model which describes the system adequately.

The initial intuitive idea to be confirmed in this problem is that in the steady state the temperature differences across the conductors (metals, used here for their mechanical strength) are much less than those across the insulators. Correspondingly, it will turn out that there is no point in considering the temperature dependence of the conductivity of the metal, but we must do so for an insulator.

## *2. Initial assumptions, orders of magnitude*

### *2.1 Analysis*

(A1) The system is geometrically and thermally (boundary conditions) **rotationally symmetric everywhere, except at the supports** (this exception will be discussed below).

(A2) The heat transfer is exclusively conductive for $R_0 < r < R_4$.

(A3) The system is assumed to be in a steady state. This behaviour will be assumed without justification (see Chapters 5–8). We assume that the two-phase medium (liquid nitrogen in equilibrium with its vapour) imposes its temperature $T_0 = 77$ K on the inner wall $r = R_0$ of the vessel, and that the surroundings are a constant temperature $T_e = 298$ K.

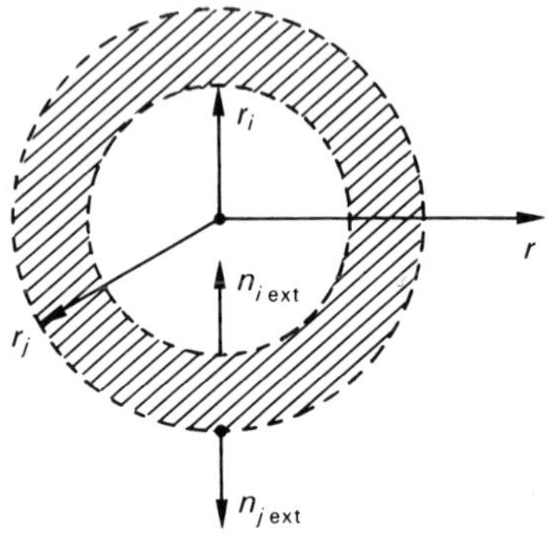


The energy balance equation applied to a system between two spheres $\Sigma_i$ and $\Sigma_j$ of radii $r_i$ and $r_j$ satisfying the conditions:

$$R_0 \leqslant r_i < r_j < R_4 + \varepsilon$$

can be used to show that for all $r_i$ and $r_j$:

$$\int_{\Sigma_i(r_i)} \mathbf{q}_i \cdot \mathbf{n}_{i\,\text{ext}}\, \mathrm{d}S - \int_{\Sigma_j(r_j)} \mathbf{q}_i \cdot \mathbf{n}_{j\,\text{ext}}\, \mathrm{d}S = 0 \tag{1}$$

which is equivalent to saying that the **total energy flux** $\Phi$ is conserved:

$$\Phi = \int_{\Sigma(r)} \mathbf{q} \cdot \mathbf{n}_{\text{ext}}\, \mathrm{d}S = \int_{\Sigma} \mathbf{q}\, \mathrm{d}S \tag{2}$$

where $\varphi$ is the algebraic energy flux density (which is not conserved) at any point $M$ in $\Sigma$.

*Notes*
1. In (2), $\varphi$ is not uniform on a sphere in the neighbourhood of the supports.
2. The energy flux is conducto-convective and radiative for $r = R_4 + \varepsilon$.

## 2.2 Calculation of the elementary thermal resistances

(a) Conduction between uniform concentric spheres. By integrating (2):

$$\forall r \in [R_i, R_j]: \quad 4\pi r^2 \left( -\lambda \frac{dT}{dr} \right) = \Phi$$

$$T_i - T_j = \frac{\Phi}{4\pi\lambda} \frac{R_j - R_i}{R_i R_j} = \mathscr{R}_{ij}\Phi$$

$$\mathscr{R}_{ij} = \frac{R_j - R_i}{4\pi\lambda R_i R_j}$$

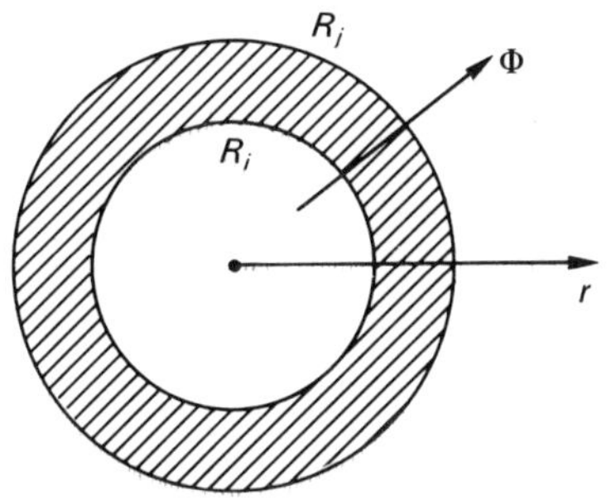


**Numerically:**

$$\mathscr{R}_{01} = 6 \times 10^{-4}\ \text{KW}^{-1}; \qquad 1.3 \times 10^{-4} < \mathscr{R}_{23} < 3.3 \times 10^{-4},$$

so long as 298 K $> T >$ 77 K.

$$\mathscr{R}_{34} = 4.35\ \text{KW}^{-1}$$

(b) Conducto-convective transfer at the walls (in this case $r = R_0$ and $r = R_4$):

$$\mathscr{R}_i^{cc} = \frac{1}{4\pi R_i^2 h_i}$$

**Numerically:** estimating $h_0$ and $h_4$:

- $h_0 \simeq 10^4\ \text{W m}^{-2}\ \text{K}^{-1}$ (see Chapter 2)
  $\mathscr{R}_0^{cc} \simeq 2 \times 10^{-4}\ \text{KW}^{-1}$
- $h_4 = 10\ \text{W m}^{-2}$; $\qquad \mathscr{R}_4^{cc} \simeq 0.088 \quad \text{KW}^{-1}$

### 2.3 Initial conclusions

At this stage, taking account of the series connections, we could take: $T_v \simeq T_0 \simeq T_1$ and $T_2 \simeq T_3$.

## 3. The supports

A more difficult problem is posed by the **supports**. Their conductivity is dependent on $T$, so the problem is non-linear, and the supports strictly break the symmetry of the system. On the other hand, the problem is greatly simplified by assuming the faces $r = R_1$ and $r = R_2$ are isothermal at $T_1$ and $T_2$, the steel producing this uniformity. The reflecting sides of the supports are assumed to be insulated.

The obvious solution compatible with the selected boundary conditions is to assume that concentric spherical surfaces within the supports are isothermal.

The energy balance, from (2), with $\omega$ as the solid angle of a support, is:

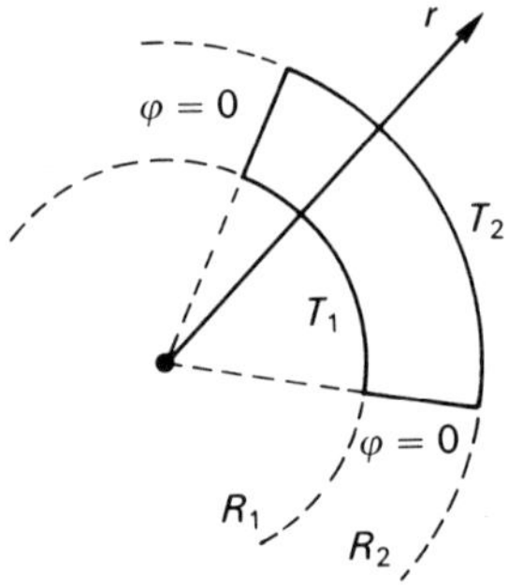


$$\frac{\Phi}{4} = \omega r^2 \left[ -\lambda(T) \frac{\mathrm{d}T}{\mathrm{d}r} \right] \qquad R_1 \leqslant r \leqslant R_2$$

Writing $\lambda = AT + B$ and integrating:

$$-\frac{\Phi}{4\omega} \frac{R_2 - R_1}{R_2 R_1} = \left[ \frac{A}{2}(T_2^2 - T_v^2) + B(T_2 - T_v) \right] \tag{3}$$

we obtain an equation in two unknowns, $T_2$ and $\Phi$.

The thermal resistance of the four supports, whose conductivity is assumed constant at $\lambda'$, is:

$$\mathscr{R}_{12}^{\text{unif}} = \frac{R_2 - R_1}{4\omega(R_1 R_2)\lambda'} \qquad \omega = 2.4 \times 10^{-2} \qquad \mathscr{R}_{12}(\lambda \text{ at } 77\ \text{K}) = 287\ \text{W}^{-1}\ \text{K}$$

$$\mathscr{R}_{12}\ (\lambda \text{ at } 298\ \text{K}) = 116\ \text{W}^{-1}\ \text{K}$$

Its true value lies between the extreme values corresponding to $\lambda'(77\ \text{K})$ and $\lambda'(298\ \text{K})$.

## 4. Energy balance

The aim of the calculation is to determine the flux $\Phi$ through the system. An exact calculation requires the solution of the following circuit:

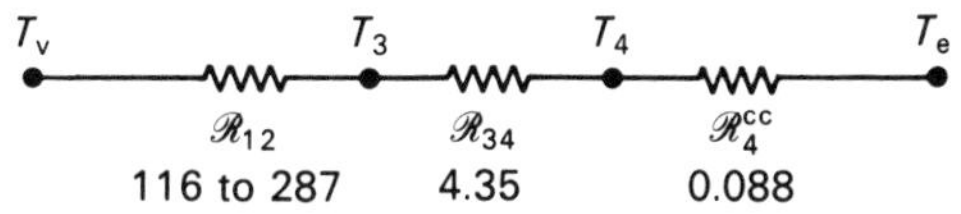


Since the problem is non-linear between $R_1$ and $R_2$, we must solve the equation:

$$-\Phi = \frac{T_e - T_3}{\mathscr{R}_{34} + \mathscr{R}_4^{cc}} = 4\omega \frac{R_2 R_1}{R_2 - R_1}\left[\frac{A}{2}(T_3^2 - 77^2) + B(T_3 - 77)\right]$$

$$\lambda = 0.03 + 4 \times 10^{-4} T \qquad \text{whence: } T_3 \simeq 292.3 \text{ K}$$

The power and duration are then given by:

$$\underline{\Phi = -1.284 \text{ W}}$$

$$\Delta t = \frac{\rho L}{2\Phi}\frac{4}{3}\pi R_0^3 \simeq 10 \text{ days}$$

## 5. Isotropy of the thermal field between $R_2$ and $R_4$

A rigorous steady-state calculation of the above system can only be carried out by a three-dimensional numerical method, because of the asymmetry introduced by the supports, particularly in the insulation. In view of the practical conclusion in the following paragraph (get rid of the insulation), this question becomes less important. The result obtained here is a good estimate of the order of magnitude sought by an engineer.

## 6. Practical conclusions

- Insulation by real insulating materials in the steady state has limited scope. There are no common high-performance thermal insulating materials such as exist for electrical insulation.
- Vacuum insulation is the most effective. (In fact, there will be a radiative heat transfer between the polished metal surfaces. This has not been taken into account here, but will be negligible. Methods of analysing such transfers are discussed in Chapter 10). The drawback is that the system requires a certain mechanical strength, which in practice requires the use of metal walls. In practice, one would not use external insulation in this application.

# Application 2

## Principles of Infra-red Long-range Detection

### *Required knowledge*

- First principles of radiation (Chapter 4 to Section 4.2).

### *Objective*

- To familiarize the student with the ideas of intensity and received flux, in a short and simple application.

### *The problem*

The atmosphere may be considered transparent in the region a few $\mu$m wide around a wavelength of 10 $\mu$m. An optical detector working in a narrow band $\Delta\lambda$ around $\lambda = 10$ $\mu$m and receiving a signal in $2\pi$ steradians is mounted on a rocket at altitude $H$. This detector may be directionally controlled either to maximize the signal detected from the earth (case 1) or to minimize this same signal (case 2). The earth is equivalent to a sphere of radius $R$, of uniform and isotropic intensity $I_\lambda$.

1. **What are the fluxes received by the detector in cases 1 and 2?** (We propose to treat the problem in its generality by two methods.)
2. Consider the second case. A spherical object of radius $r$ comes into the field of view of the detector. It is supposed to radiate in an isotropic manner with uniform intensity $I_{0\lambda}$. **What is the flux received by the detector when the object makes an**

**angle $\beta$ with the normal to the detector, at a distance $d$ from the latter?** (Assume that $d$ is large compared to $r$.)

3. **What could one detect in the first mode of operation? What use could be made of a 2 degree of freedom mounting with an accurate resolution of** $d\Omega$? Give a practical example.

**Numerical data:** $R = 6500$ km; $H = 10$ km; $r = 1$ m; $d = 2$ km; $I_\lambda = 6 \times 10^5$ W m$^{-3}$; $I_{0\lambda} = 10^5$ W m$^{-3}$; $\beta = 60°$; spectral width $\Delta\lambda = 0.1$ $\mu$m.

## The solution

**1.** The detector is equivalent to an elementary plane surface. In case 1, the normal to the detector passes through the centre of the earth, the detector facing the latter. In case 2, the detector faces in the opposite direction.

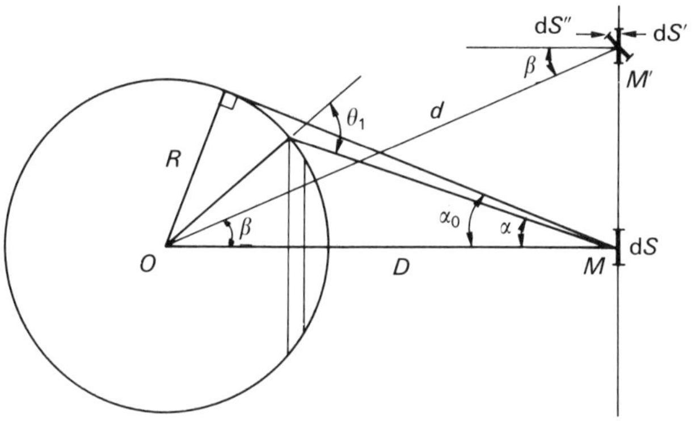


The problem to be solved is illustrated above. The plane surface element $dS$ receives radiation from part of a sphere of centre $O$ at a distance $D = R + H$.

### First calculation method

The monochromatic flux radiated by the sphere into the whole of space is given by (see Section 4.2.2):

$$d\Phi_\lambda = 4\pi R^2 (\pi I_\lambda)\, d\lambda$$

The monochromatic flux crossing a sphere of radius $D$ and containing $dS$, concentric with the source, is also $d\Phi_\lambda$ (by conservation of radiative energy). The flux density $d\varphi^i_\lambda$ on this sphere of radius $D$ is uniform:

$$d\Phi_\lambda = 4\pi D^2\, d\varphi^i_\lambda = 4\pi R^2 (\pi I_\lambda)\, d\lambda$$

$$d\varphi^i_\lambda = \left(\frac{R}{D}\right)^2 (\pi I_\lambda)\, d\lambda$$

In mode 2, the flux received $d\varphi^i_\lambda \simeq 0$. (The detector does not see the earth, but the depths of the sky. The interstellar radiation is negligible since it is characterized by a temperature of about 4 K and planetary radiation. Further, we neglect the radiation at $\lambda_0 \simeq 10\ \mu m$ which is emitted by the earth and scattered backwards from atmospheric molecules.) Actually, there will be a non-null signal at the detector.

The flux received in mode 1 corresponds to the diagram above (neglecting the attenuation of the radiation due to scattering from atmospheric molecules).

**Numerically:**

$$R = 6500 \text{ km}$$

$$D = 6510 \text{ km}$$

$$I_\lambda = 6 \times 10^5 \text{ W m}^{-3} \text{ (at } \lambda_0 = 10\ \mu\text{m)}$$

$$\Delta\lambda = 0.1\ \mu\text{m}$$

$$d\varphi^i_\lambda = \pi I_\lambda \Delta\lambda \simeq \underline{0.188 \text{ W m}^{-2}}$$

*Second calculation method (systematic, but less elegant than the above)*

If $dS_1$ is an element of the surface of the sphere, using the notation on the diagram above, the flux is (Section 4.2):

$$d^5\Phi_\lambda = I_\lambda \frac{dS_1 \cos\theta_1 \, dS \cos\alpha}{r^2} d\lambda$$

Since the intensity of the radiation concerned is isotropic, this formula is completely symmetric with regard to the elementary surface $dS$ and $dS_1$. It can be written in such a way as to simplify the calculations:

$$d^5\Phi_\lambda = I_\lambda \, dS \, d\Omega \cos\alpha \, d\lambda$$

with

$$d\Omega = \frac{dS_1 \cos\theta_1}{r^2}$$

Given the rotational symmetry of the diagram, $d\Omega$ represents the solid angle through which an elementary spherical segment of area on the axis $OM$ is seen from $M$:

$$d\Omega = 2\pi \sin\alpha \, d\alpha$$

The flux received by $dS$ is thus:

$$d^3\Phi^i_\lambda = I_\lambda \, dS \, d\lambda \, 2\pi \int_0^{\alpha \text{ limit}} \sin\alpha \cos\alpha \, d\alpha$$

$$= \pi I_\lambda \, dS \, d\lambda \frac{R^2}{D^2}$$

Whence:

$$d\varphi_\lambda^i = \pi I_\lambda \frac{R^2}{D^2} d\lambda$$

We clearly obtain the same result as before.

**2.** If a spherical object passes in front of the detector when it is turned towards the sky, the situation is still that shown in the diagram, but with the sphere as the object and the detector represented by the surface $dS'$, whose normal does not pass through the centre of the object.

The surface $dS'$ makes an angle $\beta$ with its projection $dS''$ in the plane which is tangential to a sphere of centre $O$, radius $d$ (see diagram).

The flux received by $dS'$ is the same as that received by $dS''$:

$$d\varphi_\lambda'^i(M')\, dS' = d\varphi_\lambda''^i(M')\, dS''$$

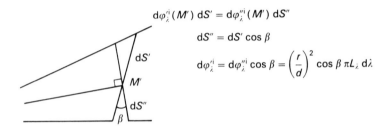

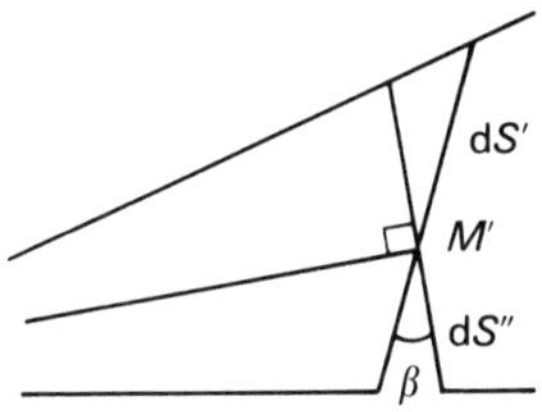


$$dS'' = dS' \cos\beta$$

$$d\varphi_\lambda'^i = d\varphi_\lambda''^i \cos\beta = \left(\frac{r}{d}\right)^2 \cos\beta\, \pi L_\lambda\, d\lambda$$

For a point source emitting the same monochromatic power isotropically, the monochromatic flux crossing the sphere of radius $d$ is the same. This flux density is:

$$\underline{d\varphi_\lambda'^i(M') \simeq 4 \times 10^{-9}\ \mathrm{W\,m^{-2}}}$$

**3.** With this procedure, the object modifies the flux undetectably. But if a directional mounting enables directional fluxes to be detected with a high resolution $d\Omega$, a significant contrast appears between the image of the object and that of the earth. One would also reconstruct images using infra-red cameras with HgCdTe detectors.

# Application 3

## Temperature of a Material exposed to Solar Radiation

### *Required knowledge*

- Steady-state energy balance (Chapters 1 and 2).
- First principles of radiation (Chapter 4).

### *Features*

- Principles of the selectivity of material radiative properties having regard to incident radiation. The choice of radiative properties for material coatings is critical, and depends strongly on the application concerned.

### *The problem*

We will consider a plane element which is part of the external face of a satellite which is exposed, at very high altitude, to direct solar radiation. (The sun is considered as a black body at temperature $T_s$.) In such conditions, heat transfer with the surroundings (the stratosphere) via conductive or conducto-convective means is negligible. The satellite's insulating structure may be thermally modelled as a simple equivalent conductance $\gamma$, between the surface of the element concerned at temperature $T$ and the fixed internal temperature $T_i$ of the satellite.

**What is the temperature** $T$ **of the plane element,** when subject to an incident solar flux density $\varphi^S$ normal to the surface, under the three following hypotheses (materials $A$, $B$ and $C$)?

Material $A$ (reference): black body

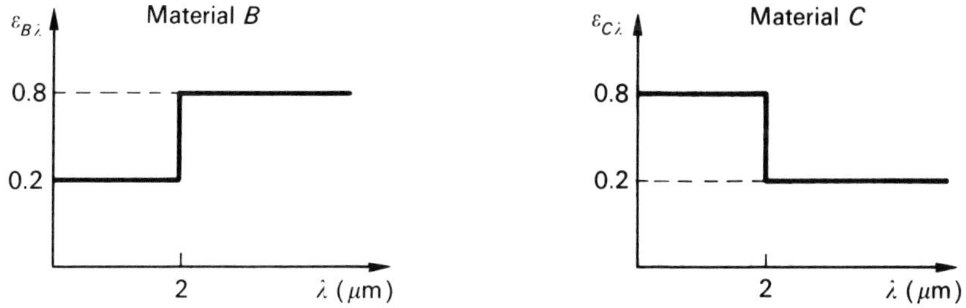


Emission and reflection from materials $B$ and $C$ are assumed to be isotropic.

**Data:** $\varphi^{\mathrm{R}} = 1400\ \mathrm{W\,m^{-2}}$; $T_{\mathrm{i}} = 273\ \mathrm{K}$; $\gamma = 0.4\ \mathrm{W\,m^{-2}\,K^{-1}}$.

## *The solution*

The required temperature $T$ may be obtained from a simple steady-state energy balance:

$$\varphi^{\mathrm{R}} S + \gamma S(T_{\mathrm{i}} - T) = 0 \tag{1}$$

where $\varphi^{\mathrm{R}}$ is the algebraic radiative flux density at the surface of the element, with inwardly directed normal:

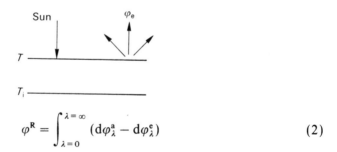

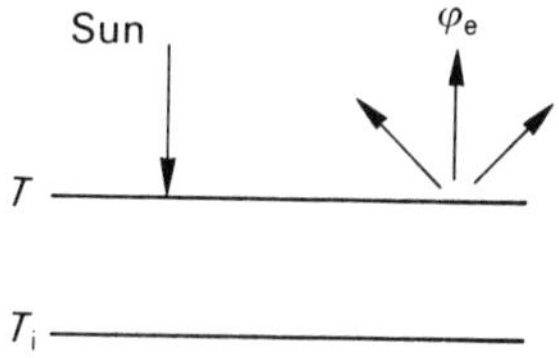


$$\varphi^{\mathrm{R}} = \int_{\lambda=0}^{\lambda=\infty} (\mathrm{d}\varphi_{\lambda}^{\mathrm{a}} - \mathrm{d}\varphi_{\lambda}^{\mathrm{e}}) \tag{2}$$

Note that:

- the monochromatic flux $\mathrm{d}\varphi_{\lambda}^{\mathrm{e}}$ is emitted in $2\pi$ steradians:

$$\mathrm{d}\varphi_{\lambda}^{\mathrm{e}} = \varepsilon_{\lambda} \pi I_{\lambda}^{0}(T)\,\mathrm{d}\lambda \tag{3}$$

  in which we have made use of the isotropy property of $I_{\lambda}^{\prime\mathrm{e}} = \varepsilon_{\lambda} I_{\lambda}^{0}(T)$.
- the absorbed monochromatic flux is **of directional origin** (bounded by the unknown solid angle through which the sun is seen, $\mathrm{d}\Omega$). If the temperature of the sun is $T_{\mathrm{S}}$:

$$\mathrm{d}\varphi_{\lambda}^{\mathrm{a}} = \varepsilon_{\lambda}\,\mathrm{d}\varphi_{\lambda}^{\mathrm{i}} = \varepsilon_{\lambda}(I_{\lambda}^{0}(T_{\mathrm{S}})\,\mathrm{d}\Omega\,\mathrm{d}\lambda) \tag{4}$$

  (the sun is assumed to be a black body, and is viewed at normal incidence, $\cos\theta = 1$).

The quantity dΩ is unknown. On the other hand, we do know the total incident flux density $\varphi^S$:

$$\varphi^S = \int_{\lambda=0}^{\lambda=\infty} d\varphi_\lambda^i \tag{5}$$

from which it follows:

$$\frac{d\varphi_\lambda^i}{\varphi^S} = \frac{I_\lambda^0(T_S)\,d\lambda}{\int_0^\infty I_\lambda^0(T_S)\,d\lambda} = \frac{\pi I_\lambda^0(T_S)\,d\lambda}{\sigma T_S^4} \tag{6}$$

Finally:

$$\varphi^R = \int_0^\infty \pi\varepsilon_\lambda \left[\frac{I_\lambda^0(T_S)}{pT_S^4}\varphi^S - I_\lambda^0(T)\right] d\lambda \tag{7}$$

Considering (1) and (7):

$$\gamma(T_i - T) + \int_0^\infty \pi\varepsilon_\lambda \left[\frac{I_\lambda^0(T_S)}{\sigma T_S^4}\varphi^S - I_\lambda^0(T)\right] d\lambda = 0 \tag{8}$$

Case 1: perfect black body: $\varepsilon_\lambda = 1$ and equation (8) becomes:

$$\gamma(T_i - T) + (\varphi^S - \sigma T^4) = 0$$

**Numerically:**

$$\underline{T_A = 393\text{ K}} \quad \text{(by iteration)}$$

Case 2: body $B$: We will use the function $z[0, (\lambda_0/\lambda_m(T))]$ (see Basic Data, Section 5.3, and Chapter 4, Section 4.4).

Equation (8) becomes:

$$\gamma(T_i - T) + 0.8(\varphi^S - \sigma T^4) - 0.6\int_0^{2\mu m} \pi\left[\frac{I_\lambda^0(T_S)}{\sigma T_S^4}\varphi^S - I_\lambda^0(T)\right] d\lambda = 0$$

If we assume that $T < 600$ K, there will only be appreciable radiation for:

$$\lambda > \frac{\lambda_m(T)}{2} \qquad \text{or} \qquad \lambda > \frac{3000}{1200} \sim 2.5\ \mu m$$

Whence:

$$\int_0^{2\mu m} \pi I_\lambda^0(T)\,d\lambda \simeq 0$$

It remains to solve:

$$\gamma(T_i - T) + 0.8(\varphi^S - \sigma T^4) - 0.6\varphi^S z\left(0, \frac{2}{\lambda_m(T_S)}\right) = 0$$

In Basic Data, Section 5.3, we find:

$$\lambda_m(T_S) \simeq \frac{3000}{5700} \simeq 0.526; \qquad x = \frac{2}{0.526} = 3.8$$

so:

$$z\left(0, \frac{2}{\lambda_m(T_S)}\right) = z(0, 3.8) = 0.932$$

and:

$$\gamma(T_i - T) + 0.24\varphi^S - 0.8\sigma T^4 = 0$$

**Numerically:**

$$\underline{T_B = 292 \text{ K}}$$

(The assumption that $T < 600$ K is satisfied.)

Case 3: body $C$.
Again, we use the function $z$, which in this case, from equation (8), gives us:

$$\gamma(T_i - T) + 0.2(\varphi^S - \sigma T^4) + 0.6 \int_0^{2\mu m} \pi \left[ \frac{I_\lambda^0(T_S)}{\sigma T_S^4} \varphi^S - I_\lambda^0(T) \right] d\lambda = 0$$

Still assuming $T < 600$ K, we find from this:

$$\gamma(T_i - T) + 0.759\varphi^S - 0.2\sigma T^4 = 0$$

**Numerically:**

$$\underline{T_C = 540 \text{ K}}$$

(The assumption that $T < 600$ K is satisfied.)

## *Practical conclusions*

The temperature of a body exposed to a radiative environment is strongly dependent on the selective character of its radiative properties. Relative to a black body, we can state the following:

- body $B$ is a selectively cold material relative to solar radiation.
- body $C$ is a selectively hot material relative to solar radiation.

The selective properties of the body depend strongly on the spectral composition of the incident radiation (which is not necessarily equilibrium radiation characterized by a single temperature).

### *Pragmatic (semi-quantitative) approach*

- The sun is at a temperature of 5700 K; its radiation corresponds in practice to the range $[(\lambda_m/2)(5700\ K), 8\lambda_m]$ or $[0.26, 4.5\ \mu m]$. The majority of the solar flux is reflected in case *B*, and similarly is absorbed in case *C*.
- The material is at a temperature 'of the order of' 300 K, corresponding to the range $[(\lambda_m/2)(300\ K), 8\lambda_m]$ or $[5, 80\ \mu m]$. As a result, emission from the material is important in case *B* but small in case *A*.

These two arguments lead to the conclusion that *B* is a 'cold' substance, relative to solar radiation, whilst *C* is a 'hot' substance for the same radiation.

# Application 4

## Heating a Metal Ingot in a Furnace

### *Required knowledge*

- Steady conditions (Chapter 1).
- Understanding of radiation (Chapter 4).

### *Relevance*

- The radiation-related concepts required to be understood for this exercise are fundamental.

### *The problem*

A spherical metal ingot of radius $R$, initially at temperature $T_0$, is supported in a rectangular furnace of sides $a$, $b$ and $c$ by a structure of negligible thermal effect. Relative to an origin at one corner, the centre of the ingot is at $a/2$, $b/3$, $c/2$.

All the walls of the furnace are isothermal at a temperature $T_F$. The uniform emissivities of the internal walls $\varepsilon'_{F\lambda}$ and of the ingot $\varepsilon'_{I\lambda}$ have been determined by a specialist laboratory, which has presented the results in terms of simple analytical expressions:

$$\varepsilon'_{F\lambda} = \frac{0.95\cos\theta}{\sqrt{\lambda}} \qquad \lambda \in [1, 100\ \mu\text{m}]$$

$$\varepsilon'_{I\lambda} = 0.6\cos\theta$$

Heat transfer is purely by radiation. **Determine the flux supplied by the furnace to the ingot, as a function of the surface temperature.**

**Numerical data:** $a = 2$ m; $b = 3$ m; $c = 2$ m; $R = 0.2$ m; $T_0 = 300$ K; $T_F = 1300$ K.

## *The solution*

A simple and careful analysis will simplify the solution considerably.

In the absence of the ingot, the radiation in the furnace at all points in space and all directions has an equilibrium intensity $I_\lambda^0(T_F)$. What disturbance to this radiation will the ingot introduce, from an energy point of view? From the point of view of emitted and absorbed radiation:

- The flux **emitted** by the furnace faces of internal area $S_F$ is (Section 4.3.3):

$$\Phi_F^e = S_F\left(\int_0^{\pi/2} \cos^2\theta\, d\Omega\right)0.95\left(\int_0^{\infty}\frac{I_\lambda^0(T_F)}{\sqrt{\lambda}}\,d\lambda\right)$$

The flux emitted by the ingot of surface $S_I$ is:

$$\Phi_I^e = S_I\left(\int_0^{\pi/2} \cos^2\theta\, d\Omega\right)0.6\,\frac{\sigma T_I^4}{\pi}$$

so that:

$$\frac{\Phi_I^e}{\Phi_F^e} = \frac{S_I}{S_F}\frac{0.6}{0.95}\frac{\sigma T_I^4}{\displaystyle\int_0^{\infty}\pi I_\lambda^0(T_F)\frac{d\lambda}{\sqrt{\lambda}}}$$

The spectral interval in which the furnace has appreciable emission is $[1, 17\ \mu\text{m}]$. We can estimate the answer by replacing $\sqrt{\lambda}$ by $\sqrt{\lambda_0}$ with $\lambda_0 = 4\ \mu$m.

Thus:

$$\frac{\Phi_I^e}{\Phi_F^e} \sim \frac{S_I}{S_F}\frac{0.6}{0.95}\,2\left(\frac{T_I}{T_F}\right)^4 \qquad S_I = 4\pi R^2 = 0.50\ \text{m}^2; \qquad S_F = 32\ \text{m}^2$$

$$\frac{\Phi_I^e}{\Phi_F^e} \sim 5.5\times 10^{-5} \qquad (T_I = T_0 = 300\ \text{K})$$

$$\frac{\Phi_I^e}{\Phi_F^e} \sim 1.88\times 10^{-2} \qquad (T_I = T_F)$$

The flux emitted by the ingot is always negligible compared to that emitted by the furnace.

- We assume that the proportion of the radiation emission by the furnace which is **absorbed** by the ingot, $\Phi_I^a/\Phi_F^e$, is negligible; this will be shown retrospectively.

We see, then, that the ingot is sufficiently small to have no effect on the equilibrium radiation in the chamber, since both emission and absorption are negligible. The ingot thus receives equilibrium radiation at the furnace temperature. Further, its surface temperature $T_I$ is clearly uniform, since the conditions around the ingot are uniform.

The radiative flux on the ingot is:

$$\Phi_I^R = \Phi_I^a - \Phi_I^e$$

The flux absorbed is:

$$\begin{aligned}\Phi_I^a &= S_I \int_0^\infty d\lambda \int_0^{\pi/2} \varepsilon'_{I\lambda} I_\lambda^0(T_F)\cos\theta \, d\Omega \\ &= 0.6 S_I \sigma T_F^4 \frac{1}{\pi} \int_0^{\pi/2} \cos^2\theta \, 2\pi \sin\theta \, d\theta \\ &= 0.4 S_I \sigma T_F^4\end{aligned}$$

Similarly, we can obtain the emitted flux by replacing $I_\lambda^0(T_F)$ by $I_\lambda^0(T_I)$ in the above equation, to obtain:

$$\Phi_I^e = 0.4 S_I \sigma T_I^4$$

The furnace thus supplies an instantaneous flux:

$$\underline{\Phi = 1.6\pi R^2 \sigma (T_F^4 - T_I^4)}$$

**Validation:**

$$\frac{\Phi_I^a}{\Phi_F^e} = \frac{2S_I}{S_F} \times \frac{0.6}{0.95} \sim 1.8 \times 10^{-2} \qquad \text{which is negligible}$$

**Discussion.** The temperature $T_I(t)$ is only the surface temperature of the ingot in unsteady conditions. Note that the boundary condition at $r = R$ can in practice be expressed in three different ways according to the respective values of $T_F$ and $T_I$:

Case 1: $T_I < 500$ K: $\Phi \simeq 1.6\pi R^2 \sigma T_F^4$
Initial phase – heating at constant flux (error <3%)

Case 2: $T_I > 1000$ K: $\Phi \simeq 1.6\pi R^2 4\sigma (1150)^3 (T_F - T_I)$
Final phase – heating with linearized flux

Case 3: $500 < T_I < 1000$ K: The non-linear flux expression must be used.

In the two first cases, the well-known linear problems resulting have simple or tabulated solutions (Schneider: see bibliography of Part 2). The third case must be solved numerically.

# Application 5

## Temperature Measurement

***Required knowledge***

*Application 5.1*

- Principles of balance and of heat transfer modes (Chapter 1).
- Non-steady conditions (Chapters 5–6).

*Application 5.2*

- The steady-state (Chapters 1 and 2).
- External convection (Chapter 13).

*Application 5.3*

- Radiative transfers (Chapters 9, 10 and 11).
- External and internal convections (Chapter 12).

*Application 5.4*

- Principles of balance and the steady state (Chapters 1 and 2).
- Radiative transfer (Chapters 9, 10 and 11).
- External laminar convection (Chapter 12).

## *Objective*

- To give a simple illustration of the precautions which must be taken when measuring temperature with a particular sensor – the thermocouple.

## *The problem*

There exist numerous methods of measuring temperature, some of which have been developed for particular applications. The two principal methods, and the most widely used in industry, are based on thermocouple and resistance thermometers. The important characteristic of these methods is that the information is given directly in the form of an electrical signal which can be transmitted, recorded or processed directly.

The simplest thermocouple consists of two conductors made of metals with different thermoelectric characteristics, welded together. In practical thermocouples, the conductors are embedded in a compressed powder of an inert mineral such as magnesia, and surrounded by a waterproof metallic sheath made, for example, from stainless steel or Inconel.

Any temperature sensor in a medium can only give an indication of **its own temperature.** From this, we have to be able to deduce the **temperature at the measuring point**, preferably in the absence of the sensor. To do this, we must take account of the response time for conduction within the sensor, of radiation and of conducto-convective exchanges between the sensor and its environment, etc.

The four following exercises are illustrations of the precautions which must be taken when measuring:

(a) response time;
(b) temperature measurement in a liquid — the thermocouple well;
(c) temperature measurement in a gas — the radiation shield;
(d) wall temperature measurement.

## *1. Response time*

**Compare the temperature given by an iron–constantan thermocouple with that given by a mercury thermometer** when the instruments are used to measure the temperature $T_f$ of a fluid which varies sinusoidally with time:

$$T_f = T_1 + T_2 \sin \omega t$$

$$\omega = \frac{2\pi}{\tau} \qquad \tau = 10\ \text{min} \quad \text{(period)}$$

The initial temperature of each instrument is $T_0$. We assume the same heat transfer coefficient $h = 28.4\ \mathrm{W\,m^{-2}\,K^{-1}}$ between each instrument and the fluid. The thermocouple has a diameter of 1 mm and its immersed length is 30 mm, and the materials of which it is made are assumed to be homogeneous. The thermometer will be considered as a mercury cylinder 6 mm diameter and 15 mm long. All the mercury is assumed to be the same temperature at any time. We ignore variations in $C_p$ and also the volume of mercury in the column above the reservoir.

**Numerical values:**

| | Thermometer | Thermocouple |
|---|---|---|
| $\rho =$ | $13.6 \times 10^3\ \mathrm{kg\,m^{-3}}$ | $7.9 \times 10^3\ \mathrm{kg\,m^{-3}}$ |
| $c_p =$ | $140\ \mathrm{J\,kg^{-1}\,K^{-1}}$ | $460\ \mathrm{J\,kg^{-1}\,K^{-1}}$ |
| $T_1 = 100\,°\mathrm{C}$ | $T_2 = 50\,°\mathrm{C}$ | $T_0 = 60\,°\mathrm{C}$ |

## 2. Temperature measurement in a liquid – the thermocouple well

Water, whose mixed fluid temperature we wish to determine, circulates at $15\ \mathrm{cm\,s^{-1}}$ in a stainless steel (chrome–nickel steel 18% Cr, 8% Ni) tube of diameter 15 mm. In order to do this, a hollow cylinder, called a thermocouple well, 4 mm diameter and of 1 mm wall thickness, also made of stainless steel, is welded onto the tube.

**How long should the thermocouple well be** in order that the difference between the indicated temperature and the actual temperature of the water should not exceed 0.5% of the difference between the water temperature and the wall temperature $T_w$ of the tube in which the water is flowing?

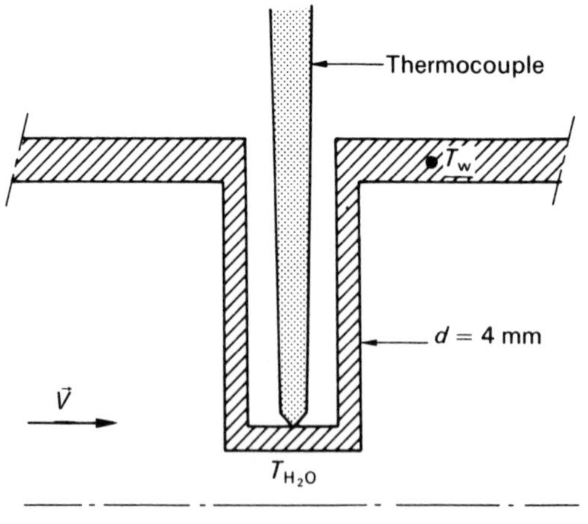

## 3. Temperature measurement in a gas

We consider a transparent gas in turbulent flow in a metallic duct at temperature $T_2 = 500$ K, of diameter $d_2 = 100$ mm, which is a grey body of emissivity $\varepsilon_2 = 0.565$.

It is proposed to measure the temperature of the gas using a thermocouple of diameter $d_3 = 0.5$ mm, on the duct axis. The thermocouple is surrounded by a thin cylindrical shield whose diameter $d_1$ is four times the thermocouple diameter $d_3$. The shield and thermocouple are taken as grey bodies of emissivities $\varepsilon_1 = 0.3$ and $\varepsilon_3 = 0.8$ respectively. The convective heat transfer coefficients between the shield and the gas and between the thermocouple and the gas are $h_1 = 100\ \mathrm{W\,m^{-2}\,K^{-1}}$ and $h_3 = 120\ \mathrm{W\,m^{-2}\,K^{-1}}$ respectively.

The temperature indicated by the thermocouple is $T_3 = 800$ K. **What is the gas temperature $T_g$? In the absence of the radiation shield, what would be the temperature $T'_3$ indicated by the thermocouple?**

## 4. Wall temperature measurement

A simple thermocouple is made of two conductors 1 and 2, whose junction, of small dimensions, is placed in contact with a wall whose temperature $T_w$ we wish to measure. The imperfect contact between the junction and the wall is sketched in Figures 1(a) and 1(b).

The junction, where the thermal EMF is generated, is assumed to be uniformly at the indicated temperature $T_j$, and can be modelled as a cylinder of radius $R$ and thickness $\eta'$. It only makes contact with the wall, which is at temperature $T_w$, through a metal ring* of conductivity $\lambda$ and rectangular cross-section with sides $\xi$ and $\eta$* (Figures 1(a) and 1(b)). Between the wall at $T_w$, which is a black body, and the junction whose emissivity is $\varepsilon_j = 0.8$ is a layer of air of thickness $\eta$ and conductivity $\lambda_a$. The sole purpose of this highly idealized model is to estimate the effect of the imperfect thermal contact. The metal conductors 1 and 2, which are of length 2 m, diameter $d_c$, conductivity $\lambda_c$, emissivity $\varepsilon_c$ and are surrounded by a body at temperature $T_a$, exchange heat by conducto-convective transfer (natural convection) with the air at $T_a$, with a convective heat transfer coefficient $h_a$.

**Evaluate the difference between the temperature $T_w$ of the wall and the measured temperature $T_j$.**

**Numerical values** (in SI units):

| | |
|---|---|
| $\eta = 10^{-4}$ m | $T_w = 50\,°\mathrm{C}$ |
| $\xi = 2 \times 10^{-5}$ m | $h_a = 5\ \mathrm{Wm^{-2}\,K^{-1}}$ |
| $R = 5 \times 10^{-4}$ m | $\lambda_c = 24.5\ \mathrm{Wm^{-1}\,K^{-1}}$ |
| $\varepsilon_j = 0.8$ | $\varepsilon_c = 0.5$ |
| $\eta' = 10^{-3}$ m | $d_c = 5 \times 10^{-4}$ m |
| $\lambda = 15\ \mathrm{Wm^{-1}\,K^{-1}}$ | $\lambda_a = 2.5 \times 10^{-2}\ \mathrm{Wm^{-1}\,K^{-1}}$ |
| | $T_a = 27\,°\mathrm{C}$ |

* The side faces of the ring are perfectly insulated.

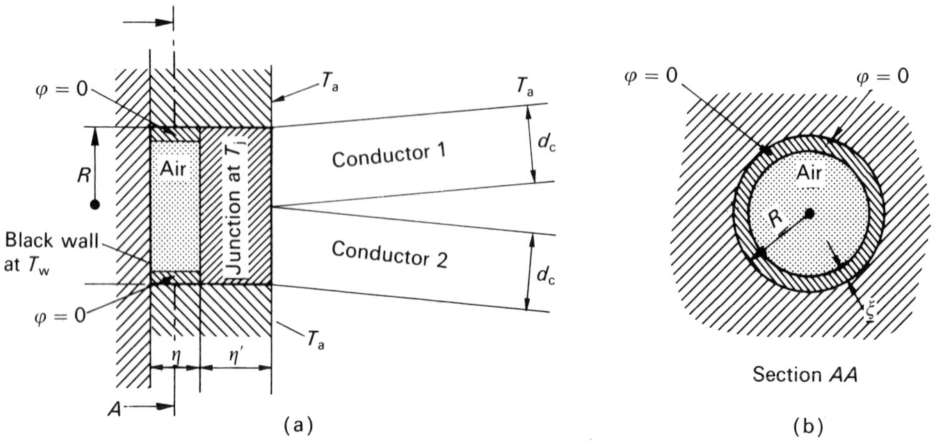


Figure 1

# The solution

## 1. Response time

The energy balance of a sensor undergoing an increase in temperature d$T$ over a time d$t$ is:

$$\rho c_p V \, dT = hS[T_f - T] \, dt$$

where $V$ is the immersed volume of the sensor, and $S$ the heat exchange area between the fluid at $T_f$ and the sensor at temperature $T$, whence:

$$\frac{dT}{dt} + kT = kT_f = k(T_1 + T_2 \sin \omega t)$$

where:

$$k = \frac{hS}{\rho c_p V}$$

write:

$$\theta = T - T_1$$

whence:

$$\frac{d\theta}{dt} + k\theta = kT_2 \sin \omega t$$

The solution of this equation is the sum of two terms:

- a transient term:

$$\theta_1 = C\,e^{-kt} = C\,e^{-t/\tau}$$

  where $\tau$ is the time constant (the response time) of the sensor;
- a harmonic term:

$$\theta_2 = \frac{T_2 k}{\sqrt{k^2 + \omega^2}} \sin(\omega t - \Psi)$$

$$\tan(\Psi) = \frac{\omega}{k}$$

The constant $C$ is determined from the initial state $T = T_0$ at $t = 0$. The temperature indicated by either one of the sensors is thus:

$$T(t) = T_1 + \left(T_0 - T_1 + \frac{T_2 \omega k}{k^2 + \omega^2}\right) e^{-kt} + \frac{T_2 k}{\sqrt{k^2 + \omega^2}} \sin(\omega t - \Psi)$$

We can see:

- a decay term,
- a phase difference between the excitation from the fluid and the response of the sensor, which increases at high frequency or low values of $k$. It is thus beneficial to choose a sensor with a small diameter (since $k$ is inversely proportional to the diameter).
- a reduction in amplitude of the oscillation by the ratio $k/\sqrt{k^2 + \omega^2}$, which becomes more significant as $k$ becomes small compared to $\omega$.

**Numerically:**

- *For the thermocouple*

$$k = 3.13 \times 10^{-2}\ \text{s}^{-1} \qquad \Psi = 0.323\ \text{rad}$$

$$T(t) = 100 - 24.94 \exp(-3.13 \times 10^{-2} t) + 47.5 \sin(\omega t - 0.323)$$

  The time constant comes out to $\tau \simeq 32$ s.
- *For the mercury thermometer*

$$k = 9.94 \times 10^{-3}\ \text{s}^{-1} \qquad \Psi = 0.811\ \text{rad}$$

$$T(t) = 100 - 15.03 \exp(-9.94 \times 10^{-3} t) + 34.4 \sin(\omega t - 0.811)$$

  The time constant is $\tau \simeq 1$ min 40 s.

The first diagram below, on which the variations of the fluid temperature and the temperatures given by the two sensors are shown, clearly shows the attenuation of amplitude and the phase lag between the measured and actual temperatures.

The second diagram shows the response of the two sensors to a step change in temperature. These curves demonstrate the response times of the sensors.

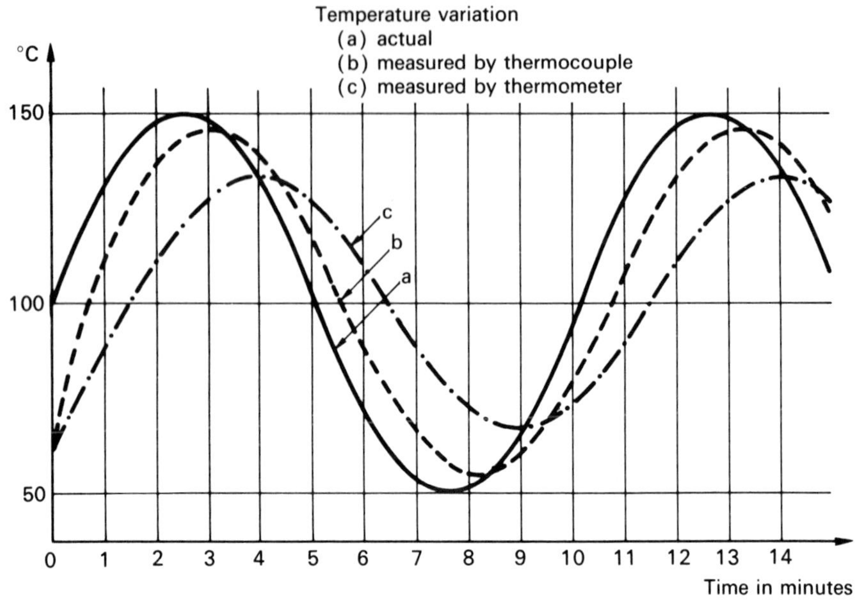
Temperature variation
(a) actual
(b) measured by thermocouple
(c) measured by thermometer
°C
150
100
50
0
1
2
3
4
5
6
7
8
9
10
11
12
13
14
c
b
a
Time in minutes

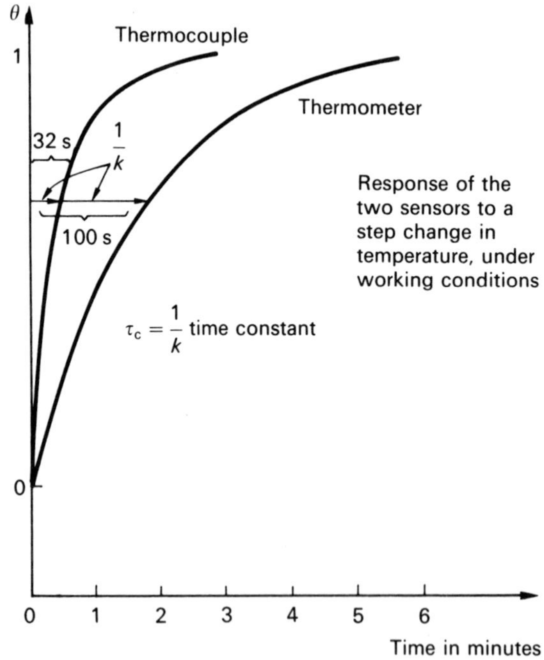
θ
1
0
Thermocouple
Thermometer
32 s
1/k
100 s
Response of the two sensors to a step change in temperature, under working conditions
$\tau_c = \frac{1}{k}$ time constant
0
1
2
3
4
5
6
Time in minutes

## *Thermocouple well*

*2(a)*

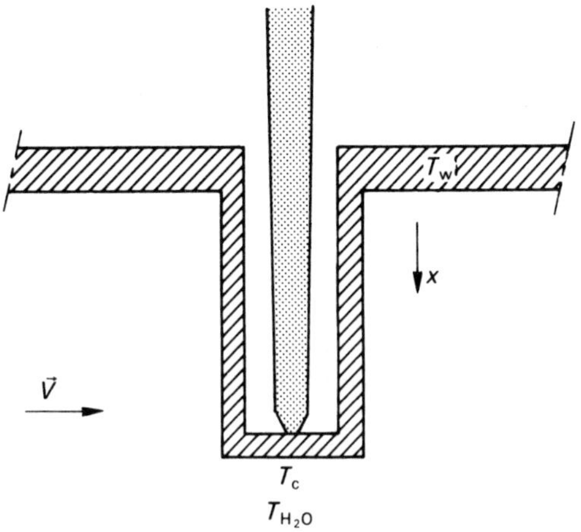


*Assumptions*

1. Perfect contact between the thermocouple and the well at the bottom of the latter.
2. The heat transfer between the tip of the well and the water is negligible compared to that between the side wall and the water. The temperature difference $T_c - T_{H_2O}$ will be the smallest possible ($\varphi = 0$ at the tip).
3. The well is a fin.
4. The base of the fin made by the well is at the tube temperature $T_w$.

Considering an element between $x$ and $x + \mathrm{d}x$, each cross-section is isothermal (the fin approximation). In the absence of flow or heat generation, the heat balance is

$$\Phi_x = \Phi_{x+\mathrm{d}x} + \Phi_{\mathrm{ext}} + \Phi_{\mathrm{int}}$$

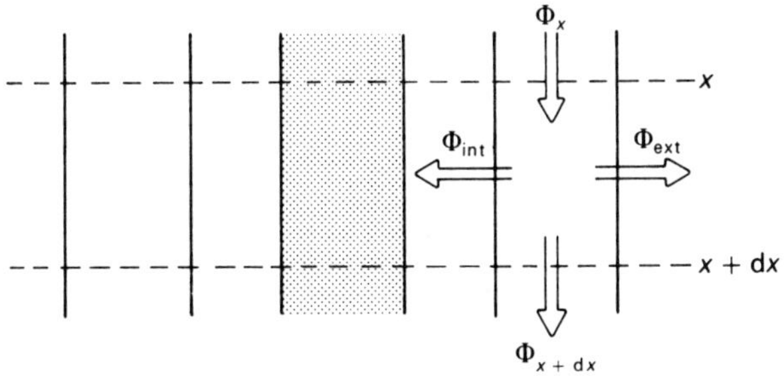

and

$$\Phi_{x+\mathrm{d}x} = \Phi_x + \frac{\mathrm{d}}{\mathrm{d}x}\Phi_x\,\mathrm{d}x$$

$$\Phi_x = -\lambda\frac{\partial T}{\partial x}A$$

$$\Phi_{\mathrm{ext}} = h_{\mathrm{CF}}[T(x) - T_{\mathrm{H_2O}}]S$$

$$\Phi_{\mathrm{int}} = h_{\mathrm{CN}}[T(x) - T_{\mathrm{air}}]S_{\mathrm{int}}$$

$$A = \frac{\pi}{4}[d^2 - (d - 2e)^2]$$

$$S = \pi d\,\mathrm{d}x$$

Heat exchange by natural convection in the air inside the well is negligible in comparison to the conducto-convective exchange by forced convection in the water:

$$\Phi_{\mathrm{int}} \ll \Phi_{\mathrm{ext}}$$

This is true for two reasons: because $h_{\mathrm{CN}} \ll h_{\mathrm{CF}}$ (by at least a factor of 1000), and because $T(x) \sim T_{\mathrm{air}}$. Then:

$$0 = \frac{\mathrm{d}}{\mathrm{d}x}\left(-\lambda\frac{\mathrm{d}T}{\mathrm{d}x}A\right) + h_{\mathrm{CF}}[T(x) - T_{\mathrm{H_2O}}]\pi d$$

Assuming the thermal conductivity of steel is constant:

$$\frac{\mathrm{d}^2T}{\mathrm{d}x^2} - \frac{\pi\,\mathrm{d}h_{\mathrm{CF}}}{\lambda A}[T(x) - T_{\mathrm{H_2O}}] = 0$$

Writing:

$$\frac{T(x) - T_{\mathrm{H_2O}}}{T_{\mathrm{w}} - T_{\mathrm{H_2O}}} = \theta(\xi) \qquad \text{and} \qquad \xi = \frac{x}{l}$$

where $l$ is the well length, we obtain:

$$\frac{\mathrm{d}^2\theta}{\mathrm{d}\xi^2} - \frac{\pi\,\mathrm{d}h_{\mathrm{CF}}l^2\theta}{\lambda A} = 0$$

$$\frac{\mathrm{d}^2\theta}{\mathrm{d}\xi^2} - \omega^2 l^2\theta = 0$$

with

$$\begin{cases} \theta(0) = 1 \\ \left.\dfrac{\mathrm{d}\theta}{\mathrm{d}\xi}\right|_{\xi-1} = 0 \end{cases}$$

This equation together with the above boundary conditions has the solution:

$$\theta(\xi) = \frac{\cosh[\omega l(1-\xi)]}{\cosh \omega l}$$

giving:

$$\frac{T(x) - T_{H_2O}}{T_w - T_{H_2O}} = \frac{\cosh \omega(l-x)}{\cosh \omega l}$$

The well length $l$ must be such that:

$$\frac{T(l) - T_{H_2O}}{T_w - T_{H_2O}} = \frac{T_c - T_{H_2O}}{T_w - T_{H_2O}} = 5 \times 10^{-3}$$

Hence:

$$\frac{1}{\cosh \omega l} = 5 \times 10^{-3} \qquad \text{or} \qquad \cosh \omega l = 200$$

$$\omega l \sim 6 \quad \Rightarrow \quad \underline{l = \frac{6}{\omega}}$$

To calculate $\omega = (\pi\, d h_{CF}/\lambda A)^{1/2}$, we need to know $h_{CF}$ for forced convection past a cylinder perpendicular to the stream. From Basic Data (Part V), for water at 60 °C:

$$\begin{cases} a = 15.5 \times 10^{-8}\ \mathrm{m^2\,s^{-1}} \\ \nu = 4.77 \times 10^{-7}\ \mathrm{m^2\,s^{-1}} \\ \lambda = 0.651\ \mathrm{W\,m^{-1}\,K^{-1}} \\ Pr = 3.08 \end{cases}$$

$$Re_d = \frac{Vd}{\nu} = \frac{0.15 \times 4 \times 10^{-3}}{4.77 \times 10^{-7}} = 1.26 \times 10^3$$

thus:

$$\begin{aligned} Nu_d &= 0.683 Pr^{1/3} Re_d^{0.406} \\ &= 0.683(3.08)^{1/3}(1.26 \times 10^3)^{0.466} \\ &= 27.7 = \frac{h_{CF} d}{\lambda_{H_2O}} \end{aligned}$$

$$\Rightarrow \quad h_{CF} = \frac{27.7 \times 0.651}{4 \times 10^{-3}} = 4.5 \times 10^3\ \mathrm{W\,m^{-2}\,K^{-1}}$$

Since heat transfer coefficients in natural convection are of order 5, our assumption that we can neglect $\Phi_{int}$ compared to $\Phi_{ext}$ is quite satisfactory.

$$\begin{aligned} A &= 3\pi\, 10^{-6}\ \mathrm{m}^2 \\ \pi d &= 4\pi\, 10^{-3}\ \mathrm{m} \end{aligned} \Rightarrow \omega^2 = \frac{4 \times 10^{-3} \times 4.5 \times 10^3}{3 \times 10^{-6} \times 17} = 3.53 \times 10^5$$

$$\omega = 594\ m^{-1}$$

The length $l$ of the well is thus

$$l \sim \frac{6}{\omega} = 1\ \mathrm{cm}$$

which is similar to the tube radius.

## 2(b)

The above calculation was made on the assumption that the flux at the tip of the well was zero. This is a first approach to a real well. A second approach is to consider an 'infinite' fin, sufficiently long for its tip temperature to be close to the ambient temperature $T_{H_2O}$.

Using the heat equation from before:

$$\frac{d^2T}{dx^2} - \omega^2(T - T_{H_2O}) = 0$$

whose solution is:

$$T - T_{H_2O} = A\, e^{-\omega x} + B\, e^{+\omega x}$$

with associated boundary conditions:

$$\begin{cases} T(0) = T_w \\ T(\infty) = T_{H_2O} \end{cases} \Rightarrow B = 0 \qquad \text{and} \qquad A = T_w - T_{H_2O}$$

so that:

$$T(x) - T_{H_2O} = (T_w - T_{H_2O})\, e^{-\omega x}$$

The distance $l$ after which the fin temperature $T(l)$ is equal to $T_{H_2O}$ to within 0.5% is given by:

$$\frac{T(l) - T_{H_2O}}{T_w - T_{H_2O}} = 0.5\% = e^{-\omega l} = 5 \times 10^{-3}$$

where

$$\omega l = 5.3 \Rightarrow \underline{l = \frac{5.3}{\omega}} = 0.9\ \mathrm{cm}$$

which result is very close to the one above. The two cases we have looked at are the two extremes; the real well will lie between them.

## 2(c)

Another approach to the thermocouple well can be made by assuming a finite fin, but with a convective heat transfer coefficient at the tip. Such a coefficient is very difficult to determine in practice: is it the same as for the side face? Surely not! Worse, the geometry at the tip is not well defined.

Nevertheless, to attempt something, we will take a plane tip with a coefficient identical to that of the sides. Starting from the fin equation:

$$\frac{d^2(T - T_{H_2O})}{dx^2} - \omega^2(T - T_{H_2O}) = 0$$

with boundary conditions:

$$\begin{cases} T(0) - T_{H_2O} = T_w - T_{H_2O} \\ -\lambda \left.\dfrac{\partial T}{\partial x}\right|_{x=l} = h_{CF}[T(l) - T_{H_2O}] \end{cases}$$

the solution is:

$$\frac{T(x) - T_{H_2O}}{T_w - T_{H_2O}} = \frac{\cosh \omega(l - x) + (h/\lambda\omega) \sinh \omega(l - x)}{\cosh \omega l + (h/\lambda\omega) \sinh \omega l}$$

The temperature at the tip of the fin is:

$$\frac{T(l) - T_{H_2O}}{T_w - T_{H_2O}} = \frac{1}{\cosh \omega l + (h/\lambda\omega) \sinh \omega l}$$

$T(l)$ tends to $T_{H_2O}$ if the denominator is large, and thus if $\omega l \gg 1$. For $\omega l = 6$, $\cosh \omega l = \sinh \omega l = 202$:

$$\frac{h_{CF}}{\lambda\omega} = \frac{h_{CF}}{\lambda}\left(\frac{\lambda A}{\pi\, d h_{CF}}\right)^{1/2} = \left(\frac{h_{CF}}{\lambda}\frac{A}{\pi d}\right)^{1/2}$$

The expression $(h_{CF}/\lambda)(A/\mathscr{P})$ is dimensionless and is called the Biot number $Bi$. The fin is said to be **thermally infinite** when $\omega l \gg 1$. In practice:

$$l_\infty = \frac{6}{\omega}$$

In this case the temperature distribution is the same as that obtained for $\varphi = 0$ at the tip.

## 3 Radiation shield

### 3(a) Balance equations

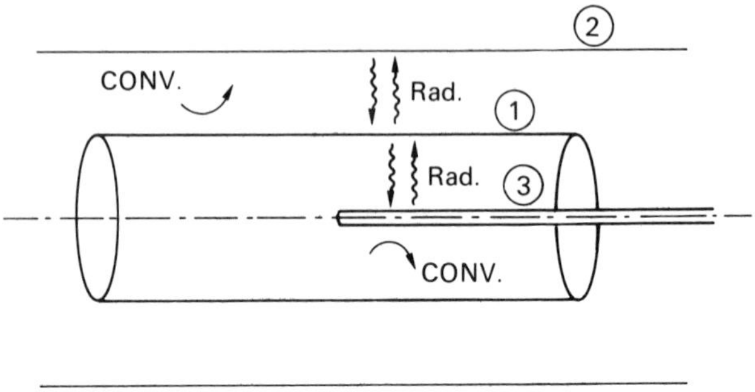


The shield ① exchanges heat

- By radiation: with the duct ②, with the thermocouple ③.
- By conducto-convection with the gas.

In steady-state conditions the balance for the shield is:

$$\sum \Phi = 0 = \Phi_R^{1\to 2} + \Phi_R^{1\to 3} + \Phi_{cc}$$

and

$$\Phi_{cc} = 2S_1 h_1 [T_1 - T_g]$$

assuming the same heat transfer coefficient for the internal and external surfaces of the shield.

Similarly for the thermocouple ③ in steady conditions:

$$\sum \Phi = 0 = \Phi_R^{3\to 1} + \Phi'_{cc}$$

and

$$\Phi'_{cc} = S_3 h_3 [T_3 - T_g]$$

### 3(b) Calculation of radiative flux

- Between shield ① and duct ②:

$$\Phi_R^{1\to 2} = S_1[\varphi_1^l - \varphi_1^i] \tag{1}$$

with:

$$\left.\begin{aligned} \varphi_1^l &= \varepsilon_1 \sigma T_1^4 + (1-\varepsilon_1)\varphi_1^i \\ S_1\varphi_1^i &= f_{21} S_2 \varphi_2^l \end{aligned}\right\} \tag{2}$$

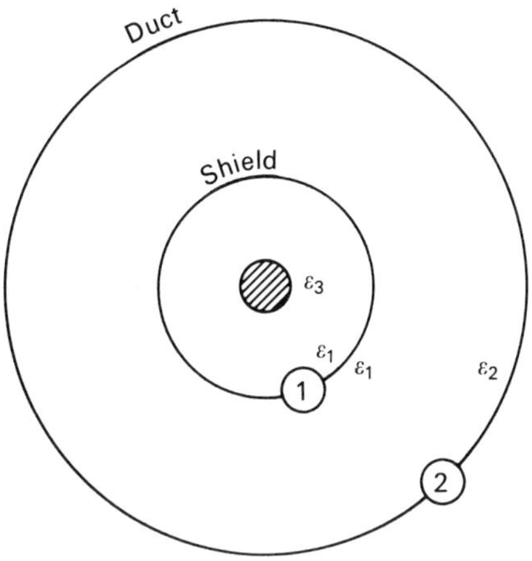


The shape factors are related by:

$$\sum_j f_{ij} = 1 \qquad f_{11} + f_{12} = 1 \Rightarrow f_{12} = 1$$

$$S_1 f_{12} = S_2 f_{21} = S_1 \Rightarrow S_1 \varphi_1^{\mathrm{i}} = S_1 \varphi_2^{\mathrm{l}}$$

$$\Rightarrow \varphi_1^{\mathrm{i}} = \varphi_2^{\mathrm{l}} \tag{3}$$

Similarly:

$$\varphi_2^{\mathrm{l}} = \varepsilon_2 \sigma T_2^4 + (1 - \varepsilon_2)\varphi_2^{\mathrm{i}} \tag{4}$$

$$S_2 \varphi_2^{\mathrm{i}} = f_{22} S_2 \varphi_2^{\mathrm{l}} + f_{12} S_1 \varphi_1^{\mathrm{l}}$$

$$f_{12} S_1 = f_{21} S_2$$

$$\Rightarrow \quad \varphi_2^{\mathrm{i}} = f_{22} \varphi_2^{\mathrm{l}} + f_{21} \varphi_1^{\mathrm{l}}$$

Shape factors:

$$f_{22} + f_{21} = 1$$

$$f_{21} = \frac{S_1}{S_2} = \frac{d_1}{d_2} \rightarrow f_{22} = 1 - \frac{d_1}{d_2}$$

$$\varphi_2^{\mathrm{i}} = \left(1 - \frac{d_1}{d_2}\right) \varphi_2^{\mathrm{l}} + \frac{d_1}{d_2} \varphi_1^{\mathrm{l}} \tag{5}$$

From equations (2), (3), (4) and (5) we obtain:

$$\varphi_2^{\mathrm{l}} = \sigma T_2^4 \frac{1 + (1 - \varepsilon_2)(\varepsilon_1 d_1/\varepsilon_2 d_2)(\sigma T_1^4/\sigma T_2^4)}{1 + (1 - \varepsilon_2)(\varepsilon_1 d_1/\varepsilon_2 d_2)}$$

This equation for the flux leaving the duct ② tends towards $\sigma T_2^4$ when:

$$\frac{\varepsilon_1 d_1}{\varepsilon_2 d_2} \frac{\sigma T_1^4}{\sigma T_2^4} \ll 1 \qquad \text{and} \qquad \frac{\varepsilon_1 d_1}{\varepsilon_2 d_2} \ll 1$$

that is, when:

$$\frac{d_1}{d_2} \ll \frac{\varepsilon_2 \sigma T_2^4}{\varepsilon_1 \sigma T_1^4} \quad \text{and} \quad \frac{d_1}{d_2} \ll \frac{\varepsilon_2}{\varepsilon_1} \tag{6}$$

These two conditions show that the flux emitted (absorbed) by the shield ① is negligible in comparison to the flux emitted (absorbed) by the duct ②. We have thus returned to the classic case of a very small body in an isothermal enclosure, which receives **equilibrium radiation** whatever the radiative properties ($\varepsilon_2$) of the enclosure. Assuming the conditions (6) are satisfied:

$$\varphi_2^l = \sigma T_2^4$$

The flux exchanged between the shield and the duct is:

$$\Phi_R^{1\to 2} = S_1 \varepsilon_1 \sigma [T_1^4 - T_2^4]$$

- Between thermocouple ③ and shield ①:

$$\Phi_R^{3\to 1} = S_3[\varphi_3^l - \varphi_3^i]$$

where:

$$\varphi_3^l = \varepsilon_3 \sigma T_3^4 + (1 - \varepsilon_3)\varphi_3^i \tag{7}$$

$$S_3 \varphi_3^i = f_{13} S_1 \varphi_1^l$$

Now:

$$f_{33} = 1 \to f_{31} = 1$$

$$S_3 f_{31} = S_1 f_{13} = S_3 \Rightarrow \varphi_3^i = \varphi_1^l \tag{8}$$

$$\varphi_1^l = \varepsilon_1 \sigma T_1^4 + (1 - \varepsilon_1)\varphi_1^i \tag{9}$$

$$S_1 \varphi_1^i = f_{11} S_1 \varphi_1^l + f_{31} S_3 \varphi_3^l$$

$$f_{31} S_3 = f_{13} S_1$$

$$\Rightarrow \quad \varphi_1^i = f_{11} \varphi_1^l + f_{13} \varphi_3^l$$

$$f_{13} = \frac{S_3}{S_1} = \frac{d_3}{d_1} \qquad f_{11} = 1 - \frac{d_3}{d_1}$$

$$\varphi_1^i = \left(1 - \frac{d_3}{d_1}\right)\varphi_1^l + \frac{d_3}{d_1}\varphi_3^l \tag{10}$$

From equations (7), (8), (9) and (10) we obtain:

$$\varphi_R^{3\to 1} = S_3 \varepsilon_1 \varepsilon_3 \frac{\sigma(T_3^4 - T_1^4)}{\varepsilon_1 + \varepsilon_3 (d_3/d_1)(1 - \varepsilon_1)}$$

### 3(c) Return to the balance equations

Having obtained expressions for the radiative fluxes, we can rewrite the thermal balance equations for the shield and the thermocouple:

$$-\Phi_R^{3\to1} + \Phi_{cc} + \Phi_R^{1\to2} = 0$$

$$\Phi_R^{3\to1} + \Phi'_{cc} = 0$$

whence:

$$\frac{d_3}{d_1}\varepsilon_1\varepsilon_3 \frac{\sigma T_3^4 - \sigma T_1^4}{\varepsilon_1 + \varepsilon_3 (d_3/d_1)(1-\varepsilon_1)} + 2h_1(T_g - T_1) - \varepsilon_1\sigma(T_1^4 - T_2^4) = 0 \qquad (11)$$

$$\varepsilon_1\varepsilon_3 \frac{\sigma T_3^4 - \sigma T_1^4}{\varepsilon_1 + \varepsilon_3(d_3/d_1)(1-\varepsilon_1)} + h_3(T_g - T_3) = 0 \qquad (12)$$

The solution of this system of two equations in two unknowns $T_1$ and $T_g$ presents no difficulty. The solutions are:

$$T_1 = 783.2\text{ K} \qquad \text{and} \qquad T_g = 808.6\text{ K}$$

The absolute error in the measurement, even with a radiation shield, is thus about 10 degrees. This takes no account of any disturbance caused to the flow field by the shield, which could lead to further errors.

### 3(d) Bare thermocouple

Thermocouple ③ is no longer protected from the radiation emitted by duct ②. For the thermocouple, in steady-state conditions:

$$\Phi''_{cc} + \Phi_R^{3\to2} = 0$$

$$S_3h_3(T'_3 - T_g) + S_3\varepsilon_3\sigma(T_3'^4 - T_2^4) = 0 \qquad (13)$$

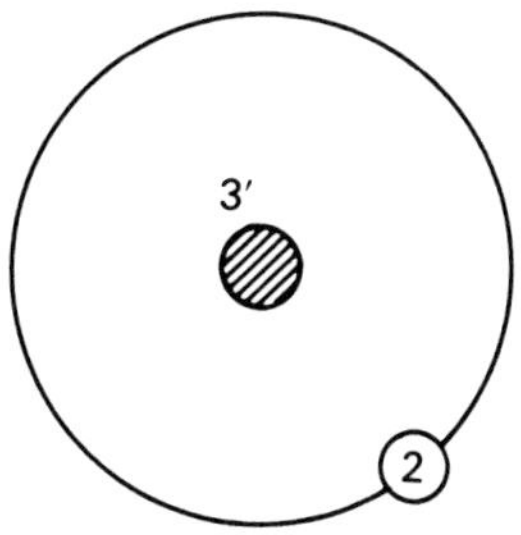

The radiative flux equation given here makes use of the fact that the flux emitted by the thermocouple is negligible in comparison to the flux emitted by duct ②. This equation gives the temperature $T'_3$ indicated by the thermocouple when it is subjected directly to the radiation emitted by the duct. After all the calculations, we obtain:

$$T'_3 = 726.8\ \text{K}$$

There is thus a major error of more than 70 °C in the measurement. Once again, we stress the principle that a thermocouple can only measure its own temperature.

## 4. *Wall temperature measurement*

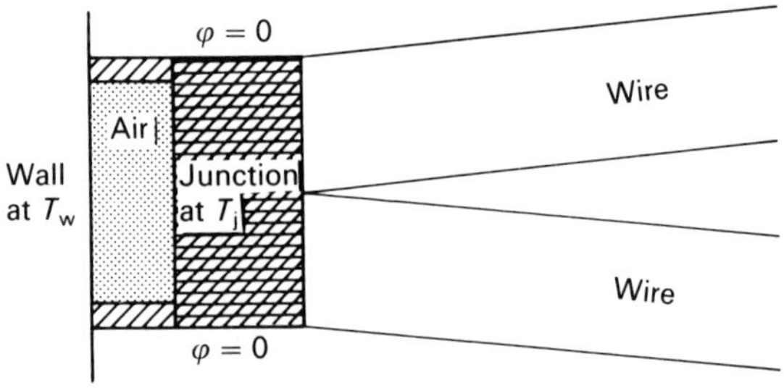


In steady-state conditions, the sum of the fluxes at the junction is zero:

$$\sum \Phi = 0$$

or

$$\Phi_{\text{entering}} = \Phi_{\text{leaving}}$$

$\Phi_{\text{entering}}$ conductive flux in the ring,
conductive flux in the air,
radiative flux in the air;
$\Phi_{\text{leaving}}$ conductive flux in the two thermocouple wires, forming fins.

- *Conductive flux in the ring*

$$\Phi_{\text{cd}}^{\text{ring}} = \frac{T_w - T_j}{R_{\text{ring}}}$$

We can use the idea of thermal resistance, since the ring is assumed to be perfectly insulated on its side faces:

$$R_{\text{ring}} = \frac{\eta}{\lambda S} = \frac{\eta}{\lambda 2\pi(R - \xi/2)\xi} = \frac{1}{9.24 \times 10^{-3}}\ \text{°C W}^{-1}$$

$$\Rightarrow \qquad \underline{\Phi_{\text{cd}}^{\text{ring}} = 9.24 \times 10^{-3}(T_w - T_j)}$$

- *Conductive flux in the air*

$$\Phi_{cd}^{air} = \frac{T_w - T_j}{R_{air}}$$

For the same reasons as for the ring, we can use the thermal resistance method:

$$R_{air} = \frac{\eta}{\lambda S'} = \frac{\eta}{\lambda\pi(R-\xi)^2} = \frac{1}{1.81 \times 10^{-4}} \ {}^\circ\mathrm{C\,W^{-1}}$$

$$\Rightarrow \qquad \underline{\Phi_{cd}^{air} = 1.81 \times 10^{-4}(T_w - T_j)}$$

⇒ it can be seen that, since air is a poor thermal conductor, the flux in the air is 200 times less than in the ring, with a cross-sectional area ten times larger.

- *Radiative flux in the air.* The total interaction between the junction at temperature $T_j$ and the wall at $T_w$ is:

$$\Phi_R^{air} = \varepsilon_j\sigma\{T_w^4 - T_j^4\}S' = h_R S'(T_w - T_j)$$

with:

$$h_R S' \simeq 4\varepsilon_j\sigma T_w^3 \pi(R-\xi)^2$$

Linearization of the radiation is possible since the whole aim of the wall temperature measurement is to have $T_j \sim T_w$, so far as is practically possible.

$$\begin{aligned} h_R S' &= 4 \times 0.8 \times 5.67 \times 10^{-8}(273+50)^3 \times \pi \times (5 \times 10^{-4} - 2 \times 10^{-5})^2 \\ &= 4.43 \times 10^{-6}\ \mathrm{W\,{}^\circ C^{-1}} \end{aligned}$$

$$\Rightarrow \qquad \underline{\Phi_R^{air} = 4.43 \times 10^{-6}(T_w - T_j)}$$

⇒ The radiative flux is negligible in comparison to the conductive fluxes.

- *Conductive flux in the wires.* The two wires comprising the thermocouple constitute two fins which dissipate energy to the ambient air by conducto-convective (natural convection) and radiative transfers through their side faces.

  Using the fin equations (see Chapter 2):

$$\frac{d^2T}{dZ^2} - m^2(T - T_a) = 0$$

$$m^2 = \frac{h\mathscr{P}_w}{\lambda S} = \frac{4h}{\lambda d}$$

$$\begin{aligned} h &= h_{CN} + h_R = h_{CN} + 4\varepsilon_c\sigma T_a^3 \\ &= 5 + 4 \times 0.5 \times 5.6 \times 10^{-8}(273+27)^3 \\ &= (5 + 3.06)\ \mathrm{W\,m^{-2}\,K^{-1}} = 8.06\ \mathrm{W\,m^{-2}\,K^{-1}} \end{aligned}$$

$$m^2 = 2.63 \times 10^3 \rightarrow m = 51.3$$

A fin is considered to be thermally infinite for a length $l_c$ so long as $\exp(-ml_e) \ll 1$, or for $l_c = 5/m$ to a precision of $10^{-2}$.

$$l_c = \frac{5}{51.3} \sim 0.1 \text{ m} \sim 10 \text{ cm}$$

Since a thermocouple will always have a length greater than $l_c$, the solution of the fin equation is:

$$T(Z) - T_a = A \exp(-mZ)$$

with boundary conditions:

$$\begin{cases} T(0) = T_j \\ T(\infty) = T_a \end{cases} \Rightarrow T(Z) - T_a = (T_j - T_a)\exp(-mZ)$$

The flux entering one wire is:

$$\Phi_{cd}^{1\text{ wire}} = \frac{\pi d_c^2}{4}\left[-\lambda \frac{dT}{dZ}\right]_{Z=0}$$

$$= \frac{\pi d_c^2}{4} \lambda m (T_j - T_a)$$

$$= \frac{\pi(5 \times 10^{-4})^2 \times 24.5 \times 51.3}{4}(T_j - T_a)$$

$$= 2.47 \times 10^{-4}(T_j - T_a)$$

So for both wires:

$$\underline{\Phi_{cd}^{2\text{ wires}} = 4.94 \times 10^{-4}(T_j - T_a)}$$

The thermal balance for the junction is:

$$\Phi_{cd}^{ring} + \Phi_{cd}^{air} + \Phi_{R}^{air} = \Phi_{cd}^{2\text{ wires}}$$

$$\Rightarrow \quad 9.43 \times 10^{-3}(T_w - T_j) = 4.94 \times 10^{-4}(T_j - T_a)$$

$$T_j(9.43 \times 10^{-3} + 4.94 \times 10^{-4}) = 9.43 \times 10^{-3} T_w + 4.94 \times 10^{-4} T_a$$

$$T_j = 48.9\,^\circ\text{C}$$

$$\Rightarrow \qquad \underline{T_w - T_j = 50 - 48.9 = 1.14\,^\circ\text{C}}$$

In the calculation of the flux leaving, we have assumed that only the two wires are effective. However, the latter do not completely cover the junction: on the exposed surface, there will be a transfer from the thermocouple to the ambient surroundings:

$$\Phi'' = hS''(T_S - T_a) = h\left(\pi R^2 - \frac{\pi}{2} d_c^2\right)(T_S - T_a)$$
$$= 3.17 \times 10^{-6}(T_S - T_a)$$

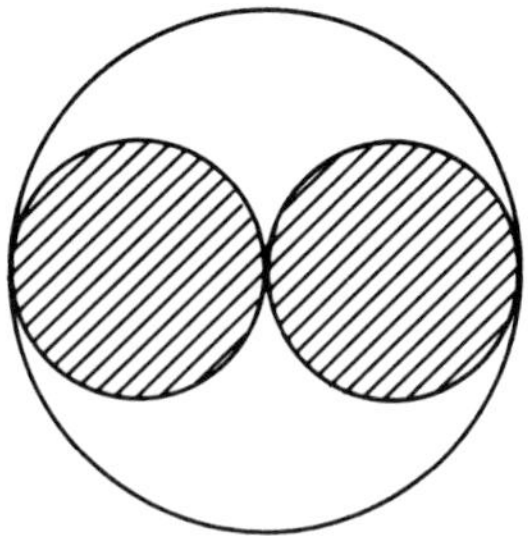

This is a hundred times less than that dissipated by the two wires ($\Phi_{cd}^{2\,\text{wires}}$), so we are justified in neglecting it.

# Application 6

## Is the Wine Cool? The Earthenware Jar

### *Required knowledge*

- Steady-state conditions (Chapter 1).
- Linear conduction models (Chapter 2).
- First principles of convection: transport theorem (Chapter 3).
- Elementary radiation (Chapter 4).

### *Objective*

- Construction of a simple physical model appropriate to an age-old application; if necessary we will make simple and realistic assumptions.

### *The problem*

A fairly porous earthenware jar, to a first approximation spherical and initially full of wine, is suspended in the shade in still air, out of direct contact with the ground. The outside temperature is $T_e = 35\,°C$. If guarded carefully, the jar will half empty itself by evaporation through the earthenware in about 30 days. We assume, however, that the outer surface of the jar remains dry.

**Determine the temperature of the wine, under (quasi-) steady conditions.**

We will use a simple thermal model, and use in it the appropriate data from Part V: Basic Data, treating the wine as water. (We could, actually, consider the problem with water directly; we could then replenish the water daily.)

**Numerical data:**

- Radius of jar: $R = 0.4$ m
- Thickness: $e = 0.01$ m
- Heat transfer coefficient, external natural convection: $h_{CN} \simeq 2.5$ W m$^{-2}$ K$^{-1}$
- Latent heat of vaporization (per unit mass): $L = 2.4 \times 10^6$ J kg$^{-1}$
- We will estimate the radiative heat transfer coefficient.

# *The solution*

## *1. Problem analysis*

We need to find the heat flows within the system.

- This system is open: vaporization of wine occurs, followed by convective transfer through the earthenware. The mass of wine is not constant, and its volume (and area) varies with time.
- Vaporization is an endothermic reaction, which requires heat to be extracted from the system.
- But conducto-convective (natural convection) and radiative heat transfers bring energy into the system.

The steady-state temperature of the wine is the result of equilibrium between these transfers. As a first approximation, we will assume it to be uniform.

## *2. Definition of the system to be studied*

This is the most difficult bit. The exact location at which evaporation takes place is ill-defined, and we cannot isolate the wine from the wall. We will therefore adopt as our open system the combination of the mass of wine at instant $t$ and the immediately adjacent part of the wall (shown shaded).

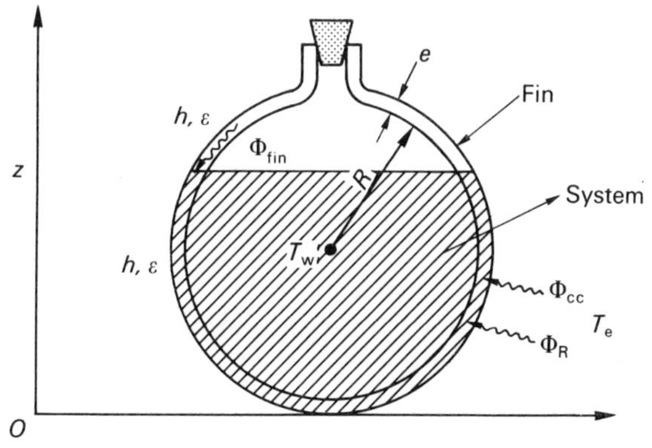

To simplify matters, we will assume that the whole system (wine + adjacent wall) is at a uniform temperature $T_w(t)$. The surface of the wine is at height $z(t)$. The part of the wall which is not in contact with the wine would appear to be a **fin**, which delivers flux into the wine, and whose importance we must determine.

### *3. Balance equations*

The simplest and most logical approach to an open system with variable volume is to use the appropriate transport theorem. Let us determine an **energy balance** of our system:

The enthalpy variation of the system arises from exchanges with its surroundings:

$$\frac{\mathrm{d}}{\mathrm{d}t}\mathscr{H} = \frac{\mathrm{d}}{\mathrm{d}t}\int_{V(t)} \rho \mathrm{h}\, \mathrm{d}\tau = \Phi^{\mathrm{CC}} + \Phi^{\mathrm{R}} + \Phi^{\mathrm{fin}} + \Phi^{\mathrm{CV}} \tag{1}$$

The fluxes appearing on the right-hand side are:

- The natural convection conducto-convective flux with the surrounding air:

  $$\Phi^{\mathrm{CC}} = h_{\mathrm{CN}} S(t)(T_e - T_W)$$

  where $S(t)$ is the area of the jar in contact with the wine.
- The radiative flux with the surroundings (with equilibrium radiation outside):

  $$\Phi^{\mathrm{R}} = S(t)\varepsilon\sigma(T_e^4 - T_W^4) \tag{2}$$

  Note that, not having access to a specialist laboratory, we will adopt a fairly high value for $\varepsilon$ for the jar walls, which are damp and porous. Take $\varepsilon \simeq 0.8$.

  It is reasonable to linearize the radiative flux (see Chapter 4):

  $$h_{\mathrm{R}} = 4\varepsilon\sigma T_e^3 \simeq 5.3 \text{ W m}^{-2}\text{ K}^{-1}$$

  if $T_e$ is around 300 K. The linearized flux equation is:

  $$\Phi^{\mathrm{R}} = S(t) h_{\mathrm{R}}(T_e - T_W) \tag{3}$$
- The flux transferred through the fin made up from the upper part of the jar (using a linear thermal model) is:

  $$\Phi^{\mathrm{fin}} = s(t)\gamma_{\mathrm{fin}}(T_e - T_W) \tag{4}$$

  where $\gamma_{\mathrm{fin}}$ is a surface conductance which will be defined below and $s(t)$ is the cross-sectional area of the fin. (The flux exchanges between the wine and the air inside the jar are ignored, which is equivalent to assuming that the internal air is at a temperature close to that of the wine.)
- The convective flux associated with the evaporation of the wine through the porous wall: a mass flux $\dot{m}(t)$, with $\dot{m} < 0$:

  $$\Phi^{\mathrm{CV}} = +\dot{m}(t) h_{\mathrm{vap}}(T_W) \tag{5}$$

This term is obviously crucial to the application: this is what gives the jar its properties. We are assuming that the wine leaves the jar as vapour.

## 4. *Application of the transport theorem*

Assuming that $\rho$ is constant throughout the system, the transport theorem allows the left-hand side of the balance equation (1) to be written:

$$\frac{\mathrm{d}}{\mathrm{d}t}\int_{V(t)} \rho \mathrm{h}\, \mathrm{d}\tau = \int_{V(t)} \rho \frac{\partial \mathrm{h}}{\partial t}\, \mathrm{d}\tau + \int_{\Sigma(t)} \rho \mathrm{h} \mathbf{v}\cdot\mathbf{n}_{\text{ext}}\, \mathrm{d}S \tag{6}$$

where $\mathbf{v}$ is the velocity of the wine/air interface $\Sigma(t)$, relative to the jar. Note that:

$$\mathscr{S}(t) = S(t) + \Sigma(t) + s(t) \tag{7}$$

The first term on the right-hand side $\int_{V(t)} \rho(\partial \mathrm{h}/\partial t)\, \mathrm{d}\tau$ corresponds to the transient state of the system which we will not consider here. Heat transfer within the wine and its adjacent wall make the temperature uniform on a time scale $\tau$ which is short in comparison to that of the motion of the surface of the wine. This assumption will be discussed at the end of the exercise.

Looking only at the second term in (6), we obtain:

$$\frac{\mathrm{d}}{\mathrm{d}t}\mathscr{H} = \Sigma(t)\rho_{\mathrm{l}}\mathbf{v}\cdot\mathbf{n}_{\text{ext}}\mathrm{h}_{\mathrm{l}}(T_{\mathrm{W}}) \tag{8}$$

where $\rho_{\mathrm{l}}$ and $\mathrm{h}_{\mathrm{l}}$ are the density and specific enthalpy of the liquid wine. Note that:

$$\dot{m} = +\Sigma(t)\rho_{\mathrm{l}}\mathbf{v}\cdot\mathbf{n}_{\text{ext}} < 0 \tag{9}$$

Equations (1), (8) and (9) give:

$$\dot{m}\mathrm{h}_{\mathrm{l}}(T_{\mathrm{W}}) = \dot{m}\mathrm{h}_{\mathrm{V}}(T_{\mathrm{W}}) + [S(t)[h_{\mathrm{R}} + h_{\mathrm{CN}}] + \gamma_{\text{fin}}\, s(t)](T_{\mathrm{e}} - T_{\mathrm{w}}) \tag{10}$$

or in terms of the latent heat of vaporization per unit mass, $L(= \mathrm{h}_{\mathrm{V}} - \mathrm{h}_{\mathrm{w}})$:

$$-\dot{m}L = [S(t)(h_{\mathrm{R}} + h_{\mathrm{CN}}) + \gamma_{\text{fin}} s(t)](T_{\mathrm{e}} - T_{\mathrm{w}}) \tag{11}$$

Complete solution of the problem requires us to be able to express $\dot{m}$ as a function of the area $S(t)$. If we neglect diffusion in that part of the jar which is not immersed, the mass evaporated will satisfy:

$$\dot{m} = -KS(t) \tag{12}$$

The phenomenological constant $K$ depends on the difference in chemical potential of the substance concerned between the inside and the outside of the jar. For water, it is proportional to the difference between the saturated vapour pressure and the partial pressure of $H_2O$ outside. It depends also on the porosity of the jar (or the actual interface area between the two phases if the evaporation actually occurs within the earthenware). We can thus deduce:

$$T_e - T_W = \frac{KL}{h_R + h_{CN} + \gamma_{fin}(s(t)/S(t))} \tag{13}$$

## 5. Effect of the fin

We would need to estimate the quantity $\gamma_{fin}$ which appears in equation (4). The non-wetted part of the jar wall makes a fin which exchanges heat with the surroundings only (with heat transfer coefficients $h_R$ and $h_{CN}$ introduced above). This assumes that the jar wall is insulated inside. The actual geometry is somewhat complex being the region between two spherical segments. To estimate the effect of the fin, we will satisfy ourselves with a rough estimate by treating the fin as a plane of thickness $e$, as in Chapter 2 (Section 2.2). Using the notation of that section, it is straightforward to calculate the fin parameter $m$:

$$m^2 = \frac{\mathscr{P}(h_{CN} + h_R)}{A\lambda} = \frac{h_{CN} + h_R}{\lambda e}$$

From Basic Data (Part V), we obtain $\lambda_{brick} \simeq 0.6\ \mathrm{W\,m^{-1}\,K^{-1}}$. In the absence of more precise information about the jar, we will assume that $\lambda \simeq 1$ SI. Furthermore:

$$h_{CN} + h_R = 2.5 + 5.3 \simeq 8\ \mathrm{W\,m^{-2}\,K^{-1}}$$

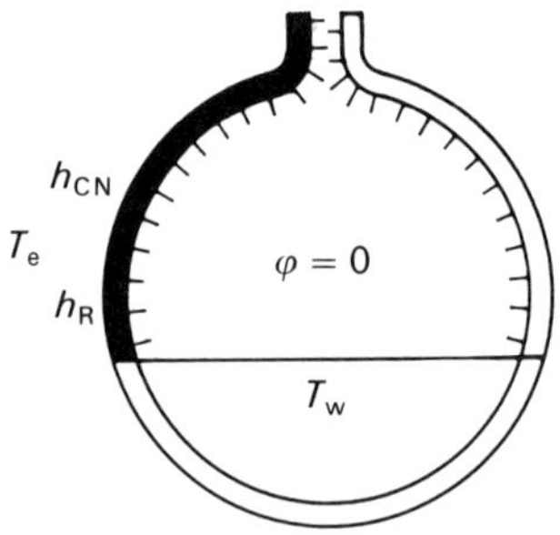


It follows that $m^2 \simeq 700\ \mathrm{m^{-2}}$. A plane fin can be considered as infinite for a length of order $3/m$ (see Section 2.2). With the rather complex geometry of this case, we will assume that the fin would be infinite for a developed length of about 15 cm. The efficiency of such a fin, compared to that of a plane surface, would be (for a plane fin – see Section 2.2):

$$\eta = \frac{\Phi_{fin}}{(h_{CN} + h_R)s(t)(T_e - T_W)} = m\frac{\lambda}{(h_{CN} + h_R)} \simeq 3.8 \tag{15}$$

It follows that:

$$\gamma_{\text{fin}} = 3.8(h_{\text{CN}} + h_{\text{R}}) \tag{16}$$

## 6. Determination of the phenomenological constant *K*

From (9) and (12), the equation of motion of the liquid surface is:

$$-KS(t) = \rho_1 \Sigma(t) \frac{\mathrm{d}z}{\mathrm{d}t} \tag{17}$$

Also:

$$S(t) = 2\pi R z(t) \qquad \text{and} \qquad \Sigma(t) = \pi(2Rz - z^2) \tag{18}$$

from which, integrating (17):

$$t = \frac{\rho_1}{2KR}\left(2R^2 - 2Rz + \frac{z^2}{2}\right)$$

Introducing the time $T$ after which the jar is half empty, we obtain:

$$K = \frac{\rho_1 R}{4T} \tag{19}$$

## 7. Conclusions

We obtain:

$$T_e - T_W = \frac{\rho_1 R}{4T} \frac{L}{(h_R + h_{CN})[1 + \eta\,(s(t)/S(t))]}$$

with:

$$\frac{s(t)}{S(t)} = \frac{e\sqrt{2Rz - z^2}}{Rz}$$

The fin has a hardly noticeable effect on the temperature of the liquid, except when the jar is almost empty. But we don't advise drinking the dregs! In fact:

$$\eta \frac{s(t)}{S(t)} \ll 1$$

We find:

$$T_e - T_W \sim \frac{\rho_1 R}{4T} \frac{L}{(h_R + h_{CN})}$$

This result is independent of time. **The wine stays cool in the jar.**

For:

$$\begin{aligned} R &= 0.4\ \text{m} \\ e &= 1\ \text{cm} \\ T &= 30\ \text{days} \\ L &= 2.4 \times 10^6\ \text{J kg}^{-1} \\ h_R + h_{CN} &\simeq 8\ \text{W m}^{-2}\ \text{K}^{-1} \end{aligned}$$

we obtain:

$$\underline{T_e - T_W \sim 12\,°\text{C}}$$

The liquid maintains a steady temperature nicely below its surroundings. This was part of the thermal *savoir faire* of the Canaanites, for example. Remains of earthenware jars several millennia old have been found to the west of Akko (St John of Acre).

The properties of leather bottles used in the past and still used today rest on the same principle. There are many applications based on the principle (and many much more complex models) such as multiphase heat transfer in porous media.

# Application 7

## Industrial Heat Recovery (Controlled Forced Ventilation with Two Flows)

### *Required knowledge*

- Steady-state conditions (Chapters 1 and 2).
- Convection (Chapters 3 and 17).

### *Features*

- Calculation of conducto-convective heat transfer coefficients.
- Calculations for counter-flow heat exchanger.
- Use of hydraulic diameter.

### *The problem*

An air conditioning system replaces stale air at temperature $T_a$ from a workshop with outside air at temperature $T_e$. The mass flow rates of air entering and leaving are the same, $\dot{m}$.

**Solution 1:** electric power $\mathscr{P}_1$ is used to raise the incoming air temperature to $T_a$.
**Solution 2:** a heat recuperator is used to heat the incoming air from the extracted air.

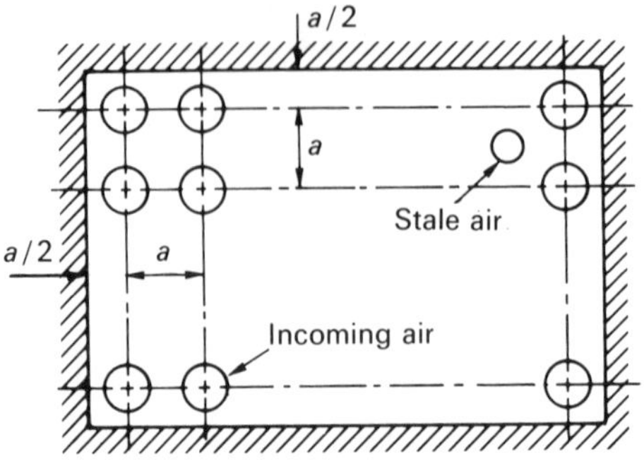


This is made up of $N$ tubes of very thin synthetic material (PVC), of thickness $e$, diameter $d$ and length $L$, assembled in a bundle in a duct. **Incoming air circulates the opposite way in the interstices between them.** We will ignore entry and exit effects arising from flow separation at the ends of the recuperator. There is assumed to be no loss to the surroundings.

To maintain the flows through the recuperator, it is necessary to supply supplementary electric power $\mathscr{P}_2$, to overcome the energy loss due to viscous friction. This will heat the fluids. To simplify the situation, we will suppose that heat is released throughout both fluids, the power per unit length d$x$ being given by:

$$P = \frac{d\mathscr{P}}{dx} = 0.025 \frac{\dot{m}}{d} \frac{V_M^2}{2} = \frac{\mathscr{P}_2}{2L} \quad \text{(in SI units)}$$

where $V_M$ is the mean velocity in one tube. (This value of $P$ calculated inside the tubes will also be used outside them.) In this method, it is also necessary to supply an additional electrical power $\mathscr{P}_3$ to raise the air coming into the workshop to $T_a$.

**Calculate the ratio of the electric power provided with the recuperator to that needed without the recuperator,** if for the same flow $\dot{m}$ the incoming air is raised to $T_a$, that is $(\mathscr{P}_2 + \mathscr{P}_3)/\mathscr{P}_1$. Show that this ratio has a minimum value for some particular recuperator length $L$, to be determined.

**Numerical data:**

$$T_a = 19\,°\text{C}; \qquad T_e = -8\,°\text{C}; \qquad \dot{m} = 0.92\ \text{kg s}^{-1}$$

Recuperator, with PVC surfaces:

$$N = 100; \qquad e = 3 \times 10^{-4}\ \text{m}; \qquad d = 3 \times 10^{-2}\ \text{m}; \qquad a = 3.76 \times 10^{-2}\ \text{m}$$

## The solution

**Solution 1:** the air entering at $T_e$ is heated electrically to $T_a$, the workshop temperature. The required power is:

$$\mathscr{P}_1 = |\dot{m}c_p|[T_a - T_e]$$

**Solution 2:** using a heat exchanger (recuperator), the incoming air is raised to a temperature $T_1^o$ ($< T_a$) by the stale air leaving the workshop, and then raised to the correct temperature electrically.

In this case, we have to provide two electrical inputs:

- $\mathcal{P}_2 = 2LP$: the power required to maintain the mass flows of the air coming in and leaving;
- $\mathcal{P}_3 = |\dot{m}c_p|[T_a - T_1^s]$; the power required to heat the fresh air from the $T_1^s$ to the workshop temperature $T_a$. We may calculate the unknown, $T_1^s$, by considering the recuperator:

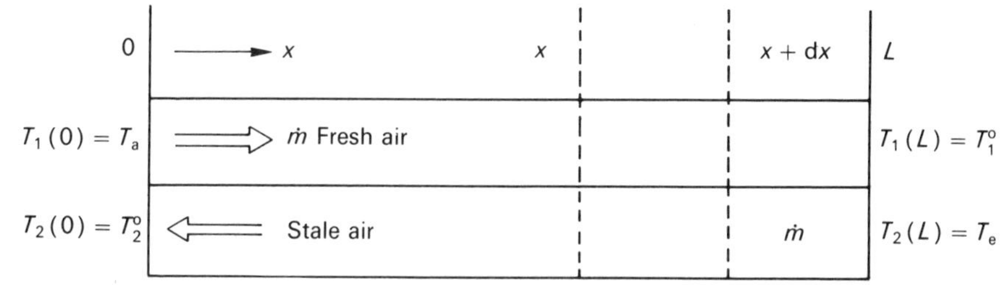


Writing a thermal balance for a slice of the recuperator between $x$ and $x + \mathrm{d}x$:

$$|\dot{m}Cc_p|\frac{\mathrm{d}T_1}{\mathrm{d}x} = (N\pi d)h_g[T_2(x) - T_1(x)] + P$$

$$-|\dot{m}c_p|\frac{\mathrm{d}T_2}{\mathrm{d}x} = (N\pi d)h_g[T_1(x) - T_2(x)] + P$$

where $h_g$ is the global heat transfer coefficient between fluid 1 and fluid 2:

$$\frac{1}{h_g S} = \frac{1}{h_i S_i} + \frac{\ln(d_e/d_i)}{2\pi\lambda l} + \frac{1}{h_e S_e}$$

For a thin walled tube:

$$S_i \sim S_e \sim S \qquad d_e = d_i + e \qquad e \ll d_i$$

$$\frac{1}{h_g S} = \frac{1}{h_i S} + \frac{e}{\lambda S} + \frac{1}{h_e S}$$

Adding the two balance equations above, term by term:

$$\frac{\mathrm{d}(T_1 - T_2)}{dx} = \frac{2P}{|\dot{m}c_p|}$$

Integrating:

$$T_1(x) - T_2(x) = \frac{2P}{|\dot{m}c_p|}x + (T_e - T_2^o)$$

This is a linear variation of temperature difference which, if substituted into one of the balance equations, will give us another relationship between $T_2^s$ and $T_1^s$. The first, putting $x = L$, gives:

$$T_2^o = T_e + T_a + \frac{2PL}{|\dot{m}c_p|} - T_1^o$$

$$\dot{m}c_p \frac{dT_1}{dx} = N\pi\, dh_g \left[ T_2^o - T_e - \frac{2Px}{\dot{m}c_p} \right] + P$$

$$\frac{dT_1}{dx} = \frac{NTU}{L}\left[ T_2^o - T_e - \frac{2Px}{|\dot{m}c_p|} \right] + \frac{P}{|\dot{m}c_p|}$$

where $NTU = (N\pi\, dL)h_g/|\dot{m}c_p| = S_{max}h_g/|\dot{m}c_p|$.

NTU is the number of transport units (Chapter 4, Section 5). Integrating:

$$T_1(x) = \frac{NTU}{L}[T_2^o - T_e]x - \frac{P \times NTU}{|\dot{m}c_p|}\frac{x^2}{L} + \frac{Px}{|\dot{m}c_p|} + T_e$$

and when $x = L$:

$$T_1(L) = T_1^o = T_e + NTU(T_2^o - T_e) - NTU\frac{PL}{|\dot{m}c_p|} + \frac{PL}{|\dot{m}c_p|}$$

Now:

$$T_2^o - T_e = \frac{2PL}{|\dot{m}c_p|} + T_a - T_1^o$$

which gives:

$$T_1^o - T_e = (T_a - T_e)\frac{NTU}{NTU + 1} + \frac{PL}{|\dot{m}c_p|}$$

$$T_1^o - T_e = (T_a - T_e)\left[\frac{NTU}{NTU + 1} + X\right]$$

where

$$X = \frac{PL}{|\dot{m}c_p|(T_a - T_e)}$$

Knowing $T_1^s$ we can obtain $\mathscr{P}_3$:

$$\mathscr{P}_3 = |\dot{m}c_p|(T_a - T_1^o) + |\dot{m}c_p|[(T_a - T_e) - (T_1^o - T_e)]$$

$$\mathscr{P}_3 = |\dot{m}c_p|(T_a - T_e)\left[1 - X - \frac{NTU}{NTU + 1}\right]$$

The ratio $\mathscr{R}$ between the power required with the recuperator and that without it is given by:

$$\mathscr{R} = \frac{\mathscr{P}_2 + \mathscr{P}_3}{\mathscr{P}_1} = \frac{2PL + |\dot{m}c_p|(T_a - T_e)(1 - X - NTU/(NTU + 1))}{|\dot{m}c_p|(T_a - T_e)}$$

$$\mathscr{R} = X + \frac{1}{NTU + 1}$$

In this equation, $X$ and $NTU$ depend on the recuperator length $L$:

$$X = \frac{PL}{|\dot{m}c_p|(T_a - T_e)} = AL$$

$$NTU = \frac{N\pi\, dh_g}{|\dot{m}c_p|} L = BL$$

$$\mathscr{R} = AL + \frac{1}{1 + BL}$$

This ratio has a turning point when its derivative is zero:

$$\frac{d\mathscr{R}}{dL} = 0 = A - \frac{B}{(BL + 1)^2}$$

$$\Rightarrow \quad \underline{L_{min} = \frac{\sqrt{(B/A)} - 1}{B}}$$

The sign of its derivative confirms that this is indeed a minimum.

**Numerically**

**Preliminary remarks.** The cross-sections of the tubes and of the interstitial spaces in one module are equal: $\Sigma = \pi(d^2/4) = a^2 - \pi(d^2/4)$. The wetted perimeters $\pi d$ are equal in the two cases, so that the hydraulic diameters $D_h$ and mean velocities $V_m$ are also equal.

The physical properties of air are taken at the mean temperature between 19 °C and −8 °C, say at 280 K:

$\lambda = 24.6 \times 10^{-3}\ \mathrm{W\,m^{-1}\,K^{-1}}$ $\quad \mu = 1.75 \times 10^{-5}\ \mathrm{kg\,m^{-1}\,s^{-1}}$

$c_p = 1006\ \mathrm{J\,kg^{-1}\,K^{-1}}$ $\quad \rho = 1.27\ \mathrm{kg\,m^{-3}}$

$Pr = 0.71$

$$P = 0.025 \frac{\dot{m}}{d} \frac{V_m^2}{2}$$

$$\dot{m} = \rho \frac{\pi d^2}{4} V_m N \Rightarrow V_m = \frac{4\dot{m}}{\rho \pi d^2 N}$$

whence:

$$P = \frac{0.2\dot{m}^3}{\rho^2 N^2 \pi^2 d^5}$$

Thus:

$$A = \frac{P}{\dot{m} c_p (T_a - T_e)} = \frac{0.2\dot{m}^3}{\rho^2 c_p N^2 \pi^2 d^5 (T_a - T_e)}$$

$$A = 1.61 \times 10^{-3}$$

To calculate $B$, we need the global heat transfer coefficient $h_g$:

$$\frac{1}{h_g} = \frac{1}{h_i} + \frac{e}{\lambda} + \frac{1}{h_e}$$

$h_i$ is the heat transfer coefficient for flow inside one tube:

$$Re_{D_h} = \frac{\rho V D_h}{\mu} = \frac{4\dot{m}/N}{\mu \mathscr{P}_w} = \frac{4\dot{m}/N}{\mu \pi d}$$

where $\mathscr{P}_w$ is the wetted perimeter of one tube.

$$Re_{D_h} = \frac{4(0.92/100)}{1.75 \times 10^{-5}\pi \times 3 \times 10^{-2}} = 2.23 \times 10^4$$

The flow is turbulent ($Re_{D_h} > 2300$).

From Basic Data (Part V):

$$Nu_{D_h} = 0.023 Re_{D_h}^{0.8} Pr^{1/3} = 61.8$$

$$= \frac{h_i D_h}{\lambda}$$

$$\Rightarrow \quad h_i = \frac{61.8 \times 24.6 \times 10^{-3}}{3 \times 10^{-2}}$$

$$= 50.7\ \mathrm{W\,m^{-2}\,K^{-1}}$$

Thus:

$$\frac{1}{h_g} = \frac{2}{50.7} + \frac{3 \times 10^{-4}}{0.15} = 4.14 \times 10^{-2}$$

$$h_g = 24.1 \text{ W m}^{-2} \text{ K}^{-1}$$

$$B = \frac{N\pi d}{\dot{m}c_p} h_g = \frac{10^2\pi \times 3 \times 10^{-2} \times 24.1}{0.92 \times 1006} = 0.246$$

$$\Rightarrow \quad L_{\text{min}} = \frac{\sqrt{(B/A)} - 1}{B} = 46.2 \text{ m}$$

$$\mathscr{R}_{\text{min}} = AL_{\text{min}} + \frac{1}{1 + BL_{\text{min}}} = 0.155$$

which represents a saving of around 85%.

# Application 8

## Supplementary Heating Using Solar Energy

### *Required knowledge*

- Steady-state conditions (Chapters 1 and 2).
- Radiative heat transfer (Chapters 4, 9 and 10).
- Internal and external convection (Chapters 12, 14 and 17).

### *The problem*

An installation to preheat water (fluid 2) before its entry into a calorifier comprises a solar absorber through which a fluid (1) circulates (driven by a pump), a heat exchanger between fluids 1 and 2 and lagged pipework between the two (Figure 1).

We will just consider the operation of this system in **steady conditions**. The entry temperature of the water $T_2'^{e}$, the external temperature $T_0$, the total solar illumination (or total solar flux density received) $E_s$ on the absorber and the flow rates $q_1$ and $q_2$ of the two fluids are all assumed to be constant.

### *Part 1 – the absorber*

As a first approximation, we will consider the absorber to be a flat and infinite plane in the $x$ and $y$ directions (Figure 2). The temperature is assumed uniform in any $xy$ plane normal to the $z$ axis; in particular, fluid 1 has a uniform temperature $T_F$ lying between $T_1'^{a}$ and $T_1'^{d}$. The absorber is exposed to atmosphere on surfaces 1 and 8 (Figure 2), and exchanges heat by forced convection (due to the wind) with ambient

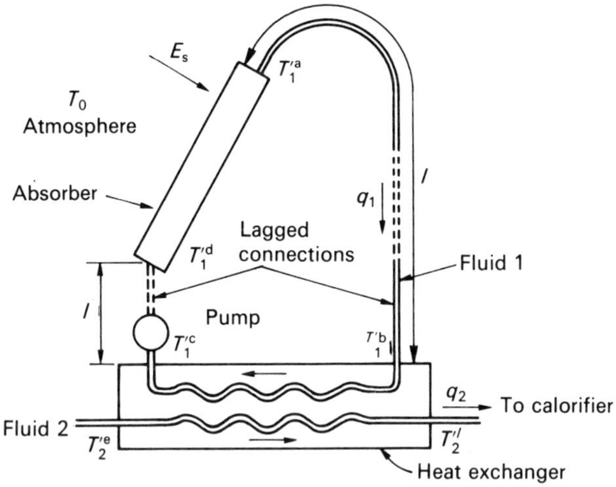


**Figure 1.** General arrangement.

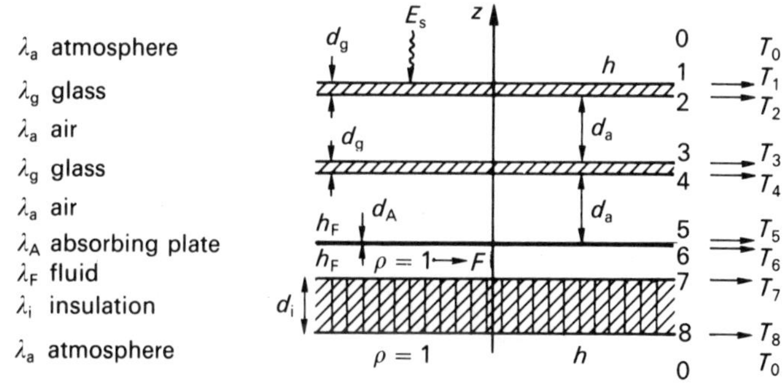


**Figure 2.** Cross-section of absorber.

air; $h$ is the relevant heat transfer coefficient. The absorber is made up of two glass plates of thickness $d_g$ and of an absorbing plate of thickness $d_A$ made of 40% nickel steel, separating two layers of air of thickness $d_a$. The air layers are thin enough to avoid natural convection. The heat transfer fluid (treated as water) circulates beneath the absorbing plate, with a characteristic mean temperature $T_F$, and exchanges heat with surfaces 6 and 7, by forced convection with heat transfer coefficient $h_F$. A glass wool insulating layer of thickness $d_i$ separates the fluid from the surroundings.

The illumination $E_s$ on face 1 of the absorber is uniform. The models used for the radiative properties of the windows and the opaque absorbing plate are given in Figures 3 and 4; they are discontinuous at $\lambda_0$. The atmosphere is, for simplicity, assumed to emit radiation as a black body at temperature $T_0$. Faces 7 and 8 of the insulation are perfectly reflecting for all radiation. Fluid 1 is transparent to radiation.

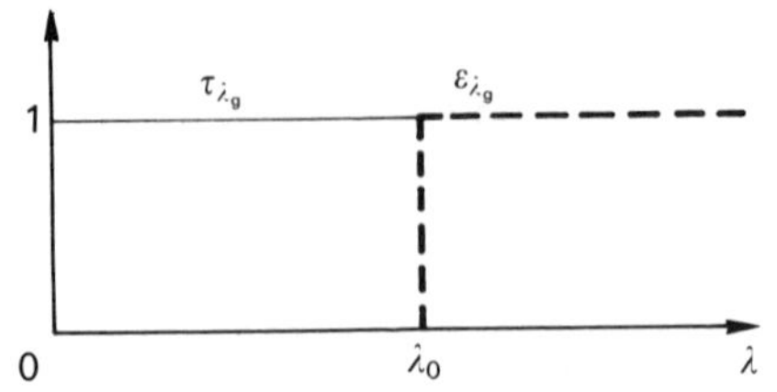


**Figure 3.** Monochromatic transmissivity $\tau_{\lambda g}$ and emissivity $\varepsilon_{\lambda g}$ (assumed independent of direction and temperature) for glass.

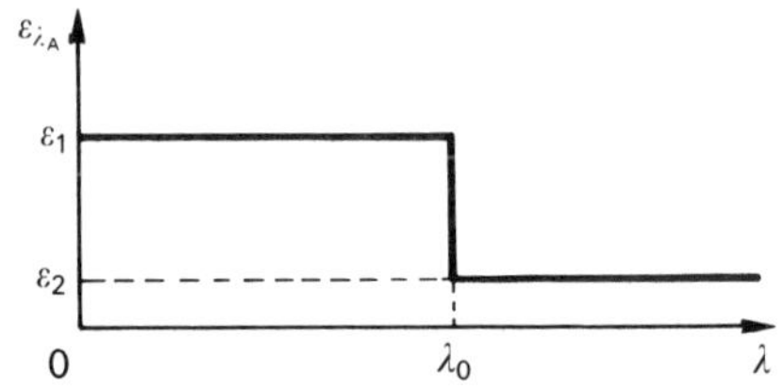


**Figure 4.** Monochromatic emissivity $\varepsilon_{\lambda A}$ for the opaque absorbing plate (assumed independent of direction and temperature).

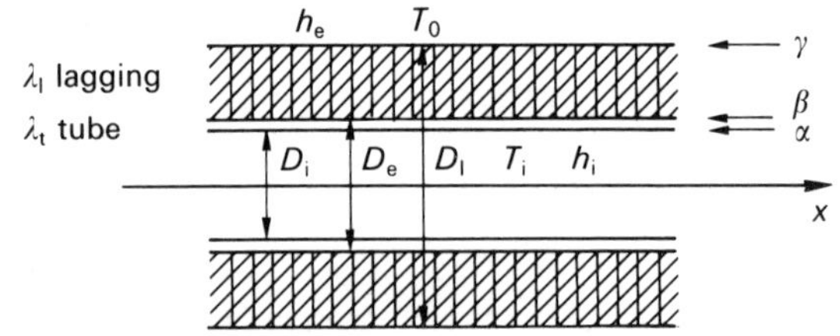


**Figure 5.** Cross-section of a lagged connection.

Temperatures $T_0$ and $T_F$ and the illumination $E_s$ are known. **What is the power $\varphi_F$ delivered to the fluid by the absorber, per unit absorber area?**

## *Part 2 – The pipework and heat exchanger*

The pipes are tubes of (20% Cr, 15% Ni) steel, of internal and external diameter $D_i$ and $D_e$ respectively. They are covered by a sleeve of glass wool lagging of outside diameter $D_l$. The heat transfer coefficient between the fluid and the tube is $h_i$, and that between the lagging and the outside air (wind) is $h_e$. Radiative heat exchange is included in $h_e$. The length $l$ of the tubes between $a$ and $b$ is the same as between $c$ and $d$ (Figure 1).

The water to be preheated (constant specific heat $c_2$) enters the heat exchanger with a mass flow rate $q_2$ at a known temperature $T_2'^{e}$, and leaves it at a known temperature $T_2'^{l}$. Fluid 1 (constant specific heat $c_1$) circulates with mass flow $q_1$. Fluid 1 leaves the absorber at $a$ at a known temperature $T_1'^{a}$. We also assume that $q_1c_1 < q_2c_2$.

1. **What is the total power $\Phi_F$ gained by the fluid in the absorber?** (this will allow the required area of the absorber to be calculated).
2. If the heat exchanger is of counter-current type, with a global heat transfer coefficient $h_g$, **what surface area of heat exchanger is needed?**

## Part 3

Carry out the numerical calculations corresponding to the above quations, using the following data (in SI units):

1. $h = 100$; $h_F = 800\ \mathrm{W\,m^{-2}\,K^{-1}}$;
   $d_a = 10^{-2}$; $d_g = 3\times10^{-3}$; $d_A = 10^{-3}$; $d_i = 10^{-1}$ m;
   $E_S = 700\ \mathrm{Wm^{-2}}$; $\varepsilon_1 = 0.9$; $\varepsilon_2 = 0.4$; $\lambda_0 = 3\ \mu\mathrm{m}$;
   $T_F = 25\,°\mathrm{C}$; $T_0 = 20\,°\mathrm{C}$.
2. $T_1'^{a} = 35\,°\mathrm{C}$; $T_2'^{e} = 15\,°\mathrm{C}$; $T_2'^{l} = 30\,°\mathrm{C}$;
   $q_1 = 0.8$; $q_2 = 1\ \mathrm{kg\,s^{-1}}$; $h_i = 1000$; $h_e = 100\ \mathrm{Wm^{-2}\,K^{-1}}$;
   $l = 20$; $D_i = 3\times10^{-2}$; $D_e = 3.4\times10^{-2}$ m; $D_l = 6\times10^{-2}$; $h_g = 10^3$ $\mathrm{Wm^{-2}\,K^{-1}}$.

# The solution

## 1. The absorber

Through the absorber, fluid 1 gains energy from the surroundings through the double glazing, and loses energy through the lagging, which is not a perfect insulator.

Treating flux in the $z$ direction as positive, the thermal power gained by the fluid is:

$$\Phi_f = \Phi_{07} - \Phi_{60}$$

where $\Phi_{07}$ is the flux through the system from 0 to 7, and $\Phi_{60}$ is that from 6 to 0. Per unit surface area,

$$\varphi_f = \varphi_{07} - \varphi_{60}$$

$\varphi_{07}$ corresponds to the loss through the lagging, and $-\varphi_{60}$ is the gain through the absorbing plate and double glazing.

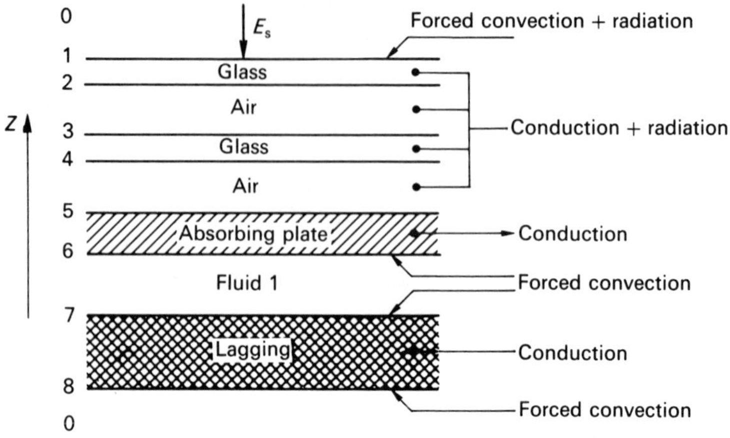


## 1.1 Calculation of $\varphi_{07}$

In steady conditions, $\varphi_{07}$ is constant, and can be calculated at various cross-sections:

- conducto-convective transfer by forced convection between surface 8 and the outside air (with a perfectly reflective wall):

$$\varphi_{07} = \mathrm{h}[T_0 - T_8]$$

- conduction through the lagging:

$$\varphi_{07} = \frac{\lambda_i}{d_i}[T_8 - T_7]$$

- conducto-convective transfer between face 7 and fluid 1:

$$\varphi_{07} = h_F[T_7 - T_F]$$

Eliminating the two unknown temperatures $T_7$ and $T_8$, we readily obtain:

$$T_0 - T_F = \varphi_{07}\left\{\frac{1}{h} + \frac{d_i}{\lambda_i} + \frac{1}{h_F}\right\}$$

## 1.2 Calculation of $\varphi_{60}$

As for the losses, $\varphi_{60}$ is spatially constant, and can be calculated at each section.

The windows are transparent to radiation of wavelength below $\lambda_0$ ($\sim 3\ \mu$m). Bodies at temperatures close to ambient or below 500 K, such as the atmosphere, the windows and the absorbing plate, will have their radiation blocked by the windows. Only that part of the solar radiation lying between 0 and 3 $\mu$m will pass through the windows. Solar radiation corresponds to a black body at 5780 K, and thus has energy lying between 0.25 $\mu$m and 3.5 $\mu$m. Thus, 98% of the solar energy $E_S$ incident on the first window passes through the windows ($Z[\lambda_0/\lambda_m(T)] = 98\%$).

- Conducto-convective transfer (forced convection between surface 6 and fluid 1):

$$\varphi_{60} = h_F(T_F - T_6)$$

- Conduction in the absorbing plate:

$$\varphi_{60} = \frac{\lambda_A}{d_A}(T_6 - T_5)$$

- Conduction through air + radiation between surfaces 5 and 4:

$$\varphi_{60} = \frac{\lambda_a}{d_a}(T_5 - T_4) + \varphi^R$$

The radiative flux is made up of two parts: the flux exchanged between surfaces 4 and 5, and the solar flux:

$$\varphi^R = \varepsilon_2\sigma(T_5^4 - T_4^4) - \varepsilon_1 E_S$$

$$\Rightarrow \quad \varphi_{60} = \frac{\lambda_a}{d_a}(T_5 - T_4) + \varepsilon_2\sigma(T_5^4 - T_4^4) - \varepsilon_1 E_S$$

- Conduction in the glass + solar radiation:

$$\varphi_{60} = \frac{\lambda_g}{d_g}(T_4 - T_3) - \varepsilon_1 E_S$$

- Conduction through air + radiation between surfaces 3 and 2:

$$\varphi_{60} = \frac{\lambda_a}{d_a}(T_3 - T_2) + \sigma(T_3^4 - T_2^4) - \varepsilon_1 E_S$$

- Conduction in the glass + solar radiation:

$$\varphi_{60} = \frac{\lambda_g}{d_g}(T_2 - T_1) - \varepsilon_1 E_S$$

- Forced conducto-convective transfer and radiation between surface 1 and the air, + solar radiation:

$$\varphi_{60} = h(T_1 - T_0) + \sigma(T_1^4 - T_0^4) - \varepsilon_1 E_S$$

The various expressions for $\varphi_{60}$ are not linear, but may be linearized by noting that the difference in temperature between fluid 1 and the air is small:

$$\frac{T_F - T_0}{T_F} = \frac{5}{298} \ll 1$$

and, by using a relationship of the form:

$$\sigma\{T_1^4 - T_2^4\} \simeq 4\sigma T_R^3(T_1 - T_2) = h_R(T_1 - T_2)$$

we can use the arithmetic mean of the temperatures of fluid 1 and ambient air as a reference temperature for the calculation of the radiative heat transfer coefficient $h_R$. The various expressions for $\varphi_{60}$ then become:

$$\begin{aligned}\varphi_{60} &= h_F(T_F - T_6)\\ &= \frac{\lambda_A}{d_A}(T_6 - T_5)\\ &= \left(\frac{\lambda_a}{d_a} + \varepsilon_2 h_R\right)(T_5 - T_4) - \varepsilon_1 E_S\\ &= \frac{\lambda_g}{d_g}(T_4 - T_3) - \varepsilon_1 E_S\\ &= \left(\frac{\lambda_a}{d_a} + h_R\right)(T_3 - T_2) - \varepsilon_1 E_S\\ &= \frac{\lambda_g}{d_g}(T_2 - T_1) - \varepsilon_1 E_S\\ &= (h + h_R)(T_1 - T_0) - \varepsilon_1 E_S\end{aligned}$$

Eliminating the unknown temperatures $T_6$ to $T_1$, we obtain:

$$T_F - T_0 = \varphi_{60}\left\{\frac{1}{h_F} + \frac{d_A}{\lambda_A} + \frac{1}{(\lambda_a/d_a) + \varepsilon_2 h_R} + 2\frac{d_g}{\lambda_g} + \frac{1}{(\lambda_a/d_a) + h_R} + \frac{1}{h + h_R}\right\}$$
$$+ \varepsilon_1 E_S\left\{\frac{1}{(\lambda_a/d_a) + \varepsilon_2 h_R} + 2\frac{d_g}{\lambda_g} + \frac{1}{(\lambda_a/d_a) + h_R} + \frac{1}{h + h_R}\right\}$$

⇒ from this expression we can calculate $\varphi_{60}$.

## 2. The pipework and heat exchanger

### 2.1 The power gained by fluid 1 in the absorber is:

$$\Phi_f = q_1 c_1\{T_1'^{a} - T_1'^{d}\}$$

If the connections were perfectly lagged, all this power would be transferred to fluid 2, so that:

$$\Phi_f = \underbrace{q_1c_1\{T_1'^{a} - T_1'^{b})}_{\text{loss in connection}} + \underbrace{q_1c_1(T_1'^{b} - T_1'^{c})}_{\text{supplied to fluid 2}} + \underbrace{q_1c_1(T_1'^{c} - T_1'^{d})}_{\text{loss in connection}}$$

In the heat exchanger, the heat given up by fluid 1 is all received by fluid 2:

$$q_1 c_1 \{T_1'^{b} - T_1'^{c}\} = q_2 c_2 \{T_2'^{l} - T_2'^{e}\}$$

This heat is known since the entry and exit temperatures and the mass flow rate of fluid 2 are known.

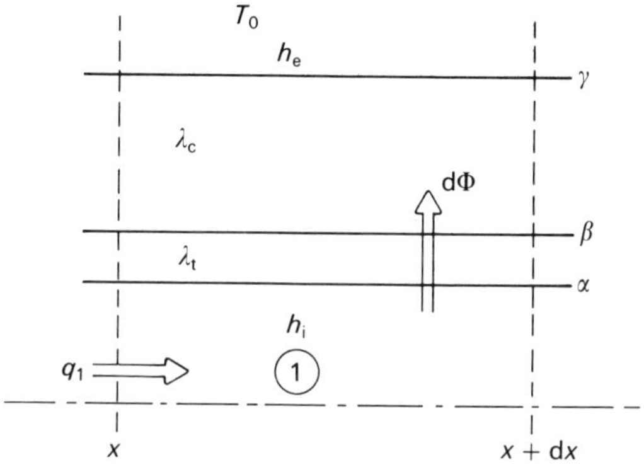


To calculate the heat losses in the pipework, let us write the heat balance between two cross-sections $dx$ apart:

$$\begin{aligned} d\Phi &= q_1 c_1 \{T_1'(x) - T_1'(x + dx)\} \\ &= -q_1 c_1 \frac{dT_1'}{dx} dx \end{aligned}$$

$$\frac{d\Phi}{dx} = -q_1 c_1 \frac{dT_1'}{dx}$$

The lost flux $d\Phi$ can be found by considering the forced convection between fluid 1 and the tube, conduction in the tube and lagging, and finally the forced convection between the lagging and the ambient air:

$$\begin{aligned} d\Phi &= \pi D_i h_i \, dx (T_1' - T_\alpha) \\ &= \frac{2\pi \, dx \lambda_t}{\ln(D_e/D_i)} (T_\alpha - T_\beta) \\ &= \frac{2\pi \, dx \lambda_l}{\ln(D_l/D_e)} (T_\beta - T_\gamma) \\ &= \pi D_l h_e \, dx (T_\gamma - T_0) \end{aligned}$$

Eliminating $T_\alpha$, $T_\beta$ and $T_\gamma$ we obtain:

$$\frac{d\Phi}{dx} = \left[\frac{1}{\pi D_i h_i} + \frac{\ln(D_e/D_i)}{2\pi\lambda_t} + \frac{\ln(D_l/D_e)}{2\pi\lambda_l} + \frac{1}{\pi D_l h_e}\right]^{-1} (T_1' - T_0) = A(T_1' - T_0)$$

Thus:

$$-q_1c_1\frac{dT'_1}{dx} = A(T'_1 - T_0) = -q_1c_1\frac{d(T'_1 - T_0)}{dx}$$

$$\frac{d(T'_1 - T_0)}{T'_1 - T_0} = -\frac{A}{q_1c_1}dx$$

$$T'_1 - T_0 = \text{const} \times \exp\left(-\frac{A}{q_1c_1}x\right)$$

so, for each of the connections:

$$T'^{b}_1 - T_0 = (T'^{a}_1 - T_0)\exp\left(-\frac{A}{q_1c_1}l\right)$$

$$T'^{d}_1 - T_0 = (T'^{c}_1 - T_0)\exp\left(-\frac{A}{q_1c_1}l\right)$$

whence:

$$T'^{a}_1 - T'^{b}_1 = (T'^{a}_1 - T_0)\left[1 - \exp\left(-\frac{A}{q_1c_1}l\right)\right]$$

$$T'^{c}_1 - T'^{d}_1 = (T'^{c}_1 - T_0)\left[1 - \exp\left(-\frac{A}{q_1c_1}l\right)\right]$$

which gives the heat losses for the connections:

$$q_1c_1(T'^{a}_1 - T'^{b}_1) = q_1c_1(T'^{a}_1 - T_0)\left[1 - \exp\left(-\frac{A}{q_1c_1}l\right)\right]$$

$$q_1c_1(T'^{c}_1 - T'^{d}_1) = q_1c_1(T'^{c}_1 - T_0)\left[1 - \exp\left(-\frac{A}{q_1c_1}l\right)\right]$$

In the last equation, $T'^{c}_1$ is unknown, but by writing the balance at the heat exchanger, we obtain:

$$T'^{c}_1 - T_0 = (T'^{a}_1 - T_0)\exp\left(-\frac{A}{q_1c_1}l\right) - \frac{q_2c_2}{q_1c_1}(T'^{l}_2 - T'^{e}_2)$$

from which we finally obtain an expression for the flux $\Phi_f$ which fluid 1 must extract from the absorber in order to produce a temperature rise from $T'^{e}_2$ to $T'^{l}_2$ in fluid 2 in the heat exchanger:

$$\Phi_f = q_1c_1[T'^{a}_1 - T_0]\left[1 - \exp\left(-\frac{Al}{q_1c_1}\right)\right]\left[1 + \exp\left(-\frac{Al}{q_1c_1}\right)\right]$$

$$+ q_2c_2\{T'^{l}_2 - T'^{e}_2\}\exp\left(-\frac{Al}{q_1c_1}\right)$$

Now the flux density in the fluid is $\varphi_f$, so that the area of the solar absorber is given by:

$$S = \frac{\Phi_f}{\varphi_f}$$

## 2.2 *The entry and exit temperatures and the mass flows are known:*

$$Cr_{min} = q_1 c_1 \qquad Cr_{max} = q_2 c_2$$

This knowledge of the entry and exit temperatures allows us to calculate the heat exchanger efficiency, which is the theoretically possible ratio of the flux transferred from the hot to the cold fluid to the maximum flux:

$$E = \frac{T_1'^b - T_1'^c}{T_1'^b - T_2'^e} \quad \text{(known)}$$

It can be readily shown (Chapter 3) that this efficiency is a function of $Cr_{min}/Cr_{max}$ and the number of transport units $NTU$:

$$E = f\left(\frac{Cr_{min}}{Cr_{max}}, NTU\right)$$

so that

$$NTU = \frac{h_g S_{max}}{Cr_{min}} = g\left(\frac{Cr_{min}}{Cr_{max}}, E\right)$$

Charts are available which show $E$ as a function of $NTU$ parameterized by $Cr_{min}/Cr_{max}$.

$$\Rightarrow \text{for } E \text{ and } \frac{Cr_{min}}{Cr_{max}} \text{ known} \Rightarrow NTU \Rightarrow S_{max} = \frac{NTU \times Cr_{min}}{h_g}$$

Knowledge of the entry and exit temperatures of the two fluids and their mass flows make the use of the preceding general method unnecessary. We may just use one of equations (3.77) or (3.78), which in the current notation gives:

$$\frac{T_2'^e - T_2'^l}{T_1'^b - T_2'^e} = \frac{q_1 c_1}{q_1 c_1 + q_2 c_2}\left\{1 - \exp\left[-h_g S_{max}\left(\frac{1}{q_1 c_1} + \frac{1}{q_2 c_2}\right)\right]\right\} \qquad (3.78)$$

with $q_2 < 0$. In this expression, only $S_{max}$ is unknown.

# 3. *Numerically*

## 3.1

$$\varphi_{07} = \frac{T_0 - T_f}{(1/h) + (d_i/\lambda_i) + (1/h_F)} = \frac{20 - 25}{10^{-2} + (10^{-1}/0.038) + (1/800)}$$

$$= -1.9 \text{ W m}^{-2}$$

To calculate $\varphi_{60}$. we need the radiative heat transfer coefficient $h_R$:

$$T_R = 295\ \text{K} \qquad h_R = 4\sigma T_R^3 = 4 \times 5.67 \times 10^{-8} \times (295)^3 = 5.8\ \text{W m}^{-2}\ \text{K}^{-1}$$

which gives:

$$\varphi_{60} = -613\ \text{W m}^{-2}$$

with:

$\lambda_g = 1\ \text{W m}^{-1}\ \text{K}^{-1}$; $\lambda_{\text{absorber}} = 10\ \text{W m}^{-1}\ \text{K}^{-1}$; $\lambda_{\text{air}} = 2.6 \times 10^{-2}\ \text{W m}^{-1}\ \text{K}^{-1}$

whence:

$$\varphi_f = \varphi_{07} - \varphi_{60} = +611\ \text{W m}^{-2}$$

The performance of the absorber may be defined as the ratio between the flux removed by fluid 1 and the incident solar energy:

$$\eta = \frac{\varphi_f}{E_S} = \frac{611}{700} = 0.87$$

⇒ energy conversion is very important since the temperature elevation is small. Even so, this performance is optimistic, since the mechanical energy necessary to circulate fluid 1 has not been taken into account.

### 3.2 *Calculation of $\Phi_F$ and the area of the absorber*

$$\frac{1}{A} = \frac{1}{\pi D_i h_i} + \frac{\ln(D_e/D_i)}{2\pi\lambda_t} + \frac{\ln(D_l/D_e)}{2\pi\lambda_l} + \frac{1}{\pi D_l h_e}$$

$$= \frac{1}{\pi 3 \times 10^{-2} \times 10^3} + \frac{\ln(3.4/3)}{2\pi \times 15.1} + \frac{\ln(6/3.4)}{2\pi \times 3.8 \times 10^{-2}} + \frac{1}{\pi 6 \times 10^{-2} \times 10^2}$$

$$= 2.44 \qquad A = 0.41$$

$$\Phi_F = 0.8 \times 4.18 \times 10^3(35 - 20)\left[1 - \exp\frac{-0.41 \times 20}{0.8 \times 4.18 \times 10^3}\right]$$

$$\times \left[1 + \exp\frac{-0.41 \times 20}{0.8 \times 4.18 \times 10^3}\right] + 4.18$$

$$\times 10^3(30 - 15)\left[\exp\frac{-0.41 \times 20}{0.8 \times 4.18 \times 10^3}\right]$$

$$= 6.28 \times 10^4\ \text{W} = 62.8\ \text{kW}$$

Absorber area:

$$S = \frac{6.28 \times 10^4}{611} = 103\ \text{m}^2$$

It is also of interest to know the entry and exit temperatures $T_1'^{b}$ and $T_1'^{c}$ of fluid 1 in the heat exchanger and also its temperature $T_1'^{d}$ at entry to the solar absorber:

$$T_1'^{a} - T_1'^{b} = (T_1'^{a} - T_0)\left[1 - \exp\left(-\frac{Al}{q_1 c_1}\right)\right]$$

$$= (35 - 20)\left[1 - \exp\left(-\frac{0.41 \times 20}{0.8 \times 4.18 \times 10^3}\right)\right.$$

$$= 15 \times 2.45 \times 10^{-3} = 4 \times 10^{-2} \Rightarrow T_1'^{a} \sim T_1'^{b} = 35\,°\mathrm{C}$$

$$T_1'^{b} - T_1'^{c} = \frac{q_2 c_2}{q_1 c_1}(T_2'^{l} - T_2'^{e}) = \frac{1}{0.8}(30 - 15) = 18.8\,°\mathrm{C}$$

$$\Rightarrow \quad T_1'^{c} = 16.3\,°\mathrm{C}$$

The temperature $T_1'^{c}$ of fluid 1 at the heat exchanger exit is below ambient temperature $T_a$. The lagging on the pipework from the heat exchanger to the absorber is therefore completely pointless.

$$T_1'^{c} - T_1'^{d} = (T_1'^{c} - T_0)\left[1 - \exp\left(-\frac{Al}{q_1 c_1}\right)\right]$$

$$= (16.3 - 20) \times 2.45 \times 10^{-3} \simeq -9 \times 10^{-3}$$

$$\Rightarrow \quad T_1'^{c} = T_1'^{d} = 16.3\,°\mathrm{C}$$

Actually, $T_1'^{d}$ is slightly higher since fluid 1 is heated by the pump.

### 3.3 Calculation of heat exchanger area $S_{\mathrm{max}}$

This can be determined either

- from charts and tables giving the efficiency $E$ as a function of $NTU$ [see reference 16], or
- using formula (3.78):

$$\frac{T_2'^{e} - T_2'^{l}}{T_1'^{b} - T_2'^{e}} = \frac{q_1 c_1}{q_1 c_1 + q_2 c_2}\left\{1 - \exp\left[-h_g S_{\mathrm{max}}\left(\frac{1}{q_1 c_1} + \frac{1}{q_2 c_2}\right)\right]\right\}$$

$$q_2 < 0$$

$$\frac{15 - 30}{35 - 30} = \frac{0.8}{0.8 - 1}\left\{1 - \exp\left[-10^{-3} S_{\mathrm{max}}\left(\frac{1}{0.8 \times 4.18 \times 10^3} - \frac{1}{4.18 \times 10^{-3}}\right)\right]\right\}$$

$$\Rightarrow \quad S_{\mathrm{max}} = 25.4\ \mathrm{m}^2$$

# Application 9

## Thermal Study of a Nuclear Reactor

### *Required knowledge*

- Steady-state conditions (Chapters 1 and 2).
- Conducto-convective heat transfer (Chapter 3).

### *Feature*

- Calculation of the efficiency of a counter-current heat exchanger.

### *The problem*

The block diagram of a nuclear power station is shown in Figure 1.

In the core, the fissile material is made into pins of radius $R_1$ enclosed in stainless steel (chrome 18%, nickel 8%) sheaths of length $L$ and radius $R_2$. The pins are assumed to occupy the whole length of the sheath (Figure 2).

These elements (pins and sheaths) are grouped into hexagonal bundles, each containing 271 elements (Figure 2), held by helicoidal wires, called swirls, which maintain their axes at 10 mm spacing. The wall thickness of the bundles is about $e = 4 \times 10^{-3}$ mm. The total of 120 adjacent assemblies make up the core (Figure 3).

The primary sodium ($Na_I$) circulates through the core, in the spaces between the sheaths. The total thermal power of the reactor is 3200 MW. The pipes between the core and the heat exchanger are perfectly insulated, and the counter-current heat exchanger is assumed to be loss free. The $Na_I$ (primary fluid) enters the core at $T_1^e = 400\,°C$ with mass flow rate $q_1$. The $Na_{II}$ (secondary fluid) leaves the heat exchanger at $T_2^1 = 540\,°C$ with a mass flow rate $q_2$ of $11\,000\ kg\,s^{-1}$. The conducto-convective heat transfer coefficient between the sheath and the sodium is $h_1 = 1.1 \times 10^5\ W\,m^{-2}\,K^{-1}$.

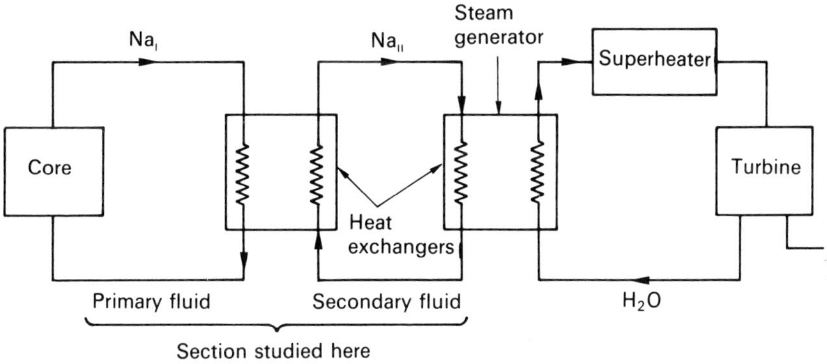


**Figure 1.** Block diagram of a nuclear power station.

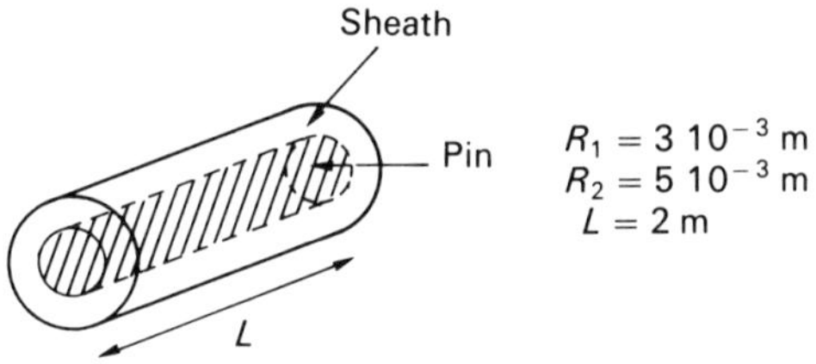


**Figure 2**

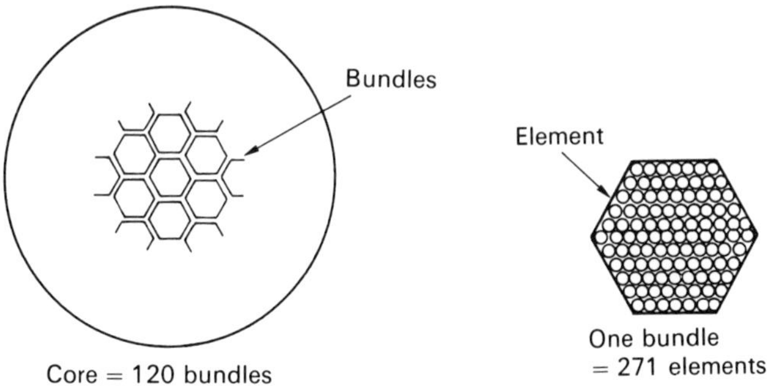


**Figure 3.** Cross-section of core and one bundle.

Given that the maximum temperature reached at any point on the fuel pin is $T_u^{max} = 2400\,°C$, **what should be the NTU of the heat exchanger for steady-state conditions?**

We will make the following approximations:

1. The power per unit volume $P$, which is only dissipated in the fissile material (fuel pins), is homogeneous and the same for all pins (a very crude approximation).

2. Axial conduction (in the $Na_I$ flow direction) is negligible equally in the pins, the sheaths, the assembly walls and the fluid.
3. The flow rate of $Na_I$ is the same in all the inter-sheath gaps.
4. The thermal conductivity of the fissile material ($UO_2$) is taken as $\lambda_u = 2.4\ W\ m^{-1}\ K^{-1}$, and is assumed independent of temperature (a gross approximation).

## Solution

The number of transport units $NTU$ depends on the heat exchanger characteristics:

$$NTU = f\left(E, \frac{Cr_{min}}{Cr_{max}}\right)$$

Between 500 °C and 700 °C, the specific heat of Na is constant, and between 250 °C and 500 °C, it only varies by 5%. We can therefore consider it constant over the temperature range concerned:

$$\frac{Cr_{min}}{Cr_{max}} = \frac{q_{min}}{q_{max}}$$

We only know $q_2$, and must therefore determine $q_1$.

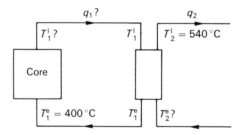

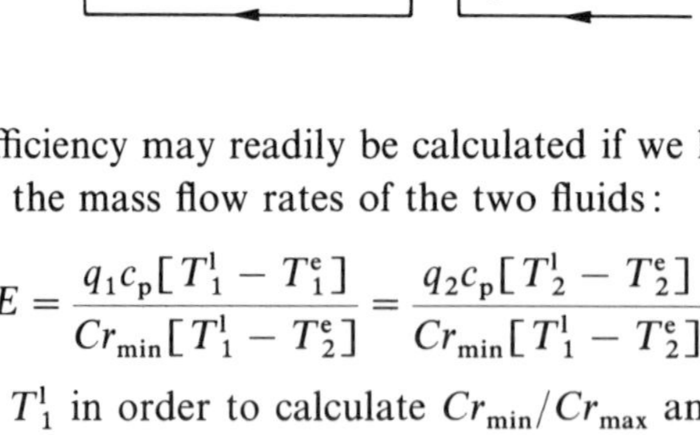


The heat exchanger efficiency may readily be calculated if we know three entry and exit temperatures and the mass flow rates of the two fluids:

$$E = \frac{q_1 c_p [T_1^l - T_1^e]}{Cr_{min}[T_1^l - T_2^e]} = \frac{q_2 c_p [T_2^l - T_2^e]}{Cr_{min}[T_1^l - T_2^e]}$$

We thus need $q_1$ and $T_1^l$ in order to calculate $Cr_{min}/Cr_{max}$ and $E$, and thus obtain $NTU$ using equation (3.85).

The total power $P_t$ of the reactor is completely removed by sodium I and transferred to sodium II (there is no heat loss in the connections, and the heat exchanger is assumed perfect). Then:

$$P_t = q_1 c_p \{T_1^l - T_1^e\} = q_2 c_p \{T_2^l - T_2^e\}$$

which gives one relationship between $q_1$ and $T_1^l$ and lets us calculate the heat exchanger entry temperature of $Na_{II}$:

$$T_2^e = T_2^l - \frac{P_t}{q_2 c_p}$$

To find a second relationship between $q_1$ and $T_1^l$, we will consider the core cooling. We know the maximum temperature $T_u^{max}$ reached by a pin. Because of the rotational symmetry of the sheath–pin combination, and because we are assuming the power per unit volume $P$ to be constant, this temperature $T_u^{max}$ is reached on the pin axis at the downstream end of the bundle.

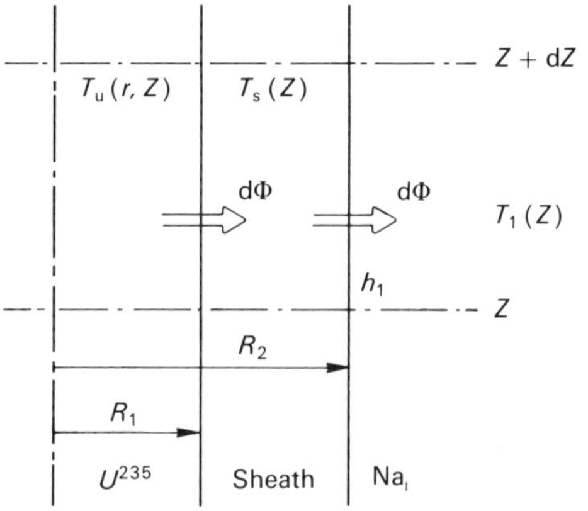


Consider a slice of the fissile pin, of thickness $dZ$. The power transferred to the fluid is:

$$\begin{aligned} d\Phi &= \pi R_1^2 \, dZ P \\ &= \frac{2\pi\lambda_s}{\ln(R_2/R_1)} [T_u(R_1, Z) - T_s(R_2, Z)] \, dZ \\ &= h_1 2\pi R_2 \, dZ [T_s(R_2, Z) - T_1(Z)] \end{aligned}$$

whence:

$$T_u(R_1, Z) - T_1(Z) = \pi R_1^2 P \left\{ \frac{1}{2\pi R_2 h_1} + \frac{\ln(R_2/R_1)}{2\pi\lambda_s} \right\}$$

$T_u(R_1, Z)$ is found from the temperature profile within the fuel:

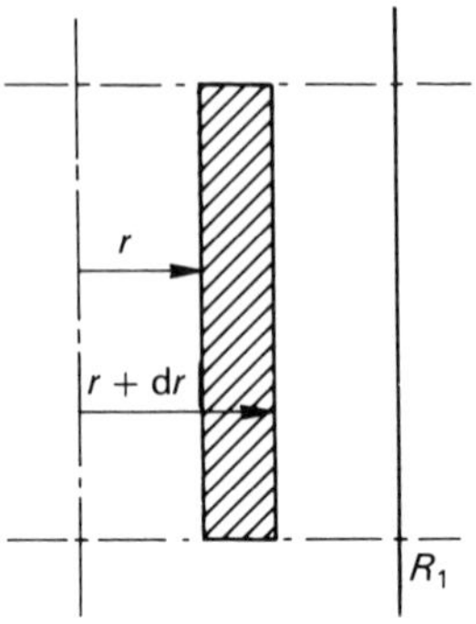


The thermal balance of the ring between $r$ and $r + \mathrm{d}r$ and of thickness $\mathrm{d}Z$ is:

$$0 = \mathrm{d}Z\{2\pi r\varphi(r) - 2\pi(r + \mathrm{d}r)\varphi(r + \mathrm{d}r)\} + 2\pi r\,\mathrm{d}r\,\mathrm{d}ZP$$

whence:

$$\frac{\partial}{\partial r}\left(-\lambda_u r\frac{\partial T}{\partial r}\right) = rP$$

After integration, and using $\partial T_u/\partial r = 0$ at $r = 0$, we obtain:

$$T_u(r, Z) - T_1(Z) = \frac{P(R_1^2 - r^2)}{4\lambda_u} + \pi R_1^2 P\left\{\frac{1}{2\pi R_2 h_1} + \frac{\ln(R_2/R_1)}{2\pi\lambda_s}\right\}$$

We have seen that $T_u$ is maximum at $r = 0$ and $Z = L$:

$$T_1(L) = T_1^1$$

so that:

$$T_u^{max} - T_1^1 = T_u(0, L) - T_1(L) = \pi PR_1^2\left\{\frac{1}{4\lambda_u\pi} + \frac{1}{2\pi R_2 h_1} + \frac{\ln(R_2/R_1)}{2\pi\lambda_s}\right\}$$

$$\Rightarrow \quad T_1^1 = T_u^{max} - \pi PR_1^2\left\{\frac{1}{4\lambda_u\pi} + \frac{1}{2\pi R_2 h_1} + \frac{\ln(R_2/R_1)}{2\pi\lambda_s}\right\}$$

The power per unit volume $P$ may be obtained from the total power of the reactor:

$$P_t = \pi R_1^2 PLn_{pin}n_{bun}$$

$$\pi R_1^2 P = \frac{P_t}{Ln_{pin}n_{bun}}$$

$$\begin{cases} n_{pin}: \text{ number of pins} \\ n_{bun}: \text{ numbers of bundles} \end{cases}$$

**Numerically:**

$$\pi R_1^2 P = \frac{3.2 \times 10^9}{2 \times 271 \times 120} = 4.92 \times 10^4 \text{ W m}^{-1}$$

$$T_1^{\mathrm{l}} = 2400 - 4.92 \times 10^4 \left\{ \frac{1}{4\pi \times 2.4} + \frac{1}{2\pi \times 5 \times 10^{-3} \times 1.1 \times 10^5} + \frac{\ln(5/3)}{2\pi \times 20} \right\} = 554\,^\circ\text{C}$$

$$q_1 = \frac{P_{\mathrm{t}}}{C_{\mathrm{p}}\{T_1^{\mathrm{l}} - T_1^{\mathrm{e}}\}} = \frac{3.2 \times 10^9}{1260[554 - 400]} = 1.64 \times 10^4 \text{ kg s}^{-1}$$

Thus $q_2 < q_1$

$$\frac{Cr_{\min}}{Cr_{\max}} = \frac{1.10 \times 10^4}{1.64 \times 10^4} = 0.67$$

$$E = \frac{q_1[T_1^{\mathrm{l}} - T_1^{\mathrm{e}}]}{q_2[T_1^{\mathrm{l}} - T_2^{\mathrm{e}}]} = \frac{1.64[554 - 400]}{1.1[554 - 309]} = 0.94$$

since:

$$T_2^{\mathrm{e}} = 540 - \frac{3.2 \times 10^9}{1.1 \times 10^4 \times 1.26 \times 10^3} = 309\,^\circ\text{C}$$

Equation (3.85) gives the efficiency of a counter current heat exchanger:

$$E = \frac{1 - \exp\left[-NTU\left(1 - \dfrac{Cr_{\min}}{Cr_{\max}}\right)\right]}{1 - \left(\dfrac{Cr_{\min}}{Cr_{\max}}\right)\exp\left[-NTU\left(1 - \dfrac{Cr_{\min}}{Cr_{\max}}\right)\right]}$$

from which we can calculate the number of transport units, $NTU$:

$$NTU = -\frac{-1}{1 - (Cr_{\min}/Cr_{\max})} \ln\left(\frac{1 - E}{1 - (Cr_{\min}/Cr_{\max})\,E}\right)$$

whence:

$$NTU = \frac{-1}{1 - 0.67} \ln\left(\frac{\Xi - 0.94}{1 - 0.67 \times 0.94}\right) = 5.51$$

This value of NTU could have been obtained directly using tables and ready reckoners (reference [16], Part I).

# Application 10

## Heat Treatment in a Tube Furnace

### *Required knowledge*

- Non-steady conduction (Chapters 5–7).
- Radiation (Chapters 4, 9 and 10).
- External laminar convection (Chapter 14).

### *Features*

- Characteristic scales in unsteady conditions.
- Calculation of conducto-convective and radiative heat transfer.
- Comparison of two heating methods (by flowing air and by radiation).

### *The problem*

Steel plates are to be heated in such a way that no point of a plate will be subject to a temperature below 327 °C. The plates are 2 m by 1 m and 10 mm thick, and are initially at 27 °C.

A tube furnace 9.25 m long and of rectangular cross-section is available. The size in the $y$ direction, perpendicular to the plane of the plates, is large enough compared to the thickness of the plates for the latter to be effectively in an infinite flow when gas circulates along the axis of the furnace. The width, in the $z$ direction, is 2.05 m.

Along the furnace axis, the spacing between plates must be a minimum of 25 cm, in order that the flow round the leading edge of a plate may be considered undisturbed by the preceding one. A similar distance is needed at the furnace entry and exit. There are thus two ways of arranging plates in the tunnel (see diagram), case 1 and case 2.

1. The plates are heated by an air current at 1300 K at a speed of $V = 10\ \mathrm{ms}^{-1}$ along the furnace axis.

   **Which is the best arrangement of plates (case 1 or case 2) to heat the maximum number of plates in 24 hours?** We will disregard the loading and unloading times: in reality, plates are loaded and unloaded continuously.

   Because of its speed, the average mixed air temperature will vary little, and may be taken as constant throughout the furnace. We will assume that expressions for Nusselt number appropriate to constant surface temperature are valid.
2. The air flow is stopped, and the walls of the tube furnace are heated to a constant and uniform temperature of 1300 K. The walls are black bodies, and the emission factor for the plates at all wavelengths is $\varepsilon = 0.5$, independent of temperature. **Answer the same question as in 1.**

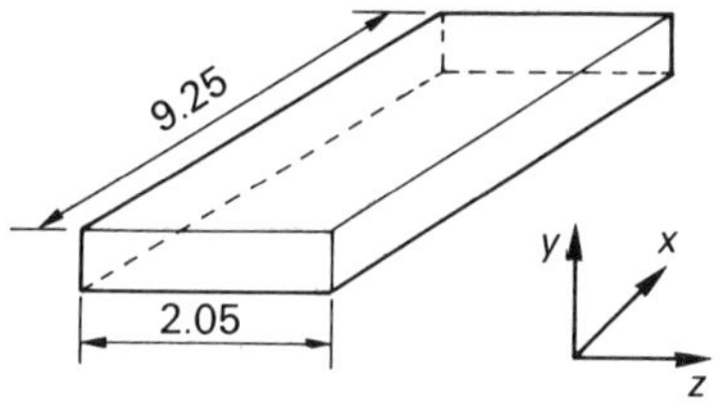


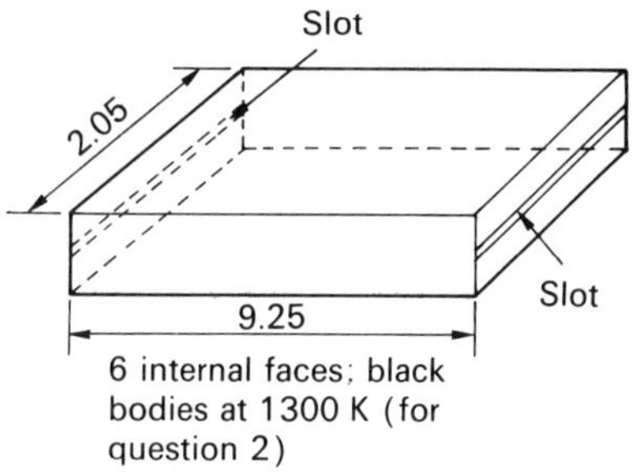


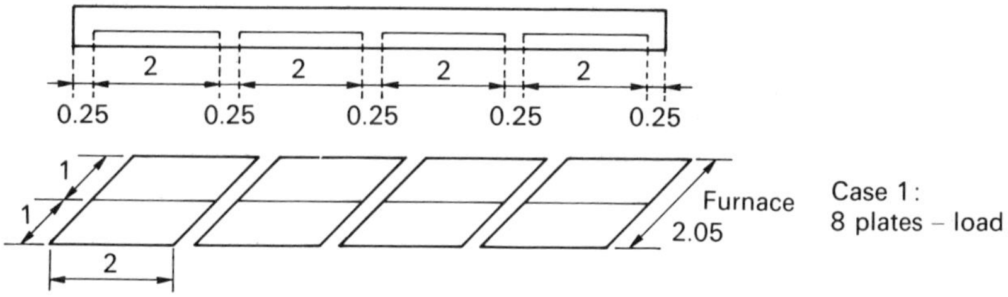


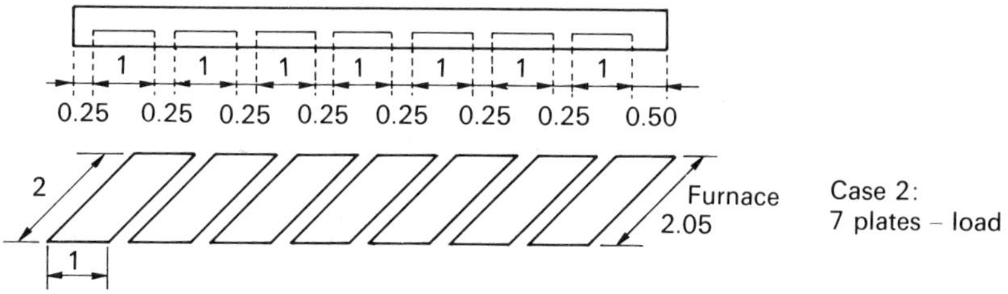


The two faces of the plates are subject to the same conducto-convective and/or radiative conditions. The supports holding the plates in the tube furnace are not shown.

# The solution

## 1 Air heating (1300 K)

For a semi-infinite flow, the heat transfer coefficient $h(x)$ falls from the leading edge and is minimum at the trailing edge. Hence, the trailing edge of each plate will be the least heated by the air current at 1300 K. If this point is at a temperature equal to or in excess of 327 °C (600 K), all other points on the plate will be at higher temperatures.

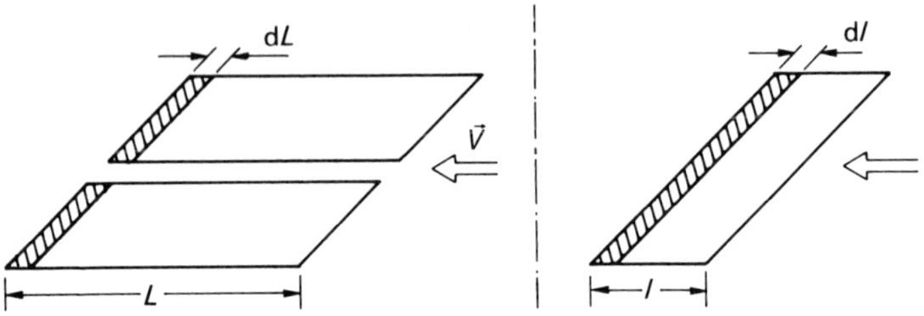


Case 1: 8 plates    Case 2: 7 plates

Air at 1300 K (from Basic Data, Part V):

$$\begin{cases} \nu = 17.9 \times 10^{-5}\ \mathrm{m^{-2}\,s^{-1}} \\ a = 24.8 \times 10^{-5}\ \mathrm{m^2\,s^{-1}} \\ Pr = 0.722 \\ \lambda = 0.0803\ \mathrm{W\,m^{-1}\,K^{-1}} \end{cases}$$

$$Re_L = \frac{LV}{\nu} = \frac{2 \times 10}{17.9 \times 10^{-5}} = 1.12 \times 10^5 < 5 \times 10^5$$

laminar flow

$$Re_l = \frac{lV}{\nu} = \frac{1 \times 10}{17.9 \times 10^{-5}} = 5.59 \times 10^4 < 5 \times 10^5$$

laminar flow

Basic Data (Part V Section 2) (uniform wall temperature):

$$Nu_x(x) = 0.324 Pr^{1/3} Re_x^{1/2} = \frac{h(x)x}{\lambda}$$

$$Nu_L(L) = \frac{h(L)L}{\lambda} = 97.2 \qquad Nu_l(l) = \frac{h(l)l}{\lambda} = 68.7$$

$$h(L) = 3.90\ \mathrm{W\,m^{-2}\,K^{-1}} \qquad h(l) = 5.52\ \mathrm{W\,m^{-2}\,K^{-1}}$$

Intuitively, the plate may be assumed to be isothermal through its thickness, since it is thin (10 mm) and the thermal conductivity of 0.5% C steel ($\lambda = 45\ \mathrm{W\,m^{-1}\,K^{-1}}$) is much higher than that of the heating air. To confirm this, we can calculate the conduction time constant:

$$\tau^{cd} = \frac{e^2}{a_{\mathrm{steel}}} = \frac{e^2}{\lambda_{\mathrm{steel}}} \times \rho c_{\mathrm{p\,steel}} = \frac{(5 \times 10^{-3})^2}{45} \times 3.64 \times 10^6 = 2.02\ \mathrm{s}$$

and the following will verify that $\tau^{cd}$ is indeed negligible in comparison to the time constant for conducto-convection: $\tau^{cd} \ll \tau^{cc}$. Let us consider a strip of the plate of width $dL$ (or $dl$) near the trailing edge:

An energy balance readily gives:

$$\rho c_p \delta V \frac{dT}{dt} = -h[T - T_{ext}]\delta S$$

so that:

$$\frac{T - T_{ext}}{T_{in} - T_{ext}} = \exp\left[-\frac{h\delta S}{\rho c_p \delta V} t\right]$$

$$= \exp - \frac{t}{\tau^{cc}}$$

where:

$$\tau^{cc} = \frac{(\rho c_p)_{steel}}{h} \frac{\delta V}{\delta S}$$

is the time constant for conducto-convection.

Thus at time $t^{cc}$ for the trailing edge to be heated from 300 K (27 °C) to 600 K (327 °C) is given by:

$$\frac{T_{final} - T_{ext}}{T_{initial} - T_{ext}} = \exp - \frac{t^{cc}}{\tau^{cc}} = \frac{600 - 1300}{300 - 1300} = 0.7$$

$$t^{cc} = -\tau^{cc} \ln 0.7 = +0.36\tau^{cc}$$

$$\tau_1^{cc} = \frac{3.64 \times 10^6 \times 5 \times 10^{-3}}{3.90} \qquad \tau_2^{cc} = \frac{3.64 \times 10^6 \times 5 \times 10^{-3}}{5.52}$$

$$= 4667 \text{ s} \qquad = 3297 \text{ s}$$

$$t_1^{cc} = 1860 \text{ s} \qquad t_2^{cc} = 1187 \text{ s}$$

$$= 28 \text{ min} \qquad \simeq 20 \text{ min}$$

The figures above speak for themselves; the characteristic times for conducto-convection and heating are very much higher than that for conduction ($\tau^{cd} = 2$ s), and the assumption of constant temperature through the thickness is quite justified.

Making use of the heating times $t_1^{cc}$ and $t_2^{cc}$ which we have just calculated, we obtain:

51 charges 72 charges

being

408 plates 504 plates

and an improved productivity of 23% using case 2 rather than case 1.

## 2. *Radiation heating*

The furnace walls are at constant temperature (1300 K); the plates enter at 300 K and leave at 600 K. The flux exchanged between the plates and the furnace is thus not a constant. The whole of the plates 'see' the furnace. Thus, initially, we have:

$$\begin{aligned}\varphi_{\text{w in}} &= \varepsilon\sigma[T^4_{\text{furn}} - T^4_{\text{initial}}] \\ &= 0.5 \times 5.67 \times 10^{-8}[1300^4 - 300^4] \\ &= 8.07 \times 10^4 \text{ W m}^{-2}\end{aligned}$$

and finally:

$$\begin{aligned}\varphi_{\text{w f}} &= \varepsilon\sigma[T^4_{\text{furn}} - T^4_{\text{final}}] \\ &= 0.5 \times 5.67 \times 10^{-8}[1300^4 - 600^4] \\ &= 7.73 \times 10^4 \text{ W m}^{-2}\end{aligned}$$

Between the initial and final state, we have $\Delta\varphi/\varphi = -(8.07 - 7.73)/8.07 = -4\%$. Without too much doubt, we can therefore assume that the radiative flux density is constant, and use an average value:

$$\varphi_{\text{w}} = \frac{\varphi_{\text{w in}} + \varphi_{\text{w f}}}{2} = 7.9 \times 10^4 \text{ W m}^{-2}$$

The plates are thus heated by a flux density which is constant in space and time, and therefore any plate is at any instant homogeneous in temperature, and its heating is simply:

$$\rho c_{\text{p}} V \frac{\text{d}T}{\text{d}t} = \varphi_{\text{w}} S$$

$$T(t) - T_{\text{in}} = \frac{\varphi_{\text{w}} S}{\rho c_{\text{p}} V} t$$

The heating is identical for cases 1 and 2. The best configuration is that which contains the most plates: case 1.

The time to heat from 300 K to 600 K is:

$$t^{\text{R}} = (T_{\text{final}} - T_{\text{initial}}) \frac{\rho c_{\text{p}} V}{\varphi_{\text{w}} S} = (600 - 300) \frac{3.64 \times 10^6 \times 5 \times 10^{-3}}{7.9 \times 10^4} = 69 \text{ s}$$

⇒ If we do not know it already, we can conclude from this that radiation heating is a lot more effective than heating with a stream of hot air.

With such a heating time, we can make 1252 charges in 24 hours, which will treat 10016 plates in the arrangement of case 1.

In conclusion, radiation heating allows us to treat nearly twenty times as many plates as hot air heating.

# Application 11

## Laser Heat Treatment of Steel

### *Required knowledge*

- Steady-state conditions. Basics of heat transfer (Part I).
- Non-steady conduction (Chapters 5–8).

### *Objective*

- Unsteady thermal analysis of a modern process. Choice of appropriate model in each operating condition.

### *The problem*

Heat treatment of steel is performed in two stages:

1. Austenitization. The item to be treated, initially at a uniform temperature $T_0$ (20 °C) is raised, at least on the surface, to a temperature above $T_a$ (1020 °C) using a laser. Austenite $\gamma$ is formed.
2. Quenching. The surface layers are cooled rapidly ($d\theta/dt \geqslant 1000\ K\ s^{-1}$) to below a critical temperature $T_c$ (720 °C). This converts the austenite into martensite, which significantly improves the hardness of the steel.

The austenitization stage is carried out with a power $CO_2$ laser, whose slightly broadened beam is scanned at very high frequency across the item. The latter, considered to be plane, is assumed to **absorb** a uniform laser flux $\varphi_L = 3 \times 10^7\ W\ m^{-2}$ during an exposure time $t_e = 130$ ms.

The item, which is in an environment at a uniform 20 °C, is assumed to have a uniform emissivity $\varepsilon$, and the heat transfer coefficient with the surrounding air is $h$.

**We wish to know the depth of the hardened region on an object of thickness $\delta$ for the two cases:**

$$\text{(a) } \delta = 2\ \text{mm} \qquad \text{(b) } \delta = 2\ \text{cm}$$

The attached chart is to be used.

**Data:** $T_0 = 20\,°\text{C}$; $T_a = 1020\,°\text{C}$; $T_c = 720\,°\text{C}$; $\varepsilon = 0.3$; $h = 10\ \text{W m}^{-2}\ \text{K}^{-1}$; $d\theta/dt = 1000\ \text{K s}^{-1}$; $t_e = 130\ \text{ms}$; $\varphi_L = 3 \times 10^7\ \text{W m}^{-2}$. We consider a 1% carbon steel (Basic Data: Part V).

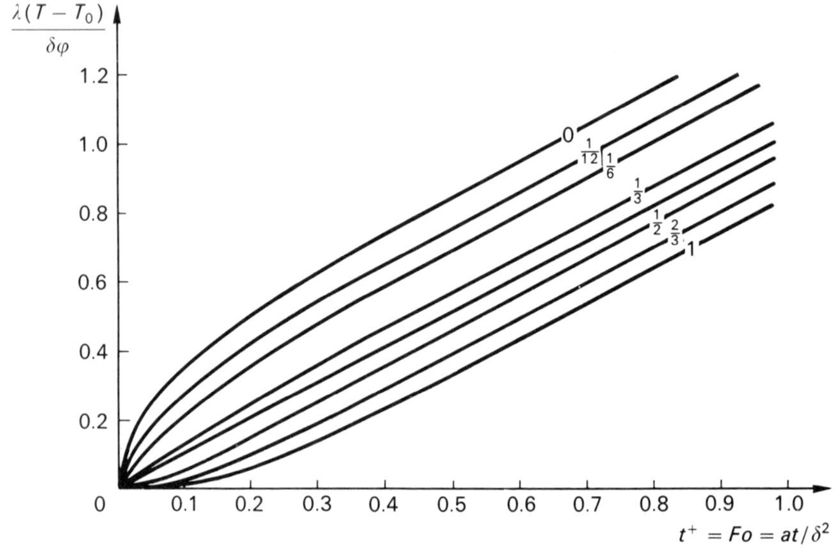


Response of a wall of thickness $\delta$, initially at $T_0$, which is exposed to a flux $\varphi$ at instant $t = 0$ on the face at $x = 0$. Face $x = \delta$ is insulated. Lines show dimensionless temperature, parameterized by $x/\delta$ (from Schneider, reference [6], Part 2).

# *The solution*

## *1. Order of magnitude analysis*

It is useful to look at the orders of magnitude of the characteristic times for the heat transfer in the first, austenitization, phase.

- exposure time: $t_e = 130$ ms,
- time associated with conducto-convective losses: $\tau^{cc}$,
- time associated with radiative losses: $\tau^{R}$,
- characteristic time for non-steady conduction in the specimen: $\tau^{cd}$.

This comparison of times will enable us to choose the appropriate model.

- The characteristic time for conduction through the whole thickness $\delta$ of the specimen is (see Chapters 6–8): $\tau^{cd} = \delta^2/a$, where $a$ is the thermal diffusivity. From Basic Data (Part V), in SI units:

  1% carbon steel at 20 °C:

$$\lambda(20\,°C) = 43 \qquad a(20\,°C) = 1.17 \times 10^{-5}$$
$$\lambda(700 - 1000\,°C) \simeq 29 \qquad a(700 - 1000\,°C) \simeq 7.86 \times 10^{-6}$$

  whence:

| $\tau^{cd}$ | *(20 °C)* | *(700–1000 °C)* |
|---|---|---|
| Case (a) | 0.34 s | 0.51 s |
| Case (b) | 34 s | 51 s |

- The Biot number represents the ratio between $\tau^{cd}$ and the characteristic time for conducto-convection $\tau^{cc}$:

$$Bi = \frac{\tau^{cd}}{\tau_{cc}} = \frac{\delta^2/a}{\rho C_p \delta/2h} = \frac{2h\delta}{\lambda} \qquad \begin{cases} \text{(a)} & Bi = 4.45 \text{ to } 6.8 \times 10^{-4} \\ \text{(b)} & Bi = 4.45 \text{ to } 6.8 \times 10^{-3} \end{cases}$$

  The conducto-convective heat transfer, being much slower than the conductive transfer, may thus be neglected.

- The radiation is characterized (so long as the plate is hot: $\bar{T}^4 \gg T_0^4$) by a time constant obtained assuming the plate to be at uniform temperature, by solving (Chapter 8):

$$\rho c_p V \frac{dT}{dt} = -S\varepsilon\sigma T^4$$

  This gives a time to cool from $T_i$ to $T_f$ of:

$$\tau^R = \frac{\rho c_p \delta}{3\varepsilon\sigma}\left(\frac{1}{T_f^3} - \frac{1}{T_i^3}\right)$$

  with:

$$\rho c_p = 3.69 \times 10^6; \qquad T_1^0 = 1400\text{ K}; \qquad T_f = 1000\text{ K}; \qquad \tau^R = \begin{cases} \text{(a)} & 89\text{ s} \\ \text{(b)} & 890\text{ s} \end{cases}$$

We can therefore also ignore radiative dissipation in both case (a) and case (b).

### Conclusion

We can consider a plate which is initially isothermal at $T_0$, cooled by conduction, and subjected to a uniform heating flux during the period $t_e$. All losses are negligible.

In case (a), the characteristic time for conduction is of the same order of magnitude as $t_e$, and we will consider a finite wall model.

In case (b), where $t_e \ll \tau^{cd}$, we can probably use a semi-infinite model, at least during the heating phase.

## 2. Derivation of equations

### 2.1 Case (a) – finite wall model

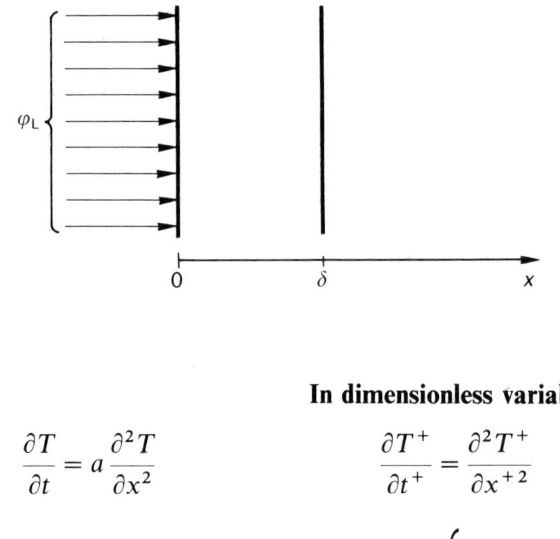


**In dimensionless variables**

$$\frac{\partial T}{\partial t} = a\frac{\partial^2 T}{\partial x^2} \qquad\qquad \frac{\partial T^+}{\partial t^+} = \frac{\partial^2 T^+}{\partial x^{+2}}$$

$$-\lambda \left.\frac{\partial T}{\partial x}\right|_0 = \begin{cases} \varphi_L & \text{if } t \leqslant t_e \\ 0 & \text{if } t > t_e \end{cases} \qquad\qquad -\left.\frac{\partial T^+}{\partial x^+}\right|_0 = \begin{cases} 1 & \text{if } t^+ \leqslant \dfrac{at_e}{\delta^2} \\ 0 & \text{if } t^+ > \dfrac{at_e}{\delta^2} \end{cases}$$

$$-\lambda \left.\frac{\partial T}{\partial x}\right|_\delta = 0 \qquad\qquad -\lambda \left.\frac{\partial T^+}{\partial x^+}\right|_1 = 0$$

$$T(x, 0) = T_0 \qquad\qquad T^+(x^+, 0) = 0$$

where we have written:

$$T^{+} = \frac{\lambda}{\delta \varphi_L}(T - T_0)$$

$$t^{+} = \frac{at}{\delta^2}; \qquad x^{+} = \frac{x}{\delta}$$

The superposition theorem allows us to write:

$$T^{+} = T_1^{+} - T_2^{+}$$

**With $T_1^+$ the solution of:**

$$\frac{\partial T_1^+}{\partial t^+} = \frac{\partial^2 T_1^+}{\partial x^{+2}}$$

$$-\left.\frac{\partial T_1^+}{\partial x}\right|_0 = 1$$

$$\left.\frac{\partial T_1^+}{\partial x}\right|_1 = 0$$

$$T_1^+(x, 0) = 0$$

**and $T_2^+$ the solution of:**

$$\frac{\partial T_2^+}{\partial t^+} = \frac{\partial^2 T_2^+}{\partial x^{+2}}$$

$$-\left.\frac{\partial T_2^+}{\partial x}\right|_0 = \begin{cases} 0 & \text{if } t^+ \leqslant \dfrac{at_e}{\delta^2} \\ 1 & \text{otherwise} \end{cases}$$

$$\left.\frac{\partial T_2^+}{\partial x}\right|_1 = 0$$

$$T_2^+(x, 0) = 0$$

The diagram given below for the finite wall gives us $T_1^+$, and the same diagram offset by $at_e/\delta^2$ gives $T_2^+$ (see diagram below for the construction giving the solution for $T^+$).

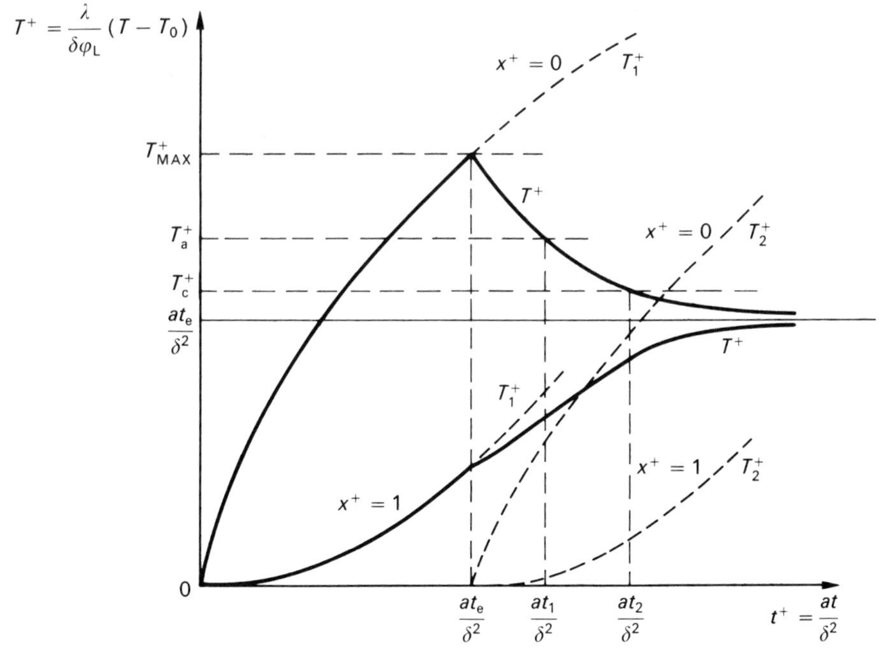

We notice that the asymptotic solution has a uniform temperature $T^+ = at_e/\delta^2$ at all points on the plate. A simple argument shows why this should be the case:

$$\rho c_p V(T_\infty - T_0) = S\varphi_L t_e$$

$$T_\infty^+ = \frac{\lambda}{\delta}\frac{(T_\infty - T_0)}{\varphi_L} = \frac{at_e}{\delta^2}$$

*Note*
This solution for $T^+$ is valid only so long as $t \ll \tau^{cc}, \tau^R$.

Hardening requirements at any section:

1. $T_{MAX}^+ > T_a^+$ (austenitization condition) ($T_{MAX}^+$ is obtained when $t^+ = Fo_e = at_e/\delta^2$).
2. $(T_a - T_c)/(t_2 - t_1) > 1000\,°C/s$.

To simplify matters in the present case, we will use the following uniform data: $a = 10^{-5}\ m^2\ s^{-1}$; $\lambda = 30\ W\ m^{-1}\ K^{-1}$:

$$Fo_e = \frac{at_e}{\delta^2} = \frac{10^{-5} \times 0.13}{(2 \times 10^{-3})^2} = 0.325$$

$$T_a^+ = \frac{\lambda}{\delta\varphi_L}(T_a - T_0) = 0.5$$

$$T_c^+ = \frac{\lambda}{\delta\varphi_L}(T_c - T_0) = 0.35$$

We can see from the diagram above that condition (1) is satisfied if $0 \leqslant x/\delta \leqslant 1/6$, i.e. $x \simeq 0.33$ mm.

It remains to verify condition (2).

The non-dimensional cooling rates are summarized in the following table, constructed using the diagram above:

| $\frac{at}{\delta^2}$ | | *0.325* | *0.350* | *0.40* | *0.50* | *0.55* |
|---|---|---|---|---|---|---|
| $x = 0$ | $T_1^+$ | 0.65 | 0.67 | 0.73 | | 0.88 |
| | $T_2^+$ | 0 | 0.17 | 0.30 | | 0.53 |
| | $T^+$ | 0.65 | $T_0^+ = 0.50$ | 0.43 | | $T_c^+ = 0.35$ |
| $x = \frac{1}{6}$ | $T_1^+$ | 0.50 | | 0.58 | 0.68 | |
| | $T_2^+$ | 0 | | 0.17 | 0.33 | |
| | $T^+$ | $T_a^+ = 0.50$ | | 0.41 | $T_c^+ = 0.35$ | |

We can see that the time taken for the insulated face $x = 0$ to cool from $T_a^+$ ($t^+ = 0.35$) to $T_c^+$ ($t^+ = 0.55$) is the longest. We therefore need to confirm the cooling rate criterion on this face only:

$$\frac{at_1}{\delta^2} = 0.35 \qquad \frac{at_2}{\delta^2} = 0.55 \Rightarrow t_2 - t_1 = 8 \times 10^{-2}\ \mathrm{s}$$

which gives a cooling rate $(T_a - T_c)/(t_2 - t_1) = 3750\ \mathrm{K\ s^{-1}}$, which far exceeds the criterion.

## 2.2 Case (b) – semi-infinite wall model

For the heating stage, we obtain (see Chapter 6, equation (6.29)) a solution for $T_1(x, t)$:

$$T_1(x, t) = T_0 + \frac{2\varphi_L\sqrt{t}}{b}\int_{x/2\sqrt{at_e}}^{\infty} \mathrm{erfc}(u)\,\mathrm{d}u$$

where the material effusivity $b = \sqrt{\lambda\rho c}$ appears (here taken as $10^4$ SI). The point which reaches temperature $T_a$ at instant $t = t_e$ is given by the above:

$$\int_{x/2\sqrt{at_e}}^{\infty} \mathrm{erfc}(u)\,\mathrm{d}u = \mathrm{ierfc}\left(\frac{x_e}{2\sqrt{at_e}}\right) = 0.4863$$

Extrapolation of the table in the Appendix to Part II gives $x_e/2\sqrt{at_e} = 7.6 \times 10^{-2}$. Whence we obtain

$$\underline{x_e = 0.17\ \mathrm{mm}}$$

For the cooling phase, we will assume that the semi-infinite wall model is still valid (to be confirmed retrospectively) and apply the superposition theorem exactly as in case (a):

$$T(x, t) = T_1(x, t) - T_2(x, t)$$

with:

$$T_2(x, t) = 0 \qquad t \leqslant t_e$$

$$T_2(x, t) = +T_1(x, t - t_e) - T_0 \qquad t > t_e$$

The face $x = 0$ cools the least rapidly (i.e. reaches $T_c$ most slowly from $T_a$). Further:

$$\sqrt{t} - \sqrt{t - t_e} = \frac{(T - T_0)\sqrt{\pi\lambda\rho c}}{2\varphi_L}$$

Face $x = 0$ reaches temperature $T_a$ after $t_1 = 0.135$ s and temperature $T_c$ after $t_2 = 0.174$ s. This corresponds to a cooling rate given by:

$$\frac{T_a - T_c}{t_2 - t_1} = 7650\ \mathrm{K\ s^{-1}} > 1000\ \mathrm{K\ s^{-1}}$$

*Note*

The time 0.174 s is still well below the conduction time of the system, which justifies the use of the semi-infinite wall model.

*Practical note*

The cooling is more effective in case (b) since it is not 'braked' by the finite size of the system as in case (a). Contrarily, the heating is less efficient, and the hardened region is thinner.

# Application 12

## Thermal Inertia of a Building

### *The problem*

Consider an apartment surrounded by eight identical apartments. The floor and ceiling of the apartment and its west and east walls are supposed to be perfectly insulated (see diagram below). The only two external faces, north and south, are exposed to outside conditions. We assume that the effects of solar radiation are to:

- vary the outside temperature according to:

$$T_e(t) = T_a + \delta T \cos^4 \omega t = T_a + \frac{3}{8}\delta T + \frac{\delta T}{2}\cos 2\omega t + \frac{\delta T}{8}\cos 4\omega t$$

  where $\omega = \pi/\tau$ and $\tau = 24$ h;
- impose an incident flux (from the sun) on the south facade (only):

$$\varphi_R = \varphi_0 \cos^4 \omega t$$

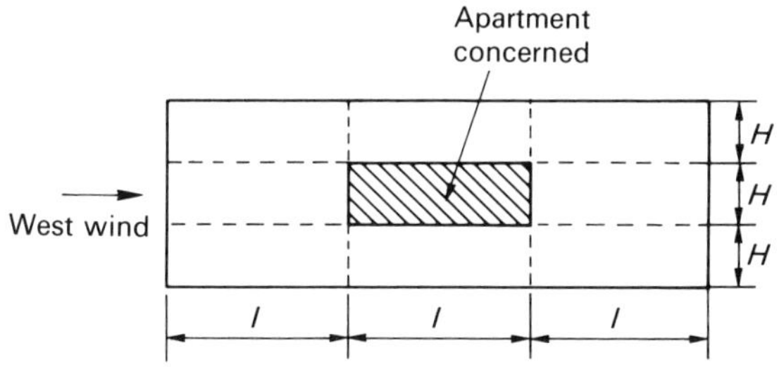


**South elevation** (the north elevation is similar).

We make the following additional assumptions:

- The atmosphere is a black body of temperature $T_e(t)$ (highly oversimplified).
- The inside air is static far from the walls, at a temperature $T_i(t)$.
- Below 5 $\mu$m, the inside air is equivalent to a black body at temperature $T_i(t)$.
- An 'average' wind of speed $V = 20$ km h$^{-1}$ blows from the west.
- The walls of the north and south facades are (for simplicity) homogeneous, of thickness $e$, emissivity $\varepsilon$, and have the same thermophysical properties as yellow pine (see Basic Data, Part V).
- For simplicity, the initial state of the problem is assumed to be a uniform internal and wall temperature of $T_0$ $(t = 0)$.

1. **Show that the walls of the north and south facades behave as 'thermal filters',** and determine their cut-off frequency $\omega_c$ in the particular conditions specified (see numerical data). *Hint: this question is simple, and is worth up to 30 mins' thought. A small numerical calculation is necessary to answer the question.*
2. **Determine the asymptotic behaviour of the inside air** $T_i(t)$, including numerical values. *Note:* the asymptotic solution of a problem is that obtained after a sufficiently long time to allow the system to have 'forgotten' the initial conditions which were in effect at the moment when the new external thermal conditions $(T_e(t), \varphi_R(t))$ were imposed.
3. **Determine the length of time needed for the asymptotic solution to become valid.** (Discuss this for a more realistic case.)

**Numerical data:**

$$T_a = 270\text{ K};\quad T_0 = 285\text{ K};\quad \delta T = 20\text{ K};\quad \varphi_0 = 300\text{ W m}^{-2};\quad H = 3\text{ m};\quad l = 10\text{ m};$$
$$e = 0.35\text{ m};\ \varepsilon = 0.8;\ V = 20\text{ km h}^{-1};\ \tau = 24\text{ h}$$

**Alternative:**

In the preceding problem, there is an implicit need to calculate the relevant conducto-convective heat transfer coefficients. This may be short circuited by taking:

$$h_e = 10.4\text{ W m}^{-2}\text{ K}^{-1} \qquad h_i = 2.9\text{ W m}^{-2}\text{ K}^{-1}$$

where $h_e$ and $h_i$ are the relevant heat transfer coefficients for external and internal convection for the system.

## *The solution*

1. The apartment is subject to a harmonic flux with circular frequency $\omega$, where

$$\omega = \frac{\pi}{\tau}$$

and $\tau$ is the period, here the length of a day.

An important question which must be answered immediately is whether this variation in outside conditions is 'seen' inside the apartment. In other words, are the components of the signal at frequencies $2\omega$, $4\omega$ transmitted or filtered out by the wall structure?

Under the highly idealized conditions of homogeneous walls (without doors or windows) assumed here, the question is to find the depth of penetration of a thermal wave* of frequency $\omega'$.

The wave attenuation law is proportional to $\exp(-x/e_p)$, where:

$$e_p = \sqrt{\frac{2a}{\omega'}}$$

**Numerically:** $a = 8.2 \times 10^{-8}\ m^2\ s^{-1}$ (Basic Data, Part V):

$$\omega' = 2\omega = \frac{2\pi}{\tau}; \qquad \tau = 24\ h; \qquad e_p = 4.7\ cm$$

For a wall of 35 cm (quite a thick wall), the signal attenuation is:

$$\exp(-35/4.7) \simeq 6 \times 10^{-4}$$

The component at $2\omega$ (even more than at $4\omega$) is totally filtered out by the apartment walls. In other words, the walls make a low-pass filter, with a cut-off frequency of:

$$e = \sqrt{\frac{2a}{\omega_c}} \Rightarrow \omega_c \simeq 1.34 \times 10^{-6}\ Hz$$

The corresponding period is $\tau_c = 54$ days.

⇒ In conclusion, the inside of the building only sees the steady-state external conditions, that is to say:

$$\bar{T}^t_e = T_a + \tfrac{3}{8}\delta T \qquad \bar{\varphi}^t_R = \tfrac{3}{8}\varphi$$

since:

$$\overline{\cos 2\omega t}^t = \overline{\cos 4\omega t}^t = 0$$

Notice that a simple physical analysis has changed a prickly unsteady problem into a straightforward steady problem.

**Numerically:**

$$T_a + \tfrac{3}{8}\delta T = 278\ K$$

An apartment (or rather, a bunker) built like this is above freezing.

*Practical note*

There are in fact many buildings made of yellow pine (chalets), but with an 'air sandwich' of 2 cm thickness between two 10 cm planks. The conclusions are similar if the openings are well sealed.

* At a single frequency, the diffusion phenomenon degenerates to a propagation phenomenon. It is therefore correct to refer to a wave (see Section 6.2).

2. In steady conditions, we obviously have:

$$T_i(t) = T_i$$

The system is summarized in the diagram shown below:

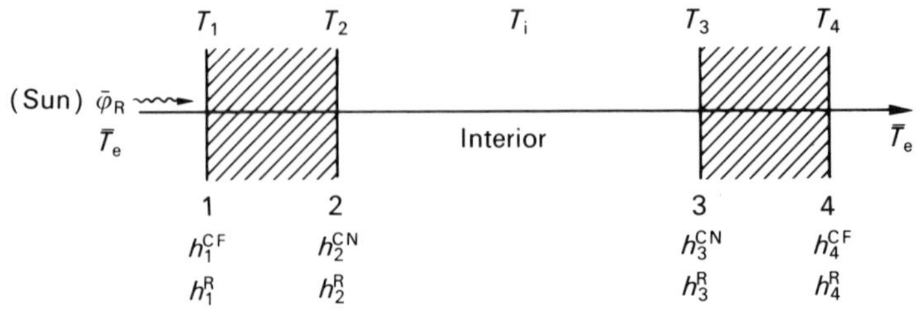


where the quantities $h^{CF}$, $h^{CN}$ and $h^R$ represent the heat transfer coefficient for forced and free convection and (linearized) radiation respectively. The justification for this assumption is obvious (Chapter 4). We must solve the following system of equations (with $\varphi > 0$ – see diagram above):

$$\begin{aligned}
\varphi &= \varepsilon\bar{\varphi}_R + (h_1^R + h_1^{CF})(\bar{T}_e - T_1) \\
&= \frac{\lambda}{e}(T_1 - T_2) \\
&= (h_2^R + h_2^{CN})(T_2 - T_i) \\
&= (h_3^R + h_3^{CN})(T_i - T_3) \\
&= \frac{\lambda}{e}(T_3 - T_4) \\
&= (h_4^R + h_4^{CF})(T_4 - \bar{T}_e)
\end{aligned}$$

If we assume:

$$h_1^{CF} = h_4^{CF}; \qquad h_1^R = h_4^R; \qquad h_2^R = h_3^R$$

and $h_2^{CN} = h_3^{CN}$, rather more doubtful in that if $T_2$ is different from $T_3$, the temperature differences driving the free convection are different.

Writing:

$$\mathcal{R} = \frac{1}{(h_1^{CF} + h_1^R)} + \frac{e}{\lambda} + \frac{1}{(h_2^{CN} + h_2^R)}$$

we obtain:

$$\bar{T}_e - T_i = \mathcal{R}\varphi - \frac{\varepsilon\varphi_R}{h_1^{CF} + h_1^R}$$

$$T_i - \bar{T}_e = \mathcal{R}\varphi$$

whence:

$$\underline{T_i - \bar{T}_e = \frac{\varepsilon \varphi_R}{2(h_1^R + h_1^{CF})}}$$

which solution is, with the assumptions made, independent of $h_2^R$ and $h_1^{CN}$ and also of $\lambda/e$!

**Numerically:**

$$h_1^R = 4\varepsilon\sigma T_e^3 \qquad h_1^{CF} = 10.4 \text{ W m}^2 \text{ K}^{-1}$$

$$\underline{T_i - \bar{T}_e = 3.2\,°\text{C}}$$

*Note*
The assumption that $h_2^{CN} + h_2^R = h_3^{CN} + h_3^R$ improves as: $e/\lambda \gg 1/(h_2^{CN} + h_2^R)$. In fact, $e/\lambda$ is about 2.4, while $h_2^R$ is of order 5!

3. The above solution is valid once the steady régime has established itself. Let us look at the characteristic times for the various transfers (see Chapter 8).

- conduction in a wall:

$$\tau^{cd} \sim \frac{e^2}{a} \sim 1.5 \times 10^6 \text{ s}$$

- linear transfer at walls 1 and 4:

$$\tau^{cc} \sim \frac{\rho c e}{h_1^{CF} + h_1^R} \sim 4.5 \times 10^4 \text{ s}$$

- transfer related to solar flux:

$$\frac{\rho c e}{\varphi_0} \sim 2040 \text{ s K}^{-1}$$

Since $\tau^{cd}$ is of the order of 17 days, the steady state will require several months to establish itself. In fact, a thickness $e$ of about 20 cm gives $\tau^{cd}$ about 4 days. Also, openings (windows, doors) change things in such a way as to shorten the characteristic time of the thermal inertia of the building.

*Note*
Ventilation has not been considered here. In the absence of heat recovery, this has the effect of introducing a variation inside the building related to the frequency $2\omega$ of the outside temperature.

# Application 13

## Temperature Measurement by Laser-induced Photothermal Effect

### *Required knowledge*

- Radiation (Chapter 4).
- Non-steady conduction (Part II).
- Boundary conditions (semi-transparent media – Chapter 1).

### *Features*

- Presentation of a method, becoming increasingly popular, of measuring temperature and, by the use of thermograms, of detecting structural faults (photothermal non-destructive testing).

### *The problem*

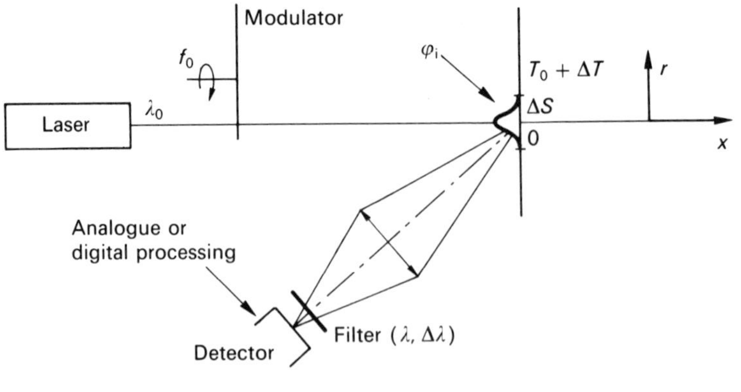

The diagram above shows the principles of a photothermal method of measuring temperature, even in the presence of an external parasitic flux $\varphi_i$ separate from that arising from the surroundings at temperature $T_a$.

The photothermal effect consists of modifying the static temperature field in a solid by supplying energy from a laser beam, of wavelength $\lambda_0$ and power $P$, which is modulated mechanically at frequency $f_0$. The monochromatic flux at wavelength $\lambda$ leaving the surface $\Delta S$ is then measured using an infra-red detector of sensitivity $D_\lambda$. The analogue or digital signal processing removes any signals which are not at frequency $f_0$.

In this example, as a first approximation we will suppose that the medium is semi-infinite and the undisturbed temperature field is uniform at temperature $T_0$. The laser beam, which is gaussian with a transverse width $\omega$ is partially reflected by the solid with a reflection factor $\rho_{\lambda_0}$.

Within a solid, we will take an exponential law for the attenuation of the power of the laser radiation. This gives (as a first approximation) laser radiation within the medium of:

$$\varphi = \frac{P(1-\rho_{\lambda_0})}{\pi\omega^2}\exp\left[-\left(\frac{r}{\omega}\right)^2\right]\exp[-\kappa\lambda_0 x]F(t) \qquad \text{where } F(t) = \exp[2\pi j f_0 t]$$

The temperature rise $\Delta T$ at point $(x, r)$ at instant $t$ is:

$$\Delta T(x, r, t) = T(x, r, t) - T_0$$

1. **Determine the signal from the detector** as a function of $T_0$, $\Delta T(x = 0, r, t)$, the sensitivity[1] $D_\lambda$ of the detector, the transmissivity $\tau_\lambda$ of the medium between the surface $\Delta S$ and the detector (including the filter), the pass band $\Delta\lambda$ of the (supposedly perfect) filter and the emissivity $\varepsilon_\lambda$ of the surface $\Delta S$.

   **Show that measurements at two wavelengths** $\lambda_1$ and $\lambda_2$ allow $T_0$ to be obtained. Discuss.
2. **Write down,** without solving them, the system of equations which give the temperature rise $\Delta T(x, r, t)$.
3. By a careful physical analysis, **simplify** the problem to be solved.

Given:

Thermal diffusivity of the material: $a = 10^{-6}\ \mathrm{m^2\,s^{-1}}$
Thermal conductivity of the material: $\lambda = 10\ \mathrm{W\,m^{-1}\,K^{-1}}$

$\omega = 0.5$ mm; $\rho_{\lambda_0} = 0.2$; $K_{\lambda_0} = 10^7\ \mathrm{m^{-1}}$; $P = 2$ W; $f_0 = 400$ Hz

[1] Defined simply as the ratio of the power detected to the power received in a given spectral interval $\Delta\lambda$.

## *The solution*

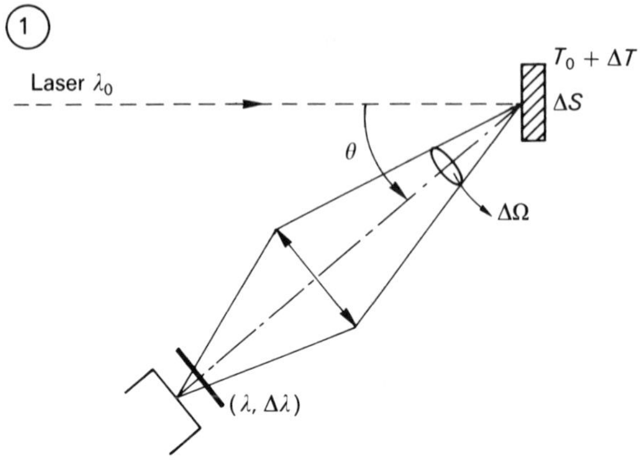


1. The signal $S_\lambda$ from the detector is proportional to the flux $\Phi_\lambda^{\rm l}$ leaving $\Delta S$ in the solid angle $\Delta\Omega$ and the wavelength interval $[\lambda, \lambda + \Delta\lambda]$ corresponding to the filter:

$$S_\lambda = D_\lambda \tau_\lambda \Phi_\lambda^{\rm l}$$

We assume that this does not affect the solution, for arbitrarily small $\Delta\lambda$ and $\Delta\Omega$:

$$\Phi_\lambda^{\rm l} = \int_{\Delta S} [\varepsilon_\lambda I_\lambda^0(T_0 + \Delta T) + (1 - \varepsilon_\lambda) I_\lambda^{\rm i}] \, {\rm d}S \cos\theta \, \Delta\lambda \, \Delta\Omega$$

The temperature rise $\Delta T$ being small ($|\Delta T| \ll T_0$), we can expand the intensity $I_\lambda^0(T_0 + \Delta T)$ as a Taylor series:

$$I_\lambda^0(T_0 + \Delta T) = I_\lambda^0(T_0) + \frac{\partial I_\lambda^0}{\partial T} \Delta T + \cdots$$

whence:

$$S_\lambda = D_\lambda \tau_\lambda \cos\theta \, \Delta\lambda \, \Delta\Omega \int_{\Delta S} \left[ \varepsilon_\lambda \left( I_\lambda^0(T_0) + \frac{\partial I_\lambda^0}{\partial T} \Delta T + \cdots \right) + (1 - \varepsilon_\lambda) I_\lambda^{\rm i} \right] {\rm d}S$$

The analogue or digital signal processing eliminates **all** signals which are not at frequency $f_0$. The signal due to $\Delta T$ is at frequency $f_0$ since it is the response to an excitation at $f_0$. Whence, assuming the emissivity is constant over $\Delta\lambda$:

$$S_\lambda = D_\lambda \tau_\lambda \cos\theta \, \Delta\lambda \, \Delta\Omega \varepsilon_\lambda \int_{\Delta S} \left. \frac{\partial I_\lambda^0}{\partial T} \right|_{T_0} \Delta T \, {\rm d}S$$

$$= \varepsilon_\lambda D_\lambda \tau_0 \cos\theta \, \Delta\lambda \, \Delta\Omega \left. \frac{\partial I_\lambda^0}{\partial T} \right|_{T_0} \int_{\Delta S} \Delta T \, {\rm d}S$$

In this expression for the signal, $\int_{\Delta S} \Delta T \, {\rm d}S$, which only depends on $f_0$ and $\Delta S$, and $\cos\theta \, \Delta\Omega$ are independent of wavelength $\lambda$. Thus if we take measurements using two different filters ($\lambda_1$ and $\lambda_2$), we have:

$$\frac{S_{\lambda_1}}{S_{\lambda_2}} = \frac{D_{\lambda_1}\tau_{\lambda_1}\,\Delta\lambda_1}{D_{\lambda_2}\tau_{\lambda_2}\,\Delta\lambda_2} \times \frac{\varepsilon_{\lambda_1}}{\varepsilon_{\lambda_2}} \times \frac{\left.\dfrac{\partial I^0_{\lambda_1}}{\partial T}\right|_{T_0}}{\left.\dfrac{\partial I^0_{\lambda_2}}{\partial T}\right|_{T_0}}$$

↑
ratio known from detector calibration

Thus knowledge of the ratio $S_{\lambda_1}/S_{\lambda_2}$ for a grey body over at least the interval $[\lambda_1, \lambda_2]$ ($\varepsilon_{\lambda_1} = \varepsilon_{\lambda_2}$) allows us to determine the temperature $T_0$.[2]

The remainder of the exercise consists of estimating the temperature rise $\Delta T$ or $\int_{\Delta S} \Delta T \, dS$, which is directly related to the signal quality.

2. We place ourselves inside the material. The temperature depends only on the coordinates $x$ and $r$ (being cylindrically symmetric) and on time.

The heat balance for a ring between $x$, $x + dx$ and $r$, $r + dr$ can be written down:

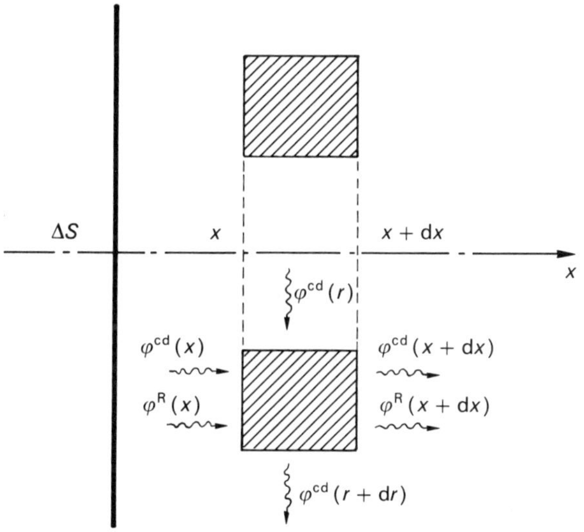


$$\rho c_p \frac{\partial T}{\partial t} 2\pi r \, dr \, dx = [\Phi^{cd}(x) - \Phi^{cd}(x + dx)] + [\Phi^{cd}(r) - \Phi^{cd}(r + dr)]$$
$$+ [\Phi^{R}(x) - \Phi^{R}(x + dx)]$$
$$= -\frac{\partial}{\partial x}\Phi^{cd}(x)\,dx - \frac{\partial}{\partial r}\Phi^{cd}(r)\,dr - \frac{\partial}{\partial x}\Phi^{R}(x)\,dx$$

[2] To confirm the hypothesis ($\varepsilon_{\lambda_1} = \varepsilon_{\lambda_2}$), a third wavelength $\lambda_3 \in [\lambda_1, \lambda_2]$ is usually used (see the discussion in Chapter 4).

where:

$$\Phi^{cd}(x) = -\lambda \left.\frac{\partial T}{\partial x}\right|_x 2\pi r \, dr$$

$$\Phi^{cd}(r) = -\lambda \left.\frac{\partial T}{\partial r}\right|_r 2\pi r \, dx$$

$$\Phi^{R}(x, r) = \frac{P_0}{\pi\omega^2} \exp\left[-\left(\frac{r}{\omega}\right)^2\right] \exp[-K_{\lambda_0} x] \exp[2\pi j f_0 t] 2\pi r \, dr$$

$$P_0 = P(1 - \rho_{\lambda_0})$$

Thus the equation giving the temperature $T(r, x, t)$ is:

$$\rho c_p \frac{\partial T}{\partial t} = \lambda \left\{ \frac{\partial^2 T}{\partial x^2} + \frac{1}{r}\frac{\partial}{\partial r}\left(r \frac{\partial T}{\partial r}\right) \right\} + \frac{P_0 K_{\lambda_0}}{\pi\omega^2} \exp\left[-\left(\frac{r}{\omega}\right)^2\right] \exp[-K_{\lambda_0} x] \exp[2\pi j f_0 t]$$

with boundary and initial conditions:

$$\left.\frac{\partial T}{\partial r}\right|_{r=0} = 0 \qquad T(\infty, x, t) = T_0$$
$$T(r, \infty, t) = T_0$$

$$-\lambda \left.\frac{\partial T}{\partial x}\right|_{x=0} = -h[T(r, 0, t) - T_a] - \varepsilon\sigma[T^4(r, 0, t) - T_a^4] + \varepsilon\varphi_i$$

$$T(r, x, 0) = T_0$$

3. The time for radial conduction (see Chapter 8):

$$\tau^{cd} = \frac{\omega^2}{a} = \frac{(5 \times 10^{-4})^2}{10^{-6}} = 0.25 \text{ s}$$

is much larger than the 'excitation time':

$$\tau_0 = \frac{1}{f_0} = 2.5 \times 10^{-3} \text{ s}$$

$$\tau^{cd} \gg \tau_0$$

We can assume that, on time scales smaller than $\tau^{cd}$, there is no thermal diffusion in the $r$ direction, and that the temperature distribution in the solid is the same as that in the laser beam:

$$\Delta T(x, r, t) = T(x, r, t) - T_0 = \theta(x, t) \exp\left[-\left(\frac{r}{\omega}\right)^2\right]$$

The laser radiation penetration length $1/K_{\lambda_0}$ ($\sim 10^{-7}$ m) is very small in comparison to the thermal diffusion length over a period $\tau_0$ ($= 1/f_0$): $\sqrt{a/f_0} \sim 5 \times 10^{-5}$ m. Thus we may consider the laser radiation to be completely absorbed at the surface ($x = 0$), so that within the solid ($x > 0$) there is pure monodirectional conduction:

$$\begin{cases} \dfrac{\partial T}{\partial t} = a \dfrac{\partial^2 T}{\partial x^2} \\ T(x, r, t) = \theta(x, t) \exp\left[-\left(\dfrac{r}{\omega}\right)^2\right] + T_0 \end{cases}$$

with boundary conditions:

$$T(\infty, r, t) = T_0$$

$$-\lambda \left.\frac{\partial T}{\partial x}\right|_{x=0} = \frac{P_0}{\pi\omega^2} \exp\left[-\left(\frac{r}{\omega}\right)^2\right] \exp[2\pi j f_0 t] - h[T(0, t, t) - T_a]$$

$$- \varepsilon\sigma[T^4(0, r, t) - T_a^4] + \varepsilon\varphi_i$$

Let us write:

$$T(x, r, t) = \theta(x) \exp\left[-\left(\frac{r}{\omega}\right)^2\right] \exp[2\pi j f_0 t] + T_0$$

then the above system becomes:

$$\begin{cases} 2\pi j f_0 \theta(x) = a \dfrac{d^2\theta}{dx^2} \\ -\lambda \left.\dfrac{d\theta}{dx}\right|_{x=0} = \dfrac{P_0}{\pi\omega^2} - h_e \theta(0) \\ \theta(\infty) = 0 \qquad h_e = h + 4\varepsilon\sigma T_0^3 \end{cases}$$

The boundary condition at $x = 0$ takes account of the fact that in the absence of the beam:

$$0 = h(T_0 - T_a) + \varepsilon\sigma(T_0^4 - T_a^4) - \varepsilon\varphi_i$$

Writing:

$$\alpha^2 = \frac{2\pi j f_0}{a} \qquad \text{and} \qquad \alpha = (1 + j)\sqrt{\frac{\pi f_0}{a}}$$

The solution of the equations is:

$$\begin{cases} \theta(x) = A\exp(\alpha x) + B\exp(-\alpha x) \\ \text{with} \quad A = 0 \qquad [\theta(\infty) = 0] \\ B = \dfrac{P_0}{\pi\omega^2(\lambda\alpha + h_e)} \end{cases}$$

$$\lambda\alpha \sim \lambda\sqrt{\frac{\pi f_0}{a}} = 10\sqrt{\frac{400\pi}{10^{-6}}} = 3.54 \times 10^5$$

$$h_e = h + 4\varepsilon\sigma T_0^3 \simeq 20 + 4 \times 5.67 \times 10^{-8}(10^3)^3$$

$$\simeq 250$$

$$\Rightarrow \quad h_e \ll \lambda\alpha$$

We can neglect the heat transfer coefficient $h_e$ in comparison to $\lambda\alpha$, so:

$$B = \frac{P_0}{\pi\omega^2\lambda\alpha} = \frac{P_0}{\lambda\pi\omega^2}\sqrt{\frac{a}{\pi f_0}}\exp\left(-j\frac{\pi}{4}\right)$$

and

$$\theta(x) = \frac{P_0}{\lambda\pi\omega^2}\sqrt{\frac{a}{\pi f_0}}\exp\left(-j\frac{\pi}{4}\right)\exp\left(-\left(\frac{\pi f_0}{a}\right)^{\frac{1}{2}}x\right)\exp\left(-\left(\frac{\pi f_0}{a}\right)^{\frac{1}{2}}jx\right)$$

which gives the temperature rise:

$\Delta T(x, r, t) =$

$$\frac{P_0}{\lambda\pi\omega^2}\sqrt{\frac{a}{\pi f_0}}(\exp\left(-\left(\frac{r}{\omega}\right)^2\right)\exp\left(-\left(\frac{\pi f_0}{a}\right)^{\frac{1}{2}}x\right)\exp\left(-\left(\frac{\pi f_0}{a}\right)^{\frac{1}{2}}jx\right)\exp\left(j\left(2\pi f_0^t\frac{-\pi}{4}\right)\right)$$

↑ skin effect ↑ thermal wave[x]

We can see straight away from this equation that we should make measurements at low frequencies.

**Numerically:**

$$\Delta T(x = 0, r = 0) = 4°\text{C}$$

# Application 14

## Window for a Domestic Oven

### *Required knowledge*

- Steady-state conditions (Chapters 1 and 2).
- Radiation (Chapters 4, 9 and 10).
- Free convection (Chapters 3 and 12).

### *Features*

- A system needing an iterative solution.
- Calculation of shape factors.

### *The problem*

We wish to find the maximum temperature $T_O$ for the oven of a domestic cooker, compatible with the temperature of the outside of the cooker window being below a critical value $T_C$ (Figure 1 and 2).

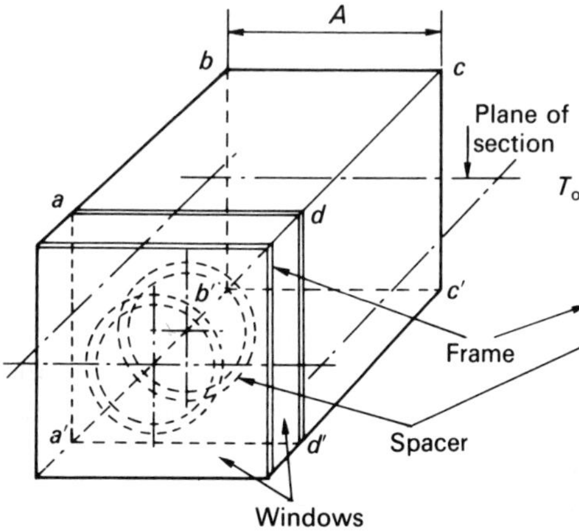


**Figure 1.** Perspective view of oven.

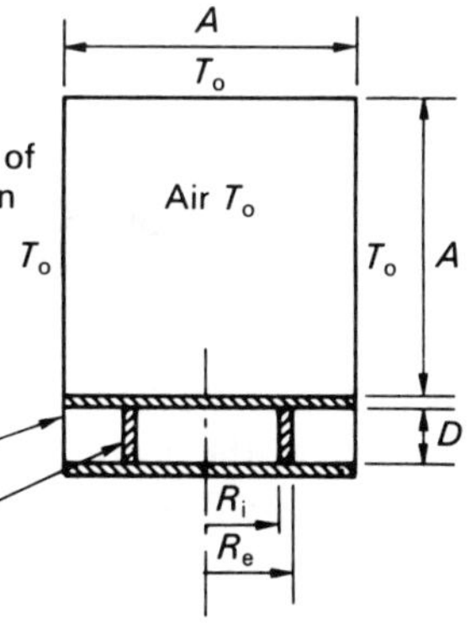


**Figure 2.** Cross-section as shown on Figure 1.

The oven is cubical; the vertical face $a\,da'\,d'$ is a double-glazed window. The five other faces are metallic, and are assumed to be at uniform temperature $T_O$. Of these five, the three vertical faces are perfectly reflecting for all radiation, or, in other words, reflection from these faces is diffuse and isotropic with a reflection coefficient $\rho = 1$. The two horizontal faces (floor and roof) are black bodies. **The various surfaces, which have the same radiative properties, are assumed to be uniformly illuminated.** All the necessary shape factors will be obtained in terms of the shape factor $f$ for two adjacent faces of a cube.

The air in the oven is assumed to be at $T_O$, except in the immediate vicinity of the glass wall. The convective heat transfer coefficient (natural convection) between the air inside the oven and the glass wall is $h_O$.

The glazing consists of two square parallel planes of thickness $e$, a distance $D$ apart (Figure 3). These panes are joined by a spacer in the form of a tube of length $D$ and radii $R_i$ and $R_e$. For simplicity, we will assume that the inside and outside surfaces of the spacer, and those of the frame at the edge of the glass, are insulating for conduction and perfectly reflecting for all radiation, the reflection being **specular**, following Descartes' law. The glass and the spacer have conductivities independent of temperature. The glass is **opaque to the infra-red radiation** which is involved in this problem, and has a monochromatic emissivity $\varepsilon_\lambda$ which is independent of direction.

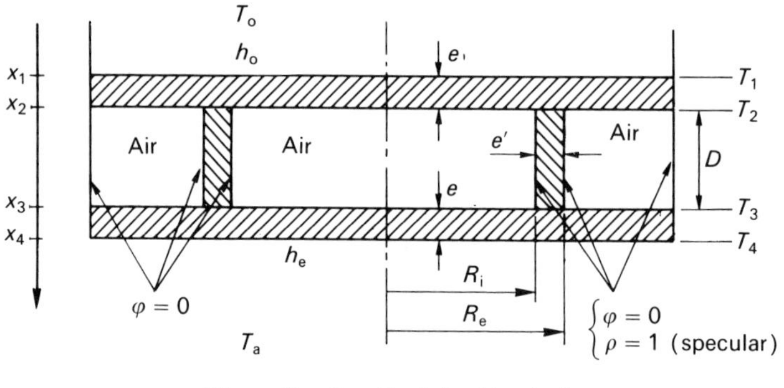


**Figure 3.** Detail of double glazing.

The kitchen is at temperature $T_a$. We assume that the coefficient of heat transfer $h_e$ between the outside of the glass and the surrounding air is known.

For simplicity, we will assume that in the glass and in the gap the isotherms are plane, and parallel to the glass. **Derive all the necessary equations to calculate:**

1. The total flux through the window;
2. The highest oven temperature $T_O$ which is compatible with an external surface temperature of the glass which is below a critical temperature $T_c$.

**Numerical data (SI units):**

$A = 0.5$ m $\quad e = 5 \times 10^{-3}$ m $\quad D = 0.01$ m

$R_i = 0.200$ m $\quad R_e = 0.202$ m $\quad f = 0.2$

$T_c = 80\,°C$ $\quad T_a = 27\,°C$

$h_e = 10$ $\quad h_o = 5$ Wm$^{-1}$ K$^{-1}$

conductivity of spacer: $\lambda_S = 0.1$ Wm$^{-1}$ K$^{-1}$

### *Radiative properties of windows*

The face of the inner pane on the oven side is partially covered by a reflecting coating. The radiative factors for this face are independent of direction and temperature, and are assumed to be independent of position on the surface.

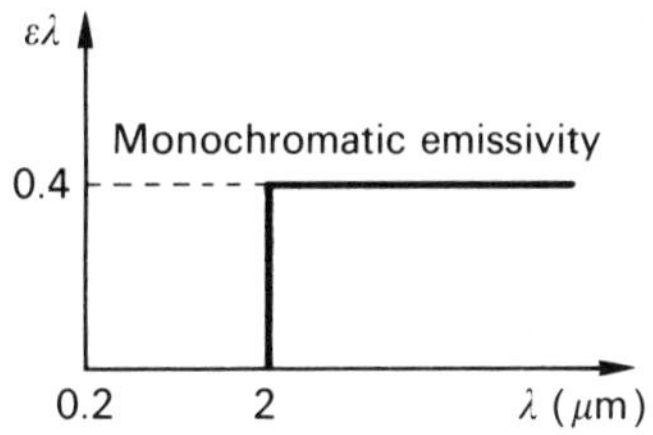


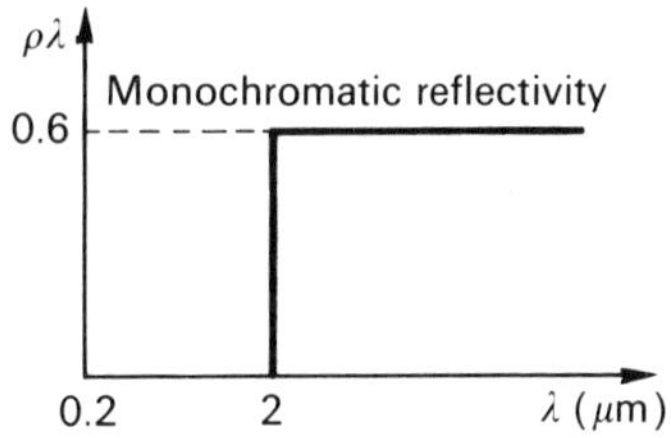


The other faces have monochromatic factors which are independent of direction.

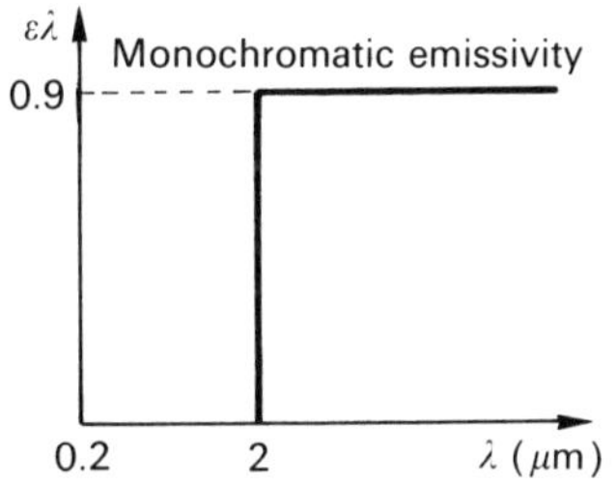


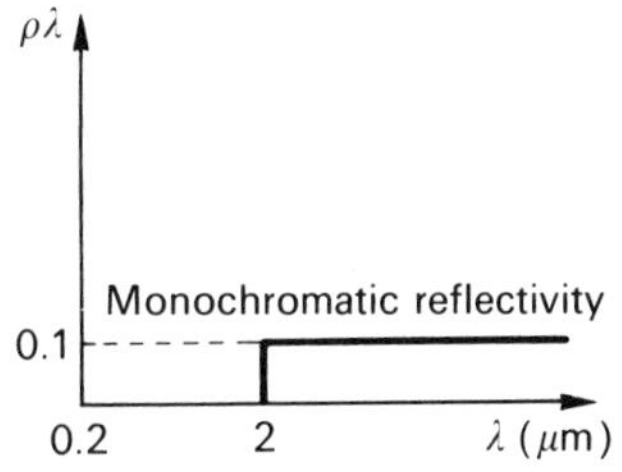


## *The solution*

### *1 Total energy flux through the window*

Under the above assumptions (lateral thermal ($\varphi = 0$) and radiative (since $\rho = 1$) insulation of the spacer), the isotherms in the spacer are parallel to the windows. We will assume the same for the windows themselves. In the steady state, each isothermal plane is crossed by the same energy flux $\Phi$, which is most easily calculated at the external face of the window. This face, which is at temperature $T_4$, exchanges

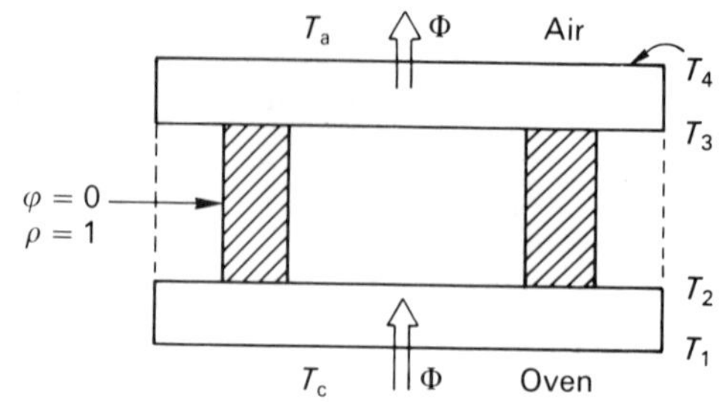


heat with its surroundings (which are in equilibrium at $T_a$) by convection and radiation:

$$\Phi = \Phi^{cc} + \Phi^{R}$$

$$\Phi^{cc} = A^2 h_e [T_4 - T_a]$$

$$\Phi^{R} = A^2 [\varphi^{\text{emitted}} - \varphi^{\text{absorbed}}]$$

In the expression for the radiative flux, we have not made use of the flux transmitted because of the radiative properties of the glass:

$$\lambda > 2\ \mu\text{m} \qquad \tau_\lambda = 0$$
$$\lambda < 2\ \mu\text{m} \qquad \tau_\lambda = 1$$

The radiation from the ambient air, which lies between 5 $\mu$m and 70 $\mu$m, is thus totally blocked by the glass. How about the radiation from the oven? In order that all or part of the radiation emitted by the oven shall be transmitted, all or part of it must have a wavelength below 2 $\mu$m. That is, $1500/T < 2\ \mu$m, or $T > 750$ K ($T > 477\,°$C). But a cooker oven can reach, at best, 350 °C, so that no radiation from the oven can cross the glass.

$$\varphi^{R} = \int_0^\infty \varepsilon_\lambda \pi I_\lambda^0(T_4)\,d\lambda - \int_0^\infty \varepsilon_\lambda \pi I_\lambda^0(T_a)\,d\lambda$$
$$= \varphi^{\text{emitted}} - \varphi^{\text{absorbed}}$$

$$\Rightarrow \qquad \Phi = A^2\left[h_e(T_4 - T_a) + \int_0^\infty \varepsilon_\lambda \pi [I_\lambda^0(T_4) - I_\lambda^0(T_a)]\,d\lambda\right]$$

If the physical properties and temperatures are known, this expression gives the total energy flux.

## 2. *Maximum oven temperature* $T_O$

By writing the expression for the flux $\Phi$ at each of the characteristic cross-sections of the double glazing, and eliminating the intermediate temperatures $T_1$, $T_2$ and $T_3$, we can obtain a relationship between $T_4$ and $T_O$:

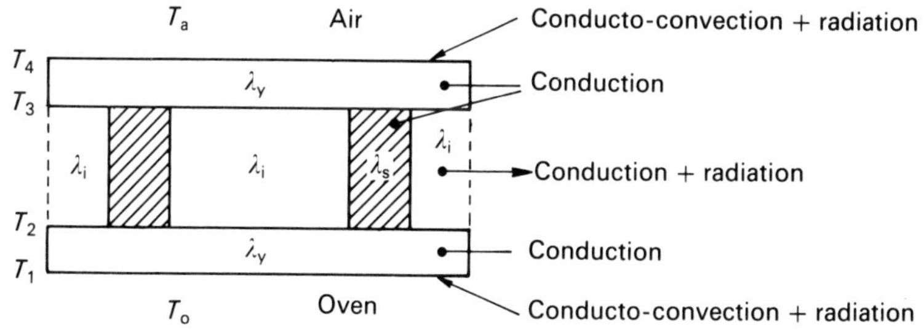


## (a) Between the outside glass and ambient air

The flux expression was written above:

$$\Phi = A^2[h_e(T_4 - T_a) + \int_0^\infty \varepsilon_\lambda \pi [I_\lambda^0(T_4) - I_\lambda^0(T_a)]\, d\lambda]$$

with

$$\begin{cases} \varepsilon_\lambda = 0.9 & \text{for } \lambda > 2\ \mu\text{m} \\ \varepsilon_\lambda = 0 & \text{for } \lambda < 2\ \mu\text{m} \end{cases}$$

Strictly, the integral for the calculation of radiative flux should be carried out from 2 $\mu$m to $\infty$. But since its temperature is below 100 °C, the glass radiates all its energy at wavelengths above 4 $\mu$m, so that integration from 2 to $\infty$ or from 0 to $\infty$ amounts to the same thing:

$$\Phi = A^2[h_e(T_4 - T_a) + 0.9\sigma(T_4^4 - T_a^4)]$$

This expression gives the flux when $T_4$ is set equal to $T_c$.

## (b) Within the outer glass

Transfer is purely conductive:

$$\Phi = \frac{\lambda_g A^2}{e}(T_3 - T_4)$$

If we know $T_4 = T_c$ and $\Phi$, we can calculate $T_3$.

### *(c) Between the two panes*

Conduction in the spacer and radiation and natural conducto-convection through the air between the two panes:

$$\Phi = \Phi_{cd}^{spacer} + \Phi_{cc}^{air} + \Phi_{R}^{2\to 3}$$

$$\Phi_{cd}^{spacer} = \frac{\lambda_S \pi (R_e^2 - R_i^2)}{D}(T_2 - T_3)$$

$$\Phi_{cc}^{air} = \frac{\lambda_e [A^2 - \pi(R_e^2 - R_i^2)]}{D}(T_2 - T_3)$$

We note that the natural convection heat transfer in the air layer is specified by an equivalent thermal conductivity (Part V, Section 4).

Since the reflection from the sides of the frame and the spacer is specular ($\rho = 1$), all radiation leaving one pane arrives at the other; from the point of view of radiative flux density, the situation is analogous to that of two infinite planes, equation (10.13):

$$\varphi_{2\to 3} = \int_0^\infty \frac{\varepsilon_{2\lambda}\varepsilon_{3\lambda}\pi[I_\lambda^0(T_2) - I_\lambda^0(T_3)]}{1 - (1 - \varepsilon_{2\lambda})(1 - \varepsilon_{3\lambda})}\,\mathrm{d}\lambda$$

$$\varepsilon_{2\lambda} = \varepsilon_{3\lambda} = 0.9 \qquad \lambda > 2\ \mu\text{m}$$

$$\varepsilon_{2\lambda} = \varepsilon_{3\lambda} = 0 \qquad \lambda < 2\ \mu\text{m}$$

$$\varphi_{2\to 3} = \int_0^\infty \frac{\varepsilon_\lambda \pi[I_\lambda^0(T_2) - I_\lambda^0(T_3)]}{(2 - \varepsilon_\lambda)}\,\mathrm{d}\lambda$$

for the same reasons as applied for transfer between the glass and the outside air, we can write:

$$\varphi_{2\to 3} = \frac{0.9}{2 - 0.9}\int_0^\infty \pi[I_\lambda^0(T_2) - I_\lambda^0(T_3)]\,\mathrm{d}\lambda$$

$$= 0.82\sigma[T_2^4 - T_3^4]$$

thus:

$$\Phi = (T_2 - T_3)\left[\frac{\lambda_S \pi(R_e^2 - R_i^2)}{D} + \frac{\lambda_e (A^2 - \pi(R_e^2 - R_i^2))}{D}\right]$$

$$+ [A^2 - \pi(R_e^2 - R_i^2)]\int_0^\infty \frac{\varepsilon_\lambda \pi[I_\lambda^0(T_2) - I_\lambda^0(T_3)]}{2 - \varepsilon_\lambda}\,\mathrm{d}\lambda$$

or:

$$\Phi = (T_2 - T_3)\left[\frac{\lambda_S \pi(R_e^2 - R_i^2)}{D} + \frac{\lambda_e (A^2 - \pi(R_e^2 - R_i^2))}{D}\right]$$

$$+ [A^2 - \pi(R_e^2 - R_i^2)] \times 0.82\sigma[T_2^4 - T_3^4]$$

from which we can calculate $T_2$.

### (d) Within the inner glass

Transfer is purely conductive:

$$\Phi = \frac{\lambda_g A^2}{e}(T_1 - T_2)$$

from which we can calculate $T_1$.

### (e) Between the inner glass and the oven

There is conducto-convective transfer between the glass and the air in the oven, and radiative transfer between the glass and the oven walls.

$$\Phi = \Phi_{cc} + \Phi_R$$

$$\Phi_{cc} = A^2 h_O (T_O - T_1)$$

$$\Phi_R = A^2 \int_0^\infty \pi[I^i_{g\lambda} - I^l_{g\lambda}]\,d\lambda$$

$I^i_{g\lambda}$ and $I^l_{g\lambda}$ are the intensities of the fluxes incident on and leaving the glass exposed to the oven radiation.

Now

$$I^l_{g\lambda} = \varepsilon_{g\lambda} I^0_\lambda(T_1) + (1 - \varepsilon_{g\lambda}) I^i_{g\lambda}$$

so that

$$\Phi_R = A^2 \int_0^\infty \varepsilon_{g\lambda}\pi[I^i_{g\lambda}(T_1, T_O) - I^0_\lambda(T_1)]\,d\lambda$$

The total flux on the internal face is:

$$\Phi = A^2\left[h_O(T_O - T_1) + \int_0^\infty \varepsilon_{g\lambda}\pi[I^i_{g\lambda}(T_1, T_O) - I^0_\lambda(T_1)]\right]d\lambda$$

Knowing $T_1$ and $\Phi$, this allows us to calculate $T_O$.

### (f) Calculation of the radiative flux between the glass and the oven walls

$$\varphi_R = \int_0^\infty \varepsilon_{g\lambda}\pi[I^i_{g\lambda}(T_1, T_O) - I^0_\lambda(T_1)]\,d\lambda$$

To calculate the intensity $I^{\mathrm{i}}_{g\lambda}$ of the flux incident on the glass, we will use the incident and leaving flux method, and divide the oven into four surfaces:

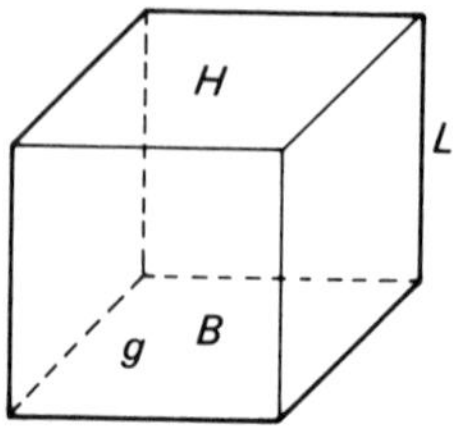


- the floor: $B$
- the ceiling: $H$
- the three metallic sides: $L$
- the glass $g$

$$I^{\mathrm{i}}_{g\lambda} = f_{gH} I^{\mathrm{l}}_{H\lambda} + f_{gB} I^{\mathrm{l}}_{B\lambda} + f_{gL} I^{\mathrm{l}}_{L\lambda}$$

$$I^{\mathrm{l}}_{H\lambda} = \varepsilon_{H\lambda} I^{0}_{\lambda}(T_{O}) + (1 - \varepsilon_{H\lambda}) I^{\mathrm{i}}_{H\lambda} = I^{0}_{\lambda}(T_{O})$$

$$I^{\mathrm{l}}_{B\lambda} = \varepsilon_{B\lambda} I^{0}_{\lambda}(T_{B}) + (1 - \varepsilon_{B\lambda}) I^{\mathrm{i}}_{B\lambda} = I^{0}_{\lambda}(T_{O})$$

since $\varepsilon_{H\lambda} = \varepsilon_{B\lambda} = 1$

then:

$$I^{\mathrm{l}}_{L\lambda} = \varepsilon_{L\lambda} I^{0}_{\lambda}(T_{O}) + (1 - \varepsilon_{L\lambda}) I^{\mathrm{i}}_{L\lambda} = I^{\mathrm{i}}_{L\lambda}$$

since $\rho_{L\lambda} = 1 \Rightarrow \varepsilon_{L\lambda} = \alpha_{L\lambda} = 0$

then:

$$I^{\mathrm{i}}_{g\lambda} = (f_{gH} + f_{gB}) I^{0}_{\lambda}(T_{O}) + f_{gL} I^{\mathrm{i}}_{L\lambda}$$

$$\begin{aligned} I^{\mathrm{i}}_{L\lambda} &= f_{Lg} I^{\mathrm{l}}_{g\lambda} + f_{LH} I^{\mathrm{l}}_{H\lambda} + f_{LB} I^{\mathrm{l}}_{B\lambda} + f_{LL} I^{\mathrm{l}}_{L\lambda} \\ &= f_{Lg} I^{\mathrm{l}}_{g\lambda} + (f_{LH} + f_{LB}) I^{0}_{\lambda}(T_{O}) + f_{LL} I^{\mathrm{i}}_{L\lambda} \\ &= \frac{f_{Lg} I^{\mathrm{l}}_{g\lambda} + (f_{LH} + f_{LB}) I^{0}_{\lambda}(T_{O})}{1 - f_{LL}} \end{aligned}$$

$$I^{\mathrm{l}}_{g\lambda} = \varepsilon_{g\lambda} L^{0}_{\lambda}(T_{1}) + (1 - \varepsilon_{g\lambda} I^{\mathrm{i}}_{g\lambda})$$

Before taking the calculation further, it is useful to write the various shape factors in terms of the shape factor $f$ appropriate to two adjacent faces of a cube:

$$f = f_{gB} = f_{gH}$$

If $f'$ is the shape factor for two opposite faces, any face of the cube is adjacent to four other faces and opposite to the other one:

$$\Rightarrow \quad f' + 4f = 1 \qquad \Rightarrow \quad f' = 1 - 4f$$

The properties of shape factors immediately lead to the following relationships:

$$f_{\mathrm{gL}} = f' + 2f = 1 - 2f$$

$$f_{\mathrm{HL}} = f_{\mathrm{BL}} = 3f$$

$$S_{\mathrm{L}} f_{\mathrm{Lg}} = S_{\mathrm{g}} f_{\mathrm{gL}} \rightarrow f_{\mathrm{Lg}} = \frac{S_{\mathrm{g}}}{S_{\mathrm{L}}} f_{\mathrm{gL}} = \frac{1 - 2f}{3}$$

$$S_{\mathrm{L}} f_{\mathrm{LB}} = S_{\mathrm{B}} f_{\mathrm{BL}} \rightarrow f_{\mathrm{LB}} = \frac{S_{\mathrm{B}}}{S_{\mathrm{L}}} f_{\mathrm{BL}} = f = f_{\mathrm{LH}}$$

$$1 = f_{\mathrm{LL}} + f_{\mathrm{LB}} + f_{\mathrm{LH}} + f_{\mathrm{Lg}} \Rightarrow f_{\mathrm{LL}} = \tfrac{2}{3}(1 - 2f)$$

Having determined the shape factors, a tedious but straightforward calculation gives:

$$I_{\mathrm{L}\lambda}^{\mathrm{i}} = \frac{(1 - 2f) I_{\mathrm{g}\lambda}^{\mathrm{i}} + 6f I_{\lambda}^{0}(T_{\mathrm{O}})}{1 + 4f}$$

and

$$I_{\mathrm{g}\lambda}^{\mathrm{i}} = \frac{4f(2 - f) I_{\lambda}^{0}(T_{\mathrm{O}}) + (1 - 2f)^2 \varepsilon_{\mathrm{g}\lambda} I_{\lambda}^{0}(T_1)}{(1 + 4f) - (1 - \varepsilon_{\mathrm{g}\lambda})(1 - 2f)^2}$$

whence:

$$\Phi^{\mathrm{R}} = A^2 \int_0^{\infty} \varepsilon_{\mathrm{g}\lambda} \pi \left\{ \frac{4f(2 - f) I_{\lambda}^{0}(T_{\mathrm{O}}) + (1 - 2f)^2 \varepsilon_{\mathrm{g}\lambda} I_{\lambda}^{0}(T_1)}{(1 + 4f) - (1 - \varepsilon_{\mathrm{g}\lambda})(1 - 2f)^2} - I_{\lambda}^{0}(T_1) \right\} \mathrm{d}\lambda$$

$$= A^2 \int_0^{\infty} \varepsilon_{\mathrm{g}\lambda} \pi \frac{4f(2 - f)\{I_{\lambda}^{0}(T_{\mathrm{O}}) - I_{\lambda}^{0}(T_1)\}}{(1 + 4f) - (1 - \varepsilon_{\mathrm{g}\lambda})(1 - 2f)^2} \mathrm{d}\lambda$$

$$\varepsilon_{\mathrm{g}\lambda} = 0 \qquad \text{for } \lambda < 2\ \mu\mathrm{m}$$

$$= 0.4 \qquad \text{for } \lambda > 2\ \mu\mathrm{m}$$

The temperatures $T_{\mathrm{O}}$ and $T_1$ of the oven and the glass are at such a level that all the radiation emitted by these two bodies is at an energy corresponding to wavelengths above 2 $\mu$m. The integral can therefore be evaluated from 0 to $\infty$ using a value of $\varepsilon_{\mathrm{g}\lambda}$ of 0.4, giving:

$$\varphi^{\mathrm{R}} = 0.4 \frac{4f(2 - f)\sigma(T_{\mathrm{O}}^4 - T_1^4)}{(1 + 4f) - 0.6(1 - 2f)^2}$$

### 3. Calculations

$$A^2 = 0.25 \text{ m}^2 \qquad T_c = 80\,°\text{C} = 353 \text{ K} \qquad T_a = 27\,°\text{C} = 300 \text{ K}$$

$$\begin{aligned}\Phi = A^2\varphi &= A^2[h_e(T_c - T_a) + 0.9\sigma(T_c^4 - T_a^4)] \\ &= 0.25[10(353 - 300) + 0.9 \times 5.67 \times 10^{-8} \times (353^4 - 300^4)]\end{aligned}$$

$$\underline{\Phi = 227 \text{ W}} \qquad \varphi = 909 \text{ W m}^{-2}$$

$$T_3 - T_c = \frac{e}{\lambda_g A^2}\Phi = \frac{e}{\lambda_g}\varphi = \frac{5 \times 10^{-3}}{1} \times 909 = 4.6 \text{ K}$$

$$\underline{T_3 = 357.6 \text{ K}}$$

$$\Phi = (T_2 - T_3)\left\{\lambda_S \frac{\pi(R_e^2 - R_i^2)}{D} + \lambda_e \frac{[A^2 - \pi(R_e^2 - R_i^2)]}{D}\right\}$$

$$+ [A^2 - \pi(R_e^2 - R_i^2)] \times \frac{0.9}{1.1} \times \sigma(T_2^4 - T_3^4)$$

To calculate the equivalent conductivity of the air layer, $\lambda_e$, we must estimate the temperature difference $T_2 - T_3$, and then carry out an iterative calculation (see Part V, Section 4). Assume:

$$\Delta T = T_2 - T_3 = 50 \text{ K}$$

$$\Rightarrow \quad Ra_D = \frac{g\beta\,\Delta T D^3}{av} = \frac{9.81 \times (1/400) \times 50 \times (0.01)^3}{3.76 \times 10^{-5} \times 2.59 \times 10^{-5}}$$

$$= 1.26 \times 10^3 < 2 \times 10^3$$

We are thus in the pseudo-conduction régime, $\lambda_e = \lambda_{air}$:

$$\lambda_{air} = 0.0337 \text{ W m}^{-1}\text{ K}^{-1}$$

Physical properties of air are taken at 400 K. Then:

$$227 = 0.858(T_2 - T_3) + 1.15 \times 10^{-8}(T_2^4 - T_3^4)$$

$$T_3 = 357.6 \text{ K} \Rightarrow 722 = 0.858T_2 + 1.15 \times 10^{-8}T_2^4$$

Iterative solution of this equation gives $T_2 = 421$ K.

*Second iteration*

$$\Delta T = T_2 - T_3 = 421 - 357.6 = 63.4 \text{ K}$$

$$Ra_D = \frac{9.81 \times (1/400) \times 63.4 \times (0.01)^3}{3.76 \times 10^{-5} \times 2.59 \times 10^{-5}} \simeq 1.6 \times 10^3$$

⇒ The natural convection régime between the two panes is always pseudo-conductive, so that $\lambda_e = \lambda_{air} = 0.0337$ W m$^{-1}$ K$^{-1}$; therefore:

$$\underline{T_2 = 421 \text{ K}}$$

In this free convection calculation, we have neglected the existence of the spacer. However, this represents a major obstacle to the development of free convection, which we have assumed to be in a vertical rectangular cavity, differentially heated. This reinforces the justification for a pseudo-conductive régime.

$$T_1 - T_2 = \frac{e}{\lambda_g}\varphi = 4.6\ \text{K}$$

$$\underline{T_1 = 425.6\ \text{K}}$$

$$f = 0.2 \qquad 4f(2-f) = 1.44$$

$$(1-2f)^2 = 0.36 \qquad 1+4f = 1.8$$

$$\varphi = h_O(T_O - T_1) + 0.4\frac{4f(2-f)\sigma(T_O^4 - T_1^4)}{(1+4f) - 0.6(1-2f)^2}$$

$$909 = 5(T_O - T_1) + 2.06 \times 10^{-8}(T_O^4 - T_1^4)$$

$$5T_O + 2.06 \times 10^{-8} T_O^4 = 3.71 \times 10^3$$

Iterative solution gives:

$$\underline{T_O = 495\ \text{K} = 222\ ^\circ\text{C}}$$

This is clearly consistent with the assumptions regarding the spectral ranges of the various radiation sources.

# Application 15

## Combustion Chamber of an Aircraft Engine (greatly simplified)

### *Required knowledge*

*First question*

- Steady-state conditions (Chapter 1).
- Radiation (Chapters 4, 9 and 10).

*Second question*

- Unsteady conditions (Chapters 5–8).

### *Features*

- Calculation of radiative heat transfer in a fairly complex axisymmetric geometry.
- Understanding of characteristic scales in unsteady conditions, and choice of suitable model for a particular problem.
- This example shows clearly that the thermal criteria for material selection in a given application are significantly different in steady and unsteady conditions.

### *The problem*

A combustion chamber is shown in simplified form in the diagram. It is rotationally symmetric. The hot gases after combustion are equivalent to a cylindrical black body

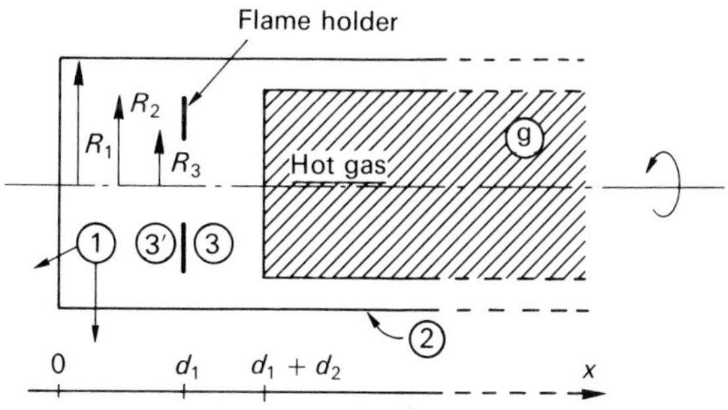


of infinite length in the $x$ direction, of radius $R_2$ (shaded in the diagram). They are separated from the opaque walls by a stream of cold gas, assumed transparent. The walls are assumed to have isotropic radiative properties and to receive a uniform flux density within each of the regions labelled ①, ②, ③ and ③'.

① consists of the base of radius $R_1$ and the sides ($0 < x < d_1$), of temperature $T_1$ and emissivity $\varepsilon_1$;
② consists of the semi-infinite side walls ($x > d_1$), at temperature $T_2$ and emissivity $\varepsilon_2 = 1$ (black body);
③ the face of the flame holder which is exposed to the hot gases. Isothermal at temperature $T_3$ and of emissivity $\varepsilon_{3\lambda}$;
③' the other face of the flame holder, of emissivity $\varepsilon_{3'}$.

I. Consider the thermal behaviour of the flame holder, assumed here to be isothermal and of zero thickness:

**What flux must be removed by conducto-convection in order that the flame holder does not exceed a given temperature $T_3$, all other temperatures being fixed?**

**Numerical data:** (we will just calculate the radiative flux $\Phi_3^R$ on face ③ of the flame holder).

| | | | |
|---|---|---|---|
| $R_1 = 0.150$ m | $R_2 = 0.149$ m | $R_3 = 0.070$ m | |
| $d_1 = 0.200$ m | $d_2 = 0.100$ m | | |
| $T_g = 2000$ K | $T_1 = 350$ K | $T_2 = 800$ K | $T_3 = 1300$ K |
| $\varepsilon_1 = 0.7$ | $\varepsilon_2 = \varepsilon_g = 1$ | $\varepsilon_{3'} = 0.7$ | |
| $\varepsilon_{3\lambda} = 0.8$ for $\lambda < 1.5\ \mu$m | | $\varepsilon_{3\lambda} = 0.4$ for $\lambda > 1.5\ \mu$m | |

II. Consider the unsteady behaviour of the flame holder, assumed to be of thickness $e$, and subjected at time $t = 0$ to a constant flux density $\varphi_0$ on one face. This flame holder, which is initially isothermal at temperature $T_0$, is cooled by conducto-convection on the other face, by a fluid whose characteristic temperature is $T_a$, with heat transfer coefficient $h$ (see diagram).

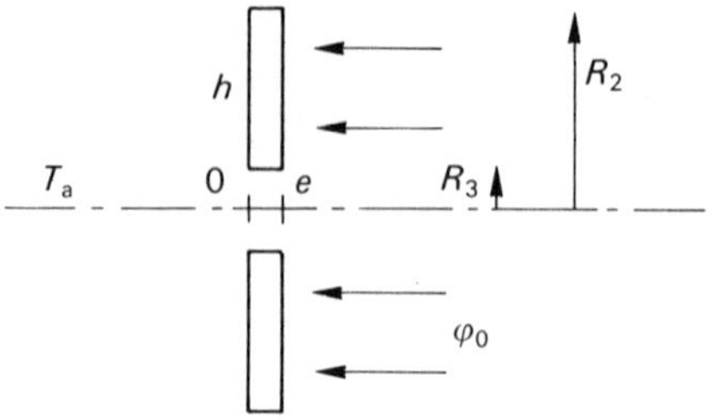


We will use the following values:

$$h = 600\ \mathrm{W\,m^{-2}\,K^{-1}}; \qquad T_a = 300\ \mathrm{K};$$
$$\varphi_0 = 5 \times 10^5\ \mathrm{W\,m^{-2}}; \qquad e = 2 \times 10^{-3}\ \mathrm{m};$$
$$T_0 = 300\ \mathrm{K}$$

1. **What practical conditions should the flame holder material withstand under steady conditions?** (quantitative answer).
2. **Determine and compare three characteristic times for the system, with the following assumptions:**

$$\text{(a)}\ \lambda = 10^{-1}\ \mathrm{W\,m^{-1}\,K^{-1}}; \qquad a = 10^{-7}\ \mathrm{m^2\,s^{-1}}$$
$$\text{(b)}\ \lambda = 50\ \mathrm{W\,m^{-1}\,K^{-1}}; \qquad a = 10^{-5}\ \mathrm{m^2\,s^{-1}}$$

3. **Study, with appropriate models, the temperature reached at $x = e$ after time $t_1 = 4$ s, for both cases (a) and (b).** Comment.

## *The solution*

1. The first question is concerned with a steady state solution. The only transfers on the flame holder are:

- a conducto-convective flux $\Phi^{cc}$;
- a radiative flux $\Phi^R_{3'}$ on face 3′;
- a radiative flux $\Phi^R_3$ on face 3.

With the usual notation, the steady energy balance equation is:

$$\Phi^{cc} = -\pi(R_2^2 - R_3^2)(\varphi_3^R + \varphi_{3'}^R) < 0 \tag{1}$$

where:

$$\varphi_3^R = \int_0^\infty \varepsilon_{3\lambda}\pi(I^i_{3\lambda} - I^0_\lambda(T_3))\,d\lambda \tag{2}$$

$I^i_{3\lambda}$ = intensity of flux incident on 3, and in so far as 3′ is grey,

$$\varphi^R_{3'} = \varepsilon_{3'}(\varphi^i_{3'} - \sigma T^4_{3'}) \tag{3}$$

Calculation of $\varphi_3^R$. The radiation incident on face ③ of the flame holder only comes from the gas or wall ②. (Since surface ② is a black body, no radiation from the rear of the flame holder is received at ③ after reflection from ②.)

It follows that:

$$I_{3\lambda}^{i} = f_{3g} I_{\lambda}^{0}(T_g) + f_{32} I_{\lambda}^{0}(T_2) \qquad (4)$$

$\varphi_3^R$ can be calculated from (2) and (4) if $f_{3g}$ and $f_{32}$ are known.

Let $f(R_i, d, R_j)$ be the shape factor between two coaxial disks of radius $R_i$ and $R_j$, a distance $d$ apart (see Part V, Section 5, in Basic Data):

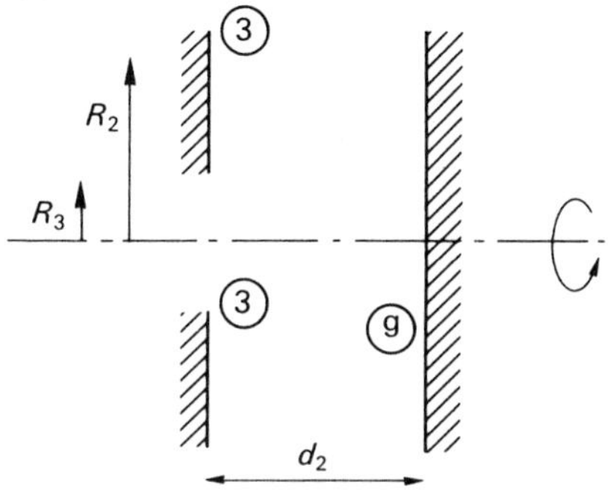


$$f_{g3} \pi R_2^2 = \pi (R_2^2 - R_3^2) f_{3g} \qquad (5)$$

$$f_{g3} = f(R_2, d_2, R_2) - f(R_2, d_2, R_3) \qquad (6)$$

(5) and (6) lead to:

$$f_{3g} = [f(R_2, d_2, R_2) - f(R_2, d_2, R_3)] \frac{R_2^2}{R_2^2 - R_3^3} \qquad (7)$$

Conservation of energy gives:

$$f_{32} = 1 - f_{3g} \qquad (7a)$$

The flux $\varphi_3^R$ may then be deduced from equations (2), (4), (7) and (7a).

$$\varphi_3^R = \int_0^{\infty} \pi \varepsilon_{3\lambda} [I_{\lambda}^{0}(T_g) f_{3g} + (1 - f_{3g}) I_{\lambda}^{0}(T_2) - I_{\lambda}^{0}(T_3)] \, d\lambda$$

Calculation of $\varphi_{3'}^R$. Equation (3) requires the calculation of $\varphi_{3'}^i$, which is entirely issued from wall ①:

$$\varphi_{3'}^i = \varphi_1^l = \varepsilon_1 \sigma T_1^4 + (1 - \varepsilon_1) \varphi_1^i$$

which gives an expression for $\varphi_1^i$:

$$\varphi_1^i = f_{11} \varphi_1^l + f_{12} \sigma T_2^4 + f_{13'} [\varepsilon_{3'} \sigma T_{3'}^4 + (1 - \varepsilon_{3'}) \varphi_1^l] + f_{1g} \sigma T_g^4 \qquad (8)$$

$$\varphi_{3'}^R = \varepsilon_{3'} \left[ \frac{\varepsilon_1 \sigma T_1^4 + (1 - \varepsilon_1)[f_{21} \sigma T_2^4 + f_{13'} \varepsilon_{3'} \sigma T_{3'}^4 + f_{1g} \sigma T_g^4]}{1 - (1 - \varepsilon_1)(f_{11} + f_{13'}(1 - \varepsilon_{3'}))} - \sigma T_{3'}^4 \right]$$

The only parameters still to be determined are $f_{11}, f_{12}, f_{13'}, f_{1g}$. Considering the following diagram, where a fictitious surface 0 is introduced, we obtain:

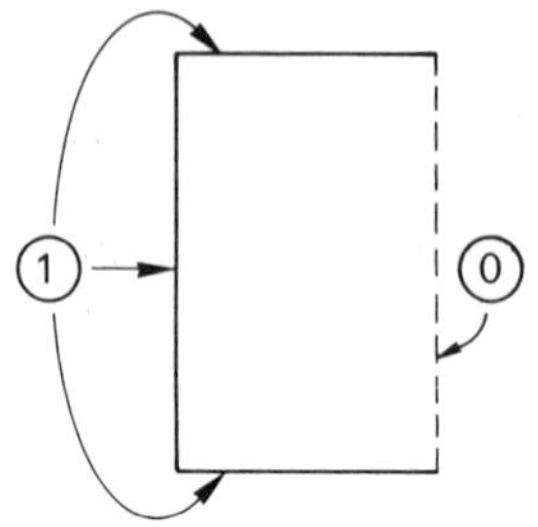


- $f_{11} = 1 - f_{10}$
- $S_1 f_{10} = S_0 f_{01} = S_0$
- $$f_{11} = 1 - \frac{S_0}{S_1} = 1 - \frac{R_1^2}{R_1^2 + 2R_1 d_1} \qquad (9)$$

Also, by reciprocity:

$$S_1 f_{13'} = S_{3'} f_{3'1} = S_{3'}$$

whence:

$$f_{13'} = \frac{R_2^2 - R_3^2}{R_1^2 + 2R_1 d_1} \qquad (10)$$

and by conservation of energy:

$$f_{12} + f_{1g} = f_{10} - f_{13'} = \frac{R_1^2 + R_3^2 - R_2^2}{R_1^2 + 2R_1 d_1} \qquad (11)$$

Assuming that:

$$f_{12} \simeq 0 \qquad (12)$$

$f_{1g}$ may be deduced from equation (11).

### 3. Numerical value of radiative flux

We find, after various calculations:

$f_{10} = 0.273$, $f_{11} = 0.727$, $f_{13'} = 0.21$, $f_{1g} = 0.063$, $f_{12} \simeq 0$

$\varphi_3^R = 1.57 \times 10^5 \text{ W m}^{-2}$ $\varphi_{3'}^R = -9.75 \times 10^4 \text{ W m}^{-2}$

$\Phi^{cc} = -3.2 \text{ kW}$

## Second question

We are interested simultaneously in the steady and unsteady behaviour of the flame holder and the specification of thermal criteria for selecting the material.

- In the steady state, and neglecting end effects (one-dimensional geometry), we obtain:

$$T(e) - T_a = \left(\frac{e}{\lambda S} + \frac{1}{hS}\right)\varphi_0 S$$

With the given data:

$$T(e) < T_3 \qquad T_3 = 1300\ \text{K} \qquad h = 600\ \text{W m}^{-2}\ \text{K}^{-1}$$
$$e = 2 \times 10^{-3}\ \text{m}$$
$$\varphi_0 = 5 \times 10^5\ \text{W m}^{-2}$$

$$\Rightarrow \quad \lambda > 6\ \text{W m}^{-1}\ \text{K}^{-1}$$

At the least, the flame holder should be a reasonable conductor.

- The characteristic times controlling the system are (see Chapter 8):
  - the conduction time: $\tau^{cd} = e^2/a$
  - the conducto-convection time: $\tau^{cc} = \rho c e/h$
  - the radiative transfer time*: $\tau^{RT} = (\rho c e/\varphi_0)\,\Delta T$

The numerical results appropriate to two typical materials (metal or ceramic) are given in the following table.

| | $\lambda$ Wm$^{-1}$ K$^{-1}$ | $a$ m$^2$ s$^{-1}$ | $\tau^{cd}$ (s) | $\tau^{cc}$ (s) | $\tau^{RT}$ $\Delta T = 1000$ K (s) |
|---|---|---|---|---|---|
| a Ceramic | $10^{-1}$ | $10^{-7}$ | 40 | 3.3 | 4 |
| b Metal | 10 | $10^{-5}$ | 0.4 | 16.6 | 20 |

- For material (a), a time period $t_1$ of 4 s satisfies $t_1 \ll \tau^{cd}$.

As a first approximation, it is the initial response which is of interest (see Chapter 6, equation (6.29)):

$$T(e) - T_0 = \frac{\varphi_0\sqrt{t}}{b} 2 \int_0^{\infty} \text{erfc}(u')\,\text{d}u' = \frac{\varphi_0\sqrt{t}}{b} 2\ \text{ierfc}(0)$$

The function 2 ierfc$(u)$ has a value of 1.128 for $u = 0$ (Appendix 2, Part 2); $b$ is the **thermal effusivity**, an important thermal quantity which represents the capacity of the temperature of a system to 'resist' short-term changes in its external

* The radiative transfer time, arising from the same model as $\tau^{cc}$ ($h \simeq \varphi_0/\Delta T$) should not be confused with the time scale for the propagation of photons.

conditions:

$$b = \sqrt{\lambda\rho c} = \lambda/\sqrt{a} = 316 \text{ WK}^{-1} \text{ S}^{-1/2}$$

**Numerically:**

$$T(e) = 3900 \text{ K}$$

This solution is unacceptable! We can see that this material, a ceramic, will not do in either unsteady or steady conditions.

- For material (b) (a metal):

$$t_1 \gg \tau^{cd}, \qquad \tau^{cd} \ll \tau^{cc}, \tau^{RT}$$

$$\Rightarrow \quad \frac{e}{\lambda} \ll \frac{1}{h} \qquad \text{or} \qquad Bi \ll 1$$

The flame holder is isothermal at any instant: $T(t)$. The energy balance becomes simply:

$$\rho c e \frac{dT}{dt} = \varphi_0 - h(T - T_a)$$

$$t = 0 \qquad T = T_a \Rightarrow \quad T - T_a = \frac{\varphi_0}{h}(1 - e^{-t/\tau^{cc}})$$

$T - T_a = \varphi_0/h$ is the steady-state solution.

For $t_1 = 4$ s, $T = 480$ K. Material (b) satisfies both the steady-state and unsteady requirements.

In conclusion, when choosing an insulator:

- not only the **conductivity** should be considered (the steady-state view),
- but also the **diffusivity** $a = \lambda/\rho c$ and the **effusivity** $b = \sqrt{\lambda\rho c}$ (different unsteady points of view).

# Application 16

## Control of an Electric Furnace

### *Required knowledge*

- Steady state (Chapter 1).
- Radiation (Chapter 10).
- Non-steady conduction (Chapters 5–8).

### *Features*

- Analysis of radiation in an axisymmetric geometry and use of the shape factor technique.
- Non-steady analysis of a system – characteristic times.

### *The problem*

We will consider a furnace whose wall, of thickness $e$, is in the form of a cylinder of internal radius $R$ and length $2L$. The outer surface of the wall is heated by $2N$ elements of length $L/N$, denoted by $i = 1, 2, \ldots, N$. Each element provides an adjustable flux density $\varphi_i$ to the wall.

At its two ends, 0 and $2N + 1$, the furnace is closed by two windows which are transparent to the radiation concerned. The windows are perfectly insulated from the furnace walls, whose ends are also insulated. The inner face of the stainless steel furnace wall is treated to have an emissivity $\varepsilon$, independent of wavelength and direction. The surroundings are at temperature $T_a$.

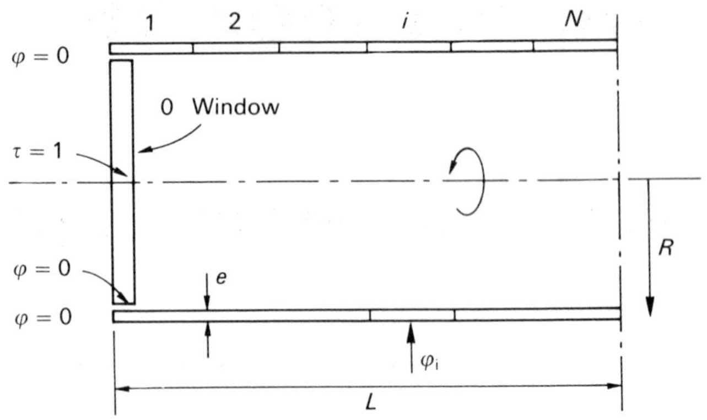


1. **Determine the distribution $\varphi_\text{i}$ of flux densities required to maintain the inner surfaces of the furnace at a uniform temperature $T$.**

1a. **Numerical values:** calculate $\varphi_1$ and $\varphi_5$ for: $N = 5$; $2L = 1.5$ m; $\varepsilon = 1$ (treatment); $e = 2$ mm; $R = 4$ cm; $T_\text{a} = 300$ K; $T = 1000$ K.

2. The mean temperature of each element is automatically controlled to between 980 and 1020 K. **Estimate the period during which an element is not heated (i.e. to fall from a supposedly uniform temperature of** 1020 K **to a mean temperature of** 980 K). Justify the model used quantitatively by reference to element $i = 1$. Discuss the case of element $i = 5$. (Neglect all heat losses from the outside of the furnace, across the surface $r = R + e$.)

## The solution

1. The radiative flux on an element $i$ is the flux temperature $\varphi_i$ supplied to that element by the heating:

$$\varphi_i = \varphi_i^l - \varphi_i^i \tag{1}$$

where $\varphi_i^l$ and $\varphi_i^i$ are the leaving and incident fluxes on $i$. In practice, other modes of heat transfer (by natural convection in the furnace) will be negligible in so far as the interior air is at a temperature very close to the temperature $T$ of the walls. Using the notation of Chapter 10:

$$i = 1, 2, \ldots, 2N \begin{cases} \varphi_i^l = \varepsilon_i \sigma T^4 + (1 - \varepsilon_i)\varphi_i^i & (2) \\ \varphi_i^i = \left( \sum_{j=1}^{2N} f_{ij}\varphi_j^l \right) + f_{i0}\varphi_0^l + f_{i2N+1}\varphi_{2N+1}^l & (3) \end{cases}$$

$$\varphi_0^l = \varphi_{2N+1}^l = \sigma T_\text{a}^4 \tag{4}$$

Note that element $i$ receives (sends) radiation from (to) the $2N$ cylindrical elements $j$ and from (to) the two end sections 0 and $2N + 1$. If the control is perfect, all elements $j$ are at the same temperature $T$, and the main exchanges will be with the ends 0 and $2N + 1$.

1a. Calculations. A large simplification is possible $\forall_i$: $\varepsilon_i = 1$. It follows that $\forall_i = 1, 2, \ldots, 2N$:

$$\varphi_i^l = \sigma T^4 \tag{5}$$

$$\sum_{j=1}^{2N} f_{ij} + f_{i0} + f_{i2N+1} = 1 \tag{6}$$

Using (1), (3). (5) and (6):

$$\varphi_i = (f_{i0} + f_{i2N+1})\sigma(T^4 - T_a^4) \tag{7}$$

We will write:

$$F_i = f_{i0} + f_{i2N+1}$$

At this stage, we need the shape factors $f_{i0}$ and $f_{i2N+1}$. In Basic Data, Part V, we find the shape factor $f^{(k)}$ between two coaxial disks of radius $R$ and a distance $kL/N$ apart (Section 5.1, case 8):

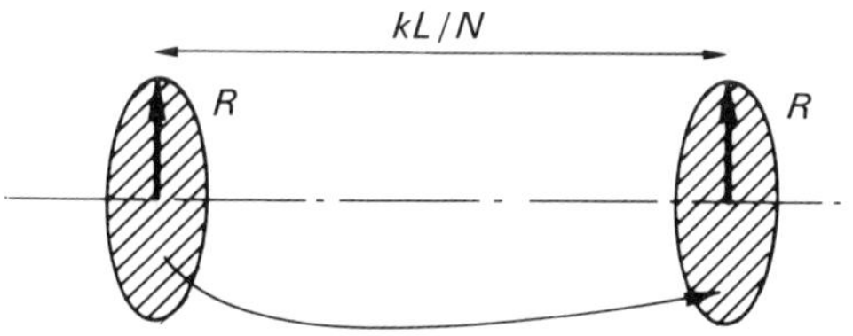


$$f^{(k)} = \tfrac{1}{2}[X - \sqrt{X^2 - 4}]$$

where $X = NR/kL$.

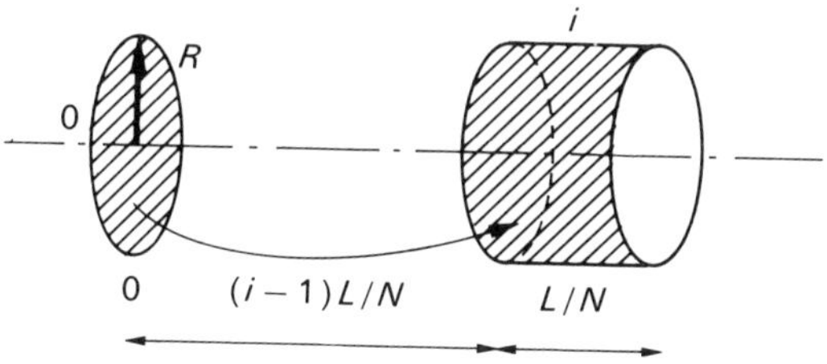


Using the definition of shape factors, it is evident that:

$$\pi R^2 f_{0i} = \pi R^2 [f^{(i-1)} - f^{(i)}]$$

with

$$f^{(0)} = 1$$

The reciprocity relationship is then:

$$2\pi R \frac{L}{N} f_{io} = \pi R^2 f_{0i}$$

**Numerically:**

$$F_1 = f_{10} + f_{1\ 2N+1} = 0.125$$
$$F_5 = f_{50} + f_{5\ 2N+1} = 9.3 \times 10^{-4}$$

The direct losses from element $i = 5$ are virtually negligible in comparison to those of element 1.

The flux densities to be supplied by the heating elements are, from (7):

$$\varphi_1 = 7.1 \times 10^3\ \text{W m}^{-2}$$

$$\varphi_5 = 53\ \text{W m}^{-2}$$

2. In this question we are concerned with the non-steady behaviour of the system. In cases of this type it is invaluable to calculate the characteristic times which control the system, in order to choose the appropriate model for each situation. Let us consider an element of length $L/N$ and thickness $e$:

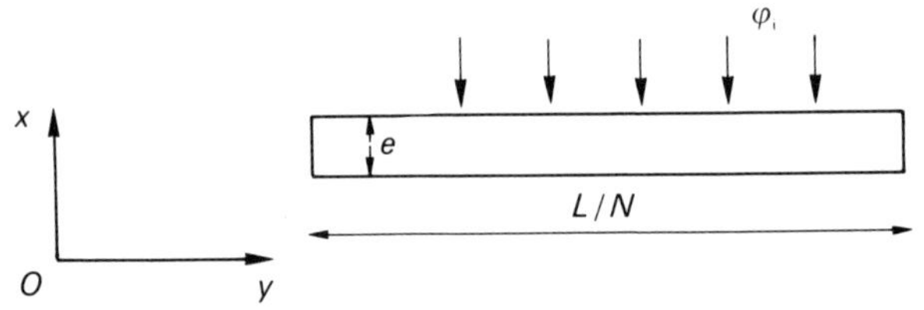


We can define two characteristic times for conduction in the $x$ and $y$ directions, respectively (see Part II):

$$\tau_{\text{cd}}^{x} = e^2/a = 4 \times 10^{-6}/5 \times 10^{-6} = 0.8\ \text{s}$$

$$\tau_{\text{cd}}^{y} = L^2/a = 2.25 \times 10^{-2}/5 \times 10^{-6} \simeq 4400\ \text{s} \simeq 1\ \text{h}\ 20\ \text{min}$$

In these, $a$ is the thermal diffusivity of the furnace element. In the absence of conducto-convective transfer, there will also be a characteristic time for radiation (not to be confused with the photon transit time). This time $\tau^{\text{R}}(\Delta T)$ can be obtained by assuming that the temperature field in any heating element is contant at any instant:

$$\rho c_{\text{p}}\left(2\pi Re \frac{L}{N}\right)\frac{\text{d}T}{\text{d}t} = F_i \sigma T^4 \left(2\pi R \frac{L}{N}\right)$$

(where we have neglected $T_{\text{a}}^4$ in comparison with $T^4$).

As a first approximation:

$$\tau^{\mathrm{R}} = \Delta t = \frac{\rho c_{\mathrm{p}} e}{F_i \sigma} \frac{\Delta T}{T_{\mathrm{m}}^4} \simeq \frac{0.2\, \Delta T}{F_i}$$

The temperature difference involved here is $\Delta T = 40\,^{\circ}\mathrm{C}$ (between 980 and 1020 K). We therefore obtain:

- *For element* $i = 1$

$$F_1 = 0.125 \Rightarrow \tau^{\mathrm{R}} = 64 \text{ s}$$

This time satisfies the inequality:

$$\tau_{\mathrm{cd}}^{x} \ll \tau^{\mathrm{R}} \ll \tau_{\mathrm{cd}}^{y}$$

which implies that at any instant element $i = 1$ can be considered as isothermal, and that only radiation transfers need be taken into account in the control of the system. **The order of magnitude of the time during which element 1 is not heated is 1 min.**

- *For element* $i = 5$

$$F_5 = 9.3 \times 10^{-4} \Rightarrow \tau^{\mathrm{R}} = 8000 \text{ s}$$

This time is of the same order of magnitude as the time for axial conduction $\tau_{\mathrm{cd}}^{y}$. Transfers by radiation and conduction should, strictly, be considered together.

In practice, the time concerned is of the order of hours, and is not important for control of the system.

# Application 17

## Insulating Windows

### *Required knowledge*

- Steady-state (Chapters 1 and 2).
- Radiation (Chapters 9, 10 and 11).
- Internal and external convection (Chapters 12, 14 and 17).

### *Features*

- A system requiring iterative solution.
- Understanding of conducto-convective heat transfer.

### *The problem*

The principle of insulating windows is based on the improvement in insulating ability introduced by an air layer between two sheets of glass.

The picture window in the south wall of a bungalow (see diagram) has two identical sheets of glass (4 mm thick, 1.5 m high and 3 m wide), separated by a layer of air at atmospheric pressure, 8 mm thick. The window is flush with the outside wall, parallel to which a wind of 5 m s$^{-1}$ is blowing. The outside temperature is $-10\,°C$ and that inside is $+20\,°C$.

**What are the thermal losses from the south façade on a dark night,** neglecting losses through the wall itself? **What is the gain compared to the case of single glazing?**

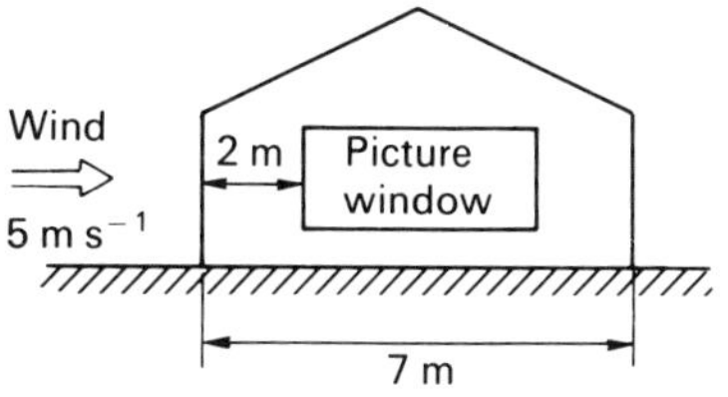


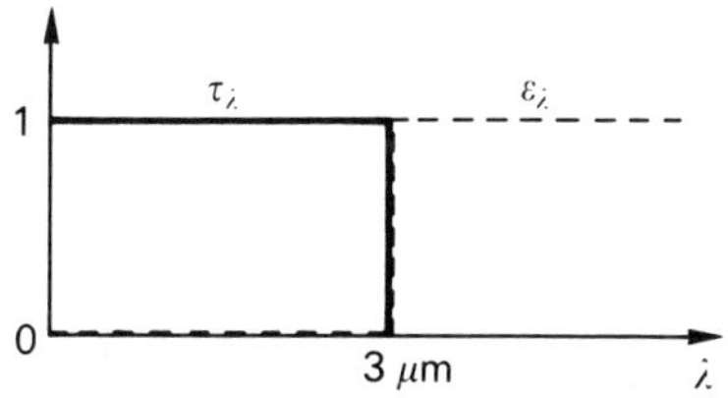


Monochromatic transmissivity $\tau_\lambda$ and emissivity $\varepsilon_\lambda$ of glass (assumed independent of direction and temperature).

Since the wall is assumed 'perfectly' insulating, the thermal losses from the south façade are solely through the picture window:

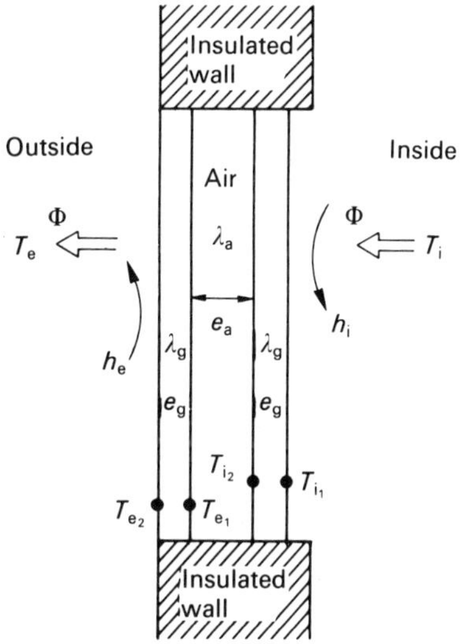


$S$: area of window

$$\begin{cases} h_i = h_i^{CN} + h_i^R \\ h_e = h_e^{CF} + h_e^R \end{cases}$$

$\lambda_g = 0.78\ \mathrm{W\ m^{-1}\ K^{-1}}$ (from Basic Data, Part V)

All three principal modes of energy transfer are involved in one or more of the four different regions:

- conducto-convective transfer (free convection) and radiation at the internal face of the window;
- conduction in the panes;

- conducto-convective transfer (free convection) and radiation in the air layer between the two panes;
- conducto-convective transfer (forced convection) and radiation at the outside face of the window.

We have not included the radiation transmitted through the two panes of glass. This is because 'terrestrial' and atmospheric radiations ($\lambda_m \sim 10\ \mu m$) are not transmitted through the glass: $\tau_\lambda = 0$ for $\lambda > 3\ \mu m$. Using the notation of the above diagram:

$$\Phi = h_i S[T_i - T_{i1}]$$

$$= S\frac{\lambda_g}{e_g}[T_{i1} - T_{i2}]$$

$$= S\frac{\lambda_e}{e_a}[T_{i2} - T_{e1}] + \Phi^R$$

$$= S\frac{\lambda_g}{e_g}[T_{e1} - T_{e2}]$$

$$= Sh_e[T_{e2} - T_e]$$

$\lambda_e$ is the equivalent thermal conductivity of the air layer between the two panes (Basic Data, Part V, Section 4):

$$\frac{\lambda_e}{\lambda} = Nu = f(Ra, Pr)$$

If there is no free convection in the air layer:

$$\lambda_e = \lambda$$

The radiation flux between the two panes can be written simply (see Chapter 10):

$$\Phi^R = S\int_0^\infty \frac{\varepsilon_\lambda \pi[I_\lambda^0(T_{i2}) - I_\lambda^0(T_{e1})]}{2 - \varepsilon_\lambda}\,d\lambda$$

$$= S\int_3^\infty \pi[I_\lambda^0(T_{i2}) - I_\lambda^0(T_{e2})]\,d\lambda$$

$$\varepsilon_\lambda = 0 \qquad \lambda < 3\ \mu m$$

$$\varepsilon_\lambda = 1 \qquad \lambda > 3\ \mu m$$

This integral is readily evaluated using the $Z$ function:

$$Z\left(\frac{\lambda_1}{\lambda_m(T)}, \frac{\lambda_2}{\lambda_m(T)}\right) = \frac{\int_{\lambda_1}^{\lambda_2} \pi I_\lambda^0(T)\,d\lambda}{\sigma T^4}$$

$$\Phi^R = S\sigma T_{i2}^4\left[1 - Z\left(0, \frac{3}{\lambda_m(T_{i2})}\right)\right] - S\sigma T_{e1}^4\left[1 - Z\left(0, \frac{3}{\lambda_m(T_{e1})}\right)\right]$$

Temperatures $T_{i2}$ and $T_{e1}$ are below 300 K. Thus:

$$\lambda_m(T) > 10\ \mu m$$

$$\frac{3}{\lambda_m(T)} < 0.3 \qquad \text{and} \qquad Z\left(0, \frac{3}{\lambda_m(T)}\right) < 5 \times 10^{-5}$$

so that function $Z$ is negligible compared to 1, and the radiation flux is:

$$\Phi^R = S\sigma(T_{i2}^4 - T_{e1}^4]$$
$$= h'^R S[T_{i2} - T_{e1}]$$

with

$$h'^R = \sigma[T_{i2}^2 + T_{e1}^2][T_{i2} + T_{e1}]$$

All the expressions for the lost flux are now linear, and we deduce:

$$\Phi = \frac{S(T_i - T_e)}{\frac{1}{h_i} + \frac{e_g}{\lambda_g} + \left(\frac{\lambda_e}{e_a} + h'^R\right)^{-1} + \frac{e_g}{\lambda_g} + \frac{1}{h_e}}$$

From this equation, we can calculate the thermal losses from the south façade if we know all the terms; $h_i$, $h_e$, $\lambda_e$ and $h'^R$ are not given, and, to calculate them, we need to know the temperatures of the various surfaces of the glass, $T_{i1}$, $T_{i2}$, $T_{e1}$ and $T_{e2}$. But these temperatures, which are intermediate between the internal temperature $T_i$ and the external temperature $T_e$, can only be calculated if we know the flux $\Phi$ and the four coefficients $h_i$, $h_e$, $\lambda_e$ and $h'^R$.
This state of affairs leads quite naturally to an iterative solution method:

We select an initial temperature distribution which allows $h_i$, $h_e$, $\lambda_e$ and $h'^R$, and then the flux $\Phi$ to be calculated, and then calculate a new temperature distribution. We continue until the variation of flux between one iteration and the next is less than the initially specified precision.

### *First iteration*

The glass panes are better conductors of heat than is air, so for our first distribution we will assume them to be isothermal:

$$T_e = -10\,°C, \qquad T_{e2} = T_{e1} = 0\,°C, \qquad T_{i2} = T_{i1} = 10\,°C. \qquad T_i = 20\,°C$$

The properties of air at the average temperature of the film are given in the following table:

| $T_m(K)$ | *268* | *278* | *288* |
|---|---|---|---|
| $\lambda$ (W/mK) | $2.37 \times 10^{-2}$ | $2.45 \times 10^{-2}$ | $2.53 \times 10^{-2}$ |
| $\nu$ ($m^2/s$) | $1.17 \times 10^{-5}$ | $1.30 \times 10^{-5}$ | $1.42 \times 10^{-5}$ |
| $a$ ($m^2/s^{-1}$) | $1.64 \times 10^{-5}$ | $1.82 \times 10^{-5}$ | $2 \times 10^{-5}$ |

- Calculation of $h_i$

$$h_i = h_i^{CN} + h_i^R$$

$$Ra_H = \frac{g\beta(T_i - T_{i1})H^3}{av}$$

$$= \frac{9.81 \times 1/288 \times 10 \times (1.5)^3}{1.42 \times 10^{-5} \times 2 \times 10^{-5}} = 4.05 \times 10^9$$

The flow down the glass inside the room is thus turbulent.

$$Nu_H = 0.13 Ra_H^{1/3} = 207.2 \Rightarrow h_i^{CN} = \frac{Nu_H \lambda}{H} = 3.5 \text{ W m}^{-2}\text{ K}^{-1}$$

$$h_i^R = \sigma(T_i^2 + T_{e1}^2)(T_i + T_{e1})$$

$$= 5.62 \times 10^{-8}[293^2 + 283^2][293 + 283]$$

$$= 5.42 \text{ W m}^{-2}\text{ K}^{-1}$$

The calculation of the radiative heat transfer coefficient has been done here using the exact expression. Given the small temperature difference between the surface and the inside air, the approximate expression

$$h^R \sim 4\varepsilon\sigma T_m^3$$

here gives exactly the same numerical value:

$$\Rightarrow \quad h_i = 8.9 \text{ W m}^{-2}\text{ K}^{-1}$$

- Calculation of $\lambda_e$ and $h'^R$. $\lambda_e$ between the two panes:

$$Ra_\delta = \frac{g\beta(T_{i2} - T_{e1})\delta^3}{av}$$

$$= \frac{9.81 \times 1/278 \times 10 \times (8 \times 10^{-3})^3}{1.3 \times 10^{-5} \times 1.82 \times 10^{-5}}$$

$$= 7.64 \times 10^2 < 2 \times 10^3$$

→ pseudo-conduction régime

$$\lambda_e = \lambda_{air} = 2.45 \times 10^{-2} \text{ W m}^{-1}\text{ K}^{-1}$$

$$h'^R = 4\sigma\varepsilon T_m^3$$

$$= 4 \times 5.67 \times 10^{-8} \times 278^3$$

$$\sim 4.9 \text{ W m}^{-2}\text{ K}^{-1}$$

Calculation of $h_e$. Outside, there is a forced flow ($v = 5 \text{ m s}^{-1}$). The momentum and thermal boundary layers do not start from the same point:

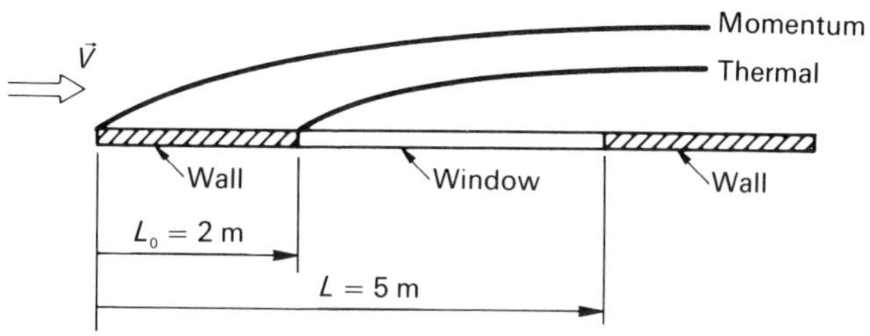


$$Re_L = \frac{vL}{\nu} = \frac{5 \times 5}{1.17 \times 10^{-5}} = 2.14 \times 10^6 > 5 \times 10^5$$

The flow is thus turbulent along the south façade of the bungalow. From Basic Data, Part V, for this particular configuration:

$$Nu_L = 0.028 Re_L^{4/5} \left[ 1 + 0.4 \left( \frac{L_0}{L} \right)^{2.75} \right] = 3.35 \times 10^3$$

$$h_e^{CF} = \frac{Nu_L \lambda}{L} = \frac{3.35 \times 10^3 \times 2.37 \times 10^{-2}}{5}$$

$$= 15.9 \text{ W m}^{-2} \text{ K}^{-1}$$

$$h_e^R = 4\varepsilon\sigma T_m^3 = 4 \times 5.67 \times 10^{-8} \times 268^3$$

$$= 4.37 \text{ W m}^{-2} \text{ K}^{-1} \simeq 4.4 \text{ W m}^{-2} \text{ K}^{-1}$$

whence:

$$h_e = 15.9 + 4.4 = 20.3 \text{ W m}^{-2} \text{ K}^{-1}$$

- Calculation of flux. Knowing all the heat transfer coefficients, we can calculate the flux density of the heat losses:

$$\varphi = \frac{\Phi}{S} = 101 \text{ W m}^{-2}$$

This flux density gives us a new temperature distribution:

$$T_i - T_{i1} = \frac{\varphi}{h_i} \sim 11.3\,^\circ\text{C}$$

$$T_{i1} - T_{i2} = \frac{e_g}{\lambda_g} \varphi \sim 0.5\,^\circ\text{C}$$

$$T_{i2} - T_{e1} = \left( \frac{\lambda_e}{e_a} + h'^R \right)^{-1} \varphi \sim 12.7\,^\circ\text{C}$$

$$T_{e1} - T_{e2} = \frac{e_g}{\lambda_g} \varphi \sim 0.5\,^\circ\text{C}$$

$$T_{e2} - T_e = \frac{\varphi}{h_e} \sim 4.97\,^\circ\text{C} \sim 5\,^\circ\text{C}$$

*Second iteration*

We will take a new distribution of temperatures:

$$T_e = -10\,°C, \qquad T_{e2} = T_{e1} = -5\,°C, \qquad T_{i2} = T_{i1} = +8\,°C, \qquad T_i = 20\,°C$$

- Calculation of coefficients. Since the average temperatures of the film have not changed by more than 1 or 2 K, we will use the same values as before for the physical properties:

$$h_i = h_i^{CN} + h_i^R$$

$$Ra_H = \frac{g\beta(T_i - T_{i1})H^3}{av} = 4.87 \times 10^9$$

$$Nu_H = 220 \qquad h_i^{CN} = 3.7\ \mathrm{W\,m^{-2}\,K^{-1}}$$

$$h_i^R = 4\varepsilon\sigma T_m^3 \simeq 5.4\ \mathrm{W\,m^{-2}\,K^{-1}}$$

$$h_i = 9.1\ \mathrm{W\,m^{-2}\,K^{-1}}$$

$$h'^R = 4\varepsilon\sigma T_m^3 \sim 4.7\ \mathrm{W\,m^{-2}\,K^{-1}}$$

$$h_e = h_e^{CF} + h_e^R$$

$$h_e^{CF} \rightarrow \text{unchanged}$$

$$h_e^R = 4\varepsilon\sigma T_m^3 \sim 4.3\ \mathrm{W\,m^{-2}\,K^{-1}}$$

$$h_e = 20.2\ \mathrm{W\,m^{-2}\,K^{-1}}$$

which gives a new flux value:

$$\varphi \sim 101\ \mathrm{W\,m^{-2}}$$

thus there is no change in the flux density of the losses, and the final temperature distribution is as follows:

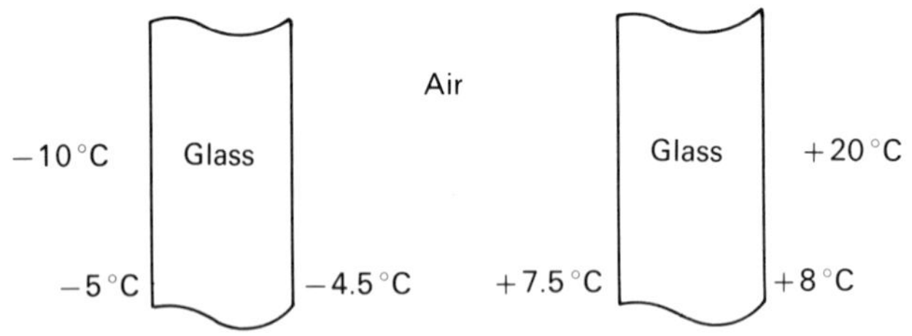


## *Comparison with single glazing*

Let us consider the same climatic conditions with a single pane of the same thickness $e_g$.

$$\varphi = \frac{\Phi}{S} = \frac{T_i - T_e}{(1/h_i) + (e_g/\lambda_g) + (1/h_e)}$$

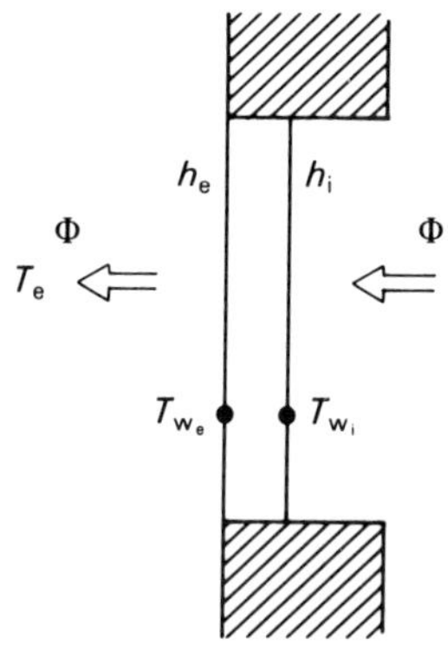


As for the double glazing, only an iterative solution is possible:

*First temperature distribution*

$$T_e = -10\,°C, \qquad T_{wi} = T_{we} = 0\,°C, \qquad T_i = +20\,°C$$

The outside film temperature is the same as for double glazing, thus the coefficient $h_e$ is the same: $h_e = 20.2\ W\ m^{-2}\ K^{-1}$.

The internal film temperature $T_m = 283$ K:

$$\lambda = 2.49 \times 10^{-2}\ W\ m^{-1}\ K^{-1}, \quad \nu = 1.36 \times 10^{-5}\ m^2\ s^{-1}, \quad a = 1.91 \times 10^{-5}\ m^2\ s^{-1}$$

$$h_i = h_i^{CN} + h_i^R$$

$$Ra_H = \frac{g\beta(T_i - T_{wi})H^3}{a\nu} = 9 \times 10^9$$

The flow is turbulent.

$$Nu_H = 0.13 Ra_H^{1/3} = 270 = \frac{h_i^{CN} H}{\lambda}$$

$$h_i^{CN} = 4.5\ W\ m^{-2}\ K^{-1}$$

$$h_i^R = 4\varepsilon\sigma T_m^3 = 4 \times 5.67 \times 10^{-8} \times (283)^3 = 5.14\ W\ m^{-2}\ K^{-1}$$

$$\Rightarrow \quad h_i = 9.6\ W\ m^{-2}\ K^{-1} \quad \text{(a well-known result!)}$$

This gives a flux density:

$$\varphi = 146\ W\ m^{-2}$$

*Second temperature distribution*

$$T_i - T_{wi} = \frac{\varphi}{h_i} = 15.2\,°C$$

$$T_{wi} - T_{we} = \varphi \frac{e_g}{\lambda_g} = 7.5\,°C$$

$$T_{we} - T_e = \frac{\varphi}{h_e} = 7.2\,°C$$

thus we can take:

$$T_{we} = -2.5\,°C \qquad \text{and} \qquad T_{wi} = 5\,°C$$

The film temperatures have not changed appreciably, so we can use the same physical properties.

$$\Rightarrow \quad h_e = 20.2 \text{ W m}^{-2}\text{ K}^{-1}$$

for $h_i$:

$$Ra_H = 6.7 \times 10^9 \qquad Nu_H = 245$$

$$\left.\begin{array}{r} h_i^{CN} = 4.1 \text{ W m}^{-2}\text{ K}^{-1} \\ h_i^{R} = 5.31 \text{ W m}^{-2}\text{ K}^{-1} \end{array}\right\} \rightarrow h_i = 9.4 \text{ W m}^{-2}\text{ K}^{-1}$$

which gives a flux density:

$$\varphi = 145 \text{ W m}^{-2}$$

This value is very close to the previous one, which corresponds to an increase in flux of about 45% compared to the flux lost through double glazing. Another important point is the feeling of comfort provided by double glazing: the wall is at 8 °C, while it is only at 5 °C with the single glazing. This produces a much bigger temperature gradient near the window.

# Application 18

## Environmental Effects on a Photovoltaic Solar Cell

### *Required knowledge*

- Steady conditions – the three modes of heat transfer (Chapters 1 and 2).
- Directional and isotropic radiation (Chapter 4; study of Chapters 9 and 10 is not essential).
- Laminar internal and external convection phenomena (Chapter 12).

### *Features*

- A system whose behaviour is strongly non-linear, requiring an iterative solution **(basic solution in free convection).**
- Simplified alternative: the problem can be solved by making *a priori* assumptions for the values of the convective heat transfer coefficients. It always requires a good understanding of basic radiative heat transfer (Chapter 4).

### *The problem*

An array of solar photovoltaic cells providing power to an isolated hamlet is mounted at the top of a mast of height $H = 10$ m. The cells cover uniformly a flat square plane of side $A = 1$ m and thickness $e' = 4$ mm, which is mounted vertically, and oriented east–west; the cells are on the south face. Thermal contact between the cells and the duralumin plate is assumed to be perfect. The cells are protected by a sheet of glass of thickness $e = 4$ mm, a distance of $E = 4$ cm from the duralumin plate. The plate/glass assembly is assumed to be perfectly thermally insulated along the four sides of length $A$, and all edge effects are neglected. The north face of the plate is coated with a selective paint with no thermal effect (but conductive). The plane faces of the various elements are labelled 1, 2, 3, 4, as in the diagram below (which is not to scale).

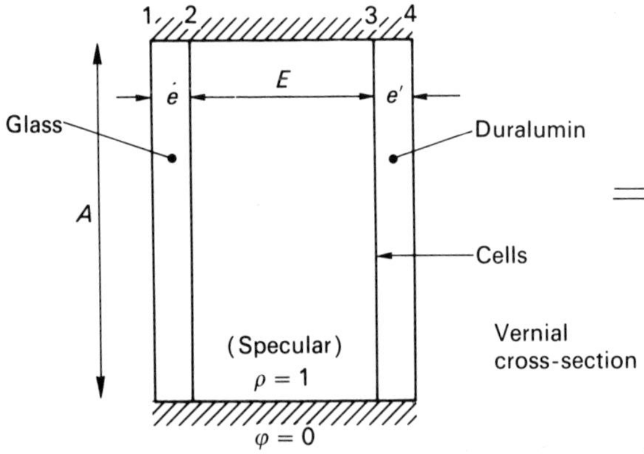


The necessary data is to be taken from Part V: Basic Data.

## *Additional data:*

*α Radiative properties (assumed independent of direction for all materials concerned)*

- Glass:

$$\alpha_\lambda = 0 \quad \lambda < 4\ \mu\text{m}; \qquad \alpha_\lambda = 1 \quad \lambda > 4\ \mu\text{m}; \qquad \tau_\lambda = 1 - \alpha_\lambda$$

- duralumin plate: *face 3 (cells):*

$$a_{3\lambda} = 1 \quad \lambda < 3\ \mu\text{m}; \qquad \alpha_{3\lambda} = 0.3 \quad \lambda > 3\ \mu\text{m}$$

*Important note*
The solar radiation **absorbed by face 3** ($\lambda < 2.5\ \mu$m) is either:

- converted to electrical energy (fraction $\eta$), or
- converted to thermal energy (fraction $(1 - \eta)$.

The dimensionless factor $\eta$ depends on $T_3$ according to:

$$\eta = 0.10 + 2 \times 10^{-3}(333 - T_3) \qquad (T_3 \text{ in K})$$

*Face 4 (selective paint)*

$$\varepsilon_{4\lambda} = 0.3 \quad \lambda < 4\ \mu\text{m}; \qquad \varepsilon_{4\lambda} = 0.7 \quad \lambda > 4\ \mu\text{m}$$

- ground:

$$\varepsilon_t = 0.8 \qquad \rho_t = 0.2$$

- air: as a first approximation, considered transparent over distances below a few tens of metres.
- the atmosphere: several kilometres thick, assumed to have the following radiative properties:

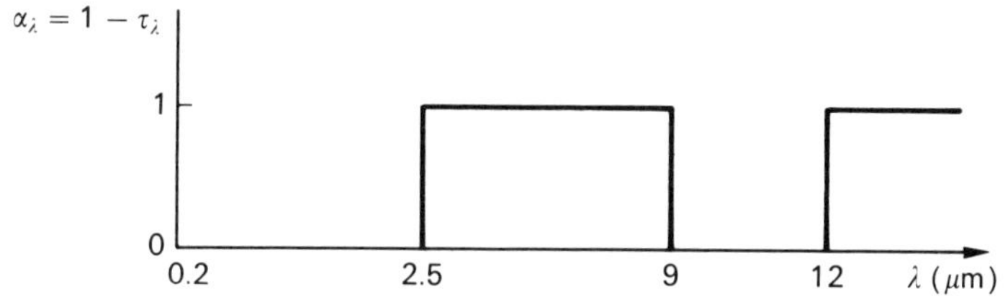


### β Other information

The **sun**: as a first estimate, the radiation reaching the ground from the sun is directional, inclined at 30° to the vertical and coming from the south. The flux density arriving at the ground on a surface normal to the solar radiation is $\varphi_s$. The spectral composition of the radiation above the atmosphere is that of a black body at 5700 K.

The **ground** is at a uniform temperature $T_t$ and the **atmosphere** and **air** are at a temperature $T_a$ far from the cells.

**Calculate the electrical power provided by the system under the specified conditions, taking account of the efficiency $\eta$ of the cells.**

**Data:** $T_a = 30\,°\text{C}$, $T_t = 40\,°\text{C}$, $T_s = 5700$ K, $\varphi_s = 900\ \text{W m}^{-2}$, $A = 1$ m, $H = 10$ m, $e = 4$ mm, $e' = 4$ mm, $E = 4$ cm.

## The solution

This example demonstrates the necessity of understanding the thermal structure of a system in an application which does not inherently involve heat transfer: photovoltaic energy conversion. The performance of the system is in fact strongly affected by an increase in absorber temperature, but this is a direct consequence of the energy source used, the sun, because 85% of solar energy may be converted to thermal energy in the most favourable case. Hence, a compromise is necessary to obtain the optimum balance between received solar flux and the maintenance of an acceptable heat level.

The actual basic application of the cells is to provide a low level of electrical energy to low power electronic systems, such as telecommunications, television, beacons, in isolated locations. They are associated with buffer batteries which allow the available power to be regulated.

The conditions specified above correspond to relatively extreme operational conditions in a very hot atmosphere, conditions used for design purposes:

- The incident solar flux is very high,
- there is no wind to assist the cooling of the system, which must be solely by radiation and very moderate free convection in the air.

An effective and cheap method of cooling will be discussed at the end of the solution.

## 1. Problem analysis

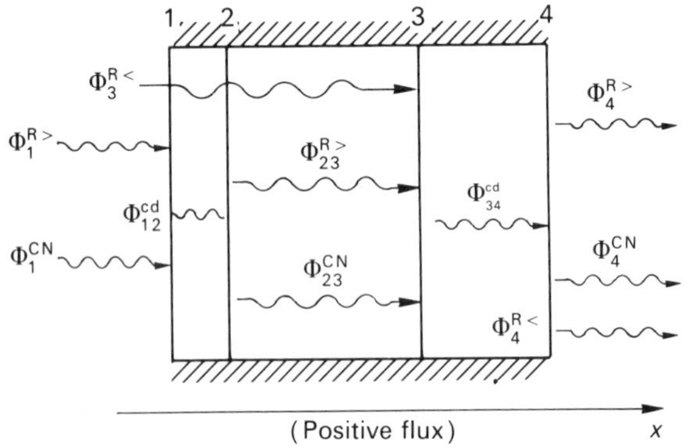


The problem is one-dimensional. The thermal sources at ambient temperature will only emit radiation between around 5–80 $\mu$m. Because of atmospheric effects, the incident solar radiation will only be in the interval (0.27–2.5 $\mu$m). We can thus separate these two types of radiation. We can then identify (see diagram):

- $\Phi_1^{CN}$, $\Phi_4^{CN}$ conducto-convective fluxes (external natural convection);
- $\Phi_{23}^{CN}$ conducto-convective flux (internal natural convection);
- $\Phi_1^{R>}$, $\Phi_4^{R>}$, $\Phi_{23}^{R>}$ radiative flux from long wavelength sources (excluding solar);
- $\Phi_3^{R<}$, $\Phi_4^{R<}$ solar radiative flux (short wavelengths);
- $\Phi_{12}^{cd}$, $\Phi_{34}^{cd}$ conductive fluxes.

The electrical power supplied will depend on the following two parameters:

- the **solar flux density received** in the interval [0.27–2.5 $\mu$m]
- the **temperature of the cells,** effectively surface 3 (assumed uniform). This is obtained by solving the thermal problem.

## 2. Energy balance of the cells

We will treat flux densities $\varphi = \Phi/S$ as positive in the positive $x$ direction.

$$\varphi + \varphi_3^{R<} - \varphi' = 0 \tag{1}$$

where:

$$\varphi' = \frac{\lambda'}{e'}(T_3 - T_4) = h_4(T_4 - T_a) + \varphi_4^{R>}(T_4, T_a, T_t) + \varphi_4^{R<}(T_s, \varphi_s) \tag{2, 3}$$

$$\varphi = \varphi_1^{R>}(T_1, T_a, T_t) + h_1(T_a - T_1) = \frac{\lambda}{e}(T_1 - T_2)$$
$$= \frac{\lambda_{23}^{CN}}{E}(T_2 - T_3) + \varphi_{23}^{R>}(T_2, T_3) \tag{4, 5, 6}$$

This system of six equations allows the six unknowns $\varphi_1$, $\varphi'$, $T_1$, $T_2$, $T_3$ and $T_4$ to be found in terms of the given $T_a$, $T_i$, $\varphi_s$. We now need to calculate the corresponding fluxes.

## 3. Radiative transfers

### 3.1 Long wavelengths (thermal sources)

We are concerned with radiation emitted by the glass, the plate, the cells, the ground and the atmosphere, and giving the fluxes $\varphi_1^{R>}$, $\varphi_4^{R>}$, and $\varphi_{23}^{R>}$ (in the spectral interval 4.5 to 80 $\mu$m). The glass is opaque, and equivalent to a black body in this spectral interval. The plate is grey.

- Calculation of $\varphi_1^{R>}$, and $\varphi_4^{R>}$:

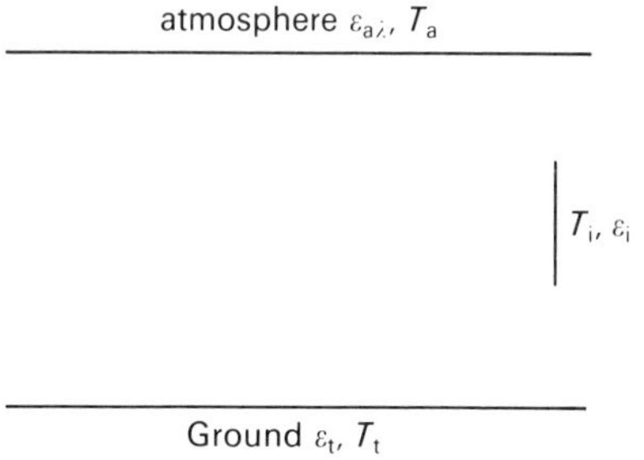


$$i = 1 \text{ or } 4$$

$$d\varphi_{i\lambda}^{R>} = (d\varphi_{i\lambda}^{i} - \pi I_\lambda^0(T_i)\, d\lambda)\varepsilon_i \tag{7}$$

$$d\varphi_{i\lambda}^{i} = f_{it}\, d\varphi_{t\lambda}^{l} + f_{ia}\, d\varphi_{a\lambda}^{l} \tag{8}$$

$$d\varphi_{a\lambda}^{l} = \varepsilon_{a\lambda}\pi I_\lambda^0(T_a)\, d\lambda \tag{9}$$

$$d\varphi_{t\lambda}^{l} = [\varepsilon_t \pi I_\lambda^0(T_t) + (1 - \varepsilon_t)\varepsilon_{a\lambda}\pi I_\lambda^0(T_a)] \tag{10}$$

By symmetry:

$$f_{it} = f_{ia} = \tfrac{1}{2} \tag{11}$$

After integration over the spectrum:

$$\varphi_i^{R>} = \tfrac{1}{2}[\varepsilon_t \sigma T_t^4 + (2-\varepsilon_t) Z \sigma T_a^4] - \sigma T_i^4 \tag{12}$$

where:

$$Z = 1 - z\left(0, \frac{12}{\lambda_m(T_a)}\right) + z\left(0, \frac{9}{\lambda_m(T_a)}\right) \tag{13}$$

$\lambda_m = 9.83\ \mu m \rightarrow Z \simeq 0.80.$

- After linearization, we thus obtain:

$$\varphi_1^{R>} = \sigma(298.0^4 - T_1^4) \overset{1^{st}\ approx}{\simeq} \underline{6.15(298.0 - T_1)} \tag{14}$$

In the same way:

$$\varphi_4^{R>} = 0.7[\sigma T_4^4 - \tfrac{1}{2}(\varepsilon_t \sigma T_t^4 + (2-\varepsilon_t) Z \sigma T_a^4)] \tag{15}$$

$$= 0.7\sigma(T_4^4 - 298.0^4) \overset{1^{st}\ approx}{\simeq} \underline{4.40(T_4 - 298.0)} \tag{16}$$

- The flux exchanged between faces 2 and 3 is:

$$\varphi_{23}^{R>} = 0.3\sigma(T_2^4 - T_3^4) \overset{1^{st}\ approx}{\simeq} \underline{2.46(T_2 - T_3)} \tag{17}$$

## *3.2 Short wavelengths ($\lambda < 2.5\ \mu m$)*

Face 1 receives direct flux:

$$\varphi_3^{i<direct} = \varphi_s \cos 60° = 450\ \mathrm{W\ m^{-2}}$$

Faces 1 and 4 receive flux reflected by the ground:

$$\varphi_3^{i<diffuse} = -\varphi_4^{i<diffuse} = \rho_t f_{1t} \cos 30° \varphi_s = 77.9\ \mathrm{W\ m^{-2}}$$

- The radiative flux on surface 4 is $\underline{\varphi_4^{R<} = -23.4\ \mathrm{W\ m^{-2}}}$
- The flux absorbed and converted to thermal energy on face 3 (cells) is, on the simple model used:

$$\varphi_3^{R<} = (1-\eta)(\phi_3^{i<direct} + \varphi_3^{i<diffuse})$$

($\eta$, the cell efficiency, is a function of $T_3$.)

Assuming $\eta = 0.12$, we obtain:

$$\underline{\varphi_3^{R<} \overset{1^{st}\ approx}{\simeq} 460\ \mathrm{W\ m^{-2}}}$$

The electric power supplied is:

$$P = \eta(\varphi_3^{i<\text{direct}} + \varphi_3^{i<\text{diffuse}})$$

The problem is to determine $T_3$. As a first approximation, $P = 63.4\ \text{W m}^{-2}$.

## 4. Conduction and conducto-convection

- Conductive flux in the glass:

$$\varphi_{12}^{cd} = \frac{0.78}{4 \times 10^{-3}}(T_1 - T_2) = 195(T_1 - T_2)$$

- Conductive flux in the plate:

$$\varphi_{34}^{cd} = \frac{164}{4 \times 10^{-3}}(T_3 - T_4) = 4.1 \times 10^4(T_3 - T_4)$$

  Taking account of the orders of magnitude of the values of $h^R$ calculated and those of $h^{CN}$ (less than 10 W m$^{-2}$ K$^{-1}$ at a guess), it is clear that $T_3 = T_4$ and, as a first approximation, $T_1 = T_2$.
- The calculation of the flux transferred by natural convection outside (at 1 and 4) and inside (between 2 and 3) requires iteration. We estimate initial values of $T_1$ $(= T_2)$ and $T_3$ $(= T_4)$.

*First iteration*

We wish to find $T_1$ and $T_3$ in the equivalent circuit:

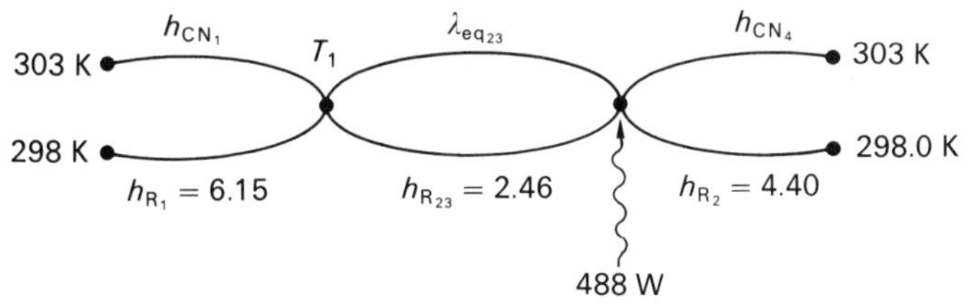


Assume (arbitrarily): $h_{CN_1} = h_{CN_4} = 3\ \text{W m}^{-2}\ \text{K}^{-1}$

$$\lambda_{eq23} = 2\lambda_{air}$$

We will introduce a global coefficient:

$$\gamma_{23} = \frac{\lambda_{eq23}}{E} = 1.4$$

Whence:

$$\underline{T_3 \simeq 350.5 \text{ K}, \ T_1 \simeq 315.7 \text{ K}, \ \eta \simeq 6.5 \times 10^{-2}}$$

$$\varphi_3^{R<} - \varphi_4^{R<} = 517 \text{ W m}^{-2}$$

whence, in SI units:

$$\begin{array}{lll} \lambda_1 = 0.0267 & \lambda_4 = 0.0281 & \lambda_{23} = 0.029 \text{ Wm}^{-1}\text{ K}^{-1} \\ a_1 = 2.33 \times 10^{-5} & a_4 = 2.60 \times 10^{-5} & a_{23} = 2.72 \times 10^{-5} \text{ m}^2\text{ s}^{-1} \\ \nu_1 = 1.64 \times 10^{-5} & \nu_4 = 1.82 \times 10^{-5} & \nu_{23} = 1.87 \times 10^{-5} \text{ m}^2\text{ s}^{-1} \\ \delta T_1 = 14.7 \text{ K} & \delta T_4 = 47.5 \text{ K} & \delta T_{23} = 34.8 \text{ K} \\ \beta_1 = 1/307 & \beta_4 = 1/326 & \beta_{23} = 1/333 \text{ K}^{-1} \end{array}$$

$$Ra_{A1} = \frac{g\beta_1 \delta T_1 A^3}{a_1 \nu_1} = 1.23 \times 10^9 \Rightarrow Nu_1 = 0.13 Ra_{A1}^{1/3} = 138$$

(turbulent régime)

$$\underline{h_{CN1} = 3.69 \text{ Wm}^{-2}\text{ K}^{-1}}$$

$$Ra_{A4} = 4.29 \times 10^9 \text{ (turbulent)} \Rightarrow Nu_4 = 210$$

$$\underline{h_{CN2} = 5.89 \text{ Wm}^{-2}\text{ K}^{-1}}$$

$$Ra_E = \frac{g\beta_3 \delta T_3 E^3}{a_{23}\nu_{23}} = 1.29 \times 10^5 \qquad \text{with} \qquad Al = 25 \Rightarrow \frac{\lambda_{eq}}{\lambda} = 0.197 Ra_E^{0.25} Al^{1/9}$$

We also introduce $\gamma_{23} = \lambda_{eq}/E = 1.89$ (the surface conductance). We obtain new estimates of $h_R$:

$$h_{R1} = 6.5; \qquad h_{R23} = 2.51; \qquad h_{R4} = 5.81$$

and we write:

$$h_{23} = \gamma_{23} = h_{R_{23}} = 4.4 \text{ (Wm}^{-2}\text{ K}^{-1})$$

We then have, in surface conductances, the following circuit:

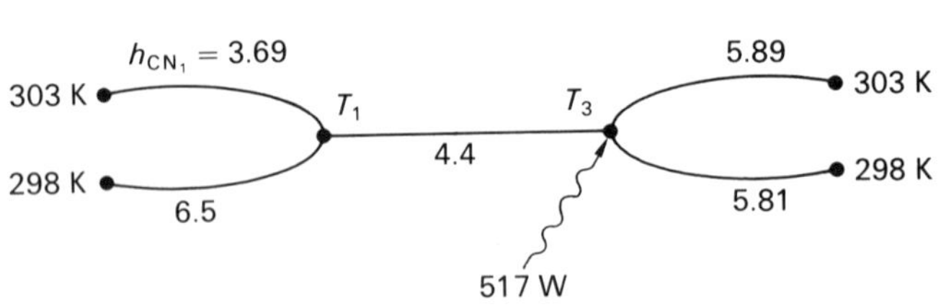


$$\Rightarrow \quad T_3 = 335.8 \text{ K}; \qquad T_1 = 311.0 \text{ K}; \qquad \eta = 9.44 \times 10^{-2}$$

and iterating again, we find:

$$\Rightarrow \quad T_3 = 342.4 \text{ K}; \qquad T_1 = 312.0 \text{ K}; \qquad \eta = 8.1 \times 10^{-2}$$

so that:

$$\boxed{P = 42.8\ \text{W}}$$

This is a lamentable output; we need to cool the panel.

A possibility (reliable, and of low cost) is to attach vertical fins to the rear face. The problem could be re-worked, with several extra iterations, for this situation.

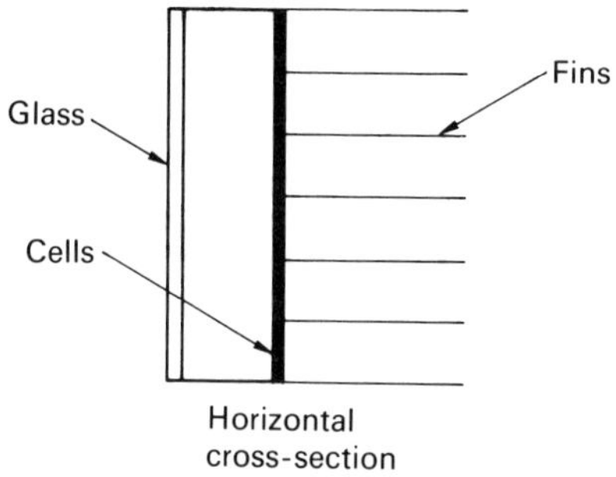

# Bibliography

### *References for Part I*


[1] Candel, S. (1990) *Mécanique des Fluides*. Dunod.
[2] Castaing, R. (1971) *Thermodynamique Statistique*. Masson.
[3] Herzfeld, K.F. and Litowitz, A. (1964) *Absorption and Dispersion of Ultrasonic Waves*. Academic Press.
[4] Hirchfelder, J.O., Curtiss, C.F. and Bird, R.B. (1964) *Molecular Theory of Gases and Liquids*. John Wiley.
[5] Sommerfeld, A. (1956) *Thermodynamics and Statistical Mechanics*. Academic Press.
[6] Ferziger, J.H. and Kaper, H.G. (1972) *Mathematical Theory of Transfer Processes in Gases*. North Holland.
[7] Prigogine, I. (1968) *Introduction à la thermodynamique des processus irréversibles*. Dunod.
[8] Lengyel, B.A. (1966) *Introduction to Laser Physics*. John Wiley.
[9] Kennards, E.H. (1938) *Kinetic Theory of Gases*. McGraw-Hill.
[10] Kittel, C. (1970) *Introduction à la physique de l'état solide* (Translated by A. Honnard). Dunod.
[11] Ziman, J.M. (1971). *Principles of the Theory of Solids*. Cambridge University Press.
[12] Touloukian, Y.S., Powell, R.W., No, C.Y. and Klemens, P.G. (1970) Vol. 1 *Thermal Conductivity – Metallic Elements and Alloys*; Vol. 2 *Thermal Conductivity – Non Metallic Solids*; Vol. 3 *Thermal Conductivity – Non Metallic Liquids and Gases*; Vol. 7 *Thermal Radiative Properties – Metallic Elements and Alloys*; Vol. 8 *Thermal Radiative Properties – Non Metallic Solids*; Vol. 9 *Thermal Properties Coatings*; Vol. 11 *Viscosity*. IFI Plenum, NY
[13] Brown, A.I. and Marco, S. (1958) *Introduction to Heat Transfer*. McGraw-Hill.
[14] Kutateladze, S.S. (1962) *Fundamentals of Heat Transfer*. E. Arnold, London
[15] Whitaker, S. (1977) *Fundamental Principles of Heat Transfer*. Pergamon Press.
[16] Kays, W.M. and London, A.L. (1964) *Compact Heat Exchangers*. McGraw-Hill.
[17] Greth-Ceng (Publications of).
[18] Huetz-Aubert, M. and Taine, J. (1978) *Revue Générale de Thermique*, **202**, 757.
[19] Soufiani, A. and Taine, J. (1988) *Applied Optics*, **27**, 3754.
[20] Reference 6, page 319 and following.

## References for Part II

[1] Baumeister, K.J. and Hamill, T.D. (1971) *J. of Heat Transfer*, **91**, 126.
[2] Kao, T. (1976) *AIAA Journal*, **14**, 818.
[3] Kazimi, M.S. and Erdwen, C.A. (1975) *J. of Heat Transfer*, **97**, 615.
[4] Abramowitz, M. and Stegun, I.A. (1965) *Handbook of Mathematical Functions*. Dover Pub. Inc., NY.
[5] Carslaw, H.S. and Jaeger, J.C. (1959) *Conduction of Heat in Solids*. Oxford Press.
[6] Schneider, P.J. (1963) *Temperature Response Charts*. John Wiley.
[7] Hladik, J. (1969) *La transformation de Laplace à plusieurs variables*. Masson.
[8] Morse, P.M. and Feshbach, H. (1953) *Methods of Theoretical Physics*. McGraw-Hill.

## References for Part III

[1] Ziman, J.M. (1964) *Principles of the Theory of Solids*. Cambridge University Press, London.
[2] Kittel, C. (1983) *Physique de l'état solide*. French edition, fifth edition. Dunod.
[3] Born, M. and Wolf, E. (1975) *Principles of Optics*. Pergamon Press.
[4] Abeles, F. (1972) *Optical Properties of Solids*. North Holland.
[5] Landau, L. and Lifchitz, E. (1969) *Physique théorique et Electrodynamique des Milieux Continus*. MIR, Moscow (2 volumes).
[6] Feynman (1970) *Cours de Physique*. Addison Wesley.
[7] Touloukian, Y.S. and DeWitt, D.P. (1970) *Thermal Radiative Properties*, Volume 7 – *Metallic Elements and Alloys*. IFI Plenum, NY.
[8] Touloukian, Y.S. and Dewitt, D.P. (1970) *Thermal Radiative Properties*, Volume 8 – *Non Metallic Solids*. IFI Plenum, NY.
[9] Touloukian, Y.S. and Dewitt, D.P. (1970) *Thermal Radiative Properties*, Volume 9 – *Coatings*. IFI Plenum, NY.
[10] Combescot, M. and Bok, J. (1985) *Journal of Luminescence*, **30**, 1-A.
[11] Beckmann, P. and Spizzichino, A. (1963) *The Scattering of Electromagnetic Waves from Rough Surfaces*. Pergamon Press.
[12] Siegel, R. and Howell, J.R. (1962) *Thermal Radiation Heat Transfer*. McGraw-Hill.
[13] Soufiani, A. and Taine, J. (1988) *Applied Optics*, **27**, 3754.
[14] Hottel, H.C. and Sarofim, A.F. (1972) *Radiative Transfer*. McGraw-Hill.
[15] Landau, L. and Lifschitz, E. (1966) *Physique statistique et Théorie du champ*. MIR, Moscow (2 volumes).
[16] Marechal, A. (1968) 'Optique géométrique générale'. *Handbuch der Physik*, Vol. XXIV, Section 3, Springer Verlag.
[17] Ferziger, J.H. and Kaper, H.G. (1965) *Mathematical Theory of Transport Processes in Gases*. McGraw-Hill.
[18] Taine, J. *Physique Statistique*, Course book, Ecole Centrale, Paris.
[19] Osizik, M.N. (1972) *Radiative Transfer and Interactions with Conduction and Convection*. McGraw-Hill.
[20] Soufiani, A., Hartmann, J.M. and Taine, J. (1985) *J.Q.S.R.T.*, **33**, 243.
[21] Zhang, L., Soufiani, A. and Taine, J. (1988) *Int. J. Heat and Mass Transfer*, **31**, 2261.

[22] Bohren, C.F. and Huffman, D.R. (1983) *Absorption and Scattering of Light by Small Particles*. John Wiley.
[23] Chandrasekhar, S. (1960) *Radiative Transfer*. Dover.
[24] Taine, J. (1983) *J.Q.S.R.T.*, **30**, 371.
[25] Hartmann, J.M., Levi di Leon, R. and Taine, J. (1984) *J.Q.S.R.T.*, **32**, 119.
[26] Bernstein, L.S. (1980) *J.Q.R.S.T.*, **23**, 157.
[27] Malkmus, W. (1963) *J. Opt. Soc. Am.*, **53**, 951.
[28] Youngs, S.J. (1975) *J.Q.S.R.T.*, **15**, 483.
[29] Soufiani, A., Hartmann, J.M. and Taine, J. (1985) *J.Q.S.R.T.*, **33**, 243.
[30] Edwards, D.W. (1976) *Advances in Heat Transfer*, **12**, 115.
[31] Goody, R.M. and Yung, Y.L. (1981) *Atmospheric Radiation* (2nd edn). Oxford University Press.
[32] Rivière, P., Soufiani, A. and Taine, J. (1992) *J.Q.S.R.T.* (in press).
[33] Godson, W.L. (1955) *J. Meteorol.*, **12**, 272.
[34] Goody, R.M. (1952) *Quart. J. Royal Meteor. Soc.*, **78**, 165.
[35] Malkmus, W. (1967) *J. Opt. Soc. Am.*, **57**, 323.
[36] Ludwig, C.B., Malkmus, W., Reardon, J.E. and Thomson, J.A.L. (1973) *Handbook of Infra-red Radiation from Combustion Gases* (Goulard, R. and Thomson, J.A.L. eds). NASA SP-3080, Washington, D.C.
[37] Young, S.J. (1977) *J.Q.S.R.T.*, **18**, 1 and **18**, 29.
[38] Levi di Leon, R. and Taine, J. (1986) *Revue de Physique Appliquée*, **21**, 825.

## References for Part IV

[1] Candel, S. (1990) *Mécanique des Fluides*. Dunod.
[2] Shade, K.W. and Maceligot, D.M. (1971) *Int. J. of Heat and Mass Transfer*, **14**, 653.
[3] Proceedings of Euromech Colloquium 237 (1988) *Influence of Density Variations on the Structure of Low Speed Turbulent Flows*, July 18–21, Marseille.
[4] Soufiani, A. and Taine, J. (1987) *Int. J. of Heat and Mass Transfer*, **30**, 437.
[5] Knudsen, J.G. and Katz, D.L. (1958) *Fluid Dynamics and Heat Transfer*. McGraw-Hill.
[6] Langhar, H.L. (1942) *Trans. ASME*, **64**, A55.
[7] Schlichting, H. (1968) *Boundary Layer Theory*. McGraw-Hill.
[8] Favre, A., Kovasnay, L.S.G., Dumas, R., Caviglio, J. and Coantic, H. (1970) *La turbulence en mécanique des fluides*. Gauthier-Villars.
[9] Laufer, J. (1954) *The Structure of Turbulent Velocity in Fully Developed Pipe Flow, NACA Report*, No. 1174.
[10] Pimont, V. (1983) Convection forcée le long d'une paroi. Application aux métaux liquides. *Rapport E11 CEA*, R.5225 (CEN Saclay 91191 Gif Cedex).
[11] Hochreiter, L.E. (1971) Turbulent Structure of Isothermal and Non Isothermal Liquid Metal Pipe Flow. PhD Thesis. Purdue University.
[12] Bremhorst, K. and Bullock, K.J. (1970) *Int. J. Heat and Mass Transfer*, **13**, 1313.
[13] Burchill, W.E. (1970) Statistical Properties of Velocity and Temperature in Isothermal and Non Isothermal Turbulent Pipe Flow. PhD Thesis, University of Illinois.
[14] Tennekes, H. and Lumley, J. (1972) *A First Course in Turbulence*. MIT Press, Cambridge.

[15] Vandromme, D., Ha-Minh, H., Viegas, J., Rubesin, M.W. and Kollman, W. (1983) *4th Turbulent Shear Flow Conf.* Karlsruhe (September).
[16] Van Driest, E.R. (1956) *J. Aero. Sci.*, **23**, 1007.
[17] Cebeci, T. (1973) *J. of Heat Transfer*, **95**, 227.
[18] Na, T.Y. and Habib, S. (1973) *Applied Science Research*, **28**, 302.
[19] Lam, C.K.G. and Bremhorst, K.A. (1981) *Journal of Fluid Engineering*, **103**, 456.
[20] Patel, V.C., Rodi, W. and Scheuerer, G. (1984) *AIAA Journal*, **23**, 1308.
[21] Sieder, E.N. and Tate, G.E. (1936) *Ind. Chem. Ing.*, **28**, 1429.
[22] Whitaker, S. (1977) *Fundamental Principles of Heat Transfer*. Pergamon.
[23] Aladyev, I.T. (1954) *NACA TM*, 1365.
[24] Deissler, R.G. (1953) *NACA TN*, 3016.
[25] Boetler, L.M.K., Young, G. and Iverson, H.W. (1948) *NACA TN*, 1451.
[26] Shade, K.W. and Maceligot, D.M. (1971) *Int. J. of Heat and Mass Transfer*, **14**, 653.
[27] Soufiani, A. and Taine, J. (1987) *Int. J. of Heat and Mass Transfer*, **30**, 437.
[28] Soufiani, A. (1987) Thèse d'Etat. Université Paris XI, Orsay.
[29] De Soto, S. (1968) *Int. J. of Heat and Mass Transfer*, **11**, 39.
[30] Kim, J., Moin, P. and Moiser, R. (1987) *J. Fluid Mech.*, **117**, 133.
[31] Examples are: FASCOD2; the ONERA code; and the code of Laboratory EM2C (ECP).
[32] AFGL and GEISA (Ecole Polytechnique, Palaiseau, France databases).
[33] Hartmann, J.M., Rosenmann, L., Perrin, M.Y. and Taine, J. (1988) *Applied Optics*, **27**, 3063.
[34] Rosenmann, L., Hartmann, J.M., Perrin, M.Y. and Taine, J. (1988) *Applied Optics*, **27**, 3902.
[35] Elsasser, W.M. (1936) *Phys. Rev.*, **54**, 128.
[36] Malkmus, W. (1963) *J. Opt. Soc. Am.*, **53**, 951
[37] Young, S.J. (1975) *J.Q.S.R.T.*, **15**, 483.
[38] Edwards, D.K. (1976) *Advances in Heat Transfer*, **12**, 115.
[39] Levi di Leon, R. and Taine, J. (1986) *Revue de Physique Appliquée*, **21**, 825.
[40] Godson, N.L. (1953) *J. Q. Met. Soc.*, **79**, 367.
[41] Linquist, G.M. and Simmons, F.S. (1972) *J.Q.S.R.T.*, **12**, 807.
[42] Soufiani, A. and Taine, J. (1986) *Proceedings of the 8th Int. Conf. Heat and Mass Transfer, San Francisco (August)*.
[43] Zhang, L., Soufiani, A. and Taine, J. (1988) *Int. J. of Heat and Mass Transfer*, **31**, 2261.
[44] Zhang, L. (1989) Thesis at ECP (20 January).
[45] Osizik, M.N. (1972) *Radiative Transfer and Interactions with Conduction and Convection*. McGraw-Hill.
[46] Chandrasekhar, S. (1960) *Radiative Transfer*. Dover.
[47] Born, M. and Wolf, E. (1975) *Principle of Optics*. Pergamon.
[48] Bohren, C.F. and Huffman, D.R. (1983) *Absorption and Scattering of Light by Small Particles*. John Wiley.
[49] Favre, A., Kovasnay, L.S.G., Dumas, R., Gaviglio, J. and Coantic, M. (1976) *La turbulence en mécanique des fluides*. Gauthier-Villars.
[50] Hanjalik, K. and Launder, B.E. (1976) *J. Fluid Mech.*, **74**, 593.
[51] Johns, W.P. and Launder, B.E. (1972) *Int. J. of Heat and Mass Transfer*, **15**, 301 and *Int. J. of Heat and Mass Transfer*, **22**, 1631.
[52] Launder, B.E. (1984) *Int. J. of Heat and Mass Transfer*, **27**, 1485.
[53] Nagano, Y. and Hishida, M. (1988) *J. of Fluids Engineering*, **109**, 156.

[54] Nagano, Y. and Kim, C. (1988) *J. of Heat Transfer*, **110**, 583.
[55] Gore, J.P., Jeng, S.M. and Faeth, G.M. (1986) *AIAA/ASME Heat Transfer and Thermophysical Conference*, Boston.
[56] Coantic, M. and Simonin, O. (1984) *J. of the Atmospheric Sciences*, **41**, 2629.
[57] Chobaut, J.P., Fotrelle, Y. and Schissler, J.M. (1986) *Physicochemical Hydrodynamics*, **4**, 67.
[58] Soufiani, A., Hartmann, J.M. and Taine, J. (1985) *J.Q.S.R.T.*, **33**, 243.
[59] Soufiani, A. and Taine, J. (1988) *Applied Optics*, **27**, 3754.
[60] Yamada, Y. (1988) *Int. J. Heat Mass Transfer*, **31**, 429.
[61] Lauriat, G. (1982) *Proceedings of the 7th Int. Heat Transfer Conf.*, **2**, NC6.

## Supplementary bibliography on convection

Two important topics in heat transfer have been covered in this work:

- Two-phase materials.
- Porous materials.

Following are some general references covering these types of heat transfer.

### *Two-phase materials*

[1] Whalley, P.B. (1987) *Boiling, Condensation and Gas–Liquid Flow*. Clarendon Press.
[2] Wallis, G.B. (1969) *One Dimensional Two-Phase Flow*. McGraw-Hill.
[3] Collier, J.G. (1981) *Convective Boiling and Condensation*. McGraw-Hill.
[4] Delhaye, J.M., Giot, M. and Riethmuller, M.L., eds (1981) *Thermalhydraulics of Two-Phase Systems for Industrial Design and Nuclear Engineering*. Hemisphere/McGraw-Hill.
[5] Bergles, A.E., Collier, J.G., Delhaye, J.M., Hewitt, G.F. and Mayinger, F. (1973) *Two-Phase and Heat Transfer in the Power and Process Industries*. Hemisphere/McGraw-Hill.
[6] Hetsroni, G. ed. (1982) *Handbook of Multiphase Systems*. Hemisphere/McGraw-Hill.

### *Porous materials*

The most complete is:

[1] Bejan, A. (1984) *Convection Heat Transfer*. John Wiley.

The basics are covered in:

[2] Bear, J. (1972) *Dynamics of Fluids in Porous Media*. American Elsevier Environmental Science.
[3] Bird, R.B., Stewart, W.E. and Lightfoot, E.M. (1960) *Transport Phenomena*. John Wiley.
[4] Klarsfeld, S. (1985) *Comptes rendus de l'école d'été du GUT*.

## References for PART V


[1] Rohsenow, W.M. and Hartnett, J.P. (1973) *Handbook of Heat Transfer*. McGraw-Hill.
[2] Knudsen, J.G. and Katz, D.L. (1958) *Fluid Dynamics and Heat Transfer*. McGraw-Hill.
[3] Jacob, M. (1949) *Heat Transfer*. John Wiley.
[4] Kutateladze, S. and Borishanskii, V.M. (1966) *A Concise Encyclopedia of Heat Transfer*. Pergamon Press.
[5] McAdams, W.H. (1961) *Transmission de la chaleur*. Dunod.
[6] Gosse, J. (1981) *Guide technique de thermique*. Dunod.
[7] Kay, J.M. (1964) *Introduction à la mécanique des fluides et la transmission de la chaleur*. Dunod.
[8] Cebeci, T. and Bradshaw, P. (1984) *Physical and Computational Aspects of Convective Heat Transfer*. Springer Verlag.
[9] Kays, W. and Crawford, M.E. (1980) *Convective Heat and Mass Transfer*. McGraw-Hill.
[10] Patten, K. and Legros, J.C. (1984) *Convection in Liquids*. Springer Verlag.
[11] Worsœ-Schmidt, P.M. (1967) 'Heat transfer in the thermal entrance region of circular tubes and annular passages with fully developed laminar flow'. *Int. J. of Heat and Mass Transfer*, **10**, 541–52.
[12] Kays, W.M. (1955) 'Numerical solutions for laminar-flow heat tranfer in circular tubes'. *Trans. ASME*, **77**, 1265–74.
[13] Graetz, L. (1883) *Annalen. d. Physik*, **18, 25** (1885).
[14] Shah, R.K. (1975) 'Thermal entry length solutions for the circular tube and parallel plates'. *Proc. Natl. Heat Mass Transfer Conf. 3rd Indian Inst. Technol. Bombay*, Vol. I, Pap., N HMT-11-75.
[15] Grigull, U. and Tratz, H. (1965). 'Thermischer einlauf in ausgebildeter laminarer Rohrströmung'. *Int. J. of Heat and Mass Transfer*, **8**, 669–78.
[16] Hornbeck, R.W. (1965) 'An all-numerical method for heat transfer in the inlet of a tube'. *Am. Soc. Mech. Eng., Pap 65*, WA/HT, 36.
[17] Heaton, H.S., Reynolds, W.C. and Kays, W.M. (1964) 'Heat transfer in annular passages: Simultaneous development of velocity and temperature fields in laminar flow'. *Int. J. of Heat and Mass Transfer*, **7**, 763–81.
[18] Dittus, F.W. and Boelter, L.M.K. (1930) *Univ. of Calif. Pubs. Eng.*, **2**, 443.
[19] Colburn, A.P. (1983) *Trans. AICHE*, **29**, 174.
[20] Lyon, R.N. (1952) *Liquid Metals Handbook*. NAVEXOS, 733.
[21] Seban, R.A. and Shimazaki, T.T. (1951) *Trans. of ASME*, **73**, 803.
[22] Sleicher, C.A. and Tribus, M. (1957) *Trans. ASME*, **79**, 798.
[23] Sparrow, E.M., Hallman, T.M. and Siegel, R. (1958) 'Steady laminar heat transfer in a circular tube with prescribed wall heat flux'. *Appl. Sci. Res., Sect.* **A7**, 386–92.
[24] Brown, G.M. (1960) 'Heat or mass transfer in a fluid in laminar flow in a circular or flat conduit'. *AIChE J.*, **6**, 179–83.
[25] Sparrow, E.M., Novotny, J.L. and Lin, S.H. (1963) 'Laminar flow of a heat-generating fluid in a parallel-plate channel'. *AIChE J.*, **9**, 797–804.
[26] Shah, R.K. and London, A.L. (1978) *Laminar Flow Forced Convection in Ducts. Advances in Heat Transfer – Supplement 1.*
[27] Wibulswas, P. (1966) Laminar Flow Heat Transfer in Non-circular Ducts. PhD Thesis, London University, London.
[28] Sparrow, E.M. and Lin, S.H. (1969) 'The developing laminar flow and pressure drop in the entrance region of annular ducts'. *J. Basics Energ.*, **91**, 345–54.

[29] Sparrow, E.M. and Lin, S.H. (1964) 'The developing laminar flow and pressure drop in the entrance region of annular ducts'. *J. Basics Eng.*, **86**, 827–34.

[30] Lundberg, R.E., Reynolds, W.C. and Kays, W.M. (1963) *Heat Transfer with Laminar Flow in Concentric Annuli with Constant and Variable Wall Temperature and Heat Flux.* NASA Tech. Note TN. D-1972.

[31] Lundberg, R.E., McCuen, P.A. and Reynolds, W.C. (1963) 'Heat transfer in annual passages. Hydrodynamically developed laminar flow with arbitrarily prescribed wall temperatures of heat fluxes'. *Int. J. of Heat and Mass Transfer*, **6**, 495–529.

[32] Shah, R.K. and London, A.L. (1971) *Laminar Flow Forced Convection Heat Transfer and Flow Friction in Straight and Curved Ducts* – A Summary of Analytical Solutions. TR No. 75 Dept. Mech. Eng., Stanford University.

[33] Shumway, R.W. and Maceligot, D.M. (1971) 'Heated gas flow in annuli with temperature dependent transport properties'. *Nucl. Sci. Eng.*, **46**, 394–407.

[34] Atkinson, B., Brocklebank, M.P., Card, C.C.H. and Smith, J.M. (1969) 'Low Reynolds number developing flows'. *AIChE J.*, **15**, 548–53.

[35] Stephan, K. (1959) 'Wärmeübergang und druckabfall beio nicht ausgebildeter Laminarströmung in Röhren und in ebenen Spalten'. *Chem. Ing. Tech.*, **31**, 773–8.

[36] Hwang, C.L. and Fan, L.T. (1964). 'Finite difference anlysis of forced convection heat transfer in entrance region of a plate rectangular duct'. *Appl. Sci. Res., Sect.* **A13**, 401–22.

[37] Schlichting, H. (1960) *Boundary Layer Theory.* McGraw-Hill.

[38] Martinelli, R.C. *et al.* (1943) *Trans. Am. Inst. Chem. Engrs.*, **38**, 943.

[39] Hilpert, R. (1932) 'VDI'. *Forshungsheft*, No. 355.

[40] Reiher, H. (1925) *Forshungsarb. a.d. Geb. d. Ingenieurives*, No. 269.

[41] Kramers, H. (1946) *Physica*, **12**, 61.

[42] Petit, J.P. (1988) *Convection Naturelle.* Cours ECP.

[43] Raithby, G.D. and Hollands, G.T. (1975) 'A general method of obtaining approximate solutions to laminar and turbulent free confection problems'. *Advances in Heat Transfer*, **11**.

[44] Catton, I. (1978) *Natural Convection in Enclosures.* 6th International Congress on Heat Transfer, Toronto, Vol. 6, p. 13–31.

[45] Siegel, R. and Howell, J.R. (1972) *Thermal Radiation Heat Transfer.* McGraw-Hill.

# Subject index


absorption 8, 79, 86
absorption coefficient 217, 218
absorptivity 8, 79, 85, 86

balance equations 61–6
  of energy 64
  of mass 61
  of momentum 62
Biot number 37–42, 136
  interpretation 141
black body 90
boundary conditions 10–20, 262
boundary layer 268–81, 283
  momentum 269
  thermal 269
  turbulent 293–4

capacity rate 73
change of state, boundary conditions involving 57–9
co-current heat exchanger 70–1
conductivity 10
continuous medium 4
contraction 109–12
counter-current heat exchanger 70–1

deformation, rate of 109–12
development lengths and régimes 282–91
diffusivity 118
  thermal 118
  momentum 253, 270
  turbulent 305
dilatation, stress of 109–12
dimensional analysis 38, 121–3, 251–9
dissipation, rate of $\varepsilon$, of $k$ 322

Eckert number 266
efficiency
  of a fin 33, 35
  of a heat exchanger 75
effusivity 126, 131–4
emission 9, 79, 87
emission coefficient 217
emissivity 87

equilibrium
  local thermodynamic 6
  medium not in 6
  thermal 4
  thermodynamic 4
equilibrium radiation 87
extinction of radiation 219

fin 27–50
  approximation 32
  infinite 36
  validity 37, 42
flux
  conductive 9–11
  conducto-convective 16, 439
  convective 13
  radiative 9, 78, 85
Fourier number 121
  interpretation 123, 139–40
friction coefficient 253
Froude number 266

global heat exchange coefficient 31, 71
Grashof number 256, 266
grey body 88

heat transfer coefficient convective 16
  linearized radiative 97

infra-red domain 81
intensity
  definition 82
  of equilibrium radiation 87, 90

$k$–$\varepsilon$ model 320
Kolmogorov, universal equilibrium law 303

material system 51
mixing
  temperature 69
  length (Prandtl) 306

Navier–Stokes equations 64
non-dimensionalization 38, 121–3, 251–9

non-steady régime 7, 113, 458, 510, 515, 528, 546, 553
NTU 75, 506
Nusselt number 252–9

opaque body 8, 78
open system 51
optical thickness 228, 231

Péclet number 254, 266
Planck's formula and constant 87
Prandtl
  model 306
  number; turbulent 252, 305

Rayleigh number 256–9
reflection 79, 86
reflectivity 79, 86
relaxation time 143
Reynolds
  decomposition 252, 296
  number 252, 257–9
  theorem 60–1
Richardson
  number 266

scattering of radiation 219
  coefficient 219
  phase function 220
semi-transparent body 9, 98, 215
shape factor 193, 197, 535, 553
shear stress 109–12, 193, 253
Stanton number 254
steady state 7
Stefan constant 91
stress 109–12
  turbulent 296–309
  viscous 109–12

thermal contact 21, 132
transition, laminar-turbulent 258–9, 293
transmission 98, 218
transmissivity 98, 218
transparent body 8
transport theorem 58

ultra-violet 81
unsteady state 7

Vaschy–Buckingham (or Π) theorem 121–3
visible domain 81
Von Karman, constant of 315

Π theorem 121–3